Lecture Notes in Operations Research

Lecture Notes in Operations Research is an interdisciplinary book series which provides a platform for the cutting-edge research and developments in both operations research and operations management field. The purview of this series is global, encompassing all nations and areas of the world.

It comprises for instance, mathematical optimization, mathematical modeling, statistical analysis, queueing theory and other stochastic-process models, Markov decision processes, econometric methods, data envelopment analysis, decision analysis, supply chain management, transportation logistics, process design, operations strategy, facilities planning, production planning and inventory control.

LNOR publishes edited conference proceedings, contributed volumes that present firsthand information on the latest research results and pioneering innovations as well as new perspectives on classical fields. The target audience of LNOR consists of students, researchers as well as industry professionals.

This series is indexed by Research Papers in Economics (RePEc), zbMATH and has been accepted by Scopus.

Ali Emrouznejad · Subhash C. Ray ·
Gokulnanda Patel
Editors

Advances in the Theory and Practice of Data Envelopment Analysis

Proceedings of the International Conference on Data Envelopment Analysis (DEA2024)

Springer

Editors
Ali Emrouznejad
Surrey Business School
University of Surrey
Guildford, UK

Subhash C. Ray
Department of Economics
University of Connecticut
Storrs, CT, USA

Gokulnanda Patel
Birla Institute of Management Technology
(BIMTECH)
Delhi, Delhi, India

ISSN 2731-040X ISSN 2731-0418 (electronic)
Lecture Notes in Operations Research
ISBN 978-3-031-98176-0 ISBN 978-3-031-98177-7 (eBook)
https://doi.org/10.1007/978-3-031-98177-7

This Springer imprint is published by the registered company Springer Nature Switzerland AG
The registered company address is: Gewerbestrasse 11, 6330 Cham, Switzerland

Contents

Introduction and Summary

Ali Emrouznejad[1(✉)], Subhash Ray[2], and Gokulnanda Patel[3]

[1] Centre for Business Analytics in Practice, Surrey Business School, University of Surrey, Guildford, Surrey, UK
A.emrouznejad@surrey.ac.uk
[2] University of Connecticut, Storrs, USA
subhash.ray@uconn.edu
[3] Birla Institute of Management Technology, Noida, India
gn.patel@bimtech.ac.in

Abstract. This volume presents a comprehensive collection of cutting-edge research on Data Envelopment Analysis (DEA), featuring contributions from the DEA2024 Conference and leading scholars in the field. Organised into five thematic sections, it highlights both methodological advancements and diverse applications across sectors such as energy, finance, healthcare, education, and industry. The book underscores usefulness of DEA in addressing current challenges, including sustainability, digital transformation, and performance optimisation under uncertainty, reflecting its evolution into a dynamic tool for efficiency analysis and strategic decision-making.

1 Introduction

Data Envelopment Analysis (DEA), introduced by Charnes, Cooper, and Rhodes (1978), is a non-parametric method for evaluating the relative efficiency of entities transforming multiple inputs into multiple outputs. Using linear programming to construct an empirical efficiency frontier, DEA enables benchmarking of Decision-Making Units (DMUs) such as firms, hospitals, schools, and public services without requiring assumptions on input-output weights or functional forms.

To address the reality that not all DMUs operate at optimal scale, Banker, Charnes, and Cooper (1984) extended the CCR model to the BCC model, incorporating variable returns to scale. This advancement allowed differentiation between pure technical inefficiency and scale inefficiency, broadening DEA's analytical scope (Emrouznejad, Podinovski, et al., 2025).

Since its inception, DEA has grown into a central research methodology and practical tool across various disciplines. With over 27,000 publications since 1978 (Scopus – see Fig. 1), DEA has been widely applied in healthcare, education, banking, energy, transportation, agriculture, and more. Research activity surged after 2000, with annual publications exceeding 1,800 since 2020, reflecting DEA's adaptability to emerging challenges like sustainability, performance management, and integration with advanced analytics.

A. Emrouznejad et al. (Eds.): DEA 2024, LNOR, pp. 1–9, 2025.
https://doi.org/10.1007/978-3-031-98177-7_1

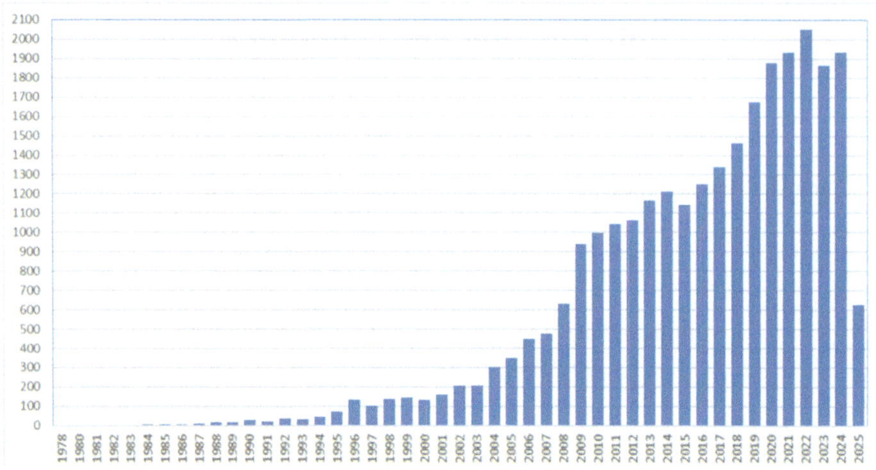

Fig. 1. Number of documents (papers, books and chapter books) published since 1978

Over five decades (Mergoni, Emrouznejad, & De Witte, 2025), DEA has evolved beyond its original models to address complexities such as uncertainty, dynamic systems, and undesirable outputs (Ray, et al, 2018). Key advancements include the Nonparametric Productivity Index (Färe et al., 1994; Ray and Desli, 1978), Network DEA (Färe and Grosskopf 1996), and Slacks-Based Measure (Tone, 2001; Emrouznejad, Brzezicki, & Lu, 2025b). DEA's flexibility has been further enhanced through integration with robust optimization, fuzzy logic (Hatami-Marbini, 2011), stochastic optimisation (Olesen et al., 2016), and machine learning techniques (Zhu et al., 2021), enabling applications in big data, predictive analytics, and decision support under uncertainty.

Today, DEA is not merely a tool for efficiency measurement but a dynamic framework for performance management, policy analysis, and strategic decision-making. This book reflects the latest developments in DEA, combining methodological innovations with practical applications across sectors. The contributions herein underscore DEA's enduring relevance and its capacity to evolve in response to contemporary operational and societal challenges.

2 Contents of this Book

This volume is collection of selected research contributions from the DEA2024 Conference, along with invited chapters from leading scholars in the field. The book showcases the latest methodological advancements and innovative applications of Data Envelopment Analysis (DEA) across a wide range of sectors. Organized into five thematic sections, it begins with theoretical developments and continues with applications in energy, finance, healthcare, education, and industry. Each chapter highlights DEA's usefulness in addressing complex efficiency and productivity challenges in diverse contexts.

2.1 Methodological and Theoretical Developments in DEA

This section showcases innovative methodologies that push the boundaries of traditional DEA models. Building on the theme of robustness, *Peykani, Emrouznejad, Mahmoodirad, Nouri, and Arabjazi (2025)* propose a generalized leader-follower network DEA model that incorporates convex uncertainty sets. This model enhances DEA's capability to assess multi-stage decision-making units (DMUs) under uncertain data conditions, particularly relevant for complex operational systems. The following chapter by Peykani, Nouri, Shojaie, Ghiyasi and Esmaeeli (2025) further expand the concept of robust DEA framework for evaluating intellectual capital efficiency under deep uncertainty. Their approach leverages conservative robust optimization to deliver stable efficiency scores in volatile environments, offering a significant contribution to DEA applications in intangible asset management.

In addressing dimensionality challenges, Dalal, Agarwal, and Mathur (2025) propose an important contribution by integrating DEA with Rough Set Theory. Their hybrid framework provides a structured method for optimal criteria selection, improving the discriminatory power of DEA in environments with large variable sets. This is followed by a chapter by Prabhash and Valiyattoor (2025) to introduce the novel concept of Imputation Propelled DEA (IPDEA), where machine learning techniques are employed to handle missing data and mitigate DEA's sensitivity to small samples. Their empirical analysis in the Indian food processing sector demonstrates how AI can effectively complement DEA for more robust performance evaluations. Following this *Yu (2025)* developed a Forest-Based resampling for confidence interval estimation of efficiencies in DEA.

A broader perspective is given by *Patel, Cvetkoska, and Kitanovikj (2025)*, who conduct a multi-technique bibliometric review of DEA applications in corporate sustainability. Their work not only maps the intellectual landscape of this field but also identifies key research clusters and outlines future directions, providing a valuable resource for scholars and practitioners. This is followed by a chapter from Ramanathan and Li (2025) to exploring the mediating role of innovation in the relationship between environmental regulations and firm performance. Using DEA, they provide critical insights into how firms can leverage innovation to balance regulatory compliance with competitive performance.

Finally, *Marques Júnior and Maciel Filho (2025)* present advancements in FRANCISCO 2.0, a specialized tool utilizing multiplicative DEA models and fractal analysis for forecasting efficiency in internet services. Their work enhances the usability and functionality of DEA in analysing modern, dynamic network infrastructures.

2.2 Energy, Environment, and Utilities

This section demonstrates DEA's utility in promoting sustainability, optimising resource use, and evaluating environmental performance. *Sufia, Singh, and Mishra (2025)* apply a two-stage DEA model combined with reliability indices to assess Indian electricity distribution utilities, offering practical recommendations for operational improvements in the energy sector. In a complementary study, Veerlapati and Perumal (2025) evaluate the efficiency of 19 Indian state-owned electricity generation utilities using a CCR DEA model. Their analysis identifies inefficiencies and benchmarks, providing strategic insights for policymakers and utility managers. This is followed by a chapter from *Kumar and Jain (2025)* who address the socio-economic impacts of decarbonization in India by employing DEA models based on by-production and weak disposability approaches. Their work quantifies the labour-related cost effects of environmental regulations, contributing to the discourse on sustainable transitions.

Turning to agriculture, Pandey, Singh, and Tiwari (2025) evaluate the environmental performance of European farms engaged in joint crop and animal husbandry production. Using DEA, they highlight best practices for sustainable farming operations, while *Pal, Chakraborty, and Ghose (2025)* investigate total factor productivity growth in Indian rice production through DEA combined with panel structural break tests, offering insights into productivity trends and agricultural efficiency. In the same area Murugesan (2025) applied DEA to measure technical, pure technical and scale efficiency of primary agriculture cooperative credit societies in Tamil Nadu.

Addressing renewable energy, Luhaniwal, Agarwal, and Mathur (2025) propose a Fuzzy DEA model with exogenously fixed variables to rank renewable energy sources. Their approach addresses uncertainty in decision-making processes related to energy policy and resource allocation.

2.3 Banking, Finance, and Insurance

This section focuses on efficiency evaluations within financial institutions, highlighting DEA's role in enhancing financial performance and stability. Muvingi (2025) applies DEA to bank asset and liability management, introducing a semi-oriented radial measure model to handle negative interest rate gaps, thus offering a novel approach to optimizing balance sheet structures, while Sinha (2025) provides a comprehensive assessment of banking efficiency in India using non-parametric DEA, revealing operational strengths and weaknesses across the sector.

Expanding on predictive analytics, Gupta, Puri, and Setia (2025) integrate network DEA with support vector regression (SVR) to forecast cost efficiency in Indian banks, showcasing a powerful hybrid methodology for financial performance forecasting, while Kushwaha and Kumar (2025) propose an integrated DEA with an entropy-based Russell Measure to evaluate the performance of Central Banks of G20 countries.

In the insurance domain, *Sinha and Amirteimoori (2025)* employ a cluster-based DEA approach to evaluate the efficiency of Indian life insurance companies, identifying peer groups and operational benchmarks.

2.4 Healthcare and Education

Efficiency in healthcare and education directly impacts societal outcomes, and this section highlights DEA's relevance in these critical sectors. Devitt, Dineen, and Zieba (2025) use a double-bootstrap DEA to evaluate cost and allocative efficiency across Irish public hospitals, identifying significant opportunities for cost reduction and efficiency improvement. In the same area, *Tigga, Sarkar, and Mor (2025)* utilize a slack-based DEA model to identify inefficiencies in India's health sector, focusing on excess resource utilization and output shortfalls.

Furthering this theme, *Sarkar and Tigga (2025)* evaluate the efficiency of public health expenditures in reducing child mortality across emerging market economies, offering valuable insights for health policy optimization. This is followed by a study on investigation and evaluation of efficiency in primary and secondary healthcare system of India by *Dar, Dar, and Gulbadin* (2025).

Turning to education institutions, Fallah and Rostamy-Malkhalifeh (2025) apply a two-stage network DEA to assess mathematical proficiency in Portuguese high schools, providing actionable insights for educational resource optimization and targeted interventions, while *GÖLCÜKCÜ (2025)* developed a fuzzy DEA model for managerial ability assessment of Turkish state universities.

Stumbriene, Ruiz, and Sirvent (2025) address bi-objective benchmarking challenges in education using DEA, offering methodological solutions for balancing multiple educational performance goals, and finally, in an analysis of educational inequality, Ghose and Choudhury (2025) investigate the impact of private tuition on the efficiency of government and aided schools in Kolkata, revealing how supplementary education influences institutional performance.

2.5 Industry-Specific Applications

The final section illustrates DEA's adaptability across various industries. In the pharmaceutical sector, Mahajan (2025) analyses the post-COVID efficiency of Indian MSMEs, shedding light on resilience factors critical to recovery and competitiveness. While Kwatra and Patel (2025) proposed a DEA model for performance mapping of startups in India. *Sithamparam and Thiagarajan (2025)* apply DEA window analysis to assess operational efficiency in the Malaysian telecommunications industry, capturing the dynamics of technological advancements and market restructuring, while *Singh, Pratap, and Laud (2025)* employ Malmquist DEA to evaluate efficiency and productivity changes in the top five global tourism destinations, highlighting the sector's recovery trajectory post-pandemic.

Turning to transportation systems, *Babazadeh, Al Raisi, and Kamran (2025)* assess container terminal efficiency in Southwest Asia, identifying operational inefficiencies and offering improvement strategies through DEA analysis. In eth same are, *Mansouri Kaleibar and Krmac (2025)* evaluate energy efficiency in Adriatic container ports, providing sustainability benchmarks for maritime logistics. Similarly, *Das, Sinha, and Mukherjee (2025)* combine stochastic frontier analysis (SFA) with the Malmquist index to examine technical efficiency and productivity trends in major Indian ports, offering a hybrid methodological perspective.

3 Conclusions

This collection demonstrates DEA's continued growth as both a research discipline and a practical framework, showcasing its adaptability to complex, real-world problems across various domains. The innovative methodologies and applications featured in this volume not only advance the theoretical boundaries of DEA but also provide actionable insights for practitioners and policymakers. As DEA continues to evolve, this book serves as a catalyst for future research and a guide for leveraging DEA in addressing emerging operational and societal challenges.

Acknowledgements. We extend our heartfelt gratitude to all the authors whose dedication, expertise, and scholarly contributions have made this volume possible. Their innovative research and commitment to advancing the field of Data Envelopment Analysis are the foundation of this collection. We also sincerely thank the reviewers for their invaluable feedback and constructive critiques, which have significantly enhanced the quality and rigor of each chapter.

We would like to express our special thanks to Springer and the production team, particularly Jialin Yan and Christian Rauscher, for their continuous support in publishing this volume within the prestigious Lecture Notes in Operations Research series.

Furthermore, we gratefully acknowledge the efforts of the organizers of DEA2024, held at BIMTECH, India, in November 2024. Their outstanding work in hosting a successful conference provided the platform for many of the contributions featured in this volume.

Finally, it is important to clarify that the responsibility for the content, accuracy, and interpretations presented within each chapter lies solely with the respective authors. As editors, we have facilitated the compilation and organization of this volume but do not assume responsibility for any errors or omissions. We encourage readers to exercise their judgment when applying the insights provided.

We trust that this collection will serve as a valuable resource and a source of inspiration for ongoing research, practical applications, and future innovations in the dynamic and ever-evolving field of Data Envelopment Analysis.

References

Babazadeh, R., Al Raisi, A., Kamran, A.M.: Efficiency Assessment of Container Terminals: The Case of Southwest Asia's Ports, Proceedings of DEA2024. Lecture Notes in Operations Research. pp. 504–519. Springer, Cham (2025). https://doi.org/10.1007/978-3-031-98177-7_34

Banker, R.D., Charnes, A., Cooper, W.W.: Some models for estimating technical and scale inefficiencies in data envelopment analysis. Manage. Sci. **30**(9), 1078–1092 (1984)

Charnes, A., Cooper, W.W., Rhodes, E.: Measuring the efficiency of decision-making units. Eur. J. Oper. Res. **2**(6), 429–444 (1978)

Dalal, I., Agarwal, S., Mathur, T.: An optimal criteria selection in efficiency assessment through integration of dea with rough set theory. In: Proceedings of DEA2024. Lecture Notes in Operations Research. Springer, Cham (2025)

Dar, Q.F., Dar, G.F., Bhat, A.: An investigation and evaluation of efficiency in primary and secondary healthcare system of india using data envelopment analysis. In: Proceedings of DEA2024. Lecture Notes in Operations Research. Springer, Cham (2025)

Das, J.K., Sinha, D., Mukherjee, S.: Analysis of technical efficiency of major Indian ports using markov chain montecarlo based stochastic frontier analysis and malmquist index. In: Proceedings of DEA2024. Lecture Notes in Operations Research. Springer, Cham (2025)

Devitt, N., Dineen, D., Zieba, M.: Cost and allocative efficiency of public hospitals in Ireland – a robust DEA approach. In: Proceedings of DEA2024. Lecture Notes in Operations Research. Springer, Cham (2025)

Emrouznejad, A., Brzezicki, Ł., Lu, C.: The development and evolution of slacks-based measure models in data envelopment analysis: a comprehensive review of the literature. J. Econom. Surv. In press, 1–28 (2025). https://doi.org/10.1111/joes.12682

Emrouznejad, A., Banker, R., Neralić, L.: Advances in data envelopment analysis: celebrating the 40th anniversary of DEA and the 100th anniversary of professor abraham charnes' birthday. Eur. J. Oper. Res. **278**(2), 365–367 (2019)

Emrouznejad, A., Podinovski, V., Lu, C., Charles, V., Moradi-Motlagh, A.: Rajiv Banker's Lasting Impact on Data Envelopment Analysis, Annals of Operations Research, **351**, 1225–1264 (2025). https://doi.org/10.1007/s10479-025-06473-3

Emrouznejad, A., Ray, S., Patel, G.: Introduction and summary. In: Proceedings of DEA2024. Lecture Notes in Operations Research. Springer, Cham (2025)

Fallah, N., Rostamy-Malkhalifeh, M.: Strategies to enhance the mathematical proficiency of underperforming high schools using data envelopment analysis. In: Proceedings of DEA2024. Lecture Notes in Operations Research. Springer, Cham (2025)

Färe, R., Grosskopf, S.: Intertemporal Production Frontiers: With Dynamic DEA (Norwell, MA: Kluwer Academic Publishers) (1996)

Färe, R., Grosskopf, S., Norris, M., Zhang, Z.: Productivity growth, technical progress, and efficiency change in industrialized countries. Am. Econ. Rev. **84**, 66–83 (1994)

Ghose, A., Choudhury, S.: Does private tuition affect efficiency of government and government aided schools? evidence of higher secondary schools based in Kolkata. In: Proceedings of DEA2024. Lecture Notes in Operations Research. Springer, Cham (2025)

Gölcükcü, A.: Managerial ability assessment of turkish state universities with DEA. In: Proceedings of DEA2024. Lecture Notes in Operations Research. Springer, Cham (2025)

Gupta, N., Puri, J., Setia, G.: Forecasting cost efficiency of banks in India with integrated DDF-based network DEA and SVR approach. In: Proceedings of DEA2024. Lecture Notes in Operations Research. Springer, Cham (2025)

Hatami-Marbini, A., Emrouznejad, A., Tavana, M.: A taxonomy and review of the fuzzy data envelopment analysis literature: two decades in the making. Eur. J. Oper. Res. **214**(3), 457–472 (2011)

Kumar, S., Jain, R.K.: Decarburization and employment: evidence from India. In: Proceedings of DEA2024. Lecture Notes in Operations Research. Springer, Cham (2025)

Kushwaha, A., Kumar, B.: Evaluating the performance of central banks of g20 countries: integrating DEA with an entropy-based russell measure. In: Proceedings of DEA2024. Lecture Notes in Operations Research. Springer, Cham (2025)

Kwatra, P., Patel, G.: Performance mapping of startups in india using data envelopment analysis. In: Proceedings of DEA2024. Lecture Notes in Operations Research. Springer, Cham (2025)

Luhaniwal, Jyoti; Agarwal, Shivi; Mathur, Trilok (2025) Fuzzy DEA Model with Exogenously Fixed Variables for Ranking of Renewable Energy Sources, Proceedings of DEA2024. Lecture Notes in Operations Research. Springer, Cham

Mansouri, K.M.,, Krmac, E.: Energy efficiency evaluation based on DEA: case study of Adriatic container ports. In: Proceedings of DEA2024. Lecture Notes in Operations Research. Springer, Cham (2025)

Marques, J., Francisco, D., Maciel, F., Franks, P.: FRANCISCO 2.0: Increasing the usability and appending multiplicative DEA models for internet services forecasting over time. In: Proceedings of DEA2024. Lecture Notes in Operations Research. Springer, Cham (2025)

Mergoni, A., Emrouznejad, A., De Witte, K.: Fifty years of Data Envelopment Analysis, European J. Opera. Res. **326**(3), 389–412 (2025). https://doi.org/10.1016/j.ejor.2024.12.049

M Raman, M.: Technical, pure technical and scale efficiency of primary agriculture cooperative credit societies in Tamil Nadu. In: Proceedings of DEA2024. Lecture Notes in Operations Research. Springer, Cham (2025)

Muvingi, J.: Application of data envelopment analysis in bank asset and liability management. In: Proceedings of DEA2024. Lecture Notes in Operations Research. Springer, Cham (2025)

Olesen, O.B., Petersen, N.C.: Stochastic data envelopment analysis - a review. Eur. J. Oper. Res. **251**(1), 2–21 (2016). https://doi.org/10.1016/j.ejor.2015.07.058

Pal, D., Chakraborty, C., Ghose, A.: Total factor productivity growth of rice production in india: an analysis based on data envelopment analysis and panel structural break test. In: Proceedings of DEA2024. Lecture Notes in Operations Research. Springer, Cham (2025)

Pandey, U., Singh, S., Tiwari, S.: Environmental performance evaluation of joint production of crop and animal husbandry in European farms. In: Proceedings of DEA2024. Lecture Notes in Operations Research. Springer, Cham (2025)

Patel, G., Cvetkoska, V., Kitanovikj, B.: Measuring corporate sustainability with data envelopment analysis: a multi-technique bibliometric review and future research agenda. In: Proceedings of DEA2024. Lecture Notes in Operations Research. Springer, Cham (2025)

Peykani, P., Emrouznejad, A., Mahmoodirad, A., Nouri, M., Arabjazi, N.

Peykani, P., Nouri, M., Shojaie, S.E., Ghiyasi, F., Esmaeeli, A.: Efficiency analysis of intellectual capital under deep uncertainty: a robust DEA approach. In: Proceedings of DEA2024. Lecture Notes in Operations Research. Springer, Cham (2025)

Peykani, P., Emrouznejad, A., Mahmoodirad, A., Nouri, M., Arabjazi, N.: A Robust Framework for Generalized Leader-Follower Network Data Envelopment Analysis under Uncertain Data, Proceedings of DEA2024. Lecture Notes in Operations Research. pp. 10–26. Springer, Cham (2025). https://doi.org/10.1007/978-3-031-98177-7_2

Prabhash, Ardhana, M., Valiyattoor, V.: Imputation propelled data envelopment analysis (IPDEA): a case of indian food processing sector. In: Proceedings of DEA2024. Lecture Notes in Operations Research. Springer, Cham (2025)

Ramanathan, R., Li, R.: The role of Innovation on the link between Environmental Regulations and Firm Performance: A DEA Approach. In: Proceedings of DEA2024. Lecture Notes in Operations Research. Springer, Cham (2025). https://doi.org/10.1007/978-3-031-98177-7_9

Sarkar, P., Nutan Shashi, T.: Efficiency and productivity of public health expenditures in reducing child mortality in emerging market economies. In: Proceedings of DEA2024. Lecture Notes in Operations Research. Springer, Cham (2025)

Singh, S., Pratap, S., Laud, S.: Evaluating efficiency and total factor productivity of top five tourism countries using malmquist DEA. In: Proceedings of DEA2024. Lecture Notes in Operations Research. Springer, Cham (2025)

Ram Pratap, S.: Bank efficiency in india: a non-parametric approach. In: Proceedings of DEA2024. Lecture Notes in Operations Research. Springer, Cham (2025)

Ram Pratap, S., Ali Reza, A.: Efficiency of indian life insurance companies-a cluster based approach. In: Proceedings of DEA2024. Lecture Notes in Operations Research. Springer, Cham (2025)

Sithamparam, A., Senthil Kumar, T.: Evaluating performance efficiency in the malaysian telecommunications industry: a data envelopment analysis approach. In: Proceedings of DEA2024. Lecture Notes in Operations Research. Springer, Cham (2025)

Stumbriene, D., José, L.R., Sirvent, I.: Benchmarking in education: data envelopment analysis approaches for solving bi-objective problems. In: Proceedings of DEA2024. Lecture Notes in Operations Research. Springer, Cham (2025)

Subhash, C.R.: Data envelopment analysis: theory and techniques for economics and operations research, Cambridge university press (2004)

Ray, S.C., Mukherjee, K., Venkatesh, A.: Nonparametric measures of efficiency in the presence of undesirable outputs: a by-production approach. Empiric. Econom. Springer **54**(1), 31–65 (2018)

Ray, S.C., Desli, E.: Productivity growth, technical progress, and efficiency change in industrialized countries: comment. Am. Econ. Rev. **87**(5), 1033–1039 (1997)

Sufia, S., Singh, A., Mishra, S.: Performance assessment of Indian electricity distribution utilities using reliability indices: a two-stage data envelopment analysis. In: Proceedings of DEA2024. Lecture Notes in Operations Research. Springer, Cham (2025)

Nutan Shashi, T., Pinak, S., Mor, N.: Identifying input excess and output shortfall in the Indian health systems using non-radial Slack Based Data Development Analysis. In: Proceedings of DEA2024. Lecture Notes in Operations Research. Springer, Cham (2025)

Tone, K.: A slacks-based measure of efficiency in data envelopment analysis. Eur. J. Oper. Res. **130**(2001), 498–509 (2001)

Varun, M.: Efficiency analysis of indian pharmaceutical msmes: a post-COVID Era. In: Proceedings of DEA2024. Lecture Notes in Operations Research. Springer, Cham (2025)

Veerlapati, R., Perumal, C.: Critical Evaluation of indian electricity generation utilities using data envelopment analysis (DEA). In: Proceedings of DEA2024. Lecture Notes in Operations Research. Springer, Cham (2025)

Yu, Z.: Forest-based resampling for confidence interval estimation of efficiencies in data envelopment analysis. In: Proceedings of DEA2024. Lecture Notes in Operations Research. Springer, Cham (2025)

Zhu, C., Zhu, N., Emrouznejad, A.: A combined machine learning algorithms and DEA method for measuring and predicting the efficiency of Chinese manufacturing listed companies. J. Manage. Sci. Eng. **6**(4), 435–448 (2021)

A Robust Framework for Generalized Leader-Follower Network Data Envelopment Analysis Under Uncertain Data

Pejman Peykani[1(✉)], Ali Emrouznejad[2], Ali Mahmoodirad[3], Mojtaba Nouri[4], and Nasim Arabjazi[5]

[1] Department of Industrial Engineering, Faculty of Engineering, Khatam University, Tehran, Iran
pejman.peykani@yahoo.com, p.peykani@khatam.ac.ir

[2] Surrey Business School, University of Surrey, Guildford, UK
a.emrouznejad@surrey.ac.uk

[3] Department of Mathematics, Bab. C., Islamic Azad University, Babol, Iran
ali.mahmoodirad@iau.ac.ir

[4] School of Industrial Engineering, Iran University of Science and Technology, Tehran, Iran
mojtaba_nouri@ind.iust.ac.ir

[5] Folkuniversitetet, Varberg, Sweden
nasimarabjazy@gmail.com

Abstract. This research focuses on assessing the efficiency of network decision-making units (DMUs) in uncertain environments using network data envelopment analysis (NDEA). Traditional NDEA models are effective in analyzing multi-stage systems but face challenges in uncertain conditions due to their dependence on precise data. To address this issue, the study introduces a leader-follower NDEA model with various returns to scale assumptions, enhancing its flexibility. A robust optimization approach utilizing convex uncertainty sets is incorporated to manage data uncertainty, ensuring reliable performance evaluations even in dynamic and imprecise scenarios. This methodology maintains the accuracy of efficiency analysis despite ambiguous or incomplete data. The proposed robust leader-follower NDEA model is validated through a real-world case study involving investment companies in the Iranian capital market. The results highlight the framework's resilience and capability to handle data ambiguity, offering valuable insights into system efficiency and identifying performance bottlenecks. This research presents a practical and dependable tool for performance evaluation in complex and uncertain operational environments.

Keywords: Network DEA · Non-Cooperative Game Theory · Data Uncertainty · Performance Measurement · Convex Robust Optimization · Budget of Robustness

A. Emrouznejad et al. (Eds.): DEA 2024, LNOR, pp. 10–26, 2025.
https://doi.org/10.1007/978-3-031-98177-7_2

1 Introduction

Network data envelopment analysis (NDEA), in the literature of efficiency analysis, is an advanced development of the traditional data envelopment analysis (DEA) model, which was developed using some sophisticated tools (Kao & Hwang, 2008, Halkos et al., 2014, Kao, 2014, Koronakos, 2019a). While DEA provides an overall efficiency assessment of the decision-making units (DMUs), considering the relation between inputs and outputs, it usually simplifies the system structure, considering it as a "black box," and does not take into consideration internal processes and interactions among subsystems or stages. NDEA overcomes this crucial shortage by providing a more refined and complete methodological structure for performance evaluation of complex systems with multi-stage or network structure. It allows, in this respect, a deeper understanding of the internal dynamics and interdependencies of these systems to be gained (Cook et al., 2010a, Kao, 2017, Koronakos et al., 2019b, Ratner et al., 2023).

NDEA is particularly relevant when dealing with processes or systems comprising several interdependent stages, where the output of one stage serves as the input for the next. For instance, in supply chains, production systems, an education sector, or banking industries, organizational operations are often divisible into sequential stages where the overall efficiency of the system is really affected by the relationships amongst them. This approach enables evaluation of the performance of each separate stage while considering the role of intermediate interactions in identifying the overall efficiency. Second, NDEA brings within the analytical framework factors that could be intermediate or unobservable, hence making results more correct and dependable. (Chen & Zhu, 2004, Chen et al., 2009, Alves & Meza, 2023, Peykani et al., 2024a).

One of the major strong points related to NDEA involves its great flexibility in terms of modeling and analyzing many network structures by accounting for different complexities typical of any real-world system. Contrary to the traditional DEA models, which usually assess the system only as black boxes, NDEA provides an opportunity to make a profound analysis of structures that can be configured either in series, parallel, mixed, or hierarchical (Kao, 2015, Zhou et al., 2018, Lin & Liu, 2022, Ding et al., 2023). This adaptability guarantees its ability to be modified to capture the operational realities of various industries and organizational setups. Additionally, NDEA incorporates exogenous factors inside its assessment model, emanating from the system, side by side with endogenous ones emanating from its external influences.

In line with that, it definitely extends the model's power in offering a holistic view that looks to efficiency, accounted for both within and outside of the system interacting. Such an integrated approach makes the NDEA an efficient tool in finding the diagnostic bottlenecks that impede the smooth flow of operations, locating the sources of inefficiency in their subsystems, and putting forward recommendations that are action-oriented for performance improvement (Tone & Tsutsui, 2009, Cook et al., 2010b, Despotis et al., 2016). Such capabilities enable NDEA not only to provide valid and meaningful efficiency assessments but also to assist managers and decision-makers in the elaboration of strategic plans and operational improvements that are evidence-based and contextually relevant. It is this rare combination of flexibility, precision, and strategic relevance that has made NDEA such a valuable methodology for optimizing complex systems.

Although the NDEA approach offers a number of advantages, it also has significant challenges with regard to data uncertainty. NDEA requires very accurate and reliable data in order to assess the efficiency of multi-stage or networked systems accurately. In real-world applications, however, data may be afflicted with various forms of uncertainty due to measurement errors, incomplete information, environmental fluctuations, or unpredictable changes in external factors (Peykani & Gheidar-Kheljani, 2022, Peykani, & Pishvaee, 2024). Such uncertainties may distort the results of the analysis, leading to misleading efficiency scores and probably wrong conclusions. As discussed in the context of NDEA, data uncertainty does become particularly problematic, considering that the approach not only focuses on the overall efficiency evaluation but rather on the interaction and the mutual relations existing among the subsystems, thus also needing to rely on the value of these interdependencies arising out of uncertain or imprecise data (Peykani et al., 2021a, 2024b, 2024c, Seyed Esmaeili & Mohammadi., 2024).

This is particularly critical in industries or systems where the external variables of market conditions, regulatory changes, or environmental factors play a huge role in shaping inputs, outputs, or intermediate variables. In this light, some researchers have developed methodologies to incorporate uncertainty into the NDEA framework using fuzzy logic, stochastic programming, and robust optimization to strengthen model resilience regarding variability in data (Emrouznejad & Tavana, 2014, Olesen & Petersen, 2016, Peykani et al., 2019, 2020a, 2022a). These kinds of approaches will let NDEA provide more realistic and reliable efficiency evaluations under uncertain scenarios; this will guarantee the obtained results are useful to decision-makers even in conditions with ambiguity in the data.

The current study proposes a robust approach to evaluate the performance of two-stage network DMUs under conditions of uncertainty. For this purpose, the leader-follower network DEA model, under various returns-to-scale assumptions, has been chosen as the foundational methodology. A robust optimization framework utilizing uncertainty sets is developed to address data uncertainty, ensuring the reliability of the evaluation process even when data is imprecise or fluctuating. This proposed approach not only extends the applicability of NDEA to uncertain environments but also provides a systematic method for incorporating robustness into network efficiency analysis. To validate the practicality and effectiveness of this methodology, a case study was conducted using real data from several investment companies operating within the Iranian capital market. The results demonstrate how the uncertain leader-follower network DEA model, combined with convex robust optimization techniques, delivers valuable insights into the performance of network DMUs despite data ambiguity. This highlights the potential of the proposed framework as a powerful decision-support tool for managers and policymakers navigating uncertain environments.

The rest of this research is organized as follows: Sect. 2 introduces the NDEA modeling framework, which employs a leader-follower approach to assess the performance of two-stage DMUs. In Sect. 3, an extension of the proposed NDEA modeling framework incorporates convex robust optimization technique for developing a robust leader-follower NDEA model to handle uncertainties in the data. Section 4 illustrates the application of the proposed robust NDEA approach by a case study on Iranian investment companies. Finally, Sect. 5 summarizes the key findings and provides possible future research directions.

2 The Generalized Leader-Follower Network DEA

This section presents the development of a generalized network DEA approach that employs the leader-follower method, a non-cooperative game theory technique, to evaluate the performance of general two-stage DMUs. Consider a general two-stage structure comprising a set of W homogeneous decision-making units $\text{DMU}_w (w = 1, ..., W)$. In this structure, each DMU processes A inputs $F_{aw} (a = 1, ..., A)$ in the first stage, produces B outputs $G_{bw} (b = 1, ..., B)$ (referred to as leakage variables) that exit the system after the first stage, and generates C intermediate variables $H_{cw} (c = 1, ..., C)$ that serve as a connection between the first and second stages. Additionally, D extra inputs $I_{dw} (d = 1, ..., D)$ are utilized in the second stage, which subsequently produces E outputs $J_{ew} (e = 1, ..., E)$.

It is important to note that the non-negative weights $\Theta_a (a = 1, ..., A)$, $\Upsilon_b (b = 1, ..., B)$, $\Omega_c (c = 1, ..., C)$, $\Psi_d (d = 1, ..., D)$, and $\Phi_e (e = 1, ..., E)$ are assigned to the corresponding elements $F_{aw} (a = 1, ..., A)$, $G_{bw} (b = 1, ..., B)$, $H_{cw} (c = 1, ..., C)$, $I_{dw} (d = 1, ..., D)$, and $J_{ew} (e = 1, ..., E)$, respectively. Moreover, the index of p is associated with the decision-making unit being evaluated. Please note that the general two-stage structure provides a comprehensive framework for analyzing the interactions and efficiency of two-stage systems while addressing interdependencies and the flow of resources between stages.

The leader-follower approach, introduced by Liang et al. (2008), is a widely used and effective method for modeling network DEA. This approach, rooted in non-cooperative game theory, assumes a hierarchical structure where one stage is prioritized as the leader (Li et al., 2014, Guo & Zhu, 2017, Chu et al., 2020, Zhou et al., 2021). In this framework, the efficiency of the leader stage is evaluated first, serving as a basis for determining the optimal solution. Subsequently, the efficiency of the follower stage is calculated in alignment with the leader's optimized performance, ensuring a structured and sequential analysis of the network. Assuming that Stage 1 holds greater importance (with the first stage designated as the leader and the second stage as the follower), the efficiency of Stage 1 for a specific DMU_p is determined using Model (1):

$$\text{Max } \Delta^{\text{S1}} = \sum_{b=1}^{B} G_{bp}\Upsilon_b + \sum_{c=1}^{C} H_{cp}\Omega_c + \omega_1 \dagger^{\text{S1}} \tag{1-1}$$

$$\text{S.t.} \sum_{a=1}^{A} F_{ap}\Theta_a = 1 \tag{1-2}$$

$$\sum_{b=1}^{B} G_{bw}\Upsilon_b + \sum_{c=1}^{C} H_{cw}\Omega_c - \sum_{a=1}^{A} F_{aw}\Theta_a + \omega_1 \dagger^{\text{S1}} \leq 0, \forall w \tag{1-3}$$

$$\omega_1\omega_2(-1)^{\omega_3}\dagger^{\text{S1}} \geq 0 \tag{1-4}$$

$$\Theta_a, \Upsilon_b, \Omega_c \geq 0, \forall a, b, c \tag{1-5}$$

With the first-stage efficiency fixed, the objective shifts to maximizing the second-stage efficiency. Accordingly, the efficiency of Stage 2 for a specific DMU$_p$ is derived from Model (2), which is formulated as follows:

$$\text{Max } \Delta^{S2} = \sum_{e=1}^{E} J_{ep}\Phi_e + \omega_1 \dagger^{S2} \tag{2-1}$$

$$\text{S.t.} \sum_{c=1}^{C} H_{cp}\Omega_c + \sum_{d=1}^{D} I_{dp}\Psi_d = 1 \tag{2-2}$$

$$\sum_{e=1}^{E} J_{ew}\Phi_e - \sum_{c=1}^{C} H_{cw}\Omega_c - \sum_{d=1}^{D} I_{dw}\Psi_d + \omega_1 \dagger^{S2} \leq 0, \forall w \tag{2-3}$$

$$\sum_{b=1}^{B} G_{bw}\Upsilon_b + \sum_{c=1}^{C} H_{cw}\Omega_c - \sum_{a=1}^{A} F_{aw}\Theta_a + \omega_1 \dagger^{S1} \leq 0, \forall w \tag{2-4}$$

$$\left(\sum_{b=1}^{B} G_{bp}\Upsilon_b + \sum_{c=1}^{C} H_{cp}\Omega_c + \omega_1 \dagger^{S1}\right) - \Delta^{S1*}\left(\sum_{a=1}^{A} F_{ap}\Theta_a\right) = 0 \tag{2-5}$$

$$\omega_1\omega_2(-1)^{\omega_3}\dagger^{S1} \geq 0 \tag{2-6}$$

$$\omega_1\omega_2(-1)^{\omega_3}\dagger^{S2} \geq 0 \tag{2-7}$$

$$\Theta_a, \Upsilon_b, \Omega_c, \Psi_d, \Phi_e \geq 0, \forall a, b, c, d, e \tag{2-8}$$

It should be noted that in Models (1) and (2), $\dagger^{S1}$ and $\dagger^{S2}$ represent decision variables associated with the returns to scale (RTS) for Stage 1 and Stage 2 of the DMU under evaluation, respectively. By configuring the binary parameters ω_1, ω_2, and ω_3 in Constraints (1–4), (2–6), and (2–7), the leader-follower network DEA approach can be adapted to different types of RTS assumptions: (1) if $\omega_1 = 0$, the model assumes constant returns to scale (CRS) (Charnes et al., 1978); (2) if $\omega_1 = 1$ and $\omega_2 = 0$, the model assumes variable returns to scale (VRS) (Banker et al., 1984); (3) if $\omega_1 = 1$, $\omega_2 = 1$, and $\omega_3 = 0$, the model assumes non-increasing returns to scale (NIRS); and (4) if $\omega_1 = 1$, $\omega_2 = 1$, and $\omega_3 = 1$, the model assumes non-decreasing returns to scale (NDRS) (Yu et al., 1996a; 1996b). Finally, the overall efficiency of the DMU under evaluation is determined using Eq. (3), as presented below:

$$\Delta^{O*} = \Delta^{S1*} * \Delta^{S2*} \tag{3}$$

It is important to emphasize that the leader-follower approach provides a distinct and non-overlapping decomposition of efficiency, ensuring a clear and unambiguous allocation of performance across stages. Alternatively, if Stage 2 is deemed more critical and acts as the leader while Stage 1 takes on the role of the follower, the efficiencies of Stage 2, Stage 1, and the overall two-stage system are evaluated using a similar methodology.

3 The Robust Leader-Follower Network DEA

This section extends the generalized two-stage network DEA framework by incorporating a convex robust optimization technique, facilitating the development of a robust leader-follower network DEA model to effectively handle data uncertainties. Now, assume that all data $\tilde{F}_{aw}(a = 1, ..., A)$, $\tilde{G}_{bw}(b = 1, ..., B)$, $\tilde{H}_{cw}(c = 1, ..., C)$, $\tilde{I}_{dw}(d = 1, ..., D)$, and $\tilde{J}_{ew}(e = 1, ..., E)$ are subject to uncertainty, each falling within an uncertainty interval defined as by $\tilde{F}_{aw} \in [F_{aw} - \hat{F}_{aw}, F_{aw} + \hat{F}_{aw}]$, $\tilde{G}_{bw} \in [G_{bw} - \hat{G}_{bw}, G_{bw} + \hat{G}_{bw}]$, $\tilde{H}_{cw} \in [H_{cw} - \hat{H}_{cw}, H_{cw} + \hat{H}_{cw}]$, $\tilde{I}_{dw} \in [I_{dw} - \hat{I}_{dw}, I_{dw} + \hat{I}_{dw}]$, and $\tilde{J}_{ew} \in [J_{ew} - \hat{J}_{ew}, J_{ew} + \hat{J}_{ew}]$, respectively.

It is worth noting that to address uncertainty and derive robust counterparts for uncertain constraints, the widely adopted robust optimization approach proposed by Bertsimas & Sim (2004) is utilized (Bertsimas et al., 2011; Gabrel et al., 2014; Peykani et al., 2020b; 2022b; 2024d). Accordingly, assuming that the first stage is of greater importance in the leader-follower approach, the robust network DEA model for evaluating the efficiency of the first stage of the DMU under evaluation is formulated as Model (4):

$$\text{Max } \Delta^{\text{S1}}_{\text{R}(\Gamma,\delta)} \tag{4-1}$$

$$\begin{aligned} \text{S.t.} & -\sum_{b=1}^{B} G_{bp}\Upsilon_b - \sum_{c=1}^{C} H_{cp}\Omega_c - \omega_1 \dagger^{\text{S1}} \\ & + Q_p^{GH}\Gamma_p^{GH} + \sum_{b=1}^{B} \xi_{bp}^{G} + \sum_{c=1}^{C} \xi_{cp}^{H} \geq -\Delta^{\text{S1}}_{\text{R}(\Gamma,\delta)} \end{aligned} \tag{4-2}$$

$$Q_p^{GH} + \xi_{bp}^{G} \geq \delta G_{bp}\Upsilon_b, \forall b \tag{4-3}$$

$$Q_p^{GH} + \xi_{cp}^{H} \geq \delta H_{cp}\Omega_c, \forall c \tag{4-4}$$

$$\sum_{a=1}^{A} F_{ap}\Theta_a + Q_p^{F}\Gamma_p^{F} + \sum_{a=1}^{A} \xi_{ap}^{F} \leq 1 \tag{4-5}$$

$$Q_p^{F} + \xi_{ap}^{F} \geq \delta F_{ap}\Theta_a, \forall a \tag{4-6}$$

$$\begin{aligned} & \sum_{b=1}^{B} G_{bw}\Upsilon_b + \sum_{c=1}^{C} H_{cw}\Omega_c - \sum_{a=1}^{A} F_{aw}\Theta_a + \omega_1 \dagger^{\text{S1}} \\ & + Q_w^{FGH}\Gamma_w^{FGH} + \sum_{a=1}^{A} \xi_{aw}^{F} + \sum_{b=1}^{B} \xi_{bw}^{G} + \sum_{c=1}^{C} \xi_{cw}^{H} \leq 0, \forall w \end{aligned} \tag{4-7}$$

$$Q_w^{FGH} + \xi_{aw}^{F} \geq \delta F_{aw}\Theta_a, \forall a \tag{4-8}$$

$$Q_w^{FGH} + \xi_{bw}^{G} \geq \delta G_{bw}\Upsilon_b, \forall b \tag{4-9}$$

$$Q_w^{FGH} + \xi_{cw}^H \geq \delta H_{cw}\Omega_c, \forall c \tag{4-10}$$

$$\omega_1\omega_2(-1)^{\omega_3}\dagger^{S1} \geq 0 \tag{4-11}$$

$$Q_p^F, Q_p^{GH}, Q_w^{FGH} \geq 0, \forall w \tag{4-12}$$

$$\xi_{ap}^F, \xi_{bp}^G, \xi_{cp}^H, \xi_{aw}^F, \xi_{bw}^G, \xi_{cw}^H \geq 0, \forall a, b, c, w \tag{4-13}$$

$$\Theta_a, \Upsilon_b, \Omega_c \geq 0, \forall a, b, c \tag{4-14}$$

Subsequently, the robust network DEA approach for calculating the efficiency of the second stage of the DMU under evaluation, based on the optimal solution obtained from Model (4), is presented as Model (5) as follows:

$$\text{Max } \Delta_{\mathrm{R}(\Gamma,\delta)}^{\mathrm{S2}} \tag{5-1}$$

$$\text{S.t.} -\sum_{e=1}^{E} J_{ep}\Phi_e - \omega_1\dagger^{S2} + Q_p^J\Gamma_p^J + \sum_{e=1}^{E}\xi_{ep}^J \geq -\Delta_{\mathrm{R}(\Gamma,\delta)}^{\mathrm{S2}} \tag{5-2}$$

$$Q_p^J + \xi_{ep}^J \geq \delta J_{ep}\Phi_e, \forall e \tag{5-3}$$

$$\sum_{c=1}^{C} H_{cp}\Omega_c + \sum_{d=1}^{D} I_{dp}\Psi_d + Q_p^{HI}\Gamma_p^{HI} + \sum_{c=1}^{C}\xi_{cp}^{HS2} + \sum_{d=1}^{D}\xi_{dp}^I \leq 1 \tag{5-4}$$

$$Q_p^{HI} + \xi_{cp}^{HS2} \geq \delta H_{cp}\Omega_c, \forall c \tag{5-5}$$

$$Q_p^{HI} + \xi_{dp}^I \geq \delta I_{dp}\Psi_d, \forall d \tag{5-6}$$

$$\begin{aligned}&\sum_{e=1}^{E} J_{ew}\Phi_e - \sum_{c=1}^{C} H_{cw}\Omega_c - \sum_{d=1}^{D} I_{dw}\Psi_d + \omega_1\dagger^{S2}\\&+ Q_w^{HIJ}\Gamma_w^{HIJ} + \sum_{c=1}^{C}\xi_{cw}^{HS2} + \sum_{d=1}^{D}\xi_{dw}^I + \sum_{e=1}^{E}\xi_{ew}^J \leq 0, \forall w\end{aligned} \tag{5-7}$$

$$Q_w^{HIJ} + \xi_{cw}^{HS2} \geq \delta H_{cw}\Omega_c, \forall c \tag{5-8}$$

$$Q_w^{HIJ} + \xi_{dw}^I \geq \delta I_{dw}\Psi_d, \forall d \tag{5-9}$$

$$Q_w^{HIJ} + \xi_{ew}^J \geq \delta J_{ew}\Phi_e, \forall e \tag{5-10}$$

$$\sum_{b=1}^{B} G_{bw}\Upsilon_b + \sum_{c=1}^{C} H_{cw}\Omega_c - \sum_{a=1}^{A} F_{aw}\Theta_a + \omega_1 \dagger^{S1} + Q_w^{FGH}\Gamma_w^{FGH} + \sum_{a=1}^{A} \xi_{aw}^F + \sum_{b=1}^{B} \xi_{bw}^G + \sum_{c=1}^{C} \xi_{cw}^{HS1} \le 0, \forall w \tag{5-11}$$

$$Q_w^{FGH} + \xi_{aw}^F \ge \delta F_{aw}\Theta_a, \forall a \tag{5-12}$$

$$Q_w^{FGH} + \xi_{bw}^G \ge \delta G_{bw}\Upsilon_b, \forall b \tag{5-13}$$

$$Q_w^{FGH} + \xi_{cw}^{HS1} \ge \delta H_{cw}\Omega_c, \forall c \tag{5-14}$$

$$\left(\sum_{b=1}^{B} G_{bp}\Upsilon_b + \sum_{c=1}^{C} H_{cp}\Omega_c + \omega_1 \dagger^{S1} + Q_p^{GH}\Gamma_p^{GH} + \sum_{b=1}^{B} \xi_{bp}^G + \sum_{c=1}^{C} \xi_{cp}^H\right) - \Delta_{R(\Gamma,\delta)}^{S1*}\left(\sum_{a=1}^{A} F_{ap}\Theta_a + Q_p^F\Gamma_p^F + \sum_{a=1}^{A} \xi_{ap}^F\right) = 0 \tag{5-15}$$

$$Q_p^F + \xi_{ap}^F \ge \delta F_{ap}\Theta_a, \forall a \tag{5-16}$$

$$Q_p^{GH} + \xi_{bp}^G \ge \delta G_{bp}\Upsilon_b, \forall b \tag{5-17}$$

$$Q_p^{GH} + \xi_{cp}^H \ge \delta H_{cp}\Omega_c, \forall c \tag{5-18}$$

$$\omega_1\omega_2(-1)^{\omega_3}\dagger^{S1} \ge 0 \tag{5-19}$$

$$\omega_1\omega_2(-1)^{\omega_3}\dagger^{S2} \ge 0 \tag{5-20}$$

$$Q_p^{FGH}, Q_p^{HI}, Q_p^J, Q_w^{FGH}, Q_w^{HIJ} \ge 0, \forall w \tag{5-21}$$

$$\xi_{ap}^F, \xi_{bp}^G, \xi_{cp}^{HS1}, \xi_{cp}^{HS2}, \xi_{dp}^I, \xi_{ep}^J \ge 0, \forall a, b, c, d, e \tag{5-22}$$

$$\xi_{aw}^F, \xi_{bw}^G, \xi_{cw}^{HS1}, \xi_{cw}^{HS2}, \xi_{dw}^I, \xi_{ew}^J \ge 0, \forall a, b, c, d, e, w \tag{5-23}$$

$$\Theta_a, \Upsilon_b, \Omega_c, \Psi_d, \Phi_e \ge 0, \forall a, b, c, d, e \tag{5-24}$$

It should be noted that in Models (4) and (5), Γ and δ represent the budget of robustness and the level of data perturbation, respectively, both of which are determined by the decision-maker. Finally, after solving Models (4) and (5) for a given Γ and δ, the overall efficiency of the DMU under evaluation is obtained using Eq. (6) as follows:

$$\Delta_{R(\Gamma,\delta)}^{O*} = \Delta_{R(\Gamma,\delta)}^{S1*} * \Delta_{R(\Gamma,\delta)}^{S2*} \tag{6}$$

It should be noted that alternatively, if Stage 2 is deemed more critical and assumes the role of the leader while Stage 1 functions as the follower under uncertain conditions, the evaluation process adjusts accordingly. In this scenario, the efficiencies of Stage 2, Stage 1, and the overall two-stage system are analyzed using a comparable methodology, ensuring methodological consistency across different setups. This adjustment allows the model to prioritize the performance metrics of Stage 2, emphasizing its leadership role in the system and its influence on overall efficiency. By applying a similar analytical framework, the model maintains the integrity of the evaluation process, even when the roles of the stages are reversed. This flexibility in the methodological approach accommodates different strategic priorities and operational realities, making it particularly useful in dynamic and uncertain environments where the importance of system components may shift over time.

4 Experimental Results

In this section, the implementation of the robust two-stage network DEA approach applied for assessing the performance of investment companies (ICs) is detailed through a real-world case study in the Iranian financial market. The methodology recognizes the operational and financial complexity of investment companies by modeling their activities as a two-stage process (Premachandra et al., 2012). In the first stage, managers focus on attracting funds from investors, an essential step for initiating operations. In the second stage, their primary objective shifts to constructing an optimal portfolio to maximize returns for stakeholders.

The inputs for the first stage include marketing and distribution costs, along with management costs. These inputs are critical in determining the efficiency with which an IC acquires and utilizes funds. The net asset value (NAV), serving as the intermediate variable, bridges the two stages. Specifically, NAV acts as the output variable for the first stage and simultaneously becomes the input variable for the second stage. This dual role highlights the importance of NAV in evaluating the operational efficiency of an IC. A company that achieves the highest NAV with minimal expenditures is deemed more efficient in operational management.

The second stage incorporates additional inputs such as fund size, the standard deviation of average return (as a proxy for risk), turnover ratio, and net expense ratio. The output in this stage is the average return, representing the financial performance of the portfolio. An IC that produces the highest average return while minimizing inputs such as NAV, fund size, risk, turnover ratio, and net expense ratio demonstrates superior portfolio management efficiency. This dual-stage assessment offers a comprehensive framework to identify operational and portfolio inefficiencies, enabling better decision-making and resource optimization for investment companies operating in dynamic financial markets.

It is worth mentioning that for implementing the robust two-stage network DEA approach, data from 15 investment companies active in the Iranian capital market were collected. The data collection process ensured that key variables relevant to the operational and financial efficiency of these companies were accurately captured. Subsequently, to execute the robust leader-follower network DEA, a 10% level of perturbation in the data was introduced to account for uncertainty, and robust budget levels of 0%, 25%, 50%,

75%, and 100% were considered. This variation in robust budget levels allowed for a comprehensive analysis of the companies' performance under differing levels of data stability and uncertainty.

In the robust leader-follower network DEA approach, greater emphasis was placed on the first stage due to its critical influence on overall efficiency. The evaluation results, presented under constant and variable returns to scale in Tables 1 and 2, provide valuable insights into the operational and financial efficiencies of investment companies across varying scenarios. A significant observation is that as the robustness budget for uncertain parameters increases from 0% to 100%, efficiency values consistently decline, reflecting the challenges posed by higher uncertainty. Among the companies assessed, IC-14 and IC-15 excel in operational management and portfolio management, respectively, with IC-14 emerging as the overall top performer. This underscores IC-14's ability to maintain superior efficiency and adaptability under uncertain conditions, distinguishing it as a model of resilience and robust performance within the evaluated group.

Table 1. Results of the Robust Two-Stage Network DEA Model under CRS Assumption

Investment Companies		Budget of Robustness				
		0%	25%	50%	75%	100%
Stage 1 (Operational Management)	IC-01	0.210623	0.184821	0.163390	0.149330	0.140996
	IC-02	0.453083	0.389215	0.335499	0.311796	0.303303
	IC-03	0.049336	0.042379	0.036527	0.033948	0.033027
	IC-04	1.000000	0.884855	0.801948	0.733351	0.669422
	IC-05	0.637291	0.549496	0.476055	0.441383	0.426616
	IC-06	0.150565	0.129326	0.111459	0.103593	0.100791
	IC-07	0.115921	0.099569	0.085814	0.079757	0.077600
	IC-08	1.000000	0.884855	0.801948	0.733351	0.669422
	IC-09	0.055156	0.047381	0.040842	0.037956	0.036923
	IC-10	1.000000	0.874034	0.771621	0.709128	0.669422
	IC-11	0.070572	0.060621	0.052251	0.048561	0.047243
	IC-12	0.011075	0.009521	0.008214	0.007631	0.007414
	IC-13	0.026309	0.022603	0.019487	0.018109	0.017612
	IC-14	0.469235	0.403984	0.349276	0.324147	0.314116
	IC-15	0.628989	0.540303	0.465708	0.432818	0.421059
Stage 2 (Portfolio Management)	IC-01	1.000000	0.841230	0.776398	0.699113	0.669422
	IC-02	1.000000	0.817637	0.744593	0.706777	0.669422

(*continued*)

Table 1. (*continued*)

Investment Companies		Budget of Robustness				
		0%	25%	50%	75%	100%
	IC-03	0.721419	0.560983	0.511461	0.496348	0.482934
	IC-04	0.296777	0.227192	0.209706	0.204188	0.198669
	IC-05	0.538043	0.423652	0.380187	0.370182	0.360177
	IC-06	0.843399	0.649308	0.595956	0.580273	0.564589
	IC-07	0.055671	0.042540	0.039375	0.038302	0.037267
	IC-08	0.954720	0.751529	0.674617	0.656863	0.639110
	IC-09	0.267660	0.205312	0.189504	0.184157	0.179178
	IC-10	1.000000	0.856098	0.772193	0.727484	0.669422
	IC-11	0.352753	0.285838	0.254370	0.242700	0.227249
	IC-12	1.000000	0.827336	0.768165	0.698353	0.669422
	IC-13	1.000000	0.856098	0.772193	0.701558	0.669422
	IC-14	0.737157	0.576092	0.520884	0.507176	0.493469
	IC-15	1.000000	0.818162	0.719829	0.693663	0.669422
Overall (Investment Companies)	IC-01	0.210623	0.155477	0.126856	0.104398	0.094386
	IC-02	0.453083	0.318237	0.249810	0.220370	0.203038
	IC-03	0.035592	0.023774	0.018682	0.016850	0.015950
	IC-04	0.296777	0.201032	0.168174	0.149741	0.132993
	IC-05	0.342890	0.232795	0.180990	0.163392	0.153657
	IC-06	0.126986	0.083972	0.066425	0.060112	0.056906
	IC-07	0.006453	0.004236	0.003379	0.003055	0.002892
	IC-08	0.954720	0.664994	0.541007	0.481711	0.427834
	IC-09	0.014763	0.009728	0.007740	0.006990	0.006616
	IC-10	1.000000	0.748258	0.595840	0.515880	0.448125
	IC-11	0.024895	0.017328	0.013291	0.011786	0.010736
	IC-12	0.011075	0.007877	0.006310	0.005329	0.004963
	IC-13	0.026309	0.019351	0.015048	0.012704	0.011790
	IC-14	0.345900	0.232732	0.181932	0.164399	0.155007
	IC-15	0.628989	0.442055	0.335230	0.300230	0.281866

Table 2. Results of the Robust Two-Stage Network DEA Model under VRS Assumption

Budget of Robustness						
Investment Companies		0%	25%	50%	75%	100%
Stage 1 (Operational Management)	IC-01	0.381604	0.335936	0.311376	0.310824	0.310824
	IC-02	0.542630	0.478117	0.444063	0.444000	0.443970
	IC-03	0.126767	0.111687	0.103730	0.103723	0.103719
	IC-04	1.000000	0.939024	0.880952	0.837209	0.818182
	IC-05	0.758633	0.624754	0.527041	0.477901	0.453408
	IC-06	0.381137	0.335767	0.311843	0.311840	0.311839
	IC-07	0.287730	0.244583	0.210375	0.192722	0.184055
	IC-08	1.000000	0.917975	0.844601	0.800368	0.797823
	IC-09	0.144163	0.127023	0.117975	0.117959	0.117951
	IC-10	1.000000	0.904007	0.843247	0.826227	0.818182
	IC-11	0.162638	0.143291	0.133084	0.133073	0.133067
	IC-12	0.027879	0.024588	0.022841	0.022820	0.022810
	IC-13	0.060678	0.053473	0.049666	0.049653	0.049646
	IC-14	0.619796	0.517648	0.439341	0.400829	0.383548
	IC-15	1.000000	0.863245	0.766091	0.696683	0.651784
Stage 2 (Portfolio Management)	IC-01	1.000000	0.895722	0.821572	0.786541	0.778269
	IC-02	1.000000	0.933333	0.870588	0.846512	0.818182
	IC-03	1.000000	0.933333	0.895238	0.822857	0.818182
	IC-04	1.000000	0.933333	0.870588	0.846512	0.818182
	IC-05	1.000000	0.933141	0.886128	0.843653	0.818182
	IC-06	1.000000	0.946342	0.895238	0.846512	0.818182
	IC-07	0.414101	0.371767	0.343411	0.338810	0.338810
	IC-08	0.974822	0.783936	0.718364	0.689523	0.674086
	IC-09	1.000000	0.904526	0.872640	0.838959	0.818182
	IC-10	1.000000	0.933333	0.870588	0.846512	0.818182
	IC-11	0.901818	0.805379	0.735959	0.731726	0.731726
	IC-12	1.000000	0.943752	0.869455	0.822813	0.818182

(*continued*)

Table 2. (*continued*)

Budget of Robustness						
Investment Companies		0%	25%	50%	75%	100%
	IC-13	1.000000	0.933333	0.895238	0.822857	0.818182
	IC-14	0.909187	0.786906	0.734685	0.727631	0.727631
	IC-15	1.000000	0.902221	0.823755	0.784045	0.776011
Overall (Investment Company)	IC-01	0.381604	0.300905	0.255818	0.244476	0.241905
	IC-02	0.542630	0.446242	0.386596	0.375851	0.363248
	IC-03	0.126767	0.104241	0.092863	0.085349	0.084861
	IC-04	1.000000	0.876423	0.766947	0.708707	0.669421
	IC-05	0.758633	0.582983	0.467026	0.403183	0.370970
	IC-06	0.381137	0.317750	0.279174	0.263976	0.255141
	IC-07	0.119149	0.090928	0.072245	0.065296	0.062360
	IC-08	0.974822	0.719634	0.606730	0.551872	0.537801
	IC-09	0.144163	0.114895	0.102950	0.098963	0.096506
	IC-10	1.000000	0.843740	0.734121	0.699411	0.669421
	IC-11	0.146670	0.115404	0.097944	0.097373	0.097369
	IC-12	0.027879	0.023205	0.019859	0.018777	0.018663
	IC-13	0.060678	0.049909	0.044463	0.040857	0.040619
	IC-14	0.563510	0.407340	0.322777	0.291655	0.279082
	IC-15	1.000000	0.778838	0.631071	0.546231	0.505791

5 Conclusions and Recommendations for Future Research

This research presented a novel approach for evaluating the performance of two-stage network DMUs in uncertain environments, addressing a critical gap in the field of efficiency analysis. By adopting the leader-follower network DEA framework and incorporating robust optimization techniques, the study provided a comprehensive methodology capable of handling data imprecision and variability. The proposed model accommodated various returns-to-scale assumptions, enhancing its flexibility and applicability to real-world scenarios. A key contribution of this work lay in the development of a robust optimization framework that ensured reliable performance evaluation, even when input and output data were uncertain. To illustrate its practical utility, the methodology was applied to a case study of investment companies in the Iranian capital market, demonstrating its effectiveness in delivering actionable insights. The results confirmed that the robust leader-follower network DEA model systematically addressed data uncertainty, offering decision-makers a powerful tool for analyzing and improving performance in dynamic and unpredictable environments.

Overall, this study not only extended the scope of network DEA methodologies but also provided a solid foundation for future research and practical applications in uncertain contexts. The proposed robust NDEA approach can be further applied to performance appraisal of two-stage network DMUs in various fields and real-life case studies, such as banks, electricity distribution companies, hotels, R&D projects, schools, universities, universities, etc. (for more details, see Avkiran, 2009, Li et al., 2012, Matthews, 2013, Huang et al., 2014, Lee & Worthington, 2016, Wanke et al., 2016, Toloo et al., 2017, Sotiros et al., 2018, Barat et al., 2019, Shi et al., 2021, Tan et al., 2021, Michali et al., 2023, Amirteimoori et al., 2024, Gupta & Kumar, 2024, Mergoni et al., 2025). Additionally, the robust NDEA approach can be extended by incorporating the Malmquist productivity index, dynamic DEA, and window DEA methodologies to facilitate dynamic performance assessment of network DMUs over a time horizon under uncertain panel data conditions (for more details, see Wanke et al., 2019, Peykani & Mohammadi, 2020, Del Barrio-Tellado et al., 2021, Peykani et al., 2021b, Lahouel et al., 2022, Fukuyama et al., 2023, Cancella et al., 2024, Ren et al., 2024, Shojaie et al., 2024a, 2024b).

References

Alves, C.G.M.D.F., Meza, L.A.: A review of network DEA models based on slacks-based measure: evolution of literature, applications, and further research direction. Int. Trans. Oper. Res. **30**(6), 2729–2760 (2023)

Amirteimoori, A., Matin, R.K., Yadollahi, A.H.: Stochastic resource reallocation in two-stage production processes with undesirable outputs: an empirical study on the power industry. Socioecon. Plann. Sci. **93**, 101894 (2024)

Avkiran, N.K.: Opening the black box of efficiency analysis: an illustration with UAE banks. Omega **37**(4), 930–941 (2009)

Banker, R.D., Charnes, A., Cooper, W.W.: Some models for estimating technical and scale inefficiencies in data envelopment analysis. Manage. Sci. **30**(9), 1078–1092 (1984)

Barat, M., Tohidi, G., Sanei, M., Razavyan, S.: Data envelopment analysis for decision making unit with nonhomogeneous internal structures: an application to the banking industry. J. Opera. Res. Soc. **70**(5), 760–769 (2019)

Bertsimas, D., Sim, M.: The price of robustness. Oper. Res. **52**(1), 35–53 (2004)

Bertsimas, D., Brown, D.B., Caramanis, C.: Theory and applications of robust optimization. SIAM Rev. **53**(3), 464–501 (2011)

Cancella, R., Nogueira, E., Silva, M., Pereira, A.: An analysis of the malmquist-DEA productivity index applied to the segment of electric power transmission in Brazil. In: Advances in the Theory and Applications of Performance Measurement and Management: Proceedings of DEA45-International Conference on Data Envelopment Analysis, pp. 233–246. Springer, Cham (2024)

Charnes, A., Cooper, W.W., Rhodes, E.: Measuring the efficiency of decision making units. Eur. J. Oper. Res. **2**(6), 429–444 (1978)

Chen, Y., Zhu, J.: Measuring information technology's indirect impact on firm performance. Inf. Technol. Manage. **5**(1–2), 9–22 (2004)

Chen, Y., Cook, W.D., Li, N., Zhu, J.: Additive efficiency decomposition in two-stage DEA. Eur. J. Oper. Res. **196**(3), 1170–1176 (2009)

Chu, J., Wu, J., Chu, C., Zhang, T.: DEA-based fixed cost allocation in two-stage systems: leader-follower and satisfaction degree bargaining game approaches. Omega **94**, 102054 (2020)

Cook, W.D., Liang, L., Zhu, J.: Measuring performance of two-stage network structures by DEA: a review and future perspective. Omega **38**(6), 423–430 (2010a)
Cook, W.D., Zhu, J., Bi, G., Yang, F.: Network DEA: Additive efficiency decomposition. Eur. J. Oper. Res. **207**(2), 1122–1129 (2010b)
Del Barrio-Tellado, M.J., Gómez-Vega, M., Gómez-Zapata, J.D., Herrero-Prieto, L.C.: Urban public libraries: performance analysis using dynamic-network-DEA. Socioecon. Plann. Sci. **74**, 100928 (2021)
Despotis, D.K., Sotiros, D., Koronakos, G.: A network DEA approach for series multi-stage processes. Omega **61**, 35–48 (2016)
Ding, T., Zhang, Y., Zhang, D., Li, F.: Performance evaluation of chinese research universities: a parallel interactive network DEA approach with shared and fixed sum inputs. Socioecon. Plann. Sci. **87**, 101582 (2023)
Emrouznejad, A., Tavana, M.: Performance Measurement with Fuzzy Data Envelopment Analysis. Springer, Berlin, Heidelberg (2014)
Fukuyama, H., Tsionas, M., Tan, Y.: Dynamic network data envelopment analysis with a sequential structure and behavioural-causal analysis: application to the Chinese banking industry. Eur. J. Oper. Res. **307**(3), 1360–1373 (2023)
Gabrel, V., Murat, C., Thiele, A.: Recent advances in robust optimization: an overview. Eur. J. Oper. Res. **235**(3), 471–483 (2014)
Guo, C., Zhu, J.: Non-cooperative two-stage network DEA model: linear vs. parametric linear. European J. Opera. Res. **258**(1), 398–400 (2017)
Gupta, M., Kumar, S.: Does income diversification affect bank efficiency, effectiveness and performance? evidence from Indian banking industry. In: Advances in the Theory and Applications of Performance Measurement and Management: Proceedings of DEA45-International Conference on Data Envelopment Analysis, pp. 165–185. Springer, Cham (2024)
Halkos, G.E., Tzeremes, N.G., Kourtzidis, S.A.: A unified classification of two-stage DEA models. Surv. Opera. Res. Manage. Sci. **19**(1), 1–16 (2014)
Huang, C.W., Ho, F.N., Chiu, Y.H.: Measurement of tourist hotels' productive efficiency, occupancy, and catering service effectiveness using a modified two-stage DEA model in Taiwan. Omega **48**, 49–59 (2014)
Kao, C.: Network data envelopment analysis: a review. Eur. J. Oper. Res. **239**(1), 1–16 (2014)
Kao, C.: Efficiency measurement for hierarchical network systems. Omega **51**, 121–127 (2015)
Kao, C.: Network Data Envelopment Analysis: Foundations and Extensions. International Series in Operations Research & Management Science. Springer (2017)
Kao, C., Hwang, S.N.: Efficiency decomposition in two-stage data envelopment analysis: an application to non-life insurance companies in Taiwan. Eur. J. Oper. Res. **185**(1), 418–429 (2008)
Koronakos, G.: A taxonomy and review of the network data envelopment analysis literature. Mach. Learn. Paradigms, 255–311 (2019a)
Koronakos, G., Sotiros, D., Despotis, D.K.: Reformulation of network data envelopment analysis models using a common modelling framework. Eur. J. Oper. Res. **278**(2), 472–480 (2019b)
Lahouel, B.B., Taleb, L., Kossai, M.: Nonlinearities between bank stability and income diversification: a dynamic network data envelopment analysis approach. Expert Syst. Appl. **207**, 117776 (2022)
Lee, B.L., Worthington, A.C.: A network DEA quantity and quality-orientated production model: an application to Australian university research services. Omega **60**, 26–33 (2016)
Li, Y., Chen, Y., Liang, L., Xie, J.: DEA models for extended two-stage network structures. Omega **40**(5), 611–618 (2012)
Li, Y., Chen, Y., Liang, L., Xie, J.: DEA models for extended two-stage network structures. Data Envelopment Analysis: A Handbook on the Modeling of Internal Structures and Networks, 261–284 (2014)

Lin, R., Liu, Q.: Directional distance based efficiency decomposition for series system in network data envelopment analysis. J. Opera. Res. Soc. **73**(8), 1873–1888 (2022)

Matthews, K.: Risk management and managerial efficiency in Chinese banks: a network DEA framework. Omega **41**(2), 207–215 (2013)

Mergoni, A., Emrouznejad, A., Witte, K.D.: Fifty years of data envelopment analysis. European J. Opera. Res. **326**(3), 389–412 (2025). https://doi.org/10.1016/j.ejor.2024.12.049

Michali, M., Emrouznejad, A., Dehnokhalaji, A., Clegg, B.: Subsampling bootstrap in network DEA. Eur. J. Oper. Res. **305**(2), 766–780 (2023)

Olesen, O.B., Petersen, N.C.: Stochastic data envelopment analysis—a review. Eur. J. Oper. Res. **251**(1), 2–21 (2016)

Peykani, P., Mohammadi, E., Emrouznejad, A., Pishvaee, M.S., Rostamy-Malkhalifeh, M.: Fuzzy data envelopment analysis: an adjustable approach. Expert Syst. Appl. **136**, 439–452 (2019)

Peykani, P., Mohammadi, E.: Window network data envelopment analysis: an application to investment companies. Int. J. Industr. Math. **12**(1), 89–99 (2020)

Peykani, P., Mohammadi, E., Farzipoor Saen, R., Sadjadi, S.J., Rostamy-Malkhalifeh, M.: Data envelopment analysis and robust optimization: a review. Expert. Syst. **37**(4), e12534 (2020a)

Peykani, P., Mohammadi, E., Jabbarzadeh, A., Rostamy-Malkhalifeh, M., Pishvaee, M.S.: A novel two-phase robust portfolio selection and optimization approach under uncertainty: a case study of Tehran stock exchange. PLoS ONE **15**(10), e0239810 (2020b)

Peykani, P., Mohammadi, E., Emrouznejad, A.: An adjustable fuzzy chance-constrained network DEA approach with application to ranking investment firms. Expert Syst. Appl. **166**, 113938 (2021a)

Peykani, P., Farzipoor Saen, R., Seyed Esmaeili, F.S., Gheidar-Kheljani, J.: Window data envelopment analysis approach: a review and bibliometric analysis. Expert. Syst. **38**(7), e12721 (2021b)

Peykani, P., Hosseinzadeh Lotfi, F., Sadjadi, S.J., Ebrahimnejad, A., Mohammadi, E.: Fuzzy chance-constrained data envelopment analysis: a structured literature review, current trends, and future directions. Fuzzy Optim. Decis. Making **21**(2), 197–261 (2022a)

Peykani, P., Gheidar-Kheljani, J., Farzipoor Saen, R., Mohammadi, E.: Generalized robust window data envelopment analysis approach for dynamic performance measurement under uncertain panel data. Oper. Res. Int. Journal **22**(5), 5529–5567 (2022b)

Peykani, P., Gheidar-Kheljani, J.: Modeling scenario-based robust data envelopment analysis approaches: an application to research and development projects. J. Oper. Res. Appl. **19**(2), 23–37 (2022)

Peykani, P., Seyed Esmaeili, F.S., Pishvaee, M.S., Rostamy-Malkhalifeh, M., Hosseinzadeh Lotfi, F.: Matrix-based network data envelopment analysis: a common set of weights approach. Socioecon. Plann. Sci. **95**, 102044 (2024a)

Peykani, P., Emrouznejad, A., Mohammadi, E., Gheidar-Kheljani, J.: A novel robust network data envelopment analysis approach for performance assessment of mutual funds under uncertainty. Ann. Oper. Res. **339**(3), 1149–1175 (2024b)

Peykani, P., Sargolzaei, M., Hamidzadeh, F., Seyed Esmaeili, F.S., Takaloo, A.: Possibilistic network DEA approach for performance evaluation of two-stage decision making units under uncertainty. In: Analytical Decision Making and Data Envelopment Analysis: Advances and Challenges, pp. 59–79. Singapore: Springer Nature Singapore (2024c)

Peykani, P., Hamidzadeh, F., Pishvaee, M.S., Memar-Masjed, E., Jabbarzadeh, A.: A robust optimization approach for estimating the most productive scale size in uncertain data envelopment analysis. In: Analytical Decision Making and Data Envelopment Analysis: Advances and Challenges, pp. 165–182. Singapore: Springer Nature Singapore (2024d)

Peykani, P., Pishvaee, M.S.: Performance evaluation of hospitals under data uncertainty: an uncertain common-weights data envelopment analysis. Healthcare **12**(6), 611 (2024)

Premachandra, I.M., Zhu, J., Watson, J., Galagedera, D.U.A.: Best-performing US mutual fund families from 1993 to 2008: evidence from a novel two-stage DEA model for efficiency decomposition. J. Bank. Finance **36**(12), 3302–3317 (2012)

Ratner, S.V., Shaposhnikov, A.M., Lychev, A.V.: Network DEA and its applications (2017–2022): a systematic literature review. Mathematics **11**(9), 2141 (2023)

Ren, T., Wang, N., Xiao, H., Zhou, Z.: Efficiency of funding to rural revitalization and regional heterogeneity of technologies in China: dynamic network nonconvex metafrontiers. Socioecon. Plann. Sci. **92**, 101825 (2024)

Seyed Esmaeili, F.S., Mohammadi, E.: Z-number network data envelopment analysis approach: a case study on the Iranian insurance industry. PLoS ONE **19**(7), e0306876 (2024)

Shi, X., Emrouznejad, A., Yu, W.: Overall efficiency of operational process with undesirable outputs containing both series and parallel processes: a SBM network DEA model. Expert Syst. Appl. **178**, 115062 (2021)

Shojaie, S.E., Sadjadi, S.J., Tavakkoli-Moghaddam, R.: Malmquist productivity index for two-stage network systems under data uncertainty: a real-world case study. PLoS ONE **19**(7), e0307277 (2024a)

Shojaie, S.E., Sadjadi, S.J., Tavakkoli-Moghaddam, R.: Malmquist productivity index under network structure and negative data: an application to banking industry. J. Syst. Manage. **10**(3), 107–118 (2024b)

Sotiros, D., Koronakos, G., Despotis, D.K.: Dominance at the divisional efficiencies level in network DEA: the case of two-stage processes. In: Data Envelopment Analysis and Performance Measurement: Recent Developments: Proceedings of the DEA40-International Conference of Data Envelopment Analysis, 63–72. Aston Business School, UK (2018)

Tan, Y., Wanke, P., Antunes, J., Emrouznejad, A.: Unveiling endogeneity between competition and efficiency in Chinese banks: a two-stage network DEA and regression analysis. Ann. Oper. Res. **306**(1), 131–171 (2021)

Toloo, M., Emrouznejad, A., Moreno, P.: A linear relational DEA model to evaluate two-stage processes with shared inputs. Comput. Appl. Math. **36**, 45–61 (2017)

Tone, K., Tsutsui, M.: Network DEA: A slacks-based measure approach. Eur. J. Oper. Res. **197**(1), 243–252 (2009)

Wanke, P., Azad, M.A.K., Emrouznejad, A., Antunes, J.: A dynamic network DEA model for accounting and financial indicators: a case of efficiency in MENA banking. Int. Rev. Econ. Financ. **61**, 52–68 (2019)

Wanke, P., Blackburn, V., Barros, C.P.: Cost and learning efficiency drivers in Australian schools: a two-stage network DEA approach. Appl. Econ. **48**(38), 3577–3604 (2016)

Yu, G., Wei, Q., Brockett, P.: A generalized data envelopment analysis model: a unification and extension of existing methods for efficiency analysis of decision making units. Ann. Oper. Res. **66**(1), 47–89 (1996a)

Yu, G., Wei, Q., Brockett, P., Zhou, L.: Construction of all DEA efficient surfaces of the production possibility set under the generalized data envelopment analysis model. Eur. J. Oper. Res. **95**(3), 491–510 (1996b)

Zhou, X., Luo, R., Yao, L., Cao, S., Wang, S., Lev, B.: Assessing integrated water use and wastewater treatment systems in China: a mixed network structure two-stage SBM DEA model. J. Clean. Prod. **185**, 533–546 (2018)

Zhou, Z., Sun, W., Xiao, H., Jin, Q., Liu, W.: Stochastic leader–follower DEA models for two-stage systems. J. Manage. Sci. Eng. **6**(4), 413–434 (2021)

Efficiency Analysis of Intellectual Capital Under Deep Uncertainty: A Robust DEA Approach

Pejman Peykani[1(✉)], Mojtaba Nouri[2], Seyed Ehsan Shojaie[3], Fatemeh Ghiyaasi[4], and Amir Esmaeli[1]

[1] Department of Industrial Engineering, Faculty of Engineering, Khatam University, Tehran, Iran
p.peykani@khatam.ac.ir

[2] School of Industrial Engineering, Iran University of Science and Technology, Tehran, Iran
mojtaba_nouri@ind.iust.ac.ir

[3] School of Industrial Engineering, Iran University of Science and Technology, Tehran, Iran
se_shojaie@mail.iust.ac.ir

[4] Department of Computer Sciences, Science and Research Branch, Islamic Azad University, Tehran, Iran

Abstract. This research introduces a robust data envelopment analysis (RDEA) methodology to evaluate intellectual capital efficiency in environments of deep uncertainty. Intellectual capital, a vital intangible asset for organizational success, is challenging to measure due to its unpredictable and variable nature. The proposed approach utilizes conservative robust optimization with a box uncertainty set to address these complexities effectively, ensuring more reliable and stable efficiency scores across diverse scenarios. The methodology accommodates variations in input and output data while exploring the interrelationships among key dimensions of intellectual capital holistically and with nuanced insights. Empirical validation using a sample of organizations demonstrates the practical applicability of the RDEA framework, highlighting its ability to handle uncertainties and account for varying returns to scale. This study contributes to the field by offering a versatile and robust tool for assessing intellectual capital efficiency, supporting more informed and resilient decision-making processes in uncertain contexts. The findings provide actionable insights and a solid foundation for advancing intellectual capital management practices.

Keywords: Intellectual Capital · Data Envelopment Analysis · Deep Uncertainty · Efficiency Assessment · Conservative Robust Optimization · Box Uncertainty Set

1 Introduction

In today's knowledge-driven economy, intellectual capital (IC) has emerged as a critical intangible asset that underpins organizational success and sustainability (Brennan & Connell, 2000, Petty & Guthrie, 2000, Marr et al., 2004). The transition from traditional

A. Emrouznejad et al. (Eds.): DEA 2024, LNOR, pp. 27–41, 2025.
https://doi.org/10.1007/978-3-031-98177-7_3

resource-based economies to knowledge-based economies has underscored the strategic importance of IC, positioning it as a critical intangible asset that drives value creation and competitive advantage. Defined as a blend of knowledge, expertise, relationships, and innovative capabilities, IC equips organizations with the tools to excel in dynamic market environments (Bontis, 2001, Inkinen, 2015, Paoloni et al., 2023). However, despite its fundamental role, the abstract and multifaceted nature of IC presents significant challenges in measurement, evaluation, and management.

IC is conventionally classified into three interrelated dimensions: human capital, structural capital, and relational capital (Chen et al., 2004, Mouritsen, 2009). Human capital encompasses the skills, expertise, and competencies of employees that directly influence organizational productivity. Structural capital refers to the infrastructure, such as systems, processes, and organizational culture, that supports and enhances employee performance. Relational capital pertains to the networks and relationships an organization maintains with external stakeholders, including customers, partners, and suppliers. The interactions among these dimensions create synergies that amplify their collective impact on organizational performance (Asiaei & Jusoh, 2015, Inkinen et al., 2017). However, the intricate and interconnected nature of these components necessitates robust methodologies capable of capturing their dynamics to evaluate their contributions effectively.

The evaluation of IC efficiency plays a pivotal role in uncovering organizational strengths and weaknesses, providing actionable insights that guide resource allocation, strategic planning, and sustainable competitiveness. It also contributes to a deeper academic understanding of the role of intangible assets in driving organizational performance. Among the available methodologies, data envelopment analysis (DEA) has emerged as a powerful and widely utilized tool for assessing IC efficiency (Leitner et al., 2005, Campisi & Costa., 2008, Wang et al., 2014). As a non-parametric technique, DEA evaluates the relative efficiency of decision-making units (DMUs) by analyzing their input-output relationships. Its adaptability to diverse performance metrics makes DEA particularly suitable for assessing IC's complex and multidimensional nature (Lu et al., 2010, Daraio et al., 2023).

Despite its utility, traditional DEA models have notable limitations. These models assume data precision and determinism, conditions that are rarely met in practice (Peykani et al., 2020a, 2020b, 2022a, 2024a). IC data are inherently uncertain due to factors such as measurement errors, environmental volatility, rapid technological changes, and qualitative aspects that resist quantification. These challenges are magnified in scenarios of deep uncertainty, where probability distributions are unknown, and variables may lack clear definitions. Under such conditions, conventional DEA approaches often fail to produce reliable or meaningful efficiency scores, limiting their applicability in real-world contexts (Despotis & Smirlis, 2002, Peykani & Pishvaee, 2024). Addressing these limitations requires advanced methodologies that integrate mechanisms to manage uncertainty effectively. Robust optimization techniques have gained prominence as a promising solution to this challenge (Ben-Tal et al., 2009, Bertsimas et al., 2011).

By explicitly accounting for uncertainty in input and output data, robust optimization approaches provide a foundation for more reliable and stable evaluations, even under highly variable conditions. Among the various frameworks available, the box uncertainty set stands out due to its simplicity, practicality, and adaptability to diverse scenarios (Soyster, 1973, Bertsimas & Sim, 2004, Gabrel et al., 2014). This framework models data variability by establishing clear bounds within which fluctuations are allowed, thereby capturing the inherent uncertainty in a structured manner. Unlike probabilistic methods, it eliminates the need for precise probability distributions, which are often unavailable or infeasible to determine in real-world settings (Li et al., 2011; 2012, Li & Floudas, 2014). By doing so, the box uncertainty set ensures that efficiency evaluations remain robust across a range of plausible scenarios, reducing sensitivity to data inaccuracies and variability. This makes it particularly well-suited for applications where data precision is unattainable but reliable decision-making is essential. Its practicality and effectiveness in handling uncertainty have made it a cornerstone in the development of advanced methodologies for efficiency assessment, including the robust extension of DEA (Sadjadi & Omrani, 2008, Toloo et al., 2022).

Accordingly, this study introduces a robust data envelopment analysis (RDEA) framework to evaluate the efficiency of intellectual capital under conditions of deep uncertainty. The proposed methodology incorporates robust optimization techniques into the DEA framework, specifically utilizing the Box Uncertainty Set to address the limitations of traditional DEA models. This approach ensures that efficiency assessments remain both reliable and meaningful, even in the presence of significant data variability and ambiguity. Furthermore, the RDEA methodology examines the interrelationships among the three dimensions of IC, providing a holistic perspective that highlights opportunities for synergy and identifies areas for improvement. Empirical validation of the proposed methodology, conducted using data from organizational case studies, highlights its effectiveness in capturing the complexities and uncertainties associated with IC evaluation. The findings underscore the practical applicability of the RDEA model, providing valuable insights for managers, policymakers, and researchers. These insights not only enhance understanding of IC dynamics but also support informed decision-making, strategic resource allocation, and long-term competitiveness.

By integrating robust optimization into the DEA framework, this study bridges the gap between conventional efficiency evaluation methods and the realities of uncertain environments. It offers both theoretical advancements and practical tools to address the complex challenges of IC assessment, highlighting the interdependencies among IC dimensions and providing a comprehensive framework for evaluation. The findings not only enrich academic discourse but also deliver actionable insights for improving IC efficiency, fostering organizational resilience, and driving success in an increasingly unpredictable global landscape. The proposed approach empowers organizations to manage uncertainty and seize opportunities in the modern knowledge economy.

The remainder of this study is organized as follows: Sect. 2 reviews the theoretical background, including the conceptual framework for evaluating intellectual capital efficiency. Section 3 describes the proposed conservative robust generalized DEA approach. Section 4 presents empirical results demonstrating the methodology's effectiveness and applicability under uncertain conditions. Section 5 concludes with key insights and recommendations for future research.

2 Generalized DEA Model

Data envelopment analysis is a non-parametric methodology designed to evaluate the relative efficiency of DMUs that convert multiple inputs into multiple outputs (Charnes et al., 1997, Cooper et al., 2011, Mergoni et al., 2025). By avoiding the need for predefined functional forms, DEA offers the flexibility to model complex and diverse systems, making it a widely utilized approach across disciplines such as operations research, management, and economics (Liu et al., 2013, Emrouznejad & Yang, 2018, Peykani et al., 2021a, 2022b). Its ability to handle multidimensional data has established DEA as a key tool for efficiency evaluation in areas where traditional parametric models are impractical. One of DEA's primary strengths lies in its capacity to construct a best-practice frontier against which the efficiency of individual DMUs is assessed.

Efficiency is expressed as the ratio of a weighted sum of outputs to a weighted sum of inputs, calculated for each DMU (Banker et al., 1989, Cook et al., 2014). A DMU is deemed efficient if it lies on the frontier, attaining an efficiency score of one, indicating optimal resource utilization. This characteristic makes DEA particularly valuable for assessing performance in scenarios with intricate input-output relationships, such as the evaluation of intellectual capital, where traditional evaluation methods often fall short. Its adaptability has led to extensions accommodating various returns to scale and contextual factors, increasing its real-world applicability (Lee & Wong, 2019).

To understand the structure of data envelopment analysis modeling, consider a scenario involving K homogeneous decision-making units $\mathrm{DMU}_k(\mathrm{k} = 1, \ldots, \mathrm{K})$, where each DMU utilizes I inputs $\mu_{ik}(\mathrm{i} = 1, \ldots, \mathrm{I})$ to generate J outputs $\lambda_{jk}(\mathrm{j} = 1, \ldots, \mathrm{J})$. The specific DMU being evaluated is denoted by the subscript p. Non-negative weights, $\Theta_i(\mathrm{i} = 1, \ldots, \mathrm{I})$ and $\Phi_j(\mathrm{j} = 1, \ldots, \mathrm{J})$, are assigned to the inputs and outputs, respectively, to reflect their relative importance in the efficiency evaluation. Additionally, the variable Ω_p, which is unrestricted in sign, facilitates adjustments to account for variations in scale. Using the introduced notation, the first DEA model, developed by Charnes et al. (1978) under the assumption of constant returns to scale (CRS), is formulated as Model (1). This pioneering model evaluates the relative efficiency of DMUs by assuming that the production scale does not impact the proportionality of input-output relationships. The CRS assumption is particularly applicable when all DMUs are considered to be operating at an optimal scale, making this model a foundational approach in DEA literature.

$$\mathrm{Max} \sum_{j=1}^{J} \lambda_{jp} \Phi_j \tag{1-1}$$

$$\mathrm{S.t.} \sum_{\mathrm{i}=1}^{\mathrm{I}} \mu_{\mathrm{ip}} \Theta_{\mathrm{i}} = 1 \tag{1-2}$$

$$\sum_{\mathrm{j}=1}^{\mathrm{J}} \lambda_{\mathrm{jk}} \Phi_{\mathrm{j}} - \sum_{\mathrm{i}=1}^{\mathrm{I}} \mu_{\mathrm{ik}} \Theta_{\mathrm{i}} \leq 0, \forall \mathrm{k} \tag{1-3}$$

$$\Theta_{\mathrm{i}}, \Phi_{\mathrm{j}} \geq 0, \forall \mathrm{i}, \mathrm{j} \tag{1-4}$$

Building on the foundational DEA model under CRS assumption, Banker et al. (1984) introduced an extension to account for variable returns to scale (VRS), formulated as Model (2). This model addresses the limitations of the CRS assumption by allowing for variations in efficiency due to scale differences among DMUs. The VRS model is particularly relevant in practical scenarios where DMUs operate under increasing or decreasing returns to scale, reflecting the reality of production environments.

$$\text{Max} \sum_{j=1}^{J} \lambda_{jp} \Phi_j + \Omega_p \tag{2-1}$$

$$\text{S.t.} \sum_{i=1}^{I} \mu_{ip} \Theta_i = 1 \tag{2-2}$$

$$\sum_{j=1}^{J} \lambda_{jk} \Phi_j - \sum_{i=1}^{I} \mu_{ik} \Theta_i + \Omega_p \leq 0, \forall k \tag{2-3}$$

$$\Theta_i, \Phi_j \geq 0, \forall i, j \tag{2-4}$$

After introducing the foundational DEA model, a more generalized and flexible framework is required to accommodate various assumptions regarding returns to scale (RTS) (Yu et al., 1996a, 1996b). These include CRS, VRS, non-increasing returns to scale (NIRS), and non-decreasing returns to scale (NDRS). To address this need, the input-oriented generalized data envelopment analysis (GDEA) model is developed, allowing for adaptability across different RTS scenarios. This comprehensive model is presented as Model (3), offering a unified approach to analyze efficiency under varying scale conditions.

$$\text{Max} \sum_{j=1}^{J} \lambda_{jp} \Phi_j + \xi_1 \Omega_p \tag{3-1}$$

$$\text{S.t.} \sum_{i=1}^{I} \mu_{ip} \Theta_i = 1 \tag{3-2}$$

$$\sum_{j=1}^{J} \lambda_{jk} \Phi_j - \sum_{i=1}^{I} \mu_{ik} \Theta_i + \xi_1 \Omega_p \leq 0, \forall k \tag{3-3}$$

$$\xi_1 \xi_2 (-1)^{\xi_3} \Omega_p \geq 0 \tag{3-4}$$

$$\Theta_i, \Phi_j \geq 0, \forall i, j \tag{3-5}$$

By modifying the binary parameters ξ_1, ξ_2, and ξ_3,, Model (3) can adapt to various returns to scale assumptions: $\xi_1 = 0$ for CRS, $\xi_1 = 1 \,\&\, \xi_2 = 0$ for VRS, $\xi_1 = 1 \,\&\, \xi_2 = 1 \,\&\, \xi_3 = 0$ for NIRS, and $\xi_1 = 1 \,\&\, \xi_2 = 1 \,\&\, \xi_3 = 1$ for NDRS. These parameter modifications provide a versatile and flexible framework, enabling the Generalized DEA model to efficiently accommodate a range of returns to scale scenarios (Peykani et al., 2022c). By adjusting the parameters, the model seamlessly adapts to varying situations, ensuring broad applicability and practicality in evaluating efficiency across diverse production possibility scales.

3 Conservative Robust Generalized DEA Model

In this section, the conservative robust generalized data envelopment analysis (CRGDEA) model is introduced, designed to address efficiency evaluation under uncertain conditions. It should be noted that the inputs $\tilde{\mu}_{ik}(\mathrm{i} = 1, ..., \mathrm{I})$ and outputs $\tilde{\lambda}_{jk}(\mathrm{j} = 1, ..., \mathrm{J})$ are subject to uncertainty and belong to interval uncertainty ranges $\tilde{\mu}_{ik} \in [\mu_{ik} - \hat{\mu}_{ik}, \mu_{\mathrm{ik}} + \hat{\mu}_{\mathrm{ik}}]$ and $\tilde{\lambda}_{jk} \in [\lambda_{jk} - \hat{\lambda}_{jk}, \lambda_{\mathrm{jk}} + \hat{\lambda}_{\mathrm{jk}}]$, respectively. Notably, the midpoint of the uncertainty intervals for inputs and outputs, represented by μ_{ik} and λ_{jk}, serves as the nominal value, while $\hat{\mu}_{ik}$ and $\hat{\lambda}_{jk}$ represent the deviations associated with the uncertain parameters. Now, to account for the uncertainty in inputs and outputs, the following model, which is equivalent to Model (3), is considered.

$$\mathrm{Max} \sum_{j=1}^{J} \tilde{\lambda}_{jp}\Phi_j + \xi_1\Omega_p \tag{4-1}$$

$$\mathrm{S.t.} \sum_{\mathrm{i}=1}^{\mathrm{I}} \tilde{\mu}_{\mathrm{ip}}\Theta_{\mathrm{i}} \leq 1 \tag{4-2}$$

$$\sum_{\mathrm{j}=1}^{\mathrm{J}} \tilde{\lambda}_{\mathrm{jk}}\Phi_{\mathrm{j}} - \sum_{\mathrm{i}=1}^{\mathrm{I}} \tilde{\mu}_{\mathrm{ik}}\Theta_{\mathrm{i}} + \xi_1\Omega_{\mathrm{p}} \leq 0, \forall \mathrm{k} \tag{4-3}$$

$$\xi_1\xi_2(-1)^{\xi_3}\Omega_{\mathrm{p}} \geq 0 \tag{4-4}$$

$$\Theta_{\mathrm{i}}, \Phi_{\mathrm{j}} \geq 0, \forall \mathrm{i}, \mathrm{j} \tag{4-5}$$

Then, to address the uncertainty present in the data in Model (4), a robust optimization approach based on a box uncertainty set is employed. This widely recognized method effectively manages variability and uncertainty in input and output data, ensuring a reliable framework for robust decision-making (Dehnokhalaji et al., 2022). Accordingly, Model (4) is transformed into Model (5) to calculate the efficiency of the DMU under evaluation in uncertain conditions.

$$\mathrm{Max} \sum_{j=1}^{J} \lambda_{jp}\Phi_j + \sum_{j\in\dagger_j} \hat{\lambda}_{jp}\nabla_j + \xi_1\Omega_p \tag{5-1}$$

$$\mathrm{S.t.} \sum_{\mathrm{i}=1}^{\mathrm{I}} \mu_{\mathrm{ip}}\Theta_{\mathrm{i}} + \sum_{\mathrm{i}\in\dagger_{\mathrm{i}}} \hat{\mu}_{\mathrm{ip}}\Delta_{\mathrm{i}} \leq 1 \tag{5-2}$$

$$\sum_{\mathrm{j}=1}^{\mathrm{J}} \lambda_{\mathrm{jk}}\Phi_{\mathrm{j}} - \sum_{\mathrm{i}=1}^{\mathrm{I}} \mu_{\mathrm{ik}}\Theta_{\mathrm{i}} + \sum_{\mathrm{j}\in\dagger_{\mathrm{j}}} \hat{\lambda}_{\mathrm{jp}}\nabla_{\mathrm{j}} + \sum_{\mathrm{i}\in\dagger_{\mathrm{i}}} \hat{\mu}_{\mathrm{ip}}\Delta_{\mathrm{i}} + \xi_1\Omega_{\mathrm{p}} \leq 0, \forall \mathrm{k} \tag{5-3}$$

$$-\Delta_{\mathrm{i}} \leq \Theta_{\mathrm{i}} \leq \Delta_{\mathrm{i}}, \forall \mathrm{i} \in \dagger_{\mathrm{i}} \tag{5-4}$$

$$-\nabla_{\mathrm{j}} \leq \Phi_{\mathrm{j}} \leq \nabla_{\mathrm{j}}, \forall \mathrm{j} \in \dagger_{\mathrm{j}} \tag{5-5}$$

$$\xi_1\xi_2(-1)^{\xi_3}\Omega_p \geq 0 \tag{5-6}$$

$$\Theta_i, \Delta_i, \Phi_j, \nabla_j \geq 0, \forall i, j \tag{5-7}$$

Model (5) represents the conservative robust generalized DEA model, designed to address the worst-case scenario in data uncertainty while accommodating different returns to scale. By incorporating robust optimization techniques, this model ensures stability and reliability in efficiency assessments, even under significant variability in input and output data. The CRGDEA model adapts to various operational scales, making it versatile for evaluating decision-making units in environments characterized by uncertainty. Its ability to consider the worst-case scenarios ensures that the analysis remains conservative and realistic, providing actionable insights for decision-makers in uncertain and dynamic contexts.

4 Case Study and Experimental Results

In this section, the efficiency of intellectual capital for several companies listed on the Tehran Stock Exchange is evaluated using the conservative robust generalized DEA model. This method is specifically designed to address data uncertainty and assess efficiency under varying scenarios of returns to scale. The CRGDEA model integrates robust optimization techniques, enabling it to provide stable and reliable assessments of efficiency even in the presence of variability in input and output data. For the evaluation, the Value-Added Intellectual Coefficient (VAIC) is selected as the input variable, representing the combined intellectual capital efficiency of each company. Complementing this input, the outputs chosen for the analysis are Annual Return (AR), Return on Equity (ROE), and Return on Assets (ROA), which are vital financial performance indicators. These outputs offer a comprehensive perspective on how effectively companies utilize their intellectual capital to achieve financial success.

It is important to note that VAIC is a composite index calculated as the sum of three distinct components, each capturing a critical dimension of intellectual capital (Pulic, 2000, Sveiby & Lloyd, 2010, Ståhle et al., 2011, Yudawisastra et al., 2018). The first component, the Value-Added Capital Employed (VACE), evaluates the effective utilization of physical and financial capital. The second component, the Value-Added Human Capital (VAHC), measures the contribution of human resources to value creation, highlighting their importance in driving innovation and growth. Lastly, the Structural Capital Value Added (SCVA) accounts for the role of organizational processes, systems, and innovations in supporting overall efficiency. This multifaceted approach ensures that all critical aspects of intellectual capital are incorporated into the analysis. By utilizing these inputs and outputs, the CRGDEA model offers a robust and comprehensive framework for evaluating how efficiently companies leverage intellectual capital to enhance financial performance.

The implementation of the CRGDEA model is conducted under two distinct returns to scale conditions, constant returns to scale and variable returns to scale, to evaluate the efficiency of 20 companies listed on the Tehran Stock Exchange. Notably, the CRGDEA model has been applied at various levels of data perturbation, including 0%, 1%, 2%, 5%, and 10%. The results under the CRS and VRS assumptions are presented in Tables 1 and 2, respectively. Additionally, to rank the companies based on intellectual capital, the average results from the five levels of data perturbation have been calculated. These rankings, under the CRS and VRS assumptions, are shown in Figs. 1 and 2, respectively.

Table 1. Results of the Conservative Robust Generalized DEA Model under CRS Assumption

INCAs	Perturbations				
	0%	1%	2%	5%	10%
INCA 01	0.943808	0.906799	0.871235	0.772595	0.631805
INCA 02	1.000000	0.960788	0.923107	0.818594	0.669422
INCA 03	0.755982	0.726338	0.697852	0.618842	0.506070
INCA 04	0.772490	0.742199	0.713091	0.632356	0.517122
INCA 05	0.753892	0.724330	0.695922	0.617131	0.504671
INCA 06	0.688458	0.661462	0.635520	0.563567	0.460868
INCA 07	1.000000	0.960788	0.923107	0.818594	0.669422
INCA 08	0.814272	0.782343	0.751660	0.666559	0.545091
INCA 09	0.846109	0.812932	0.781049	0.692620	0.566404
INCA 10	0.730928	0.702267	0.674724	0.598333	0.489299
INCA 11	0.939494	0.902655	0.867253	0.769064	0.628917
INCA 12	0.631031	0.606287	0.582509	0.516558	0.422426
INCA 13	0.701555	0.674046	0.647610	0.574289	0.469636
INCA 14	0.810938	0.779139	0.748582	0.663829	0.542859
INCA 15	0.695243	0.667982	0.641784	0.569122	0.465411
INCA 16	0.614491	0.590396	0.567241	0.503019	0.411353
INCA 17	1.000000	0.960788	0.923107	0.818594	0.669422
INCA 18	0.773002	0.742691	0.713563	0.632775	0.517464
INCA 19	1.000000	0.960788	0.923107	0.818594	0.669422
INCA 20	1.000000	0.960788	0.923107	0.818594	0.669422

Based on the results, it is observed that, as anticipated, an increase in the percentage of data perturbation results in a decline in the efficiency scores. This demonstrates the sensitivity of efficiency evaluations to varying levels of data uncertainty. Furthermore, the choice of production possibility set conditions, whether CRS or VRS, significantly influences the evaluation process. Under both conditions, companies achieving the highest efficiency scores are identified as benchmarks. These benchmark companies set the standard for best practices and serve as role models for companies exhibiting suboptimal performance in terms of intellectual capital. This highlights the importance of robust efficiency assessments in guiding improvement efforts.

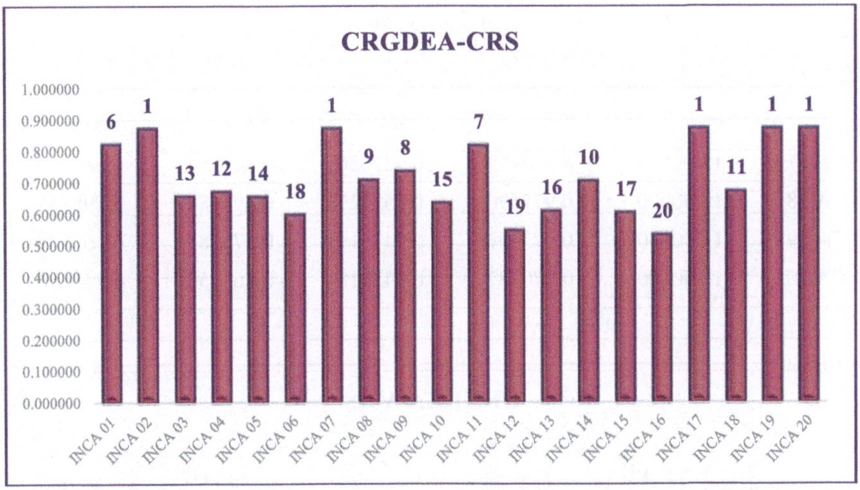

Fig. 1. Ranking Companies Based on the CRGDEA Model under CRS Assumption

Table 2. Results of the Conservative Robust Generalized DEA Model under VRS Assumption

INCA	Perturbations				
	0%	1%	2%	5%	10%
INCA 01	1.000000	0.956511	0.914804	0.799606	0.636640
INCA 02	1.000000	0.973896	0.948551	0.876784	0.769881
INCA 03	0.891489	0.868590	0.846347	0.783299	0.689199
INCA 04	0.926424	0.902877	0.879994	0.815096	0.718108
INCA 05	1.000000	0.940973	0.884642	0.730627	0.537173
INCA 06	0.817701	0.792512	0.768169	0.712665	0.644118
INCA 07	1.000000	0.971032	0.942992	0.864071	0.747933
INCA 08	1.000000	0.980198	0.960784	0.904762	0.818182
INCA 09	0.848545	0.822959	0.798651	0.730318	0.634719
INCA 10	0.787616	0.764933	0.742972	0.684657	0.600373
INCA 11	1.000000	0.952963	0.907916	0.783853	0.644666
INCA 12	0.886813	0.864626	0.843056	0.781818	0.690119
INCA 13	0.722303	0.698832	0.676180	0.612937	0.528713
INCA 14	0.869040	0.844222	0.820189	0.752494	0.656437
INCA 15	1.000000	0.942951	0.888482	0.739409	0.532716
INCA 16	0.906172	0.887040	0.868747	0.816122	0.735274

(continued)

Table 2. (*continued*)

INCA	Perturbations				
	0%	1%	2%	5%	10%
INCA 17	1.000000	0.963129	0.927650	0.828984	0.687359
INCA 18	1.000000	0.977590	0.955722	0.893184	0.798194
INCA 19	1.000000	0.973107	0.947019	0.873281	0.763834
INCA 20	1.000000	0.970737	0.942418	0.862759	0.745668

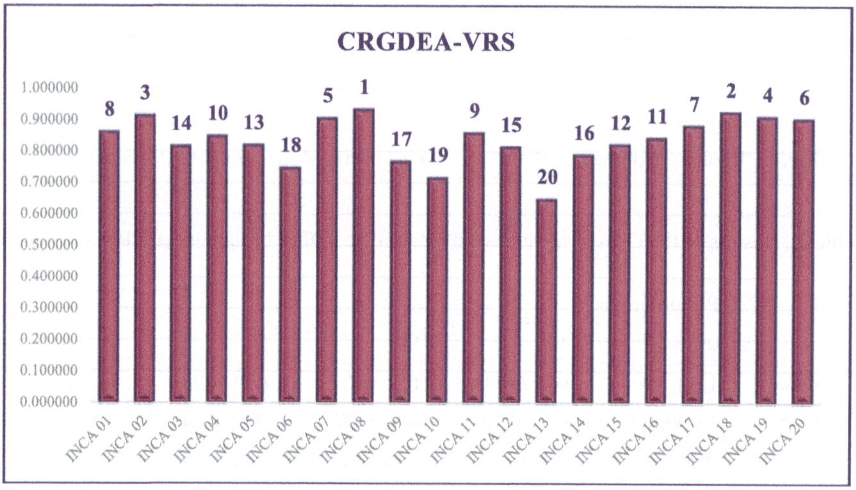

Fig. 2. Ranking Companies Based on the CRGDEA Model under VRS Assumption

5 Concluding Remarks and Future Research Directions

This study proposed a robust framework for evaluating the efficiency of intellectual capital under conditions of deep uncertainty, addressing the limitations of conventional evaluation methods. By integrating robust optimization into the DEA framework, specifically through the box uncertainty set, the proposed approach offers a practical and adaptable solution for IC efficiency assessment. The methodology not only accounts for the variability and ambiguity inherent in IC data but also provides a comprehensive perspective on the interrelationships among its key dimensions: human capital, structural capital, and relational capital. Empirical validation demonstrated its effectiveness in delivering reliable and meaningful insights, even in uncertain and dynamic contexts. By addressing the inherent complexities of real-world scenarios, where IC data are frequently affected by measurement errors, environmental fluctuations, and unpredictable dynamics, this research bridges the gap between traditional methods and the challenges of modern knowledge economies. The integration of robust optimization techniques ensures that efficiency evaluations remain reliable and stable, providing a valuable tool for both theoretical advancement and practical application. The findings have significant implications across multiple domains.

For researchers, the study contributes to the literature by introducing a robust and adaptable framework for IC efficiency evaluation, particularly under uncertain conditions. For practitioners, the methodology provides actionable recommendations for informed decision-making, strategic resource allocation, and identifying inefficiencies, thereby supporting enhanced organizational performance. Additionally, policymakers can leverage these insights to foster environments that enable organizations to optimize their intellectual capital and sustain a competitive edge. Future research could build upon this work by incorporating fuzzy mathematical programming to evaluate IC performance under fuzzy data and linguistic variables. This extension would provide greater flexibility in modeling uncertainty and qualitative factors, thereby broadening the applicability of IC efficiency evaluations (for more details, see Calabrese et al., 2013, Veltri et al., 2014, Peykani et al., 2019, 2021b, 2022d, Arabjazi et al., 2022, Pourbabagol et al., 2023, Sadeghi et al., 2024, Sonkariya & Yadav, 2024, Ghaziyani et al., 2025). Additionally, future studies could explore the evaluation of intellectual capital efficiency within a network structure, which may yield more accurate and transparent results (for more details, see Lu et al., 2004, Kao, 2014; 2017, Alves, & Meza, 2023, Gupta & Kumar, 2024, Peykani et al., 2024b, 2024c, 2024d, Shojaie et al., 2024a, 2024b, Seyed Esmaeili & Mohammadi, 2024).

References

Alves, C.G.M.D.F., Meza, L.A.: A review of network DEA models based on slacks-based measure: evolution of literature, applications, and further research direction. Int. Trans. Oper. Res. **30**(6), 2729–2760 (2023)

Arabjazi, N., Rostamy-Malkhalifeh, M., Hosseinzadeh Lotfi, F., Behzadi, M.H.: Stability analysis with general fuzzy measure: an application to social security organizations. PLoS ONE **17**(10), e0275594 (2022)

Asiaei, K., Jusoh, R.: A multidimensional view of intellectual capital: the impact on organizational performance. Manag. Decis. **53**(3), 668–697 (2015)

Banker, R.D., Charnes, A., Cooper, W.W.: Some models for estimating technical and scale inefficiencies in data envelopment analysis. Manage. Sci. **30**(9), 1078–1092 (1984)

Banker, R.D., Charnes, A., Cooper, W.W., Swarts, J., Thomas, D.: An introduction to data envelopment analysis with some of its models and their uses. Res. Government. Nonprofit Acc. **5**(1), 125–163 (1989)

Ben-Tal, A., El Ghaoui, L., Nemirovski, A.: Robust Optimization. Princeton University Press (2009)

Bertsimas, D., Sim, M.: The price of robustness. Oper. Res. **52**(1), 35–53 (2004)

Bertsimas, D., Brown, D.B., Caramanis, C.: Theory and applications of robust optimization. SIAM Rev. **53**(3), 464–501 (2011)

Bontis, N.: Assessing knowledge assets: a review of the models used to measure intellectual capital. Int. J. Manag. Rev. **3**(1), 41–60 (2001)

Brennan, N., Connell, B.: Intellectual capital: current issues and policy implications. J. Intellect. Cap. **1**(3), 206–240 (2000)

Calabrese, A., Costa, R., Menichini, T.: Using fuzzy AHP to manage intellectual capital assets: an application to the ICT service industry. Expert Syst. Appl. **40**(9), 3747–3755 (2013)

Campisi, D., Costa, R.: A DEA-based method to enhance intellectual capital management. Knowl. Process. Manag. **15**(3), 170–183 (2008)

Charnes, A., Cooper, W.W., Rhodes, E.: Measuring the efficiency of decision making units. Eur. J. Oper. Res. **2**(6), 429–444 (1978)

Charnes, A., Cooper, W., Lewin, A.Y., Seiford, L.M.: Data envelopment analysis theory, methodology and applications. J. Opera. Res. Soc. **48**(3), 332–333 (1997)

Chen, J., Zhu, Z., Xie, H.Y.: Measuring intellectual capital: a new model and empirical study. J. Intellect. Cap. **5**(1), 195–212 (2004)

Cook, W.D., Tone, K., Zhu, J.: Data envelopment analysis: prior to choosing a model. Omega **44**, 1–4 (2014)

Cooper, W.W., Seiford, L.M., Zhu, J.: Handbook on Data Envelopment Analysis. Springer, New York, NY (2011)

Daraio, C., Di Leo, S., Iazzolino, G., Laise, D.: DEA, balanced scorecard and intellectual capital including the gender dimension: a comprehensive list of indicators. Int. Trans. Oper. Res. **30**(6), 2972–3012 (2023)

Dehnokhalaji, A., Khezri, S., Emrouznejad, A.: A box-uncertainty in DEA: a robust performance measurement framework. Expert Syst. Appl. **187**, 115855 (2022)

Despotis, D.K., Smirlis, Y.G.: Data envelopment analysis with imprecise data. Eur. J. Oper. Res. **140**(1), 24–36 (2002)

Duho, K.C.T.: Intellectual capital and technical efficiency of banks in an emerging market: a slack-based measure. J. Econ. Stud. **47**(7), 1711–1732 (2020)

Emrouznejad, A., Yang, G.L.: A survey and analysis of the first 40 years of scholarly literature in DEA: 1978–2016. Socioecon. Plann. Sci. **61**, 4–8 (2018)

Gabrel, V., Murat, C., Thiele, A.: Recent advances in robust optimization: an overview. Eur. J. Oper. Res. **235**(3), 471–483 (2014)

Ghaziyani, K., Hosseinzadeh Lotfi, F., Kordrostami, S., Amirteimoori, A.: Efficiency analysis in bi-level on fuzzy input and output. Inf. Sci. **690**, 121551 (2025)

Inkinen, H.: Review of empirical research on intellectual capital and firm performance. J. Intellect. Cap. **16**(3), 518–565 (2015)

Inkinen, H., Kianto, A., Vanhala, M., Ritala, P.: Structure of intellectual capital–an international comparison. Acc. Aud. Acc. J. **30**(5), 1160–1183 (2017)

Kao, C.: Network data envelopment analysis: a review. Eur. J. Oper. Res. **239**(1), 1–16 (2014)

Kao, C.: Network Data Envelopment Analysis: Foundations and Extensions. Springer International Publishing, Switzerland (2017)

Lee, C.S., Wong, K.Y.: Advances in intellectual capital performance measurement: a state-of-the-art review. Bottom Line **32**(2), 118–134 (2019)

Leitner, K.H., Schaffhauser-Linzatti, M., Stowasser, R., Wagner, K.: Data envelopment analysis as method for evaluating intellectual capital. J. Intellect. Cap. **6**(4), 528–543 (2005)

Liu, J.S., Lu, L.Y., Lu, W.M., Lin, B.J.: A survey of DEA applications. Omega **41**(5), 893–902 (2013)

Lu, W.M., Wang, W.K., Tung, W.T., Lin, F.: Capability and efficiency of intellectual capital: the case of fabless companies in Taiwan. Expert Syst. Appl. **37**(1), 546–555 (2010)

Marr, B., Schiuma, G., Neely, A.: Intellectual capital–defining key performance indicators for organizational knowledge assets. Bus. Process. Manag. J. **10**(5), 551–569 (2004)

Mouritsen, J.: Classification, measurement and the ontology of intellectual capital entities. J. Hum. Res. Cost. Acc. **13**(2), 154–162 (2009)

Paoloni, P., Modaffari, G., Ricci, F., Della Corte, G.: Intellectual capital between measurement and reporting: a structured literature review. J. Intellect. Cap. **24**(1), 115–176 (2023)

Petty, R., Guthrie, J.: Intellectual capital literature review: measurement, reporting and management. J. Intellect. Cap. **1**(2), 155–176 (2000)

Peykani, P., Pishvaee, M.S.: Performance evaluation of hospitals under data uncertainty: an uncertain common-weights data envelopment analysis. Healthcare **12**(6), 611 (2024)

Peykani, P., Mohammadi, E., Emrouznejad, A., Pishvaee, M.S., Rostamy-Malkhalifeh, M.: Fuzzy data envelopment analysis: an adjustable approach. Expert Syst. Appl. **136**, 439–452 (2019)

Peykani, P., Mohammadi, E., Farzipoor Saen, R., Sadjadi, S.J., Rostamy-Malkhalifeh, M.: Data envelopment analysis and robust optimization: a review. Expert. Syst. **37**(4), e12534 (2020a)

Peykani, P., Mohammadi, E., Jabbarzadeh, A., Rostamy-Malkhalifeh, M., Pishvaee, M.S.: A novel two-phase robust portfolio selection and optimization approach under uncertainty: a case study of Tehran stock exchange. PLoS ONE **15**(10), e0239810 (2020b)

Peykani, P., Farzipoor Saen, R., Seyed Esmaeili, F.S., Gheidar-Kheljani, J.: Window data envelopment analysis approach: a review and bibliometric analysis. Expert. Syst. **38**(7), e12721 (2021a)

Peykani, P., Mohammadi, E., Emrouznejad, A.: An adjustable fuzzy chance-constrained network DEA approach with application to ranking investment firms. Expert Syst. Appl. **166**, 113938 (2021b)

Peykani, P., Hosseinzadeh Lotfi, F., Sadjadi, S.J., Ebrahimnejad, A., Mohammadi, E.: Fuzzy chance-constrained data envelopment analysis: a structured literature review, current trends, and future directions. Fuzzy Optim. Decis. Making **21**(2), 197–261 (2022a)

Peykani, P., Seyed Esmaeili, F.S., Mirmozaffari, M., Jabbarzadeh, A., Khamechian, M.: Input/output variables selection in data envelopment analysis: a shannon entropy approach. Mach. Learn. Knowl. Extra. **4**(3), 688–699 (2022b)

Peykani, P., Gheidar-Kheljani, J., Farzipoor Saen, R., Mohammadi, E.: Generalized robust window data envelopment analysis approach for dynamic performance measurement under uncertain panel data. Oper. Res. Int. J. **22**(5), 5529–5567 (2022c)

Peykani, P., Memar-Masjed, E., Arabjazi, N., Mirmozaffari, M.: Dynamic performance assessment of hospitals by applying credibility-based fuzzy window data envelopment analysis. Healthcare **10**(5), 876 (2022d)

Peykani, P., Esmaeili, F.S.S., Pishvaee, M.S., Rostamy-Malkhalifeh, M., Hosseinzadeh Lotfi, F.: Matrix-based network data envelopment analysis: a common set of weights approach. Socioecon. Plann. Sci. **95**, 102044 (2024c)

Peykani, P., Emrouznejad, A., Mohammadi, E., Gheidar-Kheljani, J.: A novel robust network data envelopment analysis approach for performance assessment of mutual funds under uncertainty. Ann. Oper. Res. **339**(3), 1149–1175 (2024d)

Pourbabagol, H., Amiri, M., Taghavifard, M.T., Hanafizadeh, P.: A new fuzzy DEA network based on possibility and necessity measures for agile supply chain performance evaluation: a case study. Expert Syst. Appl. **220**, 119552 (2023)

Pulic, A.: VAIC™ an accounting tool for IC management. Int. J. Technol. Manage. **20**(5–8), 702–714 (2000)

Sadjadi, S.J., Omrani, H.: Data envelopment analysis with uncertain data: an application for Iranian electricity distribution companies. Energy Policy **36**(11), 4247–4254 (2008)

Seyed Esmaeili, F.S., Mohammadi, E.: Z-number network data envelopment analysis approach: a case study on the Iranian insurance industry. PLoS ONE **19**(7), e0306876 (2024)

Shojaie, S.E., Sadjadi, S.J., Tavakkoli-Moghaddam, R.: Malmquist productivity index for two-stage network systems under data uncertainty: a real-world case study. PLoS ONE **19**(7), e0307277 (2024a)

Shojaie, S.E., Sadjadi, S.J., Tavakkoli-Moghaddam, R.: Malmquist productivity index under network structure and negative data: an application to banking industry. J. Syst. Manage. **10**(3), 107–118 (2024b)

Soyster, A.L.: Convex programming with set-inclusive constraints and applications to inexact linear programming. Oper. Res. **21**(5), 1154–1157 (1973)

Ståhle, P., Ståhle, S., Aho, S.: Value added intellectual coefficient (VAIC): a critical analysis. J. Intellect. Cap. **12**(4), 531–551 (2011)

Toloo, M., Mensah, E.K., Salahi, M.: Robust optimization and its duality in data envelopment analysis. Omega **108**, 102583 (2022)

Veltri, S., Mastroleo, G., Schaffhauser-Linzatti, M.: Measuring intellectual capital in the university sector using a fuzzy logic expert system. Knowl. Manag. Res. Pract. **12**(2), 175–192 (2014)

Wang, W.K., Lu, W.M., Kweh, Q.L., Cheng, I.T.: Does intellectual capital matter? Assessing the performance of CPA firms based on additive efficiency decomposition DEA. Knowl.-Based Syst. **65**, 38–49 (2014)

Yu, G., Wei, Q., Brockett, P.: A generalized data envelopment analysis model: a unification and extension of existing methods for efficiency analysis of decision making units. Ann. Oper. Res. **66**(1), 47–89 (1996a)

Yu, G., Wei, Q., Brockett, P., Zhou, L.: Construction of all DEA efficient surfaces of the production possibility set under the generalized data envelopment analysis model. Eur. J. Oper. Res. **95**(3), 491–510 (1996b)

Yudawisastra, H.G., Manurung, D.T., Husnatarina, F.: Relationship between value added capital employed, value added human capital, structural capital value added and financial performance. Invest. Manage. Finan. Innov. **15**(2), 222–231 (2018)

Gupta, M., Kumar, S.: Does income diversification affect bank efficiency, effectiveness and performance? evidence from Indian banking industry. In: Advances in the Theory and Applications of Performance Measurement and Management: Proceedings of DEA45-International Conference on Data Envelopment Analysis, 165–185. Springer, Cham (2024)

Li, Z., Floudas, C.A.: A comparative theoretical and computational study on robust counterpart optimization: III. improving the quality of robust solutions. Industr. Eng. Chemis. Res. **53**(33), 13112–13124 (2014)

Li, Z., Ding, R., Floudas, C.A.: A comparative theoretical and computational study on robust counterpart optimization: i. robust linear optimization and robust mixed integer linear optimization. Industr. Eng. Chemis. Res. **50**(18), 10567–10603 (2011)

Li, Z., Tang, Q., Floudas, C.A.: A comparative theoretical and computational study on robust counterpart optimization: II. Probabilistic guarantees on constraint satisfaction. Industr. Eng. Chemis. Res. **51**(19), 6769–6788 (2012)

Lu, W.M., Kweh, Q.L., Nourani, M., Wang, W.K.: (2014). The effects of intellectual capital on dynamic network bank performance. In: Proceedings of International Conference on Contemporary Economic Issues, pp. 71–78 (2014)

Mergoni, A., Emrouznejad, A., Witte, K.D.: Fifty years of data envelopment analysis. European J. Opera. Res. **326**(3), 389–412 (2025). https://doi.org/10.1016/j.ejor.2024.12.049

Peykani, P., Hamidzadeh, F., Pishvaee, M.S., Memar-Masjed, E., Jabbarzadeh, A.: A robust optimization approach for estimating the most productive scale size in uncertain data envelopment analysis. In: Analytical Decision Making and Data Envelopment Analysis: Advances and Challenges, pp. 165–182. Singapore: Springer Nature Singapore (2024a)

Peykani, P., Sargolzaei, M., Hamidzadeh, F., Seyed Esmaeili, F.S., Takaloo, A.: Possibilistic network DEA approach for performance evaluation of two-stage decision making units under uncertainty. In: Analytical Decision Making and Data Envelopment Analysis: Advances and Challenges, pp. 59–79. Singapore: Springer Nature Singapore (2024b)

Sadeghi, N., Rostamy-Malkhalifeh, M., Behzadi, M.H.: Fuzzy utility mining on DEA. In: Advances in the Theory and Applications of Performance Measurement and Management: Proceedings of DEA45-International Conference on Data Envelopment Analysis, pp. 109–121. Springer, Cham (2024)

Sonkariya, A., Yadav, S.P.: Development of Intuitionistic fuzzy revenue efficiency models in DEA: an application to the Indian public sector banks. In: Advances in the Theory and Applications of Performance Measurement and Management: Proceedings of DEA45-International Conference on Data Envelopment Analysis, pp. 137–149. Springer, Cham (2024)

Sveiby, K.E., Lloyd, T.: Methods for Measuring Intangible Assets. Programvara (2010)

An Optimal Criteria Selection in Efficiency Assessment Through Integration of DEA with Rough Set Theory

Ishu Dalal(✉), Shivi Agarwal, and Trilok Mathur

Department of Mathematics, Birla Institute of Technology and Science, Pilani 333031, India
{p20220433,shivi,tmathur}@pilani.bits-pilani.ac.in

Abstract. Data Envelopment Analysis (DEA) is a prominent nonparametric technique used to assess the efficiency of decision-making units (DMUs) by using multi criteria. However, traditional DEA models can be significantly impacted by criteria that do not contribute significantly to the efficiency analysis, thereby reducing accuracy and discriminatory power. Additionally, for DEA models to produce reliable results, the number of DMUs should be greater than the number of criteria included. This paper introduces a Rough Data Envelopment Analysis (RDEA) approach, which integrates Rough Set Theory (RST) with DEA to effectively handle this problem. RST is used by the RDEA framework to find and remove less contributing criteria from the input and output data in efficiency analysis. RST generates lower and upper approximations which helps in identifying criteria that are not significantly contributing to the efficiency analysis. Once these criteria have eliminated from the data set, the DEA models may be utilized to provide a more accurate and reliable efficiency evaluation of DMUs. This theoretical framework leverages the capabilities of RST to streamline input and output data, enhancing the effectiveness of DEA in evaluating efficiency. Also, a numerical example is provided to show implementation of this method.

Keywords: DEA · Rough set theory · Efficiency · MCDM

1 Introduction

Efficiency evaluation plays a crucial role in assessing the performance of decision-making units (DMUs) across various domains, from economics to production systems. Traditional efficiency evaluation methods, such as Data Envelopment Analysis (DEA), often require multiple input and output variables to compute efficiency scores. However, as the number of attributes increases, the complexity of analysis grows, potentially leading to redundant or insignificant attributes that

A. Emrouznejad et al. (Eds.): DEA 2024, LNOR, pp. 42–51, 2025.
https://doi.org/10.1007/978-3-031-98177-7_4

dilute the interpretability and reliability of the results. High-dimensional data often contain redundant or irrelevant attributes that do not contribute meaningfully to efficiency evaluation. This not only increases computational complexity but may also distort efficiency scores, making the evaluation process less robust. Identifying a reduced set of attributes that retains the key information is essential for simplifying the analysis while ensuring meaningful and reliable results. Reducing attributes allows for a more focused and streamlined analysis by eliminating noise and redundancy in the data. It enhances the interpretability of efficiency scores, reduces computational effort, and provides clearer insights into the core factors driving efficiency. Furthermore, it helps address overfitting issues, ensuring that the efficiency evaluation remains robust and generalizable. While existing studies have proposed methods for efficiency evaluation, limited attention has been given to systematic approaches for reducing attributes in the presence of high-dimensional data. Most studies either retain all attributes or rely on heuristic methods that lack theoretical rigor, resulting in suboptimal efficiency evaluation. This study proposes a systematic approach to attribute reduction for efficiency evaluation, ensuring that the reduced set of attributes retains the discriminatory power of the original dataset. By applying this approach, we aim to demonstrate that efficiency scores derived from reduced attributes are comparable to those obtained using the full set of attributes, thereby validating the utility and reliability of the reduction process.

Rough Set Theory (RST), introduced by Pawlak [1], has emerged as a robust tool for handling uncertainty and reducing attributes in decision-making processes. Attribute reduction is a core issue of rough set theory and also an essential preprocessing step in data mining. It is particularly effective in scenarios where data is imprecise, ambiguous, or incomplete, making it a valuable alternative to traditional methods like Principal Component Analysis (PCA) or statistical feature selection techniques. Over the years, various studies have demonstrated the applicability of RST in feature selection tasks. For instance, Skowron and Rauszer [2] extended RST with decision rules, improving its interpretability in classification problems. Greco *et al.* [3] introduced the variable precision rough set model, which enhances RST's capability to handle noisy datasets. RST has found wide application in high-dimensional datasets, especially in fields like healthcare and bioinformatics. For example, Slowinski and Stefanowski [4] utilized RST to select critical attributes for diagnosing diseases, highlighting its ability to reduce complexity while maintaining decision accuracy. Similarly, Maji *et al.* [5] developed hybrid rough-fuzzy systems for feature selection, demonstrating the adaptability of RST in uncertain environments. Tsang *et al.* [6] introduced formal concepts of attributes reduction with fuzzy rough sets. Setiawan *et al.* combined artificial neural network (ANN) with rough set theory (RST), named as ANNRST, is proposed to predict missing values of attribute. The main advantage that rough set analysis has is that it requires no additional parameters to operate other than the supplied data [7]. It works by making use of the granularity structure of the data only. This is a major difference when compared with Dempster-Shafer theory [8] and fuzzy set theory which require probability assignments and membership values, respectively. Jia *et al.* [9] proposed a generalized attribute reduct that not only considers the data but also user preference. Joulaei and Mirbolouki [10] used Quick-Reduct algorithm of the rough set theory (RST) to reduce inputs or outputs.

In the realm of industrial and financial applications, RST has been employed to optimize decision-making processes. Zhong *et al.* [11] applied RST for attribute reduction in financial risk assessment, while Komorowski *et al.* [12] showed its effectiveness in quality control processes in manufacturing. Comparative studies have also highlighted RST's advantages over other attribute reduction techniques. For instance, Jensen and Shen [13] emphasized its computational efficiency and resilience against overfitting when applied to high-dimensional data compared to PCA and genetic algorithms.

Recent advancements in RST have led to the development of hybrid models, further extending its functionality. For example, fuzzy rough sets [14] and rough set neural networks [15] have been introduced to address complex datasets with overlapping and vague boundaries. These models bridge the gap between classical RST and modern data analysis techniques, demonstrating its growing relevance in handling multidimensional and uncertain data. Jelonek *et al.* [16] presented an empirical study of the use of the rough set approach to the reduction of data for a neural network classifying objects described by quantitative and qualitative attributes. Swiniarski [17] presented an application of rough sets and statistical methods to feature reduction and pattern recognition. Renpu and Zhang [18] presented a hybrid system to efficiently extract classification rules from the decision table using a rough set. Lei [19] improved the prediction capacity of stock price trend by introducing an integrated prediction method based on Rough Set (RS) and Wavelet Neural Network (WNN).

The main objective of this paper is to utilize the discernibility matrix, which is a fundamental part of Rough Set Theory (RST) to identify and eliminate less significant input and output attributes in a DEA framework. In this study, the discernibility matrix is employed to reduce the number of attributes in the dataset. The discernibility matrix identifies the minimal subset of attributes that can distinguish between decision classes, ensuring that redundant or irrelevant attributes are effectively eliminated. To validate the effectiveness of the reduced attributes, a Rough DEA (Data Envelopment Analysis) model is utilized. The efficiency results obtained using the original set of attributes are compared with those derived from the reduced attribute set. This comparative analysis demonstrates that the reduced set of attributes retains the essential information required for meaningful efficiency evaluation, confirming the validity and practicality of the proposed attribute reduction methodology. This methodology is also practically applicable across a wide range of sectors such as manufacturing, healthcare, education, logistics, and energy, where decision-making often involves multiple uncertain or redundant variables.

2 Preliminaries

Rough set theory: A rough set theory is a mathematical approach to deal with vagueness and uncertainty in data. It was introduced by Zdzisław Pawlak [1] in 1982 and The concept revolves around approximating sets. Unlike traditional statistical methods, RST does not require additional information like probability distributions or membership functions, making it highly suitable for real-world problems where such information may not be available or is difficult to ascertain. In rough set theory, a set is written as [a b][c d] where left interval is the lower approximation and right interval is the upper approximation[20]. There is no need of detailed information about the uncertainty spread in the data as we need in other methods used for handling uncertainty. Also it's representation partitions the data in two parts in which one contain the certain values of the data which are contained in lower approximation and other values which we are not certain about but may occur are contained in the other part which is known as the upper approximation. Unlike fuzzy set theory which is subjective based approach, rough set theory is an objective based approach which makes it more simpler to use. Rough set theory doesn't need construction of membership function, uncertainty can be handled using this in a more simpler way. Qualitative data can also be handled using rough set theory as we can induce decision rules from the data present in qualitative form which will help to understand the main information that the whole dataset carries without dealing with the complexity of the data. Attribute reduction is also one of the applications of rough set theory in which reducts and core are found which are used to replace the whole set of attributes with the reduced set of attributes without compromising the critical information that the whole data set carries. These are some important points about the representation of rough set theory and its use:

Lower Approximation. The set of all elements that are definitely part of the target set based on the information available[1].

Upper Approximation The set of all elements that might possibly be included in the target set[1].

Discernibility Matrix: The Discernibility Matrix is a tabular representation used in Rough Set Theory (RST) to identify the minimal subsets of attributes (reducts) necessary to distinguish between different objects in a dataset. It plays a critical role in attribute reduction by providing the basis for generating reducts[21].

Discernibility Function: In rough set theory to find reduct and core, the discernibility function is constructed. This function which the discernibility matrix can compute, has the following form[22]:

$$f_D(C) = \prod_{(x,y)\in U^2} \left\{ \sum \delta(x,y) : (x,y) \in U^2 \text{ and } \delta(x,y) \neq \emptyset \right\}$$

where $\delta(x, y)$ represents the attributes discerning DMUs 'x' and 'y', and in this function boolean summation and product are used.

Reduct: In Rough Set Theory (RST), a reduct is defined as a minimal subset of attributes that preserves the same classification power as the original set of attributes. The primary goal of a reduct is to reduce redundancy in the attribute space without losing essential information for decision-making or classification[22]. A reduct is not necessarily unique; for a given dataset, there can be multiple reducts, each of which provides an alternative minimal representation of the data. Identifying reducts enables the simplification of data while ensuring that the essential relationships and dependencies among objects are maintained.

Core: The core in Rough Set Theory represents the set of indispensable attributes. It is derived as the intersection of all possible reducts of a dataset and contains those attributes that are essential for preserving the classification power of the data. Unlike the attributes in a reduct, attributes in the core cannot be removed without adversely affecting the classification accuracy or the overall information content of the dataset. The core serves as a foundation for attribute reduction, as it identifies the most critical elements that must be retained during the reduction process[22].

3 Numerical Example

In this study, data from seven cement companies taken from a research paper[23] are analyzed to evaluate their efficiency. The data set consisted of three inputs which are capital, cooperation experience, and cost, and the outputs are timely delivery and technology level.

Inputs and Outputs

Capital: Capital refers to the total financial investment made by the cement company, including funds used for infrastructure, machinery, and operations.

Cooperation experience: It reflects the number of cooperative projects, partnerships, or collaborations the company has been involved in.

Cost: Cost represents the total operational expenses incurred by the company during the production and delivery processes which is taken as rough variable due to inherent fluctuations in operational and logistical processes.

Timely delivery: It is considered as rough variable, as it involves a degree of uncertainty and variability. It reflects the company's general ability to meet delivery schedules, but external factors such as supply chain disruptions, weather conditions, or logistical delays may affect its consistency.

Technology level: It represents the degree of technological advancement within the company.

Below is the data table representing the values collected for 7 cement companies based on the input and output variables explained above. Among these variables, Technology Level and Timely Delivery are considered rough variables, as their values may not be precise and are subject to variation or uncertainty due to external or unmeasurable factors. These rough variables are represented using lower and upper approximations in the dataset to better capture their uncertain nature. This approach allows for a more realistic analysis when dealing with imprecise or fluctuating data in decision-making models (Tables 1,2 and 3).

Table 1. Input values

DMU	Capital	Cooperation experience	Cost
1	14	3	(8,9) (6,12)
2	12	2	(13,14), (11,17)
3	10	3	(31,32), (28,35)
4	1.6	2	(14,15), (12,18)
5	10	2	(24,5), (22,28)
6	7	3	(51,52), (49,55)
7	3	3	(36,37), (34,40)

Table 2. Output values

DMU	Timely delivery	Technology level
1	(82,85), (80,92)	86
2	(55,58), (53,65)	75
3	(78,81), (76,88)	73
4	(67,70), (65,77)	83
5	(61,64), (59,71)	90
6	(70,73), (68,80)	84
7	(74,77), (72,84)	90

Handling Rough Variables. To handle these rough variables for efficiency evaluation Rough DEA model[23][24] is used which gives efficiency interval corresponding to each DMU. The interval is considered the efficiency interval evaluated for kth DMU relative to the other DMUs. If the upper bound of this interval is 1 then the DMU is considered as efficient otherwise inefficient[23].

Table 3. Results

DMU	Efficiency interval	Relatively efficient
1	[1,1]	Yes
2	[0.9177,0.9251]	No
3	[0.7630,0.7830]	No
4	[1,1]	Yes
5	[1,1]	Yes
6	[0.6862,0.7052]	No
7	[0.7246,0.7441]	No

The discernibility matrix corresponding to the given data set is as follows in which each row and column is represented by the DMUs:

	D_1	D_2	D_3	D_4	D_5	D_6	D_7
D_1	ϕ						
D_2	B,C	ϕ					
D_3	A,C	B,C	ϕ				
D_4	A,B,C	A,C	A,B,C	ϕ			
D_5	A,B,C	C	B,C	A,C	ϕ		
D_6	A,C	A,B,C	A,C	A,B,C	A,B,C	ϕ	
D_7	A,C	A,B,C	A,C	A,C	A,B,C	A,C	ϕ

In this matrix, each entry represents the attributes discerning DMU 'i' from DMU 'j'. The diagonal entries contain no attributes as they represent the attributes that differentiate DMU 'i' from itself. Now using the discernibility function based on the above matrix we will find out the reduct and core. The discernibility function is as follows:

$$= C \cdot (B + C) \cdot (A + C) \cdot (A + B + C) \cdot (A + B)$$
$$= C \cdot (A + C) \cdot (A + B)$$
$$= C \cdot (A + B)$$
$$= C \cdot A + C \cdot B$$

From this function we get two possible reducts (C,A) and (C,B) and their intersection is C, which is core of the given attributes. Any of the possible reducts can be used instead of all set of attributes without the loss of the critical information that the whole data set carries. As the attribute 'C' is in both the reducts, which concludes that this is the most critical and if we remove this attribute, then the dataset will not be able to retain the critical information. Now comparing the efficiency results from both the original set and the reduced set of attributes the results are as follows:

Table 4. Efficiency results using original and reduced attribute sets

DMU	Reduct efficiency interval	Original efficiency
1	[1,1]	[1,1]
2	[0.9251,0.9477]	[0.9177,0.9251]
3	[0.7630,0.7722]	[0.7630,0.7830]
4	[1,1]	[1,1]
5	[1,1]	[1,1]
6	[0.6862,0.6955]	[0.6862,0.7052]
7	[0.7246,0.7338]	[0.7246,0.7441]

4 Results and Analysis

An important observation from the data Table 4 is that the efficiency results remain consistent when calculated using both the original set of attributes and the reduced set of attributes. This indicates that the reduced attribute set retains the essential discriminatory power of the original set, allowing for reliable efficiency evaluation without loss of critical information. Such consistency suggests that the eliminated attributes may be redundant or carry overlapping information with the retained ones. The use of attribute reduction not only simplifies the model but also enhances computational efficiency while maintaining analytical accuracy–an important outcome in practical decision-making scenarios.

5 Conclusion

This study demonstrates the effectiveness of employing the discernibility matrix-based reduction method, rooted in rough set theory, for minimizing the number of attributes while preserving the accuracy of efficiency evaluations. The discernibility matrix approach identified the core attributes that are essential for maintaining the classification ability of the dataset and also provided two possible reducts. The results of the efficiency evaluation using the reduced attributes were then compared to those obtained from the original, full set of attributes. This comparison revealed that the efficiency scores calculated with the reduced attributes were consistent with those derived using the original attributes, validating the effectiveness of the attribute reduction process. The findings demonstrate that the reduced set of attributes can serve as a meaningful and efficient representation of the data, allowing for accurate efficiency evaluation while reducing computational burden.

6 Scope for Future Work

Future research can build upon this approach by exploring its application in other industries and decision-making contexts, where uncertainty and data complexity play significant roles. Additionally, integrating other hybrid models combining rough set theory with advanced computational techniques could further enhance the precision and applicability of attribute reduction and efficiency evaluation processes.

References

1. Pawlak, Z.: Rough sets. Int. J. Comput. Inf. Sci. **11**, 341–356 (1982)
2. Skowron, A., Rauszer, C.: The discernibility matrices and functions in information systems. In: Intelligent Decision Support: Handbook of Applications and Advances of the Rough Sets Theory, pp. 331–362. Springer (1992)
3. Greco, S., Matarazzo, B., Slowinski, R.: Rough sets theory for multicriteria decision analysis. Eur. J. Oper. Res. **129**(1), 1–47 (2001)
4. Słowiński, R., Stefanowski, J., Greco, S., Matarazzo, B.: Rough set based processing of inconsistent information in decision analysis. Control. Cybern. **29**(1), 379–404 (2000)
5. Maji, P., Roy, S.: Rough-fuzzy clustering and unsupervised feature selection for wavelet based mr image segmentation. PLoS ONE **10**(4), e0123677 (2015)
6. Tsang, E.C., Chen, D., Yeung, D.S., Wang, X.-Z., Lee, J.W.: Attributes reduction using fuzzy rough sets. IEEE Trans. Fuzzy Syst. **16**(5), 1130–1141 (2008)
7. Kira, K., Rendell, L.A.: The feature selection problem: traditional methods and a new algorithm. In: Proceedings of the Tenth National Conference on Artificial Intelligence, pp. 129–134 (1992)
8. Almuallim, H., et al.: Learning with many irrelevant features. In: AAAI, vol. 91, pp. 547–552. Citeseer (1991)
9. Jia, X., Shang, L., Zhou, B., Yao, Y.: Generalized attribute reduct in rough set theory. Knowl. Based Syst. **91**, 204–218 (2016)
10. Joulaei, M., Mirbolouki, M.: Reduction of dea-performance factors using rough set theory: an application of companies in the iranian stock exchange. Adv. Math. Financ. Appl. **5**(1), 53–67 (2020)
11. Zhong, N., Dong, J., Ohsuga, S.: Using rough sets with heuristics for feature selection. J. Intell. Inf. Syst. **16**, 199–214 (2001)
12. Komorowski, J.: Learning rule-based models-the rough set approach. Comprehensive Biomedical Physics, Amsterdam (2014)
13. Jensen, R., Shen, Q.: Fuzzy-rough sets assisted attribute selection. IEEE Trans. Fuzzy Syst. **15**(1), 73–89 (2007)
14. Dubois, D., Prade, H.: Rough fuzzy sets and fuzzy rough sets. Int. J. Gen. Syst. **17**(2–3), 191–209 (1990)

15. Ding, S., Chen, J., Xu, X., Li, J.: Rough neural networks: a review. J. Comput. Inf. Syst. **7**(7), 2338–2346 (2011)
16. Jelonek, J., Krawiec, K., Slowiński, R.: Rough set reduction of attributes and their domains for neural networks. Comput. Intell. **11**(2), 339–347 (1995)
17. Świniarski, R.W.: Rough sets methods in feature reduction and classification (2001)
18. Li, R., Wang, Z.-O.: Mining classification rules using rough sets and neural networks. Eur. J. Oper. Res. **157**(2), 439–448 (2004)
19. Lei, L.: Wavelet neural network prediction method of stock price trend based on rough set attribute reduction. Appl. Soft Comput. **62**, 923–932 (2018)
20. Bonikowski, Z.: Algebraic structures of rough sets. In: Rough Sets, Fuzzy Sets and Knowledge Discovery: Proceedings of the International Workshop on Rough Sets and Knowledge Discovery (RSKD'93), Banff, Alberta, Canada, 12–15 October 1993, pp. 242–247. Springer (1994)
21. Pawlak, Z.: Rough set theory and its applications. J. Telecommun. Inf. Technol. 7–10 (2002)
22. Pawlak, Z.: Rough set theory and its applications to data analysis. Cybern. Syst. **29**(7), 661–688 (1998)
23. Jamal, M.A., El-Khodary, I.A., Ali, D.S.: Developing a two-stage data envelopment analysis model handling rough variables. In: 2021 International Conference on Data Analytics for Business and Industry (ICDABI), pp. 306–314. IEEE (2021)
24. Xu, J., Li, B., Wu, D.: Rough data envelopment analysis and its application to supply chain performance evaluation. Int. J. Prod. Econ. **122**(2), 628–638 (2009)

Imputation Propelled Data Envelopment Analysis (IPDEA): A Case of Indian Food Processing Sector

Ardhana M. Prabhash and Vipin Valiyattoor(✉)

Department of Economic Sciences, Indian Institute of Science Education and Research, Bhopal 462066, MP, India
vipin@iiserb.ac.in

Abstract. This study explores the complementarity of Machine Learning tools for performance analysis. We mainly focus on applying the supervised Machine Learning (ML) framework to improve the performance evaluation and data-driven goal setting of food processing companies in India (2011–21). Mathematical programming-based Data Envelopment Analysis (DEA) is the most popular tool for performance evaluation of decision-making units(DMUs) such as hospitals, companies, and local self-government institutions, etc. The relative benchmarking efficiency frontier is framed by using the given input(s)-output(s) of each DMUs. Disregarding the wider usage and applications in various fields, this tool is criticized for its sensitivity in the context of finite samples. In this study, we propose an Imputation Propelled DEA (IPDEA) as an alternative to overcome this limitation of conventional Non-Parametric DEA. In this study, we propose an Imputation Propelled DEA (IPDEA) using Machine learning algorithms to overcome the issues of the missing values in the data for the non-parametric Data Envelopment Analysis (DEA), and the prediction of the target variable. We consider Indian food processing industries to evaluate the proposed framework and presented the cases to demonstrate the proposed approach. In short, the DEA efficiency scores are sensitive to the sample size, and the proposed IPDEA can address the issue of finite samples to the best of its ability. Combining DEA with machine learning algorithms enhances our ability to accurately assess the Indian industries' performance, which we can carry out through two parallel analyses, i.e., one with only actual data and the other with imputed data (along with actual data). Our results show a significant improvement in estimating efficiency scores using IPDEA.

Keywords: Performance Prediction · Machine Learning · DEA · IPDEA

1 Introduction

This study mainly focuses on developing a supervised machine learning framework by integrating the conventional DEA for the performance prediction and missing value imputation of the data associated with the industrial sector, specifically the Food Processing Companies in India. The Indian domestic food market is expected to expand

A. Emrouznejad et al. (Eds.): DEA 2024, LNOR, pp. 52–61, 2025.
https://doi.org/10.1007/978-3-031-98177-7_5

to 344 billion USD by 2027, representing nearly a 40% increase from its current size (World of Food India, 2011 [1]). Although India's agricultural productivity is relatively strong, there are significant challenges related to wastage, while the level of food product processing remains relatively low. Some advanced countries have achieved high food processing levels to meet consumable standards, while India's overall processing capacity remains only 10%. Consequently, India's food processing sector is relatively small, and its contribution to the global trade of processed food exports remains at approximately 2.5%. With its greater forward and backward linkages, the food processing sector acts as a segment of the manufacturing sector that integrates the traditional agrarian sector and the service sector. Therefore, conducting research and developing tools to improve these processes' efficiency and ensure greater sustainability within the food processing sector is crucial. Assessing the performance of the food processing industry and examining factors contributing to its growth, such as research and development (R&D) and imports, are inevitable for its overall expansion. Numerous research efforts have scrutinized the performance of India's food processing sector. The non-availability of adequate unit-level data (i.e., input-output data) remains a notable hurdle in India's food processing sector performance evaluation during the post-reform period. Furthermore, the instances of missing data points may result in misleading outcomes, particularly when using non-parametric methods for efficiency estimation (DEA).

Moreover, from a methodological point of view, the food processing sector rightly represents the industrial sector in India. The database we use (Prowess IQ database of the Centre for Monitoring Indian Economy Pvt. Ltd.) for this study provides a considerable number of observations as a sample (with and without missing values), which facilitates us to test the appropriateness of the proposed methodology. Compared to the other industry groups (such as basic metals and textiles), the sub-segments are highly diversified. It includes diversified categories such as beer & alcohol, dairy products, sugar, tea, vegetable oils and products, and the remaining sub-sectors (which include bakery products, cocoa products, confectionery, coffee, and marine foods, etc...). As a result, an ideal performance analysis framework should conceptualize 'group frontiers' for these subsectors However, due to data limitations (aggravated by missing values), most of the studies (Bhandari and Vipin (2026) for instance)) mainly rely on the analysis based on meta frontier. The proposed IPDEA approach has the potential to retain as many observations from even small sub-segments of the food processing sector as possible, to provide a more realistic analysis of the food processing sector with severe data challenges.

In the non-parametric DEA estimation context, missing values are typically replaced with extrapolated values, which may overlook the accuracy of these values and often result in biased estimation of efficiency scores.

Given this background, we employ the non-parametric Data Envelopment Analysis (DEA) to gauge the performance of food processing enterprises in India, which assesses the relative efficiencies of the Decision-Making Units (DMUs), using alternative approaches.

DEA is a powerful frontier based tool to estimate the efficiency of DMUs. The original model for the efficiency frontier of the production units is introduced by Farrell (1957) [6]. Based on this, the very first DEA model was proposed by Charnes, Cooper, and Rhodes (1978) (the CCR Model) and was later modified by Banker, Charnes, and Cooper (BCC) in the year 1984 [2, 3] (see also Emrouznejad, et al., 2025). During the last forty-five years, DEA has evolved theoretically and has emerged as a popular non-parametric technique. Network DEA ([15] for instance), DEA with undesirable outputs [e.g., 13], and fuzzy DEA [14] are notable advancements in this regard. Disregarding the wide acceptability and application of DEA, this approach is prone to a finite sample. The estimated efficiency scores can only converge to the actual efficiency scores (against the true frontier) asymptotically [15, 16]. The proposed IPDEA approach helps the researcher to have a relatively larger sample size and move to the above-specified large sample characteristics to the extent possible.

For this purpose, we carry out two parallel analyses, i.e., one with only actual data and the other with imputed data (along with actual data). Our study proposes the integration of cutting-edge machine learning algorithms and DEA for performance prediction and to address the issue of finite sample variation, which is called Imputation Propelled DEA (IPDEA). We try to exploit this nature of Machine Learning algorithms for efficiency analysis and to minimize the limitations of finite sample bias in DEA estimates. Through this, we check it for the performance evaluation with DEA using imputed values of the food processing companies' data to provide more accurate information about the performance of the food processing sector.

Despite significant advancements in AI and ML, these technologies' long-term economic implications are still poorly understood. However, the empirical literature has yet to tap into the potential of machine learning for efficiency as well as productivity prediction. In this study, the trends and challenges confronting Indian industries are identified, with a focus on the food processing industry.

We propose the operation of cutting-edge machine learning algorithms for Missing value imputation and performance prediction. Here, we are proposing an Imputation Propelled DEA (IPDEA) using Machine learning algorithms to overcome the issues of the missing values in the data for the non-parametric analysis. We exploit substitutability nature of Machine Learning algorithms for efficiency analysis and to minimize the limitations of finite sample bias in DEA estimates.

2 Data Envelopment Analysis (DEA)

For this study, we use the conventional Data Envelopment Analysis (DEA) methodology for measuring the efficiency of the food processing companies with further refining. DEA is a nonparametric method based on linear programming problems for forming a production frontier using the observed data points and measurement of the efficiency score of decision-making units (DMUs) such as a firm or government bodies. It is a methodology for measuring the relative edge of a set of decision-making units (DMUs) that use multiple inputs to produce outputs. Crisp input and output data are indispensable in conventional DEA. Still, the observed values of the input and output data are vague and imprecise when addressing real-world problems. In DEA, we create a benchmark technology from the observed input(s)-output(s) bundles of the firms. The DEA is based on certain assumptions about the production technology frontier.

Assumption 1: All actually observed input-output combinations are feasible.
Assumption 2: The production possibility set is convex.
Assumption 3: Inputs are freely disposable.
Assumption 4: Outputs are freely disposable. The output-oriented measure of TE of the kth firm requires the solution to the following LP problem:

$$\text{Maximize } \varphi_k$$

subject to

$$\sum \lambda_j y_j \geq \varphi_k y_k \text{ for j} = 1 \text{ to N}$$

$$\sum \lambda_j \text{AT}_j \leq \text{AT}_k \text{ (Total Assets)}$$

$$\sum \lambda_j \text{SS}_j \leq \text{SS}_k \text{ (Stores \& Spares)}$$

$$\sum \lambda_j \text{PPE}_j \leq \text{PPE}_k \text{ (Packaging \& Packing Expenses)}$$

$$\sum \lambda_j \text{R}_j \leq \text{R}_k \text{ (Raw Materials)}$$

$$\sum \lambda_j \text{RLT}_j \leq \text{RLT}_k \text{ (Rent \& Lease Rent)}$$

$$\sum \lambda_j \text{RM}_j \leq \text{RM}_k \text{ (Repairs \& Maintenance)}$$

$$\sum \lambda_j \text{TC}_j \leq \text{TC}_k \text{ (Total Capital)}$$

$$\sum \lambda_j \text{PFW}_j \leq \text{PFW}_k \text{ (Power, Fuel \& Water Charges)}$$

$$\sum \lambda_j \text{ME}_j \leq \text{ME}_k \text{ (Miscellaneous Expenditures)}$$

$$\sum \lambda_j = 1 \text{ (Convexity constraint)}$$

$$\lambda_j \geq 0 \text{ for all j} = 1, \ldots, \text{N}$$

$$\varphi_k \in \mathbb{R} \text{ (Unrestricted)}$$

where y is the output, the input variables used are given in parentheses. Let φ^*_k be the optimum solution of the above linear programming problem, $1/\varphi^*_k$ will be the output-oriented technical efficiency of firm k [4].

3 Machine Learning Algorithms for Input Value Imputations

Application of Machine Learning and Deep Learning has gained attention in the recent past, and studies claim that integration of AI-ML tools can improve the estimation efficiency scores [5]. For instance, Zhu et. Al (2021) shows how machine learning algorithms can be integrated with non-parametric DEA estimation and forecasting efficiency scores [19]. As mentioned above, DEA scores are sensitive to the sample size. As a result, missing values can aggravate the bias in the estimation of efficiency scores estimated using DEA. We plug-in in the imputation techniques in performance evaluation. Following regression models are tested and trained for their best performance using various cross-validation techniques and fine-tuning. The value of cv was fixed to 10 for cross-validation over each model. Also, to evaluate the model, Root Mean Square Error (RMSE), R2 score, and Mean Square Error (MSE) are used.

The regression methods used are:

- *Decision Tree Regressor: Hyper-parameters used are: (random state = 40), rgr parameters:rgr criterion:(squared error,friedman mse,poisson), rgr max features:(None,auto, sqrt, log2), rgr max depth:(10,40,45,60), rgr ccp alpha:(0.009,0.01,0.05,0.1) Values for criterion, max feature, and max depth are found through grid search, and keeping these values constant, the MSE,RMSE and R2 is measured for different values of ccp alpha from 0 to 0.1.*
- *Random Forest Regressors: Hyperparameters used are: rgr criterion:(squared error,friedman mse,poisson), rgr n estimators:(30,50,100), 'rgr max depth':(10,20,30), 'rgr max features':(None, 'auto', 'sqrt', 'log2'), Values for criterion, max feature, and max depth are found through grid search. The values for these parameters, which gave the maximum value of R2 measure, is chosen.*
- *Adaptive Booster: Hyperparameters used are: rgr = AdaBoostRegressor(nestimators = 100)rgrparametersrgrbaseestimator: (be1, be2), rgrrandomstate: (0, 10), Values are found through grid search. The values for these parameters, which gave the maximum value of R2 measure are chosen.*
- *Ridge regression: Hyperparameters used are: rgr = Ridge(alpha = 1.0, positive = T rue) rgrparametersrgrsolver: (auto, lbf gs), The values for these parameters, which gave the maxi- mum value of R2 score-measure, are chosen. • Linear Regressor: Here MSE,RMSE and R2 measure is found for different values of n and which gave the maximum value of R2 score, is chosen. Rgrparameters rgrpositive: (True, False).*

Figure 1 shows the framework for the imputation of missing input values.

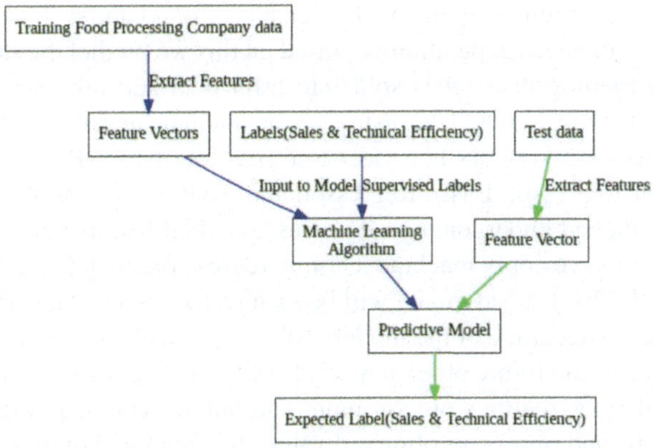

Fig. 1. Framework for Imputation/prediction of missing values

4 Imputation Propelled Data Envelopment Analysis (IPDEA)

One prominent hurdle in assessing the efficiency of the industrial sector in India (at the unit level) is the non-availability of input(s)- output(s) data. The lack of data points can result in misleading results, especially in the case of non-parametric estimation. So, crisp input and output data are crucial parts of the conventional DEA approach. Disregarding the stated advantages of DEA, it is sensitive to the sample variation. This study tries to address this issue to the best of its ability. Here, we propose the integration of cutting-edge machine learning algorithms and DEA for performance prediction and to address the issue of finite sample bias in the DEA estimates. Here, we propose an Imputation Propelled DEA (IPDEA) to address the issue related to sample variation in the industrial sector's data. We estimate the efficiency frontiers of the decision-making units(DMUs), which are relative in nature and influenced by the variation in the sample data. For estimation, the nonparametric estimator DEA can be used to estimate the efficiency of production units.. Here, the efficiency scores are estimated using the linear programming problem by considering the convex hull of the observations, which satisfies the assumptions regarding the technical efficiency of the DMUs [7–9].

5 IPDEA in Practice

Here, we mainly focus on developing a supervised machine learning framework for the performance prediction and missing value imputation of the data associated with the Food Processing Companies in India using the CMIE ProwessIQ dataset (starting from 2011 to 2021). The number of training instances of this data set is 10254 and features are 8. Therefore, the shape of the training data set is (10254,10), including the company name and year. And test instances are 1140, so the shape is (1140,10). However, there are some missing values in the data. So here, filling the values in the data set by using mean, median, and interpolation methods (Lin & Tsai 2020). The major features of this company data are Total assets, stores & spares, Packaging and packing expenses, Raw materials, Rent &lease rent, Repairs &Maintenance, Total capital, Power, fuel &water charges, and Miscellaneous expenditures. Based on this we predict the sales of the given companies. The training data set is split into train data and test data in 70:30 ratio. The missing values are imputed by using mean, median and interpolation methods. Then applied various regressors like Decision Tree Regressor, Random forest, Ridge Regressor, Linear Regressor, Lasso Regressor, and Adaptive Boosting in each imputed train data. The data had undergone some processes called feature selection and feature extraction. After that we apply machine learning regression models for training data set to train the model. This trained model will be then tested on test data set and validation data set to check the accuracy of the model [10]. Data Envelopment Analysis (DEA) is employed to forecast the future efficiency levels of food processing companies in India. While traditional DEA requires precise input and output data, real-world issues often involve ambiguous and imprecise observed values for input and output data.

Experimental Setup

Based on the data, we find a suitable model or regression model from the algorithms that we had run using the trained model, like Linear Regression, Decision Tree, Ridge Regression, Random Forest etc. using grid search cv and also by analyzing the evaluation matrices. Based on the scores of the different evaluation criteria, we select the best model and apply the model to the given test data.

Evaluation matrices of different regressors, such as MSE, RMSE, and R2 scores, serve to gauge the model's precision. The summary of experimental findings are given in Table 1. These figures consist of the efficacy of diverse regression techniques employing Median imputation [11].

Table 1 shows the performance parameters of different regressors used for median imputation.

Table 1. ML-Based Median Imputation: Performance of Regressors

Regressor	RMSE	Score
Adaptive Boosting	10717.63	0.709
Decision Tree	8274.19	0.826
Ridge Regression	8511.00	0.816
Linear Regression	8359.95	0.823
Random Forest	6512.59	0.892

The relative performance of different regression algorithms shows that the Random Forest has the lowest MSE, RMSE, and highest Score (0.8923), suggesting it performs well on the model with Median imputation, followed by Decision Tree regression and Linear Regression.

DEA Efficiency scores of the Indian food processing firms: With and without Imputation

As explained above, we estimated the efficiency scores of food processing firms using conventional DEA and newly proposed IPDEA. Efficiency of food processing firms using actual observations with all inputs and output variables for the period 2011–2021(only 2901 observations). However, the estimation using the imputed data of inputs and output variables of the firms for the same period involves 10077 observations retained in the sample using machine learning-based imputations.

To check whether the distribution of efficiency scores remain same with and without imputation, over the period under consideration, we conduct Wilcoxon matched-pairs signed-rank test [17]. The result shows that the median of efficiency scores under two approaches are different and the null hypothesis is rejected with 1% level of significance (with a P value 0.001).

This test procedure involved identifying firms with actual value and their efficicny scores estimated under DEA and IPDEA. We found a statistically significant difference in these two estimates.

The average (geometric mean) efficiency scores calculated using the above approaches are given in Fig. 2 below.

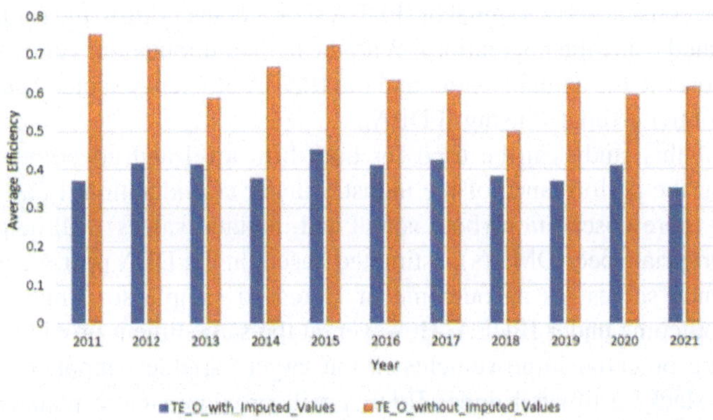

Fig. 2. Average Efficiency of Food-processing Firms: DEA Vs IPDEA

The average efficiency of the food processing sector during the period 2011–2021 is estimated 0.63 in the case of actual data. However, in the estimation using imputed data, the average efficiency of the food processing sector is 0.39, which shows greater potential for improvement in this industry under consideration.

We observed a significant difference in the average efficiency scores of the food processing firms using DEA and IPDEA. This can be considered as evidence of our estimated frontier using IPDEA reaching closer to the true frontier. In other words, by retaining observations with missing values (with the help of imputation), we try to resolve the issue of bias arising due to sampling in the case of conventional DEA—the novelty of this study.

6 Conclusion

This study involves assessing the performance of food processing enterprises in India employing a supervised machine learning algorithm. We use the proposed IPDEA approach for retaining observations with missing values in our analysis, by imputing their missing values of input-output variables. We compare various approaches to achieve this target and among different algorithms, the Random Forest model gives better predictions of the target variables. With low variance and bias, random forest algorithm provides the best imputation and, can be generalized to obtain imputed values of input-output variables used for non-parametric efficiency estimation. So, we use the imputed values of input-output variables in the estimation of efficiency scores of the food processing sector in India. By imputing the input values using machine learning algorithms based on median imputation, we could retain another 7176 observations (with a total 10077) in our sample as against only 2901 observations in the actual sample.

We employed a conventional non-parametric DEA approach in the initial analysis and the latter using imputed data. As claimed in the existing literature, DEA efficiency scores are sensitive to the sample size, and this study uses the imputed input-output data to address this issue to the greatest extent possible.

A comparison of average DEA scores estimated using actual data and IPDEA model shows significant difference through. From an empirical point of view, this observed difference can be considered as a signal of IPDEA scores converging to the actual efficiency scores(estimated using the population). With the help of imputation, our estimated frontier moves closer to the actual frontier and the IPDEA efficiency scores converge closer to the true frontier estimated using IPDEA.

Results of this study can be used for both firm level and governmental policies for improving the performance of the industry under consideration. In a firm point of view, having more observations(both actual and imputed values) will help to identify the most appropriate peer DMU(s), estimated based on the DEA peer-weights. Having lower efficiency scores (as an outcome of increased sample size) may sound as an undesirable outcome under IPDEA. However, at the same time it provides more scope for identifying potential improvements in the target variables (input or output). The input-output slacks estimated under IPDEA will provide a more realistic picture of the production framework, which can be used for firm-level input-output adjustments to achieve greater efficiency in the future. Any second stage estimation using IPDEA efficiency scores may provide consistent estimates of the coefficients associated with the factors determining firm efficiency. From a macro policy perspective, efficiency scores estimated using IPDEA can be considered as more reliable performance estimate as against the conventional DEA scores. The potential for improvement in the overall performance of the industry can be estimated using IPDEA. The impact of governmental macro level policies on firm performance can be assessed in a more realistic manner using IPDEA efficiency scores.

Acknowledgement. We appreciate the valuable feedback from the editor and the two reviewers, which helped shape our work. Additionally, we're grateful to Dr. Tanmay Basu from the Department of Data Science and Engineering, IISER Bhopal, for providing insightful comments on an earlier draft of this manuscript. Ms. Ardhana M Prabhash would like to acknowledge the funding support (Centrally Administered Contingency Grant 2024–25) received from the Indian Council of Social Sciences Research (ICSSR) for conducting this study. The usual disclaimers apply.

References

1. World of Food India, The Food Processing Industry in India (2011). https://worldfoodindia.gov.in/
2. Charnes, A., Cooper, W.W., Rhodes, E.: Measuring the efficiency of decision-making units. Eur. J. Oper. Res. **2**(6), 429–444 (1978)
3. Banker, R.D., Charnes, A., Cooper, W.W.: Some models for estimating technical and scale inefficiencies in data envelopment analysis. Manage. Sci. **30**(9), 1078–1092 (1984)
4. Ray, S.C.: Data Envelopment Analysis: Theory and Techniques for Economics and Operations Research. Cambridge University Press (2004)
5. Emrouznejad, A., Shale, E.: A combined neural network and DEA for measuring efficiency of large scale datasets. Comput. Ind. Eng. **56**(1), 249–254 (2009)
6. Farrell, M.J.: The measurement of productive efficiency. J. Roy. Stat. Soc. **120**, 253–290 (1957). https://doi.org/10.2307/2343100
7. Bargagli-Stoffi, F.J., Niederreiter, J., Riccaboni, M.: Supervised learning for the prediction of firm dynamics. In: Data science for Economics and Finance, pp. 19–41.Springer, Cham (2021). https://doi.org/10.1007/978-3-030-66891-4_2
8. Basuchoudhary, A., Bang, J.T., Sen, T.: Machine-learning techniques ineconomics: new tools for predicting economic growth. New York Springer (2017). ISBN: 978-3-319-69014-8
9. Bhandari, A.K., Vipin, V.: Efficiency and related technological aspects of the Indian food processing industry: a non-parametric analysis. J. Developing Areas **50**(6), 227–243 (2016)
10. Lago, L.P.: Imputation of household survey data using mixed models. University Of Wollongong Thesis Collection 1954–2016 (2015). https://ro.uow.edu.au/theses/4369/
11. Farhangfar, A., Kurgan, L.A., Pedrycz, W.: A novel framework for imputation of missing values in databases. IEEE Trans. Syst. Man Cybern. Part A Syst. Hum. **37**(5), 692–709 (2007)
12. Mergoni, A., Emrouznejad, A., De Witte, K.: Fifty years of data envelopment analysis. Eur. J. Oper. Res. **326**(3), 389–412 (2025). https://doi.org/10.1016/j.ejor.2024.12.049
13. Mishra, S., Singh, A., Sufia.: Environmental performance of coal-based thermal generation plants: a two-stage DEA model with undesirable outputs. In: International Conference on Data Envelopment Analysis, pp. 151–163. Springer, Cham, Switzerland (2023)
14. Sadeghi, N., Rostamy-Malkhalifeh, M., Behzadi, M.H.: Fuzzy utility mining on DEA. In: International Conference on Data Envelopment Analysis, pp. 109–121. Springer, Cham, Switzerland (2023)
15. Sotiros, D., Koronakos, G., Despotis, D.K.: Dominance at the divisional efficiencies level in network DEA: the case of two-stage processes. Omega **85**, 144–155 (2019)
16. Simar, L., Wilson, P.W.: Estimation and inference in two-stage, semi-parametric models of production processes. J. Econom. **136**(1), 31–64 (2007)
17. Wilcoxon, F., Katti, S.K., Wilcox, R.A.: Critical values and probability levels for the Wilcoxon rank sum test and the Wilcoxon signed rank test, vol. 1. Pearl River, NY: American Cyanamid (1963)
18. Zervopoulos, P.D., Kanas, A., Emrouznejad, A., Molyneux, P.: Can super-efficiencies improve bias correction? a bayesian data envelopment analysis approach. In International Conference on Data Envelopment Analysis, pp. 21–31. Springer, Cham, Switzerland (2023)
19. Zhu, N., Zhu, C., Emrouznejad, A.: A combined machine learning algorithms and DEA method for measuring and predicting the efficiency of Chinese manufacturing listed companies. J. Manag. Sci. Eng. **6**(4), 435–448 (2021)
20. Emrouznejad, A., Podinovski, V., Lu, C., Charles, V., Moradi-Motlagh, A.: Rajiv Banker's Lasting Impact on Data Envelopment Analysis, Annals of Operations Research, **351**, 1225–1264, (2025). https://doi.org/10.1007/s10479-025-06473-3

Forest-Based Resampling for Confidence Interval Estimation of Efficiencies in Data Envelopment Analysis

Yu Zhao(✉)

Tokyo University of Science, 1-3 Kagurazaka, Shinjuku-ku, Tokyo 162-8601, Japan
yu.zhao@rs.tus.ac.jp

Abstract. Standard Data Envelopment Analysis (DEA) models are deterministic, and previous studies have made numerous efforts to introduce statistical analysis into DEA, such as the bootstrapping algorithm and regression-based approaches. In this study, we consider probabilistic variations present in input-output vectors and propose a forest-based sampling procedure to handle the statistical properties of efficiencies across different orientations of the DEA model. To capture the probability distributions of the data, we classify observed decision-making units into several clusters and maximize the information gain of each cluster using a Gaussian-based entropy function. The proposed approach is illustrated using a data set used in previous studies.

Keywords: Data Envelopment Analysis (DEA) · Forest-Based Resampling · Confidence Interval · Gaussian-based entropy

1 Introduction

Data Envelopment Analysis (DEA) is a non-parametric methodology for evaluating the relative efficiency of decision-making units (DMUs) that utilize multiple inputs to produce multiple outputs. Since its inception, DEA has been extensively applied across various sectors such as economics, healthcare, education, and energy to assess performance and inform policy decisions. Traditional DEA models assume a deterministic framework where all input and output data are measured precisely, and any deviations from the efficient frontier are attributed solely to inefficiency.

However, real-world data are often subject to measurement errors, noise, and other stochastic variations that can significantly impact the accuracy of efficiency estimates. Recognizing this limitation, researchers have sought to incorporate statistical inference into DEA to account for data uncertainty and provide more robust efficiency evaluations. Simar and Wilson [23] pioneered the use of bootstrap techniques in DEA, allowing for the estimation of confidence intervals and hypothesis testing of efficiency scores. This approach has been further developed in subsequent studies. For instance, Kneip et al. [15] integrated asymptotic properties into efficiency analysis, enhancing the statistical foundation of

A. Emrouznejad et al. (Eds.): DEA 2024, LNOR, pp. 62–76, 2025.
https://doi.org/10.1007/978-3-031-98177-7_6

DEA. Moradi-Motlagh and Emrouznejad [20] examined the methodological evolution of the bootstrap approach over the past two decades, including the robust probabilistic approach [3,5] and the conditional efficiency approach [7,8], among others, highlighting the growing role of statistical inference in efficiency analysis. Michali et al. [19] further extended bootstrap methods in DEA by introducing a subsampling bootstrap approach for network structures. Their method accounts for the inner structure of the production process and incorporates undesirable factors, improving inference robustness. Other recent development of the bootstrap approach in confidence intervals estimation are Dia et al. [9], Kerstens et al. [14], Walheer [26], Ngo and Tsui [21], Lin and Lu [17], Kang et al. [13].

Another significant advancement is the development of Stochastic Nonparametric Envelopment of Data (StoNED) by Kuosmanen and Kortelainen [16]. StoNED combines the strengths of DEA and Stochastic Frontier Analysis (SFA) by incorporating a stochastic component into the non-parametric DEA framework. This allows for the separation of inefficiency and statistical noise without imposing a specific functional form on the production function. StoNED has been applied in various contexts, demonstrating its effectiveness in handling stochastic variations in efficiency analysis. However, as noted by Mergoni et al. [18], semi-parametric approaches are generally more computationally demanding than standard parametric models [24] and are particularly affected by the curse of dimensionality [16].

Advances in machine learning have also influenced DEA research. For instance, studies such as Zhu et al. [28], Aparicio et al. [2], Boubaker et al. [4], Esteve et al. [10], and Shi and Zhao [22] combine machine learning methods with DEA models to measure and predict efficiency. On the other hand, studies such as Valero-Carreras et al. [25], Aparicio and Esteve [1], Guillen et al. [12], Moragues et al. [18], and Zhao [27] focus on improving the estimation of production technologies. To the best of our knowledge, there has been limited development in estimating confidence intervals for efficiencies using an integrated approach that combines machine learning and DEA methods.

Despite these advancements, there remains a need for methods that can effectively capture probabilistic variations in input-output data while preserving the strengths of both machine learning and DEA methods. This gap exists due to the following reasons: (a) bootstrap-based approaches, such as Simar and Wilson [23], generally assume that efficiency itself is a random term, while the observed input-output data are considered free of measurement errors; (b) the production model in StoNED [16] explicitly treats outputs as random variables while considering inputs as true levels, a common issue in other parametric regression-based models such as SFA; and (c) recent efforts to integrate machine learning methods into DEA have primarily focused on improving the predictive accuracy of efficiency or the performance of estimated frontiers, with relatively little attention given to confidence interval estimation of efficiencies.

In this study, we propose a forest-based resampling approach to estimate confidence intervals of efficiencies in DEA. By classifying observed DMUs into clusters and maximizing the information gain using a Gaussian-based entropy function [6], we aim to model the underlying data distribution more accurately. This method leverages ensemble learning techniques to generate resampled datasets that reflect the stochastic characteristics of the inputs and outputs, thereby providing more reliable efficiency estimates.

The remainder of this paper is organized as follows. Section 2 introduces the preliminary theory of density trees. Section 3 presents the tree-based sampling algorithms in DEA, detailing algorithms for sampling input variables conditioned on output levels. In Sect. 4, we focus on constructing confidence intervals for efficiencies in DEA. Section 5 provides experimental results using a real-world dataset of electric utilities, comparing the efficiency scores from the conventional DEA model with those from our proposed method. Finally, Sect. 6 concludes the paper.

2 Preliminary Theory: Density Tree

Let us represent the input-output vector as $\mathbf{v} \in \mathbb{R}^d := (\mathbf{x}, \mathbf{y}) \in \mathbb{R}^m \times \mathbb{R}^s$. We can express a split function for binary trees as $h_j(\mathbf{v}, \boldsymbol{\theta}) \in \{0, 1\}$, with $\boldsymbol{\theta} \in \mathcal{T}$ representing the split parameters associated with the jth node ($j = 0, 1, \ldots, J$) and $\mathcal{T}$ is the set of all possible split parameters. The split function can be optimized in a greedy manner by using the following maximization:

$$\boldsymbol{\theta}^* = \arg\max_{\boldsymbol{\theta} \in \mathcal{T}} I(\mathcal{S}_j, \boldsymbol{\theta}), \tag{1}$$

where $\mathcal{S}_j$ is a subset of $\mathcal{S}_0 := \{\mathbf{v}\}$ at the jth node and in the context of a binary tree, we have

$$\mathcal{S}_j^L(\mathcal{S}_j, \boldsymbol{\theta}) = \{\mathbf{v} \in \mathcal{S}_j | h_j(\mathbf{v}, \boldsymbol{\theta}) = 0\}, \ \mathcal{S}_j^R(\mathcal{S}_j, \boldsymbol{\theta}) = \{\mathbf{v} \in \mathcal{S}_j | h_j(\mathbf{v}, \boldsymbol{\theta}) = 1\}. \tag{2}$$

These represent the left and right data points that arrive at the jth node. A visualization is presented in Fig. 1.

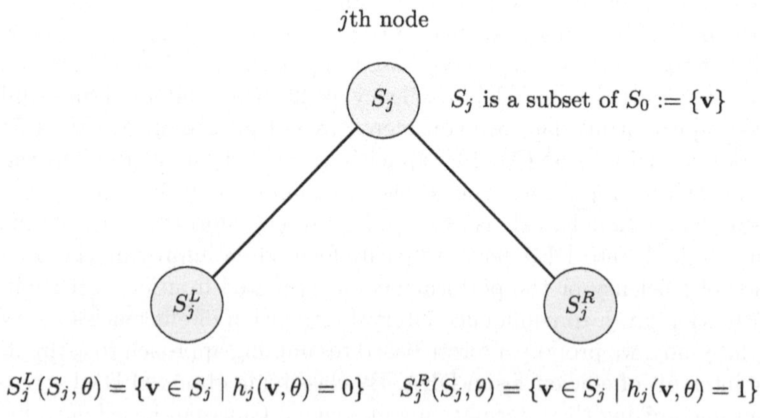

Fig. 1. Left and right Subsets

The objective function $I(\cdot)$ represents the information gain, and is defined as follows:

$$I(\mathcal{S}_j, \boldsymbol{\theta}) = H(\mathcal{S}_j) - \sum_{i \in \{L,R\}} \frac{|\mathcal{S}_j^i|}{|\mathcal{S}_j|} H(\mathcal{S}_j^i). \tag{3}$$

The symbol $|\cdot|$ indicates the cardinality of a specific set, while H denotes the entropy function which is applicable for the clustering of given data points. Following Criminisi and Shotton [6], we use a Gaussian-based entropy function to define the splitting criteria, ensuring that each resulting cluster can be well approximated by a Gaussian distribution. Now let $\mathbf{v} \in \mathcal{S}_j$ be a realization of the random variable $\mathbf{V}$. We assume that $\mathbf{V}$ follows the multivariate Gaussian distribution,

$$\mathbf{V} \sim \mathcal{N}_d(\boldsymbol{\mu}_j, \boldsymbol{\Sigma}_j), \tag{4}$$

with mean $\boldsymbol{\mu}_j \in \mathbb{R}^d$ and the covariance matrix $\boldsymbol{\Sigma}_j$ belongs to the space of symmetric positive definite $d \times d$ matrices. The corresponding probability density function is then given by

$$p(\mathbf{v}|\boldsymbol{\mu}_j, \boldsymbol{\Sigma}_j) = \frac{1}{(2\pi)^{d/2} \det(\boldsymbol{\Sigma}_j)^{1/2}} \exp\left(-\frac{1}{2}(\mathbf{v} - \boldsymbol{\mu}_j)^\top \boldsymbol{\Sigma}_j^{-1}(\mathbf{v} - \boldsymbol{\mu}_j)\right). \tag{5}$$

The entropy of a d-variate Gaussian for the jth node is defined as

$$H(\mathcal{S}_j) = \frac{1}{2} \log\left((2\pi e)^d \det(\boldsymbol{\Sigma}_j)\right), \tag{6}$$

where $\det(\cdot)$ indicates a determinant for a specific matrix. Consequently, the expression in (3) becomes

$$I(\mathcal{S}_j, \boldsymbol{\theta}) = \log\left(\det(\boldsymbol{\Sigma}_j)\right) - \sum_{i \in \{L,R\}} \frac{|\mathcal{S}_j^i|}{|\mathcal{S}_j|} \log\left(\det(\boldsymbol{\Sigma}_j^i)\right). \tag{7}$$

Thus, the splitting parameters $\boldsymbol{\theta}$ are computed by maximizing the determinants of the covariance matrices. This approach divides the entire dataset (that is, $\{\mathbf{v}\}$) into several clusters. Each cluster is characterized by a leaf node. In addition, weighting by cardinality of the subsets of $\{\mathbf{v}\}$ prevents the formation of clusters consisting of single points.

Let $l(\mathbf{v}) : \mathbb{R}^d \rightarrow \mathbb{N}$ be an index of a leaf node. The density of the tree is then represented as

$$p(\mathbf{v}) = \sum_{l(\mathbf{v})} \frac{|\mathcal{S}_{l(\mathbf{v})}|}{|\mathcal{S}_0|} p\left(\mathbf{v}|\boldsymbol{\mu}_{l(\mathbf{v})}, \boldsymbol{\Sigma}_{l(\mathbf{v})}\right), \tag{8}$$

where $\sum_{l(\mathbf{v})} \frac{|\mathcal{S}_{l(\mathbf{v})}|}{|\mathcal{S}_0|} = 1$ and the above expression defines a mixture of Gaussian density.

This approach is illustrated in Fig. 2. We generate 200 two-dimensional data points from the concave function $Y = X^{0.5} \exp(-U)$ with $U \sim \text{Exp}(\lambda = 1/\mu), \mu = 1/5$ and $X \sim \text{Uniform}(0, 1)$. The expression (7) is maximized at various depths ranging from 0 to 5. At a depth of 0, the entire data set is divided into two clusters (the covariance ellipses) with the aim of maximizing the Gaussianity of each cluster. The information gain for each cluster (see the numbers in the centra of ellipses) is 0.78. At a depth of 1, the number of clusters increases to four, and the information gain is also updated. At depths of 2, 3, 4, and 5, the clusters stabilize, and the information gain no longer increases. We observe that as the depth of the tree increases, more clusters are formed and stabilized. In addition, the centers of these clusters tend to be located in areas of high data density, while areas far from the centers are situated in regions of low density.

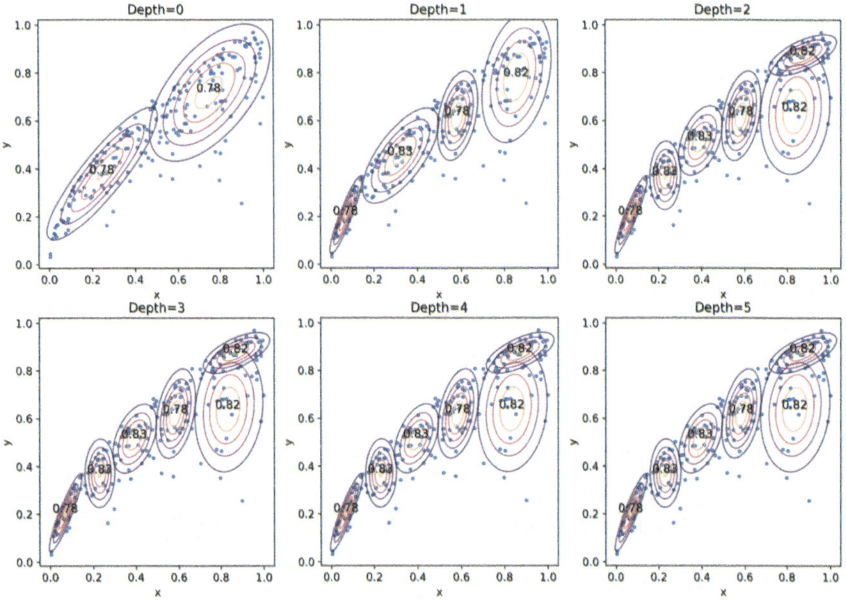

Fig. 2. Data, covariance ellipses, and information gain

It is worth noting that although we begin with the assumption that $\mathbf{V} \sim \mathcal{N}_d(\boldsymbol{\mu}_j, \boldsymbol{\Sigma}_j)$, the experiment in Fig. 2 suggests that the density tree remains effective even when the global distribution of the input-output vector deviates from Gaussianity. This is because the Gaussian assumption primarily influences the splitting criteria, leading to clusters that locally approximate Gaussian distributions. In fact, in the above experiment, it can be rigorously derived that, given $X = x$, Y follows a scaled Beta distribution, specifically $Y \mid X = x \sim x^{0.5} Z$, where $Z \sim \text{Beta}(5, 1)$.

3 Tree-Based Sampling Algorithms in DEA

In DEA, both input-oriented and output-oriented models are widely used in practical analysis. Previous studies have applied a bootstrap method to enhance statistical analysis in DEA. In this study, we propose a tree-based sampling algorithm for generating input-output vectors. For the sake of simplicity, we focus on the input orientation. The output orientation can be explained in a similar manner.

Assuming variable returns to scale, we can empirically construct the production possibility set P by using the following definition:

$$P := \left\{ (\mathbf{x}, \mathbf{y}) \in \mathbb{R}^{m+s} \;\middle|\; \sum_{h=1}^{n} \lambda_h \mathbf{x}_h \leq \mathbf{x}, \sum_{h=1}^{n} \lambda_h \mathbf{y}_h \geq \mathbf{y}, \sum_{h=1}^{n} \lambda_h = 1, \boldsymbol{\lambda} \geq \mathbf{0} \right\}. \tag{9}$$

The input-oriented efficiency relative to P is defined as

$$\hat{\delta} = \min\{\delta > 0 | (\delta \mathbf{x}, \mathbf{y}) \in P\}. \tag{10}$$

We assume that the unknown true frontier is deterministic; therefore, $\mathbf{x}$ can be considered as a realization of the random input variable $\mathbf{X}$. Since the efficiency term is calculated from the observed inputs and outputs, $\hat{\delta}$ is also considered as a realization of the random variable D with $E(D) = \mu > 0$ and $Var(D) < \infty$. To measure the uncertainty of inputs $\mathbf{X}$ given a specific level of outputs $\mathbf{y}$, we sample the input variables using the following algorithm.

Algorithm 1. Tree-based Sampling Algorithm: Multivariate Gaussian

Require: Dataset $\{(\mathbf{x}_h, \mathbf{y}_h)\}_{h=1}^{n}$
Ensure: Samples $\{\mathbf{x}_h^b\}_{h=1}^{n}$, $b = 1, \dots, B$

Construct the Density Tree:
for each node $j = 0, 1, \dots, J$ **do**
  Maximize the information gain function to find split parameters:

$$\boldsymbol{\theta}^* = \arg\max_{\boldsymbol{\theta} \in \mathcal{T}} I(\mathcal{S}_j, \boldsymbol{\theta})$$

  Split the data into left and right subsets:

$$\mathcal{S}_j^L = \{\mathbf{v} \in \mathcal{S}_j \mid h_j(\mathbf{v}, \boldsymbol{\theta}^*) = 0\},$$
$$\mathcal{S}_j^R = \{\mathbf{v} \in \mathcal{S}_j \mid h_j(\mathbf{v}, \boldsymbol{\theta}^*) = 1\}$$

end for
Collect Leaf Nodes:
for each leaf node $\mathcal{S}_l$ **do**
  Calculate mean $\boldsymbol{\mu}_l$ and covariance $\boldsymbol{\Sigma}_l$
end for
Conditional Sampling:
for each data point $(\mathbf{x}_h, \mathbf{y}_h) \in \mathcal{S}_l$ **do**

Compute conditional mean and covariance:

$$\boldsymbol{\mu}_{x|y} = \boldsymbol{\mu}_x + \boldsymbol{\Sigma}_{xy}\boldsymbol{\Sigma}_{yy}^{-1}(\mathbf{y}_h - \boldsymbol{\mu}_y)$$
$$\boldsymbol{\Sigma}_{x|y} = \boldsymbol{\Sigma}_{xx} - \boldsymbol{\Sigma}_{xy}\boldsymbol{\Sigma}_{yy}^{-1}\boldsymbol{\Sigma}_{yx}$$

Sample $\mathbf{x}_h^b$ from $\mathbf{X} \mid \mathbf{Y} = \mathbf{y}_h \sim \mathcal{N}_m(\boldsymbol{\mu}_{x|y}, \boldsymbol{\Sigma}_{x|y})$

end for

This algorithm constructs a density tree by recursively partitioning the dataset to maximize the Gaussianity of the subsets. At each leaf node, a multivariate Gaussian distribution is fitted. For each data point, conditional sampling is performed using the Gaussian parameters of the corresponding leaf node, given the observed output $\mathbf{y}_h$. For an illustrative explanation, Fig. 3 shows the conditional distribution of $\mathbf{X}$ given the mean of $\{\mathbf{y}|\mathbf{Y} = \mathbf{y}_h \in \mathcal{S}_l, \forall h\}$ for all ls.

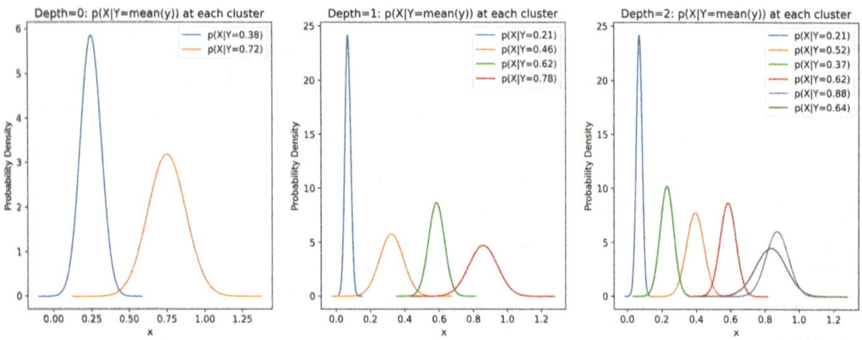

Fig. 3. Conditional distributions

It should be noted that in Algorithm 1, some data points may be influenced by multiple distributions. It is uncertain whether it is appropriate to sample inputs solely based on the distribution of their classified leaf node. To address this issue, we further propose the use of a the mixture of Gaussians.

Algorithm 2. Tree-based Sampling Algorithm: Mixture of Gaussians

Require: Dataset $\{(\mathbf{x}_h, \mathbf{y}_h)\}_{h=1}^n$
Ensure: Samples $\{\mathbf{x}_h^b\}_{h=1}^n$, $b = 1, \ldots, B$

Construct the Density Tree (as in Algorithm 1)
Collect Cluster Information:
for each leaf node $\mathcal{S}_l$ **do**
  Calculate mean $\boldsymbol{\mu}_l$, covariance $\boldsymbol{\Sigma}_l$, and prior probability $p(\mathbf{L} = l) = \frac{|\mathcal{S}_l|}{|\mathcal{S}_0|}$
end for
Conditional Sampling:
for each data point $(\mathbf{x}_h, \mathbf{y}_h)$ **do**

Compute posterior probabilities:

$$p(\mathbf{L} = l \mid \mathbf{Y} = \mathbf{y}_h) = \frac{p(\mathbf{Y} = \mathbf{y}_h \mid \mathbf{L} = l)p(\mathbf{L} = l)}{\sum_{l'} p(\mathbf{Y} = \mathbf{y}_h \mid \mathbf{L} = l')p(\mathbf{L} = l')}$$

where $p(\mathbf{Y} = \mathbf{y}_h \mid \mathbf{L} = l) = \mathcal{N}_s(\mathbf{y}_h \mid \boldsymbol{\mu}_y^l, \boldsymbol{\Sigma}_{yy}^l)$

Sample cluster l^* based on the posterior probabilities

Compute conditional mean and covariance:

$$\boldsymbol{\mu}_{x|y}^{l^*} = \boldsymbol{\mu}_x^{l^*} + \boldsymbol{\Sigma}_{xy}^{l^*}(\boldsymbol{\Sigma}_{yy}^{l^*})^{-1}(\mathbf{y}_h - \boldsymbol{\mu}_y^{l^*})$$
$$\boldsymbol{\Sigma}_{x|y}^{l^*} = \boldsymbol{\Sigma}_{xx}^{l^*} - \boldsymbol{\Sigma}_{xy}^{l^*}(\boldsymbol{\Sigma}_{yy}^{l^*})^{-1}\boldsymbol{\Sigma}_{yx}^{l^*}$$

Sample $\mathbf{x}_h^b$ from $\mathbf{X} \mid \mathbf{Y} = \mathbf{y}_h, \mathbf{L} = l^* \sim \mathcal{N}_m(\boldsymbol{\mu}_{x|y}^{l^*}, \boldsymbol{\Sigma}_{x|y}^{l^*})$

end for

The mixture of conditional Guassians can be written as

$$p(\mathbf{X}|\mathbf{Y} = \mathbf{y}) = \sum_l \frac{|\mathcal{S}_l|}{|\mathcal{S}_0|} p(\mathbf{X}|\mathbf{Y} = \mathbf{y}; \boldsymbol{\mu}_l, \boldsymbol{\Sigma}_l) \tag{11}$$

with $\sum_l \frac{|\mathcal{S}_l|}{|\mathcal{S}_0|} = 1$. However, sampling directly from (11) may lead to the occurrence of unrealistic data points. For example, $\mathbf{X}$ can be sampled from the cluster where $\mathbf{Y} = \mathbf{y}$ is not observed. To prevent from this issue, we need to determine which cluster is $\mathbf{Y} = \mathbf{y}$ from and thus we use the Bayes' Rule in the sampling algorithm. Compared with the Algorithm 1, Algorithm 2 has more variety in generating $\mathbf{X}$ as the cluster used for sampling is probabilistically determined.

For the unobserved data, the decision tree constructed from the observed data may be unstable if new observations are added. The ensemble approach can be used to improve the stability of clustering the data points. A forest-based sampling algorithm is presented below:

Algorithm 3. Forest-based Sampling Algorithm

Require: Dataset $\{(\mathbf{x}_h, \mathbf{y}_h)\}_{h=1}^n$, Number of trees T
Ensure: Samples $\{\mathbf{x}_h^b\}_{h=1}^n$, $b = 1, \ldots, B$

Construct the Density Forest:

for $t = 1$ **to** T **do**

Create bootstrap sample $\mathcal{S}_0^{(t)}$

for each node j **do**

Optimize split parameters using random feature subsets:

$$\boldsymbol{\theta}^{*(t)} = \arg\max_{\boldsymbol{\theta} \in \mathcal{T}_j} I(\mathcal{S}_j^{(t)}, \boldsymbol{\theta})$$

end for

Collect leaf nodes $\mathcal{S}_l^{(t)}$, means $\boldsymbol{\mu}_l^{(t)}$, and covariances $\boldsymbol{\Sigma}_l^{(t)}$

end for

Obtain Densities from Each Tree:
Choose to use either Algorithm 1 or Algorithm 2 for each tree:

- If using Algorithm 1: Compute $p(\mathbf{X} \mid \mathbf{Y} = \mathbf{y}; \boldsymbol{\mu}_l^{(t)}, \boldsymbol{\Sigma}_l^{(t)})$
- If using Algorithm 2: Compute $p(\mathbf{X} \mid \mathbf{Y} = \mathbf{y}, \mathbf{L} = l^{(t)}; \boldsymbol{\mu}_l^{(t)}, \boldsymbol{\Sigma}_l^{(t)})$

Build Ensemble Density:
The density of the ensemble forest is obtained by averaging over all trees:

- If using Algorithm 1:

$$p_{\text{forest}}(\mathbf{X} \mid \mathbf{Y} = \mathbf{y}) = \frac{1}{T} \sum_{t=1}^{T} p(\mathbf{X} \mid \mathbf{Y} = \mathbf{y}; \boldsymbol{\mu}_l^{(t)}, \boldsymbol{\Sigma}_l^{(t)})$$

- If using Algorithm 2:

$$p_{\text{forest}}(\mathbf{X} \mid \mathbf{Y} = \mathbf{y}, \mathbf{L} = l) = \frac{1}{T} \sum_{t=1}^{T} p(\mathbf{X} \mid \mathbf{Y} = \mathbf{y}, \mathbf{L} = l^{(t)}; \boldsymbol{\mu}_l^{(t)}, \boldsymbol{\Sigma}_l^{(t)})$$

Conditional Sampling:
for each data point $(\mathbf{x}_h, \mathbf{y}_h)$ **do**
    Predict and sample $\mathbf{x}_h^b$ using the chosen algorithm:

- If using Algorithm 1: Follow steps from Algorithm 1
- If using Algorithm 2: Follow steps from Algorithm 2

end for

Figure 4 provides an intuitive explanation of how a mixture of densities is formed in a forest consisting of three trees, given a certain level of $\mathbf{Y} = \mathbf{y}_h$. The mixed distribution is observed by averaging the densities across all three trees. This approach tends to be more robust than using a single decision tree because ensembling the randomly generated trees helps reduce the correlation between the predictions (clusters) of individual trees.

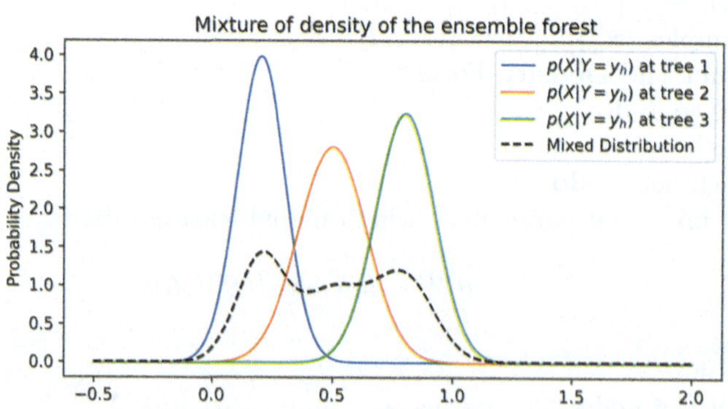

Fig. 4. Ensemble forest

4 Confidence Interval of Efficiencies in DEA

The input-oriented DEA model based on the sampled inputs is formulated as

$$\hat{\delta}^b = \min\left\{\delta^b > 0 \,\middle|\, \sum_{h=1}^{n} \lambda_h \mathbf{x}_h^b \leq \delta \mathbf{x}^b, \sum_{h=1}^{n} \lambda_h \mathbf{y}_h \geq \mathbf{y}, \sum_{h=1}^{n} \lambda_h = 1, \boldsymbol{\lambda} \geq \mathbf{0}\right\}, \quad (12)$$

for $b = 1, \ldots, B$, and $\{\mathbf{x}_h^b, h = 1, \ldots, n\}$ are drawn from the Algorithms 1, 2, or 3. The average efficiency is then

$$\bar{\delta} = \sum_{b=1}^{B} \hat{\delta}^b. \quad (13)$$

According to the Central Limit Theorem, we have $\bar{\delta} \to \delta^*$ as $B \to \infty$.

It is worth noting that DEA efficiency is generally considered to be biased. This holds true only if there are no measurement errors in $\{\mathbf{x}_h, h = 1, \ldots, n\}$ and thus $P \subset P^*$ is satisfied (P^* is the unknown, deterministic production possibility set). In contrast, our approach assumes that there are measurement errors in $\{\mathbf{x}_h, h = 1, \ldots, n\}$; therefore, there is no direct subset relation between P and P^*.

In this study, we adopt a nonparametric bootstrap method to estimate the confidence interval for $\bar{\delta}$ by resampling from the set $\{\hat{\delta}^b, b = 1, \ldots, B\}$. This is done by generating K bootstrap samples, where each sample $\{\tilde{\delta}^{b,k}, b = 1, \ldots, B\}$, for $k = 1, \ldots, K$, is created by randomly drawing B data points from $\{\hat{\delta}^b, b = 1, \ldots, B\}$ with replacement. We then compute the means for all bootstrap samples. According to the Central Limit Theorem, if the bootstrap sample size is sufficiently large (i.e., $K \to \infty$), then the mean $\bar{\delta}$ approximately follows a normal distribution. Thus, we can easily calculate the 95% confidence interval for $\bar{\delta}$ by sorting $\{\sum_{b=1}^{B} \tilde{\delta}^{b,k}, k = 1, \ldots, K\}$ in ascending order and computing the 2.5% and 97.5% percentiles of this distribution.

5 Experiments

We use the dataset in Table 1 presented in [11] (see also [23] for an analysis using the bootstrap approach) to illustrate our proposed method. This dataset contains 19 electric utilities with three inputs and one output.

The density trees constructed using Algorithm 1 and Algorithm 2 are identical, as shown in Fig. 5 and Table 1. Based on the original dataset and the resulting clusters, we find that Cluster 1 contains smaller-scale DMUs with lower input and output values, reflected in their lower means across all variables. Cluster 2 represents medium-scale DMUs, exhibiting moderate values in inputs and outputs, as evident from their intermediate mean values. Cluster 3 includes larger-scale DMUs, characterized by significantly higher values in the variables "Fuel", "Capital", and "Output", indicating higher productivity or scale. The clear distinctions in the mean and covariance matrices among the clusters demonstrate that the classification based on density trees effectively captures the underlying structure of the data, leading to meaningful segmentation (Table 2).

Table 1. Input and Output Data

No	Labor	Fuel	Capital	Output
1	179.0	3969.1	1005.5	3531.4
2	96.0	896.32	194.6	841.3
3	94.0	803.17	200.0	690.5
4	141.0	1629.9	564.1	1498.1
5	147.0	3264.4	617.4	3347.7
6	261.0	1427.5	347.1	1263.2
7	448.0	6365.3	1680.4	5734.0
8	223.0	4911.0	1319.4	4107.7
9	279.0	7779.3	1785.6	6964.2
10	214.0	2307.5	690.0	1961.0
11	367.0	4334.7	1268.9	4207.2
12	97.0	1870.7	416.5	1752.7
13	143.0	2730.6	779.8	2667.6
14	95.0	540.61	305.0	457.2
15	241.0	9509.2	1892.1	9539.9
16	147.0	1561.1	718.5	1360.6
17	306.0	1699.0	306.3	1689.7
18	182.0	881.16	182.3	750.3
19	500.0	2548.9	650.1	2439.5

Table 2. Cluster Assignments

Cluster	Number of DMU
Cluster 1	2, 3, 4, 5, 12, 13, 14, 16
Cluster 2	1, 6, 10, 17, 18, 19
Cluster 3	7, 8, 9, 11, 15

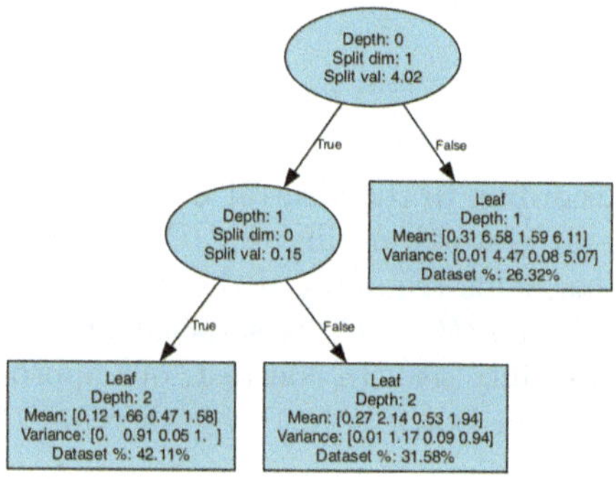

Fig. 5. Density Tree

Table 3 shows the efficiencies measured by the original DEA model (10), Algorithms 1 and 2. The bootstrap confidence intervals are calculated with $B = 2000$. It can be seen that the average standard deviation of efficiencies measured using Algorithm 1 is less than those measured using Algorithm 2 ($0.0428 < 0.0617$). This is because the sampling procedure in Algorithm 2 probabilistically selects a cluster for each data point by computing the posterior probabilities, while the sampling procedure in Algorithm 1 only uses the leaf node corresponding to the data point. Furthermore, the average width of confidence intervals in Algorithm 1 is narrower than that in Algorithm 2.

Table 4 summarizes the results using Algorithm 3 with different sampling procedures as described in Algorithms 1 and 2, respectively. For each algorithm, we construct a forest using ten density trees based on bootstrapped samples. The two forest algorithms also show that the average standard deviation of efficiencies with the sampling procedure of Algorithm 1 is less than those with the sampling procedure of Algorithm 2; however, they are very close. We also observe that the standard deviation difference ($|0.0461 - 0.0464| = 0.0003$) is smaller than that ($|0.0617 - 0.0428| = 0.0189$) in Table 3, indicating that the ensemble method is useful in reducing variance by combining predictions from multiple trees. Moreover, the average width of confidence intervals in Algorithm 3 with the sampling procedure of Algorithm 1 is the same as that with the sampling procedure of Algorithm 2. This implies that ensembling density trees is useful for improving the robustness of the sampling process.

Table 3. Results of Algorithms 1 and 2

No	$\hat{\delta}$	Algorithm 1				Algorithm 2			
		$\bar{\delta}$	Std. Dev of $\bar{\delta}$	2.5% CI	97.5% CI	$\bar{\delta}$	Std. Dev of $\bar{\delta}$	2.5% CI	97.5% CI
1	0.8691	0.9073	0.0392	0.9056	0.9091	0.9151	0.0760	0.9119	0.9183
2	1.0000	0.9538	0.0559	0.9512	0.9562	0.9505	0.0777	0.9470	0.9537
3	1.0000	0.9590	0.0584	0.9565	0.9615	0.9635	0.0715	0.9603	0.9667
4	0.9307	0.9588	0.0399	0.9570	0.9606	0.9425	0.0615	0.9398	0.9453
5	1.0000	0.9912	0.0156	0.9906	0.9919	0.9193	0.0752	0.9160	0.9226
6	0.9071	0.9310	0.0627	0.9283	0.9338	0.9410	0.0663	0.9380	0.9438
7	0.8909	0.9333	0.0526	0.9310	0.9356	0.9321	0.0544	0.9298	0.9345
8	0.8208	0.8686	0.0662	0.8657	0.8714	0.8971	0.0711	0.8940	0.9001
9	0.8885	0.9634	0.0416	0.9616	0.9652	0.9629	0.0421	0.9611	0.9646
10	0.8469	0.9117	0.0517	0.9095	0.9140	0.9377	0.0658	0.9347	0.9405
11	0.9531	0.8726	0.0652	0.8697	0.8755	0.8972	0.0696	0.8942	0.9003
12	1.0000	0.9634	0.0349	0.9619	0.9650	0.9373	0.0636	0.9346	0.9402
13	0.9602	0.9804	0.0232	0.9794	0.9814	0.9346	0.0664	0.9316	0.9374
14	1.0000	0.9998	0.0032	0.9997	0.9999	0.9953	0.0443	0.9931	0.9970
15	1.0000	1.0000	0.0000	1.0000	1.0000	1.0000	0.0000	1.0000	1.0000
16	0.8885	0.9550	0.0422	0.9530	0.9568	0.9416	0.0628	0.9388	0.9444
17	1.0000	0.9175	0.0558	0.9151	0.9200	0.9403	0.0626	0.9375	0.9430
18	1.0000	0.9699	0.0586	0.9673	0.9723	0.9566	0.0762	0.9530	0.9599
19	0.9441	0.9083	0.0458	0.9063	0.9104	0.9346	0.0652	0.9316	0.9375
Mean	0.9421	0.9445	0.0428	0.9426	0.9464	0.9420	0.0617	0.9393	0.9447

To summarize the methodological aspects, Algorithm 1 constructs a density tree by maximizing Gaussianity, ensuring that the conditional distribution $\mathbf{X} \mid \mathbf{Y} = \mathbf{y}_h$ used in the sampling procedure is Gaussian. Algorithm 2 adopts a Bayesian framework to account for uncertainty in cluster membership by leveraging a mixture model. Instead of using only the leaf node corresponding to each data point, this algorithm considers all clusters (leaf nodes) in the tree and probabilistically selects a cluster for each data point by computing posterior probabilities, thereby introducing more variability and potentially enhancing the diversity of generated samples. Algorithm 3 builds an ensemble of density trees (a forest) to improve the robustness of the sampling process. By combining predictions from multiple trees, this method reduces variance and mitigates overfitting, while bootstrap sampling and random feature selection ensure diversity among the trees.

Table 4. Results of Algorithm 3

No	$\hat{\delta}$	Algorithm 3 with Algorithm 1				Algorithm 3 with Algorithm 2			
		$\bar{\delta}$	Std. Dev of $\bar{\delta}$	2.5% CI	97.5% CI	$\bar{\delta}$	Std. Dev of $\bar{\delta}$	2.5% CI	97.5% CI
1	0.8691	0.9192	0.0690	0.9162	0.9222	0.9180	0.0674	0.9149	0.9209
2	1.0000	0.9863	0.0349	0.9848	0.9878	0.9869	0.0339	0.9853	0.9883
3	1.0000	0.9899	0.0284	0.9887	0.9911	0.9914	0.0261	0.9903	0.9925
4	0.9307	0.9213	0.0523	0.9191	0.9234	0.9228	0.0510	0.9206	0.9249
5	1.0000	0.9537	0.0598	0.9512	0.9563	0.9519	0.0606	0.9493	0.9546
6	0.9071	0.9182	0.0505	0.9160	0.9203	0.9206	0.0505	0.9185	0.9228
7	0.8909	0.9069	0.0555	0.9046	0.9093	0.9080	0.0558	0.9056	0.9104
8	0.8208	0.9220	0.0652	0.9192	0.9247	0.9224	0.0654	0.9195	0.9252
9	0.8885	0.9192	0.0437	0.9174	0.9212	0.9185	0.0435	0.9167	0.9205
10	0.8469	0.9278	0.0550	0.9254	0.9302	0.9287	0.0549	0.9263	0.9311
11	0.9531	0.9262	0.0616	0.9234	0.9288	0.9258	0.0643	0.9230	0.9286
12	1.0000	0.9408	0.0451	0.9389	0.9427	0.9424	0.0437	0.9406	0.9442
13	0.9602	0.9346	0.0633	0.9318	0.9373	0.9362	0.0642	0.9334	0.9390
14	1.0000	1.0000	0.0001	1.0000	1.0000	1.0000	0.0006	1.0000	1.0000
15	1.0000	1.0000	0.0000	1.0000	1.0000	1.0000	0.0000	1.0000	1.0000
16	0.8885	0.9188	0.0580	0.9164	0.9214	0.9149	0.0569	0.9124	0.9173
17	1.0000	0.9317	0.0516	0.9294	0.9340	0.9275	0.0508	0.9251	0.9298
18	1.0000	0.9854	0.0400	0.9837	0.9871	0.9859	0.0368	0.9843	0.9874
19	0.9441	0.9490	0.0480	0.9469	0.9512	0.9487	0.0502	0.9466	0.9508
Mean	0.9421	0.9448	0.0464	0.9428	0.9468	0.9448	0.0461	0.9428	0.9468

6 Concluding Remarks

We propose three distinct algorithms for sampling inputs conditional on a given level of outputs and estimating confidence intervals for efficiencies. These algorithms offer greater flexibility in selecting sampling procedures in practice and fundamentally differ from previous methods in the following aspects: (a) The proposed algorithms capture the statistical variation in input-output vectors during the sampling process by treating both inputs and outputs as random variables, thereby making the resulting efficiency a random variable as well. In contrast, bootstrap approaches, such as Simar and Wilson [23], assume no measurement errors in the input-output vectors during sampling. Consequently, these methods require a bias-correction process to adjust for errors in the resulting efficiency estimates. (b) Regression-based models, such as StoNED and SFA, often require additional parametric assumptions to estimate DMU-specific efficiencies, as they rely on the conditional distribution of inefficiencies given a composite random error term. The proposed algorithms and bootstrap-based approaches are more intuitive, as they primarily focus on the distribution of efficiencies, allowing for the direct application of conventional statistical confidence interval estimation methods.

One limitation of this study is the lack of empirical comparison between the proposed algorithms and existing confidence interval estimation methods using real-world data. This is primarily because the assumed data generation process differs fundamentally from that of existing methods, and the true values of inputs and outputs for observations are unknown. Moreover, we lack sufficient information to precisely define the production frontier in real-world datasets. Addressing this limitation may require comprehensive and well-designed experiments using various artificial datasets to validate the proposed algorithms. Another limitation is the potential impact of outliers, as the density tree divides the data into multiple clusters, and the presence of outliers could affect the robustness of the sampling process.

Future research directions include extending the proposed approach to other DEA models, such as non-radial or slacks-based measures, exploring alternative probabilistic models beyond the Gaussian assumption, quantifying the impact of the presence of outliers, and developing more robust and efficient algorithms for density tree construction and sampling.

References

1. Aparicio, J., Esteve, M.: How to peel a data envelopment analysis frontier: a cross-validation-based approach. J. Oper. Res. Soc. **74**(12), 2558–2572 (2023)
2. Aparicio, J., Kapelko, M., Ortiz, L.: Enhancing the measurement of firm inefficiency accounting for corporate social responsibility: a dynamic data envelopment analysis fuzzy approach. Eur. J. Oper. Res. **306**(2), 986–997 (2023)
3. Aragon, Y., Daouia, A., Thomas-Agnan, C.: Nonparametric frontier estimation: a conditional quantile-based approach. Economet. Theor. **21**(2), 358–389 (2005)
4. Boubaker, S., Le, T.D., Ngo, T., Manita, R.: Predicting the performance of MSMEs: a hybrid DEA-machine learning approach. Ann. Oper. Res. 1–23 (2023)
5. Cazals, C., Florens, J.P., Simar, L.: Nonparametric frontier estimation: a robust approach. J. Econometrics **106**(1), 1–25 (2002)
6. Criminisi, A., Shotton, J.: Decision Forests for Computer Vision and Medical Image Analysis. Springer, Cham (2013)
7. Daraio, C., Simar, L.: Directional distances and their robust versions: computational and testing issues. Eur. J. Oper. Res. **237**(1), 358–369 (2014)
8. Daraio, C., Simar, L., Wilson, P.W.: Fast and efficient computation of directional distance estimators. Ann. Oper. Res. **288**(2), 805–835 (2020)
9. Dia, M., Takouda, P.M., Golmohammadi, A.: Assessing the performance of Canadian credit unions using a three-stage network bootstrap DEA. Ann. Oper. Res. 1–33 (2022)
10. Esteve, M., Aparicio, J., Rodriguez-Sala, J.J., Zhu, J.: Random forests and the measurement of super-efficiency in the context of free disposal hull. Eur. J. Oper. Res. **304**(2), 729–744 (2023)
11. Fare, R., Grosskopf, S., Kokkelenberg, E.C.: Measuring plant capacity, utilization and technical change: a nonparametric approach. Int. Econ. Rev. 655–666 (1989)
12. Guillen, M.D., Aparicio, J., Esteve, M.: Gradient tree boosting and the estimation of production frontiers. Expert Syst. Appl. **214**, 119,134 (2023)

13. Kang, H.J., Kim, C., Choi, K.: Combining bootstrap data envelopment analysis with social networks for rank discrimination and suitable potential benchmarks. Eur. J. Oper. Res. **312**(1), 283–297 (2024)
14. Kerstens, K., Sadeghi, J., Van de Woestyne, I., Zhang, L.: Malmquist productivity indices and plant capacity utilisation: new proposals and empirical application. Ann. Oper. Res. **315**(1), 221–250 (2022)
15. Kneip, A., Simar, L., Wilson, P.W.: Asymptotics and consistent bootstraps for DEA estimators in nonparametric frontier models. Economet. Theor. **24**(6), 1663–1697 (2008)
16. Kuosmanen, T., Kortelainen, M.: Stochastic non-smooth envelopment of data: semi-parametric frontier estimation subject to shape constraints. J. Prod. Anal. **38**, 11–28 (2012)
17. Lin, S.W., Lu, W.M.: Efficiency assessment of public sector management and culture-led urban regeneration using the enhanced Russell-based directional distance function with stochastic data. J. Oper. Res. Soc. **75**(8), 1624–1642 (2024)
18. Mergoni, A., Emrouznejad, A., De Witte, K.: Fifty years of data envelopment analysis. Eur. J. Oper. Res. **326**(3), 389–412 (2025). https://doi.org/10.1016/j.ejor.2024.12.049
19. Michali, M., Emrouznejad, A., Dehnokhalaji, A., Clegg, B.: Subsampling bootstrap in network DEA. Eur. J. Oper. Res. **305**(2), 766–780 (2023)
20. Moradi-Motlagh, A., Emrouznejad, A.: The origins and development of statistical approaches in non-parametric frontier models: a survey of the first two decades of scholarly literature (1998–2020). Ann. Oper. Res. **318**(1), 713–741 (2022)
21. Ngo, T., Tsui, K.W.H.: Estimating the confidence intervals for DEA efficiency scores of Asia-pacific airlines. Oper. Res. Int. J. **22**(4), 3411–3434 (2022)
22. Shi, Y., Zhao, W.: An integrated machine learning and DEA-predefined performance outcome prediction framework with high-dimensional imbalanced data. INFOR: Inf. Syst. Oper. Res. **62**(1), 100–129 (2024)
23. Simar, L., Wilson, P.W.: Sensitivity analysis of efficiency scores: how to bootstrap in nonparametric frontier models. Manage. Sci. **44**(1), 49–61 (1998)
24. Tsionas, M.G.: Convex non-parametric least squares, causal structures and productivity. Eur. J. Oper. Res. **303**(1), 370–387 (2022)
25. Valero-Carreras, D., Aparicio, J., Guerrero, N.M.: Support vector frontiers: a new approach for estimating production functions through support vector machines. Omega **104**, 102,490 (2021)
26. Walheer, B.: Global Malmquist and cost Malmquist indexes for group comparison. J. Prod. Anal. **58**(1), 75–93 (2022)
27. Zhao, Y.: Empirical estimation of the production frontier. In: International Conference on Data Envelopment Analysis, pp. 59–69. Springer, Cham (2023)
28. Zhu, N., Zhu, C., Emrouznejad, A.: A combined machine learning algorithms and DEA method for measuring and predicting the efficiency of Chinese manufacturing listed companies. J. Manage. Sci. Eng. **6**(4), 435–448 (2021)

Measuring Corporate Sustainability with Data Envelopment Analysis: A Multi-Technique Bibliometric Review and Future Research Agenda

Gokulananda Patel[1], Violeta Cvetkoska[2](✉), and Bojan Kitanovikj[2]

[1] Birla Institute of Management Technology, Greater Noida, India
gn.patel@bimtech.ac.in

[2] Faculty of Economics, Ss. Cyril and Methodius University in Skopje, Skopje, North Macedonia
{vcvetkoska,bojan.kitanovikj}@eccf.ukim.edu.mk

Abstract. The interest in assessing companies' sustainability has been growing among both scholars and practitioners, peaking in recent years. Motivated by that, in this article, we systematically reviewed the literature on using data envelopment analysis (DEA) for measuring corporate sustainability thanks to a multi-technique bibliometric approach consisting of descriptive bibliometric analysis, co-authorships analysis, bibliographic coupling, and keyword co-occurrence analysis. Visualization maps of keyword networks were created based on data from a sample of 392 articles published in Scopus-indexed journals between 2002 – the first year an article on the topic emerged – and 2023, the last completed year at the time of writing. The data was acquired and cleared following the rigorous Preferred Reporting Items for Systematic Reviews and Meta-Analyses (PRISMA) protocol. After the bibliometric analysis, the authors conducted a systematic literature review, separating the literature into five main clusters, including sustainability performance and eco-efficiency, (supply chain) sustainability assessment, environmental implications of corporate sustainability, outcomes of measuring corporate sustainability, and ESG efforts and sustainability support. The article presents the foundational articles and most productive countries and authors, unveiling the inherent intellectual structure of the topic, its recent developmental trajectory, and six avenues for future research.

Keywords: Corporate Sustainability · Data Envelopment Analysis · Bibliometric Analysis

1 Introduction

Businesses today face increasing scrutiny regarding their environmental, social, and governance (ESG) practices, driven by societal demands, regulatory pressures, and stakeholder expectations (Gillan et al., 2021). Originating from ecology and easily finding a place in business and organizational sciences, sustainability and the triple bottom line of

A. Emrouznejad et al. (Eds.): DEA 2024, LNOR, pp. 77–94, 2025.
https://doi.org/10.1007/978-3-031-98177-7_7

profit, people, and planet has been embraced by corporations in various degrees (Ludwig & Sassen, 2022). This comes in the face of an increasing demand for resource conservation and environment protection with society allocating part of the blame to global corporations with large carbon footprints. Recent incidents of unethical corporate practices and so-called greenwashing have again reinvigorated the debate around the sustainability of corporations and properly assessing it for smarter and more efficient and effective decision-making, research and development, business intelligence and advocacy (Ab Wahab, 2021; Walker et al., 2021).

The heightened interest in assessing companies' sustainability has led to the emergence of innovative analytical methods, with data envelopment analysis (DEA) gaining prominence as a powerful non-parametric approach for measuring efficiency across multiple dimensions, which is free of the burdens regarding rigid functional forms (Shokri Kahi et al., 2017; Emrouznejad et al., 2023). DEA's flexibility, relative simplicity, ability to handle big data of inputs and outputs simultaneously, and data analysis reliability, position it as a robust technique for real-world applications (Mergoni et al., 2024). Over the past two decades, DEA has seen exponential growth in application across various fields, including corporate sustainability. Its ability to assess performance by simultaneously accounting for multiple inputs and outputs, without the need for a predefined functional form, makes it particularly well-suited for evaluating the complex relations of environmental, social, and economic factors that underpin sustainability (Ray, 2020). This methodological flexibility has attracted widespread attention from scholars and practitioners seeking quantitative, evidence-based insights into corporate sustainability performance with the ability to combine it with other techniques in the field of operations research and business analytics (Patel, 2019).

The increasing pressure for ESG compliance has further accelerated the adoption of DEA in this domain (Cheng et al., 2023). Regulatory frameworks and voluntary guidelines alike are prompting firms to integrate sustainability into their strategic objectives. However, the growing debate between voluntary corporate social responsibility (CSR) and imposed CSR highlights a critical tension: while some organizations adopt sustainability practices proactively, others respond reactively to external pressures (Gatti et al., 2019). This dichotomy has significant implications for organizational efficiency, as firms strive to balance compliance with innovation in their pursuit of sustainable growth (Jiang et al., 2021). DEA provides a valuable tool to analyse these dynamics, offering insights into how efficiently resources are being utilized to achieve sustainability goals (Fathi et al., 2022).

Motivated by this, the objective of this research endeavour is to investigate the publishing trends of the use of DEA for corporate sustainability assessment, synthesize the findings, and call for further areas which could be tackled by future research. Moreover, this study undertakes a comprehensive bibliometric approach and a systematic review of DEA applications in corporate sustainability research by analysing 392 articles published in Scopus-indexed journals between 2002 and 2023. Thus, it explores the field's intellectual structure, developmental trajectory, and thematic clusters. A multi-technique bibliometric approach—including descriptive bibliometric analysis, co-authorship analysis, bibliographic coupling, and keyword co-occurrence analysis—provides a detailed map of the field's evolution, identifying foundational works, prolific contributors, and emerging trends.

The systematic literature review categorizes existing research into five key clusters: sustainability performance and eco-efficiency, (supply chain) sustainability assessment, environmental implications of corporate sustainability, outcomes of measuring corporate sustainability, and ESG efforts and sustainability support. These clusters highlight the diverse applications of DEA, emphasizing its role in addressing the challenges of integrating sustainability into organizational strategy and operations. By synthesizing prior scholarship, identifying research gaps, and proposing six avenues for future exploration, this study contributes to the growing discourse on measuring and managing corporate sustainability. Beyond academic insights, it offers practical implications for policymakers and business leaders aiming to enhance organizational efficiency and sustainability performance in an era of heightened accountability and environmental urgency.

2 Methods and Data

To fulfil the research objective, the study presents a multi-technique bibliometric analysis, which has been proven as a research instrument for evaluating and analysing scientific literature by uncovering the research dynamics of a scientific niche, often using visualization maps (Zupic & Čater, 2015). This analysis has gained in popularity due to the increased accessibility of data for text mining and software for its analysis, as well as the more objective, bird's eye view of a certain research field, which is useful for scholars and the advancement of the discipline in question.

Considering this, a multi-technique bibliometric analysis is conducted, which consists of descriptive bibliometric analysis, co-authorship analysis, bibliographic coupling, and keyword co-occurrence analysis. While the descriptive and the co-authorship analyses are some of the first such techniques to appear, they are still widely used today to single out the main publication trends, the most influential publications, and scientific voices as well as to showcase the dynamics and collaborations between authors, countries, and contexts. Additionally, bibliographic coupling puts together the research works based on the similarity of their reference lists, yielding findings related to the domains of the broader scientific literature. To top it all off, the keyword co-occurrence analysis establishes links between main topics or keywords, based on the content of the research itself, making sure to highlight the topics that are researched together and across years (Donthu et al., 2021).

Regarding the data, a search query was performed in the Scopus database, which indexes quality research outputs from all over the world for the following keywords in the abstract, title, and keywords of the articles: "data envelopment analysis" OR "DEA" AND "sustainability". The research endeavour focused solely on data from Scopus-indexed scientific journals, meaning original, double-blind, peer-reviewed articles. There were no limitations set on the timeframe, so the data referred to the period of 2002 as the first year an article on the topic emerged until 2023 as the latest completed year at the time of writing.

Afterward, to clean the data and come up with the final sample, the Preferred Reporting Items for Systematic Reviews and Meta-Analyses (PRISMA) guidelines were followed (Page et al., 2021). The undertaken steps per the PRISMA guidelines are provided in Fig. 1.

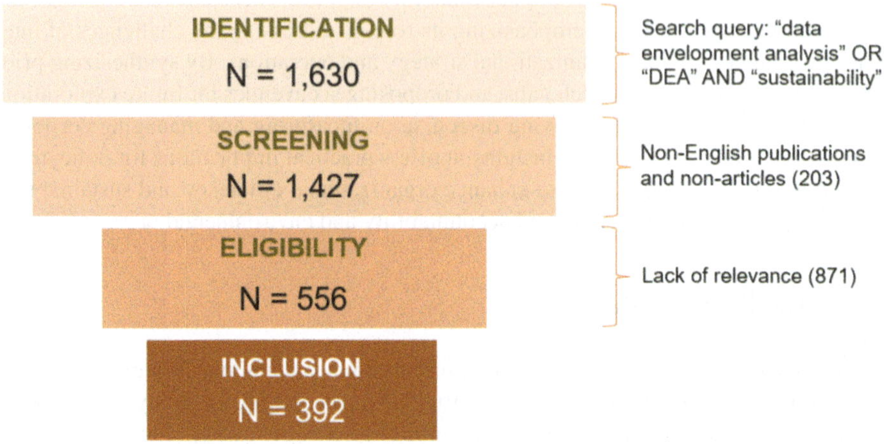

Fig. 1. Data identification and inclusion steps per PRISMA guidelines

The initial data from the search query resulted in 1,630 articles. This data underwent a detailed, two-person manual relevance check to determine the suitability for inclusion in the research. In the end, the final sample consisted of 392 articles.

3 Results

Starting from the descriptive bibliometric analysis approach, it can be observed that the interest in applying DEA methodologies for sustainability assessment at the corporate level is growing. Figure 2 presents the publication trends as an evolutionary development of the entire field as well as the trends in the last five years.

Between 2001 and 2010, research activity in this area was minimal, with only one or zero articles published annually. This reflects the nascent stage of combining DEA methods with sustainability research, where awareness and interest were limited. From 2011 onwards, there is a noticeable but modest increase in publications. This suggests a growing recognition of DEA as a useful tool for analysing sustainability, coinciding with an increase in global attention to sustainability metrics like ESG. Starting from 2019, the number of published articles rose sharply, peaking at 74 articles in 2023. This exponential growth reflects a surge in interest, likely driven by rising regulatory pressures for ESG compliance, global sustainability goals, organizational efficiency, and the increasing adoption of advanced analytical tools like DEA in academic and practical settings.

The five-year trends highlight the rapid rise in publications during the most recent years, with consistent growth from 36 articles in 2019 to 74 in 2023. The slight increase

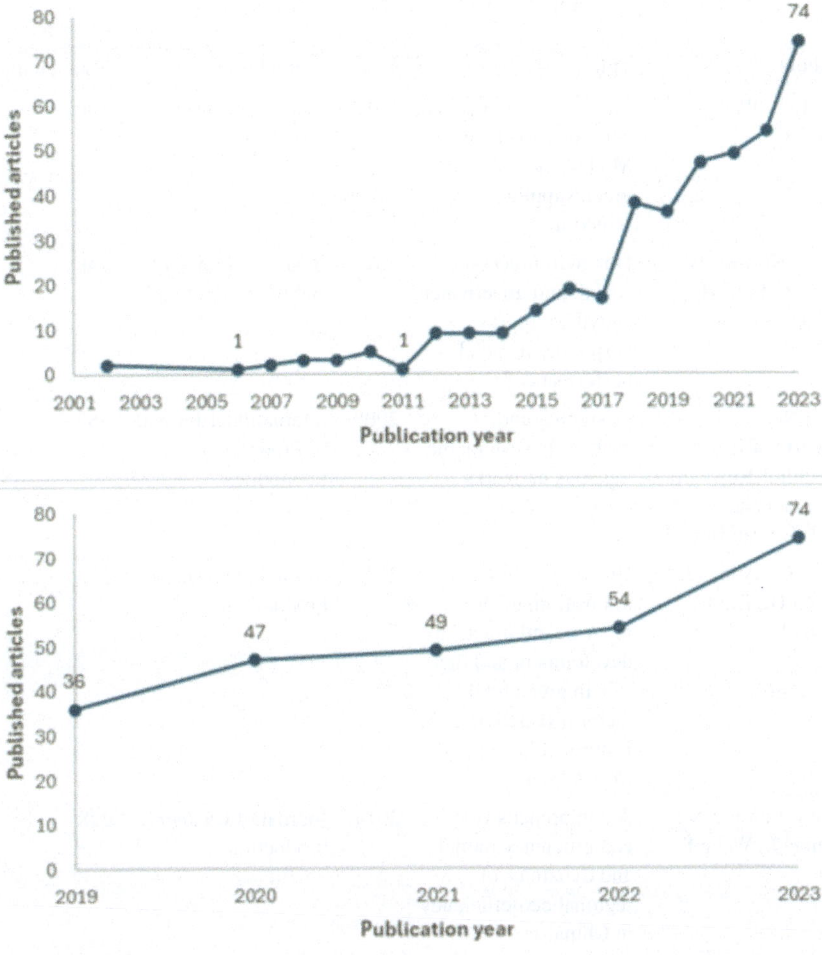

Fig. 2. Publication trends for the use of DEA for corporate sustainability assessment

from 2020 to 2021, followed by a sharper rise from 2022 to 2023, could be linked to the global pandemic's aftermath and the growing business concern for boosting efficiency where sustainability gained prominence as businesses and governments restructured priorities.

In this sense, a couple of articles and scientific voices have emerged as exceptionally influential in the field. Subsequently, the seven most cited articles are shown in Table 1 as each of them has more than two hundred citations.

When it comes to the authors, expectedly, the most cited ones are Kuo R.J., Wang Y.C., and Tien F.C. with 536 citations followed by Xie J. (466 citations) and Nozawa W., Yagi M., Fujii H., and Managi S. each with 448 citations.

Table 1. Most cited articles in the field

Authors	Title	Year	Journal title	Total citations
Kuo R.J.; Wang Y.C.; Tien F.C	Integration of artificial neural network and MADA methods for green supplier selection	2010	Journal of Cleaner Production	536
Xie J.; Nozawa W.; Yagi M.; Fujii H.; Managi S	Do environmental, social, and governance activities improve corporate financial performance?	2019	Business Strategy and the Environment	448
Frota Neto J.Quariguasi; Bloemhof-Ruwaard J.M.; van Nunen J.A.E.E.; van Heck E	Designing and evaluating sustainable logistics networks	2008	International Journal of Production Economics	357
Jiakui C.; Abbas J.; Najam H.; Liu J.; Abbas J	Green technological innovation, green finance, and financial development and their role in green total factor productivity: Empirical insights from China	2023	Journal of Cleaner Production	272
Huang J.; Yang X.; Cheng G.; Wang S	A comprehensive eco-efficiency model and dynamics of regional eco-efficiency in China	2014	Journal of Cleaner Production	228
Lee K.-H.; Farzipoor Saen R	Measuring corporate sustainability management: A data envelopment analysis approach	2012	International Journal of Production Economics	215
Kuang B.; Lu X.; Zhou M.; Chen D	Provincial cultivated land use efficiency in China: Empirical analysis based on the SBM-DEA model with carbon emissions considered	2020	Technological Forecasting and Social Change	204

The articles in this research field were published in 124 distinct, reputable scientific journals with 13 of them publishing at least five articles each. Yet, the leading five ones are the *Journal of Cleaner Production* (166 articles), *Socio-Economic Planning Sciences* (16 articles), *Technological Forecasting and Social Change* (12 articles), *International Journal of Production Economics* (10 articles), and *Business Strategy and the Environment* (7 articles).

Moving onto the bibliographic coupling, we can see the extent to which two documents cite the same sources, helping to identify thematic clusters and intellectual linkages in the literature (Zupic & Čater, 2015). The network, presented in Fig. 3 as an output of the bibliographic technique is divided into several color-coded clusters, each representing a group of articles with high bibliographic coupling. These clusters indicate distinct research themes or subfields within DEA and sustainability. Illustratively, the red cluster focuses on recent advancements in ESG assessments, as many articles from 2021–2023 are prominent in this group, with key authors including Jiakui et al. (2023), Cui et al. (2021), and Peng et al. (2021). The blue cluster is dominated by foundational works, such as Bai and Sarkis (2014). This cluster may focus on general DEA methodology and its application to sustainability performance evaluations in niche business operations like supply chain management. Besides the prominent names, some articles positioned at the periphery (e.g., Ina et al. (2020), Lapinskaitė & Skvarciany (2023)) may represent emerging or niche research areas that are less integrated into the mainstream but have the potential to influence future developments.

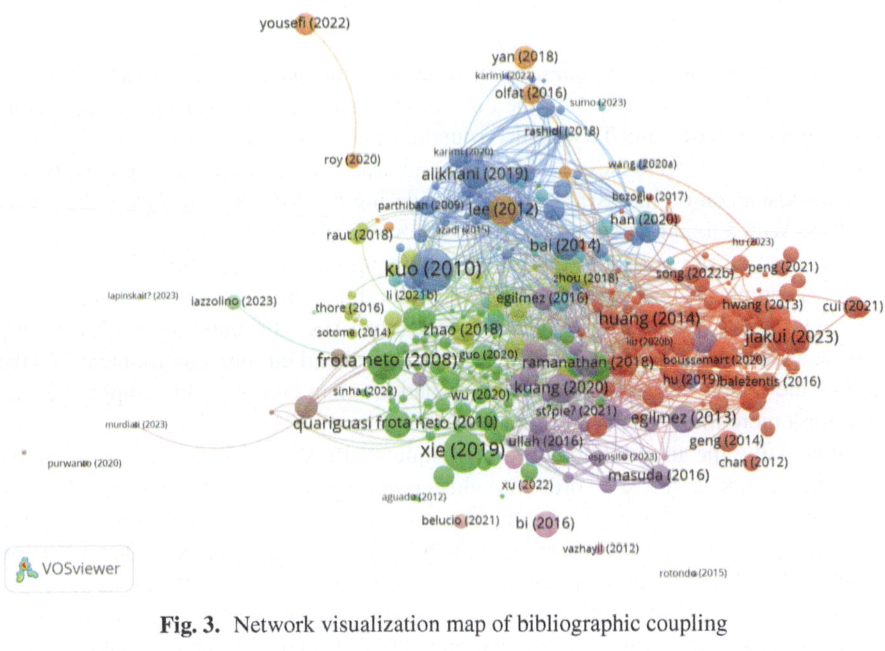

Fig. 3. Network visualization map of bibliographic coupling

The dense network of lines connecting the keywords demonstrates strong intellectual linkages between articles, reflecting a cohesive research domain. Articles within clusters are more closely related, but there are also cross-cluster connections, indicating interdisciplinary applications and overlapping methodologies.

On the other hand, the country co-authorship analysis presents a diverse landscape of DEA applications in corporate sustainability assessment. Table 2 presents the leading countries by the number of published articles and the total citations.

Table 2. Countries based on the total number of published articles and citations

Country	Number of articles	Country	Number of citations
China	143	China	6314
Iran	48	United States	2128
United States	37	Iran	1596
Australia	24	Australia	1419
India	23	Taiwan	1117
Taiwan	22	United Kingdom	1022
United Kingdom	21	Japan	788
Italy	18	Netherlands	595
Spain	18	Canada	583
Oman	16	Pakistan	566

As it can be noted, the Chinese context has been the most productive and cited one in the research field, followed by the ones in Iran and the United States. Contributors from Australia, India, and Taiwan also round up the top five in both aspects. When we look at the interconnectedness between the countries and the peers among ourselves, a network visualization map was created considering 63 distinct countries, which form 184 links with a total link strength of 302. It is shown in Fig. 4.

The intra-cluster collaboration is evident among the authors, while there are many positive examples of peers collaborating and researching different contexts. For instance, the cooperation between authors from China, Oman, Australia, and Iran, while there are also scientists who mainly stick to their geographical and cultural environment like the case of the Dutch, Italian, Spanish, Swiss, and German authors. This emphasizes the need for a more in-depth cross-regional collaboration.

Eventually, the final bibliometric technique is the keyword co-occurrence analysis, which connects the keywords into clusters based on their content and if they are researched together. Figure 5 presents the network visualization map, separating the keywords into five clusters: yellow (sustainability performance and eco-efficiency), red ((supply chain) sustainability assessment), green (environmental implications of corporate sustainability), blue (outcomes of measuring corporate sustainability), and violet (ESG efforts and sustainability support). The map is made up of 108 original keywords connected with 2,603 links and a total link strength of 19,716.

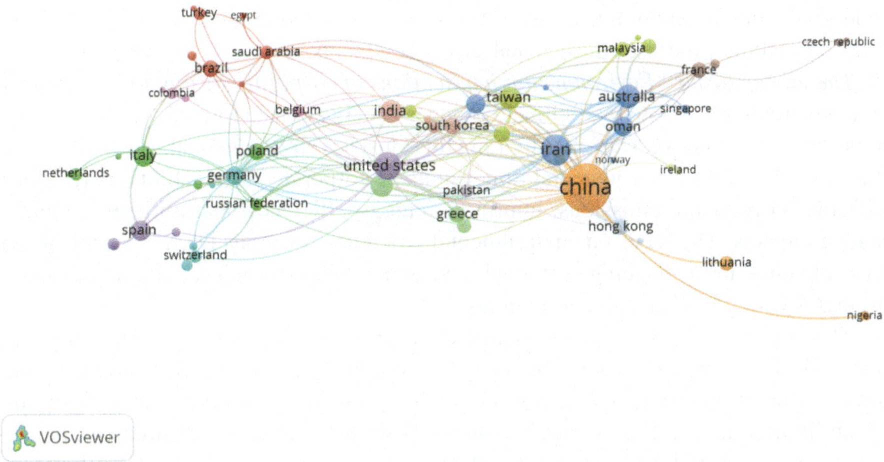

Fig. 4. Network visualization map of co-authorship analysis

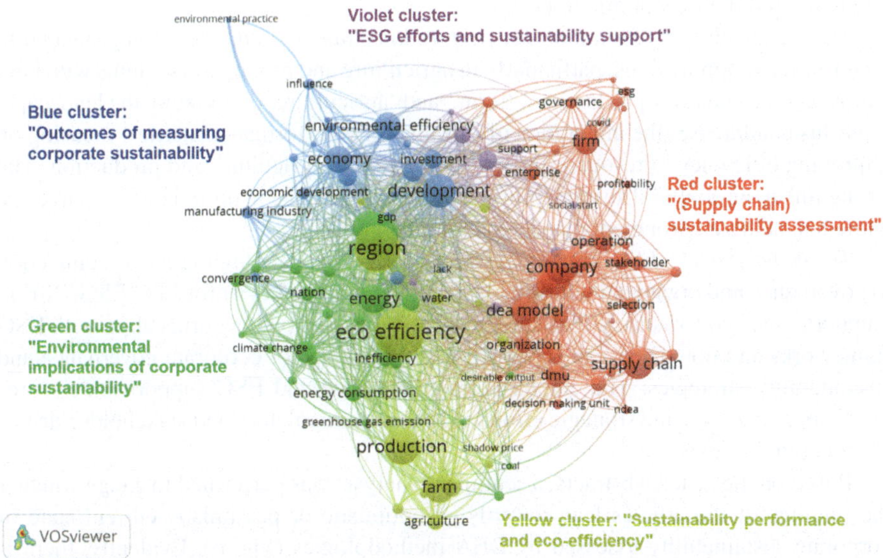

Fig. 5. Network visualization map of keyword co-occurrence

The red cluster "(Supply chain) sustainability assessment" focuses on keywords such as 'company,' 'supply chain,' 'DEA model,' 'stakeholder,' and 'operation' which dominate this cluster, suggesting a strong emphasis on the application of DEA in assessing sustainability within supply chains and corporate settings. This cluster reflects research on optimizing operations and supply chains to improve sustainability performance, stakeholder engagement, and governance structures. It ties closely to corporate social responsibility (CSR) and ESG compliance efforts. The prominence of 'DEA model' underscores the methodological role of DEA in addressing inefficiencies and improving decision-making processes in sustainability assessments.

The green cluster "Environmental implications of corporate sustainability" considers keywords such as 'energy,' 'climate change,' 'energy consumption,' 'greenhouse gas emission,' and 'convergence,' highlighting this cluster's environmental orientation. This cluster explores the environmental dimensions of corporate sustainability, particularly focusing on reducing emissions, improving energy efficiency, and mitigating climate change impacts. The focus on environmental efficiency positions DEA as a critical tool for evaluating and enhancing corporate environmental performance, aligning business strategies with global sustainability goals.

The blue cluster "Outcomes of measuring corporate sustainability" brings together keywords like 'environmental efficiency,' 'economy,' 'economic development,' and 'manufacturing industry,' indicating an emphasis on the economic outcomes of sustainability initiatives. This research explores how sustainability measurement impacts broader economic indicators such as GDP, economic growth, and industrial development. It also examines the balance between environmental and economic objectives, suggesting the use of DEA in evaluating the trade-offs and synergies between sustainability initiatives and their economic impacts.

The yellow cluster "Sustainability performance and eco-efficiency" is grounded in sector-specific applications, particularly in agriculture and production systems with keywords 'eco-efficiency,' 'production,' 'farm,' and 'agriculture' dominating the landscape. This cluster addresses the integration of sustainability performance metrics, focusing on improving efficiency in resource-intensive sectors like agriculture and production. The strong link to agricultural sustainability indicates DEA's role in optimizing resource use while balancing environmental and economic outputs.

The violet cluster "ESG efforts and sustainability support" highlights the connection to governance and organizational structures with prevailing keywords like 'ESG,' 'firm,' 'support,' and 'governance'. This cluster reflects research on how firms align with ESG frameworks and how DEA can support decision-making for corporate governance and sustainability strategies. The emphasis on governance and ESG support underscores DEA's potential to guide strategic efforts in achieving regulatory and stakeholder-driven sustainability targets.

Based on the article abstracts, a sentiment analysis was performed to gauge whether the findings of the articles lean towards an optimistic or pessimistic current state of corporate sustainability assessed by DEA methodologies (Fig. 6). Evidently, there is mostly a positive sentiment in the research landscape where different DEA variants have pointed out the efficient aspects of corporate sustainability, while more than 50 articles found negative relationships in their sustainability operations.

Building on the high-level sentiment directions in Fig. 6, Fig. 7 offers a more granular breakdown of how strongly the authors of the reviewed DEA articles articulated their findings in the sense of whether notably positive or negative terms dominate the narrative. Both figures are rooted in sentiment analysis, however, Fig. 7 highlights the extent or strength of sentiment, rather than simply the sentiment direction. It describes the sentiment analysis scores in more detail, showcasing the growing importance and the strong words used in the research landscape to determine the optimism/pessimism of

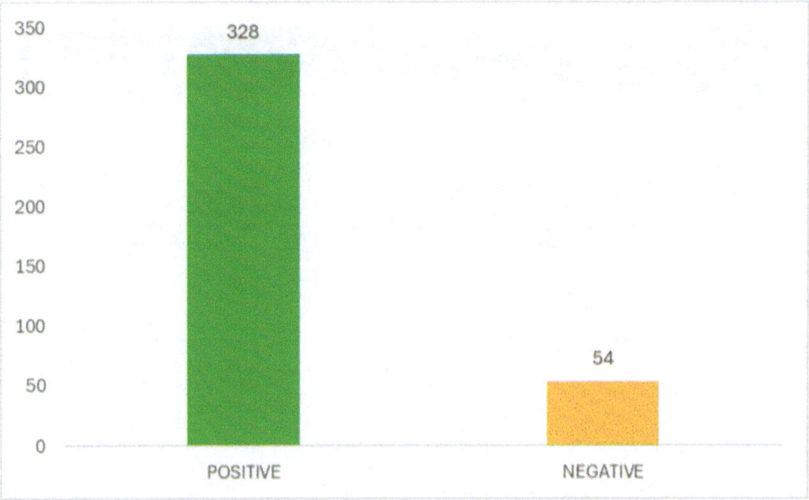

Fig. 6. Sentiment analysis output

the findings. Namely, the average score was 0.909, while the median one was 0.968. Furthermore, the range of the relevance scores went from 0.501 until 0.998.

Regarding the connection to DEA, the words identified in the sentiment analysis reflect the authors' assessments of sustainability performance as measured or discussed through DEA. Illustratively, articles that used words like "efficient," "waste reduction," "optimistic results," or "significant benefits" when describing corporate sustainability best practices thanks to DEA models, were labelled as having a positive sentiment. These words push the sentiment score closer to the upper range (above 0.9), indicating a high level of confidence or positivity about DEA's role in identifying strong sustainability performance. On the other hand, studies, that used terms such as "inefficient," "burden," "uneven progress," or "disappointing outcomes" in the context of DEA evaluations can indicate challenges or negative aspects, lowering the sentiment score.

Thus, the stronger or more extreme the word choice, the higher or lower the sentiment score becomes, helping us discern not only whether the authors' conclusions are positive or negative overall but also how emphatically they view DEA's role in corporate sustainability. This can help researchers interested in DEA applications to consider how the DEA-based studies are framed in the literature and the intensity with which findings are communicated in terms of pointing out promising pathways for improvement or highlighting bottlenecks and challenges (Wankhade et al., 2022).

4 Discussion and Review

The application of DEA in sustainability research has demonstrated significant value in addressing the complex and multidimensional challenges of corporate sustainability performance. By enabling the measurement of efficiency across diverse inputs and outputs, DEA has emerged as a robust tool for firms aiming to align their operations with ESG compliance requirements (Lu et al., 2023).

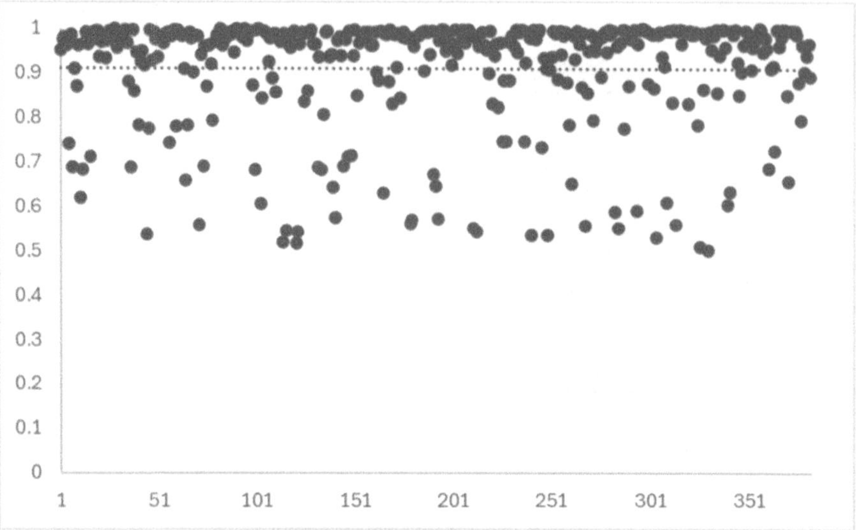

Fig. 7. Analysis of sentiment relevance scores

DEA has proven to be an effective approach for evaluating environmental performance (Mishra et al., 2023), and the efficiency of firms' sustainability activities, offering actionable insights for ESG compliance (Shu et al., 2024). As firms face mounting pressure from stakeholders, regulators, and society to prioritize sustainability, DEA provides a quantitative framework to identify inefficiencies and opportunities for improvement. By considering multiple dimensions of sustainability—such as environmental performance, social impact, and governance practices—DEA helps organizations assess their overall alignment with ESG criteria. Its ability to account for heterogeneous inputs and outputs positions it as a critical tool for firms striving to demonstrate compliance with increasingly stringent ESG standards (Xie et al., 2019).

One of DEA's most impactful contributions is its role in identifying industry benchmarks for corporate sustainability (Zaare Tajabadi & Daneshvar, 2023). By evaluating the performance of peer firms or industry leaders, DEA helps establish best practices and guides firms toward more sustainable operational strategies.

The flexibility of DEA in incorporating diverse inputs and outputs has been instrumental in evaluating sustainability performance. Inputs such as resource consumption, energy use, and greenhouse gas emissions, and outputs like reductions in carbon footprint, improvements in social welfare, and economic contributions, are seamlessly integrated into the DEA framework (Jiang et al., 2021; Mwambo et al., 2021). This ability to capture the multidimensional nature of sustainability ensures that DEA provides a holistic view of corporate performance. For instance, DEA models have been used to assess firms' ability to achieve eco-efficiency by minimizing resource use while maximizing positive environmental and social outcomes (Kuo et al., 2020).

In recent years, sector-specific applications of DEA have risen significantly, reflecting a growing recognition of the unique sustainability challenges faced by different industries. The development and application of advanced DEA models, such as Malmquist

DEA, stochastic DEA, and other variations, have opened new avenues for research and practice. Malmquist DEA, for example, allows for dynamic assessments of efficiency over time, capturing the effects of technological progress and policy changes (Voltes-Dorta et al., 2024). Stochastic DEA models have introduced probabilistic elements, enabling researchers to account for uncertainties and variability in input and output data, which are particularly relevant in sectors like agriculture, manufacturing, and energy (Jahani Sayyad Noveiri et al., 2021). Table 3 presents the most-used DEA techniques in sustainability research together with their relevance scores. Interestingly enough, it is evident that DEA is sometimes paired with AHP, ANP, ANN, regression panel analysis, and similar for more comprehensive results.

Table 3. Applied DEA methods and their supplementary techniques

DEA method	Relevance score	Supplementary method	Relevance score
NDEA	1.57	ANN	1.42
eco-DEA	1.43	multiple attitude decision making	1.23
FS DEA	1.38	ANP	1.07
fuzzy DEA	1.17	multi objective optimization	1.05
DEA TOPSIS	1.15	AHP	0.95
CCR DEA	1.11	tobit	0.93
dynamic DEA	1.01	meta frontier	0.79
multiple stage DEA	0.96	panel analysis	0.77
EM DEA	0.92		
DEA	0.89		
ZSG DEA	0.76		

One of the contemporary aspects of DEA research is treating bad outputs and inputs in empirical settings. In many real-world contexts, production processes generate not only desirable outputs (e.g., goods and services) but also undesirable outputs such as pollution, waste, or defects (Halkos & Petrou, 2019). Incorporating these "bad" outputs into the DEA methodology has gained new ground, prompting fresh modelling frameworks (Barbosa et al., 2023). Some of the frequently used approaches for tackling this type of inputs and outputs include weak disposability, by-production, and joint production (Taleb et al., 2023; Ray et al., 2018; Førsund, 2021).

The weak disposability assumption posits that undesirable outputs can only be reduced if desirable outputs are simultaneously scaled down. Illustratively, Taleb et al. (2023) proposed an improved super-efficiency DEA model under weak disposability to evaluate the environmental efficiency of port operations aiming for Net Zero targets. Similarly, Dar et al. (2021) employed weak disposability in assessing Indian banks' efficiency, demonstrating how accounting for bad outputs (e.g., non-performing assets) can alter performance rankings. While weak disposability is intuitive and relatively straightforward to implement, it can sometimes oversimplify pollution decline or waste-reduction strategies by forcing proportional reductions of all outputs. In contrast, the by-production framework separates the production of undesirable outputs into a distinct sub-technology, thereby allowing for an increased representation of pollution or defects generated by specific stages of the production process. Ray et al. (2018) showed how different abatement strategies can affect overall efficiency by treating bad outputs as separate, yet related, sub-technologies. It is demonstrated that this approach can result in fresh policy insights, especially when combined with advanced sensitivity analyses. The joint production approach recognizes that certain processes produce good and bad outputs inextricably, within a single technology. Førsund (2021) offered a comprehensive examination of performance measurement under joint production, emphasizing how intended (desirable) and unintended (undesirable) outputs can be inherently linked. Because both outputs originate from the same technology, joint production models often provide more realistic efficiency measures when those outputs cannot be separated at any production stage.

Still, as weak disposability remains accessible and widely used, by-production and joint production are said to provide more in-depth analyses of how undesirable outputs arise and might be mitigated (Jahanshahloo et al., 2005; Liu et al., 2010). Hence, it is worth noting that the labels "desirable" or "undesirable" when speaking of DEA inputs and outputs can be the result of context-dependent assumptions about bad inputs as well as bad outputs (Wojcik et al., 2017).

Thus, while DEA has made substantial contributions, challenges remain in its application to corporate sustainability. Data availability and quality are critical concerns, as the reliability of DEA results depends on accurate and comprehensive input and output metrics (Fotova Čiković et al., 2022). Moreover, the lack of standardized sustainability indicators across industries poses a challenge for benchmarking and cross-sectoral comparisons (Gerdessen & Pascucci, 2013). However, these challenges also present opportunities for further refinement and development of DEA methodologies. Incorporating emerging data sources, such as big data and remote sensing, and expanding the use of hybrid models that integrate DEA with other analytical tools can enhance its applicability and relevance.

In this sense, we propose several areas for future research grounded in the approach of text mining the abstracts, titles, and keywords, which pointed out the following keywords as those that are mostly researched from 2021 onwards 'social start,' 'circular economy,' 'ESG', 'renewable energy,' 'port,' 'investment', and more. This agenda for future research is as follows:

- There is a growing need for industry-specific longitudinal sustainability analysis which can span several years and different industries to get a more comprehensive picture of the developments of sustainability and efficiency in this regard,
- Future studies can look into combining DEA with machine learning techniques for enhancing predictive capabilities,
- Future models could explore trade-offs between environmental, social, and economic goals as they may overlap, but not always, making it harder for managers to choose a single strategic path and fulfilment of all objectives in a specific area,

- Further examination is needed into how organizational behaviour influences DEA-measured sustainability efficiency, including top management support, voluntary actions from employees, employee citizenship behaviour, leadership styles, and more,
- New research contexts are needed considering that most of the studies are centred around countries in America and Asia, often neglecting other regions,
- Further introduction of new metrics is needed like for example, those associated with the circular economy, which is becoming an increasingly large priority in the corporate world.

5 Conclusion

This study aimed to provide a comprehensive review of the application of DEA in measuring corporate sustainability, addressing a growing interest among scholars and practitioners in understanding and evaluating firms' sustainability efforts. By employing a rigorous bibliometric and systematic review approach, grounded in the PRISMA protocol, we analysed 392 Scopus-indexed articles published between 2002 and 2023. Through techniques such as bibliometric analysis, co-authorship network mapping, bibliographic coupling, and keyword co-occurrence analysis, we identified five key thematic clusters: sustainability performance and eco-efficiency, (supply chain) sustainability assessment, environmental implications of corporate sustainability, outcomes of measuring corporate sustainability, and ESG efforts and sustainability support.

The findings unveil the intellectual structure of the field, highlighting foundational articles, key contributors, and influential countries driving research on DEA and sustainability. The bibliometric analysis demonstrates the exponential growth in publications, reflecting the increasing recognition of DEA as a robust method for analysing the multidimensional nature of sustainability challenges. By integrating diverse inputs, such as resource consumption and emissions, and outputs, such as carbon footprint reduction and social impact, DEA has become instrumental in benchmarking sustainability performance, guiding firms toward greater efficiency, and achieving ESG compliance. The dense connections between clusters reflect the interdisciplinary nature of DEA in sustainability research. Themes such as eco-efficiency, ESG efforts, and supply chain assessments often overlap, highlighting the integrated approach needed to address complex sustainability challenges.

Bibliometric analysis is a powerful method for mapping and understanding research trends, but it has several limitations that should be considered when interpreting its findings. Hence, the study relied on the Scopus database, which may lead to biased results, yet this can be alleviated with the inclusion of other databases like IEEE, Google Scholar, and more. Also, highly cited articles or authors may dominate the analysis, potentially overshadowing newer, niche, or interdisciplinary research. Yet, this type of literature analysis remains the most objective and widely applied in management, operations research, business analytics, and intelligence.

By synthesizing existing knowledge and proposing avenues for future research, this review contributes to advancing the understanding of DEA's role in sustainability assessment. The identified research gaps, such as integrating new sustainability metrics, developing hybrid DEA models, and exploring under-researched regions and sectors, provide a roadmap for future inquiry. With this, this study is poised to enrich the academic dialogue on corporate sustainability and offer practical insights for policymakers and organizations seeking to align their operations with global sustainability goals.

References

Ab Wahab, M.: Is an unsustainability environmentally unethical? Ethics orientation, environmental sustainability engagement and performance. J. Clean. Prod. **294**, 126240 (2021)

Bai, C., Sarkis, J.: Determining and applying sustainable supplier key performance indicators. Supply Chain Manag. Int. J. **19**(3), 275–291 (2014)

Barbosa, A.B.R.A.B., Lopes-Ahn, A., Vilela, B., Costa, M., de Oliveira, R., Gomes, H.A.: DEA with bootstrap, weight restrictions and negative outputs: shortcomings of the proposal for the Brazilian distribution system operators regulation. In: Emrouznejad, A., Thanassoulis, E., Toloo, M. (eds.) Advances in the Theory and Applications of Performance Measurement and Management: Proceedings of DEA45—International Conference on Data Envelopment Analysis, pp. 47–58. Springer, Cham (2024). https://doi.org/10.1007/978-3-031-61597-9_5

Cheng, L.T., Lee, S.K., Li, S.K., Tsang, C.K.: Understanding resource deployment efficiency for ESG and financial performance: a DEA approach. Res. Int. Bus. Financ. **65**, 101941 (2023)

Cui, X., Zhao, T., Wang, J.: Allocation of carbon emission quotas in China's provincial power sector based on entropy method and ZSG-DEA. J. Clean. Prod. **284**, 124683 (2021)

Dar, A.H., Mathur, S.K., Mishra, S.: The efficiency of indian banks: a DEA, Malmquist and SFA analysis with bad output. J. Quant. Econ. **19**, 653–701 (2021)

Donthu, N., Kumar, S., Mukherjee, D., Pandey, N., Lim, W.M.: How to conduct a bibliometric analysis: an overview and guidelines. J. Bus. Res. **133**, 285–296 (2021)

Emrouznejad, A., Marra, M., Yang, G.L., Michali, M.: Eco-efficiency considering NetZero and data envelopment analysis: a critical literature review. IMA J. Manag. Math. **34**(4), 599–632 (2023)

Fathi, A., Karimi, B., Saen, R.F.: Sustainability assessment of supply chains by a novel robust two-stage network DEA model: a case study in the transport industry. Soft. Comput. **26**(13), 6101–6118 (2022)

Førsund, F.R.: Performance measurement and joint production of intended and unintended outputs. J. Prod. Anal. **55**, 157–175 (2021)

Fotova Čiković, K., Martinčević, I., Lozić, J.: Application of data envelopment analysis (DEA) in the selection of sustainable suppliers: a review and bibliometric analysis. Sustainability **14**(11), 6672 (2022)

Gatti, L., Vishwanath, B., Seele, P., Cottier, B.: Are we moving beyond voluntary CSR? Exploring theoretical and managerial implications of mandatory CSR resulting from the new Indian companies act. J. Bus. Ethics **160**, 961–972 (2019)

Gerdessen, J.C., Pascucci, S.: Data envelopment analysis of sustainability indicators of European agricultural systems at regional level. Agric. Syst. **118**, 78–90 (2013)

Gillan, S.L., Koch, A., Starks, L.T.: Firms and social responsibility: a review of ESG and CSR research in corporate finance. J. Corp. Finan. **66**, 101889 (2021)

Halkos, G., Petrou, K.N.: Treating undesirable outputs in DEA: a critical review. Econ. Anal. Policy **62**, 97–104 (2019)

Ina, P., Erie, F., Dian, M.: Alleviating poverty through islamic microfinance: outreach and efficiency (study on BMT and islamic cooperatives in Indonesia). Экономика региона **16**(4), 1362–1376 (2020)

Noveiri, M.J.S., Kordrostami, S., Amirteimoori, A.: Sustainability assessment and most productive scale size: a stochastic DEA approach with dual frontiers. Environ. Model. Assess. **26**(5), 723–735 (2021)

Jahanshahloo, G.R., Lotfi, F.H., Shoja, N., Tohidi, G., Razavyan, S.: Undesirable inputs and outputs in DEA models. Appl. Math. Comput. **169**(2), 917–925 (2005)
Jiakui, C., Abbas, J., Najam, H., Liu, J., Abbas, J.: Green technological innovation, green finance, and financial development and their role in green total factor productivity: empirical insights from China. J. Clean. Prod. **382**, 135131 (2023)
Jiang, T., Zhang, Y., Jin, Q.: Sustainability efficiency assessment of listed companies in China: a super-efficiency SBM-DEA model considering undesirable output. Environ. Sci. Pollut. Res. **28**(34), 47588–47604 (2021)
Kuo, K.C., Lu, W.M., Kweh, Q.L., Le, M.H.: Determinants of cargo and eco-efficiencies of global container shipping companies. Int. J. Logist. Manag. **31**(4), 753–775 (2020)
Lapinskaitė, I., Skvarciany, V.: Sustainable metamorphosis: examining sustainability transformation into value of financial institutions. J. Bus. Econ. Manag. **24**(5), 923–938 (2023)
Liu, W.B., Meng, W., Li, X.X., Zhang, D.Q.: DEA models with undesirable inputs and outputs. Ann. Oper. Res. **173**, 177–194 (2010)
Lu, W.M., Kweh, Q.L., Ting, I.W.K., Ren, C.: How does stakeholder engagement through environmental, social, and governance affect eco-efficiency and profitability efficiency? zooming into Apple Inc.'s counterparts. Bus. Strat. Environ. **32**(1), 587–601 (2023)
Ludwig, P., Sassen, R.: Which internal corporate governance mechanisms drive corporate sustainability? J. Environ. Manag. **301**, 113780 (2022)
Mergoni, A., Emrouznejad, A., De Witte, K.: Fifty years of data envelopment analysis. Eur. J. Oper. Res. **326**(3), 389–412 (2025). https://doi.org/10.1016/j.ejor.2024.12.049
Mishra, S., Singh, A., Sufia: Environmental performance of coal-based thermal generation plants: a two-stage DEA model with undesirable outputs. In: Emrouznejad, A., Thanassoulis, E., Toloo, M. (eds.) Advances in the Theory and Applications of Performance Measurement and Management: Proceedings of DEA45—International Conference on Data Envelopment Analysis, pp. 151–163. Springer, Cham (2024). https://doi.org/10.1007/978-3-031-61597-9_13
Mwambo, F.M., Fürst, C., Martius, C., Jimenez-Martinez, M., Nyarko, B.K., Borgemeister, C.: Combined application of the EM-DEA and EX-ACT approaches for integrated assessment of resource use efficiency, sustainability and carbon footprint of smallholder maize production practices in sub-Saharan Africa. J. Clean. Prod. **302**, 126132 (2021)
Page, M.J., et al.: The PRISMA 2020 statement: an updated guideline for reporting systematic reviews. bmj **372** (2021)
Patel, G.: Integrations and applications of analytic hierarchy process with data envelopment analysis–a literature review. Int. J. Anal. Hierarchy Process **11**(2), 228–268 (2019)
Peng, C., Li, B., Nan, B.: An analysis framework for the ecological security of urban agglomeration: a case study of the Beijing-Tianjin-Hebei urban agglomeration. J. Clean. Prod. **315**, 128111 (2021)
Ray, S.C.: Data envelopment analysis: a nonparametric method of production analysis. In: Handbook of Production Economics, pp. 1–62 (2020)
Ray, S.C., Mukherjee, K., Venkatesh, A.: Nonparametric measures of efficiency in the presence of undesirable outputs: a by-production approach. Empir. Econ. **54**(1), 31–65 (2018)
Shokri Kahi, V., Yousefi, S., Shabanpour, H., Farzipoor Saen, R.: How to evaluate sustainability of supply chains? A dynamic network DEA approach. Ind. Manag. Data Syst. **117**(9), 1866–1889 (2017)
Shu, T., Liao, X., Yang, S., Yu, T.: Towards sustainability: Evaluating energy efficiency with a super-efficiency SBM-DEA model across 168 economies. Appl. Energy **376**, 124254 (2024)
Taleb, M., Khalid, R., Emrouznejad, A., Ramli, R.: Environmental efficiency under weak disposability: an improved super efficiency data envelopment analysis model with application for assessment of port operations considering NetZero. Environ. Dev. Sustain. **25**(7), 6627–6656 (2023)

Voltes-Dorta, A., Britto, R., Wilson, B.: Efficiency of global airlines incorporating sustainability objectives: a malmquist-DEA approach. J. Air Transp. Manag. **119**, 102634 (2024)
Walker, A.M., Opferkuch, K., Lindgreen, E.R., Simboli, A., Vermeulen, W.J., Raggi, A.: Assessing the social sustainability of circular economy practices: industry perspectives from Italy and the Netherlands. Sustain. Prod. Consumpt. **27**, 831–844 (2021)
Wankhade, M., Rao, A.C.S., Kulkarni, C.: A survey on sentiment analysis methods, applications, and challenges. Artif. Intell. Rev. **55**(7), 5731–5780 (2022)
Wojcik, V., Dyckhoff, H., Gutgesell, S.: The desirable input of undesirable factors in data envelopment analysis. Ann. Oper. Res. **259**(1), 461–484 (2017)
Xie, J., Nozawa, W., Yagi, M., Fujii, H., Managi, S.: Do environmental, social, and governance activities improve corporate financial performance? Bus. Strateg. Environ. **28**(2), 286–300 (2019)
Zaare Tajabadi, F., Daneshvar, S.: Benchmark approach for efficiency improvement in green supply chain management with DEA models. Sustainability **15**(5), 4433 (2023)
Zupic, I., Čater, T.: Bibliometric methods in management and organization. Organ. Res. Methods **18**(3), 429–472 (2015)

Bank Efficiency in India: A Non-parametric Approach

Ram Pratap Sinha(✉)

Government College of Engineering and Leather Technology, Kolkata, India
rampratapsinha39@gmail.com

Abstract. Indian banking sector has faced several challenges in the past one decade due to a variety of factors including the macroeconomic slowdown since 2007–08, demonetisation and the pandemic caused by COVID. The buildup of excess capacity in the industrial sector led to worsening financial position of corporate firms and steep rise in the proportion of NPA s in the commercial banking sectors which remained mainly concentrated in the public sector banks. This led to the adoption of stringent asset quality norms, passage of Insolvency and Bankruptcy Code and recapitalization and consolidation of public sector banks. In the above context, the present study seeks to evaluate bank efficiency for 30 commercial banks (both public and private) for the period 2017–18 to 2023–24 using a directional DEA approach with undesirable output. Second, the study estimates both local returns to scale and scale elasticity for the period under observation. Finally, in the second stage, bank efficiency is explained by using both local likelihood and GAM regression. The second stage results indicate the importance of bank size and return on asset as determinants of bank efficiency.

Keywords: Technical efficiency · Data Envelopment Analysis · Undesirable output

1 Introduction

In India, the commercial banking sector occupies a place of prominence in the financial sector in terms of intermediation activity. About 95% of Indian household park their savings in the form of bank deposits and only 10% is invested in the capital market. Further, commercial bank deposits to GDP ratio was 72% for 2020–21. The principal reason behind the popularity of the banking system is that the commercial banks deal with over the counter (OTC) products and not with pass-through securities whose market value fluctuates with movements in the capital market and the pay off to the investor depends on the market value of the financial asset at the time of redemption.

However, the commercial banks themselves are exposed to various forms of risk (credit risk, market risk, liquidity risk, operational risk etc.) during the discharge of their business activity and thus their financial health is vulnerable to the changes in the various risk parameters and adverse developments in the macro-economy and the financial sector. It is a common knowledge that credit risk is the prime source of risk for the commercial

A. Emrouznejad et al. (Eds.): DEA 2024, LNOR, pp. 95–108, 2025.
https://doi.org/10.1007/978-3-031-98177-7_8

banks since fund based activity is the dominant form of intermediation performed by them. In the Indian context, build up of Non-performing Asset (NPA) remained a major cause of concern for various stakeholders (including the market regulator) since the onset of banking sector reform in 1991.

In the Indian context, numerous research studies have estimated the efficiency of Indian commercial banks in the aftermath of the introduction of financial and banking sector reform in India [Bhattacharyya, Lovell and Sahay (1997), Saha and Ravisankar (2000),Sathye (2003), Kumbhakar & Sarkar (2003), Ram Mohan and Ray (2004), Das, Nag and Ray (2005), Chatterjee and Sinha (2006), Kumar and Gulati (2009), Gupta and Kumar (2024)]. Some of these studies have used the undesirable output approach for evaluating bank performance [Dar, Mathur, & Mishra (2021), Khati and Mukherjee (2021), Sanati and Bhandari (2023)]. However, there is a lack of research studies relating to the estimation, analysis and explanation of bank efficiency in the recent phase which witnessed adverse macroeconomic disturbances due to the COVID outbreak and growth in cross-border conflicts. Although the Indian economy did not have any direct exposure to such conflicts, it had to face the unwarranted consequences of macroeconomic sluggishness. Secondly, the banking sector also witnessed major regulatory and structural changes in the form of tightening of asset quality norm, introduction of bankruptcy code and intra-public sector bank consolidation in the recent period. Against this backdrop, the present study seeks to reexamine the banking sector performance in India with three specific value additions. First, based on the efficiency analysis (with undesirable output), the present study classifies the in-sample banks in to three clusters with low, moderate and high levels of undesirable output (Non Performing Asset) induced risk. Second, the study computes output responsiveness to the scal of operation of the banks in the form of scale elasticity. Further, while explaining bank efficiency, the current study uses Locfit and General Additive Model for carrying out both local and global regression. The use of machine learning based techniques instead of conventional regression methods is preferred here for capturing non-linearities present in the relationships between the efficiency scores and the explanatory variables.

In view of this, the present study seeks to evaluate and explain commercial bank efficiency in India using a directional distance function model which includes both desirable and undesirable outputs. The sample data set covers 12 public sector and 18 private sector commercial banks operating in India for the period and extends over the period 2017–18 to 2023–24. The present study has five sections and proceeds as follows. Section 2 provides an overview of NPA problem in the Indian context, review the related literature and outlines the motivation for the study. Section 3 describes the methodology of efficiency evaluation. Section 4 includes description of variables, data and a brief discussion of the first and second stage estimation of results. Section 5 concludes.

2 Banking Sector Overview, Related Literature and Motivation

2.1 Indian Commercial Banks and the NPA Problem

Indian banking sector remained under public sector dominance since 1969 as 20 private sector banks were nationalized in two phases in 1969 and 1980 respectively and no new private sector banks were allowed to enter the market. During the 1970s and 1980s, the public sector banks made rapid progress in respect of expansion of branch network and fund based activity. However, due to the absence of prudential lending norms and monitoring of borrowers, there was a secular decline in asset quality during the period. In order to improve the financial health of the commercial banks and also to infuse competition and more accountability in the banking system, banking sector reform was initiated in 1991 involving deregulation of entry and introduction of prudential lending, accounting and securities valuation norms. A recurrence of of the NPA problem took place during 2014–15 and 2015–16 following the post-global crisis macroeconomic slowdown. However, this time it was concentrated in the non-priority sector lending as opposed to the priority sector centric NPA crisis that occurred during the previous instance. The rising NPA problem led to the initiation of a slew of measures taken by the Reserve Bank of India(RBI) and the Indian government. The measures taken by the RBI included the creation of a large loan database by the RBI (Central Repository Information on Large Credits) that covers all loans over '5 crores, changes in the asset disclosure norm, introduction of The Strategic Debt Restructuring (SDR) scheme in June 2015 (which aims to empower commercial banks to convert loans into equity shares of the struggling companies, facilitating a change in ownership and improving recovery prospects) and the 'Scheme for Sustainable Structuring of Stressed Assets'(S4A) which is an optional framework for the resolution of large stressed accounts without change of promoters. The government measures included improving the corporate governance of the public sector banks (the Indradhanush initiative), infusion of fresh capital and the passage of The Insolvency and Bankruptcy Code (IBC) of 2016 which amended and consolidated the then existing insolvency laws and provided a framework for the timely resolution of corporate, partnership, and individual insolvency cases.

2.2 Related Efficiency Literature

The earliest study on the efficiency evaluation of Indian commercial banks [Bhattacharyya et al. (1997)] estimated efficiency and returns to scale of public, private and foreign commercial banks operating in India for the period 1986–91 using DEA and SFA. Saha and Ravishankar (2000) found a general improvement in the performance of public sector banks for the period 1992–95. Sathye (2003) made efficiency evaluation of 94 commercial banks operating in India for 1997–98 and found that the outcome depends on the approach taken for estimation. Kumbhakar and Sarkar (2003) used SFA and found a declining trend in cost inefficiency of public and private sector banks for 1986–2000. Ray and Ram Mohan (2004) made revenue efficiency evaluation of public and private sector banks for1992–2000. Ray (2007) and Das, Ray and Nag (2005) evaluated size and labour use efficiency of commercial banks. Kumar and Gulati (2009) measured efficiency and effectiveness of Indian public sector banks. Tabak and Tecles (2010) applied Bayesian stochastic frontier for estimating cost and profit efficiency of public, private and foreign banks operating in India.

In recent times, several studies estimated efficiency of public and private sector banks with bad output. These include Puri and Yadav (2014, 2017), Dar et.al (2021), Khati and Mukherjee (2021) and Sanati and Bhandari (2023).

2.3 Motivation for the Present Study

In spite of the presence of several efficiency research studies relating to Indian commercial banks, the present study is undertaken for two reasons. First, apart from efficiency evaluation, the present study makes a classification of the Indian public and private sector banks based on their bad output (NPA) contribution and thereby provides useful information regarding their financial health.Second, the present study provides estimates of both returns to scale and scale elasticity of the observed commercial banks. Third, by using both local and global regression techniques, the present study provides additional insight in to the relationship of efficiency with other operating parameters of bank performance and the growth in the level of economic activity.

3 Estimation Methodology

Section 3.1 discusses the first stage efficiency evaluation while Sect. 3.2 discusses the second stage local and global regression analysis.

3.1 Efficiency Evaluation

In the present study, we have estimated bank efficiency, returns to scale and scale elasticity by using Data Envelopment Analysis (DEA) which was originally introduced by Charnes, Cooper and Rhodes (1978) in the context of constant returns to scale which was extended to the case of variable returns to scale by Banker, Charnes and Cooper (1984) [See also Emrouznejad, et al., 2025). Inter alia, the advantages of DEA over its econometric counterparts include its ability to handle multiple outputs in a framework which does not require a beforehand specification of the output and input variables. For an extensive account of the progress of DEA based efficiency research literature refer Mergoni et al. (2025), Mirzaei, et al. (2024) and Mahmoudabadi, et al. (2019).

The evaluation of technical efficiency in the present study is based on the use of the directional distance function [Chambers, Chung and Fare (1996) and] which is a more general approach for determining efficiency as compared to the radial or non-radial approaches. The concept of directional distance function is based on the benefit function introduced by Luenberger (1992a, 1992b).The directional distance function approach is particularly useful for the undesirable output model of efficiency evaluation because it allows for simultaneous expansion of desirable outputs and contraction of undesirable outputs.In the radial approach undesirable outputs require transformation through appropriate transformation functions.

For explaining the methodology of estimation, let us consider a production technology that converts a vector of inputs in to multiple outputs some of which are good and the remaining are bad outputs. The input vector is represented by X, the good output vector is represented by Y_G and the vector of bad outputs is denoted by Y_B The production possibility set P_T encompasses the feasible production plans of the producer (including good output, bad output and input sets) and can be presented as: P_T={(X, $Y_G, Y_B) \in R^+$, Here R^+ represents the non-negative quadrant of the real space. X can produce $Y_G and Y_B$}. The output set O(Y)$\in P_T$ includes the set of feasible outputs for each input vector while the input set I(X) includes the input vectors that are feasible with each output vector.

The production technology is assumed to exhibit free disposability of inputs and good outputs, weak disposability of undesirable outputs) and convexity. We may specify them as:

(a) Weak disposability of undesirable outputs: Good and undesirable outputs exhibit null-jointness. Thus the production of undesirable output can be reduced only by a simultaneous reduction in good outputs. Thus if $(Y_G, Y_B) \in P_T$ then $0 < \theta < 1$ imply that $(\theta Y_G, \theta Y_B) \in P_T$. Thus if $\theta = 0$ then $\theta Y_G, \theta Y_B$=0.
(b) Free disposability of good outputs:
(c) Convexity: The underlying technology is convex implying that convex combinations of feasible production processes are also feasible.

For the construction of the production frontier, the directional distance function considers the potential for increasing outputs and decreasing inputs simultaneously. Thus the directional distance function can be written as:

$$D_{DDF} = \text{Max}\ [\beta : (\{(\text{X} - \beta, (Y_G + g_{YG}), (Y_B - \beta g_{YB})\} \in P_T]$$

where g_X, g_{YG} and g_{YB} represent the direction vectors relating to input, good output and undesirable output respectively and β is a scalar. The directional distance function satisfies.

(a) representation: D_{DDF} (X, $Y_G, Y_G, Y_B, g_x, g_{YG}, g_{YB}) \geq 0 \leftrightarrow (X, Y_G, Y_B) \in P_T$
(b) translation: The translation property implies that

$$D_{DDF}(X - \beta g_X, (Y_G + \beta g_Y G), (Y_B - \beta g_Y B)|g_X, g_y G, g_Y B) = \in R^+$$

The directional distance function provides a complete characterization of production technology as it simultaneously considers the input and output sides. F ä re and Grosskopf (2000) showed that the input and output distance functions of Shephard (1953, 1970) and Farrell (1957) are special cases of the directional distance function.

3.2 Second Stage Regression

The manifest efficiency of the observed firms depends on the contextual variables which indirectly influences bank performance. In the present context, we have used both local likelihood regression and Generalised Additive Model (GAM) regression for assessing the nature of the relationship between the response and the contextual variables.

3.2.1 Local Regression

For analyzing the relationship between the response variable and each predictor variable separately, we have used the local likelihood regression approach. The local likelihood estimation procedure was introduced by Tibshirani and Hastie (1987) which fits a parametric function locally. The procedure, while trying to make a local estimation at a particular point, gives more weight to the observations which are in the nearest neighbourhood of that point. In the present context, we have used the locfit method [Loader (2012)].

3.2.2 GAM Regression

In addition, we have also estimated the relationship between efficiency and the contextual variables by using Generalised Additive Model (GAM) introduced by Hastie and Tibshirani (1986). In the GAM approach, the response variable is related to a sum of the smooth functions of the predictor variables. The smooth functions are estimated by scatterplot smoothers. The GAM approach thus maximizes the expected loglikelihood function. Generalised Additive Models perform better than the linear regression models in modelling non-linear complex relationships between the response and contextual variables thereby providing added flexibility when compared to linear regression methods.

4 Variables, Data and Results

The current section includes three Subsects. 4.1 provides a very brief discussion about the efficiency (input and output) indicators, the contextual variable used in the second stage analysis for explaining bank efficiency and data used in the estimation procedure. Subsection 4.2 describes the efficiency related results. Subsection 4.3 discusses the second stage results.

4.1 Description of Inputs, Outputs, Contextual Variables and Data

Bank efficiency evaluation in the first stage of the study needs identification of input and output indicators. However, since the banking industry deals with financial services, outputs and inputs can be defined in a number of ways. To be more specific, there are at least five approaches for choosing bank output: the production approach, the financial intermediation approach, the user cost approach, the value added approach and the risk mitigation approach. The production approach as introduced by (Benston, 1965) considers banks as customer service providers. Thus as per this approach, bank outputs reflect the services provided to the customers including the number and types of transactions, documents processed or specialized services provided over a given time period. (Sealey and Lindley, 1977) introduced the intermediation approach for defining commercial bank inputs and outputs. This approach treats commercial banks as financial intermediaries which mobilizes financial resources from depositors invest the same in business, commerce and investible securities. The user cost approach (Hancock, 1985; Hancock, 1991) identifies input or output of the banking industry depending on the positivity/negativity of the relative net revenue contribution. The value added approach introduced by (Berger et al., 1987) and (Berger & Humphrey, 1992) considers such activities as output indicators which contributes significant value added as assessed while allocating operating cost. The value added approach assumes that the insurers provide three major services: risk-pooling and risk bearing, real financial services and intermediation.

In the present study, we have followed the risk mitigation approach for selecting inputs and outputs. We have viewed banking activity from the income-expenditure perspective Thus we have included operating expenses and interest expenses as the two input indicators and interest income and other income (from off balance sheet and fee based activities) as the output indicators (Table 1). For representing undesirable output, we have included net non-performing asset. We have chosen net NPA over gross NPA as it is a better indicator of financial health of banks. Growth of net NPA reduces the ability of the banks to generate fund based income and hence needs to be minimised.

For carrying out second stage regression, contextual variables influencing bank-efficiency are to be identified. In the present case, we have chosen four important contextual variables: capital adequacy ratio, return on asset, log of total asset (indicator of bank size) and real GDP growth rate (Table 1). Capital adequacy is the standard measure of bank solvency while return on equity is a well accepted measure of bank profitability. Bank efficiency performance is also found to be dependent on the size of the commercial bank and consequently log of total asset is also considered as a contextual variable.

Table 1. Selected inputs, outputs and contextual variables

Variables	Indicator type
Operating expenses	Inputs
Interest expenses	
Interest income	Desirable Outputs
Other income	
Net Non-performing Asset	Undesirable Output
Capital adequacy ratio	Contextual variables
Return on asset	
Log of total asset	
Real GDP growth rate	

Source: Selected by the author

4.2 Efficiency Estimation Related Results

Table 2 provides the descriptive statistics of efficiency scores for the seven year observation period. During the period under consideration, mean efficiency showed secular decline during the last 5 years. Obviously, during this sub-phase, standard deviation exhibited an upward trend. Efficiency distribution exhibited negative trend during the entire period while skewness remained positive except for 2023–24. Figure 1 exhibits movements in mean efficiency scores.

4.2.1 Sources of Inefficiency

It is of interest to analyse the input and output specific inefficiencies for the observed year. Table 3 thus provides information about the mean input and output specific efficiencies for the period 2017–18 to 2023–24. The extent of input and output specific inefficiencies have been computed by comparing the observed and projected relating to the inputs

Table 2. Descriptive statistics of estimated efficiency scores

Descriptive Statistics	2018	2019	2020	2021	2022	2023	2024
Mean efficiency	0.9766	0.9844	0.9777	0.9681	0.9680	0.9646	0.9600
Standard Deviation	0.0287	0.0256	0.0278	0.0391	0.0368	0.0432	0.0430
Skewness	-1.2358	-1.6997	-1.3000	-1.0902	-0.9877	-1.0007	-0.8746
Kurtosis	1.2149	1.9407	1.1750	0.1904	0.0983	0.2666	-0.5511

Source: Calculated

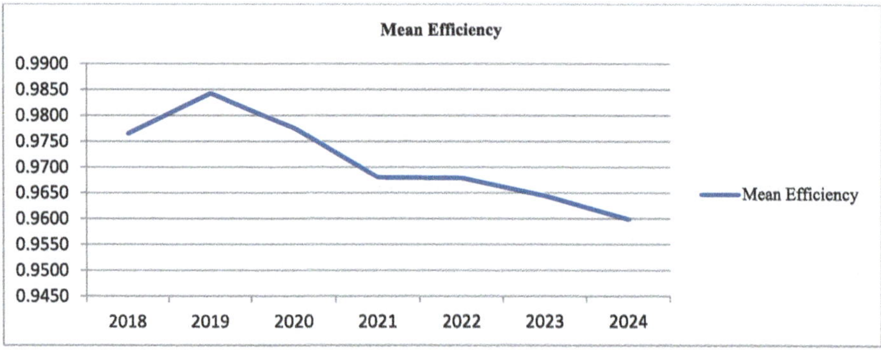

Fig. 1. Movements in mean efficiency scores

and outputs included in the study. The corresponding ratios have been computed for each bank and by deducting the ratio from unity we get the degree of input/output specific inefficiencies. Ideally in the absence of any inefficiency the ratio of observed and projected values will be the same. In order to find out the degree of inefficiency for inputs and bad output we have considered the ratio of projected and observed values and in case of good outputs we have considered the ratio of observed to projected values. Table 3 clearly indicates that inefficiency is mostly concentrated in NPA management.

Table 3. Input and output specific inefficiencies (in percentage terms)

Variables	2018	2019	2020	2021	2022	2023	2024
Interest Expenses	0.0170	0.0095	0.0191	0.0271	0.0218	0.9832	0.0164
Operating Expenses	0.0181	0.0087	0.0141	0.0167	0.0200	0.9609	0.0176
Interest Income	0.1029	0.0065	0.0128	0.0159	0.0194	0.9845	0.0174
Other Income	0.0471	0.0195	0.0302	0.0453	0.0696	0.9704	0.0891
Net Non Performing Asset	0.1854	0.0858	0.1347	0.1467	0.3113	0.7620	0.2050

Source: Calculated

Since non-performing assets constitute the most important source of technical efficiency, we have ranked the observed commercial banks on the basis of slack-data ratio relative to the undesirable output for the period under observation and classified them in to three clusters where each cluster includes 10 commercial banks. Table 4 provides the requisite information. The table shows that the mean slack-data ratio, we have applied both is 0.04 for the banks belonging to cluster A, 0.17 for cluster B and 0.37 for cluster C. Cluster A and B includes 7 private sector banks and 3 public sector banks respectively. However, cluster C has equal contribution (6 each) from private and public sector commercial banks.

Table 4. Observed bank performance relating to the undesirable output

Cluster	Included banks	Mean slack-data ratio
A	HDFC Bank, State Bank of India, Nainital Bank, Kotak Mahindra Bank, IndusInd Bank, Union Bank, Canara Bank, RBL Bank, Tamilnad Mercantile Bank and Catholic Syrian Bank	0.04
B	City Union Bank, ICICI Bank, Punjab National Bank, Axis Bank, Bank of Baroda, Federal Bank, DCB Bank, Dhanalaxmi Bank and Indian Overseas Bank	0.17
C	Yes Bank, Karnataka Bank, South Indian Bank, Bank of India, Bank of Maharashtra, UCO Bank, Central Bank of India, Karur Vysya Bank, Punjab & Sind Bank, Jammu & Kashmir Bank	0.37

Source: Calculated

4.2.2 Returns to Scale and Scale Elasticity

In order to understand the nature of bank responsiveness to changes in the scale of production, we have estimated returns to scale and scale elasticity of the commercial banks using the output oriented radial approach. See Zelenyuk (2013) for the equivalence of scale elasticity measure under the directional distance function approach with the input and output oriented scale elasticity measures. Table 5 presents summary information regarding the movement in the returns to scale during the in-sample years. Table 5 shows that the number of banks displaying constant, increasing and decreasing returns to fluctuated between 7 to18 (for CRS), 4 to 17 (for IRS) and 2 to 12 (for DRS) during the period under observation.

Table 6 provides the mean, lower bound and upper bounds of scale elasticity of the observed commercial banks for the observed period. Note that for each observed years, some banks have been identified as outliers. Thus we have excluded the outlier banks from our set of observations while calculating the scale elasticity statistics. From the table it is clearly seen that with the exception of 2019, the mean values of scale elasticity remained below unity.A detailed analysis of the 2019 scale elasticity data further shows

Table 5. Returns to scale of the observed commercial banks

Returns to Scale Type	2018	2019	2020	2021	2022	2023	2024
Constant returns to scale (CRS)	14	9	7	18	7	10	8
Increasing returns to scale (IRS)	4	9	17	10	12	13	17
Decreasing returns to scale (DRS)	12	12	6	2	11	7	5

Source: Calculated

that if we exclude one particular bank from our set of observations then the remaining banks have mean scale elasticity below 1.

Table 6. Scale elasticity statistics of the observed commercial banks

Particulars	2018	2019	2020	2021	2022	2023	2024
No of outlier banks	1	2	2	2	2	2	1
Scale elasticity lower bound	0.8384	0.8489	0.8752	0.9054	0.9225	0.8751	0.8901
Scale elasticity average	0.951	2.0614	0.9439	0.962	0.9949	0.9549	0.9787
Scale elasticity upper bound	1.0635	3.2738	1.0127	1.0187	1.0674	1.0346	1.0674

Source: Calculated

4.3 Impact of Contextual Variables on Bank

As indicated earlier, in the present study, we have used both local and global regression for exploring the relationship between the efficiency scores and the contextual variables selected for the study. The benefit of using local regression is its ability to capture the local features of the data by using weighted means of neighbourhood points. Global regression techniques use a single equation for describing the relationship between the response variable and the explanatory variables.

4.3.1 Local Likelihood Regression

For local regression, we have considered one explanatory variable at a time and used "locfit" package. The characteristics of regression are provided in Table 7. The relations are graphically presented in Figs. 2 through 5.

Figures 2 presents the relationship of log of efficiency with the four contextual variables [capital to risk weighted asset ratio (CRAR), log of total asset, return on asset (ROA) and GDP growth rate]. Note that, in this case we have considered each explanatory variable on a stand alone basis.

Table 7. Characteristics of local regression

Particulars	CRAR	LogTA	ROA	GDP Growth Rate
Number of data points	210	210	210	210
Evaluation structure	Rectangular tree	Rectangular tree	Rectangular tree	Rectangular tree
Number of evaluation points	8	7	10	9
Degree of fit	2	2	2	2
Fitted degree of freedom	5.802	5.314	6.544	6.476

Source: Calculated

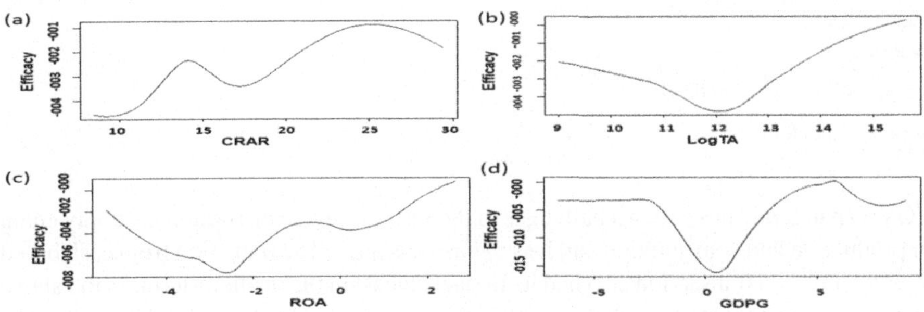

Fig. 2. Relationship of efficiency with contextual variables

4.3.2 GAM Regression

In addition to local likelihood estimation of parameters, we have performed General Additive Model (GAM) regression. The results are included in Table 8. The results indicate the coefficients of bank size and return on asset are positive and statistically significant. The coefficient of CRAR (capital adequacy ratio (CRAR) is negative but insignificant. The coefficient of (real) GDP growth rate is positive but not significant. The output shows that 37.8% is explained by the estimate.

5 Concluding Observations

As mentioned earlier, the government and the Reserve Bank of India adopted several measures for improving the financial health of Indian commercial banks. In case of the public sector banks, Rs 70,000 crores were infused for the recapitalization of the banks and during 2019 and 2020 the number of public sector banks were reduced from 19 to 12 through a series of inter-public sector bank mergers. We have already mentioned other measures initiated for course correction. While, these measures had beneficial effects on the Indian banking industry, the macro-economic environment is still far from encouraging and there are both internal and external adverse factors at play. Inter alia, the

Table 8. Results of GAM regression

Particulars	Nature of coefficient	Estimate	Standard Error	Estimate/Standard Error	t Ratio
Intercept	Parametric	−0.2888	0.0376	−7.688	~0.0000
CRAR	Parametric	−0.0061	0.0437	−0.139	0.89
Log of total asset	Smooth term	1	1	6.248	**0.0139**
ROA	Smooth term	2.143	2.582	5.556	**0.0027**
GDP growth rate	Smooth term	1	1	2.024	0.1576
Adjusted R squared	0.35				
Deviance explained	37.8%				
Generalised Cross Validation	0.0239				
Scale estimate	0.0227				

Source: Calculated

COVID pandemic posed new challenges in the form of severe disruption in supply chain of production and consumption and loss of business days. The transition from cash based to fintech-driven transaction also had its impact at least in the medium-term. In the above context, the present study seeks to identify the commercial banks with relatively higher proportion of NPA related slacks and found that at one third of the in-sample banks need due attention from the management and the market regulator for improving the financial condition. The second stage analysis shows that for efficient transformational performance, bank size and financial health are crucial. These results are important for analyzing the financial health of the banks. In specific, the above analysis shows that DEA can serve as an useful tool for taking the analysis of bank performance to a different level and thus can immensely help the decision making process both from managerial and regulatory perspectives.

However, the current study has some limitations. First, the study covers a 7 year period only. In future, a more long term study can provide more insights about bank performance. Further, the second-stage analysis could be further enriched by considering more contextual variables for improving the predictive quality of the second stage analysis.

References

Benston, G.J.: Branch banking and economies of scale. J. Financ. **20**(2), 312–331 (1965)

Berger, A.N., Humphrey, D.B.: Measurement and efficiency issues in commercial Banking. In: Griliches, Z. (ed.) Output Measurement in the Service Sectors, National Bureau of Economic Research, Studies in Income and Wealth, vol. 56, pp. 24–79. University of Chicago Press, Chicago (1992)

Bhattacharya, A., Lovell, C.K., Sahay, P.: The impact of liberalization on the productive efficiency of Indian commercial banks. Eur. J. Oper. Res. **98**, 332–345 (1997). https://doi.org/10.1016/S0377-2217(96)00351-7

Chambers, R.G., Chung, Y., F ä re, R.: Benefit and distance functions. J. Econ. Theory **70**, 407–419 (1996)

Chatterjee, B., Sinha, R.P.: Cost efficiency and commercial bank lending: some empirical results. Indian Econ. J. **54**, 145–163 (2006). https://doi.org/10.1177/0019466220060109

Chung, Y.H., Fare, R., Grosskopf, S.: Productivity and undesirable outputs: a directional distance function approach. J. Environ. Manag. **51**(3), 229–240 (1997)

Dar, A.H., Mathur, S.K., Mishra, S.: The efficiency of Indian banks: a DEA, malmquist and SFA analysis with bad output. J. Quant. Econ. **19**(4), 653–701 (2021). https://doi.org/10.1007/s40953-021-00247-x

Das, A., Nag, A., Ray, S.C.: Liberalisation, ownership and efficiency in Indian banking: a non-parametric analysis. Econ. Polit. Weekly **40**(12), 1190–1197 (2005). http://www.jstor.org/stable/4416367

Emrouznejad, A., Podinovski, V., Lu, C., Charles, V., Moradi-Motlagh, A.: Rajiv banker's lasting impact on data envelopment analysis. Ann. Oper. Res. **351**, 1225–1264, (2025). https://doi.org/10.1007/s10479-025-06473-3

Färe, R., Grosskopf, S.: Theory and application of directional distance functions. J. Prod. Anal. **13**, 93–103 (2000). https://doi.org/10.1023/A:1007844628920

Farrell, M.J.: The measurement of productive efficiency. J. R. Stat. Soc. Ser. A General **120**(3), 253–281 (1957)

Gupta, M., Kumar, S.: Does income diversification affect bank efficiency, effectiveness and performance? Evidence from Indian banking industry. In: Advances in the Theory and Applications of Performance Measurement and Management. DEA45 2023. Lecture Notes in Operations Research. Springer, Cham (2024). https://doi.org/10.1007/978-3-031-61597-9_14

Loader, C.: Smoothing: local regression techniques. In: Gentle, J., Härdle, W., Mori, Y. (eds.) Handbook of Computational Statistics. Springer Handbooks of Computational Statistics. Springer, Heidelberg (2012). https://doi.org/10.1007/978-3-642-21551-3_20

Hastie, T., Tibshirani, R.: Generalized additive models. Stat. Sci. **1**(3), 297–310 (1986). https://doi.org/10.1214/ss/1177013604

Hancock, D.: The financial firm: production with monetary and nonmonetary goods. J. Polit. Econ. **93**(5), 859–880 (1985)

Hancock, D.: A Theory of Production for the Financial Firm. Kluwer Academic Publishers, Boston (1991)

Khati, K.S., Mukherjee, D.: Productive efficiency and non-performing assets of Indian banks in the post-global financial crisis period. South Asia Econ. J. **22**(2), 186–204 (2021). https://doi.org/10.1177/13915614211029072

Kumar, S., Gulati, R.: Technical efficiency and its determinants in the Indian domestic banking industry: an application of DEA and Tobit analysis. Am. J. Finan. Account. **1**(3), 256–296 (2009). https://doi.org/10.1504/AJFA.2009.026484

Kumbhakar, S. C., & Sarkar, S. (2003). Deregulation, Ownership and Efficiency in Indian Banking. Arthaniti-Journal of Economic Theory and Practice,2(1–2),1–26(2003). https://doi.org/10.1177/0976747920030101

Luenberger, D.G.: Benefit functions and duality. J. Math. Econ. **21**, 461–481 (1992)

Luenberger, D.G.: New optimality principles for economic efficiency and equilibrium. J. Optim. Theory Appl. **75**, 221–264 (1992). https://doi.org/10.1007/BF00941466

Mahmoudabadi, M.Z., Emrouznejad, A.: Comprehensive performance evaluation of banking branches: a three-stage slacks-based measure (SBM) data envelopment analysis. Int. Rev. Econ. Finan. **64**, 359–376 (2019)

Mergoni, A., Emrouznejad, A., De Witte, K.: Fifty years of data envelopment analysis. Eur. J. Oper. Res. **326**(3), 389–412 (2025). https://doi.org/10.1016/j.ejor.2024.12.049

Mirzaei, M.S., Emrouznejad, A.: Bank stock performance during the COVID-19 crisis: does efficiency explain why Islamic banks fared relatively better? Ann. Oper. Res. **334**, 317–355 (2024)

Puri, J., Yadav, S.P.: Improved DEA models in the presence of undesirable outputs and imprecise data: an application to banking industry in India. Int. J. Syst. Assur. Eng. Manag. **8**(Suppl 2), 1608–1629 (2017). https://doi.org/10.1007/s13198-017-0634-4

Puri, J., Yadav, S.P.: A fuzzy DEA model with undesirable fuzzy outputs and its application to the banking sector in India. Expert Syst. Appl. **41**(14), 6419–6432 (2014). https://doi.org/10.1016/j.eswa.2014.04.013

Ram Mohan T .T. and S C Ray :Comparing Performances of Public and Private Sector Banks-A Revenue Maximising Approach, Economic and Political Weekly,20,1271 - 1276 (2004)

Ray, S.C.: Are some Indian banks too large? an examination of size efficiency in Indian banking. J. Prod. Anal. **27**, 41–56 (2007)

Saha, A., Ravisankar, T.S.: Rating of Indian commercial banks: a DEA approach. Eur. J. Oper. Res. **124**(1), 187–203 (2000)

Gargi, S., Bhandari, A.K.: "Operational Efficiency in the presence of undesirable byproducts: an analysis of Indian banking sector under traditional and market-based banking framework. NIBM Working Paper Series, WP24/May (2023)

Sathye, M.: Efficiency of banks in a developing economy: the case of India. Eur. J. Oper. Res. **148**(3), 662–671 (2003)

Sealey, C., Lindley, J.: Inputs, outputs, and the theory of production and cost of depository financial institutions. J. Finan. **32**, 1251–1266 (1977)

Shephard, R.W.: Cost and Productions. Princeton University Press, Princeton (1953)

Shephard, R.W.: Theory of Cost and Productions. Princeton University Press, Princeton (1970)

Tibshirani, R., Hastie, T.: Local likelihood estimation. J. Amer. Statist. Assoc. **82**, 559–567 (1987)

Zelenyuk, V.: A scale elasticity measure for directional distance function and its dual: theory and DEA estimation. Eur. J. Oper. Res. **228**(3), 592–600 (2013)

The Role of Innovation on the Link Between Environmental Regulations and Firm Performance: A DEA Approach

Ramakrishnan Ramanathan[1](✉) and Ruiqian Li[2]

[1] University of Sharjah, 27272 Sharjah, United Arab Emirates
RRamanathan@sharjah.ac.ae
[2] Heilongjiang University, Harbin 150080, China

Abstract. The nature of impact of regulatory policies on improving firm performance is a hotly debated topic based on the Porter's hypothesis. Even more debated is the role of innovation in this link. The literature has several studies that suggested a positive role of innovation while there are equal number of studies that suggested a negative and even a neutral role. In this paper, we provide a new look at this topic using data envelopment analysis (DEA). Unlike previous studies that looked at the relationships primarily using statistical principles, we first study the link between regulatory pressures on firms and their performance using DEA using the software MaxDEA. We then study the impact of innovation using both Tobit and normal OLS regressions. We use data from 36 different Chinese industries over a five-year time period, totaling to a sample size of 180. In line with previous studies, the intensity of environmental regulations is captured in our study using the level of industrial emissions. Financial performance is captured using three measures – profits, interest expenses and total assets. Innovation is captured using R&D expenditure. Our study has supported a weak influence of innovation on the link between regulatory pressures and firm performance. In this sense, this paper joins previous studies that found a weak positive or neutral link on the influence of innovation on the regulation-performance link. We provide reasons for this conclusion and highlight suggestions for future research in this direction.

Keywords: Data Envelopment Analysis · DEA · Porter hypothesis · China

1 Introduction

We present a new application of DEA in this paper to advance existing knowledge on the impact of environmental regulations (ER) on innovation and financial performance (FP) at sector level. Topics related to the environment have always been a key issue all over the world. Academics, practitioners and governments across the world have conducted a number of studies in understanding how to ensure coordinated development of economy without harming the environment. As for China, it is also urgent to balance economic development and environmental protection (Du et al. 2017; Wu et al. 2015). China

A. Emrouznejad et al. (Eds.): DEA 2024, LNOR, pp. 109–120, 2025.
https://doi.org/10.1007/978-3-031-98177-7_9

has seen tremendous economic development in recent decades, surpassing Japan as the second largest economy in 2010. Such a rapid development has resulted in significant environmental issues (Bi et al. 2015; Wu et al. 2015). For example, the total volume of industrial water discharged, the total volume of industrial waste gas emission and the industrial solid wastes generated in 2010 were 2374732 (10000 tons), 519168 (100 million cu.m) and 240944 (10000 tons), which were 1.22, 3.76 and 2.95 times that of corresponding emissions in 2000 respectively (Data source: China Statistical Yearbook on Environment 2011). There is evidence that innovation plays an important role in Chinese economy, as evidenced by the latest 2016 global innovation index (GII), where China features for the first time in the Top 25 innovative economies of the world (GII 2016). Hence, the topic linking environment, innovation and performance is of renewed interest to the case of China.

In general, economic development and environmental protection appear contradictory to each other. However, contrast to the traditional wisdom, the famous Porter Hypothesis has been proposed, focusing on the relationship between environment and economy. Porter (1991) and his subsequent studies Porter and van der Linde (1995a, 1995b) claim that properly designed ER can encourage firms to innovate their products or their production processes so that not only meet the requirements of ER but also improve their own FP. Through investing in innovation, not only the additional compliance costs can be offset, but also the efficiency in the use of resources can be improved and their competitive advantages can be created. A lot of studies have been conducted to verify Porter Hypothesis. The research topics include the relationship between ER and FP, the impact of innovation on FP, the impact of ER on innovation and the link between environmental performance and FP. Till now, the results are inconclusive (Ramanathan et al. 2010; Rennings and Rammer 2011). Some of the previous studies have used firm-level data (e.g., Zhao and Sun 2016) while some others have used aggregated industry-level data (e.g., Qi et al. 2014; Ramanathan et al. 2010).

In this paper, we investigate the impact of ER on innovation and FP in China. This study makes at least three contributions to the literature. Ours is one of the few studies that focus on all the three constructs (ER, innovation and FP) of relevance to environmental policy. Unlike previous research that used statistics based approaches (regression and/or SEM), we attempt to look at the complex interrelationships from the lens of DEA. Finally, it has been pointed out by Horváthová (2012) that most of studies on related topics are based on the U.S. and Canadian data, fewer studies deal with European data, and there are only few studies analysing emerging or Asian countries. In this sense, we can provide more empirical evidence from an emerging developing country (China).

2 Literature Review

2.1 Data Envelopment Analysis (DEA)

Data envelopment analysis (DEA) is a well-known methodology for measuring the relative efficiency of homogeneous decision making units (DMUs) with multiple inputs and multiple outputs (Charnes et al. 1978; Ramanathan 2003). DEA has been originally developed as a tool for performance measurement. It has been traditionally employed to assess efficiency of DMUs that consume certain inputs to produce certain outputs.

Accordingly, a DEA model uses a certain number of inputs and a certain number of outputs. The objective of DEA would be to identify DMUs that have maximized outputs consuming minimal inputs. Generally, DMUs can include manufacturing units, departments of big organizations such as universities, schools, hospitals, bank branches and so on (Mergoni et al., 2025).

We now focus on the application of DEA in the context of the interesting problem in environmental policy. It involves analysing the interrelationships among ER, innovation and FP. The relevant literature is discussed in the next section.

2.2 The Inter-relationship Among ER, Innovation and FP

Most of the studies that analyzed interrelationships among ER, innovation and FP stemmed from Porter Hypothesis. For convenience, we structure this section using subsections that focus on ER-FP link, Innovation-FP link, ER-Innovation link, and the ER-Innovation-FP nexus.

The Impact of ER on FP. Many studies have investigated the impact of ER on performance. As mentioned earlier, while a traditional view would see ER as detrimental to FP of firms, Porter (1991) argued that ER could have positive impact on FP. This Porter hypothesis has been tested extensively (Ambec and Barla 2002; Indrani and Das 2006). The following studies supported the positive relationship (e.g., Boiral 2007; Hamilton 1995; Majumdar and Marcus 2001; Sarkis 2001).

Majumdar and Marcus (2001) used DEA to measure productivity efficiency and demonstrated that well-designed ER had positive impact on productivity than less well-designed ER. Chintrakarn (2008) showed that increased stringency in ER had led to greater technical efficiency (computed by stochastic a frontier model) in US manufacturing industries. Zhang et al. (2011) evaluated total factor productivity growth in China and investigated the impact of ER on productivity. Their results showed that the enforcement of ER in China was, in general, far below the levels achieved in the best performing regions, and that the more stringent enforcement of ER would help to improve productivity growth in China.

The Impact of ER on Innovation. The topic on the impact of ER on innovation has also been widely studied. Innovation in firms could be stimulated using a number of drivers, including ER. Horbach (2008) considered the dynamic character of environmental innovations and found ER encouraged environmental innovation. Kammerer (2009) showed that both customer benefit and regulation played a key role for environmental product innovations in German manufacturers of electrical and electronic appliances. Cai and Zhou (2014) found that both internal and external pressures could trigger eco-innovation. ER as one of these external pressures could affect eco-innovation. Except the perspective from external and internal pressures, Kesidou and Demirel (2012) analyzed the factors affect eco-innovation from supply-side factors and demand-side mechanisms and found that the stringency of ER had different impact on firms' eco-innovation when the levels of firms' innovation were different.

Some other studies focus on the relationship between ER and innovation directly but find mixed results on the ER-innovation link. Jaffe and Palmer (1997) found lagged environmental compliance expenditures had a significant positive effect on R&D expenditures, which meant that ER had positive impact on innovation, and thus supported Porter Hypothesis. However, Brunnermeier and Cohen (2003) found that increased in pollution abatement expenditures could lead to environmental innovation, but increased monitoring and enforcement activities related to existing regulations did not provide any additional incentive to innovate in US manufacturing industries. Jiménez (2005) demonstrated that indirect regulations (voluntary agreements) could lead to SMEs in Chile innovate significantly more. Yang and Yang (2015) demonstrated that the enforcement of a series of environmental policies had accelerated the process of eco-innovations in China. Bi et al. (2017) revealed that the large enough government subsidy could lead both firms to invest in technology.

In sum, there is still no consensus on the ER-innovation link, and has been a topic of research among the academic community. In this paper, we test this relationship in Chinese context using our DEA-based methodology.

The Impact of Innovation on FP. In this subsection, we focus on the impact of innovation on FP. Some scholars have found that innovation can have positive impact on FP. For example, Boasson and MacPherson (2001) indicated both university R&D and corporate R&D had positive role in company performance. Chen et al. (2006) claimed that the performance of both green production innovation and green process innovation have significant positive correlation with the corporate competitive advantage. The investment on innovation could be benefit for business. Doran and Ryan (2012) examined the impact of regulation, customer expectations and voluntary agreements on eco-innovation first, then assessed the impact of eco-innovation on firm performance. The results showed that eco-innovation played an important role on firm performance.

In summary, it appears that the literature on innovation – FP link remain ambiguous in its conclusions.

ER, Innovation and FP Nexus. There are very few studies that focused on interrelationships among all the three constructs (ER, innovation and FP) simultaneously. Triebswetter and Wackerbauer (2008) suggested regulatory pressure could drive environmental innovation. Besides, both regulatory pushed innovations and voluntary innovations could contribute to the competitive performance of sample companies in a similar way. Rubashkina et al. (2015) used innovation activity and productivity as capture of competitiveness and found ER had positive impact on the output of innovation activity, which supported "weak" Porter Hypothesis. But ER had no impact on productivity and could not support "strong" Porter Hypothesis. Ramanathan et al. (2018) applied DEA to capture the flexibility of ERs and found that innovation capabilities could affect FP of firms significantly if the firms felt that the ERs they faced were flexible. If the firms felt that the ERs they faced were inflexible, innovation capabilities could not improve their FP significantly.

The literature focused on all relationships and investigated the intermediary role of innovation on the ERs-FP link is rather limited. Ramanathan et al. (2010) used SEM method and found that ER in the UK were significant in improving economic performance of the industrial sectors, ER negatively influenced innovation in the short run, while innovation negatively influenced economic performance in these sectors. Zhao and Sun (2016) analyzed these three constructs simultaneously on pollution-intensive corporations. The results showed that ER had a significant weak positive effect on the corporation innovation, corporation innovation had a significant strong positive effect on corporation competitiveness, and ER had an insignificant negative effect on corporation competitiveness.

As a summary of this section linking ER, innovation and FP, we conclude that the literature appears to have mixed results on the interrelationships among these three constructs. However, there is evidence that innovation could play an intermediary role on the relationship between ER and FP because innovation is influenced by ER and innovation influences FP. In this paper, we aim to examine the intermediary role of innovation on the ER-FP link (Fig. 1).

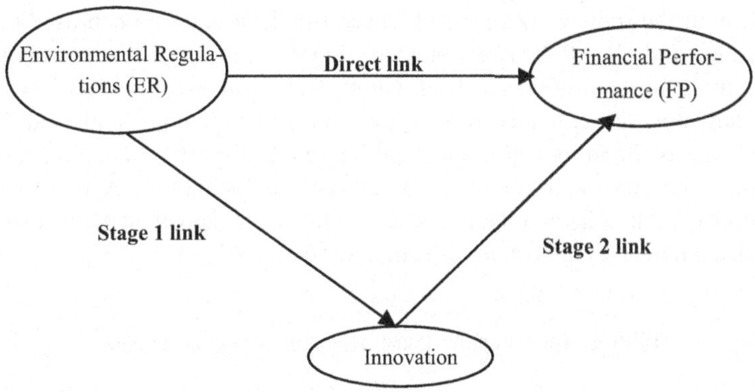

Fig. 1. Conceptual framework showing the influence of environmental regulations and innovation on performance (Source: Ramanathan et al., 2010)

Most of traditional research in this area has used statistical tools (regression or structural equations) to understand the inter-linkages among ER, innovation and FP. In this paper, we approach this research from the point of view of Operational Research modelling. Specifically, DEA has the capacity to capture the efficiency of producing a set of identical outputs by consuming a set of identical inputs. This ability of DEA has not been utilized well in the literature in the context of the nexus among ER, FP and Innovation, even though DEA has been extensively applied to the context of eco-efficiency (Emrouznejad et al., 2023). Hence, we employ data envelopment analysis (DEA) based on its multi-criteria interpretation.

2.3 Application of DEA to the Interrelationships Among ER, Innovation and FP

We first employ DEA with inputs related to environmental regulations (ER) and outputs related to financial performance (FP). This part examines the direct link between ER and FP. We employed an input-oriented CCR DEA model to compute performance scores. In the second part, we use variables related to innovation to check whether the DEA scores are significantly influenced by innovation measures.

Based on the literature and what we have discussed earlier, we propose our hypothesis on the impact of innovation on the ER-FP link.

H1: Innovation (measure using R&D expenditure) has a significant impact on the link between ER and FP (that is, in explaining the DEA scores with ER as inputs and FP as outputs).

3 Data and Methodology

The data collected in this study are based on two-digit Standard Industrial Classification in China. The required data are collected from different China statistical yearbooks. Due to the availability of data, the time periods cover from 2009 to 2013 and 36 industrial sectors are included in our analysis.

In line with the framework in Fig. 1, several indicators have been used to capture ER, innovation and FP. All data are obtained from different statistical yearbooks. Environmental indicators are collected from China Statistical Yearbook on Environment, financial indicators are collected from China Industry Economy Statistical Yearbook (name changed as China Industry Statistical Yearbook after 2013) and innovation indicators come from Statistical Yearbook on Science and Technology Activities of Industrial Enterprises. All collected data is based on the same statistical scope of large and medium-size enterprises in each industrial sector (see Table 1).

Table 1. Indicators for constructs, units and data sources.

Constructs	Indicators	Unit	Source of Data
Environmental regulations	Industrial Sulphur Dioxide Emission	Ton	China Statistical Yearbook on Environment
	Industrial Waste Gas Emission	100 Million cu.m	
Financial performance	Profits before Tax	100 Million Yuan	China Industry Economy Statistical Yearbook
	Interest Expense	100 Million Yuan	
	Total Assets	100 Million Yuan	
Innovation	R&D Expenditure	10000 Yuan	Statistical Yearbook on Science and Technology Activities of Industrial Enterprises

ROA and profits before tax are used to capture FP in this paper. ROA has been widely used as a proxy for FP in related research (e.g., Claver et al. 2007; Sharma 2001; Yu and Chen 2015; Zhao and Sun 2016). Profits before tax is similar with ROA, since they are all accounting-based profitability measures. Thus, we also use it as another proxy for FP in this research. We use the data collected from China Industry Economy Statistical Yearbook to calculate ROA first. ROA is the ratio of earnings before interest and tax

(EBIT) to average assets. EBIT equals profits before tax plus interest expense, and the average assets are the average value of last year-end assets and this year-end asset.

Based on Ramanathan et al. (2010), the intensity of ER has been captured using environmental emissions in this study. We have chosen two different environmental emissions for this construct. One is industrial sulphur dioxide emission (SO2) and the other is industrial waste gas emission. Both of these two are main sources of air pollutants caused by enterprises. As pointed out by Madsen (2009), SO2 is one of the most common industrial pollutants. There are also some studies used environmental emission to capture the intensity of ER (e.g., Gollop and Roberts 1983; Madsen 2009; Xing and Kolstad 2002).

Finally, innovation has been captured using R&D expenditure based on several previous studies. R&D expenditure means the amount of money that each sector spends on R&D activities. Some prior studies also use R&D indicators to measure innovation (e.g., Ambec and Barla 2002; Hamamoto 2006; Jaffe and Palmer 1997; Kneller and Manderson 2012; Zhao and Sun 2016).

Correlation coefficients between various variables are shown in Table 2. Most of correlation coefficients are significant. As expected, pollutant emissions have generally negative correlations with most of the variables.

Table 2. The correlation coefficients between various variables.

	1	2	3	5	6
1. Profits before Tax	1				
2. ROA	0.132*	1			
3. R&D Expenditure	0.690***	−0.222***	1		
4. Industrial Sulphur Dioxide Emission	0.158**	−0.334***	−0.005	1	
5. Industrial Waste Gas Emission	0.204***	−0.348***	0.129*	0.875***	1

Notes: * $p < 0.1$; ** $p < 0.05$; *** $p < 0.01$

4 Results

We use MaxDEA to conduct all DEA analysis. Table 3 presents the descriptive statistics of DEA results.

In order to verify our hypothesis, we use Tobit regression with DEA scores as the dependent variable and R&D Expenditure and R&D Employees (the two measures are innovation) as the independent variables. Tobit regression is usually preferred to the usual OLS regressions when DEA scores are used as the dependent variable because DEA scores are bounded by zero and one at both the ends (Hoff, 2007). Results are presented in Table 4.

The results show that innovation expenditure is weakly (at 10% level) significant in impacting the DEA scores. This supports our hypothesis that innovation (weakly) influences the link between regulatory pressures and financial performance.

Table 3. The descriptive statistics of DEA scores.

	Sample size	DEA Scores			
		Max	Min	Mean	Std. Dev
2009	36	1	0.003948	0.407481	0.403781
2010	36	1	0.002963	0.382357	0.391703
2011	36	1	0.00193	0.268835	0.29031
2012	36	1	3.00E-06	0.029697	0.16637
2013	36	1	0.000109	0.045941	0.167208
Overall	180	1	3.00E-06	0.012868	0.104979

Table 4. Regressions results with DEA scores as the Dependent Variable

Variable	
Expenditure	−0.0234*
	(−1.769)
Employee	
_cons	0.7015***
	(4.047)
Year FE	Yes
N	180
F	
Pseudo R2/Adj. R2	0.4078

t statistics in parentheses, * $p < 0.1$, ** $p < 0.05$, *** $p < 0.01$

5 Discussion

We have used DEA to help investigate the intermediary role of innovation on the link between ER and FP in this paper. Our study has shown that innovation has a weak impact on the ER-FP link in Chinese context. By proving a weak impact of innovation, this paper has joined similar studies (e.g., Ramanathan et al. 2010; Triebswetter and Wackerbauer 2008) in the literature. There could be several reasons behind such a result. These are discussed below.

Firstly, the different environmental strategies adopted by firms in each sector may be one reason for the results, when investigating the interrelationships among three constructs. Firms engaged in different environmental strategies usually lead to different economic results and hence may further affect their interrelationships. In current Chinese context, firms still lack the incentives to adopt proactive environmental strategies and prefer to adopt reactive environmental strategies, even under the background of ER become more and more stringent (Yu and Chen 2015). In this sense, with the stringer of

ER, firms are able to reduce pollutant emissions. However, it cannot stimulate innovation and the lack of innovation cannot further promote their financial performance. Hence, firms' attitudes towards environmental strategies can have impact on the interrelationships. This may the reason why innovation does not appear to have a much stronger impact on the link between ER and FP.

Another reason may lie in the type of ER. Previous studies have proved that different types of ER can have different impact on innovation and FP (e.g., Zhao & Sun, 2016). One of the commonly used methods is to classify ER into two types, namely command and control regulations and market-based regulations. In this regard, it cannot provide enough incentives for firms to innovate so that innovation has a weak impact on the relationship between ER and FP.

Lastly, innovation itself may also contribute to our results. As reported by GII (2016), China has already made a great improvement in innovation in recent years. However, we may still give priority to economic development and neglect environmental protection in some regions of current China. Therefore, firms themselves may focus much on overall technologies innovation rather than environmental innovation in the past few years. Consequently, technological innovation will improve significantly while environmental innovation remains unchanged.

6 Conclusions

Our study has contributed to literature in multiple ways. The first contribution is the use of DEA to study the links among environmental regulations, financial performance and innovation. The second contribution is the focus on China in this paper. Unlike most of previous studies that focused in the North American/European context (Horváthová 2012), our paper is one of the few attempts to investigate the applicability the ER-Innovation-FP relationship in a developing country (China) context.

Another contribution occurs in terms of our ability to focus on all the three constructs (ER, innovation and FP) of relevance to environmental policy. By doing so, our study joins very limited number of studies that looked at the trilateral relationship. We have found that innovation has a weak impact on ER-FP relationship in Chinese context, which is somewhat surprising in the context of growing innovation capabilities in China (GII 2016). As discussed in the previous section, there could be several reasons for this surprising finding. Our result shows that innovation is relatively disassociated in the context of environmental management efforts in China, and thus highlights the need to integrate environmental innovation efforts in Chinese firms more tightly. Our result could be strengthened by conducting separate analysis focusing on market-based (flexible) and command-and-control (inflexible) regulations (Black et al. 2010). Similarly, efforts that focus specifically on measures of environmental innovations (rather than general innovation) will also help (Montabon et al. 2007). These form scope for future research.

References

Ambec, S., Barla, P.: A theoretical foundation of the Porter hypothesis. Econ. Lett. **75**(3), 355–360 (2002). https://doi.org/10.1016/S0165-1765(02)00005-8

Emrouznejad, A., Marra, M., Yang, G.L., Michali, M.: Eco-efficiency considering NetZero and data envelopment analysis: a critical literature review. IMA J. Manag. Math. **34**(4), 599–632 (2023)

Banker, R.D., Charnes, A., Cooper, W.W.: Some models for estimating technical and scale inefficiencies in data envelopment analysis. Manag. Sci. **30**(9), 1078–1092 (1984). https://doi.org/10.1287/mnsc.30.9.1078

Bi, G., Jin, M., Ling, L., Yang, F.: Environmental subsidy and the choice of green technology in the presence of green consumers. Ann. Oper. Res. **255**(1–2), 547–568 (2017). https://doi.org/10.1007/s10479-016-2106-7

Bi, G., Luo, Y., Ding, J., Liang, L.: Environmental performance analysis of Chinese industry from a slacks-based perspective. Ann. Oper. Res. **228**(1), 65–80 (2015). https://doi.org/10.1007/s10479-012-1088-3

Black, A., Ramanathan, R., Nath, P.: Flexible environmental regulations and performance: the moderating role of innovation at sector level. In: The Academy of Management Annual Meeting, Montréal, Canada, 6–10 August, 2010, pp. 1–41 (2010)

Boasson, V., MacPherson, A.: The role of geographic location in the financial and innovation performance of publicly traded pharmaceutical companies: empirical evidence from the United States. Environ. Plan A **33**(8), 1431–1444 (2001). https://doi.org/10.1068/a3431

Boiral, O.: Corporate greening through ISO 14001: a rational myth? Organ. Sci. **18**(1), 127–146 (2007)

Brunnermeier, S.B., Cohen, M.A.: Determinants of environmental innovation in US manufacturing industries. J. Environ. Econ. Manag. **45**(2), 278–293 (2003). https://doi.org/10.1016/S0095-0696(02)00058-X

Cai, W., Zhou, X.: On the drivers of eco-innovation: empirical evidence from China. J. Clean. Prod. **79**, 239–248 (2014). https://doi.org/10.1016/j.jclepro.2014.05.035

Charnes, A., Cooper, W.W.: The non-archimedean CCR ratio for efficiency analysis: a rejoinder to Boyd and Färe. Eur. J. Oper. Res. **15**(3), 333–334 (1984). https://doi.org/10.1016/0377-2217(84)90102-4

Charnes, A., Cooper, W.W., Rhodes, E.: Measuring the efficiency of decision making units. Eur. J. Oper. Res. **2**(6), 429–444 (1978). https://doi.org/10.1016/0377-2217(78)90138-8

Chen, Y., Lai, S., Wen, C.: The influence of green innovation performance on corporate advantage in Taiwan. J. Bus. Ethics **67**(4), 331–339 (2006)

Chintrakarn, P.: Environmental regulation and U.S. states' technical inefficiency. Econ. Lett. **100**(3), 363–365 (2008). https://doi.org/10.1016/j.econlet.2008.02.030

Claver, E., López, M.D., Molina, J.F., Tarí, J.J.: Environmental management and firm performance: a case study. J. Environ. Manag. **84**(4), 606–619 (2007). https://doi.org/10.1016/j.jenvman.2006.09.012

Cook, W.D., Zhu, J., Bi, G., Yang, F.: Network DEA: additive efficiency decomposition. Eur. J. Oper. Res. **207**(2), 1122–1129 (2010). https://doi.org/10.1016/j.ejor.2010.05.006

Doran, J., Ryan, G.: Regulation and firm perception, eco-innovation and firm performance. Eur. J. Innov. Manag. **15**(4), 421–441 (2012). https://doi.org/10.1108/14601061211272367

Du, S., Hu, L., Wang, L.: Low-carbon supply policies and supply chain performance with carbon concerned demand. Ann. Oper. Res. **255**(1–2), 569–590 (2017). https://doi.org/10.1007/s10479-015-1988-0

GII. Global Innovation Index 2016: Switzerland, Sweden, UK, U.S., Finland, Singapore Lead; China Joins Top 25. World Intellectual Property Organization, Geneva (2016). http://www.wipo.int/pressroom/en/articles/2016/article_0008.html. Accessed 15 Aug 2016

Gollop, F.M., Roberts, M.J.: Environmental regulations and productivity growth: the case of fossil-fueled electric power generation. J. Polit. Econ. **91**(4), 654–674 (1983)

Hamamoto, M.: Environmental regulation and the productivity of Japanese manufacturing industries. Res. Energy Econ. **28**(4), 299–312 (2006). https://doi.org/10.1016/j.reseneeco.2005.11.001

Hamilton, J.T.: Pollution as news: media and stock market reactions to the toxics release inventory data. J. Environ. Econ. Manag. **28**, 98–113 (1995). https://doi.org/10.1006/jeem.1995.1007

Horbach, J.: Determinants of environmental innovation-new evidence from German panel data sources. Res. Policy **37**(1), 163–173 (2008). https://doi.org/10.1016/j.respol.2007.08.006

Hoff, A.: Second stage DEA: comparison of approaches for modelling the DEA score. Eur. J. Oper. Res. **181**(1), 425–435 (2007)

Horváthová, E.: The impact of environmental performance on firm performance: short-term costs and long-term benefits? Ecol. Econ. **84**, 91–97 (2012). https://doi.org/10.1016/j.ecolecon.2012.10.001

Indrani, R.C., Das, S.K.: Re-visiting the Porter Hypothesis (2006). https://doi.org/10.1227/01.NEU.0000349921.14519.2A

Jaffe, A.B., Palmer, K.: Environmental regulation and innovation: a panel data study. Rev. Econ. Stat. **79**(4), 610–619 (1997)

Jiménez, O.: Innovation-oriented environmental regulations: direct versus indirect regulations; an empirical analysis of small and medium-sized enterprises in Chile. Environ Plan A **37**(4), 723–750 (2005). https://doi.org/10.1068/a3736

Kammerer, D.: The effects of customer benefit and regulation on environmental product innovation: empirical evidence from appliance manufacturers in Germany. Ecol. Econ. **68**(8–9), 2285–2295 (2009). https://doi.org/10.1016/j.ecolecon.2009.02.016

Kesidou, E., Demirel, P.: On the drivers of eco-innovations: empirical evidence from the UK. Res. Policy **41**(5), 862–870 (2012). https://doi.org/10.1016/j.respol.2012.01.005

Madsen, P.M.: Does corporate investment drive a "race to the bottom" in environmental protection? A reexamination of the effect of environmental regulation on investment. Acad. Manag. J. **52**(6), 1297–1318 (2009)

Majumdar, S.K., Marcus, A.A.: Rules versus discretion: the productivity consequences of flexible regulation. Acad. Manag. J. **44**(1), 170–179 (2001). https://doi.org/10.2307/3069344

Mergoni, A., Emrouznejad, A., De Witte, K.: Fifty years of data envelopment analysis. Eur. J. Oper. Res, **326**(3), 389–412 (2025). https://doi.org/10.1016/j.ejor.2024.12.049

Montabon, F., Sroufe, R., Narasimhan, R.: An examination of corporate reporting, environmental management practices and firm performance. J. Oper. Manag. **25**(5), 998–1014 (2007). https://doi.org/10.1016/j.jom.2006.10.003

Porter, M.E.: America's green strategy. Sci. Am. **264**(4), 168 (1991). https://doi.org/10.1038/scientificamerican0491-168

Porter, M.E., van der Linde, C.: Toward a new conception of the environment-competitiveness relationship. J. Econ. Perspect. **9**(4), 97–118 (1995)

Porter, M.E., van der Linde, C.: Green and competitive: ending the stalemate. Harv. Bus. Rev. **28**(6), 128–129 (1995). https://doi.org/10.1016/0024-6301(95)99997-E

Qi, G.Y., Zeng, S.X., Shi, J.J., Meng, X.H., Lin, H., Yang, Q.X.: Revisiting the relationship between environmental and financial performance in Chinese industry. J. Environ. Manag. **145**, 349–356 (2014). https://doi.org/10.1016/j.jenvman.2014.07.010

Ramanathan, R.: An Introduction to Data Envelopment Analysis: A Tool for Performance Measurement., 1st edn SAGE Publications Pvt. Ltd. (2003)

Ramanathan, R., Black, A., Nath, P., Muyldermans, L.: Impact of environmental regulations on innovation and performance in the UK industrial sector. Manag. Decis. **48**(10), 1493–1513 (2010). https://doi.org/10.1108/00251741011090298

Ramanathan, R., Ramanathan, U., Bentley, Y.: The debate on flexibility of environmental regulations, innovation capabilities and financial performance–A novel use of DEA. Omega, **75**, 131–138 (2018)

Rubashkina, Y., Galeotti, M., Verdolini, E.: Environmental regulation and competitiveness: empirical evidence on the Porter Hypothesis from European manufacturing sectors. Energy Policy **83**, 288–300 (2015). https://doi.org/10.1016/j.enpol.2015.02.014

Sarkis, J.: Manufacturing's role in corporate environmental sustainability- Concerns for the new millennium. Int. J. Oper. Prod. Manag. **21**(5/6), 666–686 (2001)

Sharma, S.: Different strokes: regulatory styles and environmental strategy in the North-American oil and gas industry. Bus. Strateg. Environ. **10**(6), 344–364 (2001). https://doi.org/10.1002/bse.303

Triebswetter, U., Wackerbauer, J.: Integrated environmental product innovation in the region of Munich and its impact on company competitiveness. J. Clean. Prod. **16**, 1484–1493 (2008). https://doi.org/10.1016/j.jclepro.2007.09.003

Wu, J., Song, M., Yang, L.: Advances in energy and environmental issues in China: theory, models, and applications. Ann. Oper. Res. **228**(1), 1–8 (2015). https://doi.org/10.1007/s10479-014-1528-3

Xing, Y., Kolstad, C.D.: Do lax environmental regulations attract foreign investment? Environ. Res. Econ. **21**(1), 1–22 (2002). http://ideas.repec.org/a/kap/enreec/v21y2002i1p1-22.html

Yang, F., Yang, M.: Analysis on China's eco-innovations: regulation context, intertemporal change and regional differences. Eur. J. Oper. Res. **247**(3), 1003–1012 (2015). https://doi.org/10.1016/j.ejor.2015.07.029

Yu, W., Chen, Q.: Environmental regulations and industrial performance evidence from the revision of Water Pollution Prevention and Control Law in China. Problemy Ekorozwoju (Prob. Sustain. Dev.) **10**(1), 41–48 (2015)

Zhang, C.H., Liu, H.Y., Bressers, H.T.A., Buchanan, K.S.: Productivity growth and environmental regulations - accounting for undesirable outputs: analysis of China's thirty provincial regions using the Malmquist-Luenberger index. Ecol. Econ. **70**(12), 2369–2379 (2011). https://doi.org/10.1016/j.ecolecon.2011.07.019

Zhao, X., Sun, B.: The influence of Chinese environmental regulation on corporation innovation and competitiveness. J. Clean. Prod. **112**, 1528–1536 (2016). https://doi.org/10.1016/j.jclepro.2015.05.029

FRANCISCO 2.0: Increasing the Usability and Appending Multiplicative DEA Models for Internet Services Forecasting Over Time

Francisco Daladier Marques Júnior(✉) and Franks Patrício Maciel Filho

Instituto Federal de Educação da Paraíba (IFPB), Cajazeiras, Paraíba, Brazil
daladierjr@ifpb.edu.br, maciel.franks@academico.ifpb.edu.br

Abstract. Over the last thirty years, it has been demonstrated that the chaotic behaviour of Internet traffic was irregular and exhibited a contagion that radiated across the TCP/IP stack, referred to as self-similarity (SS). However, most studies examined this behaviour within a single computer network. With the rise of virtual networks, it was observed that SS could be tamed by employing fractal decision variables per alternative. Consequently, a fractal evaluation framework emerged as a specialist system employing multiplicative Data Envelopment Analysis (DEA) models to compare the virtual networks, which served as the Decision Making Units (DMUs). This led to the building of a unique global tool named FRANCISCO: FRactal Network Cloud Infrastructure Service Comparison and Optimization, which conducted comparisons using classical super-efficiency models of DEA and performed exploratory data analysis for each one of the DMUs. Nevertheless, the initial development of FRANCISCO left some gaps; therefore, the contributions of this work are: a) to enhance the usability of FRANCISCO by creating help options; b) to select public datasets to facilitate the use and understanding in the operationalisation of the tool, and; c) to implement static and dynamic multiplicative models of DEA.

Keywords: Fractal · DEA · Web tool · Internet Services Prediction · Decision-making

1 Introduction

Self-similarity (SS) in computer networks celebrates its 30th anniversary since its discovery by Mandelbrot (Leland et al. 1994). This seminal work suggested that self-similar behaviour could significantly influence the planning, control, and analysis of high-speed networks, intensifying irregular traffic bursts rather than smoothing them. Over the following decade, researchers validated the stochastic nature of self-similarity (SS) across various networks, topologies, and protocols, challenging Poisson models (Paxson & Floyd, 1995).

The foundations of self-similarity trace back to Mandelbrot's studies in 1965, revealing clustered errors in communication channels that exhibited fractal patterns. Mandelbrot demonstrated that apparent randomness harboured inherent patterns where large

A. Emrouznejad et al. (Eds.): DEA 2024, LNOR, pp. 121–134, 2025.
https://doi.org/10.1007/978-3-031-98177-7_10

numbers follow large numbers and small numbers follow small numbers (Mandelbrot & Wallis, 1968).

By 2019, most research still concentrated on a single network's time series. Marques Júnior et al. (2019a) introduced a framework for comparing multiple virtual networks using fractal decision variables, applying multicriteria decision-making (MCDM) techniques like Windows Multiplicative DEA (WMDEA) to identify optimal configurations.

Customer preferences on the Internet constantly evolve, altering fractal behaviours. Thus, it is essential to select suitable fractal patterns to provide specific services. This work revisits the evolution of SS research, introduces a web MCDM friendly tool so-called by FRActal Network Cloud Infrastructure Service Comparison and Optimisation (FRANCISCO). By using the FRANCISCO the decision-maker will have the possibility of choosing one or more sets of tools at the same time that can offer smooth traffic with maximized TCP transfer rate over time. FRANCISCO's choice of the best alternative guarantees long-term optimized services for Internet subscribers.

FRANCISCO is a tool developed using R with a web-based graphical interface. It offers the most current and advanced features and resources in research related to fractal theory and self-similarity in the computer networks. As a network evaluation method, it also makes its mark by being the first network evaluation tool to use the MCDM method, specifically DEA, and its wide range of models, especially those of super-efficiency classic and multiplicative and windows multiplicative DEA model, which allow for the ranking of DMUs.

The contributions of this work are as follows: a) revisiting the main studies on self-similarity (SS) over the past 30 years, outlining a timeline of their evolution; b) presenting a method for evaluating multiple virtual networks over time, including the provision of various public datasets; c) suggesting directions for achieving optimal Internet services based on fractals dataset for evaluating virtual networks, and outlines potential advancements in Internet service delivery, and; d) launching and explaining the FRANCISCO's use.

The rest of this paper is organised as follows. Section 2 presents a background of related works and main concepts from self-similarity, fractals, network optimization directions, and DEA models. In Sect. 3, we propose an upgrade from the performance evaluation framework of (Marques Júnior et al., 2019a) for the assessment of virtual networks. In Sect. 4, we present the FRANCISCO tool, as a promising practical application of this research for computer networks evaluation using DEA formulations, and indices related to fractal theory and self-similarity, as decision variables. Finally, the Sect. 5 presents the conclusions and the direction for future works.

2 Self-Similarity (SS) Background

In computer networks, early studies focused on the implications of SS on device buffer spaces. For instance, (Park & Crovella, 1996) reported that increasing SS levels within the TCP protocol degraded performance. Later works extended these discussions to long-range dependence's (LRD) characterisation, modelling, and performance evaluation (Park & Crovella, 1997). Agit & Popescu (2001) noted that LRD influenced network performance metrics, such as throughput, packet loss, latency, and buffer occupancy, adversely affecting quality of service (QoS). (Owezarski & Larrieu, 2006) demonstrated an inverse relationship between LRD intensity and QoS levels.

Initially, most studies focused on monofractal behaviours. However, Feldmann et al. (1998) revealed that network flows exhibit multifractal patterns akin to cascades, useful for detecting local singularities and understanding network traffic over time (Millán et al., 2014).

To illustrate some of the key studies on self-similarity (SS), refer to Fig. 1, which presents its discovery in telecommunication networks by Mandelbrot (1965) up to the launch of the network service prediction tool, named FRActal Network Cloud Infrastructure Service Comparison and Optimisation (FRANCISCO) (Andrade & Marques Junior, 2023).

We used a bibliometric web tool called Research Rabbit to correlate all articles cited here in regards to fractals via artificial intelligence. A CSV file and a graph map showing the relationships between the works were generated, and they are available at the URL: https://abre.ai/jilq.

According to Park & Crovella (1997), network performance, as captured by throughput, packet loss rate, and packet retransmission rate, gradually deteriorates with an increase in heavy-tailed traffic, while queue delay, response time, and fairness degrade more drastically. The extent to which heavy-tailed traffic affects the SS is determined by how congestion control can shape the traffic from the source into an average steady output flow while preserving information. Regarding performance, the paper showed that SS has an opposite effect on network performance. One key point is that the distribution of queue sizes with SS traffic decays more slowly than traffic exhibiting short-range dependence (SRD), using the Poisson probability distribution function.

Hohn et al. (2002) investigated how transitions in time scales cause dependencies between scale behaviours of arrival levels between layers L2 – with IP – and L3 – with TCP – across all types of time scales. Therefore, it is important to study the arrival process due to its potential impact on both the IP level and web server performance.

Terdik & Gyires (2009) classified SS studies into generations: i) the first generation (1994–2004), which addressed the shortcomings of Poisson models because these studies simplified network traffic characteristics and were not suitable for modelling bursty traffic in LANs and WANs; ii) the second generation, from 2004 onwards, challenged the adequacy of these results in networks, arguing that traditional Poisson-based models and others are still more appropriate for characterising current Internet traffic.

According to Terdik & Gyires (2009), a possible explanation is that, as the speed and volume of Internet traffic grow spectacularly, any network traffic irregularity, such as SS, may cancel out due to high-speed optical connections, new communication protocols, and the large number of multiplexed flows. However, the authors neglected to mention that the traffic in these optical networks is based on TCP/IP. Thus, even if the backbones are optical, Internet service delivery remains chaotic in bursts, flows, packet fragments, datagrams, frames, and so on, in addition to users' changing preferences and the characteristics of accessed applications.

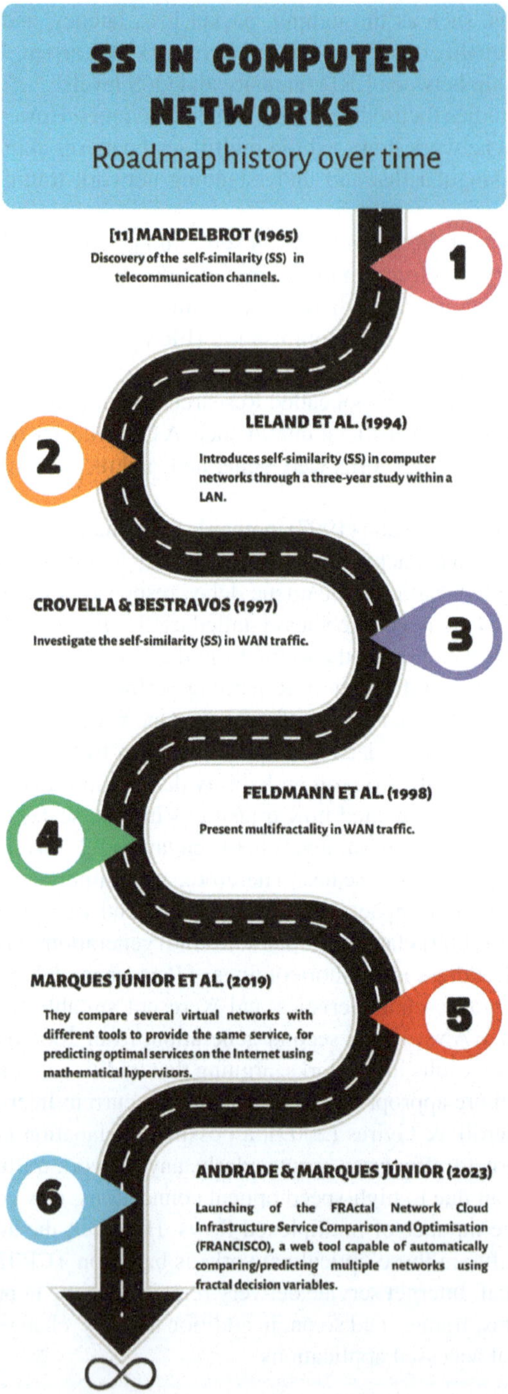

Fig. 1. Self-similarity (SS) roadmap history over time.

Mandelbrot also introduced fractal geometry, which can compute surface areas with irregular perimeters, since Euclidean geometry was incapable of such calculations (Mandelbrot & Taleb, 2012). With fractal geometry, several indices were introduced, one of which stands out: the fractal dimension, with its various methods, which have been applied for prediction in various fields such as medicine, seismology, astronomy, stock markets, information security, cryptography/entropy, blockchain, etc.

Another form of fractal evaluation, introduced in computer networks by Marques Júnior et al. (2019a), applies to both TCP traffic and other Internet protocols and is known as the fractal dimension *D* or Hausdorff dimension. This fractal index serves as a measure of irregularity in time series, with values varying in the range $[1, 2)$ (Gneiting et al., 2012). Thus, two important conclusions can be drawn regarding the value of D: 1) When *D* tends to 1, the continuous random variables under analysis exhibit a smoothing effect, i.e., the TCP bandwidth averages remain more stable over time, and; 2) When *D* tends to 2, the time series exhibits a roughness effect, meaning TCP traffic is irregular with infinite variance and high variability over time.

Xu & Feng (2014) demonstrated that TCP traffic is multifractal, and upon observing the singularity spectrum, it was shown that the multifractal characteristics of TCP traffic are weakened with increasing bandwidth and statistical multiplexing, but they cannot be eliminated.

The work of Jena & Popescu (2001), showed the influence of LRD properties on delay performance in L3 and L4 layers is reported, and network performance metrics such as throughput, packet loss, latency, and buffer occupancy levels are affected by the presence of the LRD phenomenon in many types of networks. The impact of LRD on Quality of Service (QoS) was analysed by Owezarski & Larrieu (2006), showing that the higher the LRD, the lower the QoS.

According to Rezaul & Grout (2007), experimental results showed that the CoLoRaDE algorithm is capable of reducing the LRD of received packet sequences in the router's buffer before transmission to the network core. However, the only way to compare the multiplexed client flows is by selecting the lowest value of the Hurst parameter or *H*, meaning the algorithm selects the flow with the least memory of service delivery over time, i.e., it erroneously selects flows with the least memory over time.

The research of Kaklauskas & Sakalauskas (2009) introduced the concept of fractal dimension in Internet traffic analysis. The authors argue that it is necessary to find a forecasting index that can preserve the characteristics of the time series. The goal is to help the network administrator deliver traffic with the most satisfactory service quality. However, the paper does not discuss the conclusions that can be drawn from the network traffic analysis when calculating the fractal dimension or indicate which protocols were analysed in real measurements. It is worth mentioning that all calculated fractal dimension values were below 1, which invalidates the results. However, the fractal dimension is always greater than the topological dimension, and for these time series, the topological dimension value should be 1.

One of the limiting factors of fractal traffic analysis or similar methods is that it requires the decision-maker to have in-depth knowledge of statistics and its computational tools, such as R and its respective libraries, as well as the Python programming language, among others. To address this gap, the Selfis applet was released, a Java tool that computes fractal indices in time series with vectors limited to 512 positions, in addition to using some exploratory data analysis tools. However, Selfis works with only one time series per network and does not accept files with extensions other than.txt.

Thus, the FRActal Network Cloud Infrastructure Service Comparison and Optimisation (FRANCISCO) was launched by Andrade & Marques Junior (2023) as a web application for the comparison/prediction of a time series from multiple networks, without size limitations, using either.txt files or output formats from iperf and Apache Bench to collectively analyse the L3 and L4 layers of TCP/IP. FRANCISCO employs super-efficiency DEA (Data Envelopment Analysis) models (Andersen & Petersen, 1993) to compare the networks under analysis, as well as the super multiplicative DEA model and the windows multiplicative DEA model. Thus, the decision-maker can customise their decision variables, adapting them to the characteristics of each service to be evaluated on the Internet. In conclusion, FRANCISCO is the most comprehensive fractal traffic analysis tool to date and is being improved to accept public datasets, passive measurements, multi-criteria decision-making models, and the use of other evaluation techniques such as artificial intelligence, among others.

2.1 Prediction of Optimal Services in Virtual Networks Based on Fractal Traffic

As can be observed, the vast majority of studies on SS and fractals in computer networks evaluate their effects by computing various fractal indices and performing exploratory analysis on a single network, usually using passive measurements from network devices. Passive measurements are difficult to evaluate because a range of protocols are available in all layers of the TCP/IP protocol stack to aid in service delivery. Therefore, when analysing passive traffic, a series of relevant information for the evaluation are observed; however, most of the other information is irrelevant to the decision-maker, who should discard it.

On the other hand, active measurements use benchmark tools accepted by the international community, which evaluate only the relevant layers and/or protocols. This work used active measurement tools such as iperf and apache-bench.

There are numerous models for predicting services in computer networks and in any evaluation involving large data volumes (big data), where the aim is to find the best/worst pattern to solve a specific problem. Thus, it is necessary to employ either a multicriteria decision-making (MCDM) technique, such as DEA, ELECTRE, PROMETHEE, TOPSIS, VIKOR, among others, or even a hybrid set of the main techniques by combining DEA with Fuzzy logic, analytical hierarchy process (AHP), machine learning, artificial intelligence techniques, and other combinations of strategies (Whaiduzzaman et al., 2014). The purpose of using these evaluation techniques is to provide alternatives for solving the complex multicriteria optimisation problem in computer networks.

Therefore, the decision-maker must have experience in using a multicriteria decision-making technique, or even in mixing them and, if possible, customising them to achieve the best results. However, we encourage the scientific community to always compare a large number of alternatives for providing the same types of services, as this will allow the prediction of which alternative has the best effects of LRD for service delivery on the Internet.

2.2 DEA Models

DEA is a MCDM technique that has been in use for more than 45 years to evaluate DMUs according to the concept of the relative efficiency of the same units of production appraisal in almost all areas (Mergoni et al., 2025). DEA uses linear programming for computing the relative efficiency of each DMU. DEA is based on Pareto-Koopman's efficiency or envelope frontier, i.e. trying to decrease the input without changing the output values (input-oriented), as well as trying to increase the output without modifying the input values (output-oriented), or simultaneously minimizing the input and maximizing the output in non-oriented models.

However, many DEA models were created over time under two main assumptions of constant returns to scale (CRS) (Charnes et al., 1978) or variable return to scale (VRS) (Emrouznejad et al., 2025). VRS are models characterised by three types of assumptions which are: a) increasing return to scale (IRS) – the DMUs operate far below their optimal capacity; b) constant return to scale (CRS) – the DMUs operate in their optimal capacity, and; c) decreasing return to scale (RTS) – the DMUs operate far under their optimal capacity.

Firstly, for the characterization of every DEA model, it is mandatory to know the concept of Production Possibility Set (PPS). PPS is obtained through the estimation of the distance of each unit to the border, which will define if a DMU is or is not efficient. For this reason, consider each DMU/unit uses m inputs to produce s outputs. Let a set of n DMUs be evaluated as $\mathrm{DMU_j}$ (for each $j = 1, \ldots, n$) linked to the input vector of $X_j = (x_{i,j}, \ldots, x_{m,j})$ and output vector of $Y_i = (y_{i,j}, \ldots, y_{s,j})$, then the PPS can be described as:

$$P = \{(X, Y) : X \ can \ produce \ Y\} \tag{1}$$

In the meantime, the super-efficiency DEA (SDEA) model was proposed by Marques Júnior et al. (2019a) to rank efficient DMUs on a reference expanded technology for all other units. Therefore, there are also several other research that suggested other ways to select only the most efficient unit using distinct DEA models instead SDEA formulations as can be seen ((Amin & Toloo, 2007), (Toloo & Salahi, 2018)). SDEA approach removes the under-evaluation unit called DMU_0 from observations set during the assessments of each DMU.

The difference between DEA and SDEA efficiency scores is that when a DMU is efficient on standard DEA models then its score on SDEA models is upper than the unit, even though inefficient keep themselves with the same score in both models. FRANCISCO computes the DEA's efficiency indexes by using the classic super-efficiency DEA under CRS and VRS, as well as the super multiplicative DEA CRS model and the windows multiplicative DEA CRS DEA model devised by (Marques Júnior et al., 2019a).

3 The FRANCISCO's Core

The idea for FRANCISCO originated from the fractal traffic evaluation framework developed by (Marques Júnior et al., 2019a). This framework consists of six steps, that are:

i) **Assembling virtual networks experimentation scenarios**, which can be conducted on type I and II hypervisors, containers of virtualization, cloud computing environments, software-defined networks, 5G, web servers, L3/L4 protocols, etc.;
ii) **Execution of measurements** using active traffic generation tools, e.g., iperf, apache-bench, among others;
iii) **Capturing and processing the fractal decision variables** of each set of distinct tools to provide the same network services over time, forming n DMUs, as shown in Eq. (2);
iv) **Conducting the evaluation to construct the DEA decision matrix** with all the decision variables computed for each DMU;
v) **Employing a mathematical hypervisor** based on classical and multiplicative DEA models or even other MCDM methods, artificial intelligence methods, or even hybrid approaches, and;
vi) **The decision-making**, which involves selecting the most efficient DMUs, especially those chosen by the multiplicative DEA models.

In this work, we evaluate TCP/IP layers – L3 and L4 – using the webserver benchmarking tool so-called apache-bench[1] (ab) and presenting the results in a Mendeley's public dataset[2]. It's worth to mention that we have been creating some Mendeley's public datasets that are obtained around the computer network lab discipline that in each semester create a new computer network distinct challenge to solve since 2019.

The main FRANCISCO's goal is to know which is the best way for delivering three video stream files[3](with 480, 720, and 1080 pixels) comparing the two main webservers of the world[4] – Apache2 and Nginx – inside of a Docker container running in a guest virtual machine (VM) Ubuntu 20.04 server. These web servers were varying 15 different TCP congestion control algorithms like bic, bbr, cdg, cubic, dctcp, htcp, hybla, illinois, lp, nv, scalable, vegas, veno, westwood, and yeah. On another side, a guest VM ran the ab with 5000 requisitions and 500 clients competing for the resources of the webservers under analysis. It is worth mentioning, that which guest VM running ab does not vary the TCP congestion control algorithms, i.e. this guest VM always keeps the default Linux TCP congestion control algorithm as cubic (Fig. 2).

We have been using an adapted version from the evaluation framework launched by Marques Júnior et al. (2019). The main difference is that Marques Júnior et al. (2019) only used the Windows Multiplicative DEA model as a math hypervisor, i.e. a dynamic DEA model, while in this work is employed a super Multiplicative DEA CRS model

[1] Apache-bench – Apache HTTP server benchmarking tool available at URL: https://httpd.apache.org/docs/2.4/programs/ab.html.

[2] Super Multiplicative VRS DEA input-oriented model results obtained with 660 DMUs using apache bench tool for evaluating video stream files on two main webservers from the market, that are Apache2 vs Nginx, varying 15 TCP congestion control algorithms on Xen hypervisor – Mendeley's public dataset available at URL: https://data.mendeley.com/drafts/4xn5bynbwv.

[3] The stream video files only are differing in the number of pixels and size, because their contents are the same.

[4] Usage statistics and market shares of web servers from 2024 December – Available at URL: https://w3techs.com/technologies/overview/web_server.

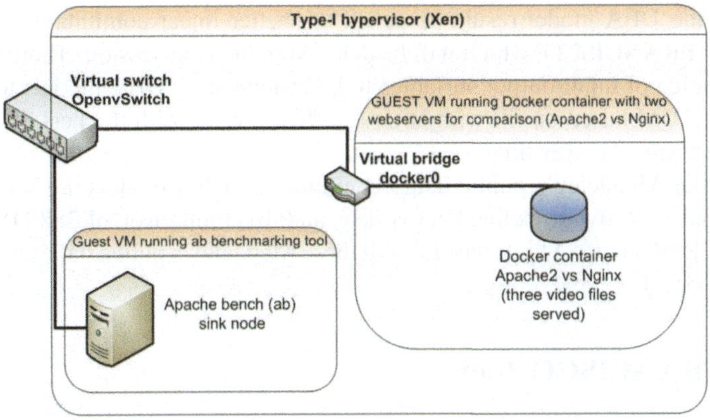

Fig. 2. Guest to Guest to Container (GGC) experimentation's topology.

and the DEA classic models, as well as the super-efficiency classic DEA formulations acting as a **forecast hypervisor**.

To assess the performance of the DMUs the topology **Guest to Guest to Container (GGC)** was created. At GGC, measurements are conducted from the Guest VM OS (sink node) through the virtual switch from type-I hypervisor (Xen (OpenvSwitch)) which passes the packets/flows to another guest VM. Finally, the Guest VM forwards the packets/flows via a virtual bridge (docker0) to the container tool running both web servers.

Both Guest VMs are running Ubuntu 20.04 server, but only in the VM that runs the webservers inside Docker's container does the amount of virtual CPU (vCPU) and virtual RAM (vRAM) is varying to assess the behaviour of the webservers delivering the three video files. The vRAM quantities are 1 and 4 GB, and for vCPU we've got 1,2,3, and 4. The experiments were made by student groups from my network lab discipline. Thus, the number of experiments or DMUs in all groups should be:

$$Number_{ofDMUs} = N_{webservers} * N_{videos} * N_{vCPU} * N_{vRAM} * N_{TCPalgorithms} \quad (2)$$

$$Number_{ofDMUs} = 2 * 3 * 4 * 2 * 15 = 720$$

It must be mentioned which one group forgot to make 60 measurements, then the total number of DMUs performed is **660**. The decision variables are separated into input and output variables, which are: I) Input-variables: 1) Fractal dimension (rodogram) from ttime obtained from ab output file; 2) Time taken for tests, and; 3) Time per request. O) Output-variables are: 1) HTTPS Transfer rate; 2) Hurst parameter from time series, and; 3) Requests per second.

In the **forecast hypervisor** occurs with the application of the DEA models of the' decision variables for a fair ranking of DMUs.

The ending layer is the **decision-making (DM)**. Here, the decision-maker has to choose the best set of settings to deliver more stable/most efficient **virtual networks** (VN) web services over time by using any DEA model on FRANCISCO. Finally, after employing the DEA model results to select the better linear combination from units obtained by FRANCISCO, which will be detailed in the next session. There has been a stepwise choice of input/output variables to be evaluated by DEA models to carry out an optimised DM based on SS with LRD to offer services with the best TCP/IP stack protocol performance over time.

Others our Mendeley's public datasets employing DEA models are: a) evaluating cloud computing software-defined networks[5], and; b) comparison of 288 DMUs of the L3 from TCP/IP on type-I hypervisors[6]. All these Mendeley's public dataset can be used by FRANCISCO.

4 The FRANCISCO Tool

The FRANCISCO[7] is a web tool developed with the aim of bridging the gap identified in the network evaluation field, as this research suggests, especially following the introduction of network analysis via methods linked to MCDM, like the DEA. Another pivotal factor was the necessity to simplify the multitude of variations concerning DEA allocation configurations, since this MCDM method allows the evaluator to make countless adjustments, among them the choice of orientation, which evidently affects the outcome.

Initially, FRANCISCO provides the capability to register DMUs through different benchmarking tools. Apache-bench is one such tool, but it also accommodates DMUs created by iperf and numerical time series. Another option is to log a DMU table with all decision variables already filled in permanently, or rather, pre-computed. In line with this study's proposal, the tool records a time series per DMU, as well as a set of time series per DMU in a windows multiplicative DEA comparison. By allowing the reading of raw network measurement files generated by these benchmarking tools – without any prior data treatment – the tool ensures greater convenience when creating data sets for the computation of fractal indices related to self-similarity and, consequently, DEA evaluation, with each DMU represented by a file. For the user convenience, this FRANCISCO's version brought a VN measurements dataset.

Indices calculated from the time series registered in the tool can also be chosen by the user. Hence, when generating their data set, users can select those most suitable for their evaluation interests. The primary ones, especially for the application of fractal theory and self-similarity in network evaluation, are fractal dimension and the Hurst parameter using the R/S method. Users can choose some of the fractal dimension methods, like the popular madogram, variogram, and rodogram, among 8 other method options. Other indices include variance, mean, Whittle's estimator, and the alpha tail shape parameter.

[5] Prediction of the best alternative for cloud network services delivering using fractals variables and super-efficiency DEA models for time-series data comparison - Available at URL: https://data.mendeley.com/drafts/29txvz235y.

[6] Comparison of 288 DMUs of the L3 from TCP/IP stack on type-I hypervisors by using the Super-Cobb-Douglas - SMDEA CCR-I model – Available at URL: https://data.mendeley.com/drafts/fx4rhhxrdm.

[7] FRANCISCO web tool – Available at URL: https://newfranciscodea.shinyapps.io/francisco-dea/.

In total, the tool can compute 6 possible indices, and users can choose between 1 and all 6, depending on their preference.

Reaching the actual DEA analysis, users can select which of the created index variables will be inputs and subsequently, the outputs. Afterwards, it's up to the user to pick their preferred DEA model. Available choices include CCR, BCC, slacked-based measure (SBM), SCCR, super multiplicative DEA CRS model, and the windows DEA multiplicative model. In addition, users can also select the orientation of the chosen method, whether input-oriented or output-oriented. Specifically, the SCCR is responsible for ranking DMUs. Often, the data set might possess multiple efficient DMUs, which is why super-efficiency methods are pivotal, allowing the ranking of top solutions, not just identifying the efficient ones.

Beyond pinpointing the most efficient network, the tool also offers two graphs related to the chosen DEA method: the efficiency frontier and the ray unbounded. Other graphs, like the autocorrelation function (ACF), histogram, periodogram, and the log-log cumulative distribution (LLCD) are also available, albeit solely for individual DMU exploratory analysis of data. In Fig. 3, FRANCISCO already has a set of IPERF-type DMUs generated, where the user has selected all 6 possible indices and set up the DEA in the SCCR model, the super-efficiency model. This will enable DMUs to be ranked and the most efficient network to be identified.

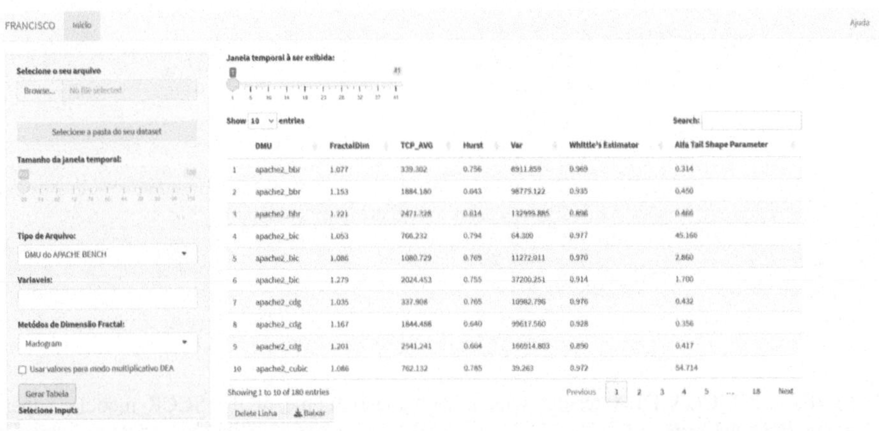

Fig. 3. FRANCISCO's home page, with the DMU dataset already generated and the configurations for DEA analysis set.

In another screen, we see in Fig. 4 the network FEDORA_2CPU_4vRAM_HTCP, identified as the most efficient with an efficiency score of 1.32.

For the first FRANCISCO's use, go to the previously mentioned FRANCISCO's URL, and click on button "Selecione a pasta do seu dataset", after a new browser windows created click in Dataset-de-teste, after that click in Select button and wait the loading of this dataset by the FRANCISCO. The next step is to click in the button "Gerar Tabela" and you will see a table like in the Fig. 3. After that, choose the input and output variables, and finally click in the button "Analyse DEA". If you want to observe the exploratory analysis of each DMU, it is necessary only click in the DMU and click in the tab DMU Analysis. Finally, to observe the DEA results click in the tab DEA results or even Graphic DEA.

FRANCISCO also allows for the creation of a DMU dataset using various file types generated by benchmarking tools. In future works must be available on FRANCISCO: i) new file types might be supported in the tool, along with an aspiration to incorporate additional MCDM methods; ii) new DEA models; iii) more fractal index options; iv) online storage of results; v) simultaneous comparison of DMU across multiple MCDM methods; vi) exporting results to create datasets, and; vii) evolve into a network orchestrator. However, an actual FRANCISCO's limitation are the support only to the Portuguese language and a better help, but we will support the English version as well.

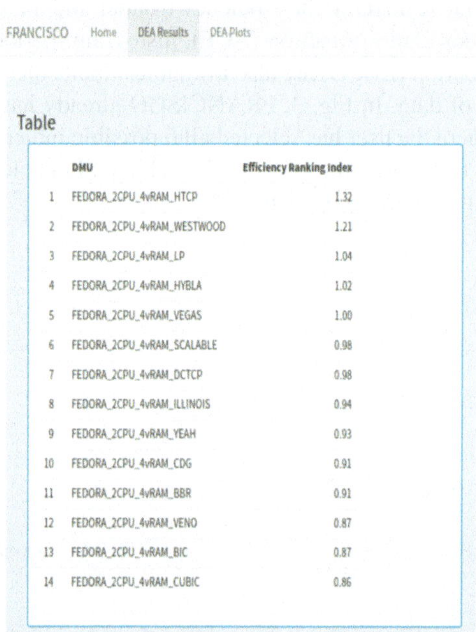

	DMU	Efficiency Ranking Index
1	FEDORA_2CPU_4vRAM_HTCP	1.32
2	FEDORA_2CPU_4vRAM_WESTWOOD	1.21
3	FEDORA_2CPU_4vRAM_LP	1.04
4	FEDORA_2CPU_4vRAM_HYBLA	1.02
5	FEDORA_2CPU_4vRAM_VEGAS	1.00
6	FEDORA_2CPU_4vRAM_SCALABLE	0.98
7	FEDORA_2CPU_4vRAM_DCTCP	0.98
8	FEDORA_2CPU_4vRAM_ILLINOIS	0.94
9	FEDORA_2CPU_4vRAM_YEAH	0.93
10	FEDORA_2CPU_4vRAM_CDG	0.91
11	FEDORA_2CPU_4vRAM_BBR	0.91
12	FEDORA_2CPU_4vRAM_VENO	0.87
13	FEDORA_2CPU_4vRAM_BIC	0.87
14	FEDORA_2CPU_4vRAM_CUBIC	0.86

Fig. 4. FRANCISCO's DEA results screen, with the outcome of the SCCR model's super-efficiency DEA analysis.

5 Conclusion and Future Works

This paper has launched one more application to DEA related to a assessment/forecasting of the better form for delivering virtual network services video stream web traffic over time. For solving this multi-criteria problem, the performance evaluation framework methodology of (Marques Júnior et al., 2019) was implemented as the FRANCISCO web tool to help the decision-making process easier on computer networks.

One can conclude that the correct use of the fractal dimension of a time series is so important to guarantee consumers a service with better, more stable, and most efficient

TCP performance over time. Only having $H > 0.5$ gives a mathematical background that there is a positive autocorrelation among the continuous random variables (CRV), but applying roughness/smoothness concept with rodogram method to measure fractal dimension allows the decision-maker to ensure choosing the super-efficient DMU which it will deliver smoother/efficient VN services traffic over time.

Researchers interested may extend this to conduct a scale-up of all virtual devices using other virtualization techniques to make a deep SS with LRD analysis using the FRANCISCO. In the FRANCISCO's future versions will be appended others DEA models, distinct MCDM methods, a testbed tool to assembly of VN to be compared, multifractals variables, and a network/cloud orchestration service to optimize the Internet services on demand.

References

Andersen, P., Petersen, N.C.: A procedure for ranking efficient units in data envelopment analysis. Manag. Sci. **39**(10), 1261–1264 (1993). https://doi.org/10.1287/mnsc.39.10.1261

Agit, J., Popescu, A.: Traffic engineering for Internet applications. In: van der Mei, R.D., Huebner-Szabo de Bucs, F. (eds.) SPIE Proceedings Series, Internet Performance and Control of Network Systems II, vol. 4523, pp. 67–78 (2001). https://doi.org/10.1117/12.434336

Amin, G.R., Toloo, M.: Finding the most efficient DMUs in DEA: an improved integrated model. Comput. Ind. Eng. **52**(2007), 71–77 (2007). https://doi.org/10.1016/j.cie.2006.10.003

Charnes, W.W.C., Rhodes, E.: Measuring the efficiency of decision making units. Eur. J. Oper. Res. **2**(1978), 429–444 (1978). https://doi.org/10.1016/0377-2217(78)90138-8

de Andrade, F.I.S., Marques Junior, F.D.: FRANCISCO: FRActal network cloud infrastructure service comparison and optimisation. In: Anais Do XXII Workshop Em Desempenho de Sistemas Computacionais e de Comunicação (WPerformance 2023), pp. 109–120 (2023). https://doi.org/10.5753/wperformance.2023.230685

Emrouznejad, A. Podinovski, V. Charles, V. Lu, C. Moradi-Motlagh, A.: Rajiv Banker's lasting impact on data envelopment analysis. Ann. Oper. Res. **351**, 1225–1264 (2025). https://doi.org/10.1007/s10479-025-06473-3

Feldmann, A., Gilbert, A.C., Willinger, W.: Data networks as cascades. ACM SIGCOMM Comput. Commun. Rev. **28**(4), 42–55 (1998). https://doi.org/10.1145/285243.285256

Gneiting, T., Sevčíková, H., Percival, D.B.: Estimators of fractal dimension: assessing the roughness of time series and spatial data. Stat. Sci. Donald B. Percival Principal Math. Appl. Phys. Lab. **27**(2), 247–277 (2012). https://doi.org/10.1214/11-STS370

Hohn, N., Veitch, D., Abry, P.: Does fractal scaling at the IP level depend on TCP flow arrival processes? In: Proceedings of the Second ACM SIGCOMM Workshop on Internet Measurment - IMW '02, vol. 63 (2002). https://doi.org/10.1145/637201.637208

lToloo, M., Salahi, M.: A powerful discriminative approach for selecting the most efficient unit in DEA, Comput. Ind. Eng. **115**(2018) 269–277 (2018). https://doi.org/10.1016/j.cie.2017.11.011

Jena, A.K., Popescu, A.: Traffic engineering for Internet applications. In: van der Mei, R.D., Huebner-Szabo de Bucs, F. (eds.), pp. 67–78 (2001). https://doi.org/10.1117/12.434336

Kaklauskas, L., Sakalauskas, L.: Application of Chaos Theory to Analysis of Computer Network Traffic (2009). https://api.semanticscholar.org/CorpusID:131234261

Leland, W.E., Taqqu, M.S., Wilson, D.V.: On the self-similar nature of ethernet traffic (extended version). IEEE/ACM Trans. Network. **2**(1), 1–15 (1994). https://doi.org/10.1109/90.282603

Mandelbrot, B.B., Wallis, J.R.: Noah, joseph, and operational hydrology. Water Res. Res. **4**(5), 909–918 (1968). https://doi.org/10.1029/WR004i005p00909

Marques Júnior, F.D., Emrouznejad, A., Dias, K.L., Cunha, P.R.F., de Castro e Silva, J.L.: Optimising virtual networks over time by using Windows Multiplicative DEA model. Expert Syst. Appl. **132**, 209–225 (2019a). https://doi.org/10.1016/J.ESWA.2019.05.005

Millán, G., Juan, E.S., Jamett, M.: A simple estimator of the Hurst exponent for self-similar traffic flows. IEEE Lat. Am. Trans. **12**(8), 1349–1354 (2014). https://doi.org/10.1109/TLA.2014.7014500

Mandelbrot, B.: Self-similar error clusters in communication systems and the concept of conditional stationarity. IEEE Trans. Commun. Technol. **13**(1), 71–90 (1965). https://doi.org/10.1109/TCOM.1965.1089090

Mandelbrot, B.B., Taleb, N.N.: Mild vs wild randomness focusing on those risks that matter. In: The Known, the Unknown and the Unknowable in Financial Risk Management : Measurement and Theory Advancing Practice (2012). https://doi.org/10.1017/CBO9781107415324.004

Mergoni, A., Emrouznejad, A., De Witte, K.: Fifty years of data envelopment analysis. Eur. J. Oper. Res. **326**(3), 389–412 (2025). https://doi.org/10.1016/j.ejor.2024.12.049

Owezarski, P., Larrieu, N.: Measurement based approach of congestion control for enforcing a robust QoS in the inter. In: International Conference on Internet Surveillance and Protection (ICISP06), p. 3 (2006). https://doi.org/10.1109/ICISP.2006.19

Paxson, V., Floyd, S.: Wide area traffic: the failure of poisson modeling. IEEE/ACM Trans. Networking **3**(3), 226–244 (1995). https://doi.org/10.1109/90.392383

Park, K., Kim, G., Crovella, M.: On the relationship between file sizes, transport protocols, and self-similar network traffic. In: Proceedings of 1996 International Conference on Network Protocols (ICNP-96), pp. 171–180 (1996). https://doi.org/10.1109/ICNP.1996.564935

Park, K., Kim, G., Crovella, M.: On the effect of traffic self-similarity on network performance. In: SPIE International Conference on Performance and Control of Netork Systems, vol. 3231, pp. 296–310 (1997). https://doi.org/10.1117/12.290419

Rezaul, M., Grout, V.: A novel algorithm for controlling long-range dependent network traffic. In: Proceedings of the Sixth International Conference on Networking (2007), Martinique, French Caribbean, 22–28 April 2007 (2007). https://doi.org/10.1109/ICN.2007.29

Terdik, G., Gyires, T.: Does the internet still demonstrate fractal nature? In: 2009 Eighth International Conference on Networks, pp. 30–34 (2009). https://doi.org/10.1109/ICN.2009.8

Whaiduzzaman, M., Gani, A., Anuar, N. B., Shiraz, M., Haque, M.N., Haque, I.T.: Cloud service selection using multicriteria decision analysis. Sci. World J. (2014). https://doi.org/10.1155/2014/459375

Xu, Y., Feng, H.: Revisiting multifractality of TCP traffic using multifractal detrended fluctuation analysis. J. Stat. Mech: Theory Exp. **2014**(2), P02007 (2014). https://doi.org/10.1088/1742-5468/2014/02/P02007

Performance Assessment of Indian Electricity Distribution Utilities Using Reliability Indices: A Two-Stage Data Envelopment Analysis

Sufia(✉), Anoop Singh, and Shambhavi Mishra

Centre for Energy Regulation (CER), Department of Management Sciences, Indian Institute of Technology Kanpur, Kanpur, India
{sufia,anoops,sbmishra}@iitk.ac.in

Abstract. Distribution utilities vary in terms of costs, quality of supply, customer segments, and network features which impacts their operational effectiveness. Quality of supply is an important determinant of operational performance. Two-stage data envelopment analysis is used to analyse operational efficiency of 19 unbundled Indian electricity distribution companies from 2016–17 to 2019–20. The first model estimates the operational efficiency of the utilities while the second model uses reliability indices as undesirable outputs in addition to the outputs and inputs used in first model. In the second stage, efficiency scores from first stage are regressed with external environmental variables such as years from unbundling, ownership, and customer structure to analyse their influence on the performance of distribution companies. Coefficients of ownership and customer structure are negative and significant in the regression results.

Keywords: Operational performance · reliability indices · Indian distribution utilities · two-stage DEA · quality of supply

1 Introduction

Indian power sector has evolved dynamically over years to support the growing energy demand of the accelerating national economy. Economic development surges the demand of reliable electricity. The Indian power sector is divided into segments of electricity generation, transmission, and distribution. The generation segment comprises of thermal, hydro, nuclear and renewable energy sources. Although the share of renewable generation is increasing, Indian power generation continues to be dominated by thermal power. India has set an ambitious target of 450 GW of renewables in its energy mix by 2030, which comes with challenges for efficient grid operation and management. The transmission and distribution segments are transmitters of electricity from electricity generator to distribution substations and then from those substations to end consumers respectively. End consumers are divided into industrial, agricultural, commercial, and residential consumers. Electricity is bought and sold as a commodity for ensuring a stable power grid by management of supply and demand. There have been significant changes

A. Emrouznejad et al. (Eds.): DEA 2024, LNOR, pp. 135–148, 2025.
https://doi.org/10.1007/978-3-031-98177-7_11

in the Indian power sector, brought about by various policies and reforms with an aim to improve efficiency, promote competition and clean energy transition. Efficient operation mainly in terms of overall technical and commercial loss reduction, network strengthening, timely tariff revisions with cost-reflective tariffs and improvement in energy access is necessary for viability of the sector.

Traditionally, the Indian power generation, transmission and distribution were vertically integrated with State Electricity Boards (SEBs) operating at state level and were responsible for production and supply of electricity. The operational and financial health of these SEBs led to reforms for restructuring of the sector. The Electricity Act (EA) of 2003 was one of the most comprehensive reforms enacted with very precise objectives set to introduce open access, create competition, and bring transparency by unbundling of generation, transmission and distribution segments and mandatory establishment of Electricity Regulatory Commissions both at central and state levels. But even after two decades of enactment of the EA, 2003, many challenges continue to hinder the performance of the sector. Distribution segment forms an important link to ensure viability of the sector as it serves the end consumers and is revenue generating point for the cost incurred in the whole process of generation, transmission, and distribution. Challenges like electricity theft, high transmission and distribution loss, cross subsidies and non-reflective tariff are affecting the viability of the distribution companies. Huge debts on most of the distribution companies is one of the reasons for delay and non-payment to the generator for the electricity purchased, thus leading to cancellation of power purchase agreements. As of October, 2024, distribution companies owe generating companies Rs 66,757.39 crore [20]. The operational and financial efficiency of the distribution companies is essential for the sustainable growth of the entire power chain and economy.

The EA, 2003 was followed by many policies and schemes over time. Accelerated Power Development Reforms Program (APDRP) introduced in 2003 aimed to enhance and strengthen the transmission and distribution network, improve the revenue collection, and minimize the aggregate technical and commercial (AT&C) losses. Restructured Accelerated Power Development and Reforms Program (R-APDRP) launched in 2008 with an objective to provide accurate baseline data and technology driven energy accounting. A major scheme launched in 2015, to revive the financial health and improve the operational efficiency of the state-owned distribution companies (DISCOMs) was Ujwal DISCOM Assurance Yojna (UDAY). A specific timeline to achieve the targeted activities for improving operational efficiencies was to be finalized, outcomes of which would be measured through indicators of reduction in AT&C loss to 15% and gap between average cost of supply and average revenue realized to zero. AT&C losses for distribution utilities decreased from 22.32% in 2020–21 to 19.4% in 2022–23 following sector revitalization efforts [21]. Another scheme, Saubhagya was introduced in 2017 with an aim to provide uninterrupted access to electricity supply to all unelectrified rural and urban households. Despite the introduction of various policies and schemes by the government, the performance improvement of DISCOMs is not satisfactory.

DISCOMs in India are regulated by State Electricity Regulatory Commissions (SERCs) under Multi-Year Tariff framework, which is a hybrid mechanism between cost-of-service and incentive regulation. Tariff for each segment of consumer is set by the SERCs to recover the cost incurred to serve them. Benchmarking or performance assessment of DISCOMs helps in examining the efficient and inefficient performers. Moreover, the efficiency or inefficiency can then be incentivized or penalized in accordance with the performance. This helps in promoting competition and better performance of the companies. There has been empirical evidence from many countries that introduction of benchmarking analysis has resulted in performance improvements of distribution companies.

Distribution utilities vary in terms of costs, supply quality, customer segments, and network features, impacting their operational effectiveness [15]. Progress in performance has been gradual, with disparities in operational, financial, and service quality performance. A comparative study on distribution utility efficiency in India should focus notably on service quality, often overlooked in prior analyses. This study aims to measure and evaluate the operational efficiency of Indian DISCOMs using reliability indices and the efficiency improvement across time.

The subsequent sections of the paper are organized in the following manner. Existing literature on the performance assessment of distribution utilities is presented in Sect. 2. The methodological framework, data and the variables used in the study are discussed in Sect. 3. Section 4 presents the key results and discussions followed by Sect. 5, which concludes the paper.

2 Review of Literature on Performance Analysis of Distribution Companies

The performance assessment of the electricity sector has been done based on various approaches, using simple methodologies of single stage efficiency studies to more complex tools like data envelopment analysis (DEA), bootstrap DEA, stochastic frontier analysis (SFA), corrected ordinary least square (COLS), three-stage virtual frontier DEA, and other such methodologies. A review of the literature revealed that the efficiency estimation studies have either focussed on segment or the entire sector using methodologies such as network DEA. Some studies have investigated effects of restructuring or regulatory changes on the performance of the sector [22].

Several studies have been carried out to analyse the technical efficiency and productivity growth of thermal generation power plants of India using methodologies such as DEA, stochastic production function, maximum likelihood estimator and directional output distance function [3, 13, 16, 18, 24]. There are many research studies in the literature that have analysed the performance of distribution companies of many countries using parametric or non-parametric methodologies [1, 9, 12]. However, only few studies have analyzed the operational efficiency of electricity distribution companies of India. This paper has reviewed the literature pertaining to efficiency analysis or benchmarking of Indian electricity distribution companies, in particular.

Thakur et al. assessed the comparative efficiencies of 26 integrated state-owned electric utilities using input-oriented Charnes, Cooper, and Rhodes (CCR) [8] and Banker, Charnes, and Cooper (BCC) [2] models of DEA [15]. The impact of scale on the efficiency was also evaluated. The study suggested the scope of significant cost reductions, restructuring and unbundling of the integrated utilities for exploitation of scale efficiencies. Thakur also analysed the productivity of State Electricity Boards using Malmquist productivity index [25]. The results showed productivity growth from 1996–97 to 2001–02 because of investments in the sector in that period but scope for efficiency improvement was also pointed out by the study. Yadav et al. evaluated the performance of 29 electricity distribution divisions of an Indian state, Uttarakhand, using input-oriented DEA [27]. The results indicated the potential for a reduction in cost and number of employees as the performance of many divisions was sub-optimal. Further, the influence of geographical area being served on efficiency was also found as plain area divisions were relatively efficient than hilly and mountain area divisions. Relative efficiency of 40 unbundled electricity distribution utilities of 17 states of India was measured by Meher and Sahu using constant returns to scale input-oriented DEA model [17]. The study found that in the Indian case, the correlation between privatization of utilities and efficiency is not necessarily positive as always assumed. The economic status of a state or the experience in adoption of reform also did not play any role in the efficiency of the utility. Major proportion of the utilities needed downsizing in their operations and exploration of alternative approaches like incentive-based regulation by the regulators to improve their overall efficiency.

Bishnoi and Gaur evaluated the relative efficiencies and benchmark of 55 Indian electricity distribution utilities for period 2014–15 using DEA along with sensitivity analysis [5]. Utilities are classified into five categories according to sensitivity impact of the variables on efficiency scores calculated by DEA. Ghosh et al. evaluated the relative performance of 28 states-owned electricity distribution utilities of India using DEA [11]. They also performed a slack analysis to identify the reasons for inefficiencies. Excessive distribution line length in comparison to the number of consumers served and the revenue realized were the main reason behind the inefficiencies. A recent study by Bobde and Tanaka used two-stage DEA with bootstrap estimation to examine the efficiencies of 13 distribution utilities of three Indian states using panel data [6]. Customer structure, population density and public ownership positively influenced the efficiency except in high population density areas where privately owned enterprises were better performers. The results found negative influence of subsidies on efficiency of the utilities. Another study by Bobde and Tanaka examined the impact of electricity reforms on technical efficiency of 21 state-level electricity distribution sector using panel data for the period 1995–2012 [7]. They used two-stage DEA and bootstrap methodology. The results indicated that partially unbundled states where generation and distribution segments were integrated with separate transmission segment and legislation of electricity reforms showed significant positive and negative influence on technical efficiency, respectively. Also, small sized electricity distribution state showed positive gain in technical efficiency when interacted with partial or full unbundling.

Khetrapal evaluated and analysed the performance of 52 Indian electricity distribution utilities post the implementation of EA, 2003, for the period from 2006 to 2013, using a translog SFA with technical inefficiency effects model [15]. Efficiency improvement was observed during that period but considerable scope for improvement was also indicated. This study found lower average technical efficiency for publicly owned utilities in comparison to privately owned ones. Environmental variables like customer density, customer structure and unbundling positively contributed to technical efficiency. The performance benchmarking in incentive-based price cap regulation and greater autonomy to the regulatory commissions were suggested as regulatory mechanisms to improve the performance of the distribution utilities. Jamsab et al. examined the effect of economic and institutional framework on the cost efficiency of 52 electricity distribution utilities of 24 Indian states using SFA [14]. Findings suggested that economic development and institutional quality positively influence the performance of the distribution utilities and the presence of both is necessary for the improvement in efficiency in a heterogenous country like India. The results also showed larger inefficiencies for larger utilities and higher total costs for private utilities.

Ghosh et al. studied the impact of three decades of Indian policy reform on techno-commercial performance of state-owned electricity utilities to find the reason behind the inefficiencies [10]. DEA improved by Shannon's entropy model was used in the study. States with vertically integrated utilities in contradiction to recommendations of EA, 2003, were fully efficient in all the three periods of study. As found in previous studies as well, most of the utilities need to downsize their scale of operation to become efficient. Region-specific policies are suggested to revive the financial health of the distribution companies. Sarangi et al. performed a two-stage DEA analysis to evaluate the efficiency of 45 electricity distribution utilities of 21 states of India for the period 2018–19 [23]. The study found that utilities with high revenue gap have high AT&C losses and a significant efficiency improvement is possible through management and financial restructuring. The second stage regression analysis found AT&C loss and agricultural consumption to strongly influence the financial health of the utilities. Patyal et al. recently compared the performance efficiencies of 48 Indian distribution companies from 24 Indian states using integrated DEA-IRP-TOSIS technique [19]. The study suggested most utilities to reduce their scale of operation to become efficient and suggested operational and financial autonomy in the governance of the utilities.

It is imperative from the literature review of efficiency analysis and benchmarking studies of Indian distribution companies that no study has considered reliability indices as determinants to measure operational efficiency. Yadav et al. has considered duration of interruption per feeder as a quality measure in his study to evaluate the performance of electricity distribution divisions of a single Indian state for 2007–08 [27]. A major gap exists in terms of examining the operational performance of the Indian distribution utilities including determinants of service quality for some recent years for better policy and regulatory interventions. Hence, the present study evaluates the performance of unbundled electricity distribution companies using reliability indices for the period 2016–17 to 2019–20.

3 Methodological Framework and Data

A two-stage DEA methodology is adopted to measure and compare the performance of Indian DISCOMs. Conventional DEA is used in the first stage to calculate the technical efficiency scores, followed by ordinary least squares (OLS) regression in the second stage.

3.1 Conventional DEA Approach and Regression Analysis

DEA is a non-parametric linear programming approach that has been widely used to measure relative efficiencies in areas of public sector, manufacturing, and service industry. Charnes, Cooper and Rhodes proposed a CCR model that is based on constant returns to scale (CRS) assumption and does not distinguish firms based on their size [2]. In simple terms, efficiency is the ratio of output to the input and the objective of any firm is to maximize its efficiency. The solution of DEA involves weight selection for inputs and outputs for each decision-making unit (DMU) under certain constraints to maximize the efficiency. DEA approach does not require an imposition of any functional form for the production function but does not account for statistical noise. Banker, Charnes and Cooper introduced an improved model to account for the variable returns to scale, which takes into consideration non-decreasing and non-increasing returns to scale [26]. A lot of methodological improvements have been made in these conventional DEA approaches.

CCR approach is argued to be more relevant for the analysis of Indian DISCOMs because they operate under similar policy and regulatory conditions [26, 23]. In the present study, input-oriented CCR DEA model is used as Indian DISCOMs operate in a sub-optimal manner [4]. So, this model is the most suitable to examine and compare the efficiency of the DISCOMs.

Consider a set of N DMUs (j = 1,…,N) and each DMU uses m number of inputs x_{ij} (i = 1,…,m) to produce s outputs y_{rj} (r = 1,…,s). The CCR linear programming model for each DMU is given by:

Min "

Subject to

$$\sum_{j=1}^{N} y_{rj}\lambda_j \geq y_{r0} \forall r = 1, \ldots\ldots, s \tag{1}$$

$$\sum_{j=1}^{N} x_{ij}\lambda_j \leq \theta\, x_{io}\ \forall i = 1, \ldots\ldots, m \tag{2}$$

$$\lambda_j \geq 0 \forall j = 1, \ldots\ldots, N \tag{3}$$

where, x_{i0} and y_{r0} are the ith input and r^{th} output for DMU 0, respectively. θ^* is the solution of this linear programming problem, which represents the efficiency score of DMU 0. Likewise, efficiency scores are calculated for each DMU. The estimated relative efficiency score lies between 0 and 1. If $\theta^* = 1$ for any DMU, it indicates that it lies on the efficiency frontier and is relatively efficient. If $\theta^* < 1$, it indicates that the DMU is relatively inefficient and there is a scope to reduce the present input levels for the same level of outputs to become fully efficient. In the BCC model, the convexity constraint is present in addition to the constraints listed above for CCR model.

Pooled-OLS regression is done to analyse the influence of external environmental variables on the efficiency scores in the second stage of the two-stage DEA.

$$Yi = \alpha + \beta Xi + \varepsilon i, i = 1, \ldots\ldots, N \tag{4}$$

where, α is the constant term and Xi are the environmental variables, and β is their respective estimated co-efficient variable. Yi represents the efficiency scores of distribution utilities estimated in the first stage, Xi is a vector representing the three independent variables, which are years from unbundling, ownership, and customer structure.

3.2 Data and Selection of Inputs and Outputs

The selection of inputs and outputs is crucial in estimating the efficiency of distribution companies. The objective of the efficiency analysis can be to assess the operational or financial performance. Based on this objective, set of inputs and outputs are selected. The selected input parameters should represent the resources being used and the output parameters should represent the production outcome. In this case, the service provided to the consumers is the measure of output. In this paper, the selection of inputs and outputs are made based on a detailed literature review of studies to estimate the efficiency of distribution utilities.

In the first stage, input-oriented DEA has been used to measure and compare relative operational efficiencies of 19 unbundled electricity distribution companies for the period 2016–17 to 2019–20, thus summing up to 76 DMUs for four years. Data used in this study is collected from PFC Reports on Performance of Power Utilities, CEA General Review Reports, Tariff Orders, Annual Reports and websites of distribution companies. This study was limited to 19 distribution companies as reliability indices for many distribution companies was unavailable. Reliability and quality of electricity being supplied is the most important outcome of electricity business. No recent study in Indian context has used quality of service parameters in the efficiency analysis of distribution utilities. The major reason behind this is the data availability constraint.

Two models have been used in this paper. First model is operational model without quality-of-service parameters and second model is operational model with quality-of-service parameters. In the first model, to measure the operational efficiency without quality-of-service parameters, five inputs and one output is used. In the second model, two quality of service determinants are used as undesirable outputs in addition to the parameters of first model. There are many methods that can be found in the literature to treat undesirable variables in DEA model. In the present study, we have followed the simplest way to treat the undesirable output, i.e., by considering it as an input in the DEA model. The total number of input and output variables are less than one-third of the total number of DMUs considered, which fulfils the criterion for a valid DEA model [10]. Table 1 summarizes the inputs, outputs and environmental variables used in the study.

Commonly used outputs include number of consumers served and electricity units sold. Following the literature, both outputs were chosen but very high correlation was found between the two variables in the preliminary observation. So, consequently only net energy sold was used as an output in the analysis. It is the energy sold annually to the end customers.

The two reliability indices considered as undesirable outputs are frequency and duration of interruption. The measure for frequency of interruption in India is represented by System Average Interruption Frequency Index (SAIFI = ΣNi*Ci/Ct). It is measured in units of number of interruptions per customer. Ni is the number of interruptions

Table 1. Model specification for first stage DEA.

Variables	Abbreviation	Model 1	Model 2
Inputs			
Low voltage power lines	LVPL	✓	✓
High voltage power lines	HVPL	✓	✓
Distribution transformation capacity	DTC	✓	✓
Operational expenditure	OPEX	✓	✓
T&D loss	TDL	✓	
Outputs			
Net energy sold	NES	✓	✓
Undesirable outputs			
Frequency of interruptions	SAIFI		✓
Duration of interruptions	SAIDI		✓

(>3 min/5 min/10 min) in the ith feeder, Ci is the number of customers in the affected feeder and Ct and is the total number of consumers served in all feeders of the area. The measure for duration of interruption in India is represented by System Average Interruption Duration Index (SAIDI = ΣTi*Ci/Ct), which is commonly used as a reliability indicator for distribution companies. SAIDI is the average outage duration for each customer served and is measured in minutes. Ti is the duration of interruptions (of > 3 min/5 min/10 min) in the ith feeder, Ci is the number of customers in the affected feeder and Ct is the total number of consumers served in all feeders of the area.

Common inputs for the assessment of electricity distribution companies include physical assets (e.g., distribution network and transformation capacity), labour and operational assets (e.g., operational expenditure) and losses in the network. Network length is a factor of geographical characteristic of the area served by the utility. In the present analysis, length of low voltage power lines, which represents the low-tension distribution lines system (upto 440 V) and length of high voltage power lines, which represents the 66 kV, 33 kV and 11 kV distribution network for each distribution utility is considered as first two inputs. The third input used is distribution transformation capacity, which is the total capacity of transformers connected to the distribution system. It characterizes the type of consumers served by the utility.

The fourth input considered is adjusted operational expenditure which includes employee cost, operation, and maintenance cost (O&M) and administrative and general expenses (A&G). The adjusted operational expenditure for all the years has been calculated by adjusting for inflation taking 2019–20 as base year. The fifth input used is transmission and distribution (T&D) losses. It represents the total technical losses which are inherent in the system and can be reduced only to a certain level and commercial losses in the process of supplying electricity to end-consumers. It is the difference between the total energy fed in the system and total energy billed.

External environmental variables may affect the efficiency of DISCOMs but are not under their direct control. The set of variables used in our study are years from unbundling, customer structure and ownership. EA, 2003 mandated the unbundling of State Electricity Boards into separate generation, transmission, and distribution entities. Most of the states accepted the regulatory change except some states, which are still operating as a single entity. Only unbundled utilities have been considered in the present study as opposed to previous studies which have considered both bundled and unbundled utilities for analysis. Years from unbundling captures the possible regulatory and internal changes in the distribution companies post unbundling, which may in turn affect the efficiency of the companies.

The second independent variable is a dummy variable representing the public or private ownership of DISCOMs. This takes value 1 if a DISCOM is state-owned and value 0 if it is private-owned. This variable is considered to assess the influence of ownership on the efficiency of DISCOMs. Previous studies have found mixed evidence regarding the effect of ownership structure on efficiency. Privately owned utilities are expected to perform better than the state-owned ones as private entities are more insulated and independent from the state decisions.

The third environmental variable is the customer structure of DISCOMs. This is measured as ratio of units sold to domestic and agricultural consumers out of total units sold by that utility. The domestic and agricultural consumers are subsidized group and are charged at a flat rate but many previous studies have shown that they are more costly to serve. This is mainly because of increased service requirements and reasons like theft of electricity, lower collection and billing efficiency, possibly resulting in lower efficiency. We expect an increase in the ratio to negatively influence the efficiency. The summary statistics for all the variables are presented in Table 2. The descriptive statistics provides an idea about the heterogeneity in the distribution utilities across the country. Also, the data related to T&D loss, SAIFI and SAIDI points out that there are utilities which are performing excellently under the same conditions, whereas, some are exhibiting below average performance.

Table 2. Descriptive statistics

Variables	Units	Maximum	Minimum	Mean	Median	Std. Dev
Low voltage power lines	km	714532	4825	115703.60	66072	150852
High voltage power lines	km	411176	2098	103827.94	85572	92612.97
Distribution transformation capacity	MVA	83244.9	1225	13585.74	12145	15258.56
OPEX	Rs Crore	8255	329	1575.72	872	1734.06

(continued)

Table 2. *(continued)*

Variables	Units	Maximum	Minimum	Mean	Median	Std. Dev
T&D loss	MU	26125	781	4651.79	2561	5436.77
Net energy sold	MU	109605	2141	19525.71	16367	21459.68
SAIFI	Nos	838	0.11	72.8	6.26	162.11
SAIDI	minutes	12518	0.21	1566.2	269	2859

4 Key Results and Discussions

The results of the analysis are presented in two stages. The first set is based on the efficiency scores derived from the DEA for both the models. In the second stage, the regression estimation results are discussed.

4.1 First Stage DEA Results

In the first model, the technical efficiency of 19 unbundled distribution companies ranges from 0.51 to 1 in CCR input-oriented DEA with 12 efficient utilities with score of 1 between 2016–17 to 2019–20. BYPL, BRPL, TPDDL and DGVCL are the most efficient DISCOMs as ranked based on average efficiency scores for four years. In the second model, when reliability indices are considered in addition to operational parameters, the technical efficiency of 19 unbundled DISCOMs ranges from 0.52 to 1 with 21 utilities with score of 1 between 2016–17 to 2019–20. BYPL, BRPL, TPDDL and UGVCL are the most efficient utilities under constant returns to scale. The average technical efficiency of all the distribution utilities is plotted in Fig. 1.

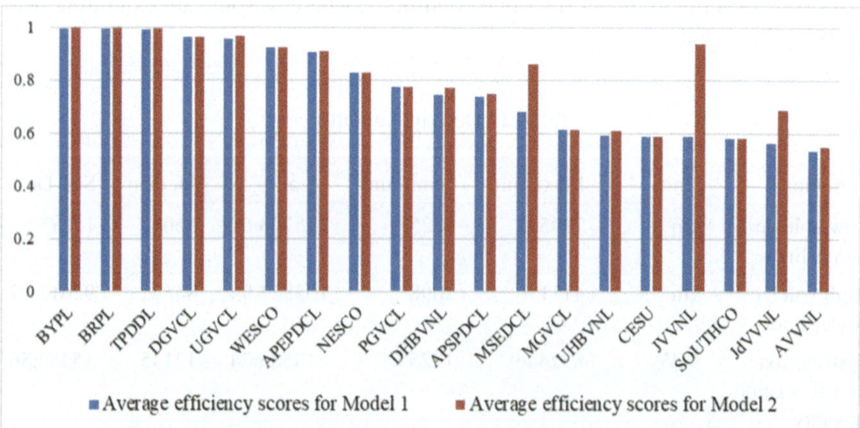

Fig. 1. Operational efficiency scores for unbundled Indian DISCOMs

The best performing DISCOMs are BYPL, BRPL and TPDDL from Delhi, and DGVCL and UGVCL from Gujarat. Private utilities are the best performers as can be seen by the efficiency scores of DISCOMs of Delhi. The efficiency scores of some DISCOMs such as MSEDCL of Maharashtra, JVVNL and JdVVNL of Rajasthan differ significantly when reliability indices are considered. They tend to be relatively more efficient among the group in Model 2 as can be seen in the above figure. This is because of their better reliability indices. This is the reason why the use of reliability indices is important when analyzing the efficiency of DISCOMs.

4.2 Second Stage Regression Results

In the second stage, the first stage efficiency scores of both the models are regressed with years from unbundling, ownership, and customer structure to analyse the effects of these variables on the efficiency of the utilities. The table below presents the results of the second-stage regression (Table 3).

Table 3. Second stage regression results

Regressor	Model 1		Model 2	
	Coef.	Std. Err.	Coef.	Std. Err.
Years from unbundling	−0.0008	0.00963	−0.0149	0.01056
Ownership	−0.2951***	0.04074	−0.2311***	0.04469
Customer structure	−0.7095***	0.13609	−0.5315**	0.14931
Constant	1.3642***	0.21048	1.5407***	0.23092
R Square	0.5147		0.3626	
Adj. R Square	0.4945		0.3359	
Std. Error	0.12502		0.13717	
Observations	76		76	

*Signif. Codes: 0 '***', 0.001 '**', 0.01 '*'.*

The coefficient of years from unbundling is negative but insignificant. The negative coefficient means that as the years from unbundling increases, the efficiency score decreases. This points to an important outcome that the newly unbundled utilities tend to be more efficient which can be attributed to many reasons like better policy design and implementation because of previous experiences. The coefficient of ownership is negative and significant in both the models which represents that the privately owned utilities are more efficient than the publicly owned ones. The privately owned utilities considered in the present study were BRPL, BYPL and TPDDL of Delhi state which were mostly efficient in both the models in all the years considered. The coefficient of customer structure is also negative and significant implying that as the proportion of sales to domestic and agricultural consumers increases, efficiency decreases. This can be due to many reasons like higher technical, commercial and distribution losses, cross-subsidy, theft of electricity, etc. because of which the cost of supplying electricity to the domestic and agricultural consumers is considered expensive. The results of this study are consistent with the expected outcome.

5 Conclusion

In this study, the efficiency of 19 Indian unbundled electricity distribution companies using panel data from 2016–17 to 2019–20 was studied using two-stage DEA. In the first stage, input-oriented CCR DEA methodology was used to calculate the efficiency scores and in the second stage, they were regressed on environmental variables. Two models were used to compare the operational efficiency after taking into consideration the reliability indices. Some major differences were seen in the efficiency scores of some utilities when reliability indices are taken into consideration. Therefore, reliability indices must be included in the assessment of operational performance.

The distribution utilities of Delhi and Gujarat are efficient in both operational model without quality-of-service parameters and operational model with quality-of-service parameters. The privately owned utilities were better performers in all the years in both the models. Years from unbundling did not have major effect on the performance of the utilities and the sign of the coefficient signifies that newly unbundled utilities perform better as compared to earlier unbundled ones. We found statistically significant and positive effect of private ownership on efficiency, which is consistent with theoretical literature. Results on the effects of ownership is mixed in the literature. Also, as expected the utilities with higher domestic and agricultural sales as compared to commercial and industrial sales have lower efficiency in both the models.

Our findings would aid policymakers, regulators, and management of utilities. Quality of supply is an important determinant of performance of DISCOMs. We suggest periodical benchmarking exercise as a part of tariff revision process for the case of India. This would promote competition, better operational and financial performance. Future studies can consider data sets for more states, for more operational and service quality parameters to better analyse the performance of Indian distribution utilities.

References

1. Arcos-Vargas, A., Núñez-Hernández, F., Villa-Caro, G.: A DEA analysis of electricity distribution in Spain: an industrial policy recommendation. Energy Policy **102**, 583–592 (2017). https://doi.org/10.1016/j.enpol.2017.01.004
2. Banker, R.D., Charnes, A., Cooper, W.W.: Some models for estimating technical and scale inefficiencies in data envelopment analysis. Manag. Sci. **30**(9), 1078–1092 (1984). https://doi.org/10.1287/mnsc.30.9.1078
3. Behera, S.K., Dash, A.P., Farooquie, J.A.: Performance analysis of coal fired power plants in India. SSRN Electron. J. 757–762 (2015). https://doi.org/10.2139/ssrn.2694154
4. Bhatt, B., Singh, A.: Power sector reforms and technology adoption in the Indian electricity distribution sector. Energy **215**, 118797 (2021). https://doi.org/10.1016/j.energy.2020.118797
5. Bishnoi, N.K., Gaur, P.: Performance evaluation and benchmarking of Indian electricity distribution utilities using data envelopment analysis along with sensitivity analysis. Int. J. Eng. Manag. Econ. **6**, 138–165 (2018). https://doi.org/10.1504/IJEME.2018.095678
6. Bobde, S.M., Tanaka, M.: Efficiency evaluation of electricity distribution utilities in India: a two-stage DEA with bootstrap estimation. J. Oper. Res. Soc. **5682**, 1–12 (2018). https://doi.org/10.1080/01605682.2017.1398202
7. Bobde, S.M., Tanaka, M.: Structural reforms and technical efficiency in the Indian electricity distribution sector. Asia-Pac. J. Oper. Res. **37**(2), 205004 (2020). https://doi.org/10.1142/S0217595920500049

8. Boyd, G., Färe, R.: Measuring the efficiency of decision making units: a comment. Eur. J. Oper. Res. **15**(3), 331–332 (1984). https://doi.org/10.1016/0377-2217(84)90101-2
9. Çelen, A., Yalçın, N.: Performance assessment of Turkish electricity distribution utilities: an application of combined FAHP/TOPSIS/DEA methodology to incorporate quality of service. Utilities Policy **23**, 59–71 (2012). https://doi.org/10.1016/j.jup.2012.05.003
10. Ghosh, S., Kumar, V., Mukherjee, V., Gupta, S.: Three decades of Indian power-sector reform: a critical assessment. Utilities Policy **68**, 101158 (2021). https://doi.org/10.1016/j.jup.2020.101158
11. Ghosh, S., Yadav, V.K., Mehta, G., Birajdar, R.: Evaluation of Indian power sector reform strategies and improvement direction though DEA. In: IEEE Power Energy Society on General Meeting, pp. 1–5 (2017). https://doi.org/10.1109/PESGM.2017.8274162
12. Ikram, M., Zahid, M., Si, K., Waheed, R., Ferraz, D.: Efficient resource utilization of the electricity distribution sector using nonparametric data envelopment analysis and influential factors. Utilities Policy **82**, 101571 (2023). https://doi.org/10.1016/j.jup.2023.101571
13. Jain, R.K., Kumar, S.: Shadow price of CO2 emissions in Indian thermal power sector. Environ. Econ. Policy Stud. **20**(4), 879–902 (2018). https://doi.org/10.1007/s10018-018-0218-9
14. Jamasb, T., Llorca, M., Khetrapal, P., Thakur, T.: Institutions and performance of regulated firms: evidence from electricity distribution in India. Econ. Anal. Policy **70**, 68–82 (2021). https://doi.org/10.1016/j.eap.2021.02.002
15. Khetrapal, P.: Performance analysis of electricity distribution sector post the implementation of electricity act 2003: empirical evidence from India. J. Adv. Manag. Res. **17**(5), 669–696 (2020). https://doi.org/10.1108/JAMR-04-2020-0060
16. Kumar, S., Jain, R.K.: Carbon-sensitive meta-productivity growth and technological gap: an empirical analysis of Indian thermal power sector. Energy Econ. **81**, 104–116 (2019). https://doi.org/10.1016/j.eneco.2019.03.015
17. Meher, S., Sahu, A.: Efficiency of electricity distribution utilities in India: a data envelopment analysis 1. Opec Energy Rev. **40**(2), 155–179 (2016). https://doi.org/10.1111/opec.12072
18. Mishra, S., Singh, A., Sufia: Environmental performance of coal-based thermal generation plants: a two-stage DEA model with undesirable outputs. In: Advances in the Theory and Applications of Performance Measurement and Management, pp. 151–163 (2024). https://doi.org/10.1007/978-3-031-61597-9_13
19. Patyal, V.S., Kumar, R., Lamba, K., Maheshwari, S.: Performance evaluation of Indian electricity distribution companies: an integrated DEA-IRP-TOPSIS approach. Energy Econ. **124**, 106796 (2023). https://doi.org/10.1016/j.eneco.2023.106796
20. PRAAPTI. https://praapti.in/. Accessed 20 Oct 2024
21. Report on Performance of Power Utilities 2022–23. https://pfcindia.com/ensite/DocumentRepository/ckfinder/files/Operations/Performance_Reports_of_State_Power_Utilities/Report Database2022-23-updated up to April 2024EntityApr.pdf
22. Rodríguez-Pomeda, J., Camacho, C.: Electricity industry regulation and innovation : benchmarking and knowledge management as appraisal tools. Universidad Autonoma de Madrid (2003). www.ecn.nl
23. Sarangi, G.K., Pradhan, A.K., Taghizadeh-Hesary, F.: Performance assessment of state-owned electricity distribution utilities in India. Econ. Anal. Policy **71**, 516–531 (2021). https://doi.org/10.1016/j.eap.2021.06.005
24. Sugathan, A., Malghan, D., Chandrashekar, S., Sinha, D.K.: Downstream electric utility restructuring and upstream generation efficiency: productivity dynamics of Indian coal and gas based electricity generators. Energy **178**, 832–852 (2019). https://doi.org/10.1016/j.energy.2019.04.107
25. Thakur, T.: Performance evaluation of indian electric power utilities based on data envelopment analysis. In: 2006 International Conference on Power Electronic, Drives and Energy Systems, New Delhi, India, pp. 1–4 (2006). https://doi.org/10.1109/PEDES.2006.344271

26. Thakur, T., Deshmukh, S.G., Kaushik, S.C.: Efficiency evaluation of the state owned electric utilities in India. Energy Policy **34**(17), 2788–2804 (2006). https://doi.org/10.1016/j.enpol.2005.03.022
27. Yadav, V.K., Padhy, N.P., Gupta, H.O.: Performance evaluation and improvement directions for an Indian electric utility. Energy Policy **39**, 7112–7120 (2011). https://doi.org/10.1016/j.enpol.2011.08.028

Critical Evaluation of Indian Electricity Generation Utilities Using Data Envelopment Analysis (DEA)

V. Ramaiah(✉) and P. Chandrasekar

Vel Tech Rangarajan Dr. Sagunthala R & D Institute of Science and Technology, Avadi, Chennai 600 062, India
ramaiahvkits@gmail.com, drchandrasekar@veltech.edu.in

Abstract. India's Electricity sector has experienced substantial transformation in recent years, propelled by a range of policy initiatives designed to foster competition, enhance efficiency, and boost the contribution of renewable energy sources within the country's overall energy portfolio. The study evaluates the relative efficiency of electricity generation utilities in 19 unbundled state-owned electricity generating utilities in India using Data Envelopment Analysis (DEA). The current investigation aims to conduct an extensive evaluation of the relative efficiency of 19 unbundled state-owned electricity generation utilities to identify inefficient generation utilities, peers, targets, DMU summary, besides providing helpful insights for intended improvements using CCR model with a constant return to scale over the years of academic years 2019–2020, 2020–2021, 2021–2022. The results include performance efficiency, gaps in inputs and outputs of inefficient utilities. Also, a peer-to-peer comparison of inefficient and efficient utilities is provided. Based on these results, inefficient utilities can develop strategic plans to improve performance.

Keywords: Electrical Generation utilities · Data Envelopment Analysis · Decision Making Unit · efficiency measurement · CCR model · efficiency scores · Targets · DMU summary · Peers

1 Introduction

1.1 Performance of Electricity Generation in India

India's complex and dynamic energy sector has evolved to meet the increasing demands of its swiftly expanding economy. In recent years, significant changes have taken place within the electricity sector, fuelled by various policy initiatives designed to enhance competition, boost efficiency, and increase the quantity of renewable energy in the nation's energy mix stated by Surya Vardhan et al. (2024). The State-owned electricity utilities predominantly dominate this sector, responsible for generating, transmitting, and distributing most of the electricity across the country. Electricity sector in India has made rapid progress and total generating capacity has increased from meager 1362 MW in 1947 to 4,70,488 MW at the end of February 2025 report by powermin.gov.in (2025).

A. Emrouznejad et al. (Eds.): DEA 2024, LNOR, pp. 149–162, 2025.
https://doi.org/10.1007/978-3-031-98177-7_12

During the fiscal year 2023–24, India ranked as the third largest electricity producer globally. The country's total electricity generation for that period amounted to 1,949 terawatt-hours (TWh), with utilities contributing 1,714 TWh to the overall generation reported at powermin.gov.in (2024). With enhanced economic activity and higher standard of living the per capita consumption of energy has increased from 16.3Kwh in 1947 to 1331 kWh in 2022–2023 presented at www.statista.com (2023) and having average annual growth of total power generation is 6–7% and where as world's demand for electricity grew by 2.2% during 2023 reported at https://www.iea.org/reports/electricity-2024/executive-summary (2024) (Fig. 1).

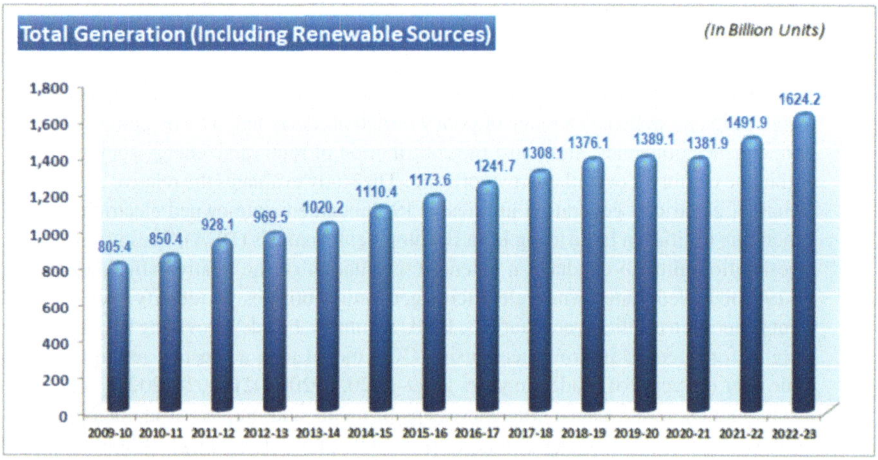

Generation Growth (%)

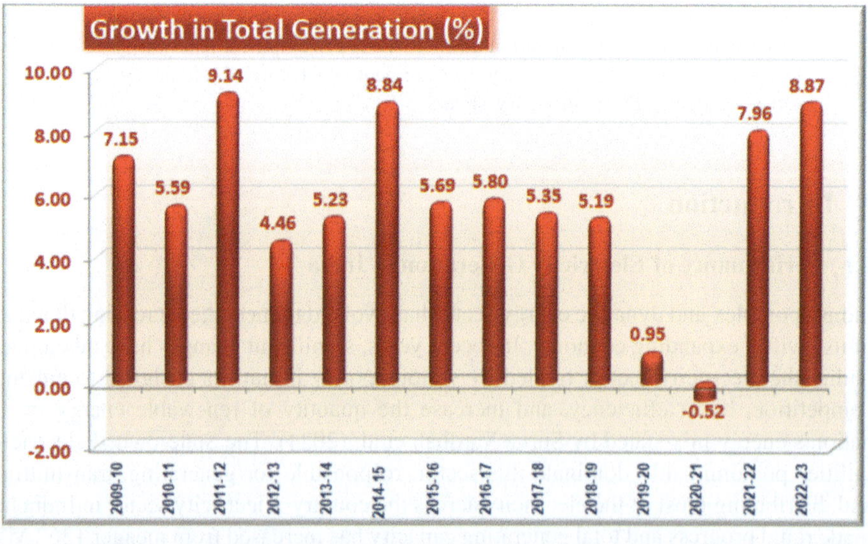

Fig. 1. Total Power generation and its growth including Renewable sources (Source: MoP, GoI https://powermin.gov.in/en/content/power-sector-glance-all-india)

1.2 Electricity Reforms in India and Challenges of Generation Utilities

To enhance the efficiency and reliability of power sector India has considered some important electricity key reforms. The Electricity Act, 2003 stands as a fundamental milestone in India's electricity reforms presented by K Narvon et al. (2004).

Many electricity generation utilities in India have faced financial challenges due to reasons like high capital costs, inefficient operations, fuel supply issues, and financial losses caused by power theft and inefficiencies in billing and collection. This financial strain has hindered investment in infrastructure upgrades and maintenance, leading to additional operational difficulties stated by H Hemi et al. (2022).

Coal dependency in electricity generation utilities introduces uncertainties, price volatility, and environmental concerns. As highlighted by Hani Aural Ibrahim et al. (2023), transitioning to cleaner alternatives and upgrading infrastructure presents significant challenges. Furthermore, high transmission and distribution losses, attributable to inadequate infrastructure (Ministry of Power, 2005), diminish system efficiency. Addressing these issues is imperative for achieving sustainability.

Integrating intermittent renewables into India's grid presents stability and storage challenges, compounded by policy, environmental, and infrastructure issues (Hassan et al., 2023). Addressing these requires complex, sustainable strategies involving policy, finance, and technology. Regulatory clarity, financial discipline, and technology adoption are crucial for long-term utility success (Archana et al., 2022).

The country's electricity sector faces various challenges, including financial constraints, fuel supply issues, transmission, and distribution losses, and integrating renewable energy sources presented by Semich Impram (2020). Addressing these challenges requires a composite approach in regulatory clarity, financial discipline, and newer technologies, which is crucial for overcoming these hurdles and ensuring the long-term success of India's electricity generation utilities report by www.ceew.in (2024).

2 Literature Review

Shafali Jain and Tripta Thakur discussed about the productivity analysis and performance evaluation of Indian Electric generation Utilities Based on DEA is being carried out for 30 Utilities and evaluated by CCR model approach from 2006–2007 presented by Shafali Jain (2010). It is found that 10 DMUs are inefficient of 30 DMUs and have a mean score of 0.76. The result of evaluation reveals significant inefficiencies and also majority of DMUs are do not operate at optimal levels. Electric utilities in developing countries are currently engaged in privatization and restructuring efforts, with a strong emphasis on improving efficiency in electricity generation, transmission, and distribution. A key contributor to the poor performance is poor technical efficiency, which is compounded by difficulties in adopting technological advancements, low plant availability (often affected by forced outages), low plant load factor (PLF), increasing cost of the fuel and high levels of auxiliary consumption. Importantly, high auxiliary consumption at the generation stage further reduces the energy available for transmission, highlighting the need for targeted improvements in technical efficiency to optimize overall generation performance.

Most of the research focuses on the wide range of performance of India's electricity distribution utilities. It is critical to investigate the performance variation of generation utilities as their scale of operation grows due to increased demand and complexity with renewable integration, as well as identify inefficient utilities and provide peers with recommendations for improvement during the enforcement of the Electricity Act 2003 and 2021 reported by K Gopal (2021).

3 Data Envelopment Analysis

Data Envelopment Analysis (DEA) is a powerful nonparametric linear programming method widely used for benchmarking and identifying best practices. It has been applied across various sectors like banks, hospitals, universities, and electricity utilities to assess technical efficiency. Many studies have focused on evaluating the relative technical efficiency of electricity generation utilities using DEA. The primary goal of these studies is to use the DEA model to pinpoint generation utilities with low efficiency scores and recommend ways to enhance their efficiency. By utilizing DEA, researchers gain insights into the resources impacting operational efficiency and design efficiency of the supply system presented by A. Charnes (1978). The CCR model, introduced by Charnes, Cooper, and Rhodes, operates under the assumption of constant returns to scale between inputs and outputs. The BCC model, an extension developed by Banker, Charnes, and Cooper, modifies the CCR framework by incorporating variable returns to scale, making it applicable to scenarios where input-output relationships exhibit varying scale efficiencies by Emrouznejad, A (2025). Mergoni (2025) conducted a comprehensive synthesis of 50 years of Data Envelopment Analysis (DEA) development, aiming to propose a standardized framework to assist researchers in streamlining their methodologies and minimizing the risk of inaccuracies when implementing DEA models. The purpose of the current research work is to analyze the performance of generating utilities using DEA methodology to identify inefficient generation utilities, peers, targets, DMU summary, besides providing helpful insights for intended improvements using CCR model with a constant return to scale over the years of years 2019–2020, 2020–2021, 2021–2022 presented by A. Charnes (1978).

3.1 Block Diagram of Data Envelopment Analysis

DEA involves multiple inputs and multiple outputs for the study. For a precise model, the number of DMUs should be three times the number of identified variables presented by S. C. Ray (2020). This requisite is met through the choice of input and output variables. These variables must be chosen to ensure they are significant to the evaluation and influence the relative performance of the DMUs. In this work, the two inputs and two outputs model is considered for investigation, as illustrated in Fig. 2. The Fig. 2 shows the collection of desirable inputs and desirable outputs. These inputs and outputs are given to Data Envelopment Analysis (DEA) presented by V. Ramaiah (2022). The inputs considered are the Total cost (Rs. Core) and Generation cost (Rs. Core). The outputs considered are net generation (MU) and the Total revenue (Rs. Crore).

Orientation and Data Collection: Efficiency can be assessed in either an input-oriented or output-oriented approach. In the present investigation, the input-orientation measure has been employed to calculate an efficiency score of 19DMUs for the years 2019–2022. The Data was collected for 19 DMUs in India from 2019 to 2022, which are managed by state-owned unbundled generation utilities. The physical data for utilities is being collected from the Power Finance Corporation of India Report 2019–2022.

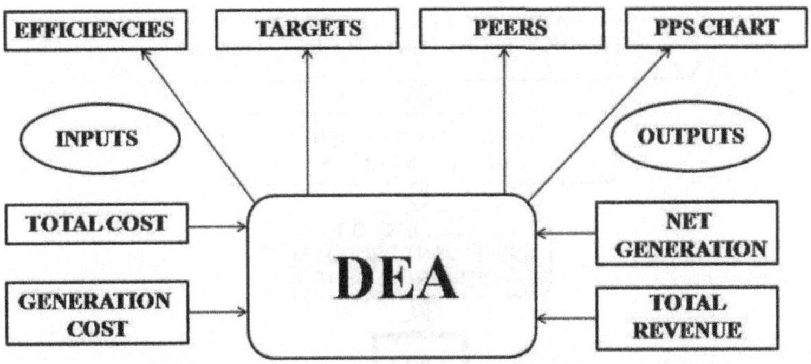

Fig. 2. Block Diagram of the Data Envelopment Analysis of both inputs and outputs considered

3.2 Procedures of DEA

Figure 3 presents the procedures of DEA - first step i.e. collecting the input and output variables for the duration from the annual reports of power Finance Corporation of India. The generation utilities are considered and selected the desired input and output variables. The input variables considered are the Total cost and the Generation Cost and the desired output variables considered are the Net Generation and Total Revenue. The application of Data Envelopment Analysis (DEA) to measure the relative efficiency of electricity generation utilities, also known as decision-making units (DMUs). DEA involves selection of a model. Subsequently, the PIM-DEA software used by A. Emrouznejad et al. (2011) is utilized to assess efficiency, productivity, set targets, identify benchmarks, and manage the performance of organizational units.

4 Mathematical Modeling and Methodology of DEA

4.1 CCR Model

Using linear programming to determine optimal weights as coefficients for inputs and outputs in Data Envelopment Analysis (DEA), assuming constant returns to scale (CRS). The data is structured into matrices X for inputs and Y for outputs, considering n decision-making units, m inputs, and s outputs. This approach provides insights into the overall technical efficiency of the decision-making units presented by Thakur Tripta (2003).

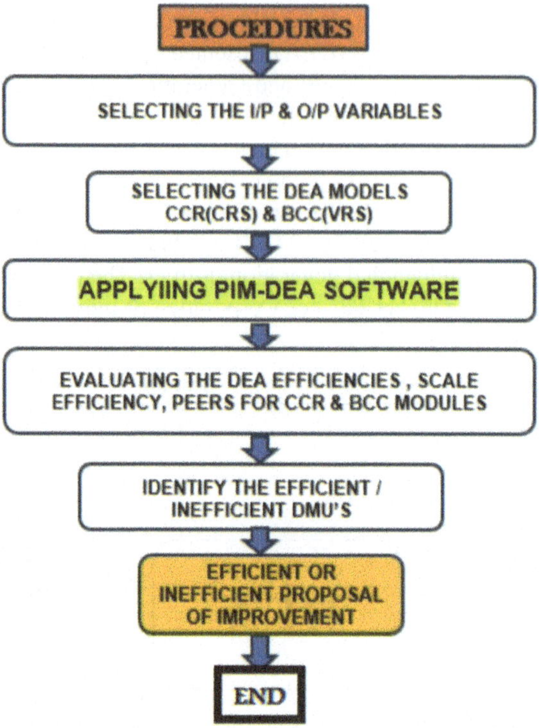

Fig. 3. Performance evaluation flow diagram

Decision Making Units DMU_j: The jth Decision Making Unit $j = 1, 2, 3, \ldots, n$

x_{ij}: The amount of the ith input of the jth Decision making Units $x_{1j}, x_{2j}, x_{3j}, \ldots, x_{nj}$
y_{ij} y_{ij}: The amount of the jth output of the jth Decision making Units $y_{1j}, y_{2j}, y_{3j}, \ldots, y_{sj}$

Initially overall efficiency is measured for every DMU and then the relative efficiency is measured from the data obtained.

The efficiency for k_o DMU is given as

$$\theta_{CCR}(k_0) = \frac{u_1 y_1 k_0 + u_2 y_2 k_0 + \ldots + u_n y_n k_0}{v_1 x_1 k_0 + v_2 x_2 k_0 + \ldots v_n x_n k_0}$$

where u_j = weight for output j, and v_i = weight for input i

The model will be subjected to the constraints.

(1) efficiency of all other DMUs will be less than or equal to 1

$$\frac{u_1 y_{1j} + u_2 y_{2j} + \cdots u_s y_{sj}}{v_1 x_{1j} + v_2 x_{2j} + \ldots v_m x_{mj}} \leq 1 (j = 1, 2, \ldots n)$$

(2) Non-negativity constraints.

$$v_1, v_2, \ldots \ \ldots \ \ldots v_m \geq 0$$
$$u_1, u_2, \ldots \ \ldots \ \ldots u_m \geq 0$$

The process of converting a fractional programming model into a linear programming model. This transformation involves scaling each input to 1 and rewriting the constraints accordingly.

Objective function

$$\text{Max } \theta_0 = u_1 y_{10} + u_2 y_{20} + \ldots + u_x y_{x0} \quad \left(\text{for o}^{\text{th}} \text{ DMU}\right)$$

Subject to

$$v_1 x_{10} + v_1 x_{10} + \ldots + v_m x_{m0}$$
$$u_1 y_{1j} + u_2 y_{2j} + \ldots u_x y_{xj} \leq v_1 x_{1j} + v_1 x_{1j} + \ldots v_m x_{mj} (j = 1, 2, \cdots n)$$
$$v_1, v_2, \ldots v_m \geq 0$$
$$u_1, u_2, \ldots u_m \geq 0$$

Assigns best weights to input variable and output variable. Maximum efficient DMU is assigned as benchmark. The DMU's form efficiency frontier. Efficiency frontier curve gives the overall technical efficiency information of every DMU.

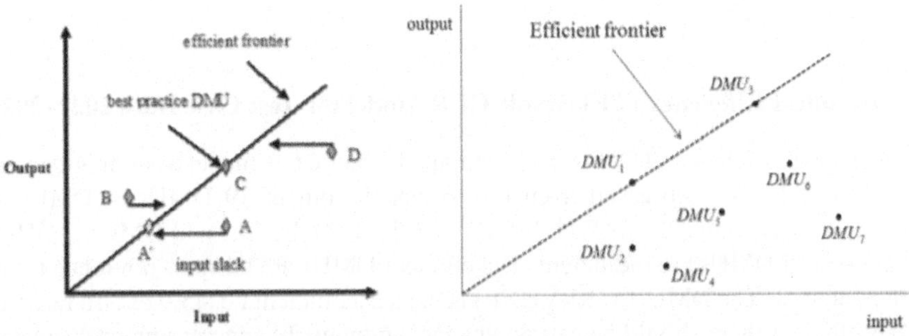

Fig. 4. Efficient Frontier Curve of CCR model 2019 (Source: Luka Vuki et al. (2020), Reenu Kumari (2019))

Thus, the CCR model (Fig. 4) in Data Envelopment Analysis (DEA) employs linear programming to assess the efficiencies of decision-making units (DMUs) by Zervopoulos, P.D (2024). The CCR model assumes constant returns to scale and determines optimal weights for inputs and outputs, providing insights into the technical efficiency of the DMUs. It measures overall efficiency for each DMU, subject to constraints, and assigns weights to input and output variables. The maximum efficient DMU serves as the benchmark, forming the efficiency frontier curve that reveals the technical efficiency of all DMUs. CCR model offers valuable insights into the efficiency and performance of decision-making units, serving as powerful tools for benchmarking and performance assessment in various sectors, including electricity generation utilities presented in Anoopsingh (2006), S. Ghosh (2021), V. K. Yadav (2008), N. K. Bishnoi (2018), A. Pahwa (2002).

5 Results and Discussions

For an extensive evaluation of the performance of Indian Electricity Generation Utilities (IEGUs), two inputs and two outputs, as shown in Fig. 5, were considered. The performance evaluation was considered with a constant return to scale with input orientation. The technical efficiency is computed using the CCR method. The technical efficiency and frequency of usage as peergroup by inefficient DMUs of the CCR model are calculated, and the results are discussed in the following sections.

Fig. 5. Block diagram of Inputs and outputs of DMU for evaluation

5.1 Technical Efficiency (TE) Result-CCR Model for Test Case Data 2021–2022

The technical efficiency (TE) result is computed using CCR model assuming constant returns to scale. It is observed from Fig. 6 that the out of 19 DMUs, 4 DMUs are efficient (DMU2_APGCL, DMU4_IPGCL, DMU11_MePGCL, and DMU12_OHPC) and the rest 15 DMUs are inefficient. The efficient DMUs are taken as reference for the inefficient ones. The DMU 15 (RRVUNL) is the least efficient i.e. 45.97%. To make the DMUs efficient there should be certain changes in inputs like increasing or decreasing the input with reference to CRS. DEA software is used to identify the efficient/inefficient DMUs and find the benchmark/peer DMUs for inefficient DMUs. It can be concluded that 4 DMUs appears with best practices by achieving technical efficiency of 1 can be taken as peers/benchmark for 15 inefficient DMUs to enhance their performance. For example, for the inefficient utility DMU-10, the technical efficiency score is 47.48%. For this utility, three peer's utilities (03) are DMUs-2, 4, 12.

5.2 DMU TARGETS and DMU Summary for 2021–2022

DMU Targets: The target and DMU summary is computed using CCR model assuming constant returns to scale and presented. It is inferred from Table 1 that the DMU1 (APGENCO) is inefficient, so the targets were given to those inefficient DMUs, such that the actual value of input parameter i.e. Total cost actual value is Rs.10202 (Cr) and the efficiency is 50.67%. So, to make DMU1 an efficient DMU the target of total cost is set to Rs.5171.63 (Cr) instead of Rs.10,202.00 (Cr). Similarly another input parameter is generation cost actual value is Rs.5727 (Cr) so the targeted value of generation cost is Rs.2903.15 (Cr) to make the DMU efficient. It is observed that only the input parameters

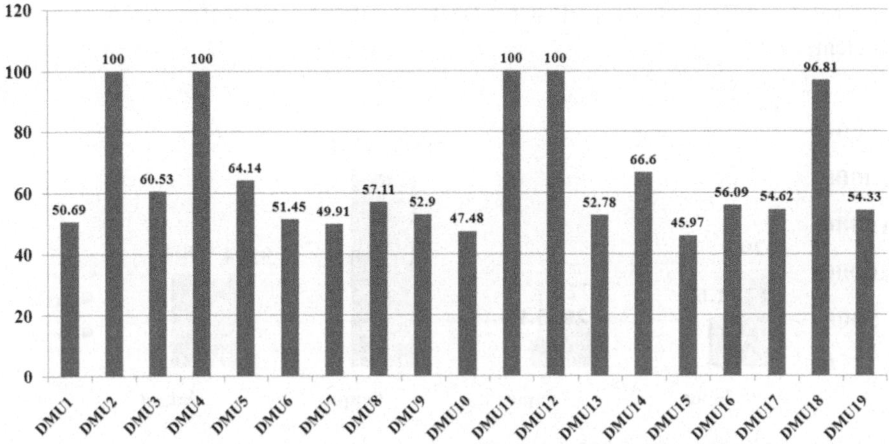

Fig. 6. Efficiency scores of Various DMUs from DEA 2021–2022

Table 1. Target values of various DMUs from PIM-DEA 2021–2022

DMU Number	TOTAL COST Value	TOTAL COST Target	GENERATION COST Value	GENERATION COST Target	NET GENERATION Value	NET GENERATION Target	TOTAL REVENUE Value	TOTAL REVENUE Target
DMU1	10202	5171.63	5727	2903.15	21111	21111	9904	9904
DMU2	284	284	252	252	1921	1921	535	535
DMU3	5947	3599.83	2566	1553.25	16582	16582	6556	6556
DMU4	327	327	187	187	208	208	703	703
DMU5	2994	1920.25	2250	1270.35	4567	4567	3983	3983
DMU6	11096	5709.04	8700	4027.89	18455	18455	11630	11630
DMU7	3910	1951.63	2971	1446.54	7661	7661	3917	3917
DMU8	9281	5300.64	4125	2355.9	25715	25715	9598	9598
DMU9	9966	5272.31	5701	3016	21707	21707	10110	10110
DMU10	25619	12164.52	15624	7418.65	50365	50365	23516	23516
DMU11	378	378	2	2	878	878	159	159
DMU12	439	439	14	14	4389	4389	562	562
DMU13	3001	1583.94	1803	951.63	9382	9382	2867	2867
DMU14	79	52.61	60	39.96	237	237	104	104
DMU15	18840	8661.08	10477	4816.46	31144	31144	16847	16847
DMU16	12857	7212.01	6578	3689.86	28734	28734	13696	13696
DMU17	9023	4928.3	5657	3089.82	24995	24995	9259	9259
DMU18	824	797.75	24	23.24	5107	7339.35	950	950
DMU19	10002	5434.12	6842	3717.28	27485	27485	10355	10355

were given the targets to make the DMUs efficient because the model considered is Input-Oriented.

DMU Summary: The graphs (Fig. 7a, b) show the actual values and targeted values to be attained to make the DMUs efficient. The red color from the figure shows the actual values of both input parameters and output parameters such as total cost, generation cost, net generation, and total revenue. The blue color from the figure shows the targeted value to be attained to make the DMUs efficient. It is observed that DMU1 (APGENCO) is inefficient so that peers are also mentioned in the Fig. 8. The peers for the DMU1 are DMU2 (APGCL), DMU4 (IPGCL), DMU12 (OHPC) and DMU2 is as it is already efficient.

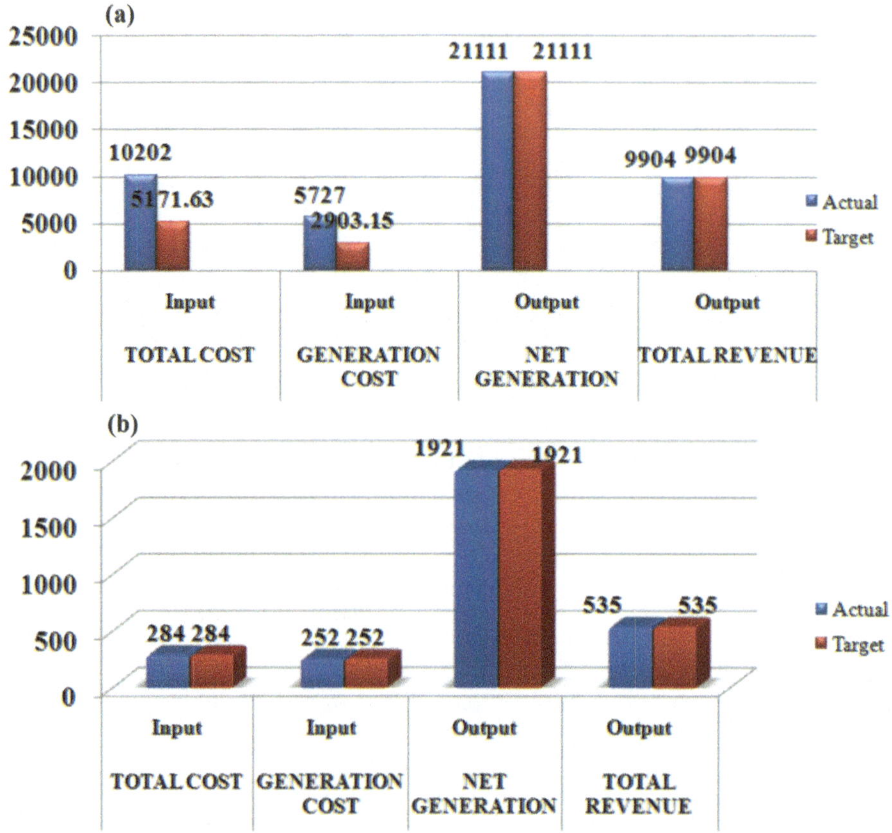

Fig. 7. **a**. DMU1 summary of various DMUs 2021–2022. **b**. DMU2 summary of various DMUs 2021–2022

5.3 DMU Peers 2021–2022

It can be noticed from the Fig. 8 that the frequency of DMU2 & 4 appearing in the peer/benchmark set of efficient DMUs is 15 using CCR model, which is relatively more frequent count. It can be regarded as a Role model in the Indian electricity generation utilities. On the other hand, DMU11 has a very low frequency (02) of appearing in the reference sets of inefficient DMUS. It is also further observed that DMU2 (APGCL), DMU4 (IPGCL), DMU11 (Mapco), DMU12 (OHPC) are taken as references because these DMUs are efficient. For Ex: DMU1 the DMU2, DMU4, DMU12 are taken as reference. For DMU2 (APGCL) no DMU has taken reference because DMU2 is itself efficient so no peers for DMU2. Similarly, for all the DMUs the peers are calculated and the efficient DMUs are taken as reference.

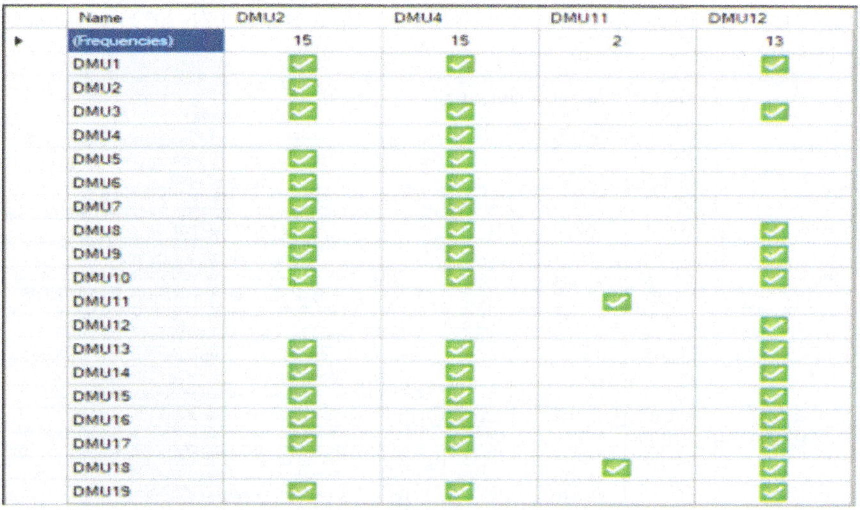

Name	DMU2	DMU4	DMU11	DMU12
(Frequencies)	15	15	2	13
DMU1	✓	✓		✓
DMU2	✓			
DMU3	✓	✓		✓
DMU4		✓		
DMU5	✓	✓		
DMU6	✓	✓		
DMU7	✓	✓		
DMU8	✓	✓		✓
DMU9	✓	✓		✓
DMU10	✓	✓		✓
DMU11			✓	
DMU12				✓
DMU13	✓	✓		✓
DMU14	✓	✓		✓
DMU15	✓	✓		✓
DMU16	✓	✓		✓
DMU17	✓	✓		✓
DMU18			✓	✓
DMU19	✓	✓		✓

Fig. 8. Peers for various DMUs 2021–2022

5.4 CCR Efficiency Analysis from 2019–2020, 2020–2021 and 2021–2022 and Their Comparison

Table 2 revealed the following observations. The DMUs display significant variations in technical efficiency scores across the country for all three assessment years 2019–2020, 2020–2021 and 2021–2022 respectively.

i. The DMUs have a national mean score of 0.71 for all three assessment years (2019–2020, 2020–2021and 2021–2022) and have 29.00% inefficiency in India.
ii. DMU11 and 12 are efficient and the rest 17DMUs (Outof19), are inefficient for the three consecutive assessment years of 2019–2020, 2020–2021and 2021–2022 respectively.

The most efficient DMUs are DMU11 and 12 for all the three assessment years (2019–2020, 2020–2021and 2021–2022).

DMU2 is efficient only for the assessment year 2021–2022. Similarly, DMU4 and 5 are efficient for only the corresponding assessment years 2020–2021 & 2019–20 respectively.

The remaining 14 DMUs (DMU1, 3, 6, 7, 8, 9, 10, 13, 14, 15, 16, 17, 18 and 19) are inefficient for the assessment year of 2021–2022.

Table 2. CCR efficiency analysis from 2019–2020, 2020–2021 and 2021–2022 (IE: inefficient, E: efficient)

DMU Number/Name	2019-2020 CCR Model TE(CRS Score)	2020-2021 CCR Model TE(CRS Score)	2021-2022 CCR Model TE(CRS Score)	Remarks
1_APGENCO	0.6313 (IE)	0.6759 (IE)	0.5069 (IE)	The DMU1 is Inefficient in all financial years of 2019-2020, 2020-2021, 2021-2022.
2_APGCL	0.6241 (IE)	0.6385 (IE)	1 (E)	The DMU2 is Inefficient in the financial year of 2019-2020, 2020-2021, and efficient in the year 2021-2022
3_CSPGCL	0.6886 (IE)	0.7133 (IE)	0.6053 (IE)	The DMU3 is Inefficient in all financial years of 2019-2020, 2020-2021, 2021-2022.
4_IPGCL	0.9045 (IE)	1 (E)	1 (E)	The DMU4 is Inefficient in the financial year of 2019-2020 and efficient in both the years of 2020-2021, and 2021-2022
5_PPCL	1 (E)	0.9268 (IE)	0.6414 (IE)	The DMU5 is Efficient only in the financial year of 2019-2020 and it is Inefficient in both the years of 2020-2021, and 2021-2022
6_GSECL	0.6252 (IE)	0.7075 (IE)	0.5145 (IE)	The DMU6 is Inefficient in all financial years of 2019-2020, 2020-2021, 2021-2022.
7_HPGCL	0.671 (IE)	0.7025 (IE)	0.4991 (IE)	The DMU7 is Inefficient in all financial years of 2019-2020, 2020-2021, 2021-2022.
8_KPCL	0.6537 (IE)	0.7406 (IE)	0.5711 (IE)	The DMU8 is Inefficient in all financial years of 2019-2020, 2020-2021, 2021-2022.
9_MPPGCL	0.6416 (IE)	0.6886 (IE)	0.529 (IE)	The DMU9 is Inefficient in all financial years of 2019-2020, 2020-2021, 2021-2022.
10_MSPGCL	0.6242 (IE)	0.6859 (IE)	0.4748 (IE)	The DMU10 is Inefficient in all financial years of 2019-2020, 2020-2021, 2021-2022.
11_MePGCL	1 (E)	1 (E)	1 (E)	The DMU11 is Efficient in all financial years of 2019-2020, 2020-2021,2021-2022.
12_OHPC	1 (E)	1 (E)	1 (E)	The DMU12 is Efficient in all financial years of 2019-2020, 2020-2021,2021-2022.
13_OPGCL	0.5501 (IE)	0.6188 (IE)	0.5278 (IE)	The DMU13 is Inefficient in all financial years of 2019-2020, 2020-2021, 2021-2022.
14_PUDUCHERRY_PCL	0.7211 (IE)	0.874 (IE)	0.666 (IE)	The DMU14 is Inefficient in all financial years of 2019-2020, 2020-2021, 2021-2022.
15_RRVUNL	0.6242 (IE)	0.6218 (IE)	0.4597 (IE)	The DMU15 is Inefficient in all financial years of 2019-2020, 2020-2021, 2021-2022.
16_TSGENCO	0.6174 (IE)	0.6892 (IE)	0.5609 (IE)	The DMU16 is Inefficient in all financial years of 2019-2020, 2020-2021, 2021-2022.
17_UPRVUNL	0.6278 (IE)	0.6729 (IE)	0.5462	The DMU17 is Inefficient in all financial years of 2019-2020, 2020-2021, 2021-2022.
18_UJVNL	0.7293 (IE)	0.8215 (IE)	0.9681	The DMU18 is Inefficient in all financial years of 2019-2020, 2020-2021, 2021-2022.
19_WBPDCL	0.6315 (IE)	0.7282 (IE)	0.5433	The DMU1 is Inefficient in all financial years of 2019-2020, 2020-2021, 2021-2022.

6 Conclusion

The current study is to determine the relative performance of 19 Indian Electricity Generation utilities (DMUs) for the period of 3 years i.e. 2019–2022. The performance of utilities has been computed by applying DEA. The CCR model of DEA has been applied to compute the overall efficiency of 19 Indian electricity Generation utilities (DMUs). The results show that 4 DMUs (DMU2, 4, 11 & 12) are found as efficient utilities across the country with CCR model during the assessment year 2021–2022. In addition, it is further inferred that the DMU2 & 4 are inefficient for the corresponding assessment years 2021–2022 and 2020–21. Therefore, only 2 DMUs (DMU11 & 12) are found as efficient utilities across the country with CCR model for the period of 3 years i.e. 2019–2022. The remaining 15 DMUs (*DMU1, 3, 6, 7, 8, 9, 10, 13, 14, 15, 16, 17, 18 and 19)* are inefficient with CCR.

To enhance the performance of such inefficient DMUs, initiatives such as.

i. Procurement of cost-efficient fuel and optimizing supply chain with establishing long term fuel contracts with national and international firms to secure at more stable prices for coal and gas.
ii. Establishing an efficient and transparent electricity market for boosting competition, downsizing the scale of operations, integrating renewable energy sources, attracting investments and optimizing operational efficiencies in order to empower generating utilities to increase energy securities.
iii. A balanced energy mix is essential in the power sector by strategically combining various energy sources like fossil fuels, renewable (solar, wind, hydro), and nuclear power to optimize electricity generation while considering factors like energy security, environmental impact, economic viability, and grid stability.

iv. Increase the plant efficiency by upgrading the technologies such as super critical boilers and combined cycle plants to enhance thermal efficiency and lower the cost of generation along with robust maintenance practices on regular basis to reduce the down time.
v. Increase the share of renewable in the generation portfolio to reduce the cost of generation.
vi. Enhance grid stability by adopting smart grid technology to improve efficiency in distribution sector and integrate renewable energy sources.
vii. Increase the energy storage system capacity to store excess energy generated from renewable during non-peak hours and supply during peak hours to reduce the demand of additional installed capacity.
viii. Collaborate with the government bodies to have strong regulatory support for the favourable policies that encourage renewable integration and price mechanism.

It is essential to find specific areas of concern and analyze the root causes of inefficiencies using sensitivity analysis of electricity generation utilities on the regional grid basis to enhance their overall effectiveness and resilience.

References

Archana, R.S., Singh, S.: Development of smart grid for the power sector in India. Cleaner Energy Syst. **2**, 100011 (2022). ISSN 2772-7831. https://doi.org/10.1016/j.cles.2022.100011

Bishnoi, N.K., Gaur, P.: Performance evaluation and benchmarking of Indian electricity distribution utilities using data envelopment analysis along with sensitivity analysis. Int. J. Eng. Manag. Econ. **6**(2–3), 138 (2018)

Charnes, A., Cooper, W.W., Rhodes, E.: Measuring the efficiency of decision-making units. Eur. J. Oper. Res. **2**(6), 429–444 (1978)

Dubash, N.K.: Electric power reform: social and environmental issues. In: Cleveland, C.J. (ed.) Encyclopedia of Energy, pp. 255–266. Elsevier, Amsterdam (2004). ISBN 9780121764807,

Emrouznejad, A., Podinovski, V., Lu, C., Charles, V., Moradi-Motlagh, A.: Rajiv Banker's lasting impact on data envelopment analysis. Ann. Oper. Res. **351**, 1225–1264 (2025). https://doi.org/10.1007/s10479-025-06473-3

Emrouznejad A., Thanassoulis, E.: Performance Improvement Management Software: PIM-DEAsoft-V3.0 User Guide, May 2011 (2011). www.DEAsoftware.co.uk. Performance Improvement Management Limited, UK ISBN: 978 1 85449 412 2

Gandhi, H.H., Hoax, B., Halim, B.J.: Strategic investment risks threatening India's renewable energy ambition. Energy Strategy Rev. **43**, 100921 (2022). ISSN 2211–467X. https://doi.org/10.1016/j.esr.2022.100921

Ghosh, S., Yadav, V.K., Mukherjee, V., Gupta, S.: Three decades of Indian power-sector reform: a critical assessment. Utilities Policy **68**, 101158 (2021)

Hassan, Q., Algburi, S., Sameen, A.Z., Salman, H.M., Jaszczur, M.: A review of hybrid renewable energy systems: Solar and wind-powered solutions: challenges, opportunities, and policy implications. Res. Eng. **20**, 101621 (2023). ISSN 2590-1230. https://doi.org/10.1016/j.rineng.2023.101621

Ibrahim, H.A., et al.: Sustainability of power generation for developing economies: a systematic review of power sources mix. Energy Strategy Rev. **47**, 101085 (2023). ISSN 2211–467X, https://doi.org/10.1016/j.esr.2023.101085

Impram, S., Nese, S.V., Oral, B.: Challenges of renewable energy penetration on power system flexibility: a survey. Energy Strategy Rev. **31**, 100539 (2020). ISSN 2211–467X. https://doi.org/10.1016/j.esr.2020.100539

Jain, S., Thakur, T., Shandilya, A.: Cost benchmarking of generation utilities using DEA: a case study of India (2010). https://doi.org/10.4236/ti.2010.14028

Kumari, R.: Data envelopment analysis recent applications in banking sector: a survey. Int. J. Res. Appl. Sci. Eng. Technol. (IJRA) **7**(XI) (2019). ISSN: 2321–9653; IC Value: 45.98; SJ Impact Factor: 7.177. www.ijraset.com

Mergoni, A., Emrouznejad, A., De Witte, K.: Fifty years of data envelopment analysis. Eur. J. Oper. Res. **326**(3), 389–412 (2025). https://doi.org/10.1016/j.ejor.2024.12.049

Pahwa, A., Feng, X., Lubkeman, D.: Performance evaluation of electric distribution utilities based on data envelopment analysis. IEEE Power Eng. Rev. **22**(11), 59 (2002)

Ray, S.C.:. Data envelopment analysis: a nonparametric method of production analysis. In: Ray, S., Chambers, R., Kumbhakar, S. (eds.) Handbook of Production Economics. Springer, Singapore (2020). https://doi.org/10.1007/978-981-10-3450-3_24-1

Sarangi, G.K., Pradhan, A.K., Taghizadeh-Hesary, F.: Performance assessment of state-owned electricity distribution utilities in India (2021). https://doi.org/10.1016/j.eap.2021.06.005

Singh, A.: Power sector reform in India: current issues and prospects. Energy policy **34**(16), 2480–2490 (2006)

Thakur, T., Deshmukh, S.G., Kaushik, S.C., Kulshrestha, M.: Impact assessment of the Electricity Act 2003 on the Indian power sector. Energy Policy **33**(9), 1187–1198 (2005)

Vardhan, B. S., Swain, A., Khedkar, M., Srivastava, I., Bokde, N.D.: An overview of Indian power sector and its Energy Management. Renew. Energy Focus **50**, 100597 (2024). ISSN 1755-0084. https://doi.org/10.1016/j.ref.2024.100597

V. Ramaiah and V. Jayasankar, Performance Assessment of Indian Electric Distribution Utilities Using Data Envelopment Analysis (DEA), https://doi.org/10.18178/ijeetc

Vukić, L., Poletan Jugović, T., Guidi, G., Oblak, R.: Model of determining the optimal, green transport route among alternatives: data envelopment analysis settings. J. Mar. Sci. Eng. **8**(10), 735 (2020)

Yadav, V.K., Padhy, N.P., Gupta, H.O.: Performance evaluation of Indian electric utility: an intercountry comparison using DEA. In: Proceedings of National Power Systems Conference (NPSC), pp. 301–304 (2008)

Zervopoulos, P.D., Kanas, A., Emrouznejad, A., Molyneux, P.: Can super-efficiencies improve bias correction? a bayesian data envelopment analysis approach. In: Advances in the Theory and Applications of Performance Measurement and Management. DEA45 2023. Lecture Notes in Operations Research. Springer, Cham (2024). https://doi.org/10.1007/978-3-031-61597-9_3

https://powermin.gov.in/en/content/power-sector-glance-all-india

https://www.statista.com/statistics/1231668/electricity-consumption-per-capita-in-india/

https://www.iea.org/reports/electricity-2024/executive-summary

https://www.ceew.in/publications/agenda-reformed-power-sector-india

https://powermin.gov.in/en/content/national-electricity-policy

Decarbonization and Employment: Evidence from India

Surender Kumar[1(✉)] and Rakesh Kumar Jain[2]

[1] Department of Economics, Delhi School of Economics, University of Delhi, Delhi 110007, India
skumar@econdse.org
[2] Indian Railways, New Delhi, India

Abstract. Potential adverse effects of decarburization in terms of output and employment losses have made countries skeptical. Higher input costs associated with decarbonization or regulatory requirements is known as the cost effect (CE). We estimate these costs in terms of labor requirements. The CE is the ratio of efficient labor requirements for a regulatory technology to a non-regulatory technology. The cost effect is split into technical change, input-mix change, and output change effects. We compute the CE using both, the by-production (BP) and the weak-disposability (WD) approaches. We use industry-level data of inputs and outputs for the Indian economy from 2000 to 2014. We find that the CE is higher for the BP approach relative to the WD approach implying the costly disposability of both CO_2 emissions and fossil fuels. We also observe that the technical change is labor-saving and the employment effect is equal for both the approaches.

Keywords: Environmental technology · Employment · Bad output · By-production · Weak disposability

1 Introduction

India is 4th largest emitter of CO_2 emissions and the most populous country in the world, with labor force participation rate below 50 %.[1] She has pledged to reduce the carbon emission intensity of GDP, by 2030 relative to 2005, by about 40–45 % and achieve the NetZero target by 2070 at the Glassgow Summit. It is claimed that regulation of fossil fuel and/or CO_2 emission intensive industries could lead to job and output losses. Concerns related to job losses also find place in the text of the Paris Agreement and ask for "*Taking into account the imperatives of a just transition of the workforce...*"[2] which

[1] Labor force participation rate was 49.8% in 2017–18 in India. http://www.mospi.gov.in/sites/default/files/publication_reports/Annual%20Report%2C%20PLFS%202017-18_31052019.pdf (as accessed on December 24, 2019).

[2] https://unfccc.int/files/meetings/paris_nov_2015/application/pdf/paris_agreement_english_.pdf (as accessed on December 24, 2019).

A. Emrouznejad et al. (Eds.): DEA 2024, LNOR, pp. 163–185, 2025.
https://doi.org/10.1007/978-3-031-98177-7_13

make countries skeptical of taking deep mitigation measures; India is not an exception.[3] It is imperative to understand the employment effects of CO_2 emission regulation. This paper estimates the employment effects of mitigating carbon emission, using industry level information in India.

The CO_2 emissions and gross energy use intensities have been declining over the period 2000 to 2014 for 44 sectors of Indian economy (Table 1).[4] Reduction in the emissions involves changes in fuel mix and/or changes in production processes, which reduces economic output or increases use of inputs. Note that though there is no formal regulatory standards for carbon emissions in India, however there are standards for controlling emission of SO_2, NO_x and particulate matter. CO_2 emissions are supplement or substitute to other local air pollutants (Kumar and Managi, 2011; Färe et.al., 2012). The declining trend in carbon intensity could be considered that Indian producers are working under a regulatory environment. We aim to estimate the employment effects of emission regulation relative to a situation in which producers of these emissions were allowed to dispose of them freely. These estimates are useful for articulating a carbon emission mitigation policy in the country.

Related Literature: The employment effects of environmental regulations on competitiveness of firms and industries have been at the core of the discussions in the literature. The literature in this area is not extensive but growing (Berman and Bui, 2001, Morgenstern et al., 2002, Gray et al., 2014, Ferris et al., 2014, Liu et al., 2017, 2019, 2021; Marin and Vona, 2021, Sheriff et al., 2019, Cole and Elliott, 2007, Gray and Shadbegian, 2013, Kahn and Mansur, 2013, Fare et al., 2018, Hille and Möbius, 2019b, Yang et al., 2024, and Fare et al., 2024).[5] In these studies the environmental regulatory impact is decomposed into three components: *output (demand or scale), cost and factor shift effects.* It is expected that, for a given technology, increased regulatory stringency reduces output of or demand for a product, resulting in lesser demand for inputs, including labor, and this effect is known as *output effect*. For a given level of output, the *cost effect* reflects increased production cost, while the *factor shift effect* is associated with change in labor cost share. The effect of factor shifts on employment hinge upon whether pollution mitigation is a supplement or substitute to labor. Overall, these studies find a small effect of environmental regulations on employment.

Morgenstern et al. (2002) measure regulatory stringency in terms of pollution abatement cost, and Hille and Möbius (2019b) use rising energy prices as a substitute of

---

[3] Apprehensions of job losses might have induced the US to withdraw from the Paris Agreement. https://www.whitehouse.gov/briefings-statements/statement-president-trump-paris-climate-accord/ (as accessed on December 24, 2019).

[4] CO_2 emitted per unit of gross output is termed as CO_2 intensity and gross energy intensity is defined as gross energy use per unit of gross output.

[5] Earlier studies measured the effects of environmental regulations on various other aspects of competitiveness such as productivity (Kumar, 2006; Kumar and Managi, 2010; Hille and Möbius, 2019a), FDI (Millimet and Roy, 2016), business performance (Lanoie et al., 2011), trade flows (Aldy and Pizer, 2015, Kumar and Prabhakar, 2016) etc.

Table 1. Carbon and energy intensities in Indian economy

Year	CO_2/Gross output	Gross energy/Gross output	Number of employees
2000	1.551	0.274	179507
2001	1.554	0.269	180985
2002	1.511	0.261	182880
2003	1.359	0.241	185234
2004	1.350	0.238	188082
2005	1.381	0.238	192055
2006	1.368	0.235	196898
2007	1.401	0.236	202449
2008	1.415	0.236	208731
2009	1.525	0.249	215817
2010	1.471	0.240	217306
2011	1.398	0.231	218597
2012	1.403	0.237	246583
2013	1.362	0.230	277234
2014	1.361	0.225	300267

Note: Gross output is measured in 2010PPP US$ millions, CO_2 emissions are measured in kilotons (hundreds), energy is measured in terra joules (TJ) and number of employees (in thousands)

regulatory stringency. Berman and Bui (2001) use binary indicators to measure environmental stringency.[6] Morgenstern et al. and Berman and Bui econometrically estimate a cost function, while Hille et al. employ a reduced form labor demand function to assess the employment effect. Cost function approach requires data on costs of pollution abatement activities (Morgenstern et al., 2002). The model specified in Berman and Bui (2001) is based on information of detailed plant level regulations. Information of pollution abatement costs and plant level regulations are generally absent in developing countries like India.

We follow a joint-production function framework to assess the employment effects of environmental regulation. These models reflect that pollution is a by-product of desired output production process and cannot be disconnected from pollution abatement activities. These models obviate the need for information on abatement costs and capture interactions among multiple pollutants in the abatement process (Färe et al., 2003). These models are used to assess the effect of restrictions in terms of reduced desired output while reducing pollutants. These models differentiate between regulated and unregulated technologies. Regulated technologies assume that reduction in pollution is

[6] Berman and Bui (2001) use the year a regulation was adopted; the year a regulation became stringent; and the year a regulation was complied with by a plant as indicators of environmental stringency.

costly, either in terms of reduced desired output or increased use of more inputs, while unregulated technologies consider that pollution can be disposed of in free.

In the present paper we build on Fare et al. (2018) that decomposes the employment effect in various components in the presence of technological change and decompose the effect into technical change, input-mix change and output change effects, and follow Ray et al., (2018) in the computation of DEA models. Emrouznejad et al. (2023) provide a structured and critical review of the literature on eco-efficiency and strategies for reducing CO_2 emissions and associated costs. Megoni et al. (2025), based on a comprehensive review of literature in the last half century, provide a state-of-the-art synthesis of DEA methodologies.[7]

For a regulatory scenario, disposability of pollutants involves two approaches: weak disposability (WD) and by-production (BP). The WD approach, propounded by Rolf Färe and his colleagues in the mid-1980s (Färe et al., 1986), is more popular in the literature. The WD approach considers that marketed output and pollutants are null-joint and pollutants are weakly disposable. This approach fails to satisfy the materials balance conditions (Førsund, 2009; Murty et al., 2012). Further, Murty et al. (2012) disagree to free disposability of polluting inputs. They also show that bad outputs are unavoidable by-products of production processes of marketed output, rather than joint products. In the present study, we follow both the approaches and compare results obtained using a common data set. We follow Ray et al. (2018) in formulation of Data Envelopment Analysis (DEA) construct for the BP approach for a regulated technology.

A variant of Färe et al. (2018) model to estimate effect of carbon emission regulation on mitigation costs in terms of labor requirements is applied in this paper. We use labor requirement function for explaining changes in employment among different sectors of Indian economy. We estimate the cost effect that compares labor requirements when CO_2 emissions are regulated with a condition when the emissions are not regulated. Opportunity cost of emission mitigation is represented by the cost effect (CE). The employment effects of regulation are split into technical change (TC), input-mix change (IC) and output change (OC) effects. The TC effect reflects change in labor requirement over time in the regulatory conditions, and the IC effect is comparable to Morgenstern et al.'s *factor shift* effect. The OC effect measures change in labor requirement due to change in emission intensity.

We find that the cost effect is higher for the BP approach in comparison to the WD approach implying that fossil-fuels which cause emissions cannot be disposed of in free. For the BP approach, Indian economy has to employ about 47% more labor for a regulatory technology relative to a non-regulatory technology, whereas the opportunity cost for the WD approach is about 27%. For both the approaches, we detect that average annual employment change effect is about 5.8% We observe a labor-saving technical progress in Indian economy for the regulatory scenario. Declining CO_2 emission intensity is positively associated with employment change for the BP approach and is negatively associated for the WD approach. The factor shift effect is labor augmenting for the BP approach, but it is labor saving for the WD approach.

[7] Proceedings of various conferences conducted on DEA which later on published as books provide recent advances in the DEA literature; for example, Emrouznejad et al. (2024) is a comprehensive compendium on the DEA methodologies.

The paper is arranged as follows: Sect. 2 designates the conceptual construct applied for estimating the employment effect of emission regulation. Details of data used in the study are given in Sect. 3. Section 4 presents and discusses the results. Section 5 offers some concluding remarks.

2 Conceptual Construct

2.1 Environmental Production Technology

Consider an industry producing marketed output vector, $y = (y_1, \ldots\ldots, y_M) \in \Re_+^M$ and bad output vector (e.g., CO_2, SO_2), $b = (b_1, \ldots\ldots, b_J) \in \Re_+^J$,, consuming emission causing (polluting) input vector (e.g., fossil fuels), $c = (c_1, \ldots\ldots, c_L) \in \Re_+^L$ and non-emission causing input vector, $x = (x_1, \ldots\ldots, x_N) \in \Re_+^N$ (e.g., labor, capital, renewable energy). Technology set, T for a given production process is demarcated as:

$$T = \left\{(c, x, y, b) \in \mathrm{R}_+^{M+J+L+N} : (c, x) can\,produce (y, b)\right\} \tag{1}$$

It is assumed that T satisfies free disposability of marketed outputs. This condition states that a marketed output can be arbitrarily reduced for a given level of both kind of inputs and bad outputs. Similarly, free disposability is assumed for both, fossil fuels and other non-fossil fuel inputs. This condition implies that use of additional quantities of all kinds of inputs is costless, i.e., no input congestion. Moreover, the production process satisfies the standard axioms of inactivity and compactness.

However, note that the condition of free disposability of pollutants is not consistent with the notion of rationality. It is not rational to produce any positive quantity of pollutants if it is possible to produce their lower quantity for a given level of inputs and marketed outputs. In the literature, the free disposability assumption of bad outputs is amended in two alternative ways: weak disposability and costly disposability. Färe et al. (1986) characterize a production technology that is consistent with the axioms of *null-jointness* between marketed output and pollutants, and good outputs and pollutants are jointly weakly disposable. The condition of *null-jointness* implies that good outputs and pollutants are jointly produced . Similarly, the condition of weak-disposability of pollutants indicates that reduction in emissions involves concurrent proportional reduction in marketed outputs . It is worth to note that Färe et al. (1986) assume free disposability of pollution causing inputs similar to free disposability of non-pollution causing inputs.

The production possibility set under weak disposability condition of pollutants and free disposability of marketed outputs and all inputs for constant returns to scale technology for $k = 1, \ldots, K$ observations, is expressed as follow:

$$S^{WD} = \left\{(y, b; c, x) : \sum_{k=1}^{K} z_k y_k \geq y; \sum_{k=1}^{K} z_k b_k = b; \sum_{k=1}^{K} z_k x_k \leq x; \sum_{k=1}^{K} z_k c_k \leq c; z_k \geq 0\right\} \tag{2}$$

where $z_k (k = 1, \ldots, K)$ are the weights assigned to each observation for construction of the production possibility set. Førsund (2009) and Murty et al. (2012) show that the weak disposability of pollutants assumption is not consistent with the material balance conditions; generation of pollution is related to use of pollution causing inputs (e.g., fossil fuels). These inputs are not completely transmitted to marketed output during the production process and some amount of these inputs is converted into pollutants.

Murty et al. (2012) assume that pollutants cannot be disposed of in free and to reduce them the polluters have to pay either in terms of reduced marketed output or higher use of conventional inputs. Generally, there is a positive association between pollution and pollution causing inputs use, and it is also possible to generate the same amount of pollution with lesser quantity of polluting inputs if they are not used efficiently. This restriction is not compatible to the condition of free disposability of inputs. It implies joint disposability of pollutants and pollution causing inputs. To reduce emissions, it is essential to reduce the use of polluting inputs, which in turn, also reduces marketed outputs.[8] Bad outputs are unavoidable by-product of the production process of marketed output rather than joint products. Murty et al., (2012) have named this production process as "by-production approach".

By-production technology (T_{BP}) is an intersection of marketed output production technology (T_1) and a production technology leading to the generation of emissions (T_2), i.e., $T_{BP} = T_1 \cap T_2$; T_1 assumes that:

$$\begin{aligned} &(c, x; y, b) \in T_1 \wedge \overline{c} > c \Rightarrow (\overline{c}, x; y, b) \in T_1 \\ &(c, x; y, b) \in T_1 \wedge \overline{x} > x \Rightarrow (c, \overline{x}; y, b) \in T_1 \\ &(c, x; y, b) \in T_1 \wedge \overline{y} < y \Rightarrow (c, x; \overline{y}, b) \in T_1 \end{aligned} \tag{3}$$

and T_2 assumes that:

$$\begin{aligned} &(c, x; y, b) \in T_2 \wedge \overline{c} \leq c \Rightarrow (\overline{c}, x; y, b) \in T_2 \\ &(c, x; y, b) \in T_2 \wedge \overline{b} \geq b \Rightarrow \left(c, x; y, \overline{b}\right) \in T_2 \\ &(c, x; y, b) \in T_2 \wedge \overline{y} \neq y \wedge \overline{x} \neq x \Rightarrow (c, \overline{x}; \overline{y}, b) \in T_2 \end{aligned} \tag{4}$$

The last condition in Eq. (4) implies that random changes in production of marketed outputs and use of standard conventional inputs have no effect on the quantity of pollutants. Production of pollutants is restricted by use of polluting inputs. The corresponding production possibility set for a by-production technology, assuming constant returns to scale, is constructed as follows:

$$S_y^{BP} = \left\{(y; c, x) : \sum_{k=1}^{K} z_k y_k \geq y; \sum_{k=1}^{K} z_k x_k \leq x; \sum_{k=1}^{K} z_k c_k \leq c; z_k \geq 0\right\} \tag{5}$$

$$S_b^{BP} = \left\{(b; c) : \sum_{k=1}^{K} \lambda_k b_k \leq b; \sum_{k=1}^{K} \lambda_k c_k \geq c; \lambda_k \geq 0\right\} \tag{6}$$

$$S^{BP} = \left\{(y, b; c, x) : (y; c, x) \in S_y^{BP} \cap (b; c) \in S_b^{BP}\right\} \tag{7}$$

[8] Other option for reducing bad outputs without reducing polluting inputs is conducting abatement activities as end-of-pipe treatment. In the end-of-pipe treatment, non-polluting inputs are diverted from main production process to abatement activities to reduce bad outputs that leads to a reduction in marketed output.

where z_k, λ_k are different intensity vectors. We follow Ray et al. (2018) in construction of S_b^{BP} that considers weak disposability of pollution and pollution causing inputs. We modify Eq. (6) to $S_b^{BP} = \left\{(b; c) : \sum_{k=1}^{K} z_k b_k = b; \sum_{k=1}^{K} z_k c_k = c; z_k \geq 0\right\}$. This expression could lead to a situation where S_b^{BP} frontier has local regions with negative slope. To avoid this problem, Murty and Russell (2017) suggest considering costly disposability of pollutants, in place of their weak disposability condition.

Murty et al. (2012) use different weights (intensity vectors) in expression of sub-technologies for producing marketed outputs and pollutants. But, following Ray et al. (2018), we use same weights in formulation of sub-technologies production possibility frontiers. Intensity variables are weights assigned to each observation for construction of production possibility frontier in DEA optimization problems. Observations that positively contribute to the expression of production possibility frontier also act as reference point or peer group for inefficient observations (Ray et al., 2018). Therefore, having separate intensity vector for sub-technologies may be logical if decision makers of these sub-technologies are working independently from each other within a firm. But in case of CO_2 emissions mitigation, marketed output production and the emissions mitigation decisions are taken simultaneously.

2.2 Employment Effect

We operationalize the model by specifying an input based environmental production technology, which is a modified version of Ray et al. (2018) and applies input-based DEA models. Mergoni et al. (2025), Emrouznejad et al. (2023) and Emrouznejad et al. (2024) provide comprehensive reviews of the related literature on DEA models and eco-efficiency. Unlike conventional input-based DEA model that seeks to proportionally reduce potential use of all inputs for a given level of good outputs and pollutants, we calculate minimum (efficient) level of labor for the given level of marketed and bad outputs, and all non-labor inputs. We classify non-emission causing inputs into two categories: labor (L) and non-labor non-emission causing inputs (x'). Labor requirement for a regulated technology for observation k' is:

$$F(c, x\prime; y, b) = \min \sum_{k=1}^{K} z_k L_k \tag{8}$$

Subject to

$$\sum_{k=1}^{K} z_k y_k \geq y_{k'}$$

$$\sum_{k=1}^{K} z_k b_k = b_{k'}$$

$$\sum_{k=1}^{K} z_k c_k = c_{k'}$$

$$\sum_{k=1}^{K} z_k x_{k'} \leq x_{k'}$$

$$z_k \geq 0$$

The objective function characterizes the minimum labor required for a certain level of good and bad outputs, and quantity of all non-labor inputs. Constraint in (i) reflects that observed desired output of an observation is less than or equal to frontier output. This implies free disposability of desired outputs. Equality constraints in (ii) and (iii) characterize joint-disposability of pollutants and pollution causing inputs, implying that an observation has to reduce use of polluting inputs to reduce production of pollutants. Free disposability of standard conventional inputs is represented in constraint (iv). Non-negativity of intensity variables in constraint (v) implies constant returns to scale.[9]

If we change the constraint in (iii) from equality to less than equal to, similar to constraint in (iv), then it implies free disposability for emission causing inputs similar to other inputs; and equality in constraint (ii) implies that marketed outputs and bad outputs are reduced jointly to zero, but bad outputs cannot be disposed of strongly.

The producers can ignore the generation of pollutants and consider free disposability of them if there are no social restriction or regulation on their generation. Similarly, emission causing inputs would be treated freely disposable like other inputs. In the absence of restrictions on generation of bad outputs, the unregulated formulation of (8) is:

$$G(c, x'; y) = \min \sum_{k=1}^{K} z_k L_k \tag{9}$$

Subject to

$$\sum_{k=1}^{K} z_k y_k \geq y_{k'}$$

$$\sum_{k=1}^{K} z_k c_k \leq c_{k'}$$

$$\sum_{k=1}^{K} z_k x_{k'} \leq x_{k'}$$

$$z_k \geq 0$$

In formulation (9), we eliminate the constraint on bad outputs that implies that these outputs are free disposable.[10] The cost effect (CE) is specified as a ratio of an optimal level of labor use for regulated and unregulated production technologies:

$$CE = F(c, x'; y, b)/G(c, x'; y) \tag{10}$$

[9] We are doing empirical analysis at the industry level, rather than firm level, therefore we are assuming constant returns to scale.

[10] Free disposability of bad outputs can be modelled in two different ways: either by treating a bad output like a good output or excluding it from the production technology. Following

The CE measures increased labor requirement due to regulation of production technology, which is an opportunity cost of emission mitigation in terms of employment. Its value is more than unity when emission reduction activities are connected with increased use of labor.

2.3 Decomposition of Employment Change

The *CE* is an index of minimum labor requirement when bad outputs are regulated to minimum labor requirement when these outputs are unregulated. Therefore, change in *CE* between two adjacent years is reflected by the index of change in *CE*, ΔCE_t^{t+1}. The ΔCE_t^{t+1} measures change in labor requirement overtime due to regulation of bad outputs. Therefore,

$$\Delta CE_t^{t+1} = \frac{\frac{F^{t+1}\left(c^{t+1},x'^{t+1};y^{t+1},b^{t+1}\right)}{G^{t+1}\left(c^{t+1},x'^{t+1};y^{t+1}\right)}}{\frac{F^t\left(c^t,x'^t;y^t,b^t\right)}{G^t\left(c^t,x'^t;y^t\right)}} = \frac{F^{t+1}\left(c^{t+1},x'^{t+1};y^{t+1},b^{t+1}\right)/F^t\left(c^t,x'^t;y^t,b^t\right)}{G^{t+1}\left(c^{t+1},x'^{t+1};y^{t+1}\right)/G^t\left(c^t,x'^t;y^t\right)} \tag{11}$$

A value of ΔCE_t^{t+1} greater than one indicates increase in labor requirement between period t and period t + 1, reflecting an increase in abatement costs. A value of one indicates that ΔCE_t^{t+1} remains unchanged overtime, and a value of less than one signifies decreased ΔCE_t^{t+1}.

The ΔCE_t^{t+1} is split into technical change (*TC*), input-mix changes (*IC*) and output changes (*OC*) (Färe et al., 2016). Färe et al. (2016) investigate causes of change in the ratio of opportunity cost of pollution abatement using an environmental production function method, while Färe et al. (2018) consider the causes of change in the employment effect for a regulatory technology using an input distance function framework. Both, Färe et al. (2016) and Färe et al. (2018), use structure of the models specified in Pasurka (2006). Pasurka (2006) examines the sources of change in pollutants only when the production of these outputs is restricted.

Following Färe et al. (2016) and Färe et al. (2018), we split the change in optimum level of labor requirement between two adjacent periods for the restricted technology using period *t + 1* as the reference period for mix-period linear programming (*LP*) problems. The decomposition is specified as:

$$\frac{F^{t+1}\left(c^{t+1},x'^{t+1};y^{t+1},b^{t+1}\right)}{F^t\left(c^t,x'^t;y^t,b^t\right)} = \frac{F^{t+1}\left(c^t,x'^t;y^t,b^t\right)}{F^t\left(c^t,x'^t;y^t,b^t\right)} \times \frac{F^{t+1}\left(c^{t+1},x'^{t+1};y^t,b^t\right)}{F^{t+1}\left(c^t,x'^t;y^t,b^t\right)} \times \frac{F^{t+1}\left(c^{t+1},x'^{t+1};y^{t+1},b^{t+1}\right)}{F^{t+1}\left(c^{t+1},x'^{t+1};y^t,b^t\right)}$$
$$= (TC) \times (IC) \times (OC) \tag{12}$$

Färe and Grosskopf (1983), numerous studies relax the constraint on bad outputs from " =" to " ≥ " to consider free disposability of these outputs. Färe et al. (2007) exclude bad outputs from the production technology to consider their free disposability. The first approach implies a downward sloping frontier between good and bad outputs; however, the later approach involves only a horizontal frontier. Therefore, following Färe et al. (2016), we exclude bad outputs in the construction of production technology to consider their free disposability.

where the change in labor requirement due to technical change is shown by TC, *IC* is the change due to changes in the mix of labor and non-labor inputs and *OC* measures the changes due to changes in emission intensity.

The *TC* component reveals the shift in labor requirement function between the adjacent periods for the regulated technology. It measures the change in efficient labor requirement due to technical progress/regress. A more than one value of TC implies that restricted technology over the period of time has become more labor intensive while a value of less than one indicates that the production technology has become less labor intensive. The *IC* component reflects changes in efficient labor requirement in the regulated labor requirement function between two adjacent periods. This component captures changes in employment requirement occurring due to changes in non-labor factors of production, which Morgenstern et al. (2002) name *factor-shift effect* of environmental regulations. A value of more than one of *IC* indicates that for a regulated technology, over the period of time, labor requirement is increasing and a value less than unity signifies decreasing demand for labor due to *factor shift effect*. Changes in labor requirement may be positive or negative depending upon whether non-labor inputs are substitute for labor or complement with labor. Finally, the *OC* indicates optimal labor requirement due to change in output mix (emission intensity) for regulated technology between two adjacent periods. A value of greater than one for the *OC* indicates increasing demand for labor due to change in output mix and a value less than one presents a case of decreasing demand for labor due to change in output mix. A value of unity of any of the components indicates that there is no change in efficient labor requirement for regulated technology.

It is preferred to use geometric mean of two adjacent periods as reference technology when specifying mixed period *LP* problems in place of selecting an arbitrary base period for decomposing of the ratio of minimum labor requirement for regulated and unregulated technology into *TC*, *IC* and *OC*. However, there are chances of occurrence of infeasible *LP* cases in mixed period problems. In our empirical application, we encountered such problems that restricted us to use $t + 1$ period technology as the reference period. Moreover, we use 2-year window to address the problem of infeasibility, while constructing the reference technology. Further, note that we carry out the empirical exercise following both the approaches: joint-disposability of pollutants and fossil fuels (i.e., the BP approach) and weak-disposability of pollutants (i.e., the WD approach).

3 Data

The required data is obtained from two sources: World Input-Output Database (WIOD) socio-Economic Account 2016[11] and WIOD Environmental Accounts.[12] WIOD socio-Economic Accounts provides information at industry level for 28 OECD and 13 other countries for the period 2000 to 2014 on labor, capital stocks, gross output and value added at current and constant prices. WIOD follows ISIC Revision 4 for industrial classification. We obtain data on CO_2 emissions, total energy use and emission relevant energy use from WIOD environmental accounts. WIOD Environmental Accounts provides data

[11] http://www.wiod.org/database/seas16 as accessed on December 08, 2019.

[12] https://ec.europa.eu/jrc/en/research-topic/economic-environmental-and-social-effects-of-globalisation as accessed on December 08, 2019.

on these variables at industry level for 41 countries following WIOD classification of industries for the period 2000 to 2016. Details of the construction of variables, followed in WIOD Socio-Economic Accounts, can be accessed in Gouma et al (2018) and on environmental variables in Corsatea et al. (2019). Both these sources provide data for 56 sectors (industries).

We apply the above-discussed framework to measure the employment effect of CO_2 emission mitigation in 44 different primary, secondary and tertiary sectors of Indian economy over the period of 2000 to 2014. The selection of time-period is limited by the availability of data. In the analysis we consider only the sectors that generate CO_2 emissions along with marketed output and use fuels that cause CO_2 emissions. The chosen sectors cover 95 and 96% of total employment and gross output produced and 93% of CO_2 emissions released by all the 56 sectors in 2014 in the country.

We consider that an industry produces gross output and CO_2 emissions using emission causing energy, renewable energy, intermediate inputs, capital stock and number of employees. Renewable energy, used at the sectoral level, is computed as gross energy use minus emission relevant energy use. Energy use is measured in Terra Joule (TJ). WIOD Environmental Accounts consider 12 different energy commodities in the computation of energy use in a sector (Corsatea et al., 2019). Out of the 44 sectors under consideration, 21 Sectors are using both emission causing and renewable energies in the production process. Remaining 23 sectors, which use emission causing energy, are primary and tertiary sectors. Use of renewable energy is generally confined to manufacturing, mining & quarrying and electricity sectors; these sectors use both emission causing and renewable energies. Gross output, capital stock and intermediate inputs are measured in value terms at 2010 prices, while the other variables are measured in quantity terms. Table 2 describes the statistics of the variables used for the commencement and terminal years, 2000 and 2014, respectively.

We observe that sectoral average gross output increased at a compound annual growth rate (CAGR) of 6.86% and CO_2 emissions have grown at a CAGR of 5.38% over the study period of 15 years, i.e., over this period the gross marketed output increased by about 180% while the growth in CO_2 emissions was about 124%, implying declining CO_2 emission intensity in the economy. During the same period, CAGR for labor and emission causing energy use was 3.43 and 4.97%, respectively. The growth rate in use of renewable energy was about 4.6% per year.

4 Results and Discussion

The LP problems are solved for regulated and unregulated technologies to generate optimal level of labor requirement. We compute efficient level of labor requirement for regulated technology for two approaches: (i) when CO_2 emissions and emission relevant energy are jointly weakly disposable, which is termed as by-production (BP) approach and (ii) when the emissions are jointly weakly disposable with marketed output, and marketed output is strongly disposable. This approach is termed as weak disposability (WD) approach. In the WD approach the emission relevant energy input is considered freely disposable. In both the approaches, when the production technology is restricted, we consider gross output (desired output), intermediate inputs, labor, capital stock and

Table 2. Summary statistics of the variables used in DEA expressions

Variable	Unit of Measurement	Mean	S.D	Maximum	Minimum
2000					
Gross Output	2010 PPP US$ mn	1518356	1982560	11100000	20144
CO_2 Emissions	Kilotons (kt)	20653	68581	446007	5
Emission causing energy	Terra Joule (TJ)	298410	905438	5959721	971
Renewable energy	Terra Joule	146205	724885	4708015	0
Capital stock	2010 PPP US$ mn	1874141	3195785	17100000	31970
Intermediate inputs	2010 PPP US$ mn	742281	959727	4179397	4703
Number of employees	Thousand	4080	12991	86061	24
2014					
Gross Output	2010 PPP US$ mn	4248203	4367750	18800000	105095
CO_2 Emissions	Kilotons (kt)	46286	145651	924741	35
Emission relevant energy	Terra Joule (TJ)	628666	1828579	12000000	1388
Renewable energy	Terra Joule	291319	1504786	9882976	0
Capital stock	2010 PPP US$ mn	5988898	7589351	43900000	198008
Intermediate inputs	2010 PPP US$ mn	2249140	2823099	12400000	10902
Number of employees	Thousand	6824	15374	90010	151

renewable energy as freely disposable in computation of LP problems. To compute optimal level of labor, when the technology is unrestricted, generation of emissions is dropped in the formulation of LP problems. Moreover, recall that for solving LP problems we follow 2-year window approach for constructing the reference technology.

Table 3, displays the geometric means of the cost effect (CE), change in CE and change in labor and its various components. The first half of the Table presents the results obtained using BP approach, while the results obtained using WD approach are given in the second half of the Table. Recall that the CE index reflects the ratio of minimum labor when production technology is regulated to when the technology is unregulated. We observe that the geometric mean of the CE for the BP approach is 1.47 and it is 1.27 for the WD approach, revealing that CO_2 emission mitigation is more costly in Indian economy for the former approach because in the latter approach the disposability of emission causing energy is considered costly. It implies that reduction in fossil fuels to reduce the CO_2 emissions is costly. For the WD approach strong disposability of fossil fuels is assumed.

Moreover, we observed that 15 sectors attained an index of one for the BP approach in year 2001 and the number of sectors fell to 12 in year 2014, while for the WD approach the number of sectors fell from 16 in 2001 to 11 in 2014; 13 sectors are common that are mitigating the emissions without incurring any mitigation cost in 2001.[13] For the BP approach, fishing & aquacultures (ISIC A03) and water transport (ISIC H50) and for the WD approach, manufacturing wood & wood products (ISIC C16), paper & paper products (ISIC C17) and electrical equipment (ISIC C27) sectors mitigate the emissions in free in 2001. In 2014, there are 10 sectors that are mitigating CO_2 emission without incurring any cost for both the approaches.[14] For the BP approach, electricity, gas, steam & air conditioning supply (ISIC D35) and manufacturers of basic metals (ISIC C24) sectors, and for the WD approach construction sector (ISIC F) also are able to freely dispose of the CO_2 emissions in Indian economy. Fishing & aquacultures (ISIC A03), manufacturing of coke & refined petroleum products (ISIC C19), pharmaceutical products & pharmaceutical preparations (ISIC C21), motor vehicles, trailers & semi-trailers (ISIC C29), water collection, treatment & supply (ISIC E36), financial services activities (ISIC K64), real estate activities (ISIC L68), education (ISIC P85) and human health & social activities (ISIC Q) sectors are disposing of the emissions freely in both initial and terminal years of the study.

Table 3 also exhibits the changes in cost effect (CCE), change in employment and its components (TC, IC and OC) for each 2-year pair from 2001 to 2014. Change in labor requirement connected with the CCE shows differences in shift of isoquants when production technology is unregulated and regulated, respectively. The CCE is associated with an average increase in labor requirement of 0.4 and 0.5% per year for the BP and WD approaches, respectively; however, the change in employment associated with CCE is not uniform over the period for both the approach.

We find that, for regulatory scenario, the employment has grown by about 5.7% annually during 2001 to 2014. For both the approaches, we observe that the geometric mean of yearly TC figure is less than one (i.e., 0.991 and 0.982) indicating labor saving TC. Note that TC is more labor saving for the WD approach as compared to the BP approach.

Recall that IC effect represents *factor shift effect*; its more than one value indicates increase in labor requirement and a less than one value reflects that use of labor is decreasing for regulatory scenarios. We obtain that average annual value of IC for the

[13] Common sectors that are freely disposing of the emissions are: fishing and aquacultures (ISIC A03), manufacturing of coke & refined petroleum products (ISIC C19), manufacturing of pharmaceutical products & pharmaceutical preparations (ISIC C21), manufacturing of motor vehicles, trailers & semi-trailers (ISIC C29), manufacturing of other transport equipment (ISIC C30), electricity, gas, steam & air conditioning supply (ISIC D35), water collection, treatment & supply (ISIC E36), construction (ISIC F), financial services activities other than insurnace & pension (ISIC K64), insurance, reinsurance & pension funding (ISIC K65), real state activities (ISIC L68), education (ISIC P85) and human health & social activities (ISIC Q).

[14] Common sectors that are disposing of the emissions in free are: fishing & aquacultures (ISIC A03), manufacturing of coke & refined petroleum products (ISIC C19), manufacturing of pharmaceutical products & pharmaceutical preparations (ISIC C21), manufacturing of computer, electronic & optical products (ISIC C26), manufacturing of motor vehicles, trailers & semi-trailers (ISIC C29), water collection, treatment & supply (ISIC E36), financial services activities other than insurnace & pension (ISIC K64), real state activities (ISIC L68), education (ISIC P85) and human health & social activities (ISIC Q).

Table 3. Cost effect and its decomposition over adjacent years

Year	CE	CCE	TC	IC	OC	Δ L
By-production approach						
2001	1.472					
2002	1.480	1.005	0.977	1.025	1.002	1.003
2003	1.478	0.999	0.994	1.027	1.061	1.083
2004	1.464	0.991	0.991	1.278	0.861	1.090
2005	1.435	0.980	0.985	1.066	0.980	1.029
2006	1.437	1.001	1.013	1.111	0.912	1.026
2007	1.452	1.011	0.973	1.008	1.037	1.018
2008	1.437	0.990	0.936	1.123	0.949	0.998
2009	1.479	1.029	0.960	1.041	1.040	1.039
2010	1.512	1.022	0.973	1.013	1.054	1.038
2011	1.456	0.963	0.922	1.116	0.941	0.968
2012	1.543	1.060	1.031	1.238	0.891	1.137
2013	1.513	0.980	1.114	0.976	1.072	1.166
2014	1.450	1.027	1.032	1.040	1.094	1.174
Geometric mean	1.473	1.004	0.991	1.078	0.989	1.057
Weak disposability approach						
2001	1.228					
2002	1.229	1.001	0.977	0.936	1.106	1.011
2003	1.219	0.992	0.988	0.898	1.219	1.082
2004	1.231	1.010	1.000	0.984	1.127	1.108
2005	1.249	1.015	0.975	0.956	1.135	1.058
2006	1.265	1.013	0.966	0.903	1.190	1.038
2007	1.293	1.022	0.967	0.923	1.155	1.030
2008	1.322	1.023	0.936	1.006	1.081	1.018
2009	1.282	0.970	0.938	0.846	1.228	0.974
2010	1.289	1.005	0.947	0.936	1.149	1.019
2011	1.288	0.999	0.967	1.049	0.983	0.996
2012	1.356	1.053	1.030	1.085	1.020	1.140
2013	1.287	0.949	1.041	0.965	1.130	1.135
2014	1.305	1.014	1.039	0.965	1.158	1.161
Geometric mean	1.272	1.005	0.982	0.956	1.127	1.058

Note: CE: cost effect; CCE: change in cost effect; TC: technical change; IC: input change; OC: output change; and Δ L: change in employment period 2001–2014 for the BP approach is greater than one but for the WD approach it is less than one. We find an average annual increase in labor requirement of 7.8% for the BP approach and an average annual decrease in labor requirement of 4.4% for the WD approach. Table 2 reveals that, in Indian economy, the use of all the inputs, including labor, has been increasing over the period of time. Increasing non-labor inputs to labor ratio is consistent with decreasing average annual employment due to mitigating activities, i.e., there is substitutability between labor and non-labor inputs in mitigating activities for the WD approach. For the BP approach, the association between increasing average annual employment due to mitigating activities and increasing non-labor to labor input ratio implies complementarity between labor and non-labor inputs in mitigating activities.

We find that the value of OC is less than one for the BP approach while it is greater than one for the WD approach. For the BP approach, annual average decrease in employment is of 1.1%, however it increases at an annual average rate of 12.7% for the WD approach. The association between declining CO_2 emissions to gross output ratio (Table 2) and OC effect hints that the CO_2 mitigation activities are labor saving for the BP approach. However, we observe that the mitigation activities are labor intensive for the WD approach.

Table 4 presents the average 2-year pair results at the sectoral level; Table 4a presents results obtained using the BP approach and Table 4b portrays results of the WD approach. For the BP approach, the CE varies from one to 8.25. Manufacturing of coke &

Table 4a. Average Cost effect and its decomposition across sectors (By Production Approach)

ISIC	CE	CCE	TC	IC	OC	Δ L	CO_2/GO
A01	2.130	1.234	0.966	1.231	1.038	1.234	0.406
A02	1.032	1.007	1.002	0.988	1.018	1.008	0.409
A03	1.010	1.000	0.992	0.973	1.064	1.027	0.283
B	1.176	1.021	0.985	1.646	0.625	1.013	3.688
C10-C12	3.718	0.954	1.011	1.038	0.975	1.023	0.104
C13-C15	1.578	0.936	0.944	0.933	1.118	0.985	0.202
C16	8.247	0.999	1.109	1.144	0.832	1.055	0.654
C17	3.977	0.971	0.926	1.352	0.825	1.033	1.521
C18	1.073	0.999	0.966	0.883	1.237	1.056	0.093
C19	1.000	1.000	0.990	2.828	0.368	1.031	0.769
C20	1.478	0.955	0.981	0.968	1.070	1.016	1.622
C21	1.000	1.000	0.957	0.983	1.090	1.025	0.018
C22	1.490	1.000	0.981	1.138	0.948	1.058	0.136
C23	5.555	0.998	1.000	0.919	1.131	1.039	8.332

(continued)

Table 4a. (*continued*)

ISIC	CE	CCE	TC	IC	OC	Δ L	CO_2/GO
C24	1.289	0.970	0.955	0.898	1.196	1.025	1.895
C25	1.271	1.002	0.994	1.104	0.964	1.057	1.005
C26	1.081	0.989	1.018	1.046	1.038	1.106	0.057
C27	1.054	0.992	0.972	0.929	1.189	1.074	0.117
C28	1.216	1.042	0.906	0.971	1.220	1.072	0.148
C29	1.002	1.000	0.966	1.117	1.011	1.090	0.082
C30	1.076	1.004	0.969	0.906	1.224	1.074	0.147
C31_C32	1.257	1.041	0.982	0.955	1.259	1.180	0.072
D35	1.003	1.000	1.002	2.197	0.477	1.051	25.619
E36	1.011	1.000	1.000	2.710	0.388	1.051	1.220
F	1.127	1.009	1.013	0.834	1.148	0.970	0.124
G45	2.162	1.013	0.991	0.605	1.681	1.007	0.136
G46	2.316	1.012	1.007	0.534	1.888	1.015	0.136
G47	2.211	1.021	0.989	0.600	1.710	1.014	0.136
H49	1.361	1.032	0.964	0.883	1.282	1.091	0.699
H50	1.025	1.008	0.989	1.075	0.970	1.032	7.790
H51	1.417	1.085	0.930	2.528	0.463	1.090	1.056
H52	1.357	1.033	1.009	1.067	0.989	1.065	0.488
I	1.842	0.997	1.021	1.015	1.013	1.050	1.029
J61	1.859	0.883	1.102	0.853	1.057	0.994	0.245
J62_J63	1.522	0.991	1.015	0.997	1.065	1.078	0.023
K64	1.000	1.000	1.000	1.020	1.055	1.076	0.040
K65	1.010	1.002	0.992	1.137	0.954	1.076	0.086
L68	1.006	1.000	1.009	0.961	1.118	1.084	0.049
M69_M70	1.498	0.964	0.991	1.311	0.855	1.111	0.157
M71	1.123	1.020	1.010	1.598	0.689	1.111	0.261
N	1.603	0.958	0.993	1.002	1.115	1.110	0.025
P85	1.002	1.000	1.002	0.888	1.200	1.067	0.074
Q	1.135	1.000	1.019	0.908	1.152	1.066	0.063
R_S	2.279	1.090	1.027	1.054	0.998	1.080	0.163
Geometric mean	1.473	1.004	0.991	1.078	0.989	1.057	0.277

Note: CE: cost effect; CCE: change in cost effect; TC: technical change; IC: input change; OC: output change; Δ L: change in employment; and CO_2/GO: CO_2 intensity of gross output refined petroleum products (ISIC C19) and basic pharmaceutical products & pharmaceutical preparations (ISIC C21) and financial services activities (ISIC K64) mitigate the emissions without incurring any cost, whereas for the industry manufacturing wood & wood products (ISIC C16), it is most expansive to mitigate CO_2 emissions in comparison to other sectors. Manufacturers of other non-metallic mineral products (ISIC C23) also find it very expansive to mitigate the emissions. For the WD approach, the index of the CE varies from one for financial services activities (ISIC K64) to 5.56 for manufacturers of other non-metallic mineral products (ISIC C23). For the most carbon intensive industry (electricity, gas, steam & air conditioning supply, ISIC D35) the CE index values are 1.003 and 1.03 for the BP and WD approaches, respectively. Manufacturing of basic pharmaceutical products & pharmaceutical preparation (ISIC C21) is least carbon intensive, and for this sector value of the CE index is one for both the approaches. In India, crop, animal production, hunting & related services activities (ISIC A01) employ about 38% of total labor force and the CE indices for the sector are 2.13 and 1.05 for the BP and WD approaches. Heterogeneity in the CE index value across sectors indicates scope for market-based instruments such as carbon taxation or carbon trading policies for cost effective mitigation of CO_2 emissions in India.

Tables 4a and 4b also present the CCE and change in employment and its components for regulatory scenario at industry level. We observe stagnation in average annual CCE for 11 and 13 sectors for the BP and WD approaches respectively; out of which 9 sectors are common.[15] In addition to common sectors for both the approaches, we find that manufacturing of basic pharmaceutical products & pharmaceutical preparation (ISIC C21) and electricity, gas, steam & air conditioning services supply (ISIC D35) for the BP approach and forestry & logging (ISIC A02), manufacturing of other non-metallic mineral products (ISIC C23) and insurance, reinsurance & pension funding (ISIC K65)

Table 4b. Average Cost effect and its decomposition across sectors (Weak Disposability Approach)

ISIC	CE	CCE	TC	IC	OC	Δ L	CO_2/GO
A01	1.051	1.005	0.969	0.990	1.039	0.997	0.406
A02	1.048	1.000	1.000	0.980	1.021	1.001	0.409
A03	1.012	1.000	0.992	0.977	1.059	1.027	0.283
B	1.194	1.022	0.984	0.975	1.056	1.013	3.688
C10-C12	1.011	1.000	1.008	0.941	1.136	1.077	0.104
C13-C15	1.002	1.000	0.972	0.867	1.249	1.052	0.202

(continued)

[15] The nine common sectors that observe stagnation in the cost effect are: fishing & aquacultures (ISIC A03), manufacturing of coke & refined petroleum products (ISIC C19), manufacturing of rubber & plastic products (ISIC C22), manufacturing of motor vehicles, trailers & semi-trailers (ISIC C29), water collection, treatment & supply (ISIC E36), financial service activities (ISIC K64), real estate activities (ISIC L68), education (ISIC P85) and human health & social work activities (ISIC Q).

Table 4b. (*continued*)

ISIC	CE	CCE	TC	IC	OC	Δ L	CO_2/GO
C16	1.052	1.001	0.967	0.912	1.194	1.053	0.654
C17	1.011	1.002	0.971	0.932	1.166	1.056	1.521
C18	1.031	0.995	0.973	0.844	1.280	1.051	0.093
C19	1.001	1.000	0.978	1.530	0.689	1.031	0.769
C20	1.479	0.956	0.981	0.968	1.070	1.016	1.622
C21	1.000	1.000	0.957	0.982	1.091	1.025	0.018
C22	1.005	0.999	0.985	0.963	1.113	1.056	0.136
C23	5.578	1.000	1.000	0.918	1.131	1.039	8.332
C24	1.352	0.980	0.955	0.875	1.227	1.025	1.895
C25	1.280	1.003	0.994	1.034	1.028	1.057	1.005
C26	1.058	0.999	1.010	1.002	1.104	1.117	0.057
C27	1.006	1.001	0.968	0.902	1.240	1.084	0.117
C28	1.216	1.042	0.906	0.968	1.224	1.072	0.148
C29	1.002	1.000	0.964	1.070	1.057	1.090	0.082
C30	1.076	1.004	0.969	0.902	1.228	1.074	0.147
C31_C32	1.089	1.010	0.961	0.855	1.387	1.140	0.072
D35	1.031	1.008	1.003	0.992	1.056	1.051	25.619
E36	1.006	1.000	1.007	1.227	0.851	1.051	1.220
F	1.060	1.000	1.014	0.814	1.165	0.961	0.124
G45	2.260	1.013	0.991	0.604	1.682	1.007	0.136
G46	2.430	1.012	1.007	0.533	1.889	1.015	0.136
G47	2.299	1.021	0.989	0.600	1.711	1.014	0.136
H49	1.520	1.032	0.964	0.865	1.309	1.091	0.699
H50	3.439	1.002	0.989	0.997	1.047	1.032	7.790
H51	1.771	1.064	0.930	1.601	0.732	1.090	1.056
H52	1.073	1.002	0.979	0.993	1.058	1.029	0.488
I	1.271	1.010	0.984	1.014	1.060	1.058	1.029
J61	1.043	0.993	0.971	0.937	1.250	1.138	0.245
J62_J63	1.481	0.968	0.966	1.000	1.089	1.052	0.023
K64	1.000	1.000	0.980	0.975	1.126	1.076	0.040
K65	1.003	1.000	0.977	0.987	1.114	1.074	0.086
L68	1.006	1.000	1.009	0.960	1.119	1.084	0.049

(*continued*)

Table 4b. (*continued*)

ISIC	CE	CCE	TC	IC	OC	Δ L	CO_2/GO
M69_M70	1.025	1.001	0.955	1.036	1.171	1.159	0.157
M71	1.124	1.020	1.010	1.520	0.723	1.111	0.261
N	1.474	0.961	0.970	0.980	1.172	1.113	0.025
P85	1.002	1.000	1.002	0.887	1.201	1.067	0.074
Q	1.135	1.000	1.019	0.908	1.152	1.066	0.063
R_S	2.281	1.090	1.027	1.054	0.998	1.080	0.163
Geometric mean	1.272	1.005	0.982	0.956	1.127	1.058	0.277

Note: CE: cost effect; CCE: change in cost effect; TC: technical change; IC: input change; OC: output change; Δ L: change in employment; and CO_2/GO: CO_2 intensity of gross output

for the WD approach also observe constancy in the CCE overtime. For the BP approach, we find that crop, animal production, hunting and related service activities (ISIC A01) had an average index value of 1.23 of CCE over the period 2001–2014, implying that the CE has increased annually by about 23%. For the BP approach, telecommunication sector, has the lowest average annual index value, showing that the cost effect has declined by 12% annually. For the WD approach, we find that, in other service activities (ISIC R_S), the CE increased by about 9% annually, but in manufacturing of chemicals and chemical products (ISIC C20), an average annual decline of about 4.5% was observed in the value of CE.

For the regulatory scenario at aggregate level, average annual 2-year pair increase in labor requirement was about 5.7% for the BP or WD approaches. But the change in labor requirement at industry level differs between these two approaches. The highest annual increase in labor requirement has been observed for crop & animal production, hunting & related service activities (ISIC A01) for the BP approach and for accommodation & food services activities (ISIC J61) for the WD approach. The highest declining labor requirement was observed for construction sector for both the approaches.

For the BP approach, 16 sectors observe labor intensive TC for the regulatory scenario and remaining 28 sectors witness labor saving or neutral TC. The highest labor-intensive TC has been observed for manufacturing wood & wood products (ISIC C16) sector and the lowest value of TC index (= 0.91) was found for manufacturing machinery & equipment n.e.c. (ISIC C28) sector. We find that for the WD approach, 12 sectors spot TC index value greater than one and remaining 32 sectors observe labor saving TC.

Recall that at aggregate level, we observe that the *factor shift effect* (IC) is labor augmenting for the BP approach and this effect is labor saving for the WD approach. At industry level, for the BP approach, we find that the index value of IC effect is greater than one for half of the sectors, implying that there is complementarity between labor and non-labor inputs in emission mitigating activities, and for remaining 22 industries there is substitutability between labor and non-labor inputs for emission abatement activities. Similarly, for the WD approach, we find that only 10 industries observe complementarity between labor and non-labor inputs in emission abatement activities.

The OC effect is related with an annual decrease in labor requirement of 1.1% for the BP approach and it is linked with an annual increase in labor requirement of 12.7% for the WD approach at aggregate level. At industry level, we find that, for the BP approach, 28 sectors observe the index value of OC greater than one, while number of industries having OC index value greater than one is 39 for the WD approach. We find that the highest average index of OC is for wholesale trade (ISIC G46) sector to the tune of 1.88 and the lowest average OC index value of 0.37 for manufacturing of coke & refined petroleum products (ISIC C19) sector for the BP approach. For the WD approach, we find that the highest and lowest average index values of OC effects are 1.89 and 0.59 for wholesale trade (ISIC G46) and manufacturing of coke & refined petroleum products (ISIC C19) sectors, respectively.

During the period of 2001–2014, overall, average labor requirement at industry level increased by about 67% (Table 2). The cumulative average employment level during the study period, at industry level increased by almost 107%. The increase in labor requirement is driven by the IC effect for the BP approach and for the WD approach it is the OC effect that drives the increase in employment. In both the approaches, the TC effect is labor saving; labor saving TC effect is higher for the WD approach in comparison to the BP approach.

5 Conclusions

There have been concerns about potential adverse effects of taking deep CO_2 emission mitigation measures. Prime focus of these concerns has been on output and employment losses resulting from CO_2 emissions mitigation, as have also been expressed in the Paris Agreement.

We estimate employment effect of CO_2 mitigation, using industry level data of Indian economy over the period of 2000 to 2014. We use joint production framework for computing emission mitigation cost in terms of additional labor requirement, which is equivalent to the cost effect estimated by Morgenstern et al. The cost effect is the ratio of efficient labor requirement to produce a given level of marketed outputs and pollutants for a regulatory technology to a non-regulatory technology. Moreover, we estimate the factors associated with change in labor requirement, i.e., the change in employment for a regulatory technology is decomposed into TC effect, IC effect and OC effect. Note that, to measure employment effect of emission regulation, we use both the BP and WD approaches.

We detect that the country has paid equivalent to 47% higher labor employment as opportunity cost for the observed declining trend in the CO_2 intensity for the BP approach, whereas the country has paid equivalent to 27% higher employment for the WD approach. This implies that for reducing carbon emissions, the disposability of emission causing inputs is not free. Moreover, it has also been observed that technical change for regulatory technology is labor saving. Declining carbon intensity in Indian economy is labor saving for the BP approach; and it is labor augmenting for the WD approach. We observe factor shift effect in favor of labor for the BP approach, but it is against labor for the WD approach.

To the best of our knowledge, this is the first study measuring employment effect of mitigating CO_2 emission at the economy level, using industry level data, for both the

BP and WD approaches. The results provide necessary inputs for a well-crafted climate policy, as they provide ex-ante estimates of labor requirement effects of mitigating the CO_2 emissions. Note that we consider each industry as a decision-making unit (DMU) due to non-availability of firm level information from individual industries. In DEA, each DMU can simulate the production technology of other DMUs. This is a limitation of the study, since production technology across industries cannot be replicated.[16] Ideally, one should carry out disaggregated industry specific analysis using firm level information, which can be a subject of future research.

Moreover, we considered that firms under both regulated and unregulated scenarios operate either on the upward slopping and horizontal portions of the good and bad outputs production frontier and rules out the possibility of downward slopping portion of the frontier. It precludes the possibilities of negative pollution abatement costs which is possible if the regulations induce firms to find new ways of pollution reduction which simultaneously enhance the production of desired output. Further, this framework considers that the specified technology is restricted to firms' observed production of pollution. It assumes that the pollution production or reduction is immobile among firms or industries, however market-based policy instruments allow the mobility. These could be the directions of future research in the decarbonization arena.

References

Aldy, J.E., Pizer, W.A.: The competitiveness impacts of climate change mitigation policies. J. Assoc. Environ. Res. Econo. **2**(4), 565–595 (2015)

Bellas, A.S.: Empirical evidence of advances in scrubber technology. Res. Energy Econo. **20**(4), 327–343 (1998)

Berman, E., Bui, L.T.: Environmental regulation and labor demand: evidence from the south coast air basin. J. Public Econ. **79**(2), 265–295 (2001)

Cole, M.A., Elliott, R.J.: Do environmental regulations cost jobs? an industry-level analysis of the UK. BE J. Econo. Analy. Policy **7**(1) (2007)

Corsatea, T.D., et al.: World input-output database environmental accounts. Update 2000–2016, Publications Office of the European Union, Luxembourg (2019). ISBN 978-92-79-64439-9. https://doi.org/10.2791/947252. JRC116234

Emrouznejad, A., Marra, M., Yang, G.L., Michali, M.: Eco-efficiency considering netzero and data envelopment analysis: a critical literature review. IMA J. Manag. Math.Manag. Math. **34**(4), 599–632 (2023)

Emrouznejad, A., Thanassoulis, E., Toloo, M.: Advances in the Theory and Applications of Performance Measurement and Management. Springer (2024)

Färe R, Grosskopf S, Lovell C.A.K.: Production frontiers. Cambridge University Press, Cambridge (1994)

Färe, R., Grosskopf, S., Pasurka, C.: Effects on relative efficiency in electric power generation due to environmental controls. Res. Energy **8**, 167–184 (1986)

Färe, R., Grosskopf, S.: Measuring output efficiency. Eur. J. Oper. Res.Oper. Res. **13**, 173–179 (1983)

Färe, R., Grosskopf, S., Pasurka, C.A., Shadbegian, R.: Pollution abatement and employment. Empiric. Econo. **54**(1), 259–285 (2018)

[16] We are very thankful to Subhash Ray for bringing this point to our notice.

Färe, R., Grosskopf, S., Pasurka, C.A.: Environmental production functions and environmental directional distance functions. Energy **32**, 1055–1066 (2007)
Färe, R., Grosskopf, S., Pasurka, C.A.: Technical change and pollution abatement costs. Eur. J. Oper. Res.Oper. Res. **248**(2), 715–724 (2016)
Färe, R., Grosskopf, S., Pasurka, C.A., Jr., Shadbegian, R.: Evaluating the employment effects of environmental regulation without abatement cost data: a nonparametric cost function approach. J. Environ. Manage. **350**, 119309 (2024)
Färe, R., Grosskopf, S., Pasurka, C.A., Weber, W.: Substitutability among undesirable outputs. Appl. Econ. **44**(1), 39–47 (2012)
Ferris, A.E., Shadbegian, R.J., Wolverton, A.: The effect of environmental regulation on power sector employment: phase I of the title IV SO_2 trading program. J. Assoc. Environ. Res. Econ. **1**(4), 521–553 (2014)
Førsund, F.R.: Good modelling of bad outputs: pollution and multiple-output production. Int. Rev. Environ. Resour. Econ.Resour. Econ. **3**(1), 1–38 (2009)
Gray, W.B.: Shadbegian, Ronald J.: Do the Job Effects of Regulation Differ with the Competitive Environment? (Chapter 3). In: Coglianese, C., Finkel, A., Carrigan, C. (eds.) Does Regulation Kill Jobs?, pp. 51–69. University of Pennsylvania Press, Philadelphia (2013)
Gray, W.B., Shadbegian, R.J., Wang, C., Meral, M.: Do EPA regulations affect labor demand? evidence from the pulp and paper industry. J. Environ. Econ. Manag. **68**(1), 188–202 (2014)
Gouma, R., Chen, W., Woltjer, P., Timmer, M.: WIOD Socio-Economic Accounts 2016: Sources and Methods. World Input Output Database (2018). http://www.wiod.org/protected3/data16/SEA/SEA16_Sources.pdf. Accessed 20 Dec 2019
Hille, E., Möbius, P.: Environmental policy, innovation, and productivity growth: controlling the effects of regulation and endogeneity. Environ. Resource Econ. **73**, 1315–1355 (2019)
Hille, E., Möbius, P.: Do energy prices affect employment? decomposed international evidence. J. Environ. Econ. Manag.Manag. **96**, 1–21 (2019)
Jain, R.K., Kumar, S.: An analysis of shadow price of CO_2 emission in Indian thermal power sector. Environ. Econ. Policy Stud. **20**(4), 879–902 (2018)
Kahn, M.: Mansur, erin: do local energy prices and regulation affect the geographic concentration of employment? J. Public Econ. **101**, 105–114 (2013)
Kumar, S., Managi, S., Jain, R.K.: CO_2 Mitigation policy for Indian thermal power sector: potential gains from emission trading. Energy Econ. **86**, 104653 (2020)
Kumar, S., Prabhakar, P.: Negative carbon leakage: evidence from South Asian countries. Working Paper No. 109–16, South Asian Network for Development and Environmental Economics (2016)
Kumar, S.: Environmentally sensitive productivity growth: a global analysis using malmquist-luenberger index. Ecol. Econ. **56**, 280–293 (2006)
Kumar, S., Managi, S.: Sulfur dioxide allowances: trading and technological progress. Ecol. Econ. **69**(10), 623–631 (2010)
Kumar, S., Managi, S.: Non-separability and substitutability among water pollutants: evidence from India. Environ. Dev. Econ. **16**, 709–733 (2011)
Lanoie, P., Laurent-Lucchetti, J., Johnstone, N., Ambec, S.: Environmental policy, innovation and performance: new insights on the porter hypothesis. J. Econo. Manage. Strategy **20**(3), 803–842 (2011)
Liu, M., Shadbegian, R., Zhang, B.: Does environmental regulation affect labor demand in china? evidence from the textile printing and dyeing industry. J. Environ. Econom. Manage. **87**, 277–294 (2017)
Liu, M., Zhang, B., Geng, Q.: Corporate pollution control strategies and labor demand: evidence from china's manufacturing sector. J. Regulatory Econo. **53**(3) (2019)
Liu, M., Tan, R., Zhang, B.: The costs of "blue sky": environmental regulation, technology upgrading, and labor demand in China. J. Dev. Econom. **150**, 102610 (2021)

Marin, G., Vona, F.: The impact of energy prices on socioeconomic and environmental performance: evidence from french manufacturing establishments, 1997–2015. Eur. Econ. Rev. **135**, 103739 (2021)

Martin, D.W., Braden, J.B., Carlson, L.J.: Estimation of process change for industrial pollution abatement. J. Air Waste Manag. Assoc.Manag. Assoc. **40**, 211–216 (1990)

Mergoni, A., Emrouznejad, A., De Witte, K.: Fifty Years of data envelopment analysis. European J. Opera. Res. **326**(3), 389–412 (2025). https://doi.org/10.1016/j.ejor.2024.12.049

Millimet, D.L., Roy, J.: Empirical tests of the pollution haven hypothesis when environmental regulation is endogenous. J. Appl. Econ. **31**(4), 652–677 (2016)

Morgenstern, R.D., Pizer, W.A., Shih, J.-S.: Jobs versus the environment: an industry-level perspective. J. Environ. Econ. Manag.Manag. **43**(2), 412–436 (2002)

Murty, S., Russell, R.: Bad Outputs. Discussion Paper in Economics, Centre for International Trade and Development, School of International Studies, Jawaharlal Nehru University, New Delhi (2017)

Murty, S., Russell, R.R., Levkoff, S.B.: On modeling pollution-generating technologies. J. Environ. Econ. Manage. **64**(1), 117–135 (2012)

Murty, M.N., Kumar, S., Dhavala, K.K.: Measuring environmental efficiency of industry: a case study of thermal power generation in India. Environ. Res. Econ. **38**, 31–50 (2007)

Pasurka, C.: Decomposing coal-fired electric power plant emissions within a joint production framework. Energy Econ **28**, 26–43 (2006)

Ray, S.C., Mukherjee, K., Venkatesh, A.: Nonparametric measures of efficiency in the presence of undesirable outputs: a by-production approach. Empirical Econo. **54**(1), 31–65 (2018)

Sheriff, G., Ferris, A.E., Shadbegian, R.J.: How did air quality standards affect employment at US power plants? the importance of timing, geography, and stringency. J. Assoc. Environ. Res. Econo. **6**, 111–149 (2019)

Yang, S., Jahanger, A., Hu, J., Awan, A.: Impact of China's carbon emissions trading scheme on firm-level pollution abatement and employment: evidence from a national panel dataset. Energy Econo. **136**, 107744 (2024)

Environmental Performance Evaluation of Joint Production of Crop and Animal Husbandry in European Farms

Utsav Pandey[1(✉)], Sanjeet Singh[1], and Saransh Tiwari[2]

[1] Decision Sciences Area, Indian Institute of Management Lucknow, Prabandh Nagar, IIM Road, Lucknow, UP 226013, India
{utsav.pandey,sanjeet}@iiml.ac.in

[2] Operations Management and Quantitative Techniques Area, Indian Institute of Management Rohtak, Rohtak, Haryana 124010, India
saransh.tiwari@iimrohtak.ac.in

Abstract. With the escalating threat of climate change, improving environmental performance across industries has become crucial. This study explores the use of Data Envelopment Analysis to measure environmental efficiency, addressing a gap in past research that overlooked the structural factors of production systems. Many studies treated production units as "black boxes," failing to identify inefficiencies accurately. To overcome this, the study incorporates emissions alongside traditional input and output parameters at each production stage to derive the overall system's environmental efficiency while decoupling stage-level efficiencies. The environmental efficiency scores are then compared with technical efficiency using separate models. The framework's applicability is demonstrated through a case study of 130 European Union farms, where 21% and 8% were efficient in technical and environmental aspects, respectively. For crop production, environmental efficiency improves significantly compared to technical efficiency, while animal husbandry shows a decline in environmental performance. The findings suggest that inefficient farms can enhance environmental performance by achieving technical competence. The study concludes that full efficiency across all stages is necessary for overall environmental efficiency. The values of marginal rate of substitution for inputs and the marginal rate of transformation for outputs hold great practical implications for policy-makers. Medium to small sized farms is environmentally more efficient. Meat and milk productions are resource intensive and attempts should be made to convert more forage land into arable land which will improve the environmental performance of agriculture.

Keyword: Network DEA · Material balance principle · Environmental efficiency · Marginal rate of substitution/transformation

1 Introduction

Several corrective actions, such as carbon taxes, Cap-and-Trade policies, and penalties, have been implemented to mitigate the impact of environmental degradation around the globe. Numerous environmental laws, such as the Clean Air Act [1] and the Clean

A. Emrouznejad et al. (Eds.): DEA 2024, LNOR, pp. 186–206, 2025.
https://doi.org/10.1007/978-3-031-98177-7_14

Water Act [2], have been established through national and international bodies such as the United Nations Framework Convention, the United Nations Environmental Programme, and the International Panel on Climate Change to ensure environmental protection. Recently concluded international summits, such as the G7 and G20, have also emphasized the adoption of sustainable development goals by participating nations.

However, growing carbon economies face a potential loss in their total production volumes if carbon-abatement measures are strictly implemented. This highlights the inherent conflict between economic growth and environmental protection, as they are interconnected through a negative trade-off [3, 4]. This interaction between the environment and the economy must be thoroughly examined to devise environmentally sustainable policies that also achieve economic gains [5]. This is the primary motivation for this study.

Agri-food products and protein-rich supplements are essential to ensuring food security, especially during pandemics. With international trade and the movement of goods restricted due to lockdown measures, any loss in food production or industrial demand for processed food due to carbon emission restrictions could cause severe food security issues. Agriculture and animal husbandry, the two main sources of food products, depend on the availability of natural resources such as land and water, as well as artificial supplements such as pesticides, fertilizers, and animal feed.

The excessive exploitation of natural resources and the overuse of chemicals play a major role in environmental degradation. Livestock farms are significant sources of greenhouse gases like methane, nitrous oxide, and ammonia, which heavily contribute to soil nitrification. Furthermore, residues from fertilizers and pesticides, including methane, ammonia, nitrous oxide, and organic compounds, cause substantial pollution [6]. Despite these challenges, the United Nations projects a roughly 15% increase in the demand for agricultural products between 2019 and 2028. Figure 1 illustrates the negligible growth in agricultural production, emphasizing the need for farms to improve productivity while considering the associated environmental costs.

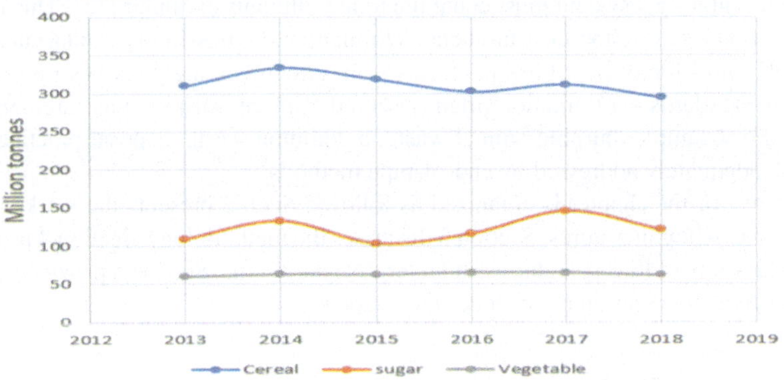

Fig. 1. Annual production trends of key crops in the EU.

While European Union (EU) member states have made strides toward implementing environmental regulations, the degree of preparedness and commitment to adhering to these laws varies considerably among countries [7]. This variation highlights the importance of analyzing the environmental productivity of the farming sector across the region. Typically, farms function in two interconnected stages: primary crop cultivation and subsequent livestock farming. Given that both agriculture and animal husbandry significantly impact environmental pollution while playing a critical role in food security, the proposed framework models them as two sequentially linked stages.

Agriculture represents a multifaceted and diverse production system involving numerous inputs and outputs. To evaluate such a multi-input, multi-output production process, a range of parametric and non-parametric methods have been developed [8]. Among these, Data Envelopment Analysis (DEA) has emerged as one of the most commonly applied non-parametric approaches. DEA offers a linear framework for assessing productivity and identifying best practices [9, 10]. One of its key advantages is that it does not rely on predefined assumptions about the structure of the production function [11, 12].

This study employs Network Data Envelopment Analysis (NDEA) to decompose the environmental efficiency of a generic system into stage-specific efficiencies using an emission-oriented approach, while simultaneously preserving economic well-being. This approach marks a notable theoretical advancement in DEA literature. Optimizing both stage-wise and overall efficiency in multi-stage production systems is inherently challenging. As noted by Despotis et al. [13], inefficient stages often receive lower priority, with resources being diverted to more efficient stages. To counter this, a method was proposed that uses input orientation for the first stage and output orientation for the second. However, such a dual-orientation approach is impractical for decision-makers in scenarios requiring a unified orientation—such as an emission-focused approach—across all stages. The proposed framework aims to enhance environmental performance for both stages equitably.

The Material Balance Condition (MBC), also known as the Material Balance Principle, is applied to ensure the conservation of pollutant flow (e.g., CO_2 and NO_x) within the system [14]. Rooted in the second law of thermodynamics, MBC mandates that the total pollutants entering a system must equal the total pollutants exiting it [15]. This method provides precise emission measurements, eliminating the possibility of inaccuracies or manipulation. Moreover, it prevents farms from circumventing penalties for violating emission standards—a limitation often observed with end-of-pipe measurement techniques. For example, burning animal waste, a common waste disposal practice in the EU, is inadequately addressed by end-of-pipe methods.

The rest of the chapter is organized as follows: Sect. 2 presents the works accomplished in existing literatures. Section 3 discusses the methodology. Section 4 presents a numerical study of European farms and interprets the results. Section 5 provides insights for policymakers. Section 6 concludes the chapter.

2 Literature Review

Relative productivity and efficiency computation gained significant popularity as decision-makers became more sensitive towards optimal resource allocation and reduce wasteful activities [16, 17]. Based on Farrell's [18] seminal work, the framework of deriving efficiency score for multiple inputs and multiple outputs were developed which later started the stream of research on Data Envelopment Analysis (DEA) [19]. Since then, DEA enabled practitioners, top-level managers, and policy makers to construct a single index for measuring the total productivity. Even though initially the use of DEA was limited to public institutions e.g., higher education [20], government authorities [21], law & civic bodies [22] but later it has been adopted by almost all the sectors for efficiency calculation: agriculture [23], bank & financial institutions [24], power plant [25], healthcare [26], etc. The detailed review of the entire body of research on DEA in the last 50 years is summarised by Mergoni et. al [27].

Traditional DEA frameworks often fall short in capturing the intricate internal structures of real-world systems. In multi-stage production systems, intermediate measures (or products) act as connectors between stages but are frequently disregarded in the simplified "Black-Box" approach [13]. DEA methods are generally applied in two ways: input-oriented (minimizing inputs) or output-oriented (maximizing outputs) [9]. To incorporate emissions as a critical aspect of production, researchers have explored approaches that either jointly minimize emissions and inputs [28] or treat emissions as fixed variables within input- or output-oriented models [29]. Non-oriented approaches aim to optimize both slack and surplus across all production factors simultaneously [30]. In the proposed model, an emission-focused approach is employed, concentrating solely on minimizing emissions.

Recently decision-makers, corporate leaders, and leaders of national and international bodies have become increasingly worried about environmental issues as it impacts social and economic development. Since authorities need to find out the units that are performing poorly to improve the present situation, DEA has been a popular choice for assessing environmental efficiency [31, 32]. Beyond the traditional input-output model, the polluting material is a part of the production function-one can refer to the rich body of literature pioneered by Färe et. al. [33–35] and Färe & Grosskopf [36]. Researchers proposed several ways to incorporate undesirable output in the tradition input-output framework of DEA. Undesirable output values can be transformed to be treated as desirable output [37]. Sometimes the nature of the variables for undesirable outputs are modified in such a way that the new variables become desirable [38]. In the most recent developments undesirable outputs are assumed to be weakly disposable [23] or using directional distance function [39].

3 Methodology

This research demonstrates the handling of undesirable outputs within a two-stage production process that adheres to the material balance principle. In conventional DEA, the technology is identified with the Production Possibility Set (PPS), defined as: $T\prime = \{(x, y)|\lambda X \leq x, \lambda Y \geq y, \lambda \geq 0\}$. The assumption of a free disposability property weakens the need for emission reduction [12, 40]. The costly disposal of emissions necessitates emission reduction such that the PPS is characterized by: $T = \{(x, y, b)|\lambda X \leq x, \lambda Y \geq y, \lambda B \leq b, \lambda \geq 0\}$. The developed DEA framework follows the frontier technology characterized as $(x, y, b) \in T$ and production possibilities defined as $(x, y, \varphi b) \in T$, where $\varphi \geq 1$.

Table 1 provides the description of the symbols used in the modelling. The framework proposed in this study deals with constructing environmental performance indicator in a two-stage production system. In the existing literature, the technical efficiency decomposition has been executed, but there are serious concerns regarding the usefulness and validity of environmental decomposition techniques. This study tries to bridge the gap where there is lack of dependable framework to compute environmental efficiency of the overall systems as well as that of individual stages (see Fig. 2).

Table 1. Notation Summary.

Category	Symbol	Description
Index		
	l	DMU Index
	i	Inputs Index
	r	Outputs Index
	d	Intermediate products Index
	o	Specific DMU under consideration
Data		
	L	DMU count
	m	Inputs count
	s	Outputs count
	D	Intermediate products count
	B	Emission vector of all DMUs
	X	Input matrix
	x_o	Input vector of o-th DMU
	Y	Output matrix
	y_o	Output vector of o-th DMU
	z_d^l	l-th DMU's d-th intermediate product
	b^l	l-th DMU's total (overall) emission
	b_k^l	l-th DMU's Stage-k emission
Variables		
	θ	Efficiency
	u_r	Weight for r-th output
	v_i	Weight for i-th input
	w_d	Weight for d-th intermediate product
	p	Weight for the pollutant emission
	p_k	Weight for the emission from stage-k
	θ_k	Stage- k efficiency, where k = 1 or 2

The following model derives inefficiencies in terms of excessive emissions:

$$\text{Min}\,\theta$$

subject to

$$\sum_{l=1}^{L}\lambda^l y_r^l \geq y_r^o, \forall r = 1,\ldots,s \tag{1}$$

$$\sum_{l=1}^{L}\lambda^l b^l \leq \theta b^o$$

$$\sum_{l=1}^{L}\lambda^l x_i^l \leq x_i^o, \forall i = 1,\ldots,m$$

$$\theta\ unrestricted, \lambda^l \geq 0, \forall l = 1,\ldots,L$$

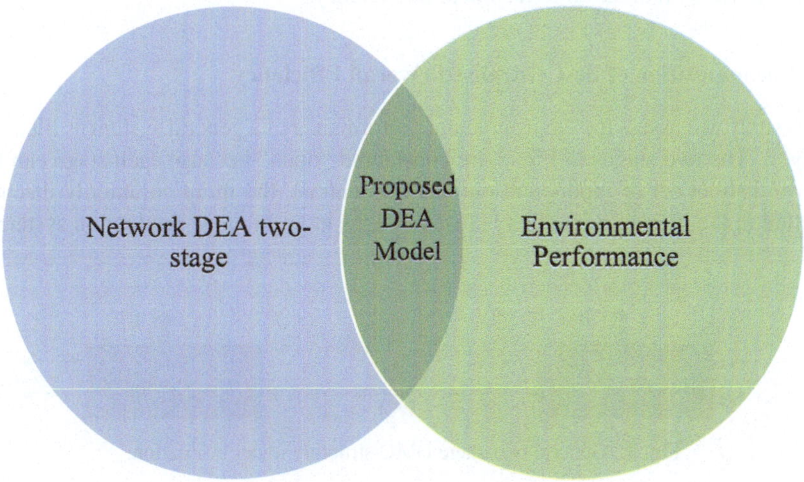

Fig. 2. Network DEA and environmental DEA research gap.

Environmental efficiency is represented by the greatest radial decrease in emissions while maintaining the current levels of inputs and outputs. The dual of model (1) can be transformed into its fractional form in model (2), which then provides an alternative definition of environmental efficiency:

$$\text{Max}\,\frac{\sum_{r=1}^{s}u_r y_r^o - \sum_{i=1}^{m}v_i x_i^o}{pb^o}$$

subject to

$$\frac{\sum_{r=1}^{s} u_r y_r^l - \sum_{i=1}^{m} v_i x_i^l}{pb^l} \leq 1, \forall l = 1, \ldots, L, \tag{2}$$

$$v_i \geq 0, u_r \geq 0, p \geq 0.$$

We have $\sum_{r=1}^{s} u_r y_r^o - \sum_{i=1}^{m} v_i x_i^o$ as benefit created in the numerator which is comparable to the virtual output in a traditional DEA. The denominator, represented as the weighted pollutant, reflects the virtual environmental cost in relation to the virtual inputs used in traditional DEA. A comparable definition of environmental efficiency is present in the eco-efficiency literature, defined as the economic value added per unit of waste [41].

Theorem 1. In model (1), an environmentally efficient DMU must be technically efficient. The environmental efficiencies of the inefficient DMUs in model (1) drop compared to technical efficiencies.

Technically inefficient DMUs generate waste in their overall production processes due to either excess inputs or insufficient outputs, leading to environmental inefficiency. Theorem 1 further demonstrates that environmental inefficiency stems from a form of mix inefficiency in addition to technical inefficiency.

3.1 Decomposition of de-Centralised Overall Efficiency

In this study, we assess the environmental performance of a general system as illustrated in Fig. 3. The two stages in Fig. 3 are considered "open" because each stage can have its own distinct set of inputs and outputs. In contrast, the more commonly discussed structure in the literature involves "closed" stages in multi-stage production systems.

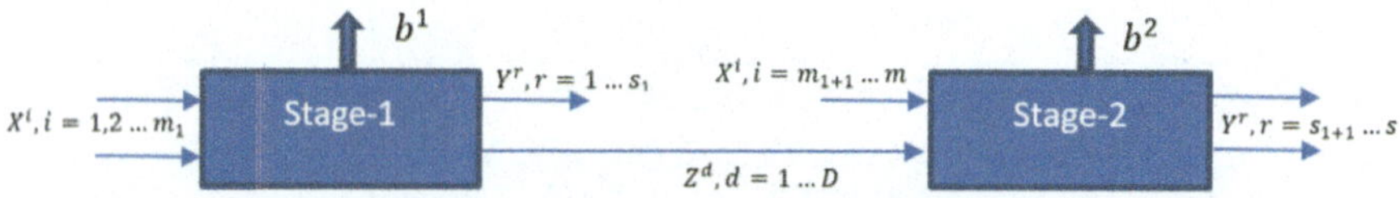

Fig. 3. Generic two-stage DMU structure under evaluation.

Note, the application of "Black-Box" DEA models would render the evaluation incorrect since the stage-1 outputs $Y^r, r = 1 \ldots s_1$ have no bearing on inputs such as $X^i, i = m_{1+1} \ldots m$. Similarly, the inputs such as $X^i, i = m_{1+1} \ldots m$ have no influence on the intermediate products $Z^d, d = 1 \ldots D$.

To apportion inputs to appropriate outputs, the decomposition of overall efficiency is performed in a top-down approach. The efficiency function of the overall system, as shown in Fig. 2 can be arithmetically converted to the sum of the efficiency scores of the stages multiplied with the corresponding emission ratio.

$$= \frac{\left(\sum_{r=1}^{s_1} u_r y_r^o + \sum_{d=1}^{D} w_d^1 z_d^o - \sum_{i=1}^{m_1} v_i x_i^o\right)}{pb^o} + \frac{\left(\sum_{r=s_2}^{s} u_r y_r^o - \sum_{d=1}^{D} w_d^2 z_d^o - \sum_{i=m_2}^{m} v_i x_i^o\right)}{pb^o}$$

$$= \frac{\left(\sum_{r=1}^{s_1} u_r y_r^o + \sum_{d=1}^{D} w_d^1 z_d^o - \sum_{i=1}^{m_1} v_i x_i^o\right)}{pb_1^o} \frac{b_1^o}{b^o} + \frac{\left(\sum_{r=s_2}^{s} u_r y_r^o - \sum_{d=1}^{D} w_d^2 z_d^o - \sum_{i=m_2}^{m} v_i x_i^o\right)}{pb_2^o} \frac{b_2^o}{b^o}.$$

The overall environmental efficiency is $(R_1\theta_1 + R_2\theta_2)$, where$\theta_k$, $k = 1 or 2$ represents the efficiency of the kth stage, while R_k where k = 1 or 2, denotes the proportion of the total emission from the kth stage. In earlier research, the shadow prices for intermediate products were set to be equal, which imposed the condition that the environmental impact of the intermediate products was consistent across the stages [42].

To enhance generalizability, the two shadow prices are not set to be equal, acknowledging that two dissimilar processes (e., agriculture and livestock) are involved. The relationships among such shadow price variables are endogenously determined as follows:

$$\text{Max } \frac{\left(\sum_{r=1}^{s_1} u_r y_r^o + \sum_{d=1}^{D} w_d^1 z_d^o - \sum_{i=1}^{m_1} v_i x_i^o\right)}{pb_1^o} \left(\frac{b_1^o}{b^o}\right) + \frac{\left(\sum_{r=s_2}^{s} u_r y_r^o - \sum_{d=1}^{D} w_d^2 z_d^o - \sum_{i=m_2}^{m} v_i x_i^o\right)}{pb_2^o} \left(\frac{b_2^o}{b^o}\right)$$

subject to

$$\frac{(\sum_{r=1}^{s_1} u_r y_r^l + \sum_{d=1}^{D} w_d^1 z_d^l - \sum_{i=1}^{m_1} v_i x_i^l)}{pb_1^l} \leq 1, \forall l = 1, \ldots, L$$

$$\frac{(\sum_{r=s_2}^{s} u_r y_r^l - \sum_{d=1}^{D} w_d^2 z_d^l - \sum_{i=m_2}^{m} v_i x_i^l)}{pb_2^l} \leq 1, \forall l = 1, \ldots, L$$

$$u_r, v_i, w_d, p \geq 0. \tag{3}$$

In traditional network DEA methods for decomposing technical efficiency using the additive approach, the stage-wise weights of the stage efficiencies are determined by optimization variables. The most popular decomposition technique is additive [43]. The input, intermediate product, and the final outputs signified as x_{io}, z_{do} and y_{ro}. The multiplier weight vectors for the three factors are: v_i, η_d u_r respectively. The overall input-oriented efficiency score e_o is computed as a convex combination of the individual efficiencies as follows:

$$e_o = W_1 \frac{\sum\limits_{d=1}^{D} \eta_d z_{do}}{\sum\limits_{i=1}^{m} v_i x_{io}} + W_2 \frac{\sum\limits_{r=1}^{s} u_r y_{ro}}{\sum\limits_{d=1}^{D} \eta_d z_{do}} \tag{4}$$

The stages are associated with a pre-defined set of weights W_1 and W_2 such that $W_1 + W_2 = 1$. This makes the efficiency decomposition subjective to the values selected for W_1 and W_2. To remove the subjectivity from the decomposition process, the stage specific weights are calculated intrinsically [13]. The same production process can now be modelled as follows:

$$e_o = W_1 \frac{\sum\limits_{d=1}^{D} \eta_d z_{do}}{\sum\limits_{i=1}^{m} v_i x_{io}} + W_2 \frac{\sum\limits_{r=1}^{s} u_r y_{ro}}{\sum\limits_{d=1}^{D} \eta_d z_{do}}, \tag{5}$$

where $W_1 = \frac{\sum_{i=1}^{m} v_i x_{io}}{\sum_{i=1}^{m} v_i x_{io} + \sum_{d=1}^{D} \eta_d z_{do}}$ and $W_2 = \frac{\sum_{d=1}^{D} \eta_d z_{do}}{\sum_{d=1}^{D} \eta_d z_{do} + \sum_{i=1}^{m} v_i x_{io}}$ are the two weights. However, there are some serious concerns regarding the selection of weight values in (5). Firstly, the weights are still intrinsically determined, and as a result the weight vectors would be chosen to optimize the value of e_o. In case of alternative optima, the weights of the two stages remain variable and depends on the chosen solution from the available optimal solutions. Thus, the entire weighing process becomes unstable and ambiguous. Secondly, the weights assigned to the stages have differential treatments. For example, both the terms used in W_1 i.e., $\sum_{i=1}^{m} v_i x_{io}$ and $\sum_{d=1}^{D} \eta_d z_{do}$ have appeared as a multiplier to W_1 in the objective function, thus their values are set in a particular way to have an optimal overall efficiency e_o. But in W_2 out of two terms $\sum_{d=1}^{D} \eta_d z_{do}$ and $\sum_{i=1}^{m} v_i x_{io}$, $\sum_{i=1}^{m} v_i x_{io}$ has no association with the second stage. This makes its association with W_2 questionable.

The proposed model overcomes all these limitations in the following manner. The stage efficiency values are: $\frac{(\sum_{r=1}^{s_1} u_r y_r^o + \sum_{d=1}^{D} w_d^1 z_d^o - \sum_{i=1}^{m_1} v_i x_i^o)}{pb_1^o}$ and $\frac{(\sum_{r=s_2}^{s} u_r y_r^o - \sum_{d=1}^{D} w_d^2 z_d^o - \sum_{i=m_2}^{m} v_i x_i^o)}{pb_2^o}$ for stage -1 and -2, respectively. The stage-wise efficiency scores follow the similar structure as that of the overall system. The overall environmental efficiency score is eventually a convex combination of the environmental efficiency scores of the underlying stages as $b^o = b_1^o + b_2^o$. The weights of stage-1 and stage-2 environmental efficiencies are: $\frac{b_1^o}{b^o}$ and $\frac{b_2^o}{b^o}$. The efficiency score of a stage is prioritized in proportion to the fraction of emission committed by the stage itself. This is a unique contribution of this paper, since all the previous research on this topic failed to express overall environmental efficiency score as a combination of the stage-wise environmental efficiency score. For a DMU under evaluation, these two weights W_1 and W_2 do not depend on the optimal solution of the model (3). That means even in the presence of alternate optima, the decomposition technique remains robust and unambiguous providing a unique pair decomposition efficiency score.

As a result, these weights can vary across different solutions in the case of alternate optimal outcomes. However, the efficiency decomposition as given in model (3) results into unique set of stage efficiencies. It is possible to transform model (3) into a linear programming model by employing Charnes-Cooper transformation technique.

The shadow price of the intermediate product gives rise to the following three conditions:

1. $\omega_d^1 > \omega_d^2$, that is, intermediate products have larger shadow prices as outputs. The production of intermediate products is environmentally more friendly than their consumption as inputs in livestock. The cumulative net environmental benefit is: $\omega_d^1 - \omega_d^2$
2. $\omega_d^1 < \omega_d^2$, that is, intermediate products have smaller shadow prices as outputs. Their consumption is environmentally more friendly than their production. The system incurs a net environmental loss of: $\omega_d^2 - \omega_d^1$
3. $\omega_d^1 = \omega_d^2$, that is, the overall system is environmentally neutral.

4 Case Study: Farm-Level Data Application

4.1 Structure of the Farms

We illustrate the applications of the developed model on 130 farms in the European Union. The schematic diagram of the firm structure, shown in Fig. 4, is carefully designed to correctly apportion inputs to relevant outputs. Inputs such as, cattle or fertiliser are not responsible to produce, say cereal or milk, respectively. The objective is to evaluate the efficiencies of production practices in executing resource allocation to reduce nitrogen emissions. As can be seen from Fig. 4, forage is produced in agriculture and consumed in livestock section.

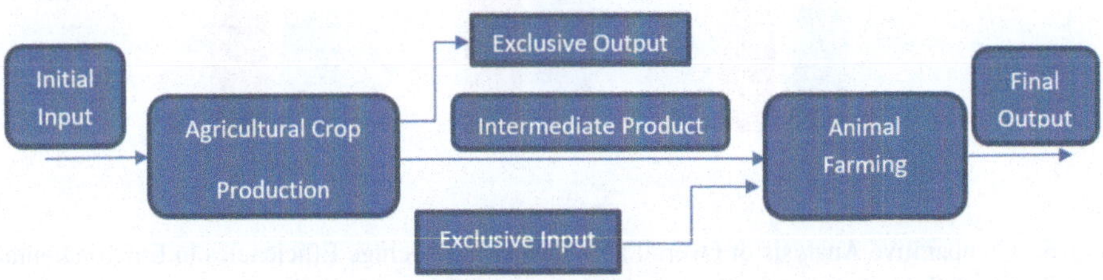

Fig. 4. Two-stage structure of European farms.

4.2 Dataset

Even though separate countries are identified as independent DMUs in practice, individual farms, acting as actual production units, are more suitable for use in this study [44, 45]. The primary data source for this study is the Farm Accountancy Data Network (https://ec.europa.eu/agriculture). The initial inputs for agriculture (stage-1) include arable land, fertilizer, and forage land. The exclusive outputs from stage-1 are potatoes, cereals, oilseeds, protein crops, industrial crops, sugar beet, and vegetables. The exclusive inputs for animal husbandry are dairy cows or livestock units, cattle feed, and grassland, with milk and meat as the final outputs (Fig. 4). Forage is considered the only intermediate product. The coefficient of variation for the input and output variables ranges from 1.3 to 1.8, indicating a comparable level of variability among the farms.

4.3 Observations and Research Findings

The difference in overall system efficiencies between the two approaches is not significantly pronounced, as illustrated in Fig. 5. There is a trend of gradual decline in the environmental efficiency of the farms, albeit by a relatively small margin, compared to the technical efficiency. The overall system demonstrates strong environmental performance, with an average efficiency of 0.9 (Median: 0.93), which further increases to 0.92 in the technical dimension (Median: 0.97) (see Fig. 5). The correlation between environmental and technical performances is positive but weak ($r = 0.25$) at a 5% significance level. Contrary to common belief, these two aspects do not necessarily exhibit a negative trade-off. This finding is supported by Theorem 1, which indicates that improvements in both efficiencies can occur simultaneously. The production systems and practices differ among the farms, as evidenced by the high standard deviations of 12% and 11% in the environmental and technical approaches, respectively.

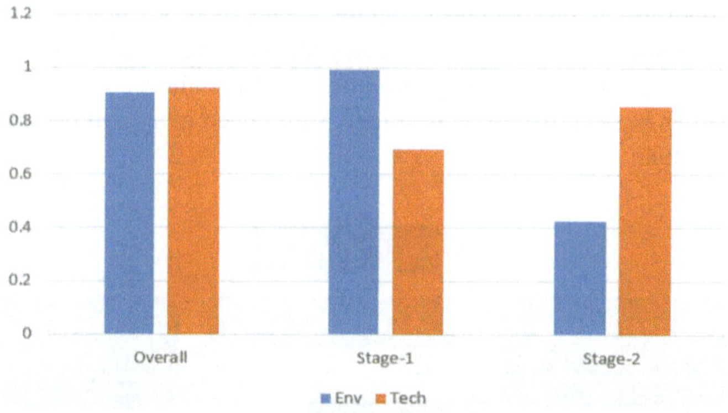

Fig. 5. Comparative Analysis of Overall and Stage-wise Average Efficiencies in Environmental and Technical Approaches.

Stage-1 (agriculture) shows a significant improvement from a low technical performance (M: 69.4%, SD: 0.24) to a high environmental performance (M: 98%, SD: 0.06) at a significance level of $p < 0.01$. In contrast, stage-2 experiences inefficiency, as its high technical efficiency (M: 85.6%, SD: 0.26) declines sharply in the environmental dimension (M: 42.8%, SD: 0.30) at the same significance level of $p < 0.01$. Consequently, without conducting a stage-wise decomposition, one might mistakenly conclude that there are only minor changes in performance between the two approaches.

Animal husbandry contributes significantly to high emission rates, mainly due to the substantial nitrogen (N) deposition rate—approximately 70% to 80%—from excreta

[46]. The technical-environmental efficiency space can be divided into four categories: (I) Advanced—where both technical and environmental efficiencies exceed 95%. The threshold of 0.95 is established based on the central tendency of efficiency scores in both approaches; (II) Environmentally Advanced—where environmental efficiency is at least 95%, but technical efficiency falls below 95%; (III) Technically Advanced—where technical efficiency is at least 95%, but environmental efficiency is below 95%; (IV) Backwards—where both efficiencies are below 95%. The distribution of farms across these categories is as follows: 25% in Region I (Advanced), 40% in Region II (Environmentally Advanced), 9% in Region III (Technically Advanced), and 26% in Region IV (Backwards). The farms are nearly evenly spread on either side of the 45° line (marked as OE in Fig. 6), suggesting no systematic bias in the design of the two approaches.

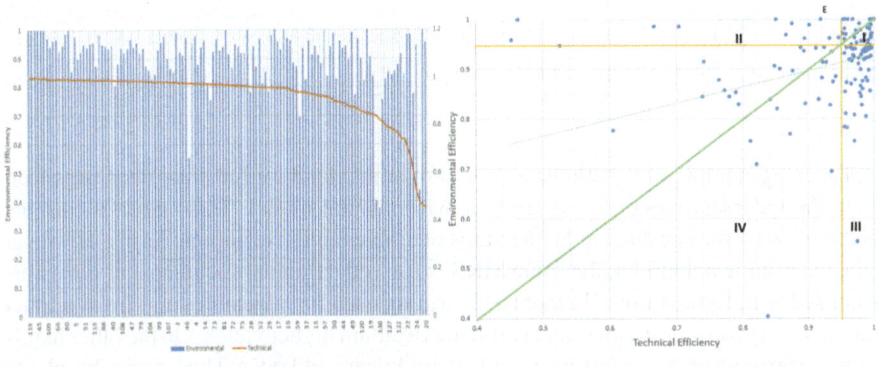

(a) Technical and environmental efficiency comparison. (b) Categorization of the DMU space.

Fig. 6. Efficiency analysis of DMUs in the Technical and Environmental approaches.

DEA also calculates weights, or multipliers, which are known as posteriori weights. These weights represent endogenously determined equilibrium prices in relative terms and can be utilized to evaluate abatement costs or the market price of emissions [47]. The differences in these weights across various approaches reflect changes in the roles of the involved factors [48]. To conduct a comparative analysis, the utilization rates of inputs and outputs are examined in relation to achieving technical and environmental objectives. This analysis aids decision-makers, such as farm owners and policymakers, in gaining a clearer understanding of the trade-offs between inputs and outputs. A reduction of 57% and 85% is noted in the number of farms that efficiently utilize forage land and fertilizer in environmental terms, compared to the technical approach (see Table 2).

In the livestock sector, the number of farms utilizing livestock efficiently decreases by 56%, underscoring the substantial greenhouse gas emissions generated throughout the lifecycle process. Cereal and meat are produced in a technically efficient manner in stage-1 and stage-2, respectively, yet they still result in high environmental costs (see Table 3). Outputs like sugar beet and industrial crops exhibit the poorest environmental performance. Conversely, potatoes and vegetables, despite their low technical efficiency, contribute positively to environmental performance.

Table 2. Count of farms with optimal input factors (by stage).

	Stage-1			Stage-2		
	Fertiliser	Arable land	Forage Land	Grassland	Dairy Cows	Cattle Feed
Technical	120.0	103.0	129.0	13.0	108.0	4.0
Environmental	18.0	52.0	56.0	11.0	47.0	4.0

Table 3. Count of farms with optimal output factors (by stage).

	Stage-1							Stage-2	
	Cereal	Protein Crops	Potato	Sugar beet	Oil-seed	Industrial	Veg	Milk	Meat
Technical	107.0	79.0	21.0	40.0	84.0	38.0	13.0	47.0	118.0
Environmental	76.0	84.0	66.0	52.0	81.0	48.0	72.0	49.0	83.0

The marginal rate of substitution (MRS) of arable land for fertilizer decreases from 5.7 in the technical model to 1.4 in the environmental model. This suggests that one quintal of fertilizer is valued at 5.7 hectares of arable land in economic terms. However, in the environmental model, the same MRS is 1.4, reflecting the social cost of emissions associated with fertilizer use. In intensive farming used for vegetables and fruits applied excessive fertilizers, pesticides, and herbicides to attain higher yields. On the other hands, extensive farming uses less fertilisers and attains lower yield rates. Thus, agricultural land requirement is very high to generate the equivalent amount of total yield as compared to intensive farming [49]. The case of intensive farming justifies the higher MRS value of land with respect to fertilizer. As shown in Fig. 7(a and b), fertilizer proves to be more effective in the technical approach. Similarly, the MRS of forage land to fertilizer (see Fig. 7(c and d)) indicates that every hundred kg of fertilizer is economically equivalent to one hectare of forage land, but this value drops to 0.2 hectares in the environmental context. Figure 8(b) illustrates that fertilizer demonstrates significantly ($p < 0.01$) higher productivity in the technical approach. Previous research has explored the relationship between the economic benefits of production systems and environmental pollution. For instance, using larger amounts of fertilizer can offset the need for more agricultural land, but this comes at the expense of increased pollution [50]. However, these studies have not quantified the extent of these impacts. Forage land contributes more in the technical approach than in the environmental approach (see Fig. 7(c) and 7(d)) at a significance level of $p < 0.01$, largely due to the high usage of fertilizer (16% of total fertilizer) on forage land [51]. The MRS of forage land to fertilizer significantly decreases ($p < 0.01$) in the environmental model. All category-A farms (22 in total) and 105 category-B farms exhibit technically optimal usage of forage land, but these figures drop to just 10 and 46, respectively, in the environmental model. The number of farms in category-A and -B with optimal fertilizer utilization is 20 and 97, respectively, which decreases to 7 and 11 in the environmental model.

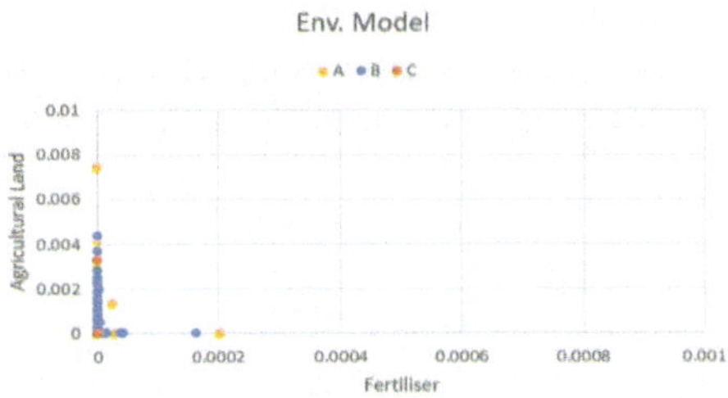

(a) Fertiliser Vs. Agri-land shadow price (Env. model).

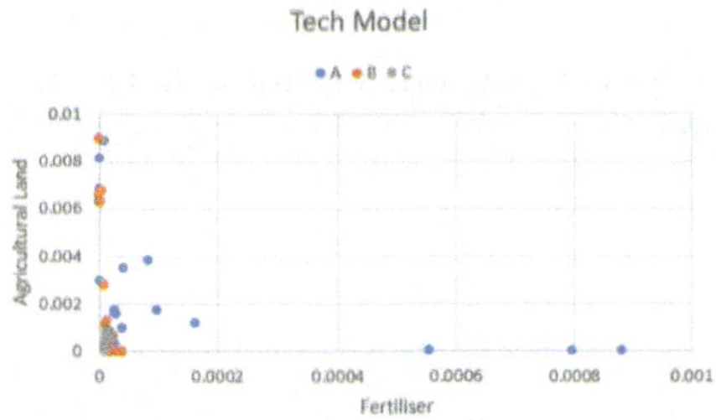

(b) Fertiliser Vs. Agri-land shadow price (Tech. model).

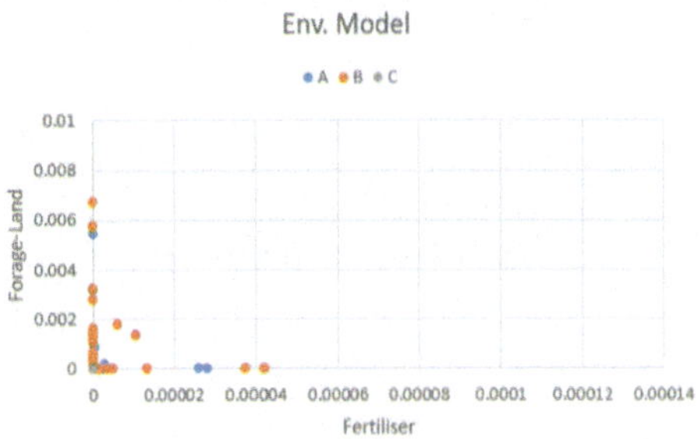

(c) Fertiliser Vs. Forage Land shadow price (Env. model).

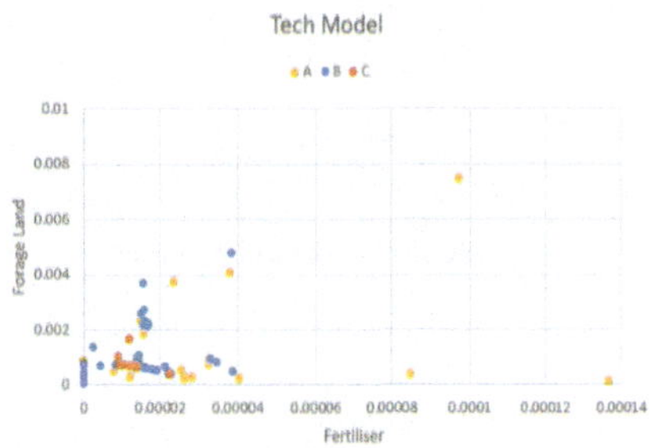

(d) Fertiliser Vs. Forage Land shadow price (Tech. model).

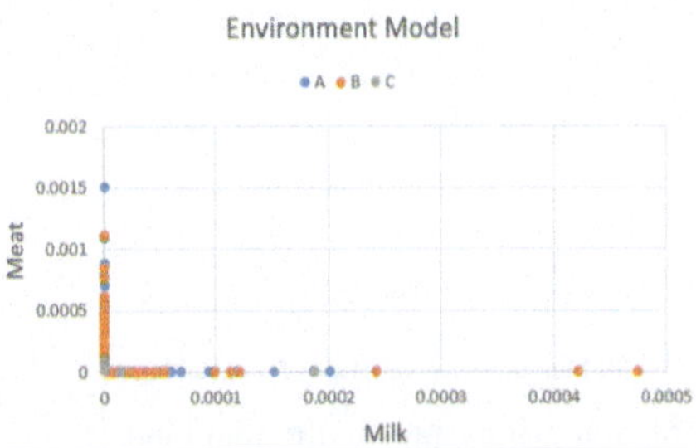

(e) Milk Vs. Meat shadow price (Env. model).

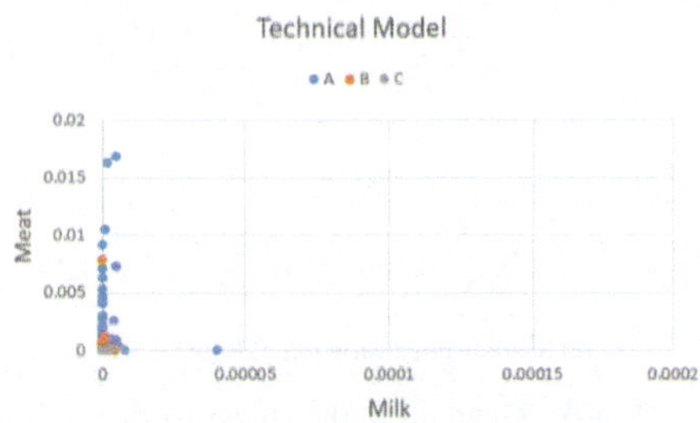

(f) Milk Vs. Meat shadow price (Tech. model).

Fig. 7. Comparison of shadow prices for production factors.

In Fig. 7(e), the distribution of weights suggests that meat and milk are considered imperfect substitutes in the environmental model. However, the marginal rate of transformation (MRT) from meat to milk in the environmental model is significantly greater than in the technical model, indicating that milk is more environmentally costly than meat. In the literature of livestock science, milk production reportedly causes the maximum amount of volatilisation of ammonia which worsens its environmental performance. Milk production in European farms is considerably more efficient in the environmental approach compared to the technical approach (see Fig. 8(c)). As long-term strategic move, some of the European farms started organic farming which requires much more natural resources than conventional farming. However, the acidification potential of milk production does not necessarily reduce in organic farming and thus, has limited impact on the environmental pollution. The number of technically efficient farms producing meat (118) far exceeds the number of environmentally efficient farms (83). There is a substantial opportunity for milk production, with only 49 farms being efficient (see Table 3).

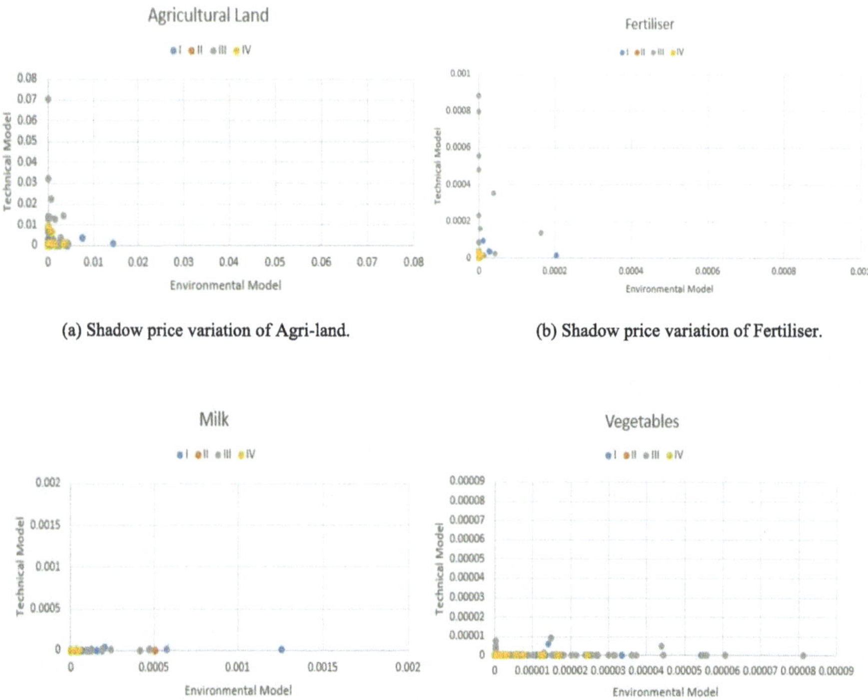

(a) Shadow price variation of Agri-land.

(b) Shadow price variation of Fertiliser.

(c) Shadow price variation of Milk.

(d) Shadow price variation of Vegetables.

Fig. 8. Changes in the shadow prices of production factors among different models.

Vegetable production is significantly ($p < 0.01$) more advantageous in the environmental approach (Fig. 8(d)). In the European countries, out of the total vegetable production 56% accounts for potatoes. Other major vegetables e.g., carrots, onions, and tomatoes account for 8%, 7%, and 5%, respectively [52]. A significant chunk of potato is grown by the commercial companies and they try to control the environmental risk factors: greenhouse gas emission, eutrophication, and soil pollution from excessive use of pesticides [53]. It results into environmentally sustainable farming of potato in these regions.

Its technical efficiency declines sharply, with seventy-two environmentally efficient farms compared to only thirteen that are technically efficient (see Table 3). The marginal rate of transformation (MRT) of vegetables to protein crops and oil-seed is 19.2 and 18.8, respectively, in the technical model, while it drops to 1.2 and 1.6 in the environmental model. This indicates that the technical yield of vegetables is considerably better than that of protein crops and oil-seed, with minimal environmental damage. Additionally, the production of oil-seed is found to be significantly (at $p < 0.01$) more beneficial from an environmental perspective than from an economic one. Among all oil-seeds, 58%, 34%, and 8% of the total oil-seed production can be attributed to rape, sunflower, and soybean. Since the EU commission for agriculture announced incentives for planting more oil-seeds in a sustainable manner, the general practice among the farmers became more environment friendly [54].

Environmentally advanced farms, regardless of their technical performance (i.e., class-I and -II), demonstrate significantly higher productivity of oil-seed compared to environmentally poor farms (class-III) (at $p < 0.05$). Thus, farms in classes IV and III have the potential to enhance their environmental performance to reach the standards of classes I and II. A similar trend is observed with protein crops. In the technical approach, protein crops and oil-seed are treated as perfect substitutes (with an MRT close to one), although protein crops show slightly better environmental impacts. Agricultural outputs such as industrial crops, and cereals exhibit a similar differentiated contribution to both environmental and technical approaches. A commonly used indicator for comparing different production factors in terms of nitrogen emissions is Nitrogen Use Efficiency (NUE) [55]. NUE is calculated as the ratio of nitrogen consumed by agricultural crops to the total nitrogen supplied through fertilizers. A lower NUE value indicates higher nitrogen leakage. The NUE values for livestock are significantly lower than those for agriculture [56], reinforcing our assertion that agriculture is more environmentally efficient than animal husbandry. The potential environmental damages associated with crops like cereals, vegetables, potatoes, and protein crops, as indicated by their respective NUE percentages (approximately 50%, 60%, 50%, and 15% according to Tiwari et al. [57]), align with the relative potentials of these crops reported in this study.

5 Policy Implications

The number of farms with near-efficient technical performance is higher than that in environment performance (see Fig. 9). Figure 9 also shows that the number of farms with efficiency till 70% is almost similar in technical and environmental approach. However, the number of farms with environmental efficiency between 70% to 90% is higher than that number in technical efficiency. For efficiency more than 90%, the scenario is reversed. Interestingly, the number by which there are environmentally more efficient farms in the range between 70% to 90%, by the same number technically efficient farms supersede environmentally efficient farms in the range between 90% to 100%.

Key implications for policy-makers and farmers can be summarised as follows:

- Large farms have higher MRS values of fertiliser than small farms with respect to arable land. It indicates larger farms experience low marginal rate of increment. Using the same analytical approach, small to medium-sized farms are environmentally more efficient. The reason behind this is the low marginal growth of yield in larger farms often force the farmers to apply excessive fertiliser.
- Almost 65% of the farms face inefficiency in milk production. It has been observed that since the farms operated under milk quota in those countries, achieving full operational efficiency became difficult.
- A negligible marginal rate of transmission (0.009) of meat to milk in technical aspect shows that a small adjustment in meat product in terms of resource consumption could lead to significant improvement in milk production.

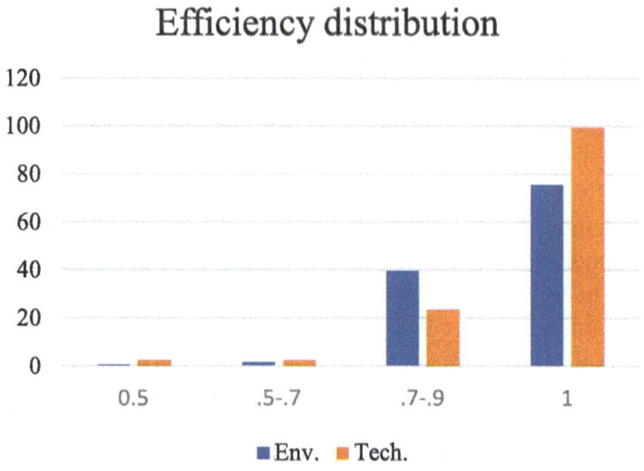

Fig. 9. Environmental and technical efficiency distribution.

- The land distribution between arable and forage land is different in different countries. Since the MRS value of the two varies significant between efficient and non-efficient farms, the policy-makers should implement land diversion schemes or quantity subsidies in the proportion of the MRS values [58].

6 Conclusion

The proposed DEA framework identifies emission sources within a complex production system and provides robust indicators for assessing the contribution of production factors to environmental damage. The case study of European farms at a micro level highlights several key points regarding potential improvements in environmental conditions and their impact on production volume (or economic performance). This framework, which integrates economic and environmental concerns, is particularly relevant for countries like those in the EU that have committed to the United Nations' Sustainable Development Goals (SDGs) by incorporating ecological considerations into national policy-making. The proposed framework suggests quantifiable measure for comparing the impact of various inputs and outputs by using the MRT and MRS values. This might help the policy-makers to set quantity for production quota or encourage production of a particular crop by giving subsidy.

The developed model could be expanded to include window analysis techniques over an extended time frame, such as five or ten years. EU countries face a variety of external factors, including weather conditions, soil nutrients, and humidity, which significantly affect farm productivity. Therefore, a region-based approach, comparing Northern and Southern European countries, could yield more equitable and reliable results. Since United Nation Framework Convention on Climate Change (UNFCC) set target for 2030, it is imperative that countries implement appropriate measures in a timely manner. DEA models can be extended to process time series data using Malmquist-Index to compute the rate of growth and expected time of achieving the set targets. A primary aim of this study is to encourage global discussions on alternative responsible farming practices while also considering economic factors.

References

1. Bryner, G.C.: Blue Skies, Green Politics: The Clean Air Act of 1990. URL: https://www.osti.gov/biblio/111385 (1995)
2. Adler, R.W., Landman, J.C., Diane, M.: the Clean Water Act 20 Years Later (1993)
3. Den Butter, F.A.G., Verbruggen, H.: Measuring the trade-off between economic growth and a clean environment. Environ. Res. Econ. **4**(2), 187–208 (1994). https://doi.org/10.1007/BF00692203
4. Marsiglio, S., Privileggi, F.: On the economic growth and environmental trade-off: a multi-objective analysis. Ann. Oper. Res. **296**(1–2), 263–289 (2021). https://doi.org/10.1007/s10479-019-03217-y
5. Baumgärtner, S., Dyckhoff, H., Faber, M., Proops, J., Schiller, J.: The concept of joint production and ecological economics. Ecol. Econ. **36**(3), 365–372 (2001). https://doi.org/10.1016/S0921-8009(00)00260-3
6. Waheed, R., Chang, D., Sarwar, S., Chen, W.: Forest, agriculture, renewable energy, and CO2 emission. J. Cleaner Prod. **172**, 4231–4238 (2018). https://doi.org/10.1016/j.jclepro.2017.10.287
7. Thonig, R., et al.: Does ideology influence the ambition level of climate and renewable energy policy? insights from four European countries. Energy Sources B **16**(1), 4–22 (2021). https://doi.org/10.1080/15567249.2020.1811806
8. Murillo-zamorano, L.R., Vega-cervera, J.A.: The use of parametric and nonparametric frontier methods to measure the productive efficiency in the industrial sector: a comparative study. Int. J. Prod. Econ. **69**(3), 265–275 (2001)
9. Charnes, A., Cooper, W.W., Rhodes, E.: Evaluating program and managerial efficiency: an application of data envelopment analysis to program follows through. Manage. Sci. **27**(6), 668–697 (1981). https://doi.org/10.1287/mnsc.27.6.668
10. Zhu, J.: DEA Under big data: Data enabled analytics and network data envelopment analysis. Ann. Opera. Res. (0123456789) (2020). https://doi.org/10.1007/s10479-020-03668-8
11. Hosseinzadeh Lotfi, F., Hatami-Marbini, A., Agrell, P.J., Aghayi, N., Gholami, K.: Allocating fixed resources and setting targets using a common-weights DEA approach. Comput. Ind. Eng. **64**(2), 631–640 (2013). https://doi.org/10.1016/j.cie.2012.12.006
12. Førsund, F.R.: Economic interpretations of DEA. Socioecon. Plann. Sci. **61**(1), 9–15 (2018). https://doi.org/10.1016/j.seps.2017.03.004
13. Despotis, D.K., Koronakos, G., Sotiros, D.: Composition versus decomposition in two-stage network DEA: a reverse approach. J. Prod. Anal. **45**(1), 71–87 (2016). https://doi.org/10.1007/s11123-014-0415-x
14. Lauwers, L., Huylenbroeck Guido, V.A.N., Coelli, T.I.M.: Materials balance-based modelling of environmental efficiency. In: Data Envelopment Analysis and Performance Management, Proceedings of the 4th International Conference of DEA (DEA2004), Birmingham, UK: 237 (2004). ISBN: 0 902683 73 X
15. Chambers, R., Melkonyan, T.: Production Technologies, Material Balance, and the Income-Environmental Quality Trade-Off. University of Exeter working paper (2012)
16. Zervopoulos, P.D., Kanas, A., Emrouznejad, A., Molyneux, P.: Can super-efficiencies improve bias correction? a bayesian data envelopment analysis approach. In: Advances in the Theory and Applications of Performance Measurement and Management. DEA45 2023. Lecture Notes in Operations Research. Springer, Cham (2024). https://doi.org/10.1007/978-3-031-61597-9_3

17. Zhu N., Zhu, C., Emrouznejad, A., Chen, L.: A combined DEA with machine learning algorithms for measuring the efficiency of Chinese listed manufacturing plants. In: Data Envelopment Analysis and Performance Measurement: Recent Developments: Proceedings of the DEA40: International Conference of Data Envelopment Analysis, Aston Business School, UK, ISBN: 978 1 85449 (2018)
18. Farrell, M.J.: The measurement of productive efficiency. J. Royal Stat. Soc. Series A (General) **120**(3), 253–290 (1957). http://goo.gl/AFhm2N
19. Charnes, A., Cooper, W.W., Rhodes, E.: Measuring the efficiency of decision making units **2**, 429–444 (1978)
20. Ahn, T., Charnes, A., Cooper, W.W.: Some statistical and DEA evaluations of relative efficiencies of public and private institutions of higher learning. Socioecon. Plann. Sci. **22**(6), 259–269 (1988). https://doi.org/10.1016/0038-0121(88)90008-0
21. Chalos, P., Cherian, J.: An application of data envelopment analysis to public sector performance measurement and accountability. J. Account. Public Policy **14**(2), 143–160 (1995). https://doi.org/10.1016/0278-4254(94)00015-S
22. Gupta, M., Bolia, N.B.: Efficiency measurement of Indian high courts using DEA: a policy perspective. J. Policy Model. **42**(6), 1372–1393 (2020). https://doi.org/10.1016/j.jpolmod.2020.06.002
23. Pandey, U., Singh, S.: Environmental performance evaluation of European farms by assessing polluting factors in joint production. J. Cleaner Prod. **328** (2021)
24. Fukuyama, H., Matousek, R.: Modelling bank performance: a network DEA approach. Eur. J. Oper. Res. **259**(2), 721–732 (2017). https://doi.org/10.1016/j.ejor.2016.10.044
25. Liu, C.H., Lin, S.J., Lewis, C.: Evaluation of thermal power plant operational performance in Taiwan by data envelopment analysis. Energy Policy **38**(2), 1049–1058 (2010). https://doi.org/10.1016/j.enpol.2009.10.057
26. Pandey, U., Singh, S.: Incorporating contextual factors into a comprehensive analysis of operational efficiency and service quality in healthcare sector. Int. J. Math. Eng. Manage. Sci. **10**(2), 300–349 (2025). https://doi.org/10.33889/IJMEMS.2025.10.2.017
27. Mergoni, A., Emrouznejad, A., De Witte, K.: Fifty years of data envelopment analysis. European J. Oper. Res. **326**(3), 389–412 (2025). https://doi.org/10.1016/j.ejor.2024.12.049
28. Førsund, F.R.: Performance measurement and joint production of intended and unintended outputs. J. Prod. Anal. **55**(3), 157–175 (2021). https://doi.org/10.1007/s11123-021-00599-9
29. Wang, K., Wei, Y.M., Zhang, X.: A comparative analysis of China's regional energy and emission performance: which is the better way to deal with undesirable outputs? Energy Policy **46**, 574–584 (2012). https://doi.org/10.1016/j.enpol.2012.04.038
30. Chen, L., Jia, G.: Environmental efficiency analysis of China's regional industry: a data envelopment analysis (DEA) based approach. J. Clean. Prod. **142**, 846–853 (2017). https://doi.org/10.1016/j.jclepro.2016.01.045
31. Song, M., An, Q., Zhang, W., Wang, Z., Wu, J.: Environmental efficiency evaluation based on data envelopment analysis: a review. Renew. Sustain. Energy Rev. **16**(7), 4465–4469 (2012). https://doi.org/10.1016/j.rser.2012.04.052
32. Xu, D.: Productivity growth in the presence of environmental regulations in chinless industry. In: Data envelopment analysis and performance management, Proceedings of the 4th International Conference of DEA (DEA2004), Birmingham, UK: 273, ISBN: 0 902683 73 X (2004)
33. Färe, R., Grosskopf, S., Tyteca, D.: An activity analysis model of the environmental performance of firms—application to fossil-fuel-fired electric utilities. Ecol. Econ. **18**(2), 161–175 (1996). https://doi.org/10.1016/0921-8009(96)00019-5
34. Färe, R., Grosskopf, S., Noh, D.W., Weber, W.: Characteristics of a polluting technology: theory and practice. J. Econ. **126**(2), 469–492 (2005). https://doi.org/10.1016/j.jeconom.2004.05.010

35. Färe, R., Grosskopf, S., Pasurka, C.A.: Environmental production functions and environmental directional distance functions. Energy **32**(7), 1055–1066 (2007). https://doi.org/10.1016/j.energy.2006.09.005
36. Färe, R., Grosskopf, S.: Modeling undesirable factors in efficiency evaluation: comment. Eur. J. Oper. Res. **157**(1), 242–245 (2004). https://doi.org/10.1016/S0377-2217(03)00191-7
37. Scheel, H.: Undesirable outputs in efficiency valuations. Eur. J. Oper. Res. **132**(2), 400–410 (2001). https://doi.org/10.1016/S0377-2217(00)00160-0
38. Zhou, H., Yang, Y., Chen, Y., Zhu, J.: Data envelopment analysis application in sustainability: the origins, development and future directions. Eur. J. Oper. Res. **264**(1), 1–16 (2018). https://doi.org/10.1016/J.EJOR.2017.06.023
39. Toloo, M., Hančlová, J.: Multi-valued measures in DEA in the presence of undesirable outputs. Omega (United Kingdom), 94 (2020). https://doi.org/10.1016/j.omega.2019.01.010
40. Zhang, L., Chen, Y.: Equivalent solutions to additive two-stage network data envelopment analysis. Euro. J. Oper. Res. **264**(3), 1189–1191 (2018). https://doi.org/10.1016/j.ejor.2017.07.042
41. Emrouznejad, A., Marra, M., Yang, G., Michali, M.: Eco-efficiency considering NetZero and data envelopment analysis: a critical literature review. IMA J. Manag. Math. **34**(4), 599–632 (2023). https://doi.org/10.1093/imaman/dpad002
42. Kao, C., Hwang, S.-N.: Efficiency decomposition in two-stage data envelopment analysis: an application to non-life insurance companies in Taiwan. Eur. J. Oper. Res. **185**(1), 418–429 (2008). https://doi.org/10.1016/J.EJOR.2006.11.041
43. Zhang, L., Chen, Y.: Equivalent solutions to additive two-stage network data envelopment analysis. Eur. J. Oper. Res. **264**(3), 1189–1191 (2018). https://doi.org/10.1016/j.ejor.2017.07.042
44. Shen, Z., Boussemart, J.P., Leleu, H.: Aggregate green productivity growth in OECD's countries. Int. J. Prod. Econ. **189**, 30–39 (2017). https://doi.org/10.1016/j.ijpe.2017.04.007
45. Grinsven, H.J.M.V., Eerdt, M.M.V., Westhoek, H., Kruitwagen, S.: Benchmarking eco-efficiency and footprints of dutch agriculture in European context and implications for policies for climate and environment. Front. Sustain. Food Syst. **3**(March), 1–17 (2019). https://doi.org/10.3389/fsufs.2019.00013
46. Haygarth, P.M., Chapman, P.J., Jarvis, S.C., Smith, R.Y.: Phosphorus budgets for two contrasting grassland farming systems in the UK. Soil Use Manage. 160–167 (1998)
47. Kortelainen, M., Kuosmanen, T.: Eco-efficiency analysis of consumer durables using absolute shadow prices. J. Product. Analy. **28**(1–2), 57–69 (2007). https://doi.org/10.1007/s11123-007-0046-6
48. Omrani, H., Alizadeh, A., Naghizadeh, F.: Incorporating decision makers' preferences into DEA and common weight DEA models based on the best–worst method (BWM). Soft. Comput. **24**(6), 3989–4002 (2020). https://doi.org/10.1007/s00500-019-04168-z
49. Mózner, Z., Tabi, A., Csutora, M.: Modifying the yield factor based on more efficient use of fertilizer - The environmental impacts of intensive and extensive agricultural practices. Ecolog. Indicat. **16**(x), 58–66 (2012). https://doi.org/10.1016/j.ecolind.2011.06.034
50. Foster, C., Audsley, E., Williams, A.G., Webster, S., Dewick, P., Green, K.: The environmental, social, and economic impacts associated with liquid milk consumption in the UK and its production. Environment (December) (2007)
51. European Commission.: Fertilisers in the EU prices, trade, and use. p. 5, (2019)
52. Frankowska, A., Jeswani, H.K., Azapagic, A.: Environmental impacts of vegetables consumption in the UK. Sci. Total. Environ. **682**, 80–105 (2019). https://doi.org/10.1016/j.scitotenv.2019.04.424
53. Haverkort, A.J., de Ruijter, F.J., van Evert, F.K., Conijn, J.G., Rutgers, B.: Worldwide sustainability hotspots in potato cultivation. 1. identification and mapping. Potato Res. **56**(4), 343–353 (2013). https://doi.org/10.1007/s11540-013-9247-8

54. Popescu, A., Stoian, E., Erban, V.S.: Eu-28 - trends and correlations, 2008-2018. Sci. Papers Seri. Manage. Econom. Eng. Agricul. Rural Dev. **19**(4), 265–272 (2019)
55. Brentrup, F.: Nitrogen use efficiency as an agro- environmental indicator nitrogen use efficiency as an agro- environmental indicator (2017)
56. Hutchings, N.J., Sørensen, P., Cláudia, M., Cordovil, S., Leip, A., Amon, B.: Measures to increase the nitrogen use efficiency of European agricultural production. Glob. Food Sec. **26**(July), 100381 (2020). https://doi.org/10.1016/j.gfs.2020.100381
57. Tiwari, J.K., Buckseth, T., Singh, R.K., Kumar, M.: Prospects of improving nitrogen use efficiency in potato: lessons from transgenics to genome editing strategies in plants. Front. Plant Sci. 11 (December), 1–6, (2020). https://doi.org/10.3389/fpls.2020.597481
58. Food and Agriculture Organization of the United Nations. The State of Food and Agriculture (2001)

Total Factor Productivity Growth of Rice Production in India: An Analysis Based on Data Envelopment Analysis and Panel Structural Break Test

Dipyaman Pal[1](✉), Chandrima Chakraborty[2], and Arpita Ghose[3]

[1] Department of Economics, Bethune College, Kolkata, West Bengal, India
dipyaman.pal@gmail.com

[2] Department of Economics, Vidyasagar University, Midnapore, West Bengal, India

[3] Department of Economics, Jadavpur University, Kolkata, India

Abstract. The objectives of this paper are i) to estimate Total Factor Productivity Growth (TFPG) of Indian rice production using Biennial Malmquist Index(BMI) of non-parametric method of Data Envelopment Analysis for 1970–71 to 2020–21, considering 11 major rice producing Indian states, ii) to decompose estimated TFPG into its different components: technical changes, efficiency changes and scale efficiency changes to find prime driving sources of TFPG and iii) to find multiple structural breaks in the TFPG series by employing panel structural break test of Ditzen et al. (2021) to check whether there exists any effect of policy change on TFPG or not. The results suggest that for Indian rice production: (i) average annual rate of change of TFPG is 1.84% for the sample period, (ii) decomposition of BMI supports that scale change is the most important factor of TFPG movement. So, there is a scope of improvement in the technical efficiency to push the states to higher growth path, (iii) existence of five structural breaks in TFPG series for the years 1978, 1986, 1994, 2001 and 2008. From the results of the panel structural break analysis it is very much prominent that the policy of green revolution and new agricultural policy 2000 has significant positive effect on TFPG of rice production. On the other hand, liberalization policy introduced in early 1990's has negative effect on the TFPG of rice production.

1 Introduction

The Total factor productivity growth (TFPG) in agriculture is one of the important components in agricultural growth because TFPG is not only the necessary condition but is also the sufficient condition for the advancement of the sector as well as the economy. In general, the term "productivity" is often misrepresented: it is used as synonyms to "labour productivity" in case of manufacturing sector, while used as synonyms to "yield productivity" in case of Agriculture. But, the consideration of yield alone as a measure of productivity provides misleading indication of the degree of productivity improvement in agriculture (Coelli, 1996).

In the existing literature mainly there exists two concepts of productivity, firstly, partial productivity which is defined as the contribution of one factor/ input (say labour

A. Emrouznejad et al. (Eds.): DEA 2024, LNOR, pp. 207–222, 2025.
https://doi.org/10.1007/978-3-031-98177-7_15

or capital) to output growth assuming that other factors are constant. Secondly, Total Factor Productivity growth (TFPG) is a variable, which explains the growth of output which is not caused by inputs but due to shift in output caused by the shift in the production function over time, holding all inputs constant (Abramovitz, 1956, Denison, 1962, 1967, 1985, Hayami et al., 1979). It has been widely recognized in the economic literature that economic growth no matter how remarkable is will not be justifiable without enhancement in total factor productivity growth. Growth in total factor productivity in agriculture is both a necessary condition as well as sufficient condition for its development. It is a necessary condition because it enables agriculture to avoid the trap of Ricardo's law of diminishing returns. It is sufficient condition because it increases production at a reduced unit cost in real terms (For example: Kahlon and Tyagi, 1983, Sidhu and Byarlee, 1992, Acharya, 1998 etc.). TFP growth can be measured by **(i)** Growth Accounting Approach (GAA) [i.e. by constructing either Solow Index (Solow, 1957), or Kendrik Index (Kendrick, 1956, 1961, 1973) or Translog-Divisia Index (Solow, 1957, Jorgenson and Griliches, 1967 etc.]. In the GAA, TFP is measured as the residual. **(ii)** Econometric (Parametric) Approach (i.e. by estimating production function or cost function). The example of parametric method can be found in Kumbhakar et al. (1999, 2000), Kumbhakar and Lovell (2000), among others. **(iii)** Non-parametric Approach (i.e. through Data Envelopment Analysis (DEA)). The most widely used measure of TFP growth is followed by constructing Malmquist Productivity Index (MPI). MPI scales output levels up or down radically with respect to the benchmark technology. Caves, Christensen and Diewert (1982) showed that for translog production function, Tornqvist output and indexes are equal to mean of two MPIs. They introduce MPI as ratio of output distance functions without any aggregation of inputs. Commonly the measurement of TFPG using MPI is done through Data Envelopment Analysis (DEA). DEA is a 'data-oriented' approach for evaluating the performance of multiple decision making units (DMUs). In DEA, TFPG is estimated by solving a mathematical programming problem given some assumptions on the production technology. Departing from existing literature sequential models was introduced by Grifell-Tatj´e & Lovell, 1999 and global frontier approach in the Malmquist literature was familiarized by Pastor & Lovell, 2005. After that, numerous developments of the Malmquist index have been suggested such as a common-weights to Malmquist index by Kao, 2010. On the other hand, Arabi et al., 2015 introduced a slacks-based measures. If we consider the decomposition of the Malmquist index many studies introduced the new concepts such as Arocena et al., 2020; Kao, 2017; Walheer, 2018 and many more. In case of two stage Malmquist index Kao & Hwang, 2014; Yu & Nguyen, 2023 introduced more sophisticated techniques. A detail literature is also available in Mergoni et al., 2024.

2 Literature Review

In recent times policy planners and research scholars are mostly concerned with the higher growth in agriculture. (Chand et al., 2007, Bhalla and Singh, 2009, Reddy and Mishra, 2009, Vaidyanathan, 2010). In fact, during 1980s at the national level, the growth performance of agriculture was splendid and due to reduction in and/or stagnation of public expenditure on agricultural infrastructure, defunct extension services and biased economic reforms it decelerated during 1990s. (Hirashima, 2000, Mahendradev, 2000, Vyas, 2001).

In the period after 1990's, there is a significant departure from the past in 1991 when Government introduced process of economic reforms, which involved deregulation, reduced government participation in economic activities, and liberalization. Though much of the reforms were not initiated to directly affect agricultural sector, the sector was affected indirectly by devaluation of exchange rate, liberalisation of external trade and disprotection to industry. Then came new international trade accord and WTO, requiring opening up of domestic market. Although initially there was positive impact of trade liberalisation on Indian agriculture but afterwards it became real threat for several commodities produced in the country. All these changes raised new challenges and provided new opportunities that required appropriate policy response. Besides, last two decades had only experienced limited price intervention, and there was a sort of policy vacuum. Because of this, there was a strong pressure on the government to come out with agriculture policy as required to provide new direction to agriculture in the new and emerging scenario. In response to this, government of India announced New Agricultural Policy in July 2000.

Since last six decades, productivity growth in agriculture has been one of the focused areas of intense research. Both development economists and agricultural economists have tried to estimate the TFPG and also tried to find out the sources of productivity growth over time. Productivity growth in the agricultural sector is considered essential if agricultural sector output is to grow at a sufficiently rapid rate to meet the demands for food and raw materials arising out of steady population growth. There are not many studies on TFPG in agricultural sector in respect of developing countries. Most of these studies found a declining TFPG in the developing countries which may be unexpected and paradoxical results. Some of the studies relating to the estimation of agricultural TFPG in developing countries are as follows: Trueblood and Coggins (2003), Alauddin, Headey and Rao (2005), Coelli and Rao (2005), Restuccia et al. (2008), Belloumi and Matoussi (2009) and many others. However, resent studies like Shahabinejad and Akbari (2010) found that during 1993–2007 total factor productivity has experienced a positive evolution in D-8 countries. Apart from these there are some studies which deals with DEA application in agriculture both in national level and international level such as Dutta (2012), Hasanov and Nomman (2011), Suresh (2015), You and Zhang (2016), Mathur and Ramnath (2018), Yang, L. and N. Zhu (2018) e Souza et al. (2020), Wysoki´nski et al. (2020), Nandy and Singh (2021), Wan and Zhou (2021), Chakraborty and Pal (2021), Pan et al. (2022), Pal et al. (2022), Shah, W.U., Zhu, N. (2024) etc.

In case of India the measurement of TFPG in agriculture involves a number of studieslike Kumar and Rosegrant (1994), Fan, Hazell and Thorat (1998), Murgai (1999), Forstner et al. (2002), Nin et al (2003), Bhushan (2005), Kumar and Mittal (2006), Bosworth and Collins (2008), Chand, Kumar and Kumar (2011) etc.

It is well recognized in empirical research that structural change is a very important analysis in case of both time series and panel data. If the time span is very long then due to major changes (in policies, events etc.) there is a high possibility to change in the likelihood and the model parameters, such as the 2007–2008 financial crisis and the 2020 COVID–19 outbreaks. So detection of the existence of breaks are therefore necessary not only for estimation purposes but also for understanding drivers of change and their effect on relationships. In time series or panel time series structural breaks (or change points) in the associations between key variables can occur. Thus the estimations and forecasts depend on knowledge about structural breaks is very essential. Structural breaks might influence interpretations and policy recommendations. Break can be unknown or known and single and multiple breaks can occur. For examples: oil price shock, COVID19 etc. So the major questions for the researchers that can we estimate the time points when the breaks occur and test the break points? **The perusal of the literature suggests that there is dearth in such studies.**

Given this background, the **objectives** of the present paper are **first of all** to find out the TFPG of Indian agricultural sector by using Biennial Malmquist Index (BMI) of non-parametric method of Data Envelopment Analysis (DEA). This study tried to find out TFPG for rice of the Indian agricultural sector for the year 1970–71 to 2020–21. Under rice 11 major producing states are considered as the multiple decision making units (DMUs). After finding out the extent of TFPG, the **second objective** is to decompose TFPG into its different components: technical changes, efficiency changes and scale efficiency changes to check which component dominates over the other while finding out the major sources of TFPG. **Thirdly**, to find out whether there exists any effect of policy changed by the Government of India or any external shock like oil price shock, COVID19 etc. in the TFPG series of rice production in Indian Agriculture by employing panel structural break analysis introduced by Ditzen et al. in 2021.

Rest of the paper is as follows:

Section 2 discusses the methodology and data source. In **Subsect. 2.1** the methodology for measuring Biennial Malmquist Index (BMI) of non-parametric method of Data Envelopment Analysis (DEA) and methodology of panel structural break analysis introduced by Ditzen et al. (2021) are discussed. Data Sources are discussed in **2.2. Section 2.3** presents the results of the analysis and some concluding remarks are made in **Sect. 2.4.**

3 Methodology and Data Source

3.1 Methodology of Measuring Biennial Malmquist Index (BMI) and Panel Structural Break Analysis Introduced by Ditzen et al. In 2021

3.2 Methodology of Measuring Biennial Malmquist Index (BMI)

This study considers the non-parametric method of Data Envelopment Analysis (DEA) introduced by Charnes, Cooper and Rhodes (1978) (CCR) and further generalized for variable returns to scale technology by Banker, Charnes and Cooper (1984) (BCC) in order to measure and decompose the Malmquist index of total factor productivity. The major advantage of using DEA is that, unlike in the parametric approach, there is no need to specify any explicit functional form for the production function (e.g., Cobb-Douglas or Translog) and mathematical programming techniques can be used to get point-wise estimates of the production function.

Departing from the common practice in the literature, some of the studies (like Forstner et al. (2002); Nin et al (2003); Bhushan (2005) among others) used the

Malmquist Index of total factor productivity growth i.e. non-parametric method of Data Envelopment Analysis (DEA). The Malmquist Productivity Index, introduced by Caves, Christensen, and Diewert (1982) and operationalized by Färe, Grosskopf, Lindgren and Roos (1992) (FGLR) to measure productivity change, is a normative measure based on a reference technology underlying observed input output data. Färe et al. (1992) (FGLR) decomposed the Malmquist Productivity Index (MPI) into technical change (TC) and technical efficiency change (TEC) considering the constant return to scale (CRS) frontier as the benchmark. However, assumption of global constant return to scale is not always a meaningful assumption about the underlying technology, so the FGLR decomposition is not particularly meaningful when CRS does not hold. In their paper Färe, Grosskopf, Norris, and Zhang (1994) (FGNZ) re-modified and extended decomposition by considering variable returns to scale (Emrouznejad, et al, 2025) and isolate specific contributions of technical efficiency change (TEC), technical change (TC), and scale efficiency change (SEC) towards the overall productivity change. According to Ray and Desli (1997), this decomposition raises a problem of internal consistency because it uses CRS and variable returns to scale (VRS) within the same decomposition. They provide a modified decomposition by using the variable returns to scale frontier as a benchmark. In that decomposition, scale efficiency change is obtained by considering both the constant returns to scale technology and the variable returns to scale technology. However, when one estimate cross-period efficiency scores (which is measured by comparing actual output of a firm in period t with the maximum producible output from period $t + 1$ input set) under a VRS technology, it may result in linear programming infeasibilities for some observations.

In 2011, Pastor, Asmild and Lovell provides a new Malmquist Index which is known as the Biennial Malmquist Index (BMI) which used the same decomposition as provided by Ray and Desli but it solved the infeasibility problem associated with the Ray-Desli decomposition of the Malmquist Index. Instead of using a contemporaneous production possibility frontier, they estimated the technical efficiency of a production unit with reference to a biennial production possibility frontier.

3.3 Methodology of Panel Structural Break Analysis Introduced by Ditzen et al (2021)

In empirical research presence of structural break and handling structural breaks is an essential step. This is particularly true in panel data because its comprised of many cross-sectional units, which are all affected by major events. The previous studies related to the testing of structural break analysis in case of the time series data only. The test of structural break analysis in case of panel data is also very important. The following methods are deals with the test and estimation of structural break analysis in case of panel data.

Wachter and Tzavalis (2012) developed a break recognition testing procedure for the linear panel data model with exogenous or pre-determined regressors in case of AR(p) series. They argued that a structural break in the slope parameters as well as in the fixed effects may be included in their method. They also suggested that breaks in the fixed effects are not forced by any type of cross-sectional similarity thus they are suggesting that breaks are correlated with all past data. They found that Monte Carlo simulations designate that the test performs adequately even in the type of panel datasets with short time-dimension often met in practice.

Hidalgo and Schafgans (2017) delivered a new Central Limit Theorem for estimators of the slope papers in case of a large dynamic panel data models, i.e. both n and T increase without bound, in the presence of strong cross-sectional dependence. They have provided two tests for breaks/homogeneity in the time dimension. The 1st test is established on the CUSUM principle whereas the second test is based on a Hausman-Durbin-Wu approach. They argued that the key features of their tests are that this test has nontrivial power, i.e. when the number of individuals is a negligible fraction or when the break happens to be towards the end of the sample, this test does not suffer from the subsidiary parameter problem. They also provided a simple bootstrap algorithm to obtain (asymptotic) valid critical values for statistics. One of the important properties of the bootstrap is that there is no need to know the underlying model of the cross-sectional dependence. A Monte-Carlo simulation analysis is used for the small sample behaviour of the tests and their bootstrap analogues.

Ditzen et al. (2021) introduced a new method for analyzing multiple structural break in time series and panel data analysis. They argued that by their method one can detect the numbers of breaks, location and provide break date confidence intervals. For these purpose they introduced a new command in STATA software called xtbreak, by which researchers can get a complete toolbox for analysing multiple structural breaks in time series and panel data. Xtbreak also detect and date an unknown number of breaks at unknown break dates. Their test is based on asymptotically valid tests for the presence of breaks, a consistent break date estimator, and a break date confidence interval with correct asymptotic coverage. They argued that their method covers three types of tests; (i) a test of no structural change against the alternative of a specific number of changes, (ii) a test the null hypothesis of no structural change against the alternative of an unknown number of structural changes, and (iii) a test of the null of s changes against the alternative of s + 1 changes. They established that this method also contains an algorithm that provides the last test successively in order to estimate the true number of breaks. They again argued that the tested break dates can be unknown or user-defined so that if one has additional information he can also examine whether there was a break in a specific point in time. After confirming that there exists some break point in the series this method estimates the locations of the breaks and delivers the related confidence intervals. They have also argued that xtbreak can deal with models of "pure" or "partial" structural change. A pure structural change model is one in which the coefficients of all explanatory variables change on the other hand in case of a partial structural change model only a subset of the coefficients changes. Xtbreak is applicable used under very general conditions. In case of panel data, xtbreak can be used even if the model errors are dependent and heteroskedastic across both time and cross-sectional units. Xtbreak can permit the endogenity in the regressors to load on the same set of factors as the errors.

In their method there are three hypotheses. Which are as follows:

(A) H_0: no breaks versus H_1: s breaks, where the number of breaks under H_1, s, is specified by the researcher.
(B) H_0: no breaks versus H_1: $1 \leq s \leq s_{max}$ breaks, where the maximum number of breaks under H_1, s_{max}, is specified by the researcher.

(C) H_0: s breaks versus H_1: s + 1 breaks, where s is specified by the researcher.

To test hypotheses (A)–(C) they employed a number of test statistics. The time series versions of these tests have appeared in Bai and Perron (1998), while the panel versions have appeared in Ditzen et al. (2021).

Karavias et al. (2022) developed a new break detection toolbox that is applicable to different sized panels. They argued that their method is easy to implement and robust to general forms of unobserved heterogeneity. Their method is almost as similar as in case of the previous methods. In the first step this method includes a structural change test. After that its employed a break date estimator and in the last step they can also find out a break date confidence interval.

4 The Data Sources

In this paper we have considered five inputs and one output data for estimating the productivity growth. The input and outputs are as follows:

Output: Production of rice; **Inputs:** (i)Seed (Kg.); (ii) Fertilizer (Kg. Nutrients); (iii)Manure (Qtl.); (iv) AREA ('000 Hectares), (v)Human Labour (Man Hrs.)[1]

All the data has been collected from the different issues of the Statistical abstract, Agriculture at a Glance, Agriculture in Brief, Handbook of Statistics on Indian Economy, www.indianstat.com (an online commercial data service), Cost of Cultivation data published by the Government of India from 1970–71 to 2020–21. The major producing states of rice are Andhra Pradesh, Assam, Bihar, Haryana, Karnataka, Madhya Pradesh, Orissa, Punjab, Tamil Nadu, Uttar Pradesh, West Bengal.

5 Results of Analysis

5.1 Results of Average Annual Rate of Changes of TFPG

The TFPG for each of the years and each of the states in case of rice are estimated. The results are then summarized to generate the information regarding the average annual rate of changes of TFPG for rice. Such estimation results for Rice are presented in Table 1.

From Table 1 in case of **Rice** it can be concluded that overall average annual rate of change of TFPG in case of rice is 1.84% for the period 1970–71 to 2020–21. Now if one considers the state wise results the average annual rate of change of TFPG is negative only in case of AS (-0.13%) for the period 1970–71 to 2020–21. This rate of changes of TFPG is highest in case of HA (9.85%) followed by UP (9.27%), MP (6.4%), KA(6.22%), OR (4.93%), AP (3.02%), PU (1.96%), WB(1.71%) and BI (1.34%). The decadal average annual rate of change of TFPG are also estimated. For decadal analysis the overall period 1970–71 to 2020–21 is broken down into four sub periods 1970–79, 1980–89, 1990–99 and 2000–2021. The result of decadal average annual rate of change of TFPG implies that this change is highest for the period 1970–79 (9.24%). This change

[1] Animal labour is not included as the determinant because most of the entries under animal labour are zero or unavailable.

may be due to the successful implementation of green revolution policies in that decade. The overall average annual rate of change of TFPG declined from 9.24% in 1970–79 to 4.72% in 1980–89 to 1.97% in 1990–99. In the period 2000–2021 the average annual rate of change of TFPG is 1.14%.

Table 1. Average Annual Growth Rate of TFPF of Rice Production.

	ALL YEAR	1970–79	1980–89	1990–99	2000–2021
AP	3.02	7.07	1.41	2.92	0.67
AS	–0.13	1.09	0.00	–0.86	–0.75
BI	1.34	1.83	1.05	1.24	1.26
HA	9.85	18.18	9.50	10.63	1.06
KA	6.22	15.07	5.23	2.07	2.49
MP	6.40	14.74	8.20	–0.41	3.09
OR	4.93	7.82	5.55	4.66	1.69
PU	1.96	3.47	1.79	1.57	1.03
TN	2.41	6.41	1.65	0.09	1.46
UP	9.27	22.23	14.22	0.21	0.42
WB	1.71	3.78	3.32	–0.43	0.15
OVER ALL	4.25	9.24	4.72	1.97	1.14

Source: Authors Own Calculations.

5.2 Results of Decomposition of TFPG

The estimated results of TFPG are then decomposed into Efficiency Changes, Scale Efficiency Changes and Technical Changes. The overall changes in the decomposition of TFPG over the period 1970–71 to 2020–21 are estimated.

In case of **Rice** overall and the decadal changes are presented in Table 2. Entries in column TEC show average annual changes in the level of technical efficiency over time for each state, a value greater than unity for this component implies that a state experienced improvement in technical efficiency over the period. Similarly, an entry with value greater (less) than unity in column TC reflects technological progress (regress) in a state over time. The change in scale efficiency over time for each state is reported in column SEC, with a value exceeding one again signaling an improvement in scale efficiency. From the results of Table 2 it can be concluded that productivity growth is mostly driven by the change in the technical efficiency for the entire sample period. The change in the scale efficiency is another major factor behind the increase in the productivity growth. The change in the technology has the lowest impact on the increase in the productivity. So it can be concluded that better utilization of factors of production and changes in the scale may pushed the sates to be on higher TFPG in case of rice for the period 1970 to 2021.

Table 2. Results of Decomposition of TFPG of Rice for the period 1970–2021

STATES	TEC	TC	SEC	MI
AP	1.003699	1.000822	1.003929	1.010126
AS	1	1	1.002755	1.002755
BI	1.013071	1.000313	0.99645	1.012403
HA	1.00304	0.99901	1.023303	1.023303
KA	1.015835	0.997989	1.013711	1.016714
MP	1.022607	1.002903	1.033512	1.040161
OR	1.043414	0.995584	1.014633	1.04531
PU	1	1	1.002303	1.002303
TN	1.008285	1.000185	1.00659	1.008523
UP	1.000471	0.999982	1.000705	1.00143
WB	1.000365	0.998302	0.998678	0.998128
AVERAGE	1.010072	0.999554	1.008779	1.014651

Source: Authors own Calculations.

5.3 Results of Panel Structural Breaks

In case of **Rice** the results of panel structural breaks are obtainable in Table 3. From the results one can conclude that there exist five structural breaks or structural changes in the TFPG series of rice. The years of structural breaks are 1978, 1986, 1994, 2001 and 2008. Thus from this results there exists six distinct periods in the TFPG of rice which are as follows:

Six distinct phases of growth were identified:

Phase I: Early-green revolution period (1970–77); Phase II: Later Phase of green revolution (1979–85); Phase III: Period of wider dissemination (1986 to 1994); Phase IV: Post-Reform Period (1995 to 2000); Phase V: Period of Recovery (2001–2008); Phase VI: Period after recovery (2009–201).

The change of TFPG for each break periods are also estimated. The results are presented in Table 4. For this analysis the overall period 1970–71 to 2020–21 is broken down into six sub periods 1970–77, 1978–83, 1994–93, 1994–2000, 2001–2007 and 2008–2021. The result of change of TFPG implies that this change is highest for the period 2001–2007 (3.1%). This change may be due to the successful implementation of new agricultural policy in 2000. TFPG change is negative in the early green revolution period however the change of TFPG has significantly increased in the latter phase of green revolution (from negative to 2.6%). Further change of TFPG declined from 2.6% in 1978–83 to 2.08% in 1984–93. In the next phase the change of TFPG is negative which indicates that the effect of liberalization on TFPG of rice is negative. In the period 2008–2021 the change of TFPG is 1.6% which is significantly lower than the previous phase.

Thus from the results of the panel structural break analysis it can be concluded that the policy of green revolution and new agricultural policy 2000 has significant positive effect on TFPG of rice production. On the other hand, liberalization policy adopted by the Government of India in early 1990's has negative effect on the TFPG of rice production.

Table 3. Results of Panel Structural Break analysis

Sequential test for multiple breaks at unknown breakpoints (Ditzen, Karavias & Westerlund 2021)				
	Test Statistic	1% Critical Value	5% Critical Value	10% Critical Value
F(1/0)	5.39	12.29	8.58	7.04
F(2/1)	22.10	13.89	10.13	8.51
F(3/2)	26.50	14.80	11.14	9.41
F(4/3)	23.98	15.28	11.83	10.04
F(5/4)	23.40	15.76	12.25	10.58
Detected number of breaks:		5	5	5
The detected number of breaks indicates the highest number of breaks for which the null hypothesis is rejected				
Estimation of break points				
#	Index	Date		
1	8	1978		
2	16	1986		
3	24	1994		
4	31	2001		
5	38	2008		

Source: Authors own Calculations.

Table 4. Results of Decomposition of TFPG of Rice for each Break Periods

	Phase-I (1971–77)				Phase-II(1978–1985)				Phase-III(1986–1993)			
	TEC	TC	SEC	MI	TEC	TC	SEC	MI	TEC	TC	SEC	MI
AP	1.0041	0.9970	1.0176	1.0178	1.0005	0.9981	0.9837	0.9829	1.0016	1.0019	1.0107	1.0142
AS	1.0000	1.0000	1.0164	1.0164	1.0000	1.0000	1.0000	1.0000	1.0000	1.0000	0.9956	0.9956
BI	1.0000	1.0000	0.9544	0.9544	0.9577	0.9921	1.0745	1.0381	1.0868	1.0090	1.0138	1.1133
HA	1.0273	1.0064	1.0026	1.0026	1.0038	0.9963	1.0551	1.0551	1.0000	1.0000	0.9607	0.9607
KA	1.0172	1.0117	1.0245	1.0312	0.9767	1.0537	0.9738	0.9738	1.0610	0.9541	1.0110	1.0110

(continued)

Table 4. *(continued)*

	Phase-I (1971–77)				Phase-II(1978–1985)				Phase-III(1986–1993)			
	TEC	TC	SEC	MI	TEC	TC	SEC	MI	TEC	TC	SEC	MI
MP	0.9866	1.0331	0.9610	0.9610	1.0200	0.9936	1.1232	1.1232	1.0462	1.0085	1.0246	1.0257
OR	1.0241	0.9971	0.9919	1.0295	1.0111	1.0103	1.0000	1.0143	1.1159	0.9981	0.9852	1.0856
PU	1.0000	1.0000	0.9957	0.9957	1.0000	1.0000	1.0066	1.0066	1.0000	1.0000	1.0055	1.0055
TN	1.0088	1.0001	1.0035	1.0231	1.0044	1.0002	0.9987	1.0025	1.0000	1.0000	0.9864	0.9864
UP	1.0029	1.0005	0.9556	0.9624	1.0000	1.0000	1.0773	1.0773	1.0000	1.0000	1.0138	1.0138
WB	1.0000	1.0000	0.9800	0.9800	1.0018	1.0006	1.0107	1.0168	1.0000	1.0004	1.0165	1.0180
AVERAGE	1.0065	1.0042	0.9912	0.9977	0.9978	1.0041	1.0276	1.0264	1.0283	0.9975	1.0022	1.0209

Source: Authors own Calculations.

Table 5. Results of Decomposition of TFPG of Rice for each Break Periods (Cond.)

	Phase-IV(1994–2000)				Phase-V(2001–2007)				Phase-VI(2008–2021)			
	TEC	TC	SEC	MI	TEC	TC	SEC	MI	TEC	TC	SEC	MI
AP	0.9796	1.0006	0.9965	0.9805	1.0188	0.9891	1.0146	1.0241	1.0115	1.0106	1.0021	1.0281
AS	1.0000	1.0000	1.0053	1.0053	1.0000	1.0000	1.0036	1.0036	1.0000	1.0000	0.9999	0.9999
BI	1.0000	1.0000	1.0182	1.0182	1.0000	1.0000	0.9996	0.9996	1.0239	1.0005	0.9502	0.9739
HA	1.0000	1.0000	1.0879	1.0879	1.0000	1.0000	1.0276	1.0276	0.9928	0.9945	1.0179	1.0179
KA	0.9568	1.0000	1.0334	0.9773	1.0178	1.0001	1.0175	1.0349	1.0420	0.9801	1.0207	1.0491
MP	1.0421	1.0000	0.9492	0.9667	1.0000	1.0000	1.0492	1.0492	1.0335	0.9897	1.0654	1.0814
OR	0.8919	1.0000	1.0943	0.9462	1.1369	1.0000	1.0107	1.1532	1.0618	0.9793	1.0150	1.0445
PU	1.0000	1.0000	0.9930	0.9930	1.0000	1.0000	1.0077	1.0077	1.0000	1.0000	1.0039	1.0039
TN	1.0010	1.0002	0.9705	0.9683	1.0037	1.0000	1.0611	1.0450	1.0218	1.0005	1.0158	1.0188
UP	1.0000	1.0000	1.0010	1.0010	1.0000	1.0000	0.9923	0.9923	1.0001	0.9996	0.9777	0.9764
WB	1.0000	1.0000	0.9975	0.9975	1.0000	1.0000	1.0044	1.0044	1.0003	0.9927	0.9893	0.9826
AVERAGE	0.9883	1.0001	1.0133	0.9947	1.0161	0.9990	1.0171	1.0311	1.0171	0.9952	1.0053	1.0160

Source: Authors own Calculations.

6 Conclusion

The present chapter estimates TFPG of Indian agriculture for Rice production over the period 1970–71 to 2020–21 considering the major producing states of these crops. Biennial Malmquist Index (BMI) of non-parametric method of Data Envelopment Analysis (DEA) is used for estimation. After estimating the TFPG we have employed panel structural break analysis introduced by Ditzen et al. (2021) to find out whether there exists any effect of policy changes by the Government of India or any external shock like oil price shock, COVID19 etc. in the TFPG series of rice production in Indian Agriculture.

The results suggest that in case of rice **first of all,** average annual rate of change of TFPG is 4.25% for the period 1970–71 to 2020–21. Not only that, the result of decadal average annual rate of change of TFPG implies that this change is highest for the period 1970–79 (9.24%). Apart from these results there exists a strong state wise variation in terms of average annual rate of change of TFPG. The overall average annual rate of change of TFPG declined from 9.24% in 1970–79 to 4.27% in 1980–89 to 1.97% in 1990–99. In the period 2000–2021 the average annual rate of change of TFPG is 1.14%.

Secondly, from the results of decomposition of Malmquist Index it can be concluded that scale changes are the most important factor causing productivity changes for rice. Among the two other alternative sources of TFPG, efficiency changes dominate over technical changes in case of rice. So, in case of rice the improvement in the technical efficiency may push the states to a higher growth path.

Thirdly, if one considers the structural changes in the TFPG series of rice, then from the results one can conclude that there exist five structural breaks in the series. The years of structural breaks are 1978, 1986, 1994, 2001 and 2008. From the results of the panel structural break analysis it is very much prominent that the policy of green revolution and new agricultural policy 2000 has significant positive effect on TFPG of rice production. On the other hand, liberalization policy introduced in early 1990's has negative effect on the TFPG of rice production.

References

Abramovitz, M.: Resource and output trends in the united states since 1870. Am. Econo. Rev. May (1956)

Acharya, S.S.: Agricultural price policy and development: some facts and emerging issues. Indian J. Agricul. Econ. **52**(1), 1–47 (1998)

Alauddin, M., Derek, H., Prasada Rao, D.S.: Explaining Agricultural Productivity Levels and Growth: An International Perspective. Working Paper, Working Paper Series No. 02/2005; Centre for Efficiency and Productivity Analysis; School of Economics University of Queensland St. Lucia, Australia (2005)

Arabi, B., Munisamy, S., Emrouznejad, A.: A new slacks-based measure of malmquist-luenberger index in the presence of undesirable outputs. Omega **51**, 29–37 (2015). https://doi.org/10.1016/j.omega.2014.08.006

Arocena, P., Saal, D.S., Urakami, T., Zschille, M.: Measuring and decomposing productivity change in the presence of mergers. Eur. J. Oper. Res.Oper. Res. **282**(1), 319–333 (2020). https://doi.org/10.1016/j.ejor.2019.08.048

Bai, J., Perron, P.: Computation and analysis of multiple structural change models. J. Appl. Economet.Economet. **18**, 1–22 (2003)

Banker, R.D., Charnes, A., Cooper, W.W.: Some models for estimating technical and scale inefficiencies in data envelopment analysis. Manage. Sci. **30**(9), 1078–1092 (1984). https://doi.org/10.1287/mnsc.30.9.1078 Banker, R.D., Charnes, A., Cooper, W.W., Swarts, J., Thomas, D.: An introduction to data envelopment analysis with some of its models and their uses. Res. Govern. Nonprofit Acc. **5**, 125–163 (1989)

Belloumi, M., Matoussi, M.S.: Measuring agricultural productivity growth in MENA countries. J. Dev. Agric. Econ. **1**(4), 103–113 (2009)

Bhalla, G.S., Singh, G.: Economic liberalisation and indian agriculture: a statewise analysis. Econ. Politic. Weekly, December 26, 2009, Vol XIIV, No 52 (2009)

Bhushan, S.: Total factor productivity growth of wheat in india: a malmquist approach. Indian J. Agricul. Econ. **60**(1) Jan-March 2005 (2005)

Bosworth, B., Collins, S.M.: Accounting for growth: comparing China and India. J Econ. Persp. **22**(1),Winter 2008, 45–66 (2008)

Caves, D., Laurits, W., Christensen, R., Diewert, W.E.: The economic theory of index numbers and the measurement of input, output and productivity. Econometrica **50**(6), 1393–1414 (1982)

Chakraborty, C., Pal, D.: Does farm size matters in determining efficiency in indian agriculture: a case study of rice production in eastern India. Int. J. Sustain. Agricul. Manage. Inf. (IJSAMI), Inderscience Publishers, (ISSN online 2054–5827; ISSN print 2054- 5819). **7**(1), February 2021; 4 issues per year (2021). https://www.inderscienceonline.com/doi/abs/https://doi.org/10.1504/IJSAMI.2021.113463

Chand, R., Kumar, P., Kumar, S.: Total factor productivity and contribution of research investment to agricultural growth in India. Policy Paper 25, National Centre For Agricultural Economics And Policy Research (2011)

Chand, R., Raju, S.S., Pandey, L.M.: Growth crisis in agriculture: severity and options at national and state levels. Econ. Pol. WklyWkly **42**(26), 2528–2533 (2007)

Charnes, A., Cooper, W.W., Rhodes, E.: Measuring the efficiency of decision making units. Eur. J. Oper. Res. **2**(6), 429–444 (1978). https://doi.org/10.1016/0377-2217(78)90138-8

Coelli, T.: A Guide to DEAP Version 2.1: A Data Envelopment Analysis (Computer) Program, CEPA. Working Paper 96/08 (1996)

Coelli, T.J., Rao, D.S.P.: Total factor productivity growth in agriculture: a malmquist index analysis of 93 countries, 1980–2000. Agric. Econ. **32**, 115–134 (2005)

Denison, E.F.: The Source of Growth in the United Sates, and Alternative before US. Committee of Economic Development, New York (1962)

Denison, E.F.: Why growth rates differ. post war experience in nine western countries. The Brooking Institution, Washington, D.C.: The Brooking Institution (1967)

Denison, E.F.: Trends in American Economic Growth: 1929–1982. The Brooking Institution, Washington, D.C. (1985)

Ditzen, J., Karavias, Y., Westerlund, J.: Testing for Multiple Structural Breaks in Panel Data (2021a)

Ditzen, J., Karavias, Y., Westerlund, J.: Testing and estimating structural breaks in time series and panel data in stata. arxiv [econ.EM] 2110.14550 (2021b). https://arxiv.org/abs/2110.14550

Ditzen, J.: Estimating long run effects and the exponent of cross-sectional dependence: an update to xtdccc2. Stata J. **21**(3), 687–707 (2021). URL https://ideas.repec.org/p/bzn/wpaper/bemps81.html

Dutta, S.: Assessment of agricultural efficiency and productivity: a study of hugli district, West Bengal, India. Int. J. Curr. Res. **4**, 190–195 (2012)

Emrouznejad, A.,V. Podinovski, C., Lu, V., Charles, A., Moradi-Motlagh: Rajiv banker's lasting impact on data envelopment analysis, annals of operations research. **351**, 1225–1264 (2025). https://doi.org/10.1007/s10479-025-06473-3

E Souza, G.D.S., Gomes, E.G., de Andrade Alves, E.R., Gasques, J.G.: Technological progress in the Brazilian agriculture. Soc. Econ. Plann. Sci. **72**, 100879 (2020)

Fan, S., Hazell, P., Thorat, S.: Government spending, growth and poverty: an analysis of interlink ages in rural India. EPTD Discussion Paper no. 33. International Food Policy Research Institute, Washington DC (1998)

Färe, R., Grosskopf, S., Lindgren, B., Roos, P.: Productivity changes in swedish pharmacies 1980 89: a non parametric malmquist approach. J. Prod. Anal. **3**(1/2), 85–101 (1992)

Färe, R., Grosskopf, S., Norris, M., Zhang, P.: Productivity growth, technical progress and efficiency changes in industrial countries. Am. Econ. Rev. **84**(1), 66–83 (1994)

Forstner, H., Anders, I.: Productivity, technology, and efficiency: An analysis of the world technology frontier when memory is infinite. SIN Working Paper Series, Working Paper no. 7, February 2002, Statistics and Information Networks branch of UNIDO (2002)

Grifell-Tatj ́e, E., Lovell, C.A.K.: Profits and productivity. Manage. Sci. **45**(9), 1177–1193 (1999). https://doi.org/10.1287/mnsc.45.9.1177

Hasanov, S., Nomman, A.: Agricultural efficiency under resources scarcity in Uzbekistan: a data envelopment analysis. Bus. Econ. Horizons (BEH) **4**, 81–87 (2011)
Hayami, Y., Kikuchi, M., Ruttan, V.W., Southworth, H.M. (eds.): Agricultural growth in Japan, Taiwan, Korea and the Philippines. The University Press of Hawaii, Honolulu (1979)
Hayami, Y., Vemon, W.R., Southword, H.M.: Agricultural Growth in Japan. Korea and Philippines, The University Press of Hawaii, Honolulu, Taiwan (1979)
Hidalgo, J., Schafgans, M.: Inference and testing breaks in large dynamic panels with strong cross sectional dependence. J. Econo. Elsevier **196**(2), 259–274 (2017)
Hirashima, S.: Issues in agricultural reforms: public investment and land market development. Econ. Pol. WklyWkly **35**(43 & 44), 3879–3884 (2000)
Jorgenson, D.W., Griliches, Z.: The Explanation of productivity change. Rev. Econ. Stud. **34**(3), 249–283 (1967)
Kahlon, A.S., Tyagi, D.S.: Agricultural Price Policy in India. Allied Publisher, New Delhi (1983)
Kao, C.: Malmquist productivity index based on common-weights DEA: the case of Taiwan forests after reorganization. Omega **38**(6), 484–491 (2010). https://doi.org/10.1016/j.omega.2009.12.005
Kao, C.: Measurement and decomposition of the Malmquist productivity index for parallel production systems. Omega **67**, 54–59 (2017). https://doi.org/10.1016/j.omega.2016.04.001
Kao, C., Hwang, S.N.: Multi-period efficiency and malmquist productivity index in two-stage production systems. Eur. J. Oper. Res.Oper. Res. **232**(3), 512–521 (2014). https://doi.org/10.1016/j.ejor.2013.07.030
Karavias, Y., Narayan, P.K., Westerlund, J.: Structural breaks in interactive effects panels and the stock market reaction to COVID-19. J. Bus. Econ. Stat. **41**(3), 653–666 (2022). https://doi.org/10.1080/07350015.2022.2053690
Kendrick, J.W.: Productivity trends: capital and labor. Rev. Econ. Stat. **38**(3), 248–257 (1956)
Kendrick, J.W.: Productivity Trends in the United States. National Bureau of Economic Research, New York (1961)
Kendrick, J.W.: Postwar Productivity Trends in the United States, 1948–1969. National Bureau of Economic Research. Columbia University Press, New York (1973)
Kumar, P., Mittal, S.: Agricultural productivity trends in india: sustainability issues. Agric. Econ. Res. Rev. **19**, 71–88 (2006)
Kumar, P., Rosegrant, M.W.: Productivity and the sources of growth of rice in India. Econ. Politic. Weekly, December-31, A183-A188 (1994)
Kumbhakar, S.C., Lovell, C.A.K.: Stochastic Frontier Analysis. Cambridge University Press, Cambridge (2000)
Kumbhakar, S.C., Heshmati, A., Hjalmarsson, L.: Parametric approaches to productivity measurement: a comparison among alternative models. Scandinavian J. Econ. **101**(3), 405-424 (1999)
Kumbhakar, S.C., Shinichiro, N., Heshmati, A.: Estimation of firm-specific technological bias, technical change and total factor productivity growth: a dual approach. Economet. Rev.. Rev. **19**(4), 493–515 (2000)
Mahendradev, S.: Economic reforms, poverty, income distribution and employment. Econom. Pol. Wkly **35**(10), 823–835 (2000)
Mathur, R.N., Ramnath, S.R.: Efficiency in food grains production in India using DEA and SFA. Central European Rev. Econom. Manage. (CEREM) **2**, 79–101 (2018)
Mergoni, A., Emrouznejad, A., De Witte, K.: Fifty years of data envelopment analysis. European J. Opera. Res. **326**(3), 389–412 (2025). https://doi.org/10.1016/j.ejor.2024.12.049
Murgai, R.: The green revolution and the productivity paradox: evidences from the Indian Punjab. Technical Report, Development Research Group, The World Bank, Washington D.C., U.S.A (1999)

Nandy, A., Singh, P.K.: Application of fuzzy DEA and machine learning algorithms in efficiency estimation of paddy producers of rural Eastern India. Benchmarking Int. J. **28**(1), 229–248 (2021)
Nin A., A. Channing, P. Paul V. Is agricultural productivity in developing countries really shrinking? new evidence using a modified nonparametric approach. J. Dev. Econom. **71**, 395-415 (2003)
Pal, D., Ghose, A., Chakraborty, C.: Measurement of efficiency loss in Indian Agricultural Sector due to COVID-19: Evidence from Rice Production. In: Rajib, B., Ananya, G.D., Soumyen, S., (Eds.), "The COVID-19 Pandemic, India and the World (2022)
Economic and Social Policy Perspectives. ISBN 9781032114965, by Routledge India. pp. 278–288
Pan, Z., Tang, D., Kong, H., He, J.: An analysis of agricultural production efficiency of yangtze river economic belt based on a three-stage DEA malmquist model. Int. J. Environ. Res. Public Health **19**(2), 958 (2022)
Pastor, J.T., Asmild, M., Lovell, C.: The Biennial malmquist productivity change index. Socioecon. Plann. Sci.. Plann. Sci. **45**, 10–15 (2011)
Pastor, J.T., Lovell, C.A.K.: A global Malmquist productivity index. Econ. Lett. **88**(2), 266–271 (2005). https://doi.org/10.1016/j.econlet.2005.02.013
Pastor, J.T., Lovell, C.A.K., Aparicio, J.: Defining a new graph inefficiency measure for the proportional directional distance function and introducing a new malmquist productivity index. Eur. J. Oper. Res.Oper. Res. **281**(1), 222–230 (2020). https://doi.org/10.1016/j.ejor.2019.08.021
Ray, S.C., Desli, E.: Productivity growth, technical progress, and efficiency change in industrialized countries: comment. Am. Econ. Rev. **87**(5), 1033–1039 (1997)
Reddy, D.N., Srijit, M.: Agriculture in the Reforms Regime. In: Narasimha Reddy, D., Srijit, M., (ed), Agrarian Crisis in India. New Delhi: Oxford University Press (2009)
Restuccia, D., Yang, D.T., Xiaodong, Z.: Agriculture and aggregate productivity: a quantitative cross-country analysis. J. Monet. Econ. **55**(2008), 234–250 (2008)
Shah, W.U., Zhu, N.: Impact of China's agricultural water policy reforms on water usage efficiency and total factor productivity change. In: Advances in the Theory and Applications of Performance Measurement and Management. DEA45 2023. Lecture Notes in Operations Research. Springer, Cham (2024). https://doi.org/10.1007/978-3-031-61597-9_21
Shahabinejad, V., Akbari, A.: Measuring agricultural productivity growth in developing eight. Iranian J. Dev. Agricul. Econom. **2**(9), 326–332 (2010)
Sidhu, D.S., Byerlee, D.: Technical changes and wheat productivity in india punjab in the post green revolution period. CIMMYT Economics Working, 92–02, CIMMYT, Mexico (1992)
Solow, R.M.: Technical change and the aggregate production function. Rev. Econ. Stat. **39**(3), 312–332 (1957)
Suresh, A.: Efficiency of agricultural production in India: an analysis using non-parametric approach. Indian J. Agricul. Econom. **70**, 471–486 (2015)
Trueblood, M.A., Coggins, J.: Inter-country Agricultural Efficiency and Productivity: A Malmquist Index Approach, mimeo. World Bank, Washington DC (2003)
Vaidyanathan, A.: Agriculture Growth in India: Role of Technology, Incentives and Institutions. Oxford University Press, New Delhi (2010)
Vyas, V.S.: Agriculture: second round of economic reforms. Econ. Pol. WklyWkly **36**(14), 829–836 (2001)
Stefan De, W., Elias, T.: Detection of structural breaks in linear dynamic panel data models. Comput. Stat. Data Anal. **56**(11), pp. 3020–3034 (2012). https://doi.org/10.1016/j.csda.2012.02.025. ISSN 0167-9473
Walheer, B.: Disaggregation of the cost Malmquist productivity index with joint and output-specific inputs. Omega **75**, 1–12 (2018). https://doi.org/10.1016/j.omega.2017.01.012
Wan, B., Zhou, E.: Research of total factor productivity and agricultural management based on malmquist-DEA modeling. Math. Problems Engi. 2021 (2021):

Wysokinski, M., Domagala, J., Gromada, A., Golonko, M., Trebska, P.: Economic and energy efficiency of agriculture. Agricul. Econom. Zemedelska Ekonomika **66**(8), 355–364 (2020)

Yang, L., Zhu, N.: On comprehensive competitiveness of enterprises using dea models: a case study of china's agricultural companies. In: Data Envelopment Analysis and Performance Measurement: Recent Developments: Proceedings of the DEA40: International Conference of Data Envelopment Analysis, Aston Business School, UK. ISBN: 978 1 85449 438 2 (2018)

You, H., Zhang, X.: Eco-efficiency of intensive agricultural production and its influencing factors in China: an application of DEA-Tobit analysis. Discrete Dynam. Nat. Soc. 2016 (2016)

Yu, M.M., Nguyen, M.A.T.: Productivity changes of Asia-Pacific airlines: A Malmquist productivity index approach for a two-stage dynamic system. Omega **115**, Article 102774 (2023). https://doi.org/10.1016/j.omega.2022.102774

Technical, Pure Technical and Scale Efficiency of Primary Agriculture Cooperative Credit Societies in Tamil Nadu

Murugesan Raman(✉)

Natesan Institute of Cooperative Management, Chennai 600 040, India
amanmurugesan60@yahoo.co.in

Abstract. This paper analyses input-oriented Data Envelopment Analysis to the measurement of technical efficiency of 110 Primary Agriculture Cooperative Credit Societies (PACS) in Tamil Nadu in India. With specification of different inputs and outputs, this study uses three approaches, namely intermediation, operating and value-added, to find out efficiency levels of PACS (known as DMUs). A number of DMUs are operating in the zone of increasing returns to scale. One-third of the DMUs are operating with the constant returns to scale. DMUs with own resources have high technical efficiency. In the second stage, Tobit model was utilized for understanding the determinants of technical efficiency, pure technical efficiency and scale efficiency of DMUs. Tobit censured regressions were estimated with maximum likelihood techniques. Self-help groups and Kissan Credit Cards hold high growth potentials for PACS in business strategy for expansion contributing to overall technical efficiency. A large portion of technical inefficiency among DMUs could be attributed to pure technical inefficiency emanating from managerial decision making on choice of inputs mix and choice of outputs produced than to the size and scale of operations undertaken by DMUs in rural areas.

Keywords: Data Envelopment Analysis · Cooperative Credit Societies · technical efficiency · Tobit maximum likelihood estimates.

1 Introduction

1.1 Cooperative Credit Movement in India

The cooperative credit movement in India since 1904, when the first Cooperative Credit Societies Act was passed, had registered an impressive growth in terms of members, borrowers, deposits collections, loans and advances to members, distribution of agriculture inputs, procurements of agriculture produces and progressive outreach to serve weaker sections of societies in rural India. Though cooperative credit societies in principle, democratic economic institutions, member-driven and member- controlled, in practice without active governments' involvements and assistance in all possible ways such as in legal, financial, managerial and in infrastructure spheres, the spread of cooperative credits societies and their strong presence in rural areas would have been limited. Realizing

A. Emrouznejad et al. (Eds.): DEA 2024, LNOR, pp. 223–250, 2025.
https://doi.org/10.1007/978-3-031-98177-7_16

the cooperative credit societies as the last mile connectivity with weaker sections of society, governments at the state and at Centre enlisted them for carrying out developmental and welfare schemes to ultimate beneficiaries. Report of the Agricultural Credit Review Committee (ACRC,1989) noted that cooperative movement was made to expand fast as an "instrument of state policy" p.176).

The significance of PACS in rural areas was eloquently discussed by the Task Force on Revival of Rural Cooperative Credit Institutions (RRCCI, 2004) in the following's words;

"A large proportion of PACS also serve as outlets for inputs and for the public distribution system for food and other essential items. Though the network of commercial banks and RRBs has spread rapidly and they now have nearly 50,000 branches, their reach in the countryside both in terms of the number of clients and accessibility to the small and marginal farmers and other poorer segments is far less than that of cooperatives. This is a compelling need to find ways for strengthening the cooperative movement and making it a well-managed and vibrant medium to serve the credit needs of rural India, especially the small and marginal farmers." (RRCCI, 2004).

Further, it went on to say:

".... in a country predominated by small or marginal land holdings, the reach of the cooperative system is much deeper than the other institutional arrangements in the rural areas". (P.25).

The RBI sponsored Expert Committee on the three-tier short-term cooperative credit in India (2012) had come to the following conclusions;

"-----having a rural penetration of over 93,000 PACS as compared to only about 50,000 rural and semi-urban branches of CBs and RRBs, the share of the cooperatives in agricultural credit have fallen to about 17% in 2011–12. (P.10). Although cooperatives are providing only 17% of agriculture credit, the share of cooperatives in total number of agricultural accounts held by the banking system is substantial. Cooperatives provided agricultural credit to 3.09 crore farmers during 2011–12 compared to only 2.55 crore farmers by Commercial banks and 82 lakhs by the RRBs. (P.15).

In 2020–21, there were 102559 PACS with membership of 13.07 crore; borrowers accounted for 5.40 crore. In the total credit flow to the agriculture, the rural cooperatives accounted for 13% while the regional rural banks and commercial banks accounted respectively 11% and 76% (RBI, 2021–22). In the case of number of Kissan Credit Cards (KCC) issued to the farmers, the cooperative banks accounted for 31.13 million followed by commercial banks with 26.87 million and regional rural banks with 13.34 million (RBI, 2021–22 P158–159).

The creation of Union Ministry of Cooperation in 2021 is the major milestone in progress towards achieving and strengthening socio-economic empowerment of rural populations and for cooperative development. To make PACS viable and sustainable in the long-run, the Union Ministry of Cooperation had identified more than 300 e-services for PACS such as in banking, insurance, Aadhaar enrolment and up-dating, health services, distribution of LPG cylinders to rural consumers, registration of PAN card, rail, bus and air tickets booking etc. for income generation. Attempts are being made to transform PACS into multipurpose business entities for viability and sustained growth. Changes involved organizational preparedness, continuous evolutions, and risk taking.

The promising venture would be development of farmers' producers' organizations societies (FPOs) especially in place of defunct societies to provide better and cost-effective services to members.

Though cooperative credit societies deal with small value accounts of depositors and borrowers of numerous small and marginal farmers and self-help groups, their advantage lies in low transaction cost due to their proximity which contribute to high operational efficiency at the ground level. The fundamental question that must be answered is whether PACS are able to allocate existing resources optimally for obtaining higher technical efficiency, given the fact that many are dependent on higher financing institutions for their business expansions and survival. The related question is: are these cooperative credit societies suffer from misallocation of resources in producing mix of outputs leading to pure technical inefficiency otherwise called managerial inefficiency? Do these PACS suffer from low volume of business leading to presence of scale inefficiency?

1.2 Organization of Paper

After a brief introduction on the primary cooperative credit societies in the Sect. 1 above, the remaining paper is organized in six sections as follows; the measurement of efficiency using DEA methodology and empirical works on the measurement of technical efficiency of the commercial banks operating in India using DEA methodology is selectively reviewed in Sect. 2. The input-oriented DEA framework with variable returns to scale is presented in Sect. 3. The data, and their description along with specification of inputs and outputs are given in Sect. 4. Section 5 analyses empirical estimates of technical, pure technical and scale efficiency of PACS obtained under alternative approaches using DEA methodology. Section 6 analyses Tobit censored maximum likelihood estimates for technical efficiency, pure technical efficiency and scale efficiency under alternative approaches. Section 7 presents the conclusion, recommendations, limitations of the study and future research.

2 Measurement of Efficiency

2.1 Empirical Studies on Data Employment Analysis

The measurement of efficiency of firms is an enduring subject of interest among academic scholars and policy makers as they provide significant insight into the allocation of resources for obtaining optimum output. Michel Farrell (1957) laid the foundation for applied frontier production function analysis and obtained firm specific efficiency and inefficiency estimates through linear programming technique. Charnes, Cooper and Rhodes (CCR, 1978) advanced Farrell's approach to measurement of efficiency through data envelopment analysis (DEA) to multiple inputs and multiple outputs. DEA approach to frontier efficiency measure involved constructing frontier production function through facets of the best performing firms from the given data set to act as benchmarks for inefficient firms in the sample for achieving either maximum level of outputs given inputs (output orientation) or for minimization of inputs given outputs (input orientation). DEA had become a standard methodological approach to measure the efficiency of firms for its flexibility in dealing with multiple inputs and multiple outputs. DEA methodology does not require a functional form for estimation of production function. Banker, Charnes and Cooper (1984) incorporated variable returns to scale (VRS) in the place of constant returns to scale assumed in CCR (1978) model so that DMU can be identified as either increasing returns to scale (IRS), or constant returns to scale (CRS) or decreasing returns to scale (DRS). Cook and Seiford (2009) provide the state of the art on data envelopment analysis and its extensions. Berger and Humphrey (1997) provide a comprehensive survey relating to application of data envelopment analysis in the financial institutions. Emrouznejad and Cabanda (2015) present a concise review on recent developments in DEA methodology, its application in manufacturing and in services sectors, likely pitfalls at different stages of research process especially not accounting for heterogeneity in inputs, neglect or avoidance of non-homogeneity of environmental factors influencing performance of DMUs, problems relating to over specification of inputs and outputs, and omission of variables just due to presence of correlations among variables biasing the efficiency scores.

2.2 Review of Literature on Banking Sector in India

In India, there is an extensive empirical works trying to capture various dimensions of performance of commercial banks by utilizing non-parametric method of data envelopment analysis. The notable works are Bhattacharyya, Lovell and Sahay (1997) for analyzing technical efficiency of commercial banks, Ray and Mohan (2004) for revenue maximizing efficiency of commercial banks, Das and Ray (2006) for technical, pure technical and scale efficiency of commercial banks, Sahoo, Das and Ghose (2006, 2009) for technical and cost efficiency of commercial banks, and Sengupta and Mandal (2007) for technical and scale efficiency of banks. These studies brought out vividly technological changes in banking sector, sources of efficiency and relative efficiency levels of public sector banks versus private sector banks and foreign banks operating in India. These works had gone on to enumerate factors contributing to convergence of efficiency levels of public sector banks, private and foreign banks in India since economic liberalization initiated in 1991.

In the second stage, a number of researchers had attempted to identify factors contributing to variations in the efficiency levels of commercial banks. The notable works are Bhattacharyya, Lovell and Sahay (1997), Gupta, Doshit and Chinubhai (2008), Das and Ghosh (2006), Ataullah and Le (2006), Aparna Bhatia and Megha (2017–18), Joyeeta Deb (2017–18), and Bhatia and Mahendru (2017–18).

There are a few studies on cooperative banks and regional rural banks at the district, state and national levels in India using data envelopment analysis. Gaurav and Krishnan (2017) studied the efficiency of district central cooperative banks at macro level; Reddy (2006) analyzed the performance of the regional rural banks at the aggregate level at India; Ganesan (2009) looked at the efficiency levels of state and district cooperative banks at the aggregate levels; Bhat and Bhat (2013) studied the performance and efficiency of district cooperative banks in the state of Jammu and Kashmir while Singh (2015) analyzed the efficiency of district cooperative banks in Punjab state.

There is conspicuous absence of empirical study relating to technical efficiency of primary cooperative credit societies in India. This study attempts to contribute to an understanding of technical, pure technical and scale efficiency of primary cooperative credit societies using primary data collected in one district in Tamil Nadu, the southern state of India.

2.3 Objectives, Research Questions and Purpose of Study

The objectives of the study are as follows:

To find out the resource- use efficiency of PACS in providing services to members in rural areas.
To find out the levels of pure technical inefficiency and scale inefficiency in delivering goods and services to members in rural areas.
To find out the determinants of technical, pure technical and scale efficiency of PACS.

The research questions are:

Do PACS suffer from misallocation of resources in producing multiple outputs and inputs leading to pure technical inefficiency otherwise called managerial inefficiency?
Do PACS suffer from low volume of business turnover leading to presence of scale inefficiency?
What are the major drivers of high performing PACS in rural areas?
How do PACS perform under alternative approaches in producing goods and services with different combinations of inputs?

Purpose
The purpose of the study is to provide information for the managers in the district central cooperative banks to decide about the scale of finance or working capital requirements for PACS so that they could determine the loanable funds for each society based on empirical findings to provide better services to farmers and other clients in rural areas. It is imperative in the context of strong competition in rural areas. The share of credit cooperatives in the rural financial market has been coming down in recent years due to the presence of multiple players such as commercial banks, private banks, non-banking finance companies, regional rural banks, and small finance banks. Despite multiple players in the rural areas, cooperative credit societies are most indispensable in serving members at low cost. The cooperative banks must optimally use available resources to serve target groups to improve their production levels and improve their economic welfare. At the same time, their sustainability depends on their volume of business and operational efficiency.

3 Data Envelopment Analysis

3.1 Input-Oriented DEA Model [CCR, 1978]

$$\begin{aligned} & \min_{\varphi,\lambda} \varphi_k \\ \text{St.} \quad & \textstyle\sum_{j=1}^{n} \lambda_j y_{rj} \geq y_{rk} \\ & \textstyle\sum_{j=1}^{n} \lambda_j x_{ij} \leq \varphi X_{ik} \\ & \lambda_j \geq 0 \end{aligned} \tag{1}$$

For

$i = 1,2, \ldots\ldots m$ (inputs); $r = 1,2 \ldots\ldots\ldots s$(outputs); $j = 1,2 \ldots\ldots\ldots n$ (firms).

The subscript k represents the DMU being evaluated, x_{ij} denotes input i used by DMU_j,

y_{rj} denotes output produced by DMU_j;

φ_k= Technical efficiency estimate for decision-making unit k under evaluation.

$$y_{rk} = \text{Observed level of output from j for DMU k;}$$

$$x_{ik} = \text{Observed level of input from j for DMU k;}$$

$$\lambda_j = \text{Weights for output and inputs from j for DMU k.}$$

In the input-oriented minimization model, the estimates of φ_k and λ are calculated by the linear programming method considering the output constraint by minimizing inputs or radial contraction of inputs to achieve the given output levels. Input oriented model specifies that a firm under reference must not use inputs exceeding inputs use of best performing DMUs in the sample while simultaneously output of the firm must not fall short of output levels of best performing DMUs.

3.2 Input-Oriented DEA Model [BCC, 1984]

Banker, Charnes and Cooper (BCC (1984) introduced variable returns to scale (VRS) input- oriented DEA model by relaxing the assumption of constant return to scale in the CCR (1978) model. Restricting the sum of weights, $\sum \lambda_j = 1$, BCC model can form convex hull in the input –output plane connecting points of sample observations more tightly leading to higher technical efficiencies or equal to those obtained for firms using constant return to scale which generate shape of conical hull connecting loosely points of sample observations (Coelli, etal, 1998, P.150).

The difference between CCR model and BCC model is the convexity condition $\sum \lambda_j = 1$ added to CCR input-oriented model.

BCC (1984) model is

$$\begin{aligned} &\min_{\varphi,\lambda} \varphi_k \\ &\sum_{j=1}^{n} \lambda_j y_{rj} \geq y_{rk} \\ \text{St.} \ &\sum_{j=1}^{n} \lambda_j x_{ij} \leq \varphi X_{ik} \\ &\lambda_j \geq 0; \\ &\sum \lambda_j = 1. \end{aligned} \tag{2}$$

The reciprocal of estimated value of φ_k is the efficiency score for a DMU. If efficiency score is one, the DMU will be on the production frontier surface and it is said to be 100% efficient. If efficiency score is less than one, the DMU is said to be inefficient and lies below the production frontier surface. In the optimization exercise, efficiency measure

φ_k is a scalar by which the firm will be scaling down inputs to reach the frontier output level. If technical efficiency obtained under CRS technology is not equal to technical efficiency obtained under VRS technology, then the difference in the efficiency is due to scale inefficiency. The scale inefficiency occurs among DMUs due to presence of either increasing returns to scale or decreasing returns to scale suggesting the lack of resources for the former or excess consumption of the resources for the latter. The technical efficiency is decomposed into pure technical efficiency under VRS and scale efficiency.

$TE_{CRS} = TE_{VRS} \times SE.$

$SE = TECRS/TEVRS.$

3.3 Relevance of CCR / BCC Model for PACS

CCR or BCC models are chosen for its versality and simplicity in explaining various components of technical efficiency. PACS are the ground level micro credit institution which uses multiple inputs to produce multiple services. BCC model is utilized for PACS as it incorporates multiple inputs and multiple outputs. PACS operate at different scales of production, though they operate in a homogenous region. As the resource-use inefficiency among PACS is widespread, due to managerial and financial constraints, along with institutional constraints, use of BCC model brings out differences in technical, pure technical and scale efficiency of PACS. Decision makers in higher financing institutions, especially in the district central cooperative banks and in apex cooperative credit institutions, can rely on empirical estimates and utilize the information on the levels of efficiency and determine level of financial assistance to PACS. At present, the fund allocation decision at the district central cooperative bank to PACS is mostly heuristic and rule of thumb rather than on scientific decision rules. BCC model provides information for cooperative leaders in higher financing institutions in formulating workable policy decision for fund allocation to PACS.

3.4 Approaches to Measurement of Efficiency

There are three approaches to the measurement of technical efficiency in banking sector; (a) Intermediation approach, (b) value-added approach and (c) operating approach. The intermediation approach suggests a bank being an intermediary between depositors and borrowers, collects deposits from savers and lend capital to borrowers. In this intermediation function, bank will be able to generate more outputs in the form of loans and advances and make investments in various securities due to its resource mobilization in the financial market (Sealy and Lindley, 1977). In the value-added approach, a bank creates more value addition to its activities by finding source of funds like deposits and borrowings in the market and involving in operations such as in provisions of loans and advances. These activities involve consumption of real resources like labor, fixed assets and incurring interest expenses (Berger, Hanweck and Humphrey (1987). In the operating approach to measure the banking efficiency, performance measures especially interest income and non-banking income are expected to be influenced by interest expenses, operating expenses and fixed assets (Leightner and Lovell, 1998). In the context of

India, the researchers have utilized all three approaches, though predominantly with input- oriented intermediation approach to the measurement of technical efficiency.

4 Data

4.1 Adequacy of Sample Data for Empirical Estimation

This study uses non-parametric method of data envelopment analysis for 110 PACS in Krishnagiri district in Tamil Nadu with reference to 2019–20. The primary data used in this paper were collected originally for the Integrated Cooperative Development Project-II phase sponsored by the Government of Tamil Nadu and National Cooperative Development Corporation (NCDC), New Delhi for assessing the requirement of financial and physical assistance to primary agriculture cooperative credit societies in the district. For this purpose, comprehensive questionnaires were administered to PACS for obtaining relevant non-financial and financial information such as members (male and females), borrowers, women self-help groups, share capital and reserves, deposits, borrowings, loans and advances, investments, recovery of principal and interest, bad debt, interest income and interest expenditure, number of employees, number of villages covered etc.

Robustness of DEA estimates are subject to number of outputs and inputs used given sample size. There are (m + s) constraints in the form of inputs (m) and outputs (s) which must be specified for technical efficiency estimates. There are n observations, one for each of $j = 1\ldots, n$, DMUs that form possible combinations from which efficiency evaluations can be secured. The number of observations should be three times higher than sum of the inputs (m) and output variables (s) i.e., $n \geq 3(m + s)$. In this study, as the number of observations is 110, the degrees of freedom are $[3 \times (4 + 3)] = 21$ for intermediation approach, 24 DF for value-added approach and 15 DF for operating approach. The specification of DEA models has enough discriminatory power in identifying efficient and inefficient DMUs in the sample.

4.2 Description of Data

Table 1 provides basic information on PACS for the year 2019–20. The members per PACS was 1910. Though there are many members, the number of borrowers per society on an average was 542. The average number of Kissan Credit Cards (KCCs) per PACS and the average number of self-help groups per PACS respectively were 66.88 and 21.57. The average number of villages per cooperative credit society were 5.48.

The average working fund of primary agriculture cooperative credit societies was Rs.403.74 lakhs in 2019–20 with 57.10% coefficient of variations among them indicating high dispersion in the resource positions of PACS. A large dispersion in PACS resources such as in share capital and reserves, deposit mobilizations, and borrowings stemmed from internal and external factors affecting performance of them.

Out of 110 PACS, number of net profit earnings societies was 61 in 2019–20. There is wide dispersion among PACS in the level of net profit earned. The average net profit per PACS was Rs. 2.96 lakhs with 126.35% coefficient of variations. The number of PACS incurring losses were 49 and the average net loss incurred by them was Rs.2.73 lakhs with coefficient of variations of 120.88%.

The average business turnover (sum of loans and advances and deposits and non-banking business income) per society was Rs. 458.04 lakhs with 15.01% coefficient of variations among societies suggesting low variability in resources mobilizations and deployments among PACS.

Average productivity of employee was Rs.114.06 lakhs with a coefficient of variation of 60.33%.

Table 1. Basic information on PACS in Krishnagiri District (values in Rs. Lakhs)

S. No	Variables	Minimum	Maximum	Mean	STD
1	Members in PACS	655	3100	1910.18	595.45
2	No. of Borrowers	136	1369	542	221
3	No of KCC Accounts	12	140	66.88	32.19
4	No of SHGs	6	35	21.57	7.37
5	Own Funds (Rs)	11.97	375.23	66.54	56.12
6	Deposits (Rs)	12.81	1153.94	130.66	161.58
7	Borrowings (Rs)	48.00	556.02	203.58	96.30
8	Total Working Fund (Rs)	138.47	1540.44	403.74	230.55
9	Total Business Turnover (Rs)	144.40	2071.22	458.40	68.82
10	Loans & Advances (Rs)	87.05	850.57	291.49	155.14
11	Investment (Rs)	7.98	676.71	67.95	82.73
12	Non- Banking Business (NBB) (Rs)	8.42	136.56	36.42	23.52
13	Fixed Assets (Rs)	9.18	47.58	23.09	9.13
14	Total Provisions (Rs)	0.00	373.30	42.26	67.03
15	Interest Income (Rs)	10.45	110.60	34.22	19.96
16	Interest Expenses (Rs)	6.73	78.66	23.98	14.87
17	Establishment costs (Rs)	4.94	35.50	14.53	6.87
18	No. of employees	2	9	4.28	1.87
19	Employee Productivity	27.28	608.87	114.06	68.82
20	(a). Net Profit (Rs) (b). No. of Profit making PACS	0.12 61	13.66	2.96	3.74
21	(a). Net Loss (Rs) (b). No of Loss making PACS	2.79 49	10.23	2.73	3.30

4.3 Intermediation Approach to Input -Output Specifications for PACS

In the intermediation approach, we have used four inputs 1). Deposits, 2). Borrowings, 3). Share capital & reserves and 4). Fixed assets and three outputs 1). Loans & advances, 2). Investments and 3). Non-banking income in Table 2.

Table 2. Specification of Inputs and Outputs for DEA in the present study

Variables	Intermediation approach (Asset-Based Approach)	Value-Added Approach (Production function Approach)	Operating Approach (Income Approach)
Inputs X1	Deposits	Labor	Interest Expenses
X2	Borrowing	Fixed Assets	Establishment Expenses
X3	Fixed Assets	Interest Expenses	Fixed Assets
X4	Share capital & Reserves	Loans &Advances	Interest Income
Outputs Y1	Loans &Advances	Investments	NBB Income
Y2	Investments	Deposits	
Y3	NBB Income	Borrowings	
Y4		NBB Income	
Y5			

Own Fund (share capital and Reserves).
We have specifically added share capital & reserves as a separate input variable along with standard variables specified in the previous empirical studies on banking efficiency. As per cooperative acts and rules, PACS must create a certain percentage of net profits as reserve fund for future contingencies. PACS, which have earned net profit consistently over the past years, would have developed large reserves. Those PACS are in a better financial position to register higher lending, higher level of investment and venture into non-banking businesses. On the other hand, PACS, which have incurred losses continually may not be able to develop reserves and might have low business turnover. Further, the growth of PACS's share capital is strongly linked with the growth of loans & advances.

Deposits and Lending

PACS are considered rural thrift and credit institutions. PACS being intermediaries between rural depositors and borrowers in rural settings transform deposits into output in the form of loans and investments.

Borrowings and Short-Term Agriculture Lending

As for input variable, borrowings are concerned, PACS depend on higher financing cooperative institutions, especially NABARD through district central cooperative banks for short-term agriculture operations. PACS acts as an intermediary between the DCCB and farmer-members at the village level for short-term agriculture production operations. For without explicitly considering the borrowings as separate input variable, the analysis will be incomplete to understand the loans & advances of PACS and its levels of technical efficiency.

Fixed Assets and Non-Banking Income

Fixed assets are specified to show the extent to which PACS is utilizing resources in generating non-banking income for PACS.

On the output side, we added non-banking income as separate variable to capture the impact of management of surplus funds which PACS deploy to produce higher volume of output which contribute to higher technical efficiency. The government of Tamil Nadu has entrusted the task of distributing essential controlled commodities to the ration

card holders in the state to the PACS. PACS obtain commission on PDS operations. A significant proportion of PACS obtain their non-banking income from PDS operations.
Investments
Own funds of PACS (share capital and reserves) are generally invested in safe investment areas. PACS do not generally use own fund for short-tern agriculture lending except for jewel loans. Generally surplus funds of PACS are invested in shares of district central cooperative bank, IFFCO and KRIBHCO and in National Saving Certificates.

4.4 Operating Approach to Input-Output Specifications for PACS

The use of operating approach to measurement of technical efficiency is to highlight variations in the performance of PACS. PACS generally suffer from lack of business diversification, poor recovery of principal and interest, mounting overdue and low business turnover. As per cooperative acts and rules, PACS is required to provide banking services to members who reside within its jurisdictional areas. These constrain the PACS to generate adequate volume of business and income to cover operational expenditures. From Table 1, it is seen that 49 of the PACS are incurring losses. The specification of inputs and outputs in the operational approach would bring out variations in technical efficiency of PACS. The operational approach shows the extent to which PACS are effectively controlling costs along with efficient use of fixed assets in producing the required income for sustained growth.

4.5 Value – Added Approach to Input- Output Specifications for PACS

Under a value-added approach, PACS are involved in transforming input into output so that the value of output is larger than value of input expenses. Three inputs namely, labor, fixed assets and interest expenses are used to produce five outputs such as loans and advances, investments, deposits, borrowings and non-banking income. It is assumed higher levels of deposits, borrowings, loans and advances, investments and non-banking incomes are resultants of high managerial inputs and ability to utilize their labor, fixed assets along with low interest costs. More efficient PACS are those who are effectively using labor and fixed assets to create non-banking income. PACS which have locational advantages can leverage their managerial and leadership positions to obtain more borrowing and deposits to increase their technical efficiency.

5 Empirical Results of Technical Efficiency

5.1 Discussion

This section provides comparative analysis of DEA estimates on technical, pure technical and scale efficiency under alternative approaches using inputs-oriented model.

Technical, pure technical and scale efficiency scores of cooperative credit societies are presented in Table 3. Under intermediation approach, the mean technical efficiency under CCR model is 83.30%; the mean value for pure technical efficiency (PTE) and scale efficiency (SE) scores are respectively 89.90% and 92.10%.

In the case of operating approach to measurement of technical efficiency, the mean technical efficiency under CCR model is 66.70%; the mean value for pure technical efficiency (PTE) and scale efficiency (SE) are respectively 78.50% and 84.40%.

Table 3. Technical, Pure technical and scale efficiency and returns to scales Under Intermediation, Operating and Value-added approaches.

Efficiency range	TE	PTE	SE	IRS	CRS	DRS
Intermediation approach						
< 0.60	10	3	3	9	1	0
0.61 - 0.90	55	39	31	40	0	15
0.91 – 1.00	45	68	76	6	37	5
Total	110	110	110	55	35	20
Mean	0.833	0.899	0.921	-	-	-
Operating approach						
< 0.60	49	12	3	47	0	2
0.61 - 0.90	41	64	65	32	0	9
0.91 – 1.00	20	34	42	1	16	3
Total	110	110	110	80	16	14
Mean	0.667	0.785	0.844	-	-	-
Value-Added approach						
< 0.60	5	0	0	5	0	0
0.61 - 0.90	52	39	29	45	3	6
0.91 – 1.00	53	71	81	14	34	3
Total	110	110	100	64	37	9
Mean	0.861	0.913	0.940	-	-	-

Note: TE = Technical efficiency; PTE = Pure technical efficiency; SE = Scale efficiency; IRS = increasing returns to scale; CRS = constant returns to scale; DRS = Decreasing returns to scale

Under value-added approach, the mean value for technical efficiency under CCR model is 86.10%; the mean values for pure technical efficiency (PTE) and scale efficiency (SE) are respectively 91.30% and 94.00%.

The differences in the means of technical, pure technical and scale efficiency scores under three methods are due to specification of different inputs and outputs and number of variables that are included in those models.

Under intermediation approach, lowest technical, pure technical and scale efficiency scores of DMUs respectively were 28.90%, 49.60% and 56.40%.

Operating approach yielded lowest technical, pure technical and scale efficiency scores respectively were 33.50%, 46.70% and 45.50%.

Under value-added approach, lowest technical, pure technical and scale efficiency scores of DMU respectively were 44.80%, 60.10% and 68.80%.

5.2 Frequency Distribution of Efficiency Scores[1]

A frequency distribution of number of DMUs under technical efficiency (TE), pure technical efficiency (PTE), and scale efficiency (SE) along with returns to scale, was grouped under three classes for easy readability and analysis.

Under intermediation approach, 35 DMUs out of 110 DMUs are technically efficient with efficiency score of one. These DMUs are technically efficient in the sense of proportionate increase in inputs result in same proportionate increase in outputs. Further these DMUs optimally combine multiple inputs in right proportions to produce multiple outputs to obtain best performance levels. Technical efficiency score under CCR model ranges from as low as less than 40% to as high as 100%. Pure technical efficiency under BCC model ranges from 41% to 100%.

Under intermediation, operating and value-added approaches to technical efficiency, a 50%, 37.27% and 47.27% of DMUs respectively are found in efficiency score ranges of 0.61 to 0.90 (Table 3).

[1] Though efficiency scores for technical, pure technical and scale efficiency under alternative approaches are grouped under eight class intervals (see histograms charts 1 to 9), Table 3 provides frequency distribution of PACS under three grouping for easy discussion. As there are lesser number of DMUs in the lower classes of efficiency scores, it is found reasonable to reduce the classes into three from eight so that a larger picture on the distribution of PACS would be obtained. Histograms for three approaches for various efficiency components are provided in eight classes.While technical efficiency scores of DMUs under intermediation approach ranged from 0.289 to 1, the pure technical efficiency and scale efficiency ranged respectively from 0.496 to 1 and 0.564 to 1. The number of class intervals is determined by applying Sturges formula ($K = 1 + 3.322 \log N$) where N is the number of DMUs in the sample. Using the formula, the number of classes obtained was approximately 8. The range which captures difference between maximum and minimum values is larger for technical efficiency estimate than for pure technical efficiency and scale efficiency of DMUs. The number of classes 8 was used to divide the range. The width obtained was 0.09 for technical efficiency, 0.06 for pure technical efficiency and 0.05 for scale efficiency. To capture the number of DMUs in the lower size classes, the number of classes at 8 is found to be optimum so that variation in the efficiency levels of DMUs could be highlighted. We have taken larger value of class width 0.09 so that at least a few DMUs could fall in lower classes across the technical, pure technical and scale efficiency scores.In the case of Operating approach to technical, pure technical and scale efficiency of DMUs, the efficiency range respectively are 0.67, 0.53 and 0.55. With class intervals of 8, the actual width is 0.08 for technical efficiency and 0.07 for both pure technical efficiency and scale efficiency. To have uniformity for class intervals for intermediation, operating and value-added approaches, class width of 0.09 is maintained. We want to include the number of DMUs in each size so that distribution of PACS could be highlighted.In the case of Value- Added approach, the actual width for technical efficiency, pure technical efficiency and scale efficiency are respectively 0.07, 0.05 and 0.04 by using the class intervals of 8. It is found even with the larger width of 0.09, we find a smaller number of DMUs in the lower classes. By reducing the width of the class interval and consequently increasing the number of classes, we could not obtain any valuable information.

Under intermediation, operating and value-added approach to pure technical efficiency, a 35.46%, 58.18% and 35.46% of DMUs respectively are found in efficiency score ranges of 0.61 to 0.90 (Table 3).

5.3 Histograms for Intermediation Approach

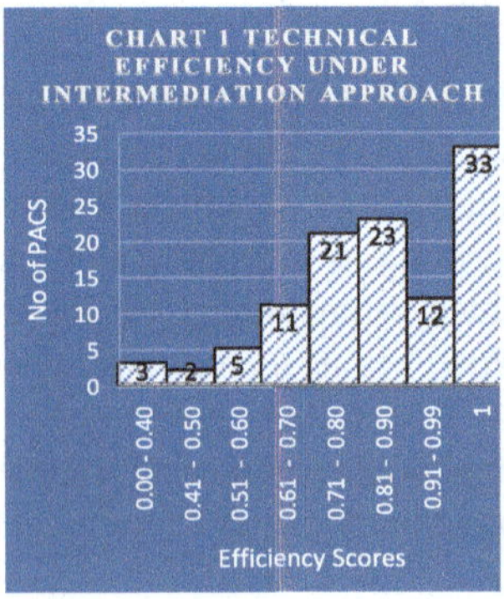

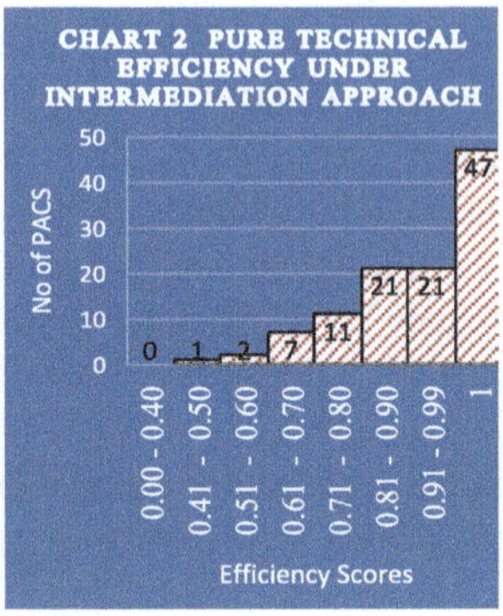

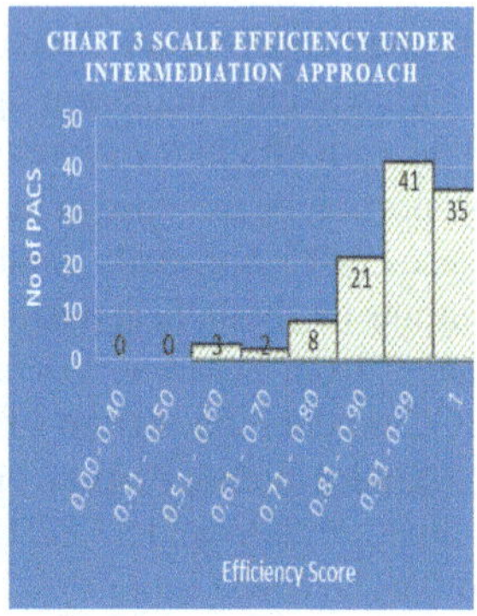

Charts 1, 2 and 3 show histograms for DMUs under intermediation approach for technical, pure technical and scale efficiency scores grouped under 8 classes. The greater number of DMUs with efficiency score of one is found for pure technical efficiency (47 DMUs) followed by DMUs with scale efficiency (35) and technical efficiency (33). In the bandwidth of 0.91 to 0.99, 41 PACS are scale efficient while 21 PACS are pure technically efficient. Only 12 PACS are found technically efficient in the second highest efficiency grouping. One may infer that it is the managerial inefficiency in relation to combination of inputs use and combination of production of output mixes that contributes to the lower level of technical efficiency of PACS.

5.4 Histograms for Operating Approach

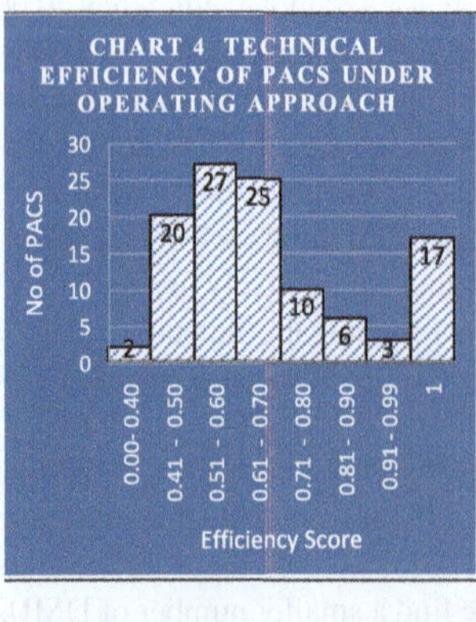

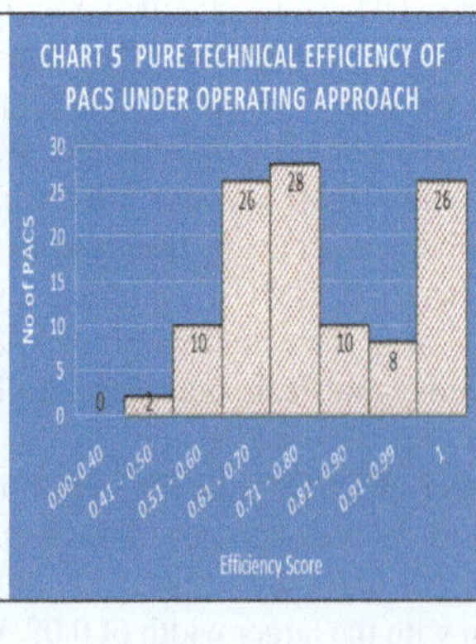

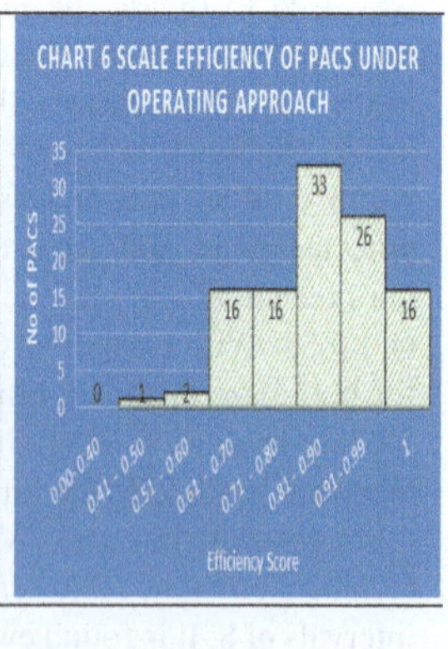

Charts 4, 5 and 6 show histograms for PACS under the operating approach to efficiency measurement. In contrast to intermediation approach to efficiency distribution, it is interesting to observe that many DMUs are concentrated in the lower classes of efficiency scores underlining PACS inability to generate adequate interest income and non-banking income in proportion to levels of expenses incurred in managing organization. Many PACS are not involved in the diversified business operations. A significant number of PACS are involved largely in the distribution of essential commodities to ration card holders in rural areas. Given institutional limitations, PACS are having low business turnover in relation to fixed assets and management expenses in running the organization. Only 17 PACS are technically efficient with perfect score of one, i.e.100% efficient.; 26 PACS are 100% pure technically efficient and 16 PACS are 100% scale efficient.

Given the resource constraints facing many PACS, efficiency in scale of operations is found to be modest compared to low pure technical efficiency. The comparison of the Chart 5 with the Chart 6 shows in contrast to number of PACS with pure technical efficiency falling on the lower side of efficiency distribution, for scale efficiency relatively many PACS fall on the higher side of the efficiency distribution; it is low pure technical efficiency of DMUs which brings down the overall technical efficiency of PACS under operating approach. However, PACS' small scale of operations also is holding back their profitability and sustainable growth.

5.5 Histograms for Value-Added Approach

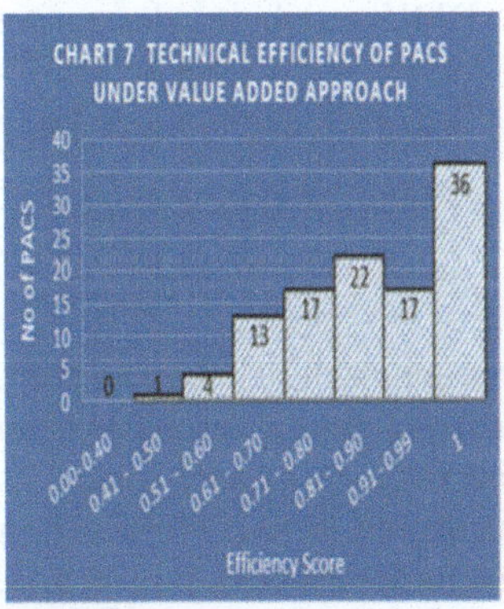

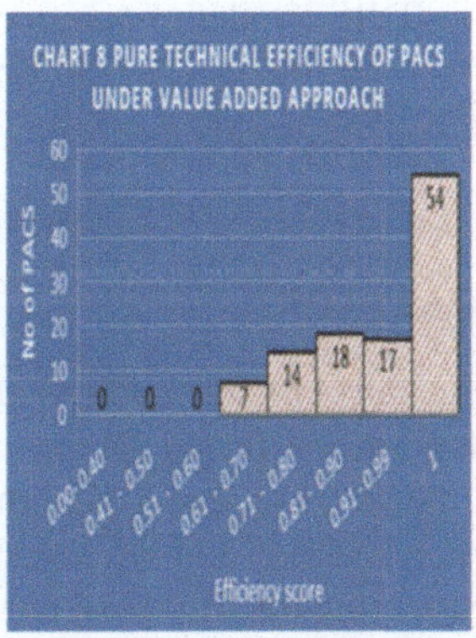

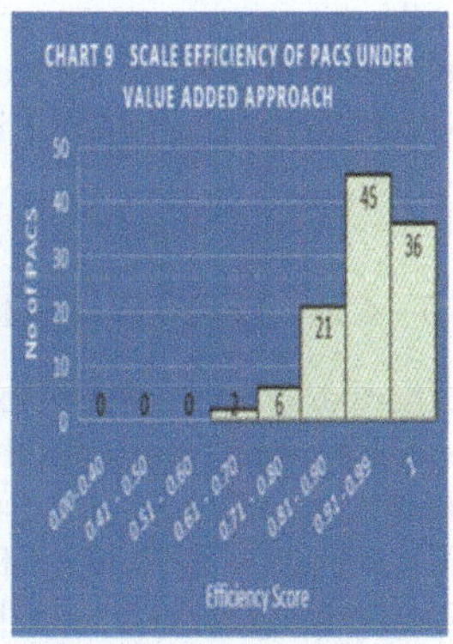

Charts 7, 8 and 9 present the histograms for PACS under value-added approach to efficiency measurement. Under the value-added approach, only 36 PACS are found to be 100% technically efficient. In the case of pure technical efficiency, 54 PACS are 100% efficient whereas 36 PACS are 100% scale efficient. In the case of value-added approach, pure technical efficiency alone cannot enhance the technical efficiency of PACS. It is the combination of scale efficiency (higher volume of business) and pure technical efficiency (effective decision making) contribute to technical efficiency of PACS. Chart 8 shows that large percentage of DMUs are not pure technically efficient, suggesting greater need for effective management practices in combining inputs to produce multiple outputs. The scale efficiency is well pronounced (45 PACS) in the efficiency class of 0.91 to 0.99. It is the presence of large number of DMUs with pure technical efficiency which fires-up overall technical efficiency to the highest level. The presence of scale efficiency alone is not enough to raise overall technical efficiency.

5.6 Returns to Scales

Returns to scale (RTS) show whether an increase in all inputs simultaneously results in more than proportionate increase in outputs (IRS) or more than proportionate decrease in output (DRS) or equal proportionate change in output (CRS). Information on the number of DMUs under IRS, CRS and DRS are provided in Table 3. Significant number of DMUs, i.e., 55 DMUs are found to be in the region of IRS under intermediation approach. Under operating and value-added approaches, 80 and 64 DMUs respectively are found in IRS region. This is to show that a large proportion of DMUs are operating in environment with greater resources constraints than they can produce outputs if resources constraints are removed. If these DMUs could manage enough resources, they would be able to scale up their production operations and their average productivity would increase to match DMUs with higher average productivity.

Out of 110 DMUs, only 33, 16 and 33 DMUs respectively are under intermediation, operating and value-added approaches characterized by constant returns to scale (CRS) with efficiency score of one; those DMUs neither use excess amounts of inputs nor face major constraints in their resource positions.

Out of 110 DMUs, 20, 16 and 9 DMUs respectively are under intermediation, operating and value-added approaches found excessively consuming inputs contributing to decreasing returns to scale (DRS) in their operations. DMUs exhibiting decreasing returns to scale can reduce their inefficiency by shifting excess inputs to those DMUs which are in the zone of increasing returns to scale (IRS) as they can produce more output.

5.7 Case Analysis of Selected PACS

Uthangari PACS is 100% technically efficient under all three approaches: intermediation, operating and value-added approaches. It was able to mobilize financial resources from members and financial institutions and transform them into loans for needy borrowers in rural areas. Uthangari PACS make profitable investments; are actively involved in non-banking services like leasing out tractor services to members under its jurisdiction. This society obtains considerable non-banking business income; from managing 17 PDS shops, operating medical store, e-service center, providing locker services to customers, renting out godown for farmers and traders for stocking of turmeric and pulses, and hiring out mini tractors to farmers in the region. In 2019–20, society registered net profit of Rs.23.00 lakhs.

Under operational approach, this DMU effectively controls establishment costs and generating enough income to the society. The fixed resources like farm equipment and buildings are utilized to the fullest capacity leading to pure technical efficiency and high overall efficiency.

The secondary occupations of members are dairying and poultry farming. The success of Uthangari PACS is due to its coverage of a large revenue village, more borrowers through self-help groups, high recovery rate of loans, diversified agriculture lending complemented by increased non-farm business operations leading to higher non-interest income, control of management and operational costs.

Another remarkable and highly diversified PACS is Bagalur PACS in the district. It is one of the well-known cooperative societies. It is technically efficient under all three approaches: intermediation, operating and value-added.

Under RRCCI (2004), Bagalur PACS has received Rs.100 lakhs; of which 60% was invested in fixed deposits at DCCB and 40% was lent for jewel loans.

Being in the main market center, it provides e-services to members and public thereby earning non-banking income to the society. In collaboration with Chettipalli PACS in this region whose banking infrastructure is on par with any commercial bank branch in the district, collect fresh vegetables from farmers -members and sort them, grade them and transport them to towns. The staff jointly put in a considerable amount of time and efforts running into late midnight till vegetables were moved to Chennai and Bangalore market centers. Their efficiency is derived from large economies of scale in business operations like agriculture lending, high recovery rate and retail sector lending due to adequate own resources, operating in large revenue villages, and locational advantages to undertake numerous non-agriculture operations.

Chettipalli PACS is efficient in TE, PTE and SE. Its social and economic benefits to members go beyond monetary valuation of costs and benefits. Its lending is largely for horticulture purposes in its jurisdiction. Society uses its own fund.

Vegetable reefer vans are provided to Chettipalli PACS to carry fresh vegetable to AMMA canteens – highly subsided low priced quality hot foods to the urban dwellers in Chennai some 280 km away daily managed by the greater Chennai Corporation. PACB has consistently declared 12% dividend to its members.

6 Tobit Regression Analysis

Tobit Model

To understand the factors influencing technical efficiency, pure technical efficiency and scale efficiency under alternative approaches, we used censured Tobit regression model and estimated it with maximum likelihood techniques. Tobit model takes following form:

$$y_i^* = x_i \prime \beta + \varepsilon_i$$

$$y_i = \begin{cases} \mathrm{y_i^*} \mathit{if}\, \mathrm{y_i^*} > 0 \\ 0 \mathit{if}\, \mathrm{y_i^*} \leq 0 \end{cases}$$

y_i^* is the unobserved variable;

y_i is the observed variable;

x_i are the explanatory variables;

β are the unknown parameters to be estimated;

ε_i is error term which is normally distributed with N $(0, \sigma^2)$.

Technical efficiency, pure technical efficiency and scale efficiency scores are regressed on independent variables such as (1) Capital-labor ratio, (2) Asset turnover ratio (3)

Employees productivity, (4) Number of Kissan Credit Cards (5) Own fund (in log form) (6) Number of self-helps groups and (7) Management expenses (in log form) as a proxy for management effectiveness and control.
Tobit estimating equation is

$$\begin{aligned}\text{TE} =& \alpha + \beta_1(\text{CapLab}) + \beta_2(\text{AssetTR}) + \beta_3(\text{Empr}) + \beta_4(\text{KCC}) \\ &+ \beta_5(\text{logOwnF}) + \beta_6(\text{ShGs}) + \beta_7(\log \text{Expens}) + e_i\end{aligned}$$

$$\begin{aligned}\text{PTE} =& \alpha + \beta_1(\text{CapLab}) + \beta_2(\text{AssetTR}) + \beta_3(\text{Empr}) + \beta_4(\text{KCC}) \\ &+ \beta_5(\text{logOwnF}) + \beta_6(\text{ShGs}) + \beta_7(\log \text{Expens}) + e_i\end{aligned}$$

$$\begin{aligned}\text{SE} =& \alpha + \beta_1(\text{CapLab}) + \beta_2(\text{AssetTR}) + \beta_3(\text{Empr}) + \beta_4(\text{KCC}) \\ &+ \beta_5(\text{logOwnF}) + \beta_6(\text{ShGs}) + \beta_7(\log \text{Expens}) + e_i\end{aligned}$$

The coefficient labeled 'intercept: $= \alpha$' is constant for the model;$\beta_1 \ldots \beta_7$ are the estimated coefficients for the variables and e_i are error terms with mean zero and positive standard deviations.
Table 4, Table 5 and Table 6 respectively provide information on the coefficient estimates of independent variables under intermediation, operating and value-added approaches for technical, pure technical and scale efficiency. The log-likelihood estimates as a summary measure for technical efficiency and pure technical efficiency under intermediation approach are larger than the log-likelihood estimates under operating and value-added approaches. In contrast, the log-likelihood estimate for scale efficiency model is higher under value-added approach followed by operating and intermediation approaches.
Capital-Labor Ratio (CapLab)
High capital-labor ratio in PACS indicates creation of more infrastructure like, new office buildings, godowns, computers and printers and furniture and fixtures. It is expected that higher capital use in relation to labor would contribute to higher level of technical efficiency of PACS. Contrary to expectations, Tobit estimate showed that for every one unit increase in capital–labor ratio, technical efficiency decreases by 0.0053195 units under intermediation approach, and decrease of 0.0102175 units under operating approach though they are statistically insignificant. The plausible explanation for a decrease in technical efficiency due to increase in capital-labor ratio is preponderance of DMUs with high capital investments. PACS must increase their volume of business operations and range of services to be provided to customers in relation to the level of capital investments and infrastructure created. Under value-added approach, the sign for capital-labor ratio is positive but again, it is not statistically significant.
Asset-Turnover Ratio (AssetTR)
The Tobit estimate showed one unit increase in asset turnover leads to 0.3603968 units increase in technical efficiency of DMUs. It is statistically significant from zero under three approaches. Greater the number of times assets are turned over the organizations could be more efficient and effective in increasing the technical efficiency scores. In other words, higher the intensity of utilization of assets, higher will be level of technical efficiency of DMUs. Under operational approach and value-added approach, asset turnover

Table 4. Dependent variable: Technical efficiency under Alternative approaches -Tobit estimates

Independent Variables	Intermediation approach		Operating approach		Value-Added Approach	
	Estimate	Pr (>t)	Estimate	Pr(>t)	Estimate	Pr(>t)
(Intercept)	–0.3645610 (–2.631)	0.00851**	0.1548698 (0.689)	0.49102	0.3027429 (1.679)	0.09313
CapLab	–0.0053195 (–1.400)	0.16145	–0.0102175 (–1.480)	0.13874	0.0074844 (1.374)	0.16949
AssetTR	0.3603968 (6.573)	4.93e-11***	0.2978448 (3.210)	0.00133**	0.1504984 (1.975)	0.04821*
EMPR	0.0003024 (1.112)	0.26613	0.0004606 (1.127)	0.25978	0.0008513 (2.019)	0.04347*
KCCs	0.0020605 (5.661)	1.51e-08***	–0.0001340 (–0.223)	0.82389	0.0015553 (3.141)	0.00169**
Log (OWNF)	0.8069359 (4.337)	1.45e-05***	0.4746285 (1.545)	0.12231	–0.0213779 (-0.087)	0.93029
SHGs	0.0109017 (7.085)	1.39e-12***	0.0067406 (2.542)	0.01103*	0.0097733 (4.616)	3.91e-06***
Expenses	–0.0491357 (–0.862)	0.38869	–0.3938157 (–4.069)	4.73e-05***	–0.0195632 (–0.245)	0.80634
Log-sigma	–2.3821562 (–28.916)	< 2e-16***	–1.7519464 (–23.021)	< 2e-16***	–2.0414075 (-24.098)	< 2e-16***
Log-likelihood	56.75444 on 9 DF		12.92952 on 9 DF		24.61213 on 9 DF	

Note: Values within brackets are t-values.
signif. codes: 0 '***' 0.001 '**' 0.01 '*' 0.05 '.' 0.1 '

ratio as an explanatory variable is statistically significant. However, the magnitude of the estimated coefficient is lower than the estimates obtained under intermediation approach.

Employees' Productivity (Empr)

An important measure of organizational effectiveness is employee productivity. Tobit estimate for employee productivity showed positive relationship between technical efficiency and employee productivity. On an average employee productivity was Rs.114.06 lakhs. Employees' productivity differences among PACS may not be attributed just only to actual efforts levels of employees, nor to their dearth of motivations and lack of trainings; rather they are influenced by various favorable factors like market centers, extent of operational areas, presence of government offices and local institutions, and unfavorable factors like presence of several competitors in the jurisdictional area etc. Under three approaches, the sign of the coefficient is on expected line. Only under value-added approach, the coefficient of the employee productivity is statistically significant. Moreover, the magnitude of the coefficient estimate is very low. For instance, one unit increase in employee productivity results in 0.0003024 units increase in technical efficiency under intermediate approach, 0.0004606 units under operating approach and 0.0008513 units under value-added approach. It all suggest there is dominance of many small DMUs contributing to low average productivity.

Kissan Credit Cards (KCC)
The total number of Kissan Credit Cards (KCC) issued by PACS in the district is 7357. The number of KCCs as an independent variable was used to find out its influence on technical efficiency of DMUs. The predicted technical efficiency increases by 0.0020605 units and 0.0015533 units for every one unit increase in the number of Kissan credit cards respectively under intermediation and value-added approaches. Though the direction of sign of estimated coefficient is as expected and it is statistically significant, it is the large number of DMUs with lower scale of operations which brings down the magnitudes of the coefficient estimates. The sign and magnitude of coefficient under operating approach is negative though it is not statistically significant. More number of Kissan credit cards contribute to increasing technical efficiency of DMUs as they indirectly influence higher volume of businesses.
Own Fund (OwnF)
DMUs with larger own resources are expected to perform better in terms of diversified business activities leading to higher overall efficiency levels; but resource positions of many PACS are unequal. PACS with strong own fund involve in jewel loans business, non-banking operations and investment activities which contribute to higher technical efficiency. In contrast, resource poor PACS must depend on higher financing institutions even for short term agriculture operations. Given the fact that resource poor PACS are numerically large and their diversification of business is limited by the extent of own resources, a disparity in resource positions determine variations in technical efficiency levels among them. Under intermediation approach, the estimate of own fund (in log form) is statistically significant and different from zero. An increase in own fund by a percentage point leads to 0.8069359 points increase in technical efficiency of PACS. It is statistically different from zero at the level of 0.001 percent. Under operational approach, an increase in own fund by one percentage point leads to 0.4746285 points increase in technical efficiency of PACS. Under value-added approach, the coefficient estimate is negative but not statistically significant. Presence of many small-value borrowers in the sample negatively influence technical efficiency scores of DMUs. PACS are serving in general a significant proportion of small and marginal farmers; PACS business transactions with these segments of village community are small value transactions potentially contributing to lower efficiency scores of PACS. This finding reinforces observations documented in various government reports that PACS are largely serving customers with smaller value transactions which are in general ignored by commercial banks in rural areas. DMUs could experience an increasing cost of serving members leading to decreasing return to scale and lower technical efficiency of PACS.
Self-help Groups (ShGs)

The number of self-help groups under the fold of PACS in the district is 2373. They act as an important bridge between group members and PACS. SHGs mobilize deposits from their members and park them in PACS; leveraging their deposits and regular repayments of previously borrowed loans, SHGs borrow more funds for their members. For several PACS, self-help groups are real resource centers for deposits and lending. The limited growth of SHGs under cooperative fold in the region is cooperative credit societies had entered very late in tapping resource potential of self-help groups compared to commercial banks in the region. The sign of the coefficient under three approaches, intermediation, operating and value-added, is on the expected line i.e., increase in the number of self-help groups increases technical efficiency of PACS. Further, coefficient estimates are statistically significant.

Management Expenses (Expense)
It is important to understand the influence of management expenses on technical efficiency of PACS. Under intermediation approach, higher the management expenses, as expected, lower is technical efficiency of PACS, though it is not statistically significant. Similar finding is found for value-added approach i.e., higher the management expenses, lower is level of technical efficiency of PACS. A unit increase in management expenses leads to 0.0491357 units decrease in the technical efficiency of DMU under intermediation approach though it is not statistically significant. Under operating approach, higher is management expenses, lower is technical efficiency. It is statistically significant at 0.001 level. A unit increase in the management expenses leads to 0.3938157 points decrease in technical efficiency of DMUs. The difference in the magnitudes in coefficient estimates is due to specifications of different inputs and outputs under alternative approaches.

Tobit Regression Analysis of Pure Technical Efficiency of PACS

Table 5 presents Tobit estimates for pure technical efficiency of PACS using same independent variables as used in technical efficiency model for intermediation, operating and value-added approaches. Pure technical efficiency capture DMUs managerial ability in combining different inputs in right proportions and producing higher levels of outputs. Of the three approaches, an intermediation approach is found to be better suited for PACS. The summary measure for the model in the form of log-likelihood estimate for pure technical efficiency under intermediation approach is larger than log-likelihood estimate under operating and value-added approaches.

Under intermediation approach, an increase in asset turnover ratio, an increase in the number of Kisan credit cards (strength of member-patronage and transactional advantages), use of society's own funds (sign of diversifications and less dependence on outside for financial assistance) and the number of self-help groups (new and strong class of customers) positively influence pure technical efficiency of PACS. These factors are statistically significant. As in the case of overall technical efficiency score, pure technical efficiency score is inversely related to capital-labor ratio suggesting decreasing productivity of capital across PACS. The productivity of employee is negatively related to pure technical efficiency score as scope for productivity gains is limited by lack of expansion of businesses which are largely conditioned by institutional and resources constraints facing DMUs in rural areas. Another negative factor which is pulling down pure technical efficiency score is management expenses. The establishment and management expenses are unavoidable despite low business turnover in many PACS. There are several PACS which must scale up their business operations to reach the stage of being viable business entity. Due to predominance of small value customers (small and marginal farmers, retailers in rural areas), the transaction cost for PACS is found to be high and thus are negatively affecting pure technical efficiency.

Table 6 show the estimated coefficients for some independent variables under operating approach are lower than coefficients estimate under intermediation approach except for management expenses. Under operating approach, a unit increase in the management

Table 5. Dependent variable: Pure Technical efficiency under Alternative approaches -Tobit estimates

Independent Variables	Intermediation approach		Operating approach		Value-Added Approach	
	Estimate	Pr (>t)	Estimate	Pr(>t)	Estimate	Pr(>t)
(Intercept)	0.0522916 (0.299)	0.765	0.6637449 (3.095)	0.00197**	0.5497725 (2.409)	0.01600*
CapLab	−0.0005501 (−0.119)	0.906	−0.0037787 (−0.573)	0.56696	0.0161123 (2.152)	0.03142*
AssetTR	0.2666334 (3.923)	8.76e-05***	0.1724259 (1.951)	0.05111	0.1040582 (1.055)	0.29160
EMPR	−0.0001338 (−0.438)	0.661	0.0002852 (0.692)	0.48900	0.0008118 (1.506)	0.13209
KCCs	0.002242 (4.753)	2.00e-06***	0.0003635 (0.626)	0.53147	0.0019665 (3.063)	0.00219**
Log (OWNF)	0.5496782 (2.327)	0.020*	0.1251133 (0.428)	0.66858	−0.1790889 (−0.586)	0.55760
SHGs	0.0076900 (3.980)	6.88e-05***	0.0045020 (1.786)	0.07412	0.0068863 (2.569)	0.01019*
Expenses	−0.0394168 (−0.563)	0.573	−0.4601016 (−4.787)	1.69e-05***	−0.0598894 (−0.602)	0.54738
Log-sigma	−2.2046868 (−23.462)	< 2e-16***	−1.8135703 (−22.160)	< 2e-16***	−1.8867208 (−18.550)	< 2e-16***
Log-likelihood	24.59258 on 9 DF		8.795762 on 9 DF		−3.554957 on 9 DF	

Note: Values within brackets are t-values.
signif. codes: 0 '***' 0.001 '**' 0.01 '*' 0.05 '.' 0.1 '

expenses, leads to 0.4601 units decrease in the pure technical efficiency score of PACS. It is statistically significant. Under value-added approach, capital-labor ratio is positively influencing pure technical efficiency. It is statistically significant at lower level. Further, the number of Kissan credit cards and the number of self-help groups are positively influencing pure technical efficiency. They are statistically significant.

Tobit Regression Analysis of Scale Efficiency of PACS
Table 6 presents coefficient estimates for scale efficiency of PACS. Scale efficiency is the measure of average productivity gains that might be obtained from large scale of operations. As a summary measure, log-likelihood estimates for value- added approach is larger than estimates under operating and intermediate approaches.

The following results are obtained: (a). a one unit increase in capital-labor ratio leads to 0.0098 units decrease in scale efficiency of DMU; (b). One unit increase in asset-turnover ratio is associated with 0.195 units and 0.291 units increase in scale efficiency scores of DMU respectively under intermediation and operating approaches; (c). Estimated coefficients are small under all three approaches implying need for higher incentives for active involvement in business decision making in the organization; (d).

Table 6. Dependent variable: Scale efficiency under Alternative approaches -Tobit estimates

Independent Variables	Intermediation approach		Operating approach		Value-Added Approach	
	Estimate	Pr (>t)	Estimate	Pr(>t)	Estimate	Pr(>t)
(Intercept)	0.1740357 (1.371)	0.170323	0.3070458 (1.932)	0.053393	0.5166703 (4.725)	2.30e-06***
CapLab	−0.0063368 (−1.804)	0.071215	−0.0098310 (−2.012)	0.044248*	−0.0017680 (−0.548)	0.5835
AssetTR	0.1955357 (3.833)	0.000127***	0.2191750 (3.332)	0.000862***	0.0760990 (1.688)	0.0915
EMPR	0.0006358 (2.330)	0.019784*	0.0004992 (1.722)	0.084993	0.0004960 (1.987)	0.0470*
KCCs	0.0007057 (2.131)	0.033125*	−0.0003734 (−0.875)	0.381452	0.0004809 (1.623)	0.1046
Log (OWNF)	0.5575708 (3.265)	0.001093**	0.4985896 (2.299)	0.021483*	0.2315023 (1.571)	0.1162
SHGs	0.0070154 (4.939)	7.85e-07***	0.0034609 (1.841)	0.065663	0.0058675 (4.623)	3.79e-06***
Expenses	−0.0100404 (-0.189)	0.850110	−0.0706272 (−1.031)	0.302578	0.0353882 (0.740)	0.4593
Log-sigma	−2.4761433 (−29.830)	< 2e-16***	−2.1034399 (−28.029)	< 2e-16***	−2.5711871 (−29.934)	< 2e-16***
Log-likelihood	24.59258 on 9 DF		49.26364 on 9 DF		62.29287 on 9 DF	

Note: Values within brackets are t-values.
signif. codes: 0 '***' 0.001 '**' 0.01 '*' 0.05 '.' 0.1 '

Coefficient estimate for a variable, number of Kissan Credit Card is found to be positive under intermediation and value- added approaches, but it is found to be negative under operating approach; (e). The coefficient estimates for scale efficiency are higher than estimates found for pure technical efficiency suggesting scale of operation is an important factor in influencing efficiency scores of PACS; (f). SHGs had come to be an integral part of PACS and as such they are recognized not just as business sub-entity, but rightly as partners in social engineering for women economic empowerment. This variable is statistically significant under intermediation and value-added approaches at the conventional level of statistical significance; (g). Coefficient estimate for managerial expenses under scale efficiency model is lower than coefficient estimate found for pure technical efficiency model indicating that scale of operations across DMUs do not exert sufficient counterforce to or neutralize the influence of low pure technical efficiency leading to overall technical inefficiency. A large portion of technical inefficiency among DMUs can be attributed to pure technical inefficiency emanating from managerial decision making on choice of inputs used and outputs produced than to the size and scale of operations undertaken by DMUs in rural areas.

7 Conclusion, Recommendations, Limitations and Future Research

Conclusion

In this paper, we made some attempts to contribute to an understanding of levels of technical, pure technical and scale efficiency of primary agriculture cooperative credit societies in Tamil Nadu in India as they were less researched subject using the methodology of data envelopment analysis. The study utilized input-oriented intermediate, value-added and operating approaches to measure the efficiency of the cooperative credit societies. The study showed that a large proportion of DMUs are operating with greater resources constraints than they can serve cooperative members. It is found PACS suffer from pure technical inefficiency which brings down overall technical efficiency of PACS. Adoption of modern management practices will go a long way in improving pure technical efficiency of PACS. Presence of high scale efficiency among DMUs is not sufficient to counter relatively low level of pure technical efficiency to neutralize and rise the overall technical efficiency of PACS.

In the second stage, technical efficiency and its components pure technical efficiency and scale efficiency scores were regressed on seven independent variables to find out their contributions to efficiency scores. They were estimated with maximum likelihood techniques. The results showed; (a) asset turnover ratio (b) number of Kissan Credit Cards (KCC) and (c) number of self-help groups (SHGs) had positive effect on components of TE, with varying degrees of statistical significance. This study brings out the significance of PACS with favorable resource endowments in diversification of businesses and increasing efficiency levels of PACS; Another interesting finding from the point of view of policy perspective is self-help groups have come to be an integral part of PACS and as such they are recognized not just as business sub-entity but rightly as partners in social engineering for women empowerment. The increase in the number of self-help groups increases technical efficiency of PACS. Women Self-help groups is a significant factor to be reckoned with in the viability of PACS. Self-help groups act as resource centers for deposit mobilization and credit dispensation to their members. Another valuable information from the perspective of the government is the need for expansion of kissan credit cards in rural areas. The government efforts to provide working capital to small and marginal farmers through the implementation of Kissan Credit cards has been yielding positive results. This scheme indirectly influenced higher volume of businesses and contributed to increasing technical efficiency of PACS. There is immense scope for financing in rural areas in the light of government focus on doublings of farmers income. PACS must venture out beyond conventional financing molds to diversify their business operations into food processing units, horticulture development, floriculture and animal husbandry for attaining higher technical efficiency, profitability and sustainability in the long-run.

Recommendations

1. PACS must actively involve in organizing and developing women self-help groups (SHGs) in rural areas as they are going to be most significant stakeholders in the cooperative credit system in the years to come. SHGs operate as bullwork against male migration. The increasing migration of male members from villages in recent

years in search of better employment opportunities in urban areas, have reduced the male members in the cooperative societies.

2. The scale of finance per Kissan Credit Card must be increased so that farmer can have high working capital for his varied production needs which would indirectly increase the business turnover of the PACS.
3. NABARD must provide larger loans to PACS through district central cooperative bank for agriculture jewel loans to the farmers. This could increase their volume of business.
4. At present, PACS are largely involved in providing credit for short-term agriculture operation which limited their growth potential in rural areas; capacity building and managerial training are required so that they must be venturing into project financing and diversifying their loan portfolios for long term sustainable growth.
5. In the context of wider use of banking technology, restricting PACS to operate within certain specified geographical areas, sounds odd with the need for large scale of business operation. Sound business principle requires economics of scale. The geographical limitations for PACS limits their expansion plans for resource mobilization and deployment. Cooperative acts and rules must be amended so that contiguous PACS must be merged or weak PACS must be merged with the stronger ones to reap advantages of economics of scale in operations to provide all round services to the farmers.
6. The Union Ministry of Cooperation has advocated repositioning of primary cooperative credit societies in India. The policy guidelines aimed at transforming PACS into multipurpose cooperatives societies so that they can provides services to members and rural population and thereby earn non-banking income for long run sustainability. Managerial assistance and IT-related services are provided to the PACS to enable them to undertake multiple services. Activities identified are Aadhaar registration to voter registration; LPG booking to railway booking to Air and Bus booking; PACS are encouraged to act as distributing agency for petrol and diesel in rural areas. The aim is that rural populations can avail varied services at lesser cost without travelling to urban centers. PACS can undertake those activities for their business expansion and serve their members more effectively.

Limitations of this Study

The data set largely depended on the financial variables. Physical variables such as number of deposit accounts, loans accounts, and total number of banking transactions accounts in a year would have provided more insights into the performance of PACS.

1. This study does not consider X-efficiency factors, especially psychological factors like individual inertia and work efforts of employees, which play an important role in the efficiency levels of PACS.
2. This study has not considered the impact of incentive systems for members and employees in the organization along with bureaucratic decision making and cooperative acts and rules on the performance of the organization.

Future Research Agenda

1. Future research must investigate the optimum size of rural credit society. There are numerous unviable PACS operating in rural areas. The amalgamations and mergers of weak PACS would contribute to serving the interest of members more effectively.
2. There is need for detailed research into the effectiveness of cooperative governance which is less researched area in PACS. Though cooperative acts and rules are known for more than 100 years, their effectiveness and hindrances to the working of PACS is less known. Board members are not aware of their duties and responsibilities and their role in the management of credit societies. This results in low resource mobilization and low business turnover

Acknowledgement. I am highly thankful to the Editors and two anonymous referees for valuable comments and suggestions which had significantly contributed to an improvement in the quality and presentation of the paper.

References

Ara, S., Akter, H.: Technical and scale efficiency of scheduled banks of bangladesh in terms of SME finance using data envelopment analysis (DEA). In: Emrouznejad, A., Banker, R., Ahn, H., Afsharian, M., (eds.) 2016, Data Envelopment Analysis and Applications: Proceedings of the 13th International Conference of DEA, August, pp. 159–165. Braunschweig, Germany (2015). ISBN: 978 1 85449 497 9

Ara, S.: Comparison between conventional banking and islamic banking in terms of X-efficiency using data envelopment analysis and malmquist productivity index analysis. In: Emrouznejad, A., Banker, S.C. Ray, Chen, L., (eds) 2016, Recent Applications of Data Envelopment Analysis: Proceedings of the 14th International Conference of DEA, pp. 19–25. Jiangham University, Wuhan, China (2016). ISBN:978 1 85449 4139

Ataullah, A., Le, H.: Economic reforms and bank efficiency in developing countries: the case of indian banking industry. Appl. Financ. Econom. **16**, 653–663 (2006)

Bakshi, A., Sinha, D.: A DEA based framework for proposed merger of public sector banks in India. In: Emrouznejad, A., Jablonsky, J., Banker and Toloo, M., (eds) 2017, Recent Applications of Data Envelopment Analysis: Proceedings of the 15th International Conference of DEA, pp.1–8. University of Economics, Prague, Czech Republic (2017). ISBN:978 1 85449 4337

Bhat, M.S., Bhat, S.A.: Financial performance and efficiency of cooperative banks in Jammu & Kashmir (India). J. Coopera. Acc. Rep. **2**(1). Summer, 16–36 (2013)

Berger, A.N., Humphrey, D.B.: Efficiency of financial Institutions: International survey and directions for future research. Eur. J. Oper. Res. **98**, 175–212 (1997)

Berger, A.N., Hanweck, G.A., Humphrey, D.B.: Competitive viability in banking: scale, scope, and product mix economies. J. Monetary Econom. **20**, 501–520 (1987)

Bhatia, A., Mahendru, M.: Factors affecting profit efficiency of scheduled commercial banks in India: a panel tobit regression analysis. Prajnan **XLVI**(3), 253–280 (2017)

Bhatia, A., Megha:, Revenue and cost efficiency analysis: empirical study of indian scheduled commercial banks. Prajnan **XLVI**(2), 155–183 (2017)

Banker, R.D., Charnes, A., Cooper, W.: Some models for estimating technical and scale inefficiencies in data envelopment analysis. Manage. Sci. **30**(9). September, 1078–1092 (1984)

Bhattacharyya, A., Lovell, C.A.K., Sahay, P.: The impact of liberalization on the productive efficiency of Indian commercial Banks. Eur. J. Oper. Res. **98**, 332–345 (1997)

Charnes, A., Cooper, W.W., Rhodes, E.: Measuring the efficiency of decision-making units. Eur. J. Oper. Res. **2**, 429–444 (1978)

Coelli, T., Rao, D.S., Prasada, Battese, G.E.: An Introduction to Efficiency and Productivity Analysis, Kluwer Academic Publishers, USA (1998)

Coelli, T.: A Guide to DEAP version 2.1: Data Envelopment Analysis (Computer) Programe, 8/96 CEPA Working Paper, Department of Econometrics, University of New England, Armidale, NSW 2351, Australia (1997)

Cook, W. D., Seiford, L.M.: Data envelopment analysis DEA – thirty years on. European J. Opera. Res. (192) 1-17 (2009)

Das, A., Nag, A., Ray, Subhash, C.: Liberalization, ownership and efficiency in indian banking: a nonparametric analysis. Econom. Politic. Weekly **40**(12), 19–25 (2005)

Das, A., Ghose, S.: Financial deregulation and profit efficiency: a non-parametric analysis of Indian Banks. J. Econ. Bus. **66**, 509–528 (2009)

Das, A., Ghose, S.: Financial deregulation and efficiency: an empirical analysis of indian banks during the post reform period. Rev. Financ. Econom. **15**, 193–221 (2006)

Deb, J.: Diversification impact on bank efficiency: a panel data analysis. Prajnan **XLVI**(3). 231–252 (2017)

Domingo, E., Cabanda, E.: An analysis of micro finance efficiency: asset and intermediation approaches using the DEA model. In: Emrouznejad, A., Banker, R,S.M. Doraisamy, Arabi, B., (eds) 2014, Recent Developments in Data Envelopment Analysis and Applications: Proceedings of the 12th International Conference of DEA, pp. 56–63. Kula Lumpur, Malaysia (2014). ISBN: 978 1 85449 487 0

Emrouznejad, A., Cabanda, E.: Introduction to data envelopment analysis and its applications. In: Osman et al (Eds). Handbook of Research on Strategic Performance Management and Measurement using Data Envelopment Analysis, pp. 235–255, IGI Global USA (2015)

Emrouznejad, A., Yang, G.L.: A survey and analysis of the first 40 years of scholarly literature in DEA: 1978–2016, socio-economic planning sciences **61**(2018), 4–8

Farrell, M.J.: The measurement of productive efficiency. J. Royal Stat. Soc. Seri. A **120**, 253–290 (1957)

Ganesan, N.: Data envelopment analysis of state and district cooperative banks in India: exploratory results. IUP J. Bank Manage. Aug-Nov.11, **III**(3 & 4), 37–53 (2009)

Gaurav, S., Krishnan, J.: How efficient are india's cooperative banks? evidence from DCCBs. Econom. Politic. Weekly **52**(12), 115–124 (2017)

Golcukcu, A, Bal, H.: Customer efficiency versus firm efficiency: a banking example. In: Baker R., Emrouznejad, A., Bal, H., Alp, I., Ali Cengiz, M., (eds) 2013, Data Envelopment Analysis and Performance Measurement: Proceedings of the 11th International Conference of DEA, pp. 93–100. Samsun, Turkey (2013). ISBN: 978 1 85449 477 1

Gupta, O.K., Doshit, Y., Chinubhai, A.: Dynamics of productive efficiency of indian banks. Int. J. Oper. Res. **5**(2), 78–90 (2008)

Ismail, M.: A DEA analysis of bank performance in Malaysia. In: Emrouznejad, A., Podinovski, V., (eds) 2004, Data Envelopment Analysis and Performance Management: Proceedings of the 4th International Conference on DEA, pp. 15–21. Birmingham, UK (2024). ISBN:0 902683 73X

Leightner, J.E., Lovell, C.A.K.: The impact of financial liberalization on the performance of thai banks. J. Econ. Bus. **50**, 115–131 (1998)

Mergoni, A., Emrouznejad, A., Witte, K.D.: Fifty years of data envelopment analysis. Eur. J. Oper. Res. **326**(3), 389–412 (2025). https://doi.org/10.1016/j.ejor.2024.12.049

Patel, G.N., Pande, S.: Assessing performance of organized pharmacy retail stores using data envelopment analysis. In: Banker, R., Emrouznejad, A, L.M. Lopes, M.R.d. Almeida, (eds)

2012, Data Envelopment Analysis: Theory and Applications: Proceedings of the 10th International Conference on DEA, August 2012, pp. 41–47. Natal, Brazil (2012). ISBN: 978 185449 437 5

Ram Mohan, T.T., Ray, Subhash, C.: Comparing performance of public and private sector banks- a revenue maximization efficiency approach. Econom. Politic. Weekly **39**(12). March, 20–26 (2004)

Reda, M.: Bank efficiency and liberalization: a case study of egyptian commercial banks. In: Emrouznejad, A., Green, R., Krovonozhko, V., (eds) 2002, Efficiency and Productivity Analysis in 21st Century: Proceedings of the 3rd International Conference on DEA, June 2002, pp.68–81. Moscow, Russia (2002). ISBN:5–93467–007–7

Reddy, A.A.: Productivity Growth in Regional Rural Banks. Econom. Politic. Weekly **41**(11), Money, Banking and Finance, March 18–24, 1079–1086 (2006)

Reserve Bank of India. Report of the Agricultural Credit Review Committee - A Review of Agricultural Credit System in India. *Reserve Bank of India,* Mumbai (1989)

Reserve Bank of India. Report of the Expert Committee to Examine Three-Tier Short-Term Cooperative Credit Structure (Prakash Bakshi, Chairman of the Committee), Reserve Bank of India, Mumbai (2012)

Reserve Bank of India.: Banking Statistics. Reserve Bank of India, Mumbai (2021–22)

Sahoo, B.K., Sengupta, J.K., Mandal, A.: Productive performance evaluation of the banking sector in india using data envelopment analysis. Int. J. Opera. Res. **4**(2), 63–79 (2007)

Sealey, C.W., Jr., Lindley, J.T.: Inputs, outputs, and a theory of production and cost at depository financial institutions. J. Finan. **32**(4), 1251–1266 (1977)

Sampaio, B., Silva, L.: On the measurement of social efficiency in microfinance institutions. In: Banker, R., Emrouznejad, A, L.M. Lopes, M.R.d. Almeida, (eds) 2012, Data Envelopment Analysis: Theory and Applications: Proceedings of the 10th International Conference on DEA, August 2012, pp.201–208. Natal, Brazil (2012). ISBN: 978 185449 437 5

Sharma, G., Sharma, D.: X-Efficiency of indian commercial banks and their determinants of service quality: a study of post global financial services in Emrouznejad, Banker, A.R., Ahn, H., Afsharian, M., (eds) 2016, Data Envelopment Analysis and Applications: Proceedings of the 13th International Conference of DEA, August 2015, pp.181–186. Braunschweig, Germany (2015). ISBN: 978 1 85449 497 9

Ravi Inder, S.: efficiency analysis of central cooperative banks in punjab: data envelopment analysis approach. Prajnan **XLIV**(1), 53–66 (2015)

The Task Force on Revival of Rural Cooperative Credit Institutions (RRCCI, 2004). National Bank for Agriculture and Rural Development (NABARD), Mumbai

Fuzzy DEA Model with Exogenously Fixed Variables for Ranking of Renewable Energy Sources

Jyoti Luhaniwal(✉), Shivi Agarwal, and Trilok Mathur

Birla Institute of Technology and Science, Pilani, Pilani Campus,
Pilani 333031, Rajasthan, India
{p20200065,shivi,tmathur}@pilani.bits-pilani.ac.in

Abstract. As the global population grows, so does the demand for energy. India, with its fast growth, industrialization, and urbanization, is struggling to meet energy needs using traditional sources. To tackle energy shortages, pollution, and climate change, it's important to find cost-effective and environment friendly alternatives. Renewable energy sources (RESs) offer a promising solution, making it important to prioritize them. India has strong potential in technologies like solar, geothermal, hydro, biomass, wave energy, and onshore and offshore wind energy. However, prioritizing these energy options involves considering many factors, often with conflicting priorities. This study proposed a fuzzy Data Envelopment Analysis (DEA) method to prioritize renewable energy sources in India, considering exogenously fixed variables that can't be controlled, and handling undesirable variables. The proposed model ranks RESs effectively. It is revealed from results that Offshore wind energy is found to be the most efficient, followed by onshore wind and hydro energy, while geothermal energy ranks the lowest. The proposed methodology and findings can help developing nations and policymakers make better decisions when adopting renewable energy sources.

Keywords: Sustainable development · Renewable energy sources · Fuzzy Data envelopment analysis

1 Introduction

Sustainable development is crucial for societal and national well-being, focusing on progress that meets current needs without compromising future generations [17,21,42]. Energy is vital for sustainable progress, and with global population and industrial growth, the demand for natural resources and energy continues to grow. Since 1950, the global population has more than doubled, and energy consumption has increased sevenfold, with a projected 30–40% rise by 2030 [19,53]. In India, recent energy demand increases stem from societal progress, population growth, and economic development. Government initiatives like "The Saubhagya Scheme" and "24*7 Power for All", along with policies like Make in India and

A. Emrouznejad et al. (Eds.): DEA 2024, LNOR, pp. 251–266, 2025.
https://doi.org/10.1007/978-3-031-98177-7_17

Atmanirbhar Bharat, will further boost energy demand [55,56]. India's installed power capacity has grown significantly, from 1,713 MW in 1950 to 426,131 MW by 2023, representing annually an average growth rate of 10.2%. As a rapidly developing nation, India is projected to require an additional power capacity of 600-1,200 GW by 2050 [28]. India's energy sector faces challenges like supply shortages, greenhouse gas emissions, fuel scarcity, reliability issues, efficiency concerns, and significant land needs. Global environmental issues, especially climate change, making these challenges worse. Renewable energy resources offer a promising solution, meeting growing electricity demands while reducing emissions and land use. As of December 2023, India's energy mix includes 56% thermal, 11% hydro, 2% nuclear, and 31% other renewable sources [2]. India's commitment to reach net zero emissions by 2070, announced at COP 26, highlights the need to shift to renewable energy sources [1]. These initiatives support global climate objectives and promote sustainable development by cutting carbon emissions and providing fair access to clean energy. Embracing renewable energy will help India tackle environmental issues while stimulating economic growth and creating jobs [1]. But the high costs of plant construction and operation, coupled with regulations requiring productivity assessments, make efficiency a crucial managerial consideration. Assessing the efficiency and priority of renewable energy alternatives involves various criteria [22]. The 2002 World Summit on Sustainable Development identified social, environmental, and economic dimensions as fundamental for sustainability assessment. Managing infrastructure, land needs, and source fluctuations make it challenging to find out the optimal energy source for India. This study takes into account a range of factors categorized into socio-political, environmental, technical, and economic aspects. These factors are essential for evaluating the sustainability of renewable sources for energy production. Decision-making involves uncertainty and conflicting criteria for policymakers and investors. Researchers have explored various methods to evaluate renewable energy sources (RESs). Mergoni et al. (2024) [40] highlighted the evolution of DEA over the past five decades, incorporating hybrid models, network DEA, and AI-driven approaches. The COOPER-framework synthesized DEA methodologies, improving reliability and robustness in efficiency evaluations. As DEA adapts to complex decision-making environments, its role in performance analysis remains crucial across various sectors. Many existing DEA models assume crisp input and output values, often overlooking the uncertainty and variability in these factors, leading to potential inaccuracies [45,46,48]. Emrouznejad et al. (2013) [20] reviewed fuzzy DEA methods to address imprecise and ambiguous data in efficiency measurement. They classified existing approaches into six categories, including the tolerance approach, α-level based approach, and fuzzy ranking, providing a comprehensive taxonomy of FDEA research over the past 30 years. Agarwal and Mathur (2020) [13] introduced a DEA model incorporating exogenously fixed fuzzy variables to evaluate the performance of alternatives with fixed external variables in the fuzzy context. Traditional models also tend to ignore exogenously fixed variable due to regulatory standards and technological constraints, as well as undesirable outputs such as pollutant emissions, which are critical for assessing environmental sustainability. To address these gaps, this study proposed a Fuzzy DEA method that incorporates undesirable criteria and exogenously fixed variables, enhancing decision-making accuracy in prioritizing RESs. It also includes wave and offshore wind energy as additional alternatives for a more comprehensive evaluation.

1.1 Literature Review

Numerous studies have utilized MCDM and DEA methods to evaluate RESs. Kongar and Rosentrater (2009) [30] introduced a DEA-based method to evaluate and rank RESs by efficiency and effectiveness. Their model analyzed various inputs and outputs, helping decision-makers to choose the best technology. This study addressed the rising demand for sustainable energy amid growing global needs and environmental concerns. Baysal et al. (2011) proposed a DEA model to prioritize six different renewable energy technologies, considering two groups of variables: economic and technical in fuzzy environment. Cristóbal et al. (2011) [48] applied an extended DEA model with minsum and minimax criteria to assess the performance of 13 sustainable energy sources in Spain. The analysis identified Windpower $10 < P < 50$ MW as the only efficient and non-dominated technology among those evaluated. Kolagar et al. (2020) [29] combined DEA and the fuzzy best-worst method (FBWM) to rank sustainable energy alternatives, considering economic, technical, political, social, and environmental aspects. A sensitivity analysis validated the technique's reliability in Iran. Lozano et al. (2020) [45] utilized DEA to analyze the performance of electricity production from both renewable and traditional sources across 126 countries for the period 2000 to 2016. The Variable Returns to Scale (VRS) model was employed to analyze various energy capacities and outputs, revealed that efficiency typically ranges between 60.0% and 79.9%. Saudi Arabia, Canada, and U.S. were consistently identified as efficient throughout the study period. Zhou et al. (2008) [59] carried out a comprehensive survey of literature on applying DEA to energy and environment (E&E) studies. They classified 100 studies and highlighted some problems within this field using DEA. Luhaniwal et al. (2025) [36] employed an integrated DEA-fuzzy AHP approach to rank renewable energy sources in India and offered insights into their comparative efficiency and feasibility. Sarkis and Talluri [50] evaluated eco-efficiency using DEA to assess environmental performance. Jha et al. (2017) [25] concentrated on the ranking of renewable energy alternatives in India, utilizing an Energy Index parameter. This index was derived by aggregating scores across various alternatives, which were assessed based on eleven environmental and techno-economic criteria using the Fuzzy AHP. Saglam (2018) [46] conducted an extensive examination of the effectiveness of seven key renewable energy technologies in U.S. using DEA, providing valuable insights into their comparative efficacy. By employing DEA methodologies, the research offered policymakers a robust framework to assess and prioritize RESs based on their effectiveness and performance. Saraswat et al. (2021) [49] undertook an evaluation of both traditional and renewable energy sources to assess their role in fostering development in India's energy sector. This assessment covered political, environmental, technical, societal, economic, and adaptable criteria, employing an hybrid MCDM approach. Haldar et al. (2023) [24] in the review article, explored the energy shift process, the integration of sustainable sources to reduce pollutant emissions and the roadmap for achieving net-zero emissions in India by 2030. They identified and ranks various obstacles based on their influence, aiming to expedite sustainable development in the Indian context. Its relevance extended to other emerging economies such as Indonesia, Ghana, Kenya, Brazil, Egypt, Bangladesh, Nepal, and beyond. This transition will lead to reduced carbon footprints by using renewable sources, thereby enhancing energy security. Le et al. (2023) [32] developed an ideal approach for formulating and managing energy storage setups. The simulation outcomes indicated that systems are situated in equatorial regions consistently offer greater advantages than those in moderate climates, mainly attributed to ample solar assets. The research emphasized the significance of utilizing hydrogen in energy retention and heightened dependence on deploying renewable energy sources. Ghoushchi et al. (2023) [23] evaluated the challenges hindering renewable energy adoption by considering severity, detectability, and frequency criteria merging a spherical fuzzy set with SWARA method. These obstacles were classified into legal, and political, economic, social, technical, administrative categories. Their ranking was established using CODAS. Noteworthy obstacles identified encompass market deficiencies and administrative modifications. Kaya and Kahraman (2010) [27] pursued twin goals within the article. This study utilized an combined fuzzy AHP-VIKOR methodology to find out the most suitable sustainable energy choice for Istanbul, after that they selected energy generation sites in Istanbul employing the same approach. Criteria weights were established using the AHP technique, while alternative ranking was conducted through the TOPSIS method. The deployment of wind and solar energy faces challenges such as resource intermittency, land and sea use conflicts, and infrastructure limitations, requiring strategic planning and policy support to overcome. A study by Luhaniwal et al. (2024) [38] on the strategic deployment of hybrid offshore solar and wind power plants in India provided a framework to tackle these challenges, integrating GIS and MCDM methods to identify suitable locations. Additionally, the assessment of offshore wind farm site suitability highlighted the importance of efficient site selection in maximizing energy potential and minimizing conflicts [37]. These studies contributed to better decision-making for the sustainable expansion of offshore renewable energy. Mishra et al. (2024) [41] evaluated the operational and environmental performance of 20 coal/lignite-based thermal power plants in India using a two-stage DEA approach. Their study highlighted the impact of factors like unit size, vintage, and plant type on efficiency, emphasizing the role of regulatory benchmarks in performance assessment. Existing DEA models for evaluating the ranking of RESs often rely on crisp data, assuming that all input and output variables have precise and fixed values [45,46,48]. However, these methods fails to capture the inherent uncertainty in some factors, which can fluctuate and are not always well-represented by exact numbers. As a result, traditional DEA models may lead to inaccurate assessments due to their inability to account for this variability. Moreover, these models often overlook exogenously fixed variables affected by regulatory standards and technological constraints that are beyond the control of energy operators yet significantly influence RES performance. Additionally, existing studies typically do not address undesirable outputs, like pollutant emissions, which are critical for evaluating the environmental sustainability of energy sources. To overcome these limitations, this study proposes a novel fuzzy DEA model that better reflects the complexities of evaluating RESs efficiency in India. By representing uncertain factors as fuzzy numbers, the model effectively captures the inherent variability and uncertainty in the data, resulting in a more accurate and realistic assessment. Furthermore, the model integrates exogenously fixed variables and undesirable outputs within a fuzzy environment, providing a more realistic approach for efficiency evaluation. This comprehensive model addresses the gaps in traditional DEA methods, offering a significant advancement in accurately assessing the performance and sustainability of renewable energy sources. The remainder of this paper is organized as follows: Sect. 2 presents an overview of available RES. Section 3 outlines the proposed methodology. Section 4 lists parameters for RES evaluation and presents the findings, sensitivity analysis, and implications for policy and management. Finally, Sect. 5 provides conclusions and recommendations based on the study's results.

2 Renewable Energy Sources and Literature Review

2.1 Renewable Energy Sources

India, one of Asia's largest countries, spans from $8°40'$ to $37°6'$ North latitude with $68°7'$ to $97°25'$ East longitude, covering around 3.287 million km^2. Rapid population rise and urbanization have led growth in energy demand, with consumption projected to increase from 650 Mtoe in 2022 to 1,200 Mtoe by 2030 [3]. As of December 2023, India's energy capacity includes 56% thermal, 11% hydro, 2% nuclear, and 31% renewable sources, shown in Fig. 1 [2]. The government is working to increase renewable energy with different initiatives. Even with economic growth, energy supply hasn't matched the demand, showing the need for sustainable sources. India, ranking fourth globally in installed renewable, solar, and wind energy capacities, is working to address this issue [5].

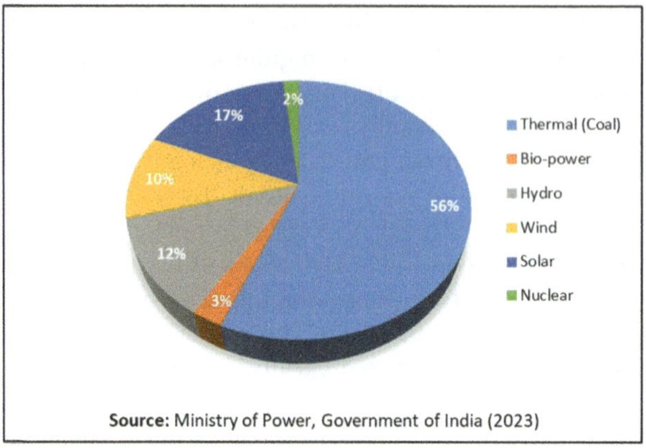

Fig. 1. Installed power capacity of India.

Solar Energy: Solar power is an abundant and limitless renewable energy source. India's strategic location results in high solar radiation, from 1200 to 2300 kWh/m^2/year, and around 3000 sun hours annually, indicating immense solar potential. The Jawaharlal Nehru National Solar Mission, initiated in 2010, committed to increase solar energy production, targeting 2000 MW of off-grid and 20,000 MW of grid solar power [47]. Although solar energy is highly sustainable, it involves some emissions due to materials like mercury and cadmium in solar cell manufacturing [58].

Hydro Energy: Hydroelectric power plants convert water's potential energy into electricity, leveraging the natural water cycle. These plants offer quick response times, on/off capabilities, and energy storage for various demands. India ranks fifth globally in hydroelectric potential, with its first plant in Darjeeling starting in 1879 [15]. By 2022, India's hydro capacity reached 46,512 MW, up from 39,788.40 MW in 2013 [31]. Though hydropower projects require significant investment, expanding mini- and micro-hydro systems can provide cost-effective electricity [44]. While emitting fewer pollutants, these projects can disrupt agriculture, land use, and communities [18].

Biomass Energy: India, with its agricultural focus, has ample biomass resources like agricultural residue, animal waste, and organic waste [52]. Various technologies convert biomass into energy, with a significant potential of 17,538 MW from agricultural residue, 5,000 MW from bagasse, and 2,556 MW from organic waste [25]. Biomass has historically been a key energy source due to its environmental benefits, abundant supply, renewability, and job creation in rural areas. Driven by cost reduction and efficiency improvements, India has over 10,205.61 MW of biomass energy production facilities [2]. Leading states in biomass energy production include Gujarat, Madhya Pradesh, Rajasthan, and Punjab. However, biomass energy emits greenhouse gases, NOx, and SOx [58].

Geothermal Energy: Geothermal energy, sourced from Earth's heat and visible in hot springs and geysers, is used in countries like the USA, Indonesia,

Turkey, Philippines, and New Zealand. In India, the Geological Survey of India (GSI) estimates a potential of 10 GW [26]. This reliable renewable energy is optimal for both off-grid and grid-connected setups, benefiting rural areas with power, industrial applications, space heating, and more [43]. Exploration in India began in 1862, with the GSI identifying 340 thermal springs and promising areas like Puga Valley in Ladakh. Despite a potential of 10,600 MW, much remains untapped [54].

Onshore Wind Energy: Onshore wind power provides electricity for both residential and industrial uses, offering a mature, cost-effective technology with a well-established supply chain [33]. It generates power by transforming the kinetic energy of air through wind turbines, which result from the Earth's uneven heating. As a renewable energy source, onshore wind reduces reliance on fossil fuels. By the end of 2023, India had nearly 44.96 GW of onshore wind capacity and is set to play an important role in meeting its energy goals [4]. To assess wind potential, 794 specialized platforms have been installed at elevations from 20 to 120 m [2].

Offshore Wind Energy: Offshore wind has been used globally for about 20 years, starting with Denmark's first turbine in 1991. Now, over 18 countries, including the UK, the Netherlands, Germany, China, and Denmark, have offshore wind projects. India has no offshore wind capacity yet, but assessments suggest a potential of 70 GW off Tamil Nadu and Gujarat [2]. Offshore turbines, which are 5 to 10 MW each, are larger than onshore ones (2 to 3 MW each). Despite higher costs due to marine infrastructure, offshore wind offers potential for improved tariffs as technology advances [2].

Wave Energy: Wave energy is derived from the movement of water surfaces caused by wind interacting with the sea or ocean. It is usually captured from waves with average heights of 2 to 3 m. Though considered more efficient than solar and wind energy, wave energy's adoption is limited by the high costs of conversion [16]. A December 2014 survey estimated the theoretical wave energy capacity at 41,300 MW [6]. Wave energy generation relies on wave amplitude and duration. Initial evaluations suggest that harnessing wave energy along India's coastline could yield 5–15 MW per meter, with a theoretical potential of 40–60 GW.

3 Methodology

3.1 Proposed Fuzzy DEA Model with Exogenously Fixed Variables

Data Envelopment Analysis is a non-parametric method in Operations Research for evaluating the efficiency of DMUs with multiple outputs and inputs. Unlike parametric methods, DEA uses linear programming to create an efficiency frontier, identifying efficient and inefficient units. The Charnes, Cooper, Rhodes

(CCR) model, introduced in 1978 by Charnes, Cooper, and Rhodes, was the first DEA model. The Banker, Charnes, Cooper (BCC) model, developed in 1984, allows for variable returns to scale. Further advancements include the Cone Ratio Model, Slack Based Model (SBM) [57], Slack Adjusted Model, and New Slack Model (NSM) [14]. Traditional models don't incorporate slacks, while the NSM addresses this by maintaining monotonicity and radial characteristics, proposed by Agarwal et al. (2011) [14]. The dual of the NSM model ensures all multipliers are positive, indicating full input and output utilization. DEA models evaluate the weightage of output and input variables to get the highest relative efficiency of a DMU while ensuring that the efficiency of other DMUs remains less than or equal to 1 with the similar weights. Efficiency evaluation in DEA can be approached in two directions: maximization of outputs with a fixed set of inputs, known as the input-oriented DEA model, or minimizing inputs for a given set of outputs, known as the output-oriented DEA model. The output-oriented NSM model with variable returns to scale (VRS) is used in this study. Uncertainty is an inherent aspect of real-life problems. This issue can be addressed using a fuzzy approach. This model handles exogenously fixed variables, which are affected by regulatory standards and technological constraints that are beyond the control of energy operators. Numerous fuzzy DEA models have been developed and implemented in DEA literature to manage such uncertainty [11,12,39]. Consider R DMUs denoted as $(r = 1, 2, \ldots, R)$ which produce n outputs defined by fuzzy numbers $\tilde{y}_{pr}; (p = 1, 2, \ldots, n)$ using m inputs $\tilde{x}_{ir}; (i = 1, 2, \ldots, m)$. The output-oriented fuzzy NSM-VRS DEA model with exogenously fixed outputs is given as follows.

$$
\begin{aligned}
&\text{Max}\, \bar{\phi}_k = \phi_k + \frac{1}{m+n}\left(\sum_{p=1}^{m} \frac{s_{pk}^{+}}{\tilde{y}_{pk}} + \sum_{i=1}^{n} \frac{s_{ik}^{-}}{\tilde{x}_{ik}}\right) \\
&\text{subject to} \\
&\sum_{r=1}^{R} \lambda_{rk}\tilde{y}_{pr} - s_{pk}^{+} = \phi_k \tilde{y}_{pk} \quad \forall p \in O_D \\
&\sum_{r=1}^{R} \lambda_{rk}\tilde{y}_{pr} - s_{pk}^{+} = \tilde{y}_{pk} \quad \forall p \in O_F \\
&\sum_{r=1}^{R} \lambda_{rk}\tilde{x}_{ir} + s_{ik}^{-} = \tilde{x}_{ik} \quad \forall i = 1, 2, \ldots, n, \\
&\sum_{r=1}^{R} \lambda_{rk} = 1, \\
&\phi_k \text{ unrestricted}, \ \lambda_{rk} \geq 0, \quad \forall r = 1, 2, \ldots, R \\
&s_{ik}^{-}, s_{pk}^{+} \geq 0, \quad \forall i = 1, 2, \ldots, n; \ \ p = 1, 2, \ldots, m.
\end{aligned}
\tag{1}
$$

where $\{\tilde{y}_{pr} : p \in O_D\}$ is discretionary outputs set and $\{\tilde{y}_{pr} : p \in O_F\}$ denotes exogenously fixed outputs set of rth DMU. λ_{rk} represents the intensity variable, s^+_{pk} signifies the slack in the pth output of the kth DMU, s^-_{ik} indicates the slack in the ith input of the kth DMU. θ_k, free in sign, represents the proportional adjustment employed to all outputs of the kth DMU to enhance efficiency. Moreover, Model (1) is converted into a fuzzy DEA version, solved using the possibility approach introduced by Lertworasirikul [34] in 2003. Output variables are expressed as triangular fuzzy numbers, denoted by $\tilde{y}_{pr}$ for each rth DMU. The possibility DEA-NSM model is then developed using five distinct levels of α (0, 0.25, 0.50, 0.75, and 1) as follows:

$$
\begin{aligned}
&\text{Max}\, \bar{\phi}_k = \left(\phi_k + \frac{1}{m+n}\left(\sum_{p=1}^{m}\frac{s^+_{pk}}{y_{pk}} + \sum_{i=1}^{n}\frac{s^-_{ik}}{x_{ik}}\right)\right)^U_\alpha \\
&\text{subject to} \\
&\left(\sum_{r=1}^{R}\lambda_{rk}y_{pr} - s^+_{pk} - \phi_k y_{pk}\right)^L_\alpha \le 0 \quad \forall p \in O_D. \\
&\left(\sum_{r=1}^{R}\lambda_{rk}y_{pr} - s^+_{pk} - \phi_k y_{pk}\right)^U_\alpha \ge 0 \quad \forall p \in O_D, \\
&\left(\sum_{r=1}^{R}\lambda_{rk}y_{pr} - s^+_{pk} - y_{pk}\right)^L_\alpha \le 0 \quad \forall p \in O_F, \\
&\left(\sum_{r=1}^{R}\lambda_{rk}y_{pr} - s^+_{pk} - y_{pk}\right)^U_\alpha \ge 0 \quad \forall p \in O_F, \\
&\left(\sum_{r=1}^{R}\lambda_{rk}x_{ir} + s^-_{ik} - x_{ik}\right)^L_\alpha \le 0 \quad \forall i = 1, 2, \ldots, n, \\
&\left(\sum_{r=1}^{R}\lambda_{rk}x_{ir} + s^-_{ik} - x_{ik}\right)^U_\alpha \ge 0 \quad \forall i = 1, 2, \ldots, n, \\
&\sum_{r=1}^{R}\lambda_{rk} = 1, \\
&\phi_k \text{ unrestricted}, \ \lambda_{rk} \ge 0, \quad \forall r = 1, 2, \ldots, R \\
&s^-_{ik}, s^+_{pk} \ge 0, \quad \forall i = 1, 2, \ldots, n; \ \ p = 1, 2, \ldots, m.
\end{aligned}
\tag{2}
$$

Model (2) assesses the relative effectiveness of DMU k. According to Lertworasirikul (2003) [34], if ϕ_k is greater than or equal to 1 for any α level, the kth DMU is deemed possibilistically efficient. Otherwise, it is considered possibilistically inefficient.

4 Results and Discussion

This study proposes a fuzzy DEA model that utilizes triangular fuzzy sets and accounts for exogenously fixed undesirable variables to rank renewable energy sources in India. For efficiency evaluation of sustainable energy sources for India, a hierarchical framework is utilized, consisting of single input, three outputs and seven RESs are used as alternatives, namely solar, onshore wind, offshore wind, hydro, biomass, geothermal, wave energy. These input-output variables and alternatives are derived from a comprehensive review of decision-making studies on energy in the existing literature [45,46,48,49]. The input-output variables utilized for the decision-making process are discussed in the lines below.

In order to comprehensively evaluate various technologies from an economic perspective, it is essential to consider cost-related factors. To ensure a comprehensive analysis without excessive complexity, this study have chosen to utilize an aggregated cost indicator known as the "Levelized Cost of Electricity (LCOE)". This indicator enables us to encompass all relevant financial aspects within the analysis. LCOE stands for the cost of energy generated per unit. It is expressed in Rs/kWh or $/kWh. This variable is considered as single input variable. The capacity factor, an output variable, defined as the ratio of actual electricity generated to the maximum possible output of a power plant, is influenced by varying environmental conditions, maintenance schedules, and operational strategies. These factors create uncertainty and imprecision in its measurement, making it suitable for treatment in a fuzzy environment. By using fuzzy logic, the inherent variability and uncertainties, capacity factor is converted into a triangular fuzzy number, allowing for a more nuanced and accurate evaluation in performance assessments. Next output variable is Job creation, which refers to the capacity of an energy project to generate potential employment opportunities. The implementation of renewable energy should consider the impact on local residents by evaluating improvements in quality of life and the potential for job creation. Pollutant Emission is a crucial indicator of the ecological footprint of renewable energy sources, inherently tied to the technology used. Since it reflects negative environmental consequences, it is considered an undesirable output. The level of emissions is primarily influenced by external factors, such as regulatory standards and technological choices made during installation, which are beyond the control of the operators. Therefore, this variable should be treated as both undesirable and exogenously fixed, highlighting its negative impact and the lack of control over its reduction.

Present DEA models often have limited use for crisp input and output variables. However, in cases where qualitative data is involved in performance evaluation and cannot be measured precisely, this data can be represented by using fuzzy numbers. Additionally, some outputs are fixed and beyond management's control. This paper extends the NSM-DEA model to work with fuzzy data to better assess the efficiency of RESs with undesirable and exogenously fixed variables. There are various ways to handle fuzzy DEA models, and this study proposes a model based on the possibility approach of the fuzzy NSM-DEA model [11,12,39]. This approach treats fuzzy data as events and fuzzy DEA models are transformed into possibility DEA models, similar to chance-constrained programming deals with uncertainty. Further, to handle the undesirable output variable, this study applied linear monotonic decreasing transformation approach, which is implemented by Seiford and Zhu (2002) [51]. A positive constant β_i is summed to the reciprocal additive transformation of the undesirable output i for each DMU to ensure that the resulting values are positive. The environmental DEA technology is then expressed as $f(u_i^k) = -u_i^k + \beta_i$. Lu and Lo (2007b) [35] applied this technique to assess the performance of various areas in China, considering both environmental and economic factors and considered GDP as the desirable output, while dust, soot, and sulfur dioxide are considered as undesirable factors.

Table 1. Data for implementing efficiency analysis.

RESs	Levelized Cost of Electricity	Capacity factor	Job Creation	Pollutant Emission
Solar	0.0447	(18, 20, 22)	10600	49.174
Onshore Wind	0.049	(36, 38, 40)	2620	17.652
Offshore Wind	0.075	(41, 43, 45)	8643.6	13
Hydro	0.048	(43, 45, 47)	2500	22.696
Biogas	0.067	(66, 68, 70)	24310	58
Geothermal	0.056	(75, 77, 79)	27050	18.913
Wave	0.08	(15, 17.5, 20)	5000	35

All the data is presented in Table 1. To conduct the efficiency analysis, data have been collected from various secondary sources [1–3,6]. Quantitative data regarding to input and output variables are extracted from multiple annual reports and balance sheets of India's renewable energy organizations [5,7–10]. After modifying the undesirable output, data is organized accordingly to apply the Fuzzy NSM-VRS output-oriented DEA model to calculate the efficiency scores for each of the seven RESs. These computations are performed using MATLAB version R2020b software and the results are tabulated in Table 2.

Table 2. Efficiency scores at 5 possibility levels and ranking of RESs.

RESs	$\alpha = 0$	$\alpha = 0.25$	$\alpha = 0.50$	$\alpha = 0.75$	$\alpha = 1$	Average	Rank
Solar	0.7801	0.7513	0.721	0.6924	0.6627	0.7215	4
Onshore Wind	1.0122	1	1	0.964	0.9056	0.97636	2
Offshore Wind	1.0833	1.0599	1.0312	1.0157	1	1.03802	1
Hydro	1.016	0.9925	0.9792	0.961	0.9321	0.97616	3
Biomass	0.7151	0.6835	0.661	0.6527	0.6413	0.67072	6
Geothermal	0.4951	0.4823	0.4551	0.4232	0.4069	0.45252	7
Wave	0.7402	0.7251	0.7116	0.6998	0.6753	0.7104	5

It displays the efficiency scores for each of the alternative RESs across different possibility levels. At a possibility level of $\alpha = 0$, offshore wind, onshore wind, and hydro energy show high efficiency scores, all greater than 1. In contrast, solar, biomass, and wave energy have slightly lower efficiency scores of 0.7801, 0.7151, and 0.7402, respectively. Geothermal energy exhibits the lowest efficiency score at 0.4951. When the possibility level is increased to $\alpha = 0.50$, only wind energy–both offshore and onshore–remains efficient. Hydro energy becomes slightly inefficient with an efficiency score of 0.9792, while all other energy sources are inefficient at this level. As α further progresses towards 1, the only efficient source is offshore wind energy. Onshore wind is slightly inefficient with value of 0.97636, followed by hydro, wave, solar, and biomass, while geothermal performs poorly with a score of 0.019. At all these possibility levels, this study identifies offshore wind as the best suitable RES and Geothermal as the worst. Consequently, the efficiencies at different possibility levels are averaged to determine the overall efficiencies and rankings of each renewable energy source, as illustrated in Table 2. The ranking of energy sources using the Fuzzy DEA method is as follows: Offshore Wind > Onshore Wind > Hydro > Solar > Wave > Biomass > Geothermal. These rankings are based on the overall efficiency scores, with Offshore Wind Energy emerging as the most suitable renewable energy source for India's future, boasting the highest overall efficiency value of 1.03802 among the available sources. Onshore energy achieved the second rank with an efficiency score of 0.97636, followed by hydro energy in third place with an efficiency of 0.97616. Solar, Wave, Biomass, and Geothermal energy obtained the fourth, fifth, sixth, and seventh ranks respectively, as shown in Table 2.

5 Conclusion

Renewable energy is crucial for achieving long-term environmental stability, particularly as India's energy needs rise due to industrialization, urban growth, and population increase. Prioritizing renewable energy is key to combating climate change and reducing reliance on diminishing fossil fuels while addressing issues like energy shortages and pollution. These technologies not only lower carbon emissions but also promote economic growth and create jobs. However, deploying wind and solar energy comes with challenges such as resource intermittency, land and sea use conflicts, and infrastructure constraints, which must be addressed through strategic planning and policy support. This study focused on assessing the ranking of different energy sources, considering seven types of renewable energy for production. A Fuzzy DEA model with exogenously fixed variables is developed to prioritize these renewable energy sources in India, specifically addressing undesirable outputs in a fuzzy environment. Based on the results of Fuzzy NSM-DEA model, offshore wind, onshore wind, and hydro energy sources are ranked as the top three among RESs in India. Additionally, geothermal and biomass exhibit comparatively lower allure compared to other RESs. Offshore wind energy emerges as the top choice, consistently securing the highest efficiency value across various possibility levels discussed above. Employing an Fuzzy DEA approach, this article ranks renewable energy sources based on efficiency scores. Future studies could explore advanced DEA models capable of handling undesirable factors more effectively. Additionally, different types of fuzzy sets, such as trapezoidal and interval type-2 fuzzy numbers, could be considered based on specific contexts. Since this study considers only cost, excluding factors like coal, fuel, and water consumption, the results may have some limitations. Future studies can incorporate these aspects for a more comprehensive analysis. This study focuses on India, and its findings may not be globally applicable due to varying RESs and parameters. Future research could explore similar issues in other countries or apply this methodology to different industries.

References

1. Ministry of Environment, Forest and Climate Change. https://moef.gov.in/moef/index.html. Accessed 19 Mar 2024
2. Ministry of New and Renewable Energy. https://mnre.gov.in. Accessed 19 Mar 2024
3. International Energy Agency. https://www.iea.org/. Accessed 20 Mar 2024
4. National Power Portal. https://npp.gov.in. Accessed 20 Mar 2024
5. Renewables 2022 Global Status Report. https://www.ren21.net/gsr-2022/. Accessed 20 Mar 2024
6. Tidal & Waves Energy in India: Survey on the Potential & Proposition of a Roadmap. https://www.ireda.in/doc/publications/sarve2.pdf. Accessed 20 Mar 2024
7. Renewable Capacity Statistics 2023. https://www.irena.org/. Accessed 22 Mar 2024
8. World Energy Transitions Outlook 1.5 °C Pathway. https://www.irena.org/. Accessed 22 Mar 2024
9. The cost of financing for renewable power. https://www.irena.org/. Accessed 23 Mar 2024
10. Renewable Power Generation Costs in 2021. https://www.irena.org/. Accessed 23 Mar 2024
11. Agarwal, S.: Efficiency measure by fuzzy data envelopment analysis model. Fuzzy Inf. Eng. **6**(1), 59–70 (2014)
12. Agarwal, S.: Fuzzy slack based measure of data envelopment analysis: a possibility approach. In: Pant, M., Deep, K., Nagar, A., Bansal, J.C. (eds.) Proceedings of the Third International Conference on Soft Computing for Problem Solving. AISC, vol. 258, pp. 733–740. Springer, New Delhi (2014). https://doi.org/10.1007/978-81-322-1771-8_63
13. Agarwal, S., Mathur, T.: DEA model for exogenously fixed variables under fuzzy environment. In: Choudhury, S., Mishra, R., Mishra, R.G., Kumar, A. (eds.) Intelligent Communication, Control and Devices. AISC, vol. 989, pp. 895–901. Springer, Singapore (2020). https://doi.org/10.1007/978-981-13-8618-3_93
14. Agarwal, S., Yadav, S.P., Singh, S.P.: A new slack DEA model to estimate the impact of slacks on the efficiencies. Int. J. Oper. Res. **12**(3), 241–256 (2011)
15. Agrawal, A., Kumar, A., Rao, T.J.: Future of Indian power sector reforms: electricity amendment bill 2014. Energy Policy **107**, 491–497 (2017)
16. Alkan, M.A.: Education and training of renewable energy resources in Turkey. Master's thesis, Fen Bilimleri Enstitüsü (2009)

17. Atilgan, B., Azapagic, A.: An integrated life cycle sustainability assessment of electricity generation in Turkey. Energy Policy **93**, 168–186 (2016)
18. Büyüközkan, G., Güleryüz, S.: Evaluation of renewable energy resources in Turkey using an integrated MCDM approach with linguistic interval fuzzy preference relations. Energy **123**, 149–163 (2017)
19. Ebrahimi, M., Rahmani, D.: A five-dimensional approach to sustainability for prioritizing energy production systems using a revised GRA method: a case study. Renew. Energy **135**, 345–354 (2019)
20. Emrouznejad, A., Tavana, M., Hatami-Marbini, A.: The state of the art in fuzzy data envelopment analysis. In: Performance Measurement with Fuzzy Data Envelopment Analysis, pp. 1–45 (2014)
21. Eren, B.M., Taspinar, N., Gokmenoglu, K.K.: The impact of financial development and economic growth on renewable energy consumption: empirical analysis of India. Sci. Total Environ. **663**, 189–197 (2019)
22. Ervural, B.C., Zaim, S., Demirel, O.F., Aydin, Z., Delen, D.: An ANP and fuzzy TOPSIS-based SWOT analysis for Turkey's energy planning. Renew. Sustain. Energy Rev. **82**, 1538–1550 (2018)
23. Ghoushchi, S.J., Garg, H., Bonab, S.R., Rahimi, A.: An integrated SWARA-CODAS decision-making algorithm with spherical fuzzy information for clean energy barriers evaluation. Expert Syst. Appl. **223**, 119884 (2023)
24. Haldar, S., Peddibhotla, A., Bazaz, A.: Analysing intersections of justice with energy transitions in India-a systematic literature review. Energy Res. Soc. Sci. **98**, 103010 (2023)
25. Jha, S.K., Puppala, H.: Prospects of renewable energy sources in India: prioritization of alternative sources in terms of energy index. Energy **127**, 116–127 (2017)
26. Kahraman, C., Kaya, İ: A fuzzy multicriteria methodology for selection among energy alternatives. Expert Syst. Appl. **37**(9), 6270–6281 (2010)
27. Kaya, T., Kahraman, C.: Multicriteria renewable energy planning using an integrated fuzzy VIKOR & AHP methodology: the case of Istanbul. Energy **35**(6), 2517–2527 (2010)
28. Khare, V., Nema, S., Baredar, P.: Status of solar wind renewable energy in India. Renew. Sustain. Energy Rev. **27**, 1–10 (2013)
29. Kolagar, M., Hosseini, S.M.H., Felegari, R., Fattahi, P.: Policy-making for renewable energy sources in search of sustainable development: a hybrid DEA-FBWM approach. Environ. Syst. Decis. **40**, 485–509 (2020)
30. Kongar, E., Kurt, A., Rosentrater, A.: Data envelopment analysis model for renewable energy technology selection. In: Northeast Decision Sciences Institute Proceedings, pp. 579–584 (2009)
31. Kumar, D., Katoch, S.: Sustainability indicators for run of the river (RoR) hydropower projects in hydro rich regions of India. Renew. Sustain. Energy Rev. **35**, 101–108 (2014)
32. Le, T.S., Nguyen, T.N., Bui, D.K., Ngo, T.D.: Optimal sizing of renewable energy storage: a techno-economic analysis of hydrogen, battery and hybrid systems considering degradation and seasonal storage. Appl. Energy **336**, 120817 (2023)
33. Lee, H.C., Chang, C.T.: Comparative analysis of MCDM methods for ranking renewable energy sources in Taiwan. Renew. Sustain. Energy Rev. **92**, 883–896 (2018)
34. Lertworasirikul, S., Fang, S.C., Joines, J.A., Nuttle, H.L.: Fuzzy data envelopment analysis (DEA): a possibility approach. Fuzzy Sets Syst. **139**(2), 379–394 (2003)
35. Lu, W.M., Lo, S.F.: A closer look at the economic-environmental disparities for regional development in china. Eur. J. Oper. Res. **183**(2), 882–894 (2007)

36. Luhaniwal, J., Agarwal, S., Mathur, T.: An integrated DEA-fuzzy AHP method for prioritization of renewable energy sources in India. Sci. Rep. **15**(1), 838 (2025)
37. Luhaniwal, J., Agarwal, S., Mathur, T.: Assessment of offshore wind farm site suitability in India using GIS and MCDM methods. Oper. Res. Int. J. **25**(2), 31 (2025)
38. Luhaniwal, J., Puppala, H., Agarwal, S., Mathur, T.: Framework for strategic deployment of hybrid offshore solar and wind power plants: a case study of India. J. Clean. Prod. **479**, 144009 (2024)
39. Mahla, D., Agarwal, S.: A credibility approach on fuzzy slacks based measure (SBM) DEA model. arXiv preprint arXiv:2102.01933 (2021)
40. Mergoni, A., Emrouznejad, A., De Witte, K.: Fifty years of data envelopment analysis. Eur. J. Oper. Res. **326**(3), 389–412 (2025). https://doi.org/10.1016/j.ejor.2024.12.049
41. Mishra, S., Singh, A., Sufia: Environmental performance of coal-based thermal generation plants: a two-stage DEA model with undesirable outputs. In: International Conference on Data Envelopment Analysis, pp. 151–163. Springer, Cham (2023)
42. Moldan, B., Janoušková, S., Hák, T.: How to understand and measure environmental sustainability: indicators and targets. Ecol. Ind. **17**, 4–13 (2012)
43. Pillai, I.R., Banerjee, R.: Renewable energy in India: status and potential. Energy **34**(8), 970–980 (2009)
44. Reddy, V.S., Kaushik, S., Panwar, N.: Review on power generation scenario of India. Renew. Sustain. Energy Rev. **18**, 43–48 (2013)
45. Rodríguez-Lozano, G., Cifuentes-Yate, M.: Efficiency assessment of electricity generation from renewable and non-renewable energy sources using data envelopment analysis. Int. J. Energy Res. **45**(13), 19597–19610 (2021)
46. Saglam, Ü.: The efficiency assessment of renewable energy sources with data envelopment analysis. In: 2018 Annual Meeting of the Decision Sciences Institute Proceedings, Chicago IL (2018)
47. Sahoo, S.K.: Renewable and sustainable energy reviews solar photovoltaic energy progress in India: a review. Renew. Sustain. Energy Rev. **59**, 927–939 (2016)
48. San Cristóbal, J.R.: A multi criteria data envelopment analysis model to evaluate the efficiency of the renewable energy technologies. Renew. Energy **36**(10), 2742–2746 (2011)
49. Saraswat, S., Digalwar, A.K.: Evaluation of energy alternatives for sustainable development of energy sector in India: an integrated Shannon's entropy fuzzy multi-criteria decision approach. Renew. Energy **171**, 58–74 (2021)
50. Sarkis, J., Talluri, S.: Ecoefficiency measurement using data envelopment analysis: research and practitioner issues. JEAPM **6**(01), 91–123 (2004)
51. Seiford, L.M., Zhu, J.: Modeling undesirable factors in efficiency evaluation. Eur. J. Oper. Res. **142**(1), 16–20 (2002)
52. Sen, S., Ganguly, S., Das, A., Sen, J., Dey, S.: Renewable energy scenario in India: opportunities and challenges. J. Afr. Earth Sci. **122**, 25–31 (2016)
53. Şengül, Ü., Eren, M., Shiraz, S.E., Gezder, V., Şengül, A.B.: Fuzzy TOPSIS method for ranking renewable energy supply systems in Turkey. Renew. Energy **75**, 617–625 (2015)
54. Shanker, R., Geological Survey of India: Geothermal atlas of India. Geological Survey of India (1991)
55. Singh, R.K., Murty, H.R., Gupta, S.K., Dikshit, A.K.: An overview of sustainability assessment methodologies. Ecol. Ind. **15**(1), 281–299 (2012)
56. Srikanth, R.: India's sustainable development goals-glide path for India's power sector. Energy Policy **123**, 325–336 (2018)

57. Tone, K.: A slacks-based measure of efficiency in data envelopment analysis. Eur. J. Oper. Res. **130**(3), 498–509 (2001)
58. Tripathi, L., Mishra, A., Dubey, A.K., Tripathi, C., Baredar, P.: Renewable energy: an overview on its contribution in current energy scenario of India. Renew. Sustain. Energy Rev. **60**, 226–233 (2016)
59. Zhou, P., Ang, B.W., Poh, K.L.: A survey of data envelopment analysis in energy and environmental studies. Eur. J. Oper. Res. **189**(1), 1–18 (2008)

Application of Data Envelopment Analysis on Bank Asset and Liability Management

Jacob Muvingi(✉)

University of Kurdistan Hewler, Erbil, Kurdistan Region of Iraq, Iraq
jacob.muvingi@ukh.edu.krd
http://www.ukh.edu.krd/faculty/13772

Abstract. Data envelopment analysis (DEA) is a method for identifying the best practices among peer decision-making units (DMUs). Bank profitability is a function of a bank's asset and liability management. The sensitivity of a bank's assets and liabilities to the movement of interest rates is a key aspect of a bank asset and liability management. The difference between rate-sensitive assets (RSA) and liabilities (RSL) represents a bank's income gap (IG). The nature of a bank's IG affects the bank's interest-bearing income. Banks generally face a decision-making challenge associated with choosing the optimum RSL, RSA, and IG structure. The current study proposed a IG efficiency analysis for a hypothetical banking sector based on three scenarios of a bank's IG; negative IG, positive IG, and a mixture of positive and negative IG. The semi-oriented radial measure model was used in the study to handle negative IG values. Targets for the outputs in each scenario were provided. Overally it was discovered that of the three scenarios, the negative IG scenario yield a higher number of efficient banks.

Keywords: Data envelopment analysis · Bank asset and liability management · Gap · Semi-oriented radial measure

1 Introduction

A banking institution's main activities are deposit-taking and lending. To attract deposits, banks offer deposit rates to customers through various accounts, including savings accounts, fixed deposit accounts, call accounts, money market deposits accounts, among many other types of investment accounts. The deposit rates paid by the bank are a form of an expense to the bank, since deposits represent a major proportion of a bank's liabilities. To balance the liabilities side, banks offer different types of loans to customers. Banks receive interest income from loan interest rates charged to customers. The difference between the deposit interest rates and loans interest rates is the net interest income of the bank. In the day-to-day operations of a bank, there is a need to match the assets and liabilities of the bank. Any mismatch between the assets and liabilities of the

A. Emrouznejad et al. (Eds.): DEA 2024, LNOR, pp. 267–279, 2025.
https://doi.org/10.1007/978-3-031-98177-7_18

bank may positively or negatively expose the bank's income to movements in the market interest rates. To proactively manage the exposure of a bank's income to changes in interest rates, bank managers proactively manage the interest-bearing assets and liabilities through asset and liability management (ALM) techniques, like income gap analysis and duration gap analysis. In the aftermath of the collapse of Silicon Valley Bank in the USA, commentators pointed to poor management and a lack of ALM discipline as the key contributors to the collapse of the bank [10].

A number of studies [15,22,23,30] were focused on bank efficiency and productivity through the non-parametric approach of data envelopment analysis (DEA). However very few studies on bank efficiency and productivity focused on bank strategic areas like asset and ALM [31], which is a key driver of bank profitability.

In order to optimise their net interest income and overall profitability while maintaining acceptable risk levels, banks implement asset and liability management strategy on their asset and liabilities portfolio [2]. A number of classical models are used by a bank to measure interest rate risk, and they include; gap analysis (repricing model), maturity model, and the duration model. Among the models for measuring and managing interest rate risk, the repricing gap is certainly the best known and most widely used [26]. In bank risk management, DEA was used in [16,19,29].

. The application of operations research in bank asset and liability management has been done through a number of models which include linear programming; [1,7,9,11,14], and [24]. The focus in a number of previous studies was based on the maximisation and minimisation of funding cost, bank costs. In my own opinion, none of the studies were focussed on the structure of the rate sensitive assets (RSA) or liabilities (RSL) and the income gap (IG) of the bank.

Outside the discipline of operations research, [13,18,25] applied IG analysis to investigate the interest risk of commercial banks in India. The basic IG analysis approach is inadequate of identifying the targets for improvements of banks within a sector, or for a single bank when the maturity buckets are treated as the decision-making units (DMUs).

In literature there is no consensus with regard to the ideal IG of a bank, for instance in [6,27], it was observed that in conditions of declining interest rates, a positive gap between RSA and RSL has decreased banks' net interest income, and the opposite was also observed. A negative IG is a situation were a bank has more RSL than RSA the assets, whereas in a positive gap there are more RSA than RSL. Handling the negative value of the IG as an output through traditional DEA models like [8] is problematic. A number of approaches in literature have been proposed to deal with negative data, [4]. The approaches include data transformation [28], changing the type of returns to scale i.e. [constant returns to scale (CRS), variable returns to scale (VRS)] [3], use of super-slacks measure model [17], and the use of the semi-oriented radial measure model (SORM) proposed in [12,20]. The existing literature focused on the bank ALM modeling is focused on positive data set, and furthermore none has applied the DEA methodology. The DEA approach allows an analysis of DMUs (banks or numerous maturity buckets) with a bigger size. It is this gap, which our current study is trying to address.

The remaining sections of the study are ordered as follows; in the second section, a background of the gap analysis and the DEA models for negative data is provided. In Sect. 3, we present the methodology to be used in generating bank efficiency ratings for the three scenarios of bank's IG. The fourth section provides a discussion of the results. The conclusion is found in the last section.

2 Background

2.1 Gap Analysis

To perform bank IG analysis which is part of bank asset and liability management, there is needy to classify assets and liabilities in terms of their interest rate sensitivity. The basic rule normally applied is that all assets and liabilities with maturities less than a year are considered rate sensitive, furthermore all assets and liabilities with variable interest rates are also considered rate sensitive. All money market securities are classified as rate sensitive, together with funding from the central bank. In [21], the single-period IG of a bank is given by the following formula;

$$RSA_t - RSL_t = IG_t \tag{1}$$

In cases where a cumulative IG is desired within a single bank over a given time horizon, i.e. one year, the cumulative IG is given as follows;

$$\sum RSA_t - \sum RSL_t = \sum IG_t. \tag{2}$$

Once the IG of a bank is established the next step will be to establish how movement in interests will affect the income of a bank, and the following formula used in [21] is used,

$$\Delta_I = \Delta_i \times IG_t, \tag{3}$$

where I is the bank income, and Δ_i is the change in interest rates, for instance if interest rates move from 10% to 15%, Δ_i will be 5%.

2.2 Standard DEA Model

The standard output-oriented DEA model with constant returns to scale of [8] is presented as follows:

$$\begin{aligned} & max\ h \\ s.t. & \sum_{j=1}^{n} \lambda_j x_{ij} \leq x_{io}\ \forall i \in I, \\ & \sum_{j=1}^{n} \lambda_j y_{rj} \geq h y_{ro}\ \forall r \in R, \\ & \sum_{j=1}^{n} \lambda_j = 1, \\ & \lambda_j \geq 0\ \forall j, \end{aligned} \tag{4}$$

where h is the objective value of the model, and $\frac{1}{h}$ is the efficiency of DMUo. The inputs and outputs are respectively represented by x_{ij} and y_{rj}, whereas the intensity variable is denoted by λ_j.

2.3 Dealing with Negative Data Through SORM

To deal with negative data i.e. negative IG, [12] proposed a partitioning approach for the input and output data [20]. Output variables having both positive and negative values for some DMUs were represented by a sum of two variables described as follows:

$$y_{kj}^1 = \begin{cases} y_{kj}, & \text{if } y_{kj} \geq 0, \\ 0, & \text{if } y_{kj} < 0, \end{cases} \quad \& \quad y_{kj}^2 = \begin{cases} 0, & \text{if } y_{kj} \geq 0, \\ -y_{kj}, & \text{if } y_{kj} < 0, \end{cases}$$

$y_k, k \in K$ are outputs which take positive values for some DMUs and negative for others.

Input variables having both positive and negative values for some DMUs were represented by a sum of two variables described as follows:

$$x_{lj}^1 = \begin{cases} x_{lj}, & \text{if } x_{lj} \geq 0, \\ 0, & \text{if } x_{lj} < 0, \end{cases} \quad \& \quad x_{lj}^2 = \begin{cases} 0, & \text{if } x_{lj} \geq 0, \\ -x_{lj}, & \text{if } x_{lj} < 0. \end{cases}$$

$x_l, l \in L$ are inputs which take positive values for some DMUs and negative for others.

[12] proposed an input-oriented SORM with VRS for some outputs with positive and negative values as follows:

$$
\begin{aligned}
& \min h \\
s.t. & \sum_{j=1}^{n} \lambda_j x_{ij} \leq h x_{io} \ \forall i, \\
& \sum_{j=1}^{n} \lambda_j y_{rj} \geq y_{ro} \quad \forall r \neq k, \\
& \sum_{j=1}^{n} \lambda_j y_{kj}^{1} \geq y_{ko}^{1}, \\
& \sum_{j=1}^{n} \lambda_j y_{kj}^{2} \leq y_{ko}^{2}, \\
& \sum_{j=1}^{n} \lambda_j = 1, \\
& \lambda_j \geq 0 \ \forall j,
\end{aligned}
\tag{5}
$$

where the optimal value of h is DMU_o efficiency, λ_j is the intensity variable.

Model (5) applies to situations where outputs are composed of both positive and negative values. [12] provided the following extension of model (5) to cater to a situation where DMUs have both positive and negative values in input and output variables.

$$
\begin{aligned}
& \min h \\
s.t. & \sum_{j=1}^{n} \lambda_j x_{ij} \leq h x_{io} \ \forall i \in I, \\
& \sum_{j=1}^{n} \lambda_j x_{lj}^{1} \leq h x_{lo}^{1} \quad \forall l \in L, \\
& \sum_{j=1}^{n} \lambda_j x_{lj}^{2} \geq h x_{lo}^{2} \quad \forall l \in L, \\
& \sum_{j=1}^{n} \lambda_j y_{rj} \geq y_{ro} \ \forall r \in R, \\
& \sum_{j=1}^{n} \lambda_j y_{kj}^{1} \geq y_{ko}^{1} \ \forall k \in K, \\
& \sum_{j=1}^{n} \lambda_j y_{kj}^{2} \leq y_{ko}^{2} \ \forall k \in K, \\
& \sum_{j=1}^{n} \lambda_j = 1, \\
& \lambda_j \geq 0 \ \forall j
\end{aligned}
\tag{6}
$$

When DMUs include input and output variables with both positive and negative values, the output-oriented VRS SORM is outlined as follows:

$$
\begin{aligned}
& max\ h \\
s.t. \sum_{j=1}^{n} \lambda_j x_{ij} &\leq x_{io}\ \forall i \in I, \\
\sum_{j=1}^{n} \lambda_j x_{lj}^1 &\leq x_{lo}^1 \quad \forall l \in L, \\
\sum_{j=1}^{n} \lambda_j x_{lj}^2 &\geq x_{lo}^2 \quad \forall l \in L, \\
\sum_{j=1}^{n} \lambda_j y_{rj} &\geq h y_{ro}\ \forall r \in R, \\
\sum_{j=1}^{n} \lambda_j y_{kj}^1 &\geq h y_{ko}^1\ \forall k \in K, \\
\sum_{j=1}^{n} \lambda_j y_{kj}^2 &\leq h y_{ko}^2\ \forall k \in K, \\
\sum_{j=1}^{n} \lambda_j &= 1, \\
\lambda_j &\geq 0\ \forall j.
\end{aligned} \tag{7}
$$

3 Research Methodology

To cater for the three scenarios of the bank's IG, the output-oriented SORM model with VRS was utilised. There are two outputs (IG and RSA) and one input (RSL) in this study. In the first instance, the IG value which is an output are all positive $RSA_j \geq RSL_j$, in that situation model (4) based on data in Table 1 is used.

In the second scenario, the IG value which is an output are all negative $RSA_j \leq RSL_j$, model (7) is adjusted as follows:

Table 1. Input and output data (first scenario)

Bank	RSL	IG	RSA
1	5,000.00	2,000.00	7,000.00
2	7,000.00	3,000.00	10,000.00
3	7,500.00	2,500.00	10,000.00
4	9,000.00	3,000.00	12,000.00
5	10,000.00	2,000.00	12,000.00
6	8,500.00	1,500.00	10,000.00
7	9,500.00	3,000.00	12,500.00
8	6,500.00	2,000.00	8,500.00
9	9,000.00	4,000.00	13,000.00
10	11,000.00	4,000.00	15,000.00
11	12,000.00	4,000.00	16,000.00
12	9,200.00	4,300.00	13,500.00
13	12,500.00	3,000.00	15,500.00
14	5,200.00	1,600.00	6,800.00
15	4,300.00	900.00	5,200.00
16	3,800.00	1,200.00	5,000.00
17	9,800.00	3,000.00	12,800.00
18	10,300.00	2,900.00	13,200.00

$$
\begin{aligned}
& max\ h \\
s.t. & \sum_{j=1}^{n} \lambda_j x_{ij} \leq x_{io}\ \forall i \in I, \\
& \sum_{j=1}^{n} \lambda_j y_{rj} \geq h y_{ro}\ \forall r \in R, \\
& \sum_{j=1}^{n} \lambda_j y_{kj}^1 \geq h y_{ko}^1\ \forall k \in K, \\
& \sum_{j=1}^{n} \lambda_j y_{kj}^2 \leq h y_{ko}^2\ \forall k \in K, \\
& \sum_{j=1}^{n} \lambda_j = 1, \\
& \lambda_j \geq 0\ \forall j.
\end{aligned} \tag{8}
$$

The data to be used in the second scenario is presented in Table 2.

Table 2. Input and output data (second scenario)

Bank	RSL	IG	RSA
1	5,000.00	−2,000.00	3,000.00
2	7,000.00	−3,000.00	4,000.00
3	7,500.00	−2,500.00	5,000.00
4	9,000.00	−3,000.00	6,000.00
5	10,000.00	−2,000.00	8,000.00
6	8,500.00	−1,500.00	7,000.00
7	9,500.00	−3,000.00	6,500.00
8	6,500.00	−2,000.00	4,500.00
9	9,000.00	−4,000.00	5,000.00
10	11,000.00	−4,000.00	7,000.00
11	12,000.00	−4,000.00	8,000.00
12	9,200.00	−4,300.00	4,900.00
13	12,500.00	−3,000.00	9,500.00
14	5,200.00	−1,600.00	3,600.00
15	4,300.00	-900.00	3,400.00
16	3,800.00	−1,200.00	2,600.00
17	9,800.00	−3,000.00	6,800.00
18	10,300.00	−2,900.00	7,400.00

In the last scenario, where both positive IG, $RSA_j \geq RSL_j$ and negative IG, $RSA_j \leq RSL_j$ gap values are observed, model (8) based on data in Table 3 is used.

4 Results and Discussion

The efficiency ratings for the three scenarios are highlighted in Table 4. The application of model (4) for scenario 1 identifies four banks (bank 1, bank 11, bank 12, and bank 16) as efficient, whilst banks (bank 2, bank 3, bank 4, bank 5, bank 6, bank 7, bank 8, bank 9, bank 10, bank 13, bank 14, bank 15, bank 17, and bank 18) are identified as inefficient. Incremental positive targets are suggested for the inefficiency DMUs (banks) as reflected in Table 5. The targets for the income gap are higher than the targets for assets.

Considering scenario 2 with a negative IG based on model (8), five banks (bank 5, bank 6, bank 13, bank 15, and bank 16) are efficient, whilst banks (bank 1, bank 2, bank 3, bank 4, bank 7, bank 8, bank 9, bank 10, bank 11, bank 12, bank 14, bank 17, and bank 18) are identified as inefficient. Decreasing positive asset targets are suggested and decreasing income gap negative targets are suggested for the inefficient DMUs (banks) as reflected in Table 6.

Table 3. Input and output data (third scenario)

Bank	RSL	IG	RSA
1	5,000.00	−2,000.00	3,000.00
2	7,000.00	−3,000.00	4,000.00
3	7,500.00	−2,500.00	5,000.00
4	9,000.00	3,000.00	12,000.00
5	10,000.00	2,000.00	12,000.00
6	8,500.00	1,500.00	10,000.00
7	9,500.00	−3,000.00	6,500.00
8	6,500.00	−2,000.00	4,500.00
9	9,000.00	−4,000.00	5,000.00
10	11,000.00	4,000.00	15,000.00
11	12,000.00	4,000.00	16,000.00
12	9,200.00	4,300.00	13,500.00
13	12,500.00	−3,000.00	9,500.00
14	5,200.00	−1,600.00	3,600.00
15	4,300.00	−900.00	3,400.00
16	3,800.00	1,200.00	5,000.00
17	9,800.00	3,000.00	12,800.00
18	10,300.00	2,900.00	13,200.00

Table 4. Efficiency ratings

Bank	Scenario 1	Scenario 2	Scenario 3
1	1.0000	0.7500	0.4355
2	0.9906	0.7000	0.3985
3	0.9200	0.8140	0.4619
4	0.9097	0.8182	0.9101
5	0.8442	1.0000	0.8442
6	0.8054	1.0000	0.8066
7	0.9079	0.8478	0.4721
8	0.9119	0.8514	0.4865
9	0.9856	0.6818	0.3792
10	0.9929	0.8140	0.9929
11	1.0000	0.8696	1.0000
12	1.0000	0.6563	1.0000
13	0.9688	1.0000	0.5938
14	0.9303	0.8630	0.4997
15	0.8914	1.0000	0.5875
16	1.0000	1.0000	1.0000
17	0.9120	0.8644	0.9120
18	0.9115	0.9046	0.9115

Lastly, using model (8) for the mixed IG scenario, three banks (bank 11, bank 12, and bank 16) are efficient, whilst banks (bank 1, bank 2, bank 3, bank 4, bank 5, bank 6, bank 7, bank 8, bank 9, bank 10, bank 13, bank 14, bank 15, bank 17, and bank 18) are identified as inefficient. Incremental positive targets are

Table 5. Observed and target values: Positive IG scenario

Bank	Observed Assets	Observed IG	Target Assets	Target IG
2	3,000.00	10,000.00	3095.24	10095.24
3	2,500.00	10,000.00	3369.05	10869.05
4	3,000.00	12,000.00	4190.48	13190.48
5	2,000.00	12,000.00	4214.29	14214.29
6	1,500.00	10,000.00	3916.67	12416.67
7	3,000.00	12,500.00	4267.86	13767.86
8	2,000.00	8,500.00	2821.43	9321.43
9	4,000.00	13,000.00	4190.48	13190.48
10	4,000.00	15,000.00	4107.14	15107.14
13	3,000.00	15,500.00	4000.00	16000.00
14	1,600.00	6,800.00	2109.52	7309.52
15	900.00	5200.00	1533.33	5833.33
17	3,000.00	12,800.00	4235.71	14035.71
18	2,900.00	13,200.00	4182.14	14482.14

Table 6. Observed and target values: Negative IG scenario

Bank	Observed Assets	Observed IG	Target Assets	Target IG
1	5000.00	−2000.00	4000.00	−1000.00
2	7000.00	−3000.00	5714.29	−1285.71
3	7500.00	−2500.00	6142.86	−1357.14
4	9000.00	−3000.00	7333.33	−1666.67
7	9500.00	−3000.00	7666.67	−1833.33
8	6500.00	−2000.00	5285.71	−1214.29
9	9000.00	−4000.00	7333.33	−1666.67
10	11000.00	−4000.00	8600.00	−2400.00
11	12000.00	−4000.00	9200.00	−2800.00
12	9200.00	−4300.00	7466.67	−1733.33
14	5200.00	−1600.00	4171.43	−1028.57
17	9800.00	−3000.00	7866.67	−1933.33
18	10300.00	−2900.00	8180.00	−2120.00

Table 7. Observed and target values: Mixed IG scenario

Bank	Observed Assets	Observed IG	Target Assets	Target IG
1	3000.00	−2000.00	6888.89	1888.89
2	4000.00	−3000.00	10037.04	3037.04
3	5000.00	−2500.00	10824.07	3324.07
4	12000.00	3000.00	13185.19	4185.19
5	12000.00	2000.00	14214.29	4214.29
6	10000.00	1500.00	12398.15	3898.15
7	6500.00	−3000.00	13767.86	4267.86
8	4500.00	−2000.00	9250.00	2750.00
9	5000.00	−4000.00	13185.19	4185.19
10	15000.00	4000.00	15107.14	4107.14
13	9500.00	−3000.00	16000.00	4000.00
14	3600.00	−1600.00	7203.70	2003.70
15	3400.00	−900.00	5787.04	1487.04
17	12800.00	3000.00	14035.71	4235.71
18	13200.00	2900.00	14482.14	4182.14

suggested for the inefficient DMUs (banks) as reflected in Table 7. The targets for the assets are higher than the targets for the income gap.

The three scenarios had a consensus on bank 16, identified as efficient. In identifying inefficient DMUs (banks), there is a consensus between scenarios 1 and 2 for banks 5, 10, 17, and 18.

5 Summary and Conclusions

A higher number of studies [13,18,25] used the IG analysis approach in the study of banks, however the models adopted are not capable of identifying targets for the variables used in the study, which may limit the banks included in the study to benchmark with the performance of banks identified as best performing. The study adopted the SORM model of [12] to handle negative values of IG in two scenarios (negative IG and mixed IG), for the positive IG scenario, a standard DEA model was used.

The models (4) and (8) reveal more efficient DMUs compared to model (7) based on a mixed IG. Overally, the second scenario yields more efficient banks, which may potray that keeping a negative IG is ideal for a bank, which is aligned with the needy for a bank to be highly liquid [5].

Future studies on applying the SORM DEA model on bank asset and liability management may need to be extended to incorporate the stochastic nature of the inputs and outputs.

References

1. Abdollahi, H.: Multi-objective programming for asset-liability management: the case of Iranian banking industry. Int. J. Ind. Eng. Prod. Res. **31**, 75–85 (2020). https://doi.org/10.22068/ijiepr.31.1.75
2. Abor, J.Y., Gyeke-Dako, A., Fiador, V.O., Agbloyor, E.K., Amidu, M., Mensah, L.: Money and Banking in Africa. AAESPD, Springer, Cham (2019). https://doi.org/10.1007/978-3-319-77458-9
3. Allahyar, M., Rostamy-malkhalifeh, M.: Negative data in data envelopment analysis: efficiency analysis and estimating returns to scale. Comput. Ind. Eng. **82**, 78–81 (2001). https://doi.org/10.1016/j.cie.2015.01.022
4. Barbosa, A.B.R.A.B., Lopes-Ahn, A., Vilela, B., Costa, M., de Oliveira Gomes, R., Ahn, H.: DEA with bootstrap, weight restrictions and negative outputs: shortcomings of the proposal for the Brazilian distribution system operators regulation. In: Emrouznejad, A., Thanassoulis, E., Toloo, M. (eds) Advances in the Theory and Applications of Performance Measurement and Management. DEA45 2023. LNOR, pp. 47–58. Springer, Cham (2024). https://doi.org/10.1007/978-3-031-61597-9_5
5. BIS: Principles for sound liquidity risk management and supervision. Basel Committee on Banking Supervision International, pp. 1–44 (2008). https://www.bis.org/publ/bcbs144.pdf
6. Brewer, E.: Bank gap management and the use of financial futures. Federal Reserve Bank of Chicago. Econ. Perspect. **9**, 12–21 (1985)
7. Chakroun, F., Abid, F.: A multiobjective model for bank asset liability management: the case of a Tunisian bank. IUP J. Financ. Risk Manag. **10**, 35–56 (2013)
8. Charnes, A., Cooper, W., Rhodes, A.: Measuring the efficiency of decision making units. Eur. J. Oper. Res. **2**, 429–444 (1978). https://doi.org/10.1016/0377-2217(78)90138-8
9. Cohen, K.J., Hammer, F.C.: Linear programming and optimal bank asset management decisions. J. Financ. **22**, 147–165 (1967)
10. Coyne, S.: Why asset and liability management should be the nerve center of modern banks. International Banker (2024). https://shorturl.at/04RS8
11. Dash, M., Pathak, R.: A linear programming model for assessing asset-liability management in banks. IUP J. Financ. Risk Manage. **8**, 50–67 (2011)
12. Emrounzejad, A., Anouze, A.L., Thanassoulis, E.: Semi-oriented radial measure for measuring the efficiency of decision making units with negative data, using DEA. Eur. J. Oper. Res. **200**, 297–304 (2010). https://doi.org/10.1016/j.ejor.2009.01.001
13. Gang, A.: Gap analysis: asset and liability management at a bank. J. Commer. Trade **16**, 14–17 (2021). https://doi.org/10.26703/jct.v16i1.32
14. Guven, S., Persentili, E.: A linear programming model for bank balance sheet management. Omega **25**, 449–459 (1997). https://doi.org/10.1016/S0305-0483(97)00008-X
15. Henriquesa, I.C., Sobreiroa, V.A., Kimuraa, H., Mariano, E.B.: Efficiency in the Brazilian banking system using data envelopment analysis. Future Bus. J. **4**, 157–178 (2018). https://doi.org/10.1016/j.fbj.2018.05.001
16. Lartey, T., James, G.A., Danso, A.: Interbank funding, bank risk exposure and performance in the UK: a three-stage network DEA approach. Int. Rev. Financ. Anal. **75**, 101753 (2021). https://doi.org/10.1016/j.irfa.2021.101753
17. Lin, R., Yang, W., Huang, H.: A modified slacks-based super-efficiency measure in the presence of negative data. Comput. Ind. Eng. **135**, 39–52 (2019). https://doi.org/10.1016/j.cie.2019.05.030

18. Makkar, A., Singh, S.: Interest rate risk of selected Indian commercial banks: an analysis of gap analysis model. IUP J. Bank Manage. **12**, 58–69 (2013)
19. Mathews, K.: Risk management and managerial efficiency in Chinese banks: a network DEA framework. Omega **41**, 207–215 (2012). https://doi.org/10.1016/j.omega.2012.06.003
20. Matin, R.K., Azizi, R.: A two-phase approach for setting targets in DEA with negative data. Appl. Math. Model. **35**, 5794–5803 (2011). https://doi.org/10.1016/j.apm.2011.05.002
21. Mishkin, F., Eakins, S.: Financial Markets and Institutions. Pearson Prentice Hall, India (2009)
22. Othman, F.M., Mohd-Zamil, N.A., Rasid, S.Z.A., Vakilbashi, A., Mokhber, M.: Data envelopment analysis: a tool of measuring efficiency in banking sector. Int. J. Econ. Financ. Issues **6**, 911–916 (2016)
23. Paradi, J.C., Zhu, H.: A survey on bank branch efficiency and performance research with data envelopment. Omega **41**, 61–79 (2013). https://doi.org/10.1016/j.omega.2011.08.010
24. Peykani, P., Sargolzaei, M., Botshekan, M. H., Oprean-Stan, C., Takaloo, A.: Optimization of asset and liability management of banks with minimum possible changes. Mathematics **11**, 2761, 1–24 (2023)
25. Raju, M.A.: Asset-liability management to control the volatility in net interest income and economic value through gap analysis of selected public and private sector banks. Int. J. Acad. Res. Bus. Soc. Sci. **6**, 123–142 (2016). https://doi.org/10.6007/IJARBSS/v6-i1/1983
26. Resti, A., Sironi, A.: Risk management and shareholders' value in banking: From risk measurement models to capital allocation policies. Wiley (2012). https://doi.org/10.1002/9781118371886
27. Rokhmawati, A.: Interest rate risk of banking sector: the effect of maturity gap on net interest income in Indonesia. J. Econ. Financ. Account. **6**, 19–31 (2019). https://doi.org/10.17261/Pressacademia.2019.1025
28. Scheel, H.: Undesirable outputs in efficiency valuations. Eur. J. Oper. Res. **132**, 400–410 (2001). https://doi.org/10.1016/S0377-2217(00)00160-0
29. Shi, X., Yu, W.: Analysis of Chinese commercial banks' risk management efficiency based on the PCA-DEA approach. Omega **41**, 1–11 (2021). https://doi.org/10.1155/2021/7306322
30. Tamatam, R., Dutta, P., Dutta, G., Lessmann: Efficiency analysis of Indian banking industry over the period 2008–2017 using data envelopment analysis. Benchmarking: Int. J. **26**, 2417–2442 (2019). https://doi.org/10.1108/BIJ-12-2018-0422
31. Vij, M.: Managing gap: a case study approach to asset-liability management of banks. Vision: J. Bus. Perspect. **9**, 49–58 (2019). https://doi.org/10.1177/097226290500900

Forecasting Cost Efficiency of Banks in India with Integrated DDF-Based Network DEA and SVR Approach

Nishtha Gupta[1(✉)], Jolly Puri[1], and Gautam Setia[2]

[1] Department of Mathematics, Thapar Institute of Engineering and Technology, Patiala 147004, Punjab, India
nnishtha_phd20@thapar.edu

[2] Department of Mechanical Engineering, Thapar Institute of Engineering and Technology, Patiala 147004, Punjab, India
gautam.setia@thapar.edu

Abstract. Cost efficiency is a significant measure to assess an organization's ability to utilize the resources to achieve the desired output at the lowest possible cost. The estimation and prediction of the cost efficiency are vital for decision-makers to strategically allocate resources, optimize operations, and forecast future financial health. This enables the identification of inefficiencies, the facilitation of strategic planning, and the enhancement of overall organizational performance. Addressing this necessity, the present study introduces a methodology to estimate the cost efficiency of decision-making units (DMUs) by employing the directional distance function (DDF)-based network data envelopment analysis (DEA) approach. To enhance the predictive capabilities, the methodology integrates a support vector machine for regression (SVR), which significantly reduces the computational resources required for re-executing the network DEA model when new units are added, especially in large and expanding datasets. The study applied this methodology to the Indian banking sector to measure the cost efficiency across two distinct divisions, 'productivity and profitability' within the network architecture. The cost efficiency scores were estimated for the banks operational in the financial period 2012–22 and predicted for the latest year 2021–22. The proposed hybrid approach demonstrates exceptional precision and accuracy in predicting cost efficiency scores when compared to actual efficiency scores, thus confirming the robustness and reliability of the integrated proposed methodology.

Keywords: Cost efficiency · Directional distance function (DDF) · Indian banking sector · Network data envelopment analysis (DEA) · Support vector regression (SVR)

1 Introduction

Cost efficiency is a metric that evaluates the effectiveness of the organizations in utilizing their resources to achieve desired outcomes at minimal costs. Estimat-

A. Emrouznejad et al. (Eds.): DEA 2024, LNOR, pp. 280–292, 2025.
https://doi.org/10.1007/978-3-031-98177-7_19

ing and predicting cost efficiency holds significant importance for the decision-makers across financial/non-financial institutions. Firstly, estimating cost efficiency assists decision makers to identify the areas of improvement, optimizes resource allocation, and enhances overall financial management. Additionally, forecasting cost efficiency helps in early identification of potential inefficiencies, allowing for proactive risk management. A well-established approach to achieving this is the integration of Data Envelopment Analysis (DEA) and Machine Learning (ML) methodologies.

Developed by Charnes et al. [5], DEA is a non-parametric linear programming technique, creates an empirical production frontier encompassing all data points, and aids measuring the efficiency of a decision-making unit (DMU) relative to other DMUs operating under similar conditions. However, traditional DEA models [7] treat DMUs as black boxes, ignoring internal operational structures. To address this, Network DEA, introduced by Färe et al. [9], extends DEA models by capturing inter-dependencies between different divisions within a network framework. Traditional DEA and network DEA models are typically either input-oriented or output-oriented. The directional distance function (DDF) approach pioneered by Chambers et al. [4], expands both DEA and network DEA models by allowing simultaneous adjustments in multiple directions, reducing inputs and increasing outputs together. With known unit-input costs, the DEA and/or network DEA models, along with DDF-based models, have been extensively employed to estimate cost efficiency scores considering both desirable and undesirable outputs, as documented in the literature [10,13,14,19,21–23]. However, to the best of our knowledge, existing literature have not explored the estimation of cost efficiency by employing the DDF-based approach within a network framework that accommodates heterogeneous input costs (varying input costs from one DMU to other), negative values of variables, and undesirable outputs in one framework. This study aims to address this gap by developing two-stage network DEA model based on DDF approach that incorporates these complexities.

Predicting cost efficiency scores for new units using DEA and network DEA requires complete re-execution of the models, which is computationally expensive, especially with exponentially growing data. To address this, the ML algorithms that automatically identify patterns in data can be integrated with the DEA methodology, enabling the predictive analysis. Numerous studies in the literature have employed DEA and ML approach to predict the efficiency scores applied in different applications [1,8,11,24]. Support vector machine for regression (SVR), has been widely utilized ML algorithm in conjunction with DEA models for forecasting the future efficiency scores [6,15]. Invented by Vapnik [25], SVR algorithm is designed to predict continuous values by finding the hyperplane that best fits the data, minimizing the prediction error within a specified margin. It fits a hyperplane that approximates the relationship between input variables and the output variable, forming a minimal-width tube (ε) around it [2,20]. Its ability to handle high-dimensional data and robustness to outliers made it an effective tool for enhancing the predictive capabilities of DEA models. To enhance the capabilities of the combined DEA and SVR approach, this study proposes a novel methodology for predicting cost efficiency within a network framework. The two-stage DDF-based network DEA model is developed to handle undesirable outputs and negative data, and further integrated with SVR algorithm to enable efficient cost efficiency prediction. The practical utility of this innovative integrated approach is demonstrated using a real dataset of the Indian banking sector, for the financial period 2012 to 2022.

The next sections of the paper are organized in the following order. Section 2 outlines the proposed approach for estimating cost efficiency using a DDF-based two-stage network DEA model. Section 3 presents the dataset of Indian banks to evaluate their cost efficiency. Section 4 summarize the key findings and discussions from the current study. In the final Sects. 5 and 6, theoretical and practical implications of the proposed approach and some concluding remarks are mentioned respectively.

2 Proposed DDF-Based Two-Stage Network DEA Cost Efficiency Model

To define the DDF-based two-stage network DEA model for estimating the cost efficiency, consider the set of n homogeneous DMUs, where each DMU_j is involved in two-stage ($k = 1, 2$) network system. Let I_1 and I_2 denote the number of exogenous inputs, while O_1^g and O_2^g represent the number of desirable outputs, and O_1^b and O_2^b indicate the number of undesirable outputs produced by divisions 1 and 2, respectively. Let $X_1 = [X_{j1}]_{1\times n} = [x_{ij1}]_{I_1\times n}$ and $X_2 = [X_{j2}]_{1\times n} = [x_{ij2}]_{I_2\times n}$ be the matrices representing the inputs utilized by the divisions 1 and 2 respectively, where X_{jk} is the input vector and x_{ijk} is the ith input of DMU_j in division k. Assume, $Y_1^g = [Y_{j1}^g]_{1\times n} = [y_{rj1}^g]_{O_1^g\times n}$ and $Y_2^g = [Y_{j2}^g]_{1\times n} = [y_{rj2}^g]_{O_2^g\times n}$ be the matrices representing the outputs produced by the divisions 1 and 2 respectively, where Y_{jk}^g is the desirable output vector and y_{rjk} be the rth desirable output of DMU_j in division k. Additionally, let $Y_1^b = [Y_{j1}^b]_{1\times n} = [y_{rj1}^b]_{O_1^b\times n}$ and $Y_2^b = [Y_{j2}^b]_{1\times n} = [y_{rj1}^b]_{O_1^b\times n}$ denote the matrices of undesirable outputs for divisions 1 and 2, respectively where Y_{jk}^b is the undesirable output vector and y_{sjk} be the sth undesirable output of DMU_j in division k. Let $Z_1^{(2)} = \left[Z_{j1}^{(2)}\right]_{1\times n} = \left[z_{bj1}^{(2)}\right]_{L_1\times n}$ be the matrix of links, where L_1 denotes the number of intermediate products from division 1 to 2, $Z_{j1}^{(2)}$ is the intermediate products' vector and $z_{bj1}^{(2)}$ is b^{th} intermediate product linked from division 1 to 2 for DMU_j. Consider c_{ij1}^{I}, c_{sj1}^{O} and c_{bj1}^{Z} be the cost corresponding to the ith variable input, s^{th} undesirable output and b^{th} intermediate product for DMU_j in division 1. This research considers homogeneous DMUs with varying input prices for each DMU and assumes that the costs associated with intermediate products, when they serve as inputs, play a crucial role in determining the cost efficiency of the network system. Therefore, total spending of inputs

$(\bar{X}_{j1})$, undesirable outputs $(\bar{Y}_{j1}^{b})$ and intermediate products $(\bar{Z}_{j1}^{(2)})$ for DMU$_j$ in division 1 are given as:

$$\begin{aligned}\bar{X}_{j1} &= [\bar{x}_{ij1}]_{I_1\times 1}, \text{ where } \bar{x}_{ij1} = x_{ij1} * c_{ij1}^{\mathrm{I}}\ \forall i,\\ \bar{Y}_{j1}^{b} &= [\bar{y}_{sj1}^{b}]_{O_1^b\times 1}, \text{ where } \bar{y}_{sj1}^{b} = y_{sj1}^{b} * c_{sj1}^{\mathrm{O}}\ \forall s, \text{ and}\\ \bar{Z}_{j1}^{(2)} &= \left[\bar{z}_{bj1}^{(2)}\right]_{L_1\times 1}, \text{ where } \bar{z}_{bj1}^{(2)} = z_{bj1}^{(2)} * c_{bj1}^{\mathrm{Z}}\ \forall b.\end{aligned}$$

It is implicit that analogous parameters can be considered for division 2. The production possibility set, denoted by PPS^N, under variable returns-to-scale assumption in the cost space is given as:

$$\begin{aligned}\mathrm{PPS}^N = \Bigg\{ & \bar{X}_1, Y_1^g, \bar{Y}_1^b, \bar{X}_2, Y_2^g, \bar{Y}_2^b, Z_1^{(2)}, \bar{Z}_1^{(2)} \ \Bigg|\ \bar{X}_1 \geq \sum_j \lambda_{j1}\bar{X}_{j1},\ Y_1^g \leq \sum_j \lambda_{j1} Y_{j1}^g,\\ & \bar{Y}_1^b = \sum_j \lambda_{j1}\bar{Y}_{j1}^b, \bar{X}_2 \geq \sum_j \lambda_{j2}\bar{X}_{j2},\ Y_2^g \leq \sum_j \lambda_{j2} Y_{j2}^g,\ \bar{Y}_2^b = \sum_j \lambda_{j2}\bar{Y}_{j2}^b,\\ & Z_1^{(2)} \leq \sum_j \lambda_{j1} Z_{j1}^{(2)},\ \bar{Z}_1^{(2)} \geq \sum_j \lambda_{j2}\bar{Z}_{j1}^{(2)}, \sum_j \lambda_{j1} Z_{j1}^{(2)} = \sum_j \lambda_{j2} Z_{j1}^{(2)}, \qquad (1)\\ & \sum_j \lambda_{j1} = 1, \lambda_{j1} \geq 0, \sum_j \lambda_{j2} = 1, \lambda_{j2} \geq 0, \forall j = 1, 2, \ldots, n. \Bigg\}\end{aligned}$$

The inequalities in Eq. (1), corresponding to the inputs and desirable outputs of both divisions represents strong disposability conditions [3]. The undesirable outputs in the system are assumed to follow weak disposability conditions, hence designated by the equality constraint. Two constraints which relate to the links between divisions, termed as intermediate products, represents the output produced from the 1st division and input-spending transmitted to division 2 respectively. The equality constraint, $(\sum_{j=1}^{n} \lambda_{j1} Z_{j1}^{2} = \sum_{j=1}^{n} \lambda_{j2} Z_{j1}^{2})$ represents the continuity between intermediate products ensuring that these products are transferred and utilized across the divisions without any wastage. The last constraint, $\sum_{j=1}^{n} \lambda_{jk} = 1$ indicates that technology set is convex and allows for VRS. By employing the input-oriented DDF approach, we seek simultaneous reduction in total-cost of inputs $(\bar{X}_1, \bar{X}_2)$ and undesirable outputs $(\bar{Y}_1^b, \bar{Y}_2^b)$, along with the total cost reduction of intermediate products serving as an input $(\bar{Z}_1^2)$ to the successive division. To directly address the presence of negative data, [16] provided a set of direction vectors. Analogously, the following set of direction vectors are established for divisions 1 and 2:

$$\begin{aligned}&\mu_{i1} = \min_j \{\bar{x}_{ij1}\}\ \forall i = 1, \ldots, I_1; \quad \mu_{i2} = \min_j \{\bar{x}_{ij2}\}\ \forall i = 1, \ldots, I_2;\\ &\nu_{s1} = \max_j \{\bar{y}_{sj1}^b\}\ \forall s = 1, \ldots, O_1^b; \quad \text{and} \quad \nu_{s2} = \max_j \{\bar{y}_{sj2}^b\}\ \forall s = 1, \ldots, O_2^b.\end{aligned} \qquad (2)$$

The input oriented DDF-based network DEA model to assess the system cost efficiency is given below:

Model 1.

$$\textbf{Min CE}_o^N = 1 - \frac{1}{2}\left[\frac{\beta_{o1}^I \sum_{i=1}^{I_1} (\bar{x}_{io1} - \mu_{i1}) + \beta_{o1}^O \sum_{s=1}^{O_1^b} (\bar{y}_{so1}^b - \nu_{s1})}{\sum_{i=1}^{I_1} \bar{x}_{io1} + \sum_{s=1}^{O_1^b} \bar{y}_{so1}^b} + \frac{\beta_{o2}^I \sum_{i=1}^{I_2} (\bar{x}_{io2} - \mu_{i2}) + \beta_{o2}^O \sum_{s=1}^{O_2^b} (\bar{y}_{so2}^b - \nu_{s2})}{\sum_{i=1}^{I_2} \bar{x}_{io2} + \sum_{s=1}^{O_2^b} \bar{y}_{so2}^b}\right]$$

subject to constraints

Stage 1: $$\sum_{j=1}^{n} \lambda_{j1}\, \bar{x}_{ij1} \le \bar{x}_{io1} - \beta_{o1}^I (\bar{x}_{io1} - \mu_{i1}), \forall i = 1, \ldots, I_1;$$

$$\sum_{j=1}^{n} \lambda_{j1}\, y_{rj1}^g \ge y_{ro1}^g \forall r = 1, \ldots, O_1^g,$$

$$\sum_{j=1}^{n} \lambda_{j1}\, \bar{y}_{sj1}^b = \bar{y}_{so1}^b - \beta_{o1}^O (\bar{y}_{so1}^b - \nu_{s1}) \forall s = 1, \ldots, O_1^b,$$

$$\sum_{j=1}^{n} \lambda_{j1}\, z_{bj1}^2 \ge z_{bo1}^2 \forall b = 1, \ldots, L_1,$$

$$\sum_{j=1}^{n} \lambda_{j1} = 1, \ \lambda_{j1} \ge 0 \, \forall j = 1, 2, \ldots, n,$$

Stage 2: $$\sum_{j=1}^{n} \lambda_{j2}\, \bar{x}_{ij2} \le \bar{x}_{io2} - \beta_{o2}^I (\bar{x}_{io2} - \mu_{i2}) \forall i = 1, \ldots, I_2,$$

$$\sum_{j=1}^{n} \lambda_{j2}\, y_{rj2}^g \ge y_{ro2}^g \forall r = 1, \ldots, O_2^g,$$

$$\sum_{j=1}^{n} \lambda_{j2}\, \bar{y}_{sj2}^b = \bar{y}_{so2}^b - \beta_{o2}^O (\bar{y}_{so2}^b - \nu_{s2}) \forall s = 1, \ldots, O_2^b,$$

$$\sum_{j=1}^{n} \lambda_{j2}\, \bar{z}_{bj1}^2 \le \bar{z}_{bo1}^2 \forall b = 1, \ldots, L_1,$$

$$\sum_{j=1}^{n} \lambda_{j1}\, z_{bj1}^2 = \lambda_{j2}\, z_{bj1}^2 \forall b = 1, \ldots, L_1,$$

$$\sum_{j=1}^{n} \lambda_{j2} = 1, \ \lambda_{j2} \ge 0 \, \forall j = 1, 2, \ldots, n,$$

where $\beta^I_{o1}, \beta^I_{o2}$ and $\beta^O_{o1}, \beta^O_{o2}$ calculates the proportion of inefficiency in the inputs and undesirable outputs, respectively of DMU_o in the both divisions.

Definition 1 (Cost-efficient DMU). *The overall two-stage network system for DMU_o under evaluation can be considered as cost efficient only if the optimal value CE^{N*}_o obtained from Model 1 is equal to 1.*

Definition 2 (Cost efficiency of divisions). *The cost efficiency of each division, denoted by CE^{N*}_{o1}, CE^{N*}_{o2}, of the DMU_o can be evaluated using the solution vector $(\beta^{*I}_{o1}, \beta^{*I}_{o2}, \beta^{*O}_{o1}, \beta^{*O}_{o2})$ optimized from the Model 1, defined as:*

$$CE^{N*}_{o1} = 1 - \left(\frac{\beta^{*I}_{o1} \sum_{i=1}^{I_1} (\bar{x}_{io1} - \mu_{i1}) + \beta^{*O}_{o1} \sum_{s=1}^{O^b_1} (\bar{y}^b_{so1} - \nu_{s1})}{\sum_{i=1}^{I_1} \bar{x}_{io1} + \sum_{s=1}^{O^b_1} \bar{y}^b_{so1}} \right) \quad \textit{and} \tag{3}$$

$$CE^{N*}_{o2} = 1 - \left(\frac{\beta^{*I}_{o2} \sum_{i=1}^{I_2} (\bar{x}_{io2} - \mu_{i2}) + \beta^{*O}_{o2} \sum_{s=1}^{O^b_2} (\bar{y}^b_{so2} - \nu_{s2})}{\sum_{i=1}^{I_2} \bar{x}_{io2} + \sum_{s=1}^{O^b_2} \bar{y}^b_{so2}} \right). \tag{4}$$

3 Application to Indian Banking Sector

The panel dataset is compiled from statistical tables relating to banks in India, an annual publication of the Reserve Bank of India (RBI) (2012–2022) database. The present work employed a panel dataset comprising of 41 Indian scheduled banks, including both Public sector banks (PuSBs) and Private sector banks (PrSBs), as listed in Table 1. Over the 11-year period (2012–2022), the dataset theoretically accounts for 451 bank-year observations. However, due to mergers and consolidations in the banking sector, the actual number of observations is reduced to 427. RBI has facilitated several mergers, particularly among public sector banks, which have led to a reduction in the total number of banks.

This study aims to comprehensively assess the bank's performance by evaluating cost efficiencies within a network architecture comprising two interconnected divisions: productivity and profitability as shown in Fig. 1. Input-output data variables are selected based on previous research to assess these operations carried out by the banks [12,17,18]. Productivity division represents operational efficiency, where banks utilize resources such as employees and fixed assets to generate deposits. Profitability division measures financial performance, where deposits and borrowings are used to generate loans, investments, and profits while also accounting for undesirable outputs like Gross NPAs. Profit (loss) after tax can take negative values, capturing financial losses incurred by banks. Deposits act as an intermediate measure, linking both divisions, emphasizing the influence of operational productivity on financial profitability. In the model,

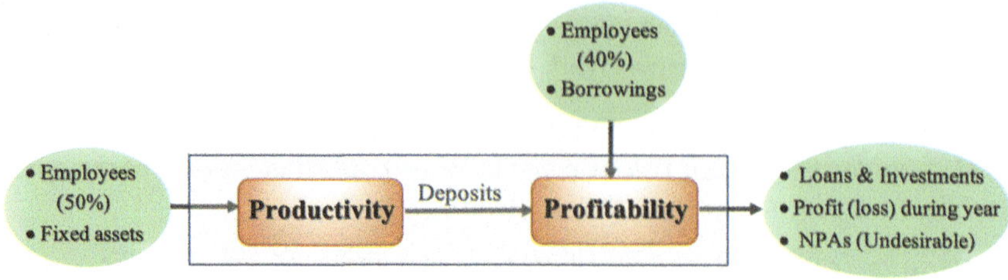

Fig. 1. Two-stage performance model of a bank.

50% of employees are engaged in productivity-related operations, such as transaction processing and customer service, which drive deposit generation. The profitability division, relying on productivity efficiency, involves 40% of employees in financial decision-making and revenue generation, while the remaining 10% are engaged in other miscellaneous activities not directly linked to these divisions. The detailed descriptions and statistical metrics for the variables used in the bank's production model can be found in Table 2. The cost associated with the inputs, intermediates and undesirable outputs are defined and explained concisely in Table 3.

Table 1. List of the Indian banks and corresponding bank codes analyzed in the study.

Bank Code	Bank Name (PuSBs)	Bank Code	Bank Name (PuSBs)	Bank Code	Bank Name (PrSBs)	Bank Code	Bank Name (PrSBs)
ALB	Allahabad Bank	ANB	Andhra Bank	AXB	Axis Bank Ltd.	BBL	Bandhan Bank Ltd.
BOB	Bank of Baroda	BOI	Bank of India	CSB	Catholic Syrian Bank Ltd.	CUB	City Union Bank Ltd.
BOM	Bank of Maharashtra	CNB	Canara Bank	DCB	Development Credit Bank	DLB	Dhanlakshmi Bank Ltd.
CBI	Central Bank of India	CRB	Corporation Bank	FDB	Federal Bank Ltd.	HDFC	HDFC Bank Ltd.
DNB	Dena Bank	INB	Indian Bank	ICICI	ICICI Bank Ltd.	IDBI	IDBI Bank Ltd.
IOB	Indian Overseas Bank	OBC	Oriental Bank of Commerce	IDFC	IDFC First Bank Ltd.	IIB	Indusind Bank Ltd.
PSB	Punjab and Sind Bank	PNB	Punjab National Bank	ING	ING Vysya Bank	JKB	Jammu & Kashmir Bank
SBI	State Bank of India	SYB	Syndicate Bank	KBL	Karnataka Bank Ltd.	KVB	Karur Vysya Bank Ltd.
UCO	United Commercial Bank	UOB	Union Bank of India	KMB	Kotak Mahindra Bank	RBL	Ratnakar Bank Ltd.
UNB	United Bank of India	VJB	Vijaya Bank Ltd.	SIB	South Indian Bank Ltd.	TMB	Tamilnadu Mercantile Bank
						YBL	Yes Bank Ltd.

Note: The dataset covers the maximum 41 banks for each year during the period 2012 to 2022. Due to mergers, the dataset has 427 observations instead of the expected 451 over the complete study period.

4 Results

Building on the practical application of our hybrid methodology, the following section presents the results obtained from integrating DDF-based network DEA with SVR to measure and predict the cost efficiency of Indian banks.

4.1 Cost Efficiency Analysis Using DDF-Based Network DEA Approach

In the current analysis, the DDF-based two-stage network DEA model have been improvised to calculate cost efficiency scores for the overall system and two

Table 2. Description and statistics of the inputs and outputs included.

Inputs	Description	Statistics (2012–2022			
		Average	Min	Max	Std. Dev.
Employees	Total number of staff across various positions and ranks.	15,949	359	1,47,707	22,956
Fixed assets	Tangible properties such as buildings, land contributing to the bank's activities.	2908	59	49,884	5259
Borrowings	External funds raised from sources like RBI, other banks, financial institutions.	31,751	0.00	4,26,043	59,801
Intermediates					
Deposits	Funds entrusted to a bank by customers and other entities.	2,79,385	4739	40,51,534	4,63,099
Desirable outputs					
Profit (loss) during year	Net earnings (or losses) generated from bank's operations over the year.	1242	−16,418	36,961	5186
Loans & investments	Financial loans provided to customers and domestic/international investments.	2,83,690	6433	4,10,3389	47,5620
Undesirable output					
Gross NPAs	Total value of non-performing loans, generating no income (Undesirable output).	15,631	19	22,3427	26,216

Note: All values, except the number of employees, are expressed in crores.

Table 3. Details of unit-costs associated with the variables.

Variable	Per-unit cost	Description	Statistics (2012–2022)			
			Average	Min	Max	Std. Dev.
Employees	$\frac{\text{Employee benefit expenses (EBE)}}{\text{Employees}}$	EBE is the amount incurred to provide health insurance, salaries & wages etc. to the employees.	2520	49	51562	5876
Fixed assets	$\frac{\text{80\% of Operating expenses (OE)}}{\text{Fixed assets}}$	OE refers to amount incurred for daily operations, like rent, utilities, marketing, insurance etc.	2256	16.34	28,668	3840
Deposits	$\frac{\text{Interest on deposits}}{\text{Deposits}}$	Interest on deposits refers to earnings generated from the funds held in savings accounts, fixed deposits etc.	13,492	33.04	14,7399	1412.243
Borrowings	$\frac{\text{Interest on borrowings}}{\text{Borrowings}}$	Interest on borrowings refers to the interest payments made to the other banks or financial entities.	687	0.00	9835	1412
Gross NPAs	$\frac{\text{Loss due to NPAs}}{\text{NPAs}}$	Loss due to NPAs is the loss from non-performing loans/assets, including unpaid principal and interest.	4033	-2571	71,374	7336

divisions, productivity and profitability of Indian banks operational during the period of study 2012–2022. Using MATLAB programs, the linear programming Model 1 was executed annually over the period from 2012 to 2022 to derive the optimal value (CE^{N*}), indicating the banks' overall cost efficiency. Further, the cost efficiency scores for each division ($\text{CE}_1^{N*}, \text{CE}_2^{N*}$) were determined utilizing the defined efficiency given in Definition 2. The average overall and divisions' cost efficiency values of the banks operational during the period of study have been demonstrated below in Fig. 2.

The graph highlights the cost efficiency scores of various banks across three categories: overall system efficiency, productivity efficiency (division 1), and profitability efficiency (division 2). It is evident that the overall efficiency for most banks seems heavily influenced by profitability efficiency, as banks with strong profitability scores demonstrated higher overall system efficiency. Banks like PNB, PSB, KMB, HDFC, and ICICI perform well across all three categories, particularly in profitability efficiency, which strongly influences their overall system efficiency. In contrast, banks such as ING and YBL show consistently low efficiency scores in all divisions, especially in productivity, leading to low overall efficiency. While, CBI shows moderate scores, performing relatively better in profitability, reflecting inefficiency in resource use. The specific banks, SBI and DLB stands out as the best-performing bank, with a perfect efficiency score of 1 in all three categories, indicating optimal performance in both productivity and profitability.

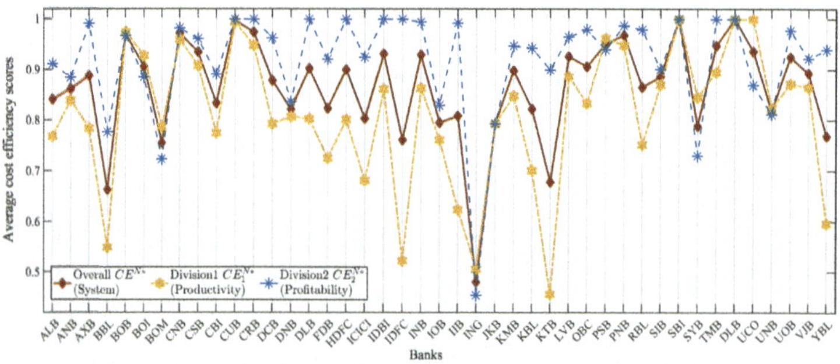

Fig. 2. Average cost efficiency of the banks across all the years 2012–2022.

To assess the robustness of a model by examining the impact of key variables on the results, sensitivity analysis has been performed. In this study, we conducted a sensitivity analysis by analyzing the influence of fixed assets on the average overall cost efficiency scores of PuSBs and PrSBs for the year 2022. The sensitivity analysis results indicate that the change in average overall cost efficiency scores after removing undesirable outputs is insignificant, suggesting that the model remains stable and robust. The removal of fixed assets leads to a 12% decline in cost efficiency for public sector banks and a 2% decline for private sector banks, suggesting greater impact of fixed assets on public sector banks, likely due to their higher operational costs. Further, a basic sensitivity analysis to assess the stability of cost efficiency outcomes has been conducted for the year 2022. We systematically eliminated the most efficient banks and analyzed the impact of their removal on the average cost efficiency scores of the remaining banks. There are seven banks (BOM, PNB, SBI, HDFC, CUB, DLB, RBL) that form the efficient frontier, and the overall average cost efficiency score across all banks was found to be 0.8667. The average cost efficiency scores, after removing each efficient bank one by one, ranged between 0.8035 and 0.8271–values that remain relatively close to the original score of 0.8667. This indicates that the overall cost efficiency results are not significantly influenced by individual top-performing banks.

4.2 Prediction of Cost Efficiency Using Integrated Network DEA-SVR Approach

To analyze the robustness of the integrated DDF-based network DEA and SVR approach, the value-based inputs/outputs and intermediates data matrices, $(\bar{X}_1, Y_1^g, Y_1^b; \bar{X}_2, Y_2^g, Y_2^b; Z_1^{(2)}, \bar{Z}_1^{(2)})$, along with the optimal system's cost efficiency scores (CE^{N*}), have been used as a feature space $\mathcal{F}$ and the targeted outputs. The study utilized a historical dataset spanning 10 years (2012–2021) with 395 operational banks as the training dataset and 2022 data set of 32 banks for testing to predict efficiency scores. Training a non-linear SVR requires selecting the appropriate kernel function, setting the penalty parameter, and determining the threshold deviation error. In this study, the *fitrsvm* toolbox in MATLAB, using the Radial Basis Function (RBF) kernel was employed for the SVR training. This toolbox facilitates automatic hyper-parameter tuning via Bayesian optimization. Then, the accuracy of the SVR method is validated by comparing the predicted cost efficiency scores (ce^{N*}) with the actual cost efficiency scores (CE^{N*}), and both set of scores are visualized in Fig. 3. The plot clearly illustrate how closely the estimated cost efficiency scores from SVR align with the actual scores for year 2022. Additionally, the obtained mean square error of **0.0056** further, highlights the high performance and accuracy of the proposed combined framework.

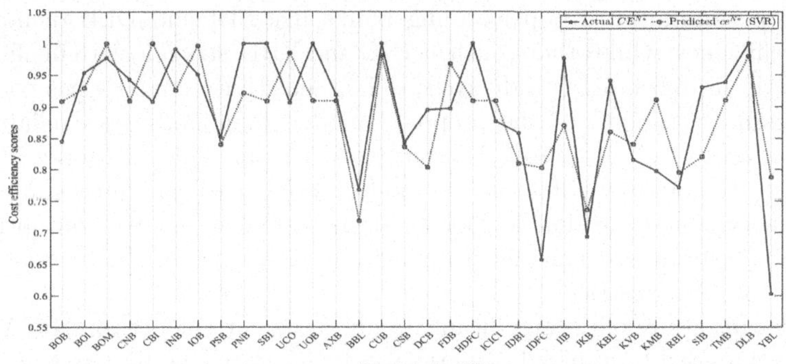

Fig. 3. Prediction results on integrated DDF-based network DEA and SVR approach.

5 Implications

In terms of **theoretical implications**, the study advances existing literature by integrating a DDF-based network DEA with SVR to estimate and predict cost efficiency by incorporating undesirable outputs, negative data, and heterogeneous input costs. The use of SVR for predictive purposes significantly reduces the computational demands associated with re-executing DEA models when handling large and dynamic (time-dependent or cross-sectional) datasets. By applying this model to Indian banks, the study provides insights into the cost efficiency of banks over a eleven-year period (2012–2022). The **practical implications** of the analysis suggests that: (i) Division 2 efficiency (profitability) plays a critical role in driving overall system's cost efficiency for banks, emphasizing the need for decision-makers to focus on improving efficiency within this division. (ii) Banks with lower division 1 efficiency (productivity), such as ING and YBL, should address resource utilization inefficiencies to enhance overall cost efficiency. (iii) To enhance their division 1 efficiency, banks like BOM, BBL, IDFC, IIB, ING, KMB, and YBL should consider training programs, invest in technology, to effectively utilize employees and fixed assets. (iv) Banks like BOM, BBL and SYB should focus on better optimization of human resources, cost-effective borrowing strategies, enhance capital efficiently to improve their division 2 efficiency (profitability). Moreover, UNB bank could benefit from cost reduction associated with fixed assets to enhance both division 1 and division 2 efficiency.

6 Conclusion

Building on previous studies and addressing identified research gaps, this study introduces a robust methodology that integrates DDF-based two-stage network DEA model with SVR algorithm for assessing and predicting the cost efficiency of Indian banks. The study reveals significant variation in the cost efficiency of Indian banks, with overall system's cost efficiency scores are primarily influenced by the efficiency of the division 2 (profitability) within the network structure. Improving profitability can enhance overall system efficiency, thus towards better financial performance. Top-performing banks like SBI and DLB demonstrate optimal efficiency, while others, such as ING and YBL, struggle with inefficiencies particularly in division 1 (productivity). The sensitivity analysis has confirmed the robustness stability of the proposed model, highlighting its reliability in evaluating cost efficiency across different scenarios and bank categories. Further, the integration notably enhances the model's predictive capabilities. Moreover, it can emerge as an invaluable tool for decision-makers focused on optimizing resource allocation and enhancing financial performance of the banking sector as well as other industries.

While this study offers significant insights, the current network DEA model could be extended to a mixed or parallel-network structure to assess additional efficiency metrics, such as revenue or profit.

References

1. An, Y., Zhai, X.: SVR-DEA model of carbon tax pricing for China's thermal power industry. Sci. Total Environ. **734**, 139438 (2020)
2. Awad, M., Khanna, R.: Support vector regression. In: Efficient Learning Machines, pp. 67–80. Springer, Cham (2015)
3. Bansal, P., Kumar, S., Mehra, A., Gulati, R.: Developing two dynamic Malmquist-Luenberger productivity indices: An illustrated application for assessing productivity performance of Indian banks. Omega **107**, 102538 (2022)
4. Chambers, R.G., Chung, Y., Färe, R.: Benefit and distance functions. J. Econ. Theory **70**, 407–419 (1996)
5. Charnes, A., Cooper, W.W., Rhodes, E.: Measuring the efficiency of decision making units. Eur. J. Oper. Res. **2**, 429–444 (1978)
6. Cheng, Y., et al.: An intelligent supplier evaluation model based on data-driven support vector regression in global supply chain. Comput. Ind. Eng. **139**, 105834 (2020)
7. Cooper, W., Seiford, L., Zhu, J.: Handbook on Data Envelopment Analysis. Springer, New York (2011)
8. Çakır, S.: Best output prediction in OECD railways using DEA in conjunction with machine learning algorithms. Ann. Oper. Res. **335**, 59–77 (2024)
9. Färe, R., Grosskopf, S., Whittaker, G.: Network DEA. In: Zhu, J., Cook, W.D. (eds.) Modeling Data Irregularities and Structural Complexities in Data Envelopment Analysis, pp. 209–240. Springer, Boston (2007)
10. Fukuyama, H., Matousek, R., Tzeremes, N.G.: A Nerlovian cost inefficiency two-stage DEA model for modeling banks' production process: evidence from the Turkish banking system. Omega **95**, 102198 (2020)
11. Kwon, H.-B., Lee, J.: Two-stage production modeling of large US banks: a DEA-neural network approach. Expert Syst. Appl. **42**, 6758–6766 (2015)
12. Kumar, S.: Banking reforms and the evolution of cost efficiency in Indian public sector banks. Econ. Chang. Restruct. **46**, 143–182 (2013)
13. Lozano, S., Gutiérrez, E., Moreno, P.: Network DEA approach to airports performance assessment considering undesirable outputs. Appl. Math. Model. **37**, 1665–1676 (2013)
14. Mishra, S., Singh, A., Sufia: Environmental performance of coal-based thermal generation plants: a two-stage DEA model with undesirable outputs. In: Advances in the Theory and Applications of Performance Measurement and Management, pp. 151–163. Springer, Cham (2024)
15. Nishtha, N., Puri, J., Setia, G.: Performance prediction of DMUs using integrated DEA-SVR approach with imprecise data: application on Indian banks. Soft. Comput. **27**, 5325–5355 (2023)
16. Portela, M.S., Thanassoulis, E., Simpson, G.: Negative data in DEA: a directional distance approach applied to bank branches. J. Oper. Res. Soc. **55**, 1111–1121 (2004)
17. Puri, J., Yadav, S.P.: Performance evaluation of public and private sector banks in India using DEA approach. Int. J. Oper. Res. **18**, 91–121 (2013)
18. Ray, S.: Cost efficiency in an Indian bank branch network: a centralized resource allocation model. Omega **65**, 69–81 (2016)
19. Sahoo, B.K., Mehdiloozad, M., Tone, K.: Cost, revenue and profit efficiency measurement in DEA: a directional distance function approach. Eur. J. Oper. Res. **237**, 921–931 (2014)

20. Smola, A.J., Schölkopf, B.: A tutorial on support vector regression. Stat. Comput. **14**, 199–222 (2004)
21. Tohidnia, S., Tohidi, G.: Estimating multi-period global cost efficiency and productivity change of systems with network structures. J. Ind. Eng. Int. **15**, 171–179 (2019)
22. Tone, K.: A strange case of the cost and allocative efficiencies in DEA. J. Oper. Res. Soc. **53**, 1225–1231 (2002)
23. Tone, K., Tsutsui, M.: Decomposition of cost efficiency and its application to Japanese-US electric utility comparisons. Socioecon. Plann. Sci. **41**, 91–106 (2007)
24. Valero-Carreras, D., Aparicio, J., Guerrero, N.M.: Support vector frontiers: a new approach for estimating production functions through support vector machines. Omega **104**, 102490 (2021)
25. Vapnik, V.: The Nature of Statistical Learning Theory. Springer, New York (1995)
26. Zhu, N., Zhu, C., Emrouznejad, A.: A combined machine learning algorithms and DEA method for measuring and predicting the efficiency of Chinese manufacturing listed companies. J. Manag. Sci. Eng. **6**, 435–448 (2020)

Evaluating the Performance of Central Banks of G20 Countries: Integrating DEA with an Entropy-Based Russell Measure

Archila Kushwaha[1(✉)] and Brajesh Kumar[2]

[1] School of Management Sciences, Varanasi, UP 221011, India
archilakushwaha25@gmail.com
[2] Central University of South Bihar, Gaya, Bihar 824236, India
brajesh@cusb.ac.in

Abstract. The evaluation of central banks' performance is a complex task, involving the assessment of their efficiency and effectiveness in achieving multiple objectives, such as price stability, financial stability, and economic growth. Existing evaluation methods have limitations, highlighting the need for a more comprehensive and robust approach to assess the performance of G20 central banks. This study aims to address this gap by using an integrated framework that combines Data Envelopment Analysis (DEA) with the Entropy-Based Russell Measure (ERM) to provide a more accurate and nuanced assessment of central banks' performance. The efficiency of financial statements of central banks is a critical indicator of economic growth and stability. This research focuses on evaluating the efficiency of the financial statements of central banks of the G20 countries for the financial year 2022. By incorporating entropy, this study will account for the use of input, output, and system weighting, offering a comprehensive analysis of performance. This approach provides a novel perspective on the financial efficiency of central banks, considering the complexity and multifaceted nature of financial data. The findings of this research will contribute valuable insights into the operational efficiency and performance evaluation of central banks within the G20, informing policy and decision-making processes. Future research could extend this framework to include additional years of data and explore the impact of macroeconomic factors on central bank efficiency, further refining the assessment tools for broader applicability.

Keywords: Data Envelopment Analysis (DEA) · Entropy-Based Russell Measure (ERM) · Central Bank Efficiency · G20 Countries · Financial Statement

1 Introduction

The efficiency of financial institutions has garnered considerable attention in economic studies, given their vital role in ensuring growth and economic stability (Rahman, 2023). Among these, banks are crucial for monetary policy implementation, effective resource

A. Emrouznejad et al. (Eds.): DEA 2024, LNOR, pp. 293–309, 2025.
https://doi.org/10.1007/978-3-031-98177-7_20

distribution, and advancing financial inclusion (Ononiwu et al., 2024). Although numerous studies have focused on the efficiency of commercial banks, central banks key players in maintaining economic and monetary stability have not received as much scrutiny (McKinley & Banaian, 2005). Central banks perform diverse roles, such as controlling inflation, managing exchange rates, stabilizing prices, and overseeing financial institutions, which complicates their efficiency assessment. Institutional efficiency reflects how effectively objectives are achieved with minimal resource wastage. For central banks, this includes two aspects: operational efficiency, which emphasizes resource optimization, and managerial efficiency, which focuses on achieving goals without redundancy (Ononiwu et al., 2024). Evaluating the efficiency of central banks, particularly in less developed economies, is challenging due to varying financial structures, market conditions, and regulatory environments (Fandl, 2018). Previous research has applied techniques such as Data Envelopment Analysis (DEA) and Stochastic Frontier Analysis (SFA) to measure banking sector efficiency. Studies have shown that DEA is particularly effective for analyzing central banks, as it accommodates multiple inputs and outputs in its framework (Mester, 2004). Additionally, the interaction between central bank efficiency and external variables, such as controlling inflation and stabilizing exchange rates, highlights the dynamic nature of their operations (Rahman, 2023). These external factors further underscore the need for robust models capable of addressing the complexities inherent in evaluating central bank efficiency (McKinley & Banaian, 2005).

The primary objective of this study is to evaluate the financial efficiency of G20 central banks for the year 2022 using a novel approach that integrates an entropy-based Russell measure with Data Envelopment Analysis (DEA). This method addresses the inherent limitations of traditional efficiency evaluation models, particularly the assumptions of equal-weight contributions from all inputs and outputs. By incorporating entropy, the study ensures a more objective and accurate determination of variable weights, reflecting their true significance in the efficiency measurement process. Various approaches have been proposed to address this equal-weight limitation (Seifert & Zhu, 1998; Thanassoulis & Dyson, 1992; Pastor et al., 1999; Ruggiero & Bretschneider, 1998). Thanassoulis & Dyson (1992) introduced a model that allows for unequal weights based on target levels for inputs and outputs, enabling DMUs to allocate weights according to their preferences. Alternatively, Zhu (1996) employed the Analytic Hierarchy Process (AHP), and Seifert & Zhu (1998) utilized the Delphi technique to determine weights, although these methods still rely on subjective expert judgment. Pastor et al. (1999) addressed this issue by introducing a weighted measure that calculates technical efficiency as the ratio of inputs to outputs using averages. This concept was extended by Fukuyama & Weber (2009), who proposed a half-weighting system that integrates average inputs and outputs into efficient computations. Although these approaches reduce the equal-weight bias, the extent of improvement remains limited. Furthermore, the dependence on subjective preferences or the number of variables in these studies often leads to weights that do not reflect the true significance of the variables.

To resolve these challenges, Cooper et al. (2007) introduced the weighted slack-based measure (WSBM), which assigns weights based on the relative importance of variables by dividing each variable's value by the sum of the corresponding variables across all DMUs. While traditional WSBM identifies significant variables, it fails to account for disorder or variables uncertainty. To address this, the entropy concept has been integrated with the Russell measure. By employing entropy to compute technical efficiency weights, this approach minimizes the equal-weight effect and enhances evaluation accuracy. The current study builds on this innovation, combining entropy with the Russell measure in DEA to improve efficiency measurement and address the shortcomings of previous methods.

The research is composed of five sections. The first section contains an introduction and provides background information and the importance of financial efficiency on the central bank of G20 Countries, the second section describes the summary of previous research. The third section describes the Integration of the Entropy-Based Russell Measure with Data Envelopment Analysis. The fourth section is based on the Case Study of G20 Central Banks Efficiency for 2022. The last section focuses on the Conclusion and Discussion.

2 Literature Review

DEA has not been applied extensively yet in the context of financial statement analysis, in spite of the fact that it avoids many of the problems usually encountered with parametric methods, like the difficulty of determining the functional form or statistical distribution of ratios. Pioneering work was undertaken by Smith (1990) who discusses the strengths and weaknesses of DEA applied to financial statement analysis. Other authors in the management-science and operations-research literature, like Thanassoulis et al. (1996), Berger & Humphrey (1997), and Feroz et al. (2003), have suggested that DEA might be superior to traditional financial ratios analysis or at least complement it. Notably, Halkos & Salamouris (2004) are one of the few studies applying DEA in this context, demonstrating its potential as a substitute or complement for conventional ratio analysis in the banking sector. Recent contributions highlight the COOPER framework as a structured synthesis of DEA methodologies in Operational Research, enhancing decision-making by incorporating feedback loops for greater reliability and robustness. It serves as a valuable resource for researchers and practitioners, promoting collaboration and innovation (Mergoni et al., 2024). Building upon the initial applications of DEA in Financial Statement Analysis, its adoption has significantly expanded within the banking sector, where its ability to assess operational efficiency and benchmark performance has been widely acknowledged. DEA was first developed by Charnes et al. (1978) and is based on the concept of technical efficiency introduced by Farrell (1957). DEA operates as a linear programming technique that converts various inputs and outputs into a single efficiency score, allowing for the comparison of banks against a constructed benchmark of efficient banks (Kamarudin et al., 2019). This non-parametric approach does not require prior assumptions about the production function, thus minimizing the risk of misspecification. The application of DEA in measuring bank efficiency has gained attraction since the 1980s, particularly in national banking sectors. It allows for the identification of efficient banks and quantifies inefficiencies in others, providing a comprehensive view of performance across the banking industry (Wang et al., 2014). Various studies have highlighted the effectiveness of DEA in evaluating bank efficiency, noting its precision and lower magnitude of inefficiency compared to other methods. In the context of Latvian banks, the DEA methodology has been tailored to address specific variables relevant to the banking sector. The research emphasizes the importance of selecting appropriate input-output combinations to avoid biases in efficiency measurement (Titko et al., 2014). For instance, the intermediation and profitability approaches have been suggested as frameworks for variable selection in DEA models. The study also indicates that the choice of variables significantly impacts the results, necessitating careful consideration during model specification (Fukuyama et al., 2023). As the application of DEA in the banking sector gained prominence, researchers have continuously refined the methodology to address the limitations of conventional models. Initially introduced by Charnes et al. (1978), the foundational DEA model has been expanded to include semi-oriented radial measures and other adaptations, allowing for variables with negative values (Srivastava et al., 2024; Portela et al., 2004). To account for undesirable outputs, models like the Enhanced Russell Measure (ERM) have been applied effectively in agricultural productivity evaluations (Arana-Jiménez et al., 2022). Further advancements include slacks-based models for multi-stage processes, such as banking operations, which incorporate interest costs and non-performing loans as outputs (Omrani et al., 2023). Network DEA models, enhanced by Egalitarian Bargaining game theory and Slack-based measures, have been employed for operational and service efficiency analyses, notably in Iranian airlines (Mahmoudi & Emrouznejad, 2023).

Subsampling bootstrap techniques have also been explored to estimate the asymptotic distribution of DEA efficiency scores, particularly in two-stage series structures, with evidence from railway systems in European countries (Michali et al., 2023). Robust optimization and fuzzy logic approaches have been integrated into DEA to handle uncertain and imprecise data, enhancing its applicability in fields like mutual fund performance and merger analyses (Peykani et al., 2024; Hatami-Marbini et al., 2011; Soltanifar et al., 2024). Directional DEA models further address eco-efficiency by optimizing desirable and undesirable outputs, as demonstrated in NetZero-focused studies across OECD countries (Taleb et al., 2024). Collectively, these advancements highlight DEA's growing versatility in complex, real-world decision-making scenarios.

This study focuses on evaluating the efficiency of central banks across G20 countries, analyzing their performance through a standardized data envelopment analysis (DEA) framework. By considering variables such as assets, liabilities, equity, income, and expenses. Unlike the European Central Bank (ECB) and the National Central Banks (NCBs) in the Euro system, which share a homogeneous monetary policy environment (Scheller, 2006), central banks in G20 nations operate under diverse economic and financial conditions, making comparative efficiency analysis a challenging yet essential endeavor. The application of Data Envelopment Analysis (DEA) to measure the efficiency of central banks has gained attention in recent years, with significant methodological advancements and findings contributing to the discourse. Lou et al. (2024) study introduces a novel methodology combining DEA with super-efficiency and the Malmquist Productivity Index (MPI). The study reveals that productivity changes in central banks are largely driven by technological shifts rather than efficiency changes, with export levels exerting a greater impact on efficiency compared to GDP, import levels, and average exchange rates. Khan et al. (2024) employ stochastic frontier analysis (SFA) to estimate the efficiency of 17 top Asian central banks using three inputs and one output. The findings highlight significant efficiency variations across the years 2016–2018, with efficiency values ranging from 0.2368 to 0.8219. Furthermore, structural equation modeling reveals a stronger positive impact of international trade on central bank efficiency compared to economic growth. Another notable contribution comes from Ihaddaden's study, which employs a radial DEA model to evaluate the efficiency of Eurosystem central banks. Results indicate an average technical efficiency of 58.37%, with notable disparities among national central banks (NCBs). Efficient units include the central banks of Spain, Finland, the Netherlands, and Lithuania, while Malta and Greece demonstrate the lowest efficiency levels at 19.76% and 14.20%, respectively. The study emphasizes the role of peer units like Spain and Finland as benchmarks for improving NCB efficiency. Taleb et al. (2024) examine the cost-benefit efficiency of banks in seven Central and Eastern European (CEE) states during periods of economic crisis and recovery (2006–2011) using DEA models. Their findings underscore that banking systems in Lithuania and Slovenia operate most efficiently, whereas Latvia's system is the least efficient. The study underscores the importance of adapting strategies to enhance efficiency amidst competitive dynamics in the banking sector.

3 Integration of Entropy-Based Russell Measure with Data Envelopment Analysis

3.1 Entropy-Based Russell Measure

The entropy-based Russell measure is employed to solve the issue of giving subjective weights to variables. Utilizing the entropy method, objective weights are generated based on the variability of each variable within the dataset. This ensures that:

- Variables with higher divergence (significance) receive higher weights (Hsiao et al., 2011).
- The DEA model reflects the true importance of each variable without bias (Hsiao et al., 2011).

The entropy-based Russell measure provides a comprehensive framework for evaluating efficiency by accounting for both slacks and proportional efficiency changes within the DEA model. Thanassoulis and Dyson (1992) proposed an objective mechanism for weighting variables that avoids assuming equal factor weights. Their approach replaces subjectively preferred targets by setting and averaging concepts for assessing input and output efficiency. This method aligns with the objectives of efficiency evaluation without relying on subjective assumptions. Similarly, Ruggiero & Bretschneider (1998), through their study with simulated data, emphasized the utility of the weighted Russell measure. However, their approach, which involved deriving weights using canonical regression analysis, presented practical challenges when applied to real case data. Further advancements in this area were made by Pastor et al. (1999) and Fukuyama & Weber (2009), who contributed to refining the conceptual framework for technical efficiency and variable weighting. Entropy, in this context, captures the "potential for disorder" within a system, illustrating that systems with greater degrees of freedom and constituents possess a wider range of possible states. While a few states may display significant order, the majority tend to exhibit disorder. Based on this concept, the following steps outline the process for deriving weights from the entropy measure, transforming the data into weighted values, and subsequently applying DEA for efficiency evaluation:

Step 1: Find each cell [Pij] (Hsiao et al., 2011).

$$[P_{ij}]_{n\times m} = \left[x_{ij}/\sum_{i=1}^{n} x_{ij}\right]_{n\times m} \quad (1)$$

where Pij is the normalized value of the i-th DMU for the j-th variable.

Step 2: Calculate entropy (Hsiao et al., 2011).

$$e_j = -k\sum_{i=1}^{n} p_{ij}\ln p_{ij}\ \forall \mathrm{j} \in R_{+}^{m} \quad (2)$$

where k is Boltzmann's constant, which is equal to k =1/ln n such that $0 \le \mathrm{ej} \le 1$

Step 3: Calculate divergence (Hsiao et al., 2011)

$$\overline{e}_{j=1-e_j} \quad (3)$$

Step 4: Calculate weights (Hsiao et al., 2011)

$$r_{j=}\overline{e}_j/\sum\nolimits_{j=1}^{m} \overline{e}_{j\forall j\in R_{+}^{m}} \quad (4)$$

where $\sum_{j=1}^{m} r_j$ =1, This allows us to find the weight of the input variables ([r1 r2.................................rm]1*m)

Step 5: Similarly, repeat steps 1-4 to calculate the weight of the output variables.

Step 6: The calculated weights were then multiplied with their respective input and output variables to ensure that the contributions of each variable accurately reflect their significance in the evaluation process.

Step 7: The weighted data was subsequently utilized as input for the Super SBM model, which was executed to calculate the efficiency scores, providing a comprehensive assessment of performance by accounting for slacks in both inputs and outputs Table 4.

3.2 Super-SBM Model: Data Envelopment Analysis

The Super-SBM (Slacks-Based Measure) DEA model, introduced by Tone (2001), extends traditional DEA by offering a mechanism to evaluate the super-efficiency of efficient Decision-Making Units (DMUs) while retaining the ability to identify the inefficiency of others. However, Tone's original model is limited in its ability to evaluate both efficient and inefficient DMUs simultaneously. Tran et al. (2019) addressed this limitation by proposing a binary linear programming model that integrates SBM and Super-SBM, achieving efficiency evaluation for both types of DMUs in a single stage. Despite its utility, their model had significant computational demands, which were later optimized by subsequent studies, reducing computation time by one-third (Tone, 2001; Tran et al., 2019).

Further applications of Super-SBM DEA have demonstrated its versatility across various domains (Emrouznejad et al., 2025). Gan et al. (2020) employed the model alongside Meta-frontier analysis and the Malmquist–Luenberger Index (MLI) to assess energy efficiency and technological heterogeneity in G20 countries. Their findings revealed a significant efficiency gap between developed and developing nations, highlighting technology as a crucial determinant of energy efficiency. Similarly, Gronwald et al. (2011) utilized the Super SBM Zero Sum Gains DEA model to examine emission allowance allocations across 24 European Union member countries. Likewise, Trnovská & Halická (2024) extended input/output radial models, hyperbolic distance function models, and directional distance function models for super-efficiency measurement, demonstrating their effectiveness through numerical applications. In the context of environmental efficiency, Ma et al. (2018) applied the Super-SBM model to measure regional environmental efficiency in China. Their research highlighted stark regional disparities, with eastern provinces outperforming central and western areas. Another study by Yasmeen et al. (2020) utilized the Super-SBM DEA methodology to evaluate water usage efficiency in China. The study underscored the positive impact of water-saving technologies like recycling and irrigation modernization on efficiency, particularly in regions facing severe water scarcity.

Building on the robust capabilities of the Super-SBM DEA model, this study applies it to evaluate the financial efficiency of the central banks of G20 countries for the year 2022. By employing the Super-SBM framework, this research provides a nuanced understanding of the efficiency levels among central banks, offering insights into the relative performance of these critical institutions. Below is the equation given by Tone (2002) used to calculate Super-SBM efficiency of the Decision-Making Units.

$$\boldsymbol{minp}_{2=}\frac{\frac{1}{m}\sum_{i=1}^{m}x_{i/x_{ik}}^{2}}{\frac{1}{s}\sum_{r=1}^{s}y_{r/y_{rk}}^{2}}$$

$$s.t.x_i^2 = \sum_{j=1,j\neq k}^{n} x_{ij}\lambda_{j,}i = 1,\ldots..m$$

$$y_r^{2\leq} \sum_{j=1,j\neq k}^{n} y_{rj}\lambda_{j,}r = 1,\ldots\ldots..s$$

$$\lambda_j \geq 0,\ j = 1\ldots n,\ j \neq k$$

$$x_{ik} \leq x_i^2, i = 1,\ldots\ldots\ldots.m$$

$$O \leq y_r^2 \leq y_{rk},\ r = 1\ \ldots \text{s}. \tag{5}$$

4 The Case Study of G20 Central Banks Efficiency Evaluation

4.1 Data Sources

The data for this study is aggregated from central bank reports and financial statements of banks functioning in G20 nations for the year 2022. These financial accounts correspond to the reporting periods unique to each country, which differ among the G20 countries. In the majority of nations, such as Argentina, Brazil, Canada, China, France, Germany, Indonesia, Italy, Mexico, Russia, Turkey, and the United States, the reporting period corresponds with the calendar year, extending from January to December. Conversely, nations such as India, Japan, Korea, and South Africa adhere to a fiscal year spanning from April to March. The United Kingdom employs a distinct financial year from March to February, while Saudi Arabia and Australia present their financial results for the July to June interval. The study meticulously accounts for these discrepancies in reporting periods to guarantee data consistency and facilitate meaningful comparisons among countries. The Saudi Central Bank (SAMA) was excluded from the research due to the unavailability of data pertaining to Interest Expenses and Interest Income, as this information is not disclosed in their financial system. We have excluded three additional countries from the analysis due to their negative equity, rendering the use of the logarithmic function unfeasible (Bagwell, 2005). The central banks are the Reserve Bank of Australia, Deutsche Bundesbank (The Central Bank of Germany), and the Bank of Mexico. Below is the List of DMUs in Table 1.

Table 1. List of DMUs

Name of the Central Bank	List of DMUs
Central Bank of the Argentina Republic	DMU1
Central Bank of Brazil	DMU 2
Bank of Canada	DMU 3
People's Bank of China	DMU 4
Banque De France	DMU 5
Reserve Bank of India	DMU 6
Bank of Indonesia	DMU 7
Bank of Japan	DMU 8
Banc d'Italia	DMU 9
The Bank of Korea	DMU 10
Bank of Russia	DMU 11
South Africa Reserve Bank	DMU 12
Central Bank of the Republic of Turkey	DMU 13
Bank of England United Kingdom	DMU 14
Federal Reserve System	DMU 15
Source: Author's	

4.2 Variables Used

Inputs are resources or efforts the bank uses to generate desirable financial outcomes.

- **Equity:** Represents the bank's capital, a foundational resource enabling operations (Sufian & Shah Habibullah, 2010).

- **Interest Expenses:** Reflect the cost incurred for attracting funds through deposits or borrowing (Titko et al., 2014).
- **Other Expenses:** Includes operational and administrative costs essential for running the bank (Lin et al., 2009).

Outputs are the desirable results or outcomes of the bank's management of its inputs.

- **Assets:** Reflect the bank's ability to accumulate resources (e.g., loans, investments) through its operations (Titko et al., 2014).
- **Liabilities:** Represent the bank's capacity to attract deposits and external funding (Titko et al., 2014).
- **Interest Income:** The revenue generated from loans and other interest-bearing activities (Taylor et al., 1997).
- **Other Income:** Income from non-interest sources, such as fees, commissions, or investments (Taylor et al., 1997).

4.3 Currency Conversion Method for Standardizing into USD

To ensure uniformity and comparability of financial data across banks in G20 countries, all monetary values were converted into USD. This standardization minimizes distortions arising from currency differences, facilitating robust cross-country analysis. The conversion process relied on average annual exchange rates provided by the International Monetary Fund (IMF), recognized for their reliability, and calculated using daily exchange rate data over the respective financial year (IMF, 2022). Each bank's financial figures, initially reported in their local currency, were converted into USD using the formula:

Value in USD = Value in Local Currency ÷ Average Annual Exchange Rate.

4.4 Result of the Analysis

This section analyses the central bank efficiency and results obtained using the proposed DEA methodology. We have used the Entropy-Based Russell measure to derive weights in Table 2 and then applied the Super SBM DEA model to obtain the financial efficiency score of G20 Countries listed in Table 3 for 2022.

Table 2 showcases the entropy-derived weights assigned to input and output variables in the Super-SBM DEA model for evaluating the financial efficiency of G20 central banks. The weights highlight the relative importance of each variable, ensuring an unbiased assessment. Among the outputs, Equity and Interest Income generally hold higher weights for central banks like the Reserve Bank of India (Equity: 0.0693) and the Bank of Korea (Interest Income: 0.0691), indicating their significant contribution to performance evaluation. Conversely, lower weights for Other Income in banks such as the Central Bank of Turkey (0.0608) reflect a reduced impact on their efficiency scores. Inputs like Total Assets and Total Liabilities exhibit uniform weights across most central banks, suggesting their foundational role in the financial framework. However, variations in Interest Expense and Other Expense weights, such as the higher emphasis for the South Africa Reserve Bank (0.0690) compared to the Bank of England (0.0603), underline differences

Table 2. The Variable weights derived from Entropy-Based Russell Measure

DMUS	(O) Equity	(O) Interest Income	(O) Other Income	(I) Total Assets	(I) Total Liabilities	(I) Interest Expense	(I) Other Expense
Central Bank of The Argentina Republic	0.0664	0.0676	0.0650	0.0673	0.0670	0.0615	0.0685
Central Bank of Brazil	0.0682	0.0691	0.0685	0.0671	0.0669	0.0690	0.0698
Bank of Canada	0.0681	0.0685	0.0683	0.0674	0.0671	0.0681	0.0692
People's Bank of China	0.0621	0.0636	0.0637	0.0663	0.0664	0.0654	0.0615
European Central Bank	0.0685	0.0669	0.0681	0.0667	0.0671	0.0672	0.0676
Reserve Bank of India	0.0693	0.0667	0.0650	0.0671	0.0669	0.0691	0.0683
Bank of Indonesia	0.0679	0.0683	0.0677	0.0674	0.0671	0.0686	0.0612
Bank of Japan	0.0664	0.0675	0.0638	0.0655	0.0658	0.0686	0.0658
Bank of Italy	0.0673	0.0673	0.0671	0.0669	0.0668	0.0675	0.0682
Bank of Korea	0.0694	0.0691	0.0685	0.0676	0.0672	0.0691	0.0698
Central Bank of Russia	0.0622	0.0679	0.0684	0.0672	0.0670	0.0677	0.0637
South Africa Reserve Bank	0.0692	0.0690	0.0684	0.0675	0.0672	0.0690	0.0691

(*continued*)

Table 2. *(continued)*

DMUS	(O) Equity	(O) Interest Income	(O) Other Income	(I) Total Assets	(I) Total Liabilities	(I) Interest Expense	(I) Other Expense
Central Bank of The Republic of Turkey	0.0608	0.0676	0.0670	0.0650	0.0658	0.0686	0.0652
Bank of England United Kingdom	0.0686	0.0604	0.0680	0.0669	0.0668	0.0603	0.0688
Federal Reserve System	0.0657	0.0604	0.0625	0.0642	0.0649	0.0604	0.0632

Source: Author's Calculation

in cost structures. These disparities reflect the diverse economic environments and operational focuses among G20 central banks, underscoring the importance of tailoring DEA models to capture such heterogeneity. The balanced weighting mechanism provided by the entropy method ensures an equitable evaluation, aligning the variables with their contextual relevance while maintaining comparability across Decision-Making Units (DMUs).

Table 3. Descriptive Statistics of Variables (Amt in Million USD)

	Equity	Interest Expenses	Other Expenses	Total Assets	Total Liabilities	Interest Income	Other Income
Max	2859.77	976.62	181.20	140305.90	248701.61	1497.02	1020.41
Min	18.81	0.017	0.0226	19.83	16.99	0.0291	0.0637
Average	4423.33	1770.09	445.79	105709.98	177043.51	2600.78	989.60
SD	7392.41	3110.80	569.37	347225.81	617654.89	4712.63	2497.53

Source: Author's Calculation

Table 3 presents the descriptive statistics for the input and output variables used to evaluate the efficiency of G20 central banks for the year 2022 (measured in million USD). The input variables include Equity, Interest Expenses, and Other Expenses, while the output variables consist of Total Assets, Total Liabilities, Interest Income, and Other Income. Among the input variables, Equity demonstrates the highest variability, with a standard deviation of 7,392.41 million USD and values ranging from 18.813 million USD to 28,597.734 million USD, reflecting significant differences in capital levels among central banks. Interest Expenses and Other Expenses also show considerable variation, with averages of 1,770.09 million USD and 445.79 million USD, respectively, indicating diverse cost structures. For the output variables, Total Assets and Total Liabilities exhibit extensive ranges, with averages of 1,057,099.82 million USD and 1,770,435.09 million USD, respectively, and high standard deviations exceeding trillions, highlighting disparities in financial scale. Interest Income (average of 2,600.78 million USD) and Other Income (average of 989.60 million USD) also vary widely, underlining diverse revenue generation strategies. These statistics underscore the heterogeneity in the financial operations and resource allocations of central banks, emphasizing the importance of employing a robust DEA model to effectively capture and analyze the efficiency of these entities.

Using the Super Slacks-Based Measure (Super SBM) Data Envelopment Analysis (DEA), the financial efficiency scores of 15 central banks were evaluated, revealing a wide range of performance levels (see Table 4). The Central Bank of the Republic of Turkey (DMU 13) and Central Bank of Brazil (DMU 2) emerged as super-efficient, with scores of 1.9592 and 1.9511, respectively, indicating exceptional resource utilization beyond the efficiency frontier. In contrast, central banks such as the People's Bank of China (DMU 4) and Bank of Russia (DMU 11) displayed significantly lower efficiency scores of 0.1656 and 0.0082, respectively, highlighting considerable inefficiencies. While some institutions, like the Reserve Bank of India (DMU 6) with a score of 1.0000, were on the efficiency frontier, others demonstrated inefficiencies in their financial operations. These findings suggest that more efficient central banks could serve as benchmarks, while less efficient ones should review their operational strategies to optimize resource management. The results emphasize the potential for policy adjustments and operational improvements, particularly for lower-ranking institutions

Table 4. Results of Efficiency Score by Integrating Entropy-Based Russell Measure with Super SBM DEA

Central Bank	*DMUs*	*Financial Efficiency Score*	*Rank*
Central Bank of the Republic of Turkey	DMU 13	1.9592	1
Central Bank of Brazil	DMU 2	1.9511	2
Federal Reserve System	DMU 15	1.3326	3
The Bank of England	DMU 14	1.3068	4
Reserve Bank of India	DMU 6	1.0000	5
Bank of Japan	DMU 8	0.4684	6
People's Bank of China	DMU 4	0.1656	7
The Bank of Korea	DMU 10	0.1432	8
Banc d'Italia	DMU 9	0.1338	9
Central Bank of the Argentina Republic	DMU1	0.0635	10

(continued)

Table 4. (*continued*)

Central Bank	*DMUs*	*Financial Efficiency Score*	*Rank*
Banque De France	DMU 5	0.0590	11
Bank of Canada	DMU 3	0.0580	12
South Africa Reserve Bank	DMU 12	0.0407	13
Bank of Indonesia	DMU 7	0.0329	14
Bank of Russia	DMU 11	0.0082	15

Source: Author's Calculation

5 Conclusion and Discussion

This study evaluates the financial efficiency of G20 central banks using the Super SBM DEA model, addressing the limitations of equal and subjective weight assignments through weights derived from the entropy-based Russell measure. This approach ensures a more objective and robust efficiency analysis by capturing system weights. The results reveal significant efficiency variations, with central banks like the Central Bank of Turkey and the Central Bank of Brazil exhibiting super-efficiency, while others, such as the Bank of Russia and the People's Bank of China show inefficiencies. The methodology retains the strengths of the Russell measure by allowing input contractions and output expansions, while its simplicity avoids the complexities of regression-based weight determination. These results are significant for policymakers as they can now find the low-scoring institutions to be benchmarked against the best-performing central banks. Resource utilization was improved by efficient banks in the allocation of equity and interest income while poorly efficient banks were constrained in capital structure and cost optimization. Such insights can help decision makers to design and implement comprehensive financial reforms, better allocate certain scarce resources, and enhance operational and financial resilience. This research emphasizes the need for customizing financial approaches to particular regions using the differing weights of inputs and outputs calculated through the entropy method. The effectiveness of central banks with high-interest income and substantial equity was better, which suggests that these underperforming institutions require strategic spending and additional income diversification. Moreover, the study points out the need for ongoing tracking of performance. Central banks can evaluate progress and detect new inefficiencies through regular use of the Super SBM DEA model. This model would allow for more flexible policy-making. Future studies might build upon this one by adding some political and economic variables or by applying it to different political systems to discover various efficiency distributions.

References

Arana-Jiménez, M., Sánchez-Gil, M.C., Lozano, S., Younesi, A.: Efficiency assessment using fuzzy production possibility set and enhanced russell graph measure. Comput. Appl. Math.. Appl. Math. **41**(2), 79 (2022). https://doi.org/10.1007/s40314-022-01780-y

Bagwell, C.B.: Hyperlog—a flexible log-like transform for negative, zero, and positive valued data Cytometry Part A, **64A**(1), 34–42 (2005). https://doi.org/10.1002/cyto.a.20114
Banco central do brasil. (n.d.). Retrieved 2 January 2025, from https://www.bcb.gov.br/en
Banca d'Italia. (n.d.). Homepage. Retrieved December 23, 2024, from https://www.bancaditalia.it/homepage/index.html?com.dotmarketing.htmlpage.language=1
Bank, E.C.: (n.d.). European central bank. European Central Bank. Retrieved 2 January (2025). https://www.ecb.europa.eu/home/html/index.en.html
Bank Indonesia. (n.d.). Retrieved 2 January 2025, from https://www.bi.go.id/en/default.aspx
Bank of Canada. (n.d.). Retrieved 2 January 2025, from https://www.bankofcanada.ca/
Bank of Korea. (n.d.). Retrieved 2 January 2025, from https://www.bok.or.kr/eng/main/main.do
Banxico, central bank, Banco de México. (n.d.). Retrieved 2 January 2025, from https://www.banxico.org.mx/indexen.html
Berger, A.N., Humphrey, D.B.: Efficiency of financial institutions: international survey and directions for future research. Eur. J. Oper. Res.Oper. Res. **98**(2), 175–212 (1997). https://doi.org/10.1016/S0377-2217(96)00342-6
Central Bank of the Argentine Republic. (n.d.). English information. Retrieved December 23, (2024). https://www.bcra.gob.ar/varios/English_information.asp
Charnes, A., Cooper, W.W., Rhodes, E.: Measuring the efficiency of decision-making units. Eur. J. Oper. Res.Oper. Res. **2**(6), 429–444 (1978). https://doi.org/10.1016/0377-2217(78)90138-8
Chen, F., Zhao, T., Xia, H., Cui, X., Li, Z.: Allocation of carbon emission quotas in Chinese provinces based on Super-SBM model and ZSG-DEA model. Clean Technol. Environ. Policy **23**(8), 2285–2301 (2021). https://doi.org/10.1007/s10098-021-02137-5
Cooper, W.W., Seiford, L.M., Tone, K., Zhu, J.: Some models and measures for evaluating performances with DEA: past accomplishments and future prospects. J. Prod. Anal. **28**(3), 151–163 (2007). https://doi.org/10.1007/s11123-007-0056-4
Deutsche Bundesbank. (n.d.). Home. Retrieved December 23, 2024. https://www.bundesbank.de/en
Dyson, R.G., Thanassoulis, E.: Reducing weight flexibility in data envelopment analysis. J. Oper. Res. Soc.Oper. Res. Soc. **39**(6), 563–576 (1988). https://doi.org/10.1057/jors.1988.96
Emrouznejad, A., Brzezicki, Ł., Lu, C.: Development and evolution of slacks-based measure models in data envelopment analysis: a comprehensive review of the literature. J. Econom. Surv. joes.12682 (2025). https://doi.org/10.1111/joes.12682
Fandl, M.: Monetary and financial policy in the euro area. Springer Int. Publ. (2018). https://doi.org/10.1007/978-3-319-72643-4
Farrell, M.J: The measurement of productive efficiency. J. Royal Stat. Soc. Series A (General), **120**(3), 253 (1957). https://doi.org/10.2307/2343100
Feroz, E.H., Kim, S., Raab, R.L.: Financial statement analysis: a data envelopment analysis approach. J. Opera. Res. Soc. **54**(1), 48–58 (2003). https://doi.org/10.1057/palgrave.jors.2601475
Fukuyama, H., Tsionas, M., Tan, Y.: Dynamic network data envelopment analysis with a sequential structure and behavioral-causal analysis: application to the Chinese banking industry. Eur. J. Oper. Res.Oper. Res. **307**(3), 1360–1373 (2023). https://doi.org/10.1016/j.ejor.2022.09.028
Fukuyama, H., Weber, W.L.: A directional slacks-based measure of technical inefficiency. Socioecon. Plann. Sci.. Plann. Sci. **43**(4), 274–287 (2009). https://doi.org/10.1016/j.seps.2008.12.001
Gan, G., Lee, H.-S., Lee, L., Wang, X., Wang, Q.: Network hierarchical DEA with an application to the international shipping industry in Taiwan. J. Opera. Res. Soc. **71**(6), 991–1002 (2020). https://doi.org/10.1080/01605682.2019.1603792
Gronwald, M., Ketterer, J., Trueck, S.: The dependence structure between carbon emission allowances and financial markets—a copula analysis. SSRN Electron. J. (2011). https://doi.org/10.2139/ssrn.1824224

Halkos, G.E., Salamouris, D.S.: Efficiency measurement of the Greek commercial banks with the use of financial ratios: a data envelopment analysis approach. Manag. Account. Res.. Account. Res. **15**(2), 201–224 (2004). https://doi.org/10.1016/j.mar.2004.02.001

Hatami-Marbini, A., Emrouznejad, A., Tavana, M.: A taxonomy and review of the fuzzy data envelopment analysis literature: two decades in the making. Eur. J. Oper. Res.Oper. Res. **214**(3), 457–472 (2011). https://doi.org/10.1016/j.ejor.2011.02.001

Home. (n.d.-a). Retrieved 2 January 2025, from https://www.federalreserve.gov/default.htm

Home. (n.d.-b). Retrieved 2 January 2025, from https://www.resbank.co.za/

Home. (2025, January 2). https://www.bankofengland.co.uk/

Home: 日本銀行Bank of Japan. (2024, December 30). Bank of Japan. https://www.boj.or.jp/en/index.htm

Hsiao, B., Chern, C.-C., Chiu, C.-R.: Performance evaluation with the entropy-based weighted Russell measure in data envelopment analysis. Expert Syst. Appl. **38**(8), 9965–9972 (2011). https://doi.org/10.1016/j.eswa.2011.02.033

Kamarudin, F., Sufian, F., Nassir, A., Anwar, N.A.M., Hussain, H.I.: Bank efficiency in Malaysia a DEA approach. J. Central Banking Theory and Practice **8**(1), 133–162 (2019). https://doi.org/10.2478/jcbtp-2019-0007

Khan, A.S., Akhtar, S., Alam, M.: How efficient are Indian banks? An application of stochastic frontier analysis. J. Financ. Econ.Policy (2024). https://doi.org/10.1108/JFEP-11-2023-0338

Lin, T.T., Lee, C.-C., Chiu, T.-F.: Application of DEA in analyzing a bank's operating performance. Expert Syst. Appl. **36**(5), 8883–8891 (2009). https://doi.org/10.1016/j.eswa.2008.11.018

Lou, Y., et al.: A parallel data envelopment analysis and Malmquist productivity index model of virtual frontier for evaluating scientific and technological innovation efficiency at universities. Decis. Analy. J. **10**, 100384 (2024). https://doi.org/10.1016/j.dajour.2023.100384

Ma, F., Wang, W., Sun, Q., Liu, F., Li, X.: Integrated transport efficiency and its spatial convergence in China's provinces: a super-SBM DEA model considering undesirable outputs. Appl. Sci. **8**(9), 1698 (2018). https://doi.org/10.3390/app8091698

Mahmoudi, R., Emrouznejad, A.: A multi-period performance analysis of airlines: a game-SBM-NDEA and malmquist Index approach. Res. Transp. Bus. Manag.Manag. **46**, 100801 (2023). https://doi.org/10.1016/j.rtbm.2022.100801

McKinley, V., Banaian, K.: Central bank operational efficiency: Meaning and measurement. Central Banking Publications (2005) https://ssrn.com/abstract=2508461

Mergoni, A., Emrouznejad, A., De Witte, K.: Fifty years of data envelopment analysis. Eur. J. Opera. Res. **326**(3), 389–412 (2025). https://doi.org/10.1016/j.ejor.2024.12.049

Mester, L.J.: Applying efficiency measurement techniques to central banks. SSRN Electron. J. (2004). https://doi.org/10.2139/ssrn.572442

Michali, M., Emrouznejad, A., Dehnokhalaji, A., Clegg, B.: Subsampling bootstrap in network DEA. Eur. J. Oper. Res.Oper. Res. **305**(2), 766–780 (2023). https://doi.org/10.1016/j.ejor.2022.06.022

Omrani, H., Shamsi, M., Emrouznejad, A., Teplova, T.: A robust DEA model under discrete scenarios for assessing bank branches. Expert Syst. Appl. **219**, 119694 (2023). https://doi.org/10.1016/j.eswa.2023.119694

Ononiwu, M.I., Onwuzulike, O.C., Shitu, K., Ojo, O.O.: Operational risk management in emerging markets: a case study of nigerian banking institutions. World J. Adv. Res. Rev. **23**(3), 446–459 (2024). https://doi.org/10.30574/wjarr.2024.23.3.2705

Pastor, J.T., Ruiz, J.L., Sirvent, I.: A statistical test for detecting influential observations in DEA. Eur. J. Oper. Res.Oper. Res. **115**(3), 542–554 (1999). https://doi.org/10.1016/S0377-2217(98)00153-2

Peykani, P., Emrouznejad, A., Mohammadi, E., Gheidar-Kheljani, J.: A novel robust network data envelopment analysis approach for performance assessment of mutual funds under uncertainty.

Ann. Oper. Res.Oper. Res. **339**(3), 1149–1175 (2024). https://doi.org/10.1007/s10479-022-04625-3

Portela, M.C.A.S., Thanassoulis, E., Simpson, G.: Negative data in DEA: a directional distance approach applied to bank branches. J. Opera. Res. Soc. **55**(10), 1111–1121 (2004). https://doi.org/10.1057/palgrave.jors.2601768

Rahman, M.: The effect of business intelligence on bank operational efficiency and perceptions of profitability. FinTech **2**(1), 99–119 (2023). https://doi.org/10.3390/fintech2010008

Reserve Bank of Australia. (2024, December 24). Reserve Bank of Australia. https://www.rba.gov.au/index.html

Reserve Bank of India. (n.d.). Retrieved 2 January (2025). https://www.rbi.org.in/

Ruggiero, J., Bretschneider, S.: The weighted Russell measure of technical efficiency. Eur. J. Oper. Res.Oper. Res. **108**(2), 438–451 (1998). https://doi.org/10.1016/S0377-2217(97)00150-1

Saudi central bank. (n.d.). Retrieved 2 January (2025). https://www.sama.gov.sa/en-US/Pages/default.aspx

Scheller, H.K.: The European Central Bank (Second Revised) (2006).

Seifert, L.M., Zhu, J.: Identifying excesses and deficits in Chinese industrial productivity (1953–1990): a weighted data envelopment analysis approach. Omega **26**(2), 279–296 (1998). https://doi.org/10.1016/S0305-0483(98)00011-5

Smith, P.: The use of performance indicators in the public sector. J. Royal Stat. Soc. Seri. A (Statistics in Society), **153**(1), 53 (1990). https://doi.org/10.2307/2983096

Soltanifar, M., Ghiyasi, M., Emrouznejad, A., Sharafi, H.: A novel model for merger analysis and target setting: A CSW-Inverse DEA approach. Expert Syst. Appl. **249**, 123326 (2024). https://doi.org/10.1016/j.eswa.2024.123326

Srivastava, S., Aggarwal, A., Bansal, P.: Non-homogeneous DEA approach in the presence of negative data: A promising prospective approach to enhance decision-making. Comput. Appl. Math.. Appl. Math. **43**(4), 255 (2024). https://doi.org/10.1007/s40314-024-02769-5

Sufian, F., Shah Habibullah, M.: Developments in the efficiency of the Thailand banking sector: A DEA approach. Int. J. Dev. Issu. **9**(3), 226–245 (2010). https://doi.org/10.1108/14468951011073316

Taleb, M., Emrouznejad, A., Charles, V., Khalid, R., Ramli, R.: An extended-directional mix-efficiency measure: Performance evaluation of OECD countries considering NetZero. Comput. Ind. Eng.. Ind. Eng. **189**, 109967 (2024). https://doi.org/10.1016/j.cie.2024.109967

Taylor, W.M., Thompson, R.G., Thrall, R.M., Dharmapala, P.S.: DEA/AR efficiency and profitability of Mexican banks a total income model. Eur. J. Oper. Res.Oper. Res. **98**(2), 346–363 (1997). https://doi.org/10.1016/S0377-2217(96)00352-9

Tcmb. (n.d.). Retrieved 2 January 2025. https://www.tcmb.gov.tr/wps/wcm/connect/en/tcmb+en

Thanassoulis, E., Boussofiane, A., Dyson, R.G.: A comparison of data envelopment analysis and ratio analysis as tools for performance assessment. Omega **24**(3), 229–244 (1996). https://doi.org/10.1016/0305-0483(95)00060-7

Thanassoulis, E., Dyson, R.G.: Estimating preferred target input-output levels using data envelopment analysis. Eur. J. Oper. Res.Oper. Res. **56**(1), 80–97 (1992). https://doi.org/10.1016/0377-2217(92)90294-J

The central bank of Russian federation | bank of Russia. (n.d.). Retrieved 2 January 2025. https://www.cbr.ru/eng/

The people's bank of China. (n.d.). Retrieved 2 January 2025. http://www.pbc.gov.cn/english/130437/index.html

Titko, J., Stankevičienė, J., Lāce, N.: Measuring bank efficiency: dea application. Technol. Econ. Dev. Econ. **20**(4), 739–757 (2014). https://doi.org/10.3846/20294913.2014.984255

Tone, K.: A slacks-based measure of efficiency in data envelopment analysis. Eur. J. Oper. Res.Oper. Res. **130**(3), 498–509 (2001). https://doi.org/10.1016/S0377-2217(99)00407-5

Tone, K.: A strange case of the cost and allocative efficiencies in DEA. J. Opera. Res. Soc. **53**(11), 1225–1231 (2002). https://doi.org/10.1057/palgrave.jors.2601438

Tran, T.H., Mao, Y., Nathanail, P., Siebers, P.-O., Robinson, D.: Integrating slacks-based measure of efficiency and super-efficiency in data envelopment analysis. Omega **85**, 156–165 (2019). https://doi.org/10.1016/j.omega.2018.06.008

Trnovská, M., Halická, M.: Applying path-based models to negative data with a focus on super-efficiency. In: A. Emrouznejad, E. Thanassoulis, & M. Toloo (Eds.), Advances in the Theory and Applications of Performance Measurement and Management (pp. 95–107). Springer Nature Switzerland (2024). https://doi.org/10.1007/978-3-031-61597-9_9

Wang, K., Huang, W., Wu, J., Liu, Y.-N.: Efficiency measures of the Chinese commercial banking system using an additive two-stage DEA. Omega **44**, 5–20 (2014). https://doi.org/10.1016/j.omega.2013.09.005

Yasmeen, R., Hao, G., Ye, Y., Shah, W.U.H., Tang, C.: The synergy of water resource agglomeration and innovative conservation technologies on provincial and regional water usage efficiency in China: a super SBM- DEA approach. Water **15**(19), 3524 (2023). https://doi.org/10.3390/w15193524

Zhu, J.: Data envelopment analysis with preference structure. J. Opera. Res. Soc. **47**(1), 136–150 (1996). https://doi.org/10.1057/jors.1996.12(N.d.)

Efficiency of Indian Life Insurance Companies-A Cluster Based Approach

Ram Pratap Sinha[1(✉)] and Alireza Amirteimoori[2]

[1] Government College of Engineering and Leather Technology, Kolkata, India
rampratapsinha39@gmail.com
[2] Istinye University, Istanbul, Turkey

Abstract. The extant DEA estimation of efficiency presupposes the existence of a monotone strictly convex production function. The presence of increasing or decreasing returns to scale are two cases of scale inefficiency where the firm are producing at sub-optimal scales. The convex DEA framework is, however, in conflict with the S-shaped production function introduced by Ragnar Frisch. The present study provides an empirical investigation of life insurer efficiency using the cluster-based approach introduced by Tone and Tsutsui. The present study thus estimates and explains the cluster adjusted efficiency scores of Indian life insurers.

1 Introduction

In the neo-classical framework of microeconomics, the production function is assumed to satisfy the property of homotheticity. Shephard (1953) defines a production function to be homothetic provided it can be represented as a continuous positive increasing monotonous function of a homogeneous production function. The possibility of productive inefficiency was excluded by assuming that the observed output (input usage) equals the potential output (input usage). Thus, productive firms are assumed to operate at the optimum level in the lon-run.

Shephard (1953, 1970, 1974) introduced the concepts of input and output distance functions for characterization of technology and estimation of efficiency which formed the basis of efficiency evaluation. Like Shephard, Farrell (1957) also introduced technical efficiency measures which relate observed performed with frontier performance depending the requirement of radial contraction of input or radial expansion of output. For the purpose of empirical estimation, Farrell specified the frontier as the as the most pessimistic linear envelopment of data in the setting of single output technology exhibiting constant returns to scale. The approach introduced by Farrell was extended by Charnes, Cooper and Rhodes (CCR) in 1978 in the context of multiple output and multiple input technology (exhibiting constant returns to scale) by providing a ratio-based measure of efficiency generated from the weighted sum of inputs and outputs. Farrell and

---

The original version of the chapter has been revised. To update the author's name to Alireza Amirteimoori. A correction to this chapter can be found at
https://doi.org/10.1007/978-3-031-98177-7_37

A. Emrouznejad et al. (Eds.): DEA 2024, LNOR, pp. 310–322, 2025.
https://doi.org/10.1007/978-3-031-98177-7_21

Fieldhouse (1962) pointed out that production technology often exhibits non-convexity and suggested two alternative methods for efficiency evaluation under increasing returns to scale. However, Banker, Charnes and Cooper (BCC) in 1984 (Emrouznejad, A., V. Podinovski, et al., 2025) extended the linear programming-based envelopment approach introduced by Charnes, Cooper and Rhodes to the case of variable returns to scale (allowing the technology to exhibit increasing and decreasing returns to scale over some ranges of outputs) through the addition of convexity constraint in the CCR framework.

The BCC approach towards efficiency evaluation focuses on the estimation of production efficiency without considering the aspect of scale efficiency and this is accomplished by removing the ray unboundedness property contained in the CCR approach. The estimated production possibility set is a polyhedral set which accommodate the returns to scale to be variable [Olesen and Ruggiero (2014)]. This conflicts with the U shaped average cost curve which implies an S-shaped technology displaying non-convexity. Hanoch (1975) pointed out that the assumptions made about scale elasticity along the ray through the origin do not necessarily correspond to an U shaped cost average curve and vice versa. The moot point to be noted here is that if the technology is non-convex is nature, the imposition of convexity can lead to biased estimates of technical efficiency and scale properties.

Given the background, the objective of the present study is to estimate and explain efficiency of Indian life insurers adopting a cluster-based Data Envelopment Analysis (DEA) approach introduced by Tone and Tsutsui (2015). In this approach, based on the character of technology and returns to scale the decision-making units are classified in to several clusters. The slacks exhibited by the DMUs are decomposed into two components-scale dependent and in-cluster efficiency. Finally, we compute scale adjusted efficiency which is the product of scale dependent and in-cluster efficiency. In the second stage, we have regressed the scale dependent and in-cluster efficiency scores on several contextual variables.

The present study has five sections and proceeds as follows. Section 2 provides a review of the related literature. Section 3 outlines the research methodology and provides information about variables and data. Section 4 includes the results and discussion. Section 5 concludes.

2 Review of Related Literature

As indicated in the outset, the present study uses the Data Envelopment Analysis approach for efficiency and scale evaluation. Since the original formulations of Charnes, Cooper and Rhodes (CCR) in 1978 and that of Banker, Charnes and Cooper (1984), the methodology has an exponential growth in numerous directions. Interested readers may consult Mergoni et al. (2025), Mirzaei, et al. (2024) and Mahmoudabadi, et al. (2019) in order get an idea about the progress of DEA literature.

Since our study utilizes the cluster based DEA approach, we now provide a very brief account of the prior studies involving cluster-based DEA. One of the earliest studies was by Meimand et al. (2002) which clustered DEA results to find common performance themes which were then used to measure performance against a Farrell frontier. Samoilenko and Osei-Bryson (2008) applied clustering to improve the discriminatory power of DEA in presence of heterogeneity. Po et al. (2009) adopted DEA-based clustering approach for clustering the data with input and output items. Amin et al. (2011) clarified the role of alternative optimal solutions in the clustering of multidimensional observations using DEA and pointed out that an observed DMU may be included in different clusters when the corresponding DEA estimation model has multiple optimal solutions. Paradi et al. (2012) introduced a new grouping approach in the context of DEA for identifying management groups and group performance leaders. Emrouznejad et al. (2014) provided a literature review on fuzzy DEA technique. Chen et al. (2020) introduced a new clustering approach based on DEA cross efficiency which not only clusters DMUs on the basis of production relationships but also reflects the preference of the decision maker. Sakouvogui et al. (2020) deals with the problems of extreme data points and heterogeneity in terms of a cluster-adjusted DEA model in which clusters are determined using Gap Statistic and Elbow methods.

As indicated in the outset, our study adopts a completely different approach with the objective of efficiency and scale estimation in the context of a non-convex technology.

3 Methodology, Variables and Data

3.1 Cluster Based DEA Estimation

We have followed Tone and Tsutsui (2015) for cluster-based estimation of DEA. The methodology is described briefly:

We consider a technology that converts a set of inputs (x_1, ……, x_s) into outputs ($y_{1,}$……., y_m). We denote the input and output vectors as X and Y respectively. The radial output-oriented efficiency measures for the CCR and BCC approaches can be written as:

$$E^{O}_{CCR} = Max\left[E^{O} | x_{obs} \geq \lambda X,\ E^{O}\, y_{obs} \leq \lambda Y, \lambda \geq 0\right] \tag{1}$$

$$E^{O}_{BCC} = Max\left[E^{O} | x_{obs} \geq \lambda X,\ E^{O}\, y_{obs} \leq \lambda Y,\ \lambda \geq 0,\ e\lambda = 1\right] \tag{2}$$

We now define the output slack in the CCR model as: $S^{+} = y_{obs}\left(E^{o}_{CCR} - 1\right)$

We can decompose S^{+} in to scale dependent and scale independent parts as:

$$S^{+} = (1-\omega)\, S^{+} + \omega\, S^{+} = (1-\omega) y_{obs}\left(E^{O}_{CCR} - 1\right) + \omega\, y_{obs}\left(E^{O}_{CCR} - 1\right)$$

Since this is an output-oriented model, we have the presence of output slack. The scale adjusted output and input can be written as:

$$y_s = y_{obs} + \omega\, y_{obs}\left(E^{O}_{CCR} - 1\right) \text{ and } x_s = x_{obs}$$

We can how find out scale adjusted efficiency by using the scale adjusted outputs and inputs. Further, we have computed efficiency with the clusters by taking the the data for the DMUs included in a specific cluster for the purpose of forming reference sets for the relative cluster.

Finally, the scale dependent efficiency part can also be computed based on the information available about scale dependent slacks.

3.2 Description of Variables and Data

Estimation of frontier based technical efficiency requires the identification of input and output indicators for the decision making units under consideration. However, deciding on the input and output indicators (variables) in the context of the life insurance sector is a difficult proposition due to the complexities involved in defining the indicator variables.

Eling and Luhnen (2010) identified three major types of inputs used in the insurance industry: labour (including agents and office staff), business services (including items such as travel, communications and advertisement) and capital (including debt and equity capital).

The measurement of life insurance output is equally challenging. O'Brien (1991) lists various approaches of output measurement which include both the income and expenditure approaches. Inter alia, the income approach encompasses the number of policies sold [Burgess and Walker (1982)] and the premium mobilised by the insurer [Colenutt (1977) and Praetz (1983) as output indicators for the life insurance industry. There are other studies, however, which consider unit cost weighted average of various services rendered by the life insurance companies [Geehan (1977)].

The present study seeks to construct revenue frontier. Therefore, we need to identify the major activities which contribute to insurer revenue as well as the inputs of production of service. Since we do not very detailed information on various inputs separately, we have considered two inputs- total operating expenses and net worth. The first input is considered as the proxy input for capturing operational activities which is required to generate premium. The second input captures the capital contribution of the equity shareholders. On the output side, we have considered net premium income and income from investments as the two outputs. For all the outputs, price is taken as unity. Consequently, revenue efficiency in the present context is synonymous with technical efficiency. In the second stage, four contextual variables (relative market share of the observed insurers, solvency ratio, return on equity and insurer age) have been included. The relative market share is an indicator of market dominance enjoyed by a life insurer and can influence insurer performance significantly. Insurer age is an indicator of the duration of market presence by the concerned life insurer. Solvency ratio and return on equity, on the other hand, are indicator of financial health and profitability of the life insurers. Table 2 presents a brief statement of the input indicators, output indicators and explanatory variables.

Table 1. Inputs, outputs and contextual variables

Description	Input indicators	Output indicators	Contextual variables
First stage (efficiency estimation)	Operating expenses, Net worth	- Total Income = Net Premium Income + Investment income	-
Second stage (regression of efficiency scores on contextual variables)	Not Applicable	Not Applicable	Herfindahl-Hirschman index of concentration, Solvency ratio, Return on equity and Insurer age

Author's selection.

The data relating to the input(s), output(s) and contextual variables have been collected from the Handbook of Indian Insurance Statistics published by the IRDA for various years. The period of analysis is 2013–14 to 2020–21. Efficiency estimation has been made using the radial approach.

4 Results and Discussion

The present section presents both first stage and second stage results relating to efficiency evaluation and explanation respectively.

4.1 Evaluation of Technology Character, Efficiency and Returns to Scale

Table 1 though 6 present the first stage summary relating to the global character of technology, local returns to scale, DMU clusters, scale adjusted efficiency scores and their two components (scale dependent efficiency and in cluster efficiency scores). The results are now discussed in brief. Please note that for the tables (1–6), 2013–14 is written as 2014, 2014–15 is written as 2015 and so on. The financial years begin from the 1st of April of a particular year and terminates on the 31st of March of the next year.

4.1.1 Global Character of Technology

Table 1 relates to the identification of the global character of technology based on the year-wise input-output relationship estimates corresponding to the 22 decision making units (life insurers). The table shows that majority of life insurance companies do not exhibit convexity for 5 out of the 7 years under observation. The LIC exhibited horizontal technology for the entire period under observation. Two life insurers exhibited vertical shape of technology for 6 in-sample years while only one life insurer showed vertical technology for 2019–20. The number of life insurers exhibiting convexity varies between 11 and 5 during the in-sample period. The number of insurers exhibiting convex technology was 11 for both 2014–15 and 2015–16 while it reached its lowest value (5) in 2018–19. Figure 1 plots the proportion of observed life insurers displaying convex technology.

Table 2. Global Character of Technology

Technology Character	2014	2015	2016	2017	2018	2019	2020	2021
Convex	7	11	11	7	7	5	7	6
Non-convex	12	8	8	12	12	14	13	13
Horizontal	1	1	1	1	1	1	1	1
Vertical	2	2	2	2	2	2	1	2
Total	22	22	22	22	22	22	22	22

Source: Calculated

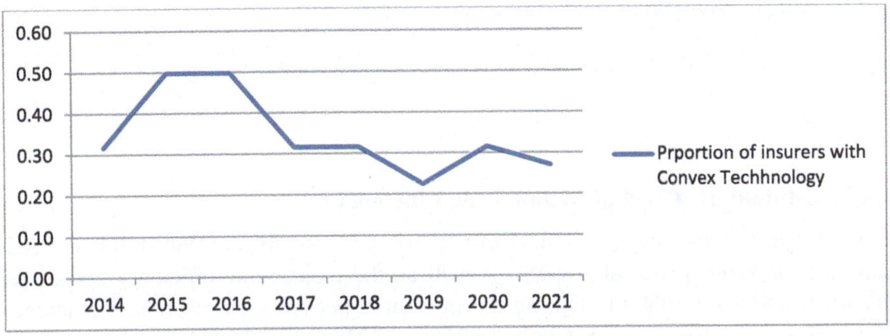

Fig. 1. Proportion of insurers with convex technology

4.1.2 Estimation of Local Returns to Scale

Table 2 provides information regarding the year-wise decomposition of the observed life insurers based on their local returns to scale. The table shows that except 2014–15, most of the in-sample life insurers were away from the optimal scale of operation. Thus, the total number of insurers displaying constant returns to scale (CRS) varied between 3 (minimum) and 12 (maximum) for the in-sample phase. The corresponding figures for increasing (IRS) and decreasing returns to scale (DRS) varied between 0 and 15 and 0 and 16 respectively. Figure 2 provides a graphical presentation regarding the proportion of life insurers exhibiting optimal scale.

Table 3. Local returns to scale

RTS	2014	2015	2016	2017	2018	2019	2020	2021
Constant	5	12	8	7	6	3	5	6
Increasing	8	7	14	15	8	7	14	0
Decreasing	9	3	0	0	8	12	3	16

Source: Calculated

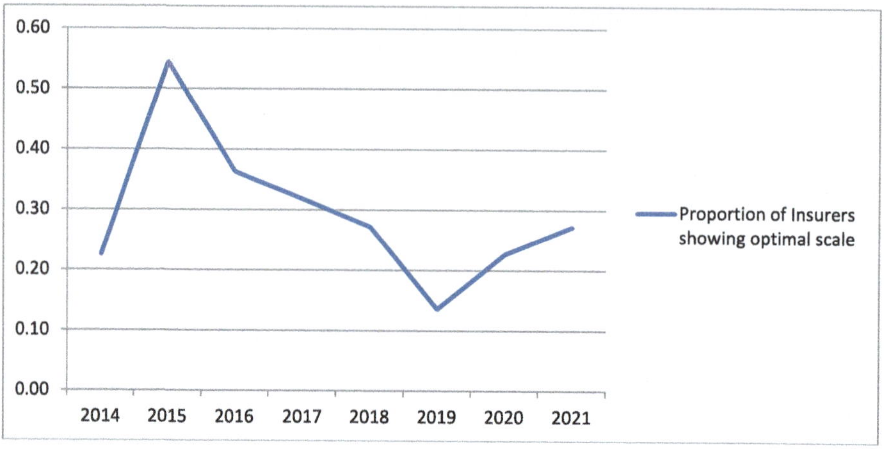

Fig. 2. Proportion of insurers exhibiting optimal scale

4.1.3 Estimation of Technical and Scale Efficiency

Table 3 presents the mean technical efficiency scores under constant returns (CRS) and variable returns to scale (VRS) as well as the mean scale efficiency scores for the years under observation. During the in-sample period, mean technical efficiency under CRS varied between 0.5363 (2020–21) to 0.8078 (2014–15) while the same under VRS varied between 0.7073 (2021) and 0.8503 (2017–18).The table further shows that mean scale efficiency remained more than 90% for the phase 2014–15 to 2017–18 but remained between 0.7621 and 0.8605 for the remaining observed years. The difference between technical efficiency scores under constant and variable returns to scale remained particularly sharp during 2020–21 in which the impact of COVID was the maximum. Figure 3 graphically presents the mean technical efficiency scores (under both CRS and VRS) and the mean scale efficiency scores for the period under observation.

Table 4. Mean technical and scale efficiency scores

Mean Efficiency Scores	2014	2015	2016	2017	2018	2019	2020	2021
Technical efficiency (CRS)	0.5800	0.8078	0.7135	0.7537	0.7675	0.6048	0.6528	0.5363
Technical efficiency (VRS)	0.6760	0.8476	0.7883	0.8206	0.8503	0.7682	0.7815	0.7073
Scale efficiency	0.8605	0.9544	0.9099	0.9205	0.9032	0.8098	0.8472	0.7621

Source: Calculated

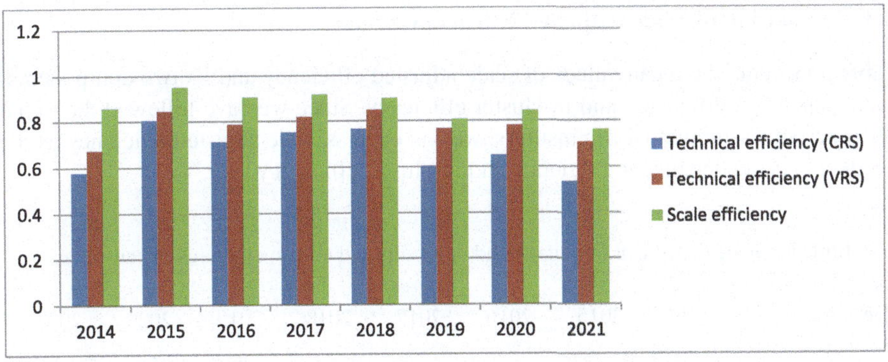

Fig. 3. Mean technical and scale efficiency scores of the observed life insurers

4.2 Formation of Decision-Making Unit (DMU) Clusters

Table 4 provides information regarding the number of clusters formed on the basis of returns to scale and scale efficiency. Cluster formation has been undertaken using a 2-metric approach involving the local returns to scale and insurer output size. The table indicates that out of the eight years under observation, the decision-making units (DUs) exhibiting increasing returns to scale have been decomposed into two separate clusters for 2015–16, 2016–17 and 2019–20. Similar procedure has been followed for the DMUs exhibiting decreasing returns to scale for 2018–19 and 2020–21. For 5 out of the 7 years under observation, the DMUs displaying a particular type of returns to scale (e.g. increasing or returns to scale) are sub-divided into more than one cluster. For the period under observation, the total number DMU clusters remained 3 for 2013–14 through 2017–18 and for 2020–21.

Table 5. Year wise DMU clusters

DMU Clusters	2014	2015	2016	2017	2018	2019	2020	2021
Constant	1	1	1	1	1	1	1	1
Increasing	1	1	**2**	**2**	1	1	**2**	0
Decreasing	1	1	0	0	1	**2**	1	**2**
Total Clusters	3	3	3	3	3	4	4	3

Source: Calculated

4.3 Estimation of Scale Adjusted Efficiency Scores

Table 5 presents the mean values of scale adjusted efficiency and its two components-scale dependent efficiency and in-cluster efficiency. Since we have followed the radial approach so for individual life insurers, we can express scale adjusted efficiency as the product of scale dependent efficiency and in-cluster efficiency (Fig. 4).

Table 6. Scale dependent, in-cluster and scale adjusted efficiency scores (mean values)

Descriptive Statistics	2014	2015	2016	2017	2018	2019	2020	2021
Scale dependent efficiency	0.8688	0.9806	0.9422	0.9505	0.9521	0.8706	0.9014	0.7750
In cluster efficiency	0.8512	0.9438	0.8350	0.9005	0.9253	0.9432	0.9144	0.9014
Scale adjusted efficiency	0.7353	0.9250	0.7837	0.8538	0.8815	0.8788	0.8212	0.6961

Source: Calculated

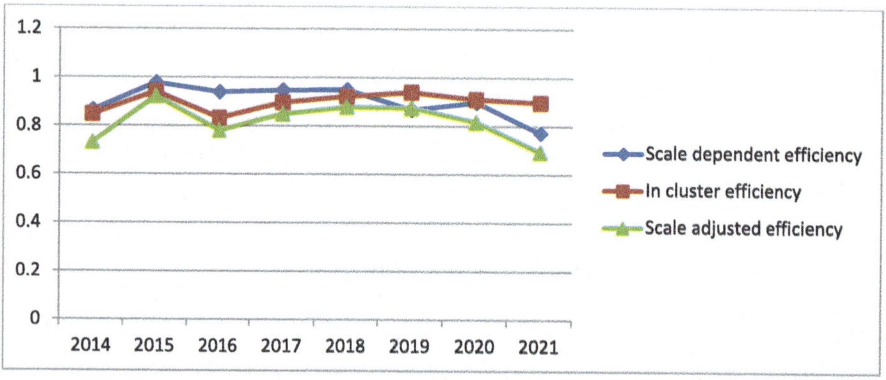

Fig. 4. Scale dependent, in cluster and scale adjusted (mean) efficiency scores

4.4 Estimation of Scale Elasticity

Table 6 presents the descriptive statistics of the scale elasticity scores for the in-sample life insurers for the period 2013–14 through 2020–21. The table indicates that except for 2020–21 mean scale elasticity remained above unity for the remaining years. Standard deviation of scale elasticity exhibited less variability during the first phase (2013–14 to 2016–17) under observation compared to the second phase (2017–18 to 2020–21). Both skewness and Kurtosis remained positive during the in-sample period.

Table 7. Descriptive statistics of scale elasticity of life insurers

Descriptive Statistics	2014	2015	2016	2017	2018	2019	2020	2021
Mean	1.0300	1.0675	1.1041	1.0538	1.2993	1.1387	1.1364	0.9994
Standard Deviation	0.2971	0.1840	0.2731	0.2134	1.0731	0.5034	0.3453	0.5286
Skewness	1.0181	2.4589	3.8649	4.3452	4.3995	3.5500	2.0967	3.2641
Kurtosis	1.7010	6.6352	16.7392	19.8375	20.1036	14.8593	5.3468	13.4105

Source: Calculated

4.5 Second Stage Regression Results

Tables 7 though 9 present the results of regressions carried out in the second stage. Table 7 presents the regression outcome for scale dependent efficiency as the dependent variable. We have applied fixed effects panel data model here based on the results of the Hausman test. We have considered 6 contextual variables (solvency ratio, insurer age, return on total asset, GDP growth rate, Herfindahi-Hirschman index of market concentration and log of total asset. Of these 6 variables, the coefficients of return on total asset and GDP growth rate are found to positive and significant at 95% level of significance.

Table 8. Regression of Scale Dependent Efficiency on the contextual variables

Explanatory variables	*Coefficient*	*Std. Error*	*t-ratio*	*p-value*
Intercept	−0.1342	1.2255	−0.1095	0.9138
Solvency	−0.0173	0.0274	−0.6297	0.5357
Age	0.01311	0.0263	0.4987	0.6232
ROTA	1.6921	0.141049	12.00	**<0.0001**
GDP Growth	0.0181	0.0051	3.530	**0.0020**
HHI	0.3796	0.7349	0.5164	0.6109
LnTA	−0.0290	0.0983	−0.2949	0.7710
Standard Error of Regression	0.1525			

Source: Calculated

Table 8 presents the regression results for in cluster efficiency (as the dependent variable).In this case, we applied random effects panel data model on the basis of the outcome of Hausman test. The regression results indicate that out of the 6 contextual variables, the coefficients of return on total asset and Hirschman-Herfindahl index are significant at 95% level. While the coefficient of return on total asset is found to be positive, the coefficient of H-H index is (expectedly) negative.

Table 9. Regression of In-cluster Efficiency on the contextual variables

Explanatory variables	*Coefficient*	*Std. Error*	*t-ratio*	*p-value*
Intercept	−0.1096	0.2877	−0.3807	0.7034
Solvency	0.0051	0.0096	0.5363	0.5917
Age	−0.0010	0.0022	−0.4712	0.6375
ROTA	0.2783	0.1359	2.048	**0.0405**
GDP Growth	0.0067	0.0049	1.371	0.1703
HHI	−0.8552	0.3940	−2.171	**0.0299**
LnTA	0.0266	0.0172	1.541	0.1233
Standard Error of Regression	0.1728			

Source: Calculated

Finally, Table 9 presents the regression estimates for scale adjusted efficiency. The table shows that the coefficients of three contextual variables (return on total asset, GDP growth rate and log of total asset) are statistically significant and positive (Table 10).

Table 10. Regression of Scale Adjusted Efficiency on the contextual variables

Explanatory variables	*Coefficient*	*Std. Error*	*t-ratio*	*p-value*
Intercept	−0.9273	0.4895	−1.895	0.0582
Solvency	−0.0017	0.0216	−0.0805	0.9358
Age	−0.0033	0.0039	−0.8424	0.3996
ROTA	1.8335	0.1839	9.970	**<0.0001**
GDP Growth	0.0251	0.0051	4.937	**<0.0001**
HHI	−0.8644	0.5268	−1.641	0.1009
LnTA	0.0718	0.0324	2.213	**0.0269**

Source: Calculated

5 Concluding Observations

The present study makes several important value additions to the extant efficiency literature relating to the life insurance industry. First, by decomposing the radial output slacks in to scale dependent and scale independent parts, the study computes scale adjusted efficiency which lies between the CCR and BCC efficiency measures respectively. Secondly, the study also provides in-cluster efficiency of observed life insurers based on the clustering of DMUs. Third, the study provides scale elasticity measures for observed insurers which gives valuable cardinal information regarding the scale optimality of the life insurers. The second stage analysis explores the relationship of scale dependent and in-cluster efficiency scores with several contextual variables. The study has immense practical implications because while benchmarking efficiency across firms or comparing different branches/sub-units of a firm, one comes across performance units of various sizes and the problem of heterogeneity cannot be overcome without resorting to cluster-based analysis.

The study is, however, based on eight years observations and can be extended for a longer period. Further, the number of contextual variables can be extended in the second stage to improve predictive and analytical ability.

References

Banker, R.D., Charnes, A., Cooper, W.W.: Some models for estimating technical and scale inefficiencies in data envelopment analysis. Manage. Sci. **30**(9), 1078–1092 (1984)

Burgess, J., Walker, G.: Returns to scale in the australian life insurance industry: a comment. Econ. Rec. **58**(3), 290–292 (1982)

Charnes, A., Cooper, W.W., Rhodes, E.: Measuring the efficiency of decision making units. Eur. J. Oper. Res. **2**(6), 429–444 (1978)

Chen, L., Wang, S.H., Wang, Y.M.: A new data envelopment analysis clustering approach within cross-efficiency framework. J. Opera. Res. Soc. **73**(3), 664–673 (2020). https://doi.org/10.1080/01605682.2020.1857667

Colenuit, D.W.: Economies of scale in the United Kingdom ordinary life assurance industry. Appl. Econ. **9**(3), 219–225 (1977)

Emrouznejad, A., Podinovski, V., Lu, C., Charles, V., Moradi-Motlagh, A.: Rajiv banker's lasting impact on data envelopment analysis. Ann. Oper. Res. **351**, 1225–1264 (2025). https://doi.org/10.1007/s10479-025-06473-3

Eling, M., Luhnen, M.: Frontier efficiency methodologies to measure performance in the insurance industry: Overview, systematization, and recent developments. Geneva Pap Risk Insur: Issues Pract. **35**(2), 217–265 (2010)

Emrouznejad, A., Tavana, M., Hatami-Marbini, A.: The state of the art in fuzzy data envelopment analysis. In: Performance Measurement with Fuzzy Data Envelopment Analysis published in Studies in Fuzziness and Soft Computing 309, pp. 1- 48, Springer-Verlag (2014)

Farrell, M.J.: The measurement of productive efficiency. J. Royal Stat. Soc. Seri. A **120**(3), 253–281 (1957)

Farrell, M.J., Fieldhouse, M.: Estimating efficient productions functions under increasing returns to scale. J. Roy. Stat. Soc. **125**, 252–267 (1962)

Gholam, R., Amin, A.E., Rezaei, S.: Some clarifications on the DEA clustering approach. Eur. J. Opera. Res. **215**(2), 498–501 (2011). https://doi.org/10.1016/j.ejor.2011.06.043

Geehan, R.: Returns to scale in the life insurance industry. Bell J. Econom. **8**(2), 497–514 (1977). https://doi.org/10.2307/3003299

Hirshhorn, R., Geehan, R.: Measuring the real output of the life insurance industry. Rev. Econ. Stat. **49**(2), 211–219 (1977)

Mahmoudabadi, M.Z., Emrouznejad, A.: Comprehensive performance evaluation of banking branches: a three-stage slacks-based measure (sbm) data envelopment analysis. Int. Rev. Econ. Financ. **64**, 359–376 (2019)

Meimand, R., Cavana, Y., Laking, R.: Using DEA and survival analysis for measuring performance of branches in New Zealand's accident compensation corporation. J. Opera. Res. Soc. **53**(3), 303–313 (2002). https://doi.org/10.1057/palgrave.jors.2601291

Mergoni, A.A. Emrouznejad, K.: De Witte: Fifty years of Data Envelopment Analysis. Eur. J. Oper. Res. **326**(3), 389–412 (2025). https://doi.org/10.1016/j.ejor.2024.12.049

Mirzaei, A., Saad, M., Emrouznejad, A.: Bank stock performance during the COVID-19 crisis: Does efficiency explain why Islamic banks fared relatively better?. Ann. Opera. Res. **334**, 317–355 (2024)

Olesen, O.B., Ruggiero, J.: Maintaining the regular ultra passum law in data envelopment analysis. Eur. J. Oper. Res. **235**(3), 798–809 (2014)

Paradi, J., Zhu, H., Edelstein, B.: Identifying managerial groups in a large Canadian bank branch network with a DEA approach. Eur. J. Oper. Res. **219**, 178–187 (2012). https://doi.org/10.1016/j.ejor.2011.12.022

Po, Y., Guh, Y., Yang, M.: A new clustering approach using data envelopment analysis. Eur. J. Oper. Res. **199**, 276–284 (2009). https://doi.org/10.1016/j.ejor.2008.10.022

Praetz, P.D.: The effect of size and other economic factors on the Expense Rate Behaviour of U. K. Life Insurance Companies, J. Inst. Actuaries **110**, 383–388 (1983)

Sakouvogui, K., Shaik, S., Addey, K.A.: Cluster-adjusted DEA Efficiency in the presence of heterogeneity: an application to banking sector. Open Econom. **3**(1), 50–69 (2020). https://doi.org/10.1515/openec-2020-0004

Samoilenko, S., Osei-Bryson, K.: Increasing the discriminatory power of DEA in the presence of the sample heterogeneity with cluster analysis and decision trees. Expert Syst. Appl. **34**(2), 1568–1581 (2008). https://doi.org/10.1016/j.eswa.2007.01.039

Shephard, R.W.: Cost and Productions. Princeton University Press, Princeton (1953)

Shephard, R.W.: Theory of Cost and Productions. Princeton University Press, Princeton (1970)

Shephard, R.W.: Indirect Production Functions. Verlag Anton Hain, Meisenheim am Glan (1974)

Tone, K., Tsutsui, M.: How to deal with non-convex frontiers in data envelopment analysis. J. Optim. Theory Appl.Optim. Theory Appl. **166**(3), 1002–1028 (2015). https://doi.org/10.1007/s10957-014-0626-3

Cost and Allocative Efficiency of Public Hospitals in Ireland – A Robust DEA Approach

Niall Devitt, Declan Dineen, and Marta Zieba(✉)

Department of Economics, University of Limerick, Limerick, Ireland
{devitt.niall,declan.dineen,marta.zieba}@ul.ie

Abstract. This study evaluates the cost and allocative efficiencies of 37 Irish public hospitals (IPH) from July 2017 to June 2018, using a unique dataset containing in-depth information on inputs, outputs, and input prices. Conventional DEA and two-stage double-bootstrap (DB) DEA models are applied to derive robust bias-corrected efficiency scores for IPH. The DB DEA results show that Irish hospitals operate with an average bias-corrected cost efficiency score of 0.756. Thus, IPH are considerably cost-inefficient as they could reduce their costs by almost 25% and still produce the same output level. Moreover, the average DB DEA allocative efficiency is derived at 0.965, whilst the mean bias-corrected technical efficiency is 0.783, indicating that technical (productive) inefficiencies are the primary driver of cost inefficiencies in IPH. The DB DEA results also suggest that reducing private practice in public hospitals negatively influences cost efficiency. Furthermore, local frontier results are generated by truncating the data sample into various hospital groups by type and size. These results demonstrate that large Type-4 hospitals are the most cost-efficient hospital units, which deal with the highest volume and the most complex cases. There is also consistency in the results, where the smaller Type-2 and Type-3 hospitals, which provide emergency care in small urban or regional settings, are the least cost-efficient. Altogether, the findings indicate that, whether applying the full-sample DB DEA frontier or the local DB DEA frontier approach, significant cost savings and reductions in input use can be achieved in IPH.

Keywords: Hospitals · Cost and Allocative Efficiency · Technical Efficiency · Double-Bootstrap DEA · Efficiency Determinants · Ireland

1 Introduction

The spiraling cost of healthcare is a well-established trend in developed countries and has highlighted the escalating healthcare expenditure burden placed on government finances. In this respect, Ireland is an interesting locale. Between 2000 and 2018, per capita annual spending on healthcare in Ireland increased by 167%, compared to 132% in the USA and 86% in France. At the same time, bed capacity decreased in Ireland from 6.13 per 1,000 population to 2.97, while in the USA and France, it decreased by a much smaller amount, from 3.49 to 2.83 and from 7.97 to 5.9 beds, respectively [1]. Ireland is in the bottom tier among European countries regarding hospital bed capacity, and a lack of investment

A. Emrouznejad et al. (Eds.): DEA 2024, LNOR, pp. 323–335, 2025.
https://doi.org/10.1007/978-3-031-98177-7_22

in Irish public hospitals (IPH) can explain the decrease in total bed provision over time. Moreover, it is projected that the total nominal gross expenditure requirements of IPH to 2035 will need to grow by between 3.6% and 5.4% per annum on average [2]. In the face of rising healthcare spending, cost increases, and underinvestment, assessing and improving efficiencies is especially relevant for the IPH sector. Thus, appropriate cost management becomes imperative for the hospital sector in Ireland, and the challenge is to reduce hospital costs and use the cost savings to expand capacity appropriately. To our knowledge, this is the first study in the Irish setting to evaluate hospitals' cost and allocative efficiencies by applying DEA and robust DB DEA models and using detailed monthly data on outputs, inputs, and input prices of IPH.

The study is structured as follows: Sect. 2 provides the context for the analysis by briefly reviewing the literature, emphasizing cost and allocative efficiency measurement in the hospital sector. Section. 3 presents the methodology and empirical approach applied. The dataset and variables are outlined in Sect. 4, while Sect. 5 presents the findings and Sect. 6 concludes.

2 Background

Following the efficiency concepts of Farrell [3], the cost efficiency (CE) of hospitals is defined in this study as choosing the optimal input sets, given the corresponding input prices and the produced hospital outputs, to minimize the hospital's total costs. The corresponding technical efficiency (TE) is then measured from the input-oriented perspective, which relates to the minimization of resources given the produced output. The difference between cost and technical efficiency defines allocative efficiency (AE), which reflects the optimal allocation of resources based solely on the input prices (i.e., even when the inputs are minimized).

Data Envelopment Analysis (DEA), first introduced by [4–6], has been widely applied in many sectors and from various perspectives (see, e.g., [7] for a comprehensive overview of fifty years of DEA). Moreover, hospitals' productive (technical) efficiency has been studied in many countries using the DEA approach for various concepts and data [8–12]. However, in contrast to estimating TE, the literature about cost and allocative efficiency measurement in the hospital sector is relatively sparse - mainly due to the lack of adequate data on input prices [13]. Cost and allocative efficiency have been investigated using DEA for hospitals in Greece [14], Norway [15], Spain [16], and the USA [13, 17]. Though, in these studies, only [15] and [17] applied bootstrap DEA methods to evaluate the CE of hospitals. In an Irish context, [18] used the parametric Stochastic Frontier Analysis (SFA) to obtain the first TE estimates for IPH – these range from 0.96–0.97 for the late 1990s. More recently, [19] conducted a panel data SFA analysis of TE in Irish public hospitals using a much richer dataset and estimating an output-distance function that accounts for multiple outputs in contrast to the standard single-output production function. Their findings show that Irish hospitals have an average short-run TE score of 0.93–0.94, whereas the persistent TE is 0.90. Furthermore, the study found that the less specialized Type-2 hospitals produce lower output, and the medium-sized hospitals are the most technically efficient units. The study also found that the share of emergency patients negatively affects hospitals' production, and the length of stay in a hospital contributes to inefficiency in IPH.

The current study extends the previous research and estimates cost and allocative efficiencies for IPH using robust DB DEA analysis. DEA has an advantage over SFA efficiency measurement methods, as it does not require prior information on the underlying production technology. However, the conventional DEA is deterministic and does not account for either sampling or measurement error [15, 20]. Moreover, to evaluate the technical and cost efficiency determinants, the standard regression analysis in the two-stage DEA approach fails to give valid inference as the true unobserved efficiency is replaced with DEA estimates, which are serially correlated by construction and are biased [21]. Therefore, this study applies the two-stage double-bootstrap (DB) DEA of [22, 23], which accounts not only for sampling variability to estimate bias-corrected efficiencies of IPH but also incorporates exogenous efficiency determinants into the bias-corrected TE and CE scores (see Subsect. 3.3 for details). Moreover, we truncate the sample into several sub-samples by hospital size and type and derive separate DB DEA scores using these local frontiers.

3 Empirical Model for IPH

In this section, we first introduce the conventional input-oriented DEA model used to obtain the non-parametric TE scores based on the input-output data for IPH. Then, we introduce the cost-efficiency DEA model to derive the corresponding CE and AE, which includes the cost-minimization problem given the corresponding input quantities and their input prices. Furthermore, we present the two-stage double-bootstrap (DB) DEA to derive robust bias-corrected efficiency scores and examine the effect of public-patient share on the cost and technical efficiencies of IPH.

3.1 Input-Oriented TE for IPH

To provide estimates of cost efficiency (CE) and allocative efficiency (AE) for IPH, a corresponding input-oriented technical efficiency (TE) is first obtained. The hospital unit aims to minimize input usage to produce a given output level. Thus, the study tries to determine whether the sample hospitals are producing a given output level with the minimum input utilization based on the actual input-output data. We use the variable returns to scale (VRS) assumption in line with [6, 24] by solving the following DEA linear programming problem:

$$\begin{aligned} \min_{\theta,\lambda} \quad & \theta_{it}(\theta,\lambda) \\ s.t.: \quad & -q_{it} + Q\lambda \geq 0 \\ & \theta_{it}x_{it} - X\lambda \geq 0 \\ & I1'\lambda = 1 \\ & \lambda \geq 0 \end{aligned} \tag{1}$$

where θ_{it} is the obtained technical efficiency (TE) score for the *i*-th hospital in month *t*, which is bounded between 0 and 1, while λ is a vector of constants; $q$it denotes a vector of outputs of the *i*-th hospital unit on a *t*-monthly basis, and x_{it} is a vector of labour and capital inputs, respectively. The constraint I1'λ creates a convex hull around a sample observation, which more closely resembles the productive activity of the hospitals of similar scale. Removing this constraint provides the constant returns to scale (CRS) DEA model of [6], and the difference between VRS and CRS results indicates the existence of possible scale inefficiencies.

3.2 Cost and Allocative Efficiency for IPH

After obtaining TE using the input-oriented DEA model in the Subsect. 3.1, the cost efficiency (CE) and allocative efficiency (AE) are further calculated in line with [24] using the cost minimization VRS DEA model and solving the following linear programming problem:

$$\begin{aligned} \min_{\lambda, x_{it}^*} \quad & w_{it} x_{it}^*, \\ s.t.: \quad & -q_{it} + Q\lambda \geq 0, \\ & x_{it}^* - X\lambda \geq 0, \\ & I1'\lambda = 1, \\ & \lambda \geq 0 \end{aligned} \tag{2}$$

where all the symbols have the same meaning as in the Subsect. 3.1. Now, x_{it}^* is the calculated cost-minimizing vector of input quantities for hospital i in month t, given the vector of two input prices, w_{it}, related to the total staff (wage rate) and the number of beds (non-pay unit costs). The cost efficiency of hospital i in month t is then calculated as the ratio of minimum costs to its actual costs as $CE_{it} = w_{it}x_{it}^*/w_{it}x_{it}$.

An important aspect of the cost-minimization DEA problem is the appropriate choice of the corresponding labour and capital input prices [13]. To obtain these prices for IPH, this study uses information from the relevant hospital reports on total hospital expenditures and the consolidated pay scales of numerous hospital staff categories for the IPH as follows: Let $(p_1, \ldots, p_j)$ represent the midpoint of the consolidated monthly salary scales for all IPH associated with the employee numbers for each staff category $(s_{1it}, \ldots, s_{jit})$. Then, the monthly wage rate of the aggregated labour (staff input-mix) for each hospital i and month t is calculated as a weighted average of the underlying wage costs associated with the corresponding staff inputs as follows:

$$w_{1it} = \frac{s_{1it}p_1 + \ldots\ldots + s_{jit}p_j}{s_{1it} + \ldots\ldots + s_{jit}} \tag{3}$$

Once the wage rate for all hospital staff is derived, the total labour costs for each hospital can be obtained as $Lab_{it} = w_{1it} * x_{1it}$, where x_{1it} is the total number of all hospital employees. Also, the capital input price, as non-pay unit costs, is derived as the difference between the total expenditures for each hospital and the total labour costs, divided by the corresponding number of hospital beds (x_{2it}):

$$w_{2it} = \frac{(TC_{it} - Lab_{it})}{x_{2it}} \tag{4}$$

After estimating CE, allocative efficiency (AE) can be derived residually as the ratio between CE and TE so that: $AE_{it} = CE_{it}/TE_{it}$. Like TE, the CE and AE scores are bounded between 0 and 1, with a score of 1 indicating a cost and allocative-efficient hospital unit.

3.3 Double-Bootstrap DEA

According to [15], bootstrapping DEA is a procedure that randomly draws with replacement from the primary data sample, mimicking the data-generating process of the underlying true model and producing multiple pseudo-estimates that allow for the calculation of the sampling error of the DEA scores and their variance, as well as their confidence intervals. In this study, to obtain both bias-corrected DEA TE and CE scores and unbiased estimates of the parameters of our posited determining variables, we adopt Algorithm 2 of the double-bootstrapping (DB) DEA method set out by Simar and Wilson [22, pp. 42–43].

The initial truncated regression model is estimated using the reciprocals of the conventional DEA TE and CE scores, $(1/(TE_{it}))$ and $(1/(CE_{it}))$, as the dependent variable and the efficiency Z-determinants as the explanatory variables. Firstly, the bootstrap is applied to correct the bias problem in the original DEA TE and CE scores. Thus, empirical sampling distributions are obtained by taking L1 = 100 drawings of residuals from a truncated normal distribution. The truncated regression model is re-estimated for each drawing, and the DEA TE and CE scores are re-calculated (performing L1 additional DEA analyses – one for each drawing) after also adjusting the input values for the ratio of the original DEA TE and CE estimates to the bias-corrected DEA TE and CE scores. The second objective is to correct the serial correlation problem of the DEA estimates. The truncated regression models are re-run again – this time with the reciprocals of the bias-corrected efficiency scores as the dependent variables as follows:

$$1/TE^{*}_{it} = f(z_{it}, \beta) + \varepsilon_{it} \tag{5}$$

$$1/CE^{*}_{it} = f(z_{it}, \beta) + \varepsilon_{it} \tag{6}$$

where z_{it} is the vector of efficiency determining variables, β is the vector of their corresponding coefficients, and ε_{it} is the error term. Similar to before, L = 2,000 drawings of residuals aaken from a truncated normal distribution. Thus, after re-estimating the truncated regression model for each drawing, we obtain the bias-corrected TE and CE scores, unbiased parameter estimates of the z-variables, and their confidence intervals as lower and upper bounds. The robust TE and CE scores are further used to derive equally robust AE scores as per Subsect. 3.2.

It is important to note that the second-stage regression on the efficiency scores will only be valid when a 'separability' condition holds; that is, for the given level of output, the minimum input or cost level does not vary with the z-variables [23, p.207]. Hence, under the separability assumption, the environmental z-variables only affect the technical or cost efficiency distribution, but they do not affect the production possibilities. To check if this condition is satisfied, the separability tests on the z-variables are also conducted in this study in line with [25] by comparing the unconditional (original) DEA scores with DEA scores that are conditioned on the z-variables.

4 Dataset and Variables

The data sample of Irish public hospitals (IPH) was collected for 12 discrete months, from July 2017 to June 2018, using publicly available information from the Health Service Executive (HSE) in Ireland and drawing on three primary data sources: (1) monthly management data reports (MMDR), (2) monthly management employment reports (MMER), and (3) consolidated pay scales (CPS) reports. The MMDRs are primarily used for the output variables and the total staff input data (FTE for all hospital employees) - and for total expenditure (costs of hospitals), which was used to calculate capital unit costs as the price of capital. The CPS reports contain the salaries and salary scales for the individual staff categories used to calculate each hospital's average monthly wage rate, weighted by the different staff categories as explained in Sect. 3. In addition, the HSE Business Information Unit (BIU) generously provided data specifically for this research on the inpatient and day-case bed capacity for each hospital in the sample.

The original sample included all acute hospitals, including maternity, children's, and specialist orthopedic hospitals. In line with [19], the initial sample of 52 hospitals was reduced to 37 traditional general and regional hospitals, which have similar structures. When combined with a time series of 12 consecutive months, this smaller sample formed a balanced dataset of 444 observations.

Table 1 presents the descriptive statistics for the output and input variables, the input prices, and z-factors for the sample of IPH. The mean number of inpatients treated monthly by each hospital is 1,211, with the largest hospital seeing 4,341 inpatients in one month. Outpatient numbers represent the greatest volume of outputs, with each IPH seeing an average of 5,630 patients per month. Similarly, on average, day-case patients account for 1,965 monthly outputs, with one hospital seeing up to 7,155 cases. In total, the three output categories are representative of all the output generated in the IPH in a monthly period. The mean wage rate, weighted salary scale of the staff hospital mix, is €3,733, and the average non-pay (clinical and non-clinical) costs per bed are €13,832.

The primary and only Z-determinant of both technical and cost (in) efficiency used in the DB DEA model is the 'Public-Patient Share' (PPS). This is the share of inpatients whose treatment costs are covered by the state and measures the degree of private market activity of IPH. The mean PPS is 83% in Table 1, but the share varies greatly between the hospitals. It is important to note that Ireland's share of private health insurance on public hospital costs stands at 30%, and after the USA, this is the second largest share among OECD countries [19]. Hence, public hospitals in Ireland provide a considerable amount of care to private patients.

Furthermore, hospital size and type are used to truncate the sample of IPH and to estimate separate local frontiers for these various hospital groups. The hospitals were divided into three discrete size classes – small, medium, and large (see Table 2). While hospital size may affect the production technology and, hence, the optimal frontier, the hospital type is also an important characteristic of hospital activity in Ireland, representing the heterogeneous nature of the services offered. IPH can be categorized into four generic hospital model types ranging from Type-1 to Type-4, in line with the degree of complexity of the care provided. Type-1 hospitals are community or district facilities with the least complexity, and without acute care (i.e., short-term, emergency, or sudden

Table 1. Summary statistics of relevant variables for 37 IPH over 12 months.

Variable and Description	Mean	St. Dev	Min	Max
Outputs (in units)				
Casemix-Adjusted Inpatients	1,211	1,058	2	4,341
Casemix-Adjusted Day Cases	1,965	1,709	88	7,155
Outpatients	5,630	4,635	549	20,104
Inputs (in units)				
Labour (Total hospital staff)	1,340	1,090	219	3,922
Capital (Number of beds)	306	227	27	832
Input prices (in EUR)				
Average wage rate of staff mix	3,733	157	3,096	3,992
Non-pay expenses per bed	13,832	4,457	2,482	36,000
Z - (In)efficiency factors				
Public Patient Share	0.8297	0.079	0.575	1.000

Table 2. Hospital *type* and *size* variables

Variable	Description	% Share
Type-2	The hospitals provide a range of surgeries and diagnostic services, but do not operate a 24-h accident and emergency department	29.7
Type-3	The hospitals provide 24-h A&E and acute surgery, along with critical care	45.9
Type-4	They encompass all activities of Type-2 and Type-3 hospitals and provide more critical services and tertiary care	24.3
Small	Hospitals with fewer than 200 beds	42.7
Medium	Hospitals with more than 200 and fewer than 400 beds	33.4
Large	Hospitals with more than 400 beds	23.9

illness care), and therefore, they are excluded from the analysis. Thus, this study partitions the dataset for three hospital-type sub-samples: Type-2, Type-3, and Type-4 (see also Table 2).

5 Findings

Subsect. 5.1 presents the TE, CE, and AE scores obtained for the whole sample of IPH using both the conventional DEA and DB DEA models while Subsect. 5.2 discusses the estimated effects of a single efficiency determinant using DB DEA. Subsection 5.3 examines the summary statistics of the estimated DB DEA efficiency scores for the truncated subsamples, referring to the different hospital types and sizes. This allows for a more granular analysis of the hospital efficiency scores.

5.1 TE, CE, and AE Scores for IPH

Table 3 presents the estimated efficiency scores of IPH for both VRS and CRS DEA models. The asterisk beside the original DEA score denotes that it is outside the bootstrapped 95% confidence interval and, hence, it is significantly different (i.e., higher or lower) from the bias-corrected DB DEA efficiency score. The results show that the conventional DEA estimates are higher than the bias-corrected DB DEA scores, except for AE in the CRS DEA model. The bias is statistically insignificant for TE and CE under VRS technology. The VRS TE estimates are also higher than the CRS scores, indicating that hospitals are not fully scale efficient, as found previously in [19] for IPH. Given these findings, the VRS DB DEA results will be used for further analysis.

Table 3. Average TE, CE, and AE scores for IPH.

Mean Efficiency Score	Original DEA	DB DEA	DB LB	DB UB
VRS DEA				
Technical Efficiency (TE)	0.8173	0.7834	0.7565	0.8226
Cost Efficiency (CE)	0.7911	0.7565	0.7322	0.7918
Allocative Efficiency (AE)	0.9679*	0.9641	0.9625	0.9678
CRS DEA				
Technical Efficiency (TE)	0.7705*	0.7352	0.7110	0.7665
Cost Efficiency (CE)	0.7411*	0.7085	0.6872	0.7380
Allocative Efficiency (AE)	0.9606*	0.9636	0.9628	0.9665

DB (double bootstrap) 95% confidence interval with LB (lower bound) and UB (upper bound).

Based on the VRS DB DEA results, the IPH are, on average, only 78% technically efficient (TE = 0.783) and only 75.6% cost efficient (CE = 0.756), indicating that they could reduce their costs by almost 25% and still produce the same amounts of output with the given level of inputs and input prices. The lower cost efficiency is also the result of the allocative *inefficiency* of IPH, with an average AE score of 0.964. Cost efficiency estimates in international studies vary across countries (see also Sect. 2). Still, a similar level of cost *inefficiency* (about 20%) has been observed in DEA research on Greek [14] and Catalan hospitals [16]. Moreover, similar to this study, [17] and [15] controlled for the bias in the original DEA scores using the double-bootstrap DEA approach and found lower cost efficiency levels of 61–64% in US hospitals and 70–74% in Norwegian hospitals, respectively.

Figure 1 also presents the VRS DB DEA efficiency results over time. There is a seasonal variation in the average TE and CE scores during the examined period from July 2017 to June 2018, with decreased performance observed throughout the winter period. This result aligns with the fact that waiting lists and the number of patients without a hospital bed typically escalate considerably in IPH during this time.

However, allocative efficiency shows minimal variation for IPH in Fig. 1, indicating that the extent of adjusting the inputs according to their prices does not change essentially. Over time, cost inefficiencies result from low productive efficiency (i.e., technical inefficiency), which can be explained by inappropriate allocation of physical resources in the hospital units.

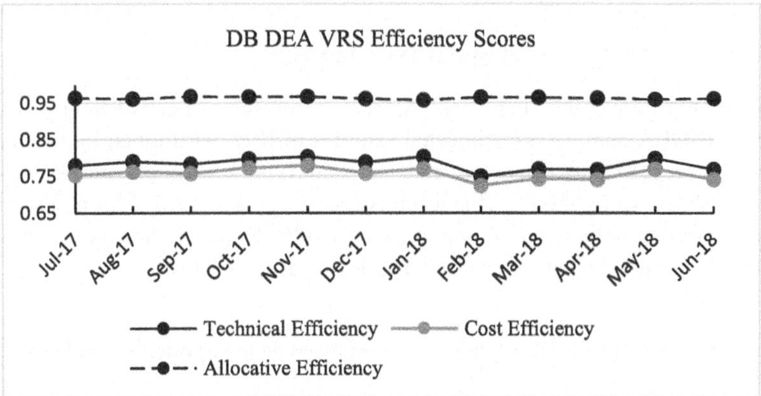

Fig. 1. Estimated average TE, CE, and AE scores monthly 2017–2018.

5.2 Efficiency Determinants

Table 4 presents the estimated coefficients of the TE and CE determinants using the DB DEA algorithm of [22] for the full sample of 37 hospitals, as outlined in Subsect. 3.3. The estimated coefficient of the single posited TE determinant, PPS (public-patient share), indicates a direct effect on the reciprocals of the TE^* and CE^* scores and, hence, an inverse effect on technical and cost efficiency, respectively. The separability tests of [25], which test the null hypothesis that the z-variables influence neither the shape nor the level of the optimal frontier (i.e., the boundary of the attainable set), were applied for our dataset (see also Subsect. 3.3). The results are available upon request and show that the separability conditions hold, indicating that the effect of the single Z_1-variable on the production process is only through the distribution of the inefficiencies.

The results in Table 4 strongly indicate that the sole Z_1-determinant, public-patient share (PPS), negatively affects the bias-corrected technical and cost efficiency scores. Importantly, this implies a difference in terms of efficiencies between public and private patients in Irish hospitals. Hence, the findings indicate that private patients increase hospital efficiencies due to the financial incentives for public hospitals in Ireland to provide care to those patients. This contributes to the current debate about reducing or increasing private practice in Irish public hospitals.

Table 4. The DB VRS DEA estimates of efficiency determinants

	Effect on $(1/TE^*_{it})$			Effect on $(1/CE^*_{it})$		
Variable	Coefficient	LB	UB	Coefficient	LB	UB
Intercept (Z_0)	0.733*	0.511	0.980	0.736*	0.491	0.993
PPS (Z_1)	0.690*	0.354	0.907	0.690*	0.382	0.969

LB = lower CI bound UB = upper CI bound. *Significant at 5% of bootstrapped CIs.

5.3 Truncated Analysis by Hospital Type and Size

Table 5 presents the average TE, CE, and AE scores and their respective rankings obtained from the DB DEA analysis for truncated samples of IPH, partitioned by hospital type and size. The local frontier analysis indicates that Type-4 hospitals are the most cost-efficient, with CE score of 0.88. This result is encouraging as Type-4 hospitals deal with the highest volume and the most complex cases. Higher levels of efficiency in these units translate into greater system-wide efficiency levels in the IPH sector.

Table 5. The DB VRS DEA efficiency scores based on the *type* and *size* of hospitals.

Hospital Group	Technical Efficiency		Cost Efficiency		Allocative Efficiency	
	Score	Ranking	Score	Ranking	Score	Ranking
Hospital Type						
Type-2	0.860	2	0.823	2	0.957	3
Type-3	0.813	3	0.787	3	0.968	2
Type-4	0.909	1	0.884	1	0.972	1
Hospital Size						
Small	0.843	3	0.809	3	0.960	3
Medium	0.930	1	0.906	1	0.974	2
Large	0.880	2	0.863	2	0.981	1

The results also show that 'Small' hospitals (units with less than 200 beds) are the least efficient, with their average TE, CE, and AE displaying the lowest rankings. In contrast, medium-sized hospitals (units with 200–400 beds) have the highest TE and CE scores, followed by large hospitals (with more than 400 beds). This result corresponds with the finding of [19] for IPH, where medium-sized hospitals had the highest effect on TE scores in the SFA output-distance function model.

Our finding that 'Small' hospitals are the least efficient in Ireland can be explained by the fact that 84% of small hospitals belong to Type-2 hospitals (see Fig. 2). Type-2 hospitals usually treat low-risk patients and are less specialized than other hospital types [26]. At the same time, Type-2 and Type-3 hospitals, which generally provide care

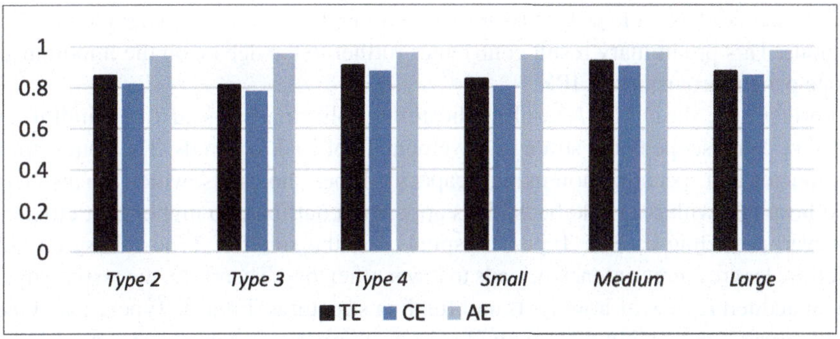

Fig. 2. Estimated average TE, CE, and AE by hospital type and size.

in a small urban or regional setting, are less technically and cost-efficient than Type-4 hospitals, which are the most efficient hospital type.

The provision of a 24-h Accident & Emergency department in Type-3 hospitals, which differentiates them from model Type-2 hospitals, may reduce their efficiency levels even further. Thus, our findings imply that the emergency care resources in Type-3 hospitals are spread too thin, so emergency care has neither the desired critical mass nor effectiveness while at the same time being less cost-efficient.

The truncated results also confirm that whether applying the DB DEA full-sample frontier or the local DB DEA frontier approach (subsamples for hospital type and size), the findings indicate that significant savings in costs and input use in IPH can be achieved. It also follows that investing public funds in specialist high-capacity emergency treatment provision in a Type-4 hospital setting would be more efficient and efficacious for the Irish public hospital sector. In contrast, the small, mostly Type-2, and the less specialized regional Type-3 hospitals require efficiency improvements.

6 Conclusions

The study provides empirical and methodological contributions to measuring cost and allocative efficiencies in publicly funded hospitals in Ireland (IPH). Using a detailed HSE dataset of IPH, this study applies conventional DEA and robust DB DEA methods, and truncates the sample into various subsets based on hospital size and type. The results present several findings regarding cost and allocative efficiencies in the IPH sector. The bias-corrected cost efficiency is estimated to be 75.6%, indicating substantial scope for improving costs in these hospitals. Moreover, the decomposition of CE into its constituent parts shows that technical efficiency is more a cause of cost efficiency than allocative efficiency, with mean bias-corrected TE and AE scores of 78% and 96%, respectively. Hence, as the primary driver of cost inefficiency in IPH is technical (productive) inefficiency rather than allocative inefficiency, cost savings are possible through the reduction of inputs rather than through the optimal allocation of inputs given the prices of those inputs.

In addition, the two-stage DB DEA analysis shows a strong negative effect on cost and technical efficiency when the share of public patients increases. As noted earlier, this could be explained by the private sector's role in funding and incentivizing these public hospitals. This preliminary result contributes further to the debate on the importance of private patient provisions in IPH.

Furthermore, the DB DEA results indicate that policymakers need to be mindful of the role of size and scope in the strategic development of Irish hospitals. The concentration of resources and specialization in high-capacity Type-4 hospitals, which are also large-sized hospitals with over 400 beds, does appear to contribute to higher cost efficiency. However, providing care in Type-3 hospitals and the smallest Type-2 hospital units, which are less resource-intensive, leads to lower efficiency. Therefore, the results suggest that an audited review of how costs and funding structures differ in Type-2 and Type-3 hospitals and how to align them with Type-4 hospitals may produce substantial savings for taxpayers.

In summary, the findings from this research imply that cost incentives should be implemented in the IPH sector through better management of available resources and that the best practices should be widely shared amongst the weaker efficiency-performing hospital types and sizes. Alternatively, more elective activity could be channeled towards the more efficient hospital types identified to take advantage of their superior cost performance. Further research could expand on this analysis to include more detailed labour and capital inputs and their corresponding prices, examining the allocation of the input mix more closely rather than relying solely on total labour and capital. Additionally, a more comprehensive analysis of the efficiency determinants could be conducted, and other potential efficiency factors, such as the quality of care, could be explored. Their impact could be further examined using local frontiers for various hospital types and sizes to derive relevant policy recommendations for IPH.

References

1. OECD: Health Statistics, OECD, Paris (2022). https://www.oecd-ilibrary.org/social-issues-migration-health/data/oecd-health-statistics_health-data-en. Accessed 22 Oct 2024
2. Keegan, C., Brick, A., Bergin, A., Wren, M.-A., Henry, E., Whyte, R.: Projections of expenditure for public hospitals in Ireland, 2018–2035, based on the Hippocrates model. Econ. Soc. Res. Inst., 117 (2020)
3. Farrell, M.J.: The measurement of the productive efficiency. J. Roy. Stat. Soc., Ser. A, CXX, Part **3**, 253–290 (1957)
4. Charnes, A., Cooper, E., Rhodes, E.: Measuring the efficiency of decision-making units. Eur. J. Oper. Res. **2**(6), 429–444 (1978)
5. Färe, R., Grosskopf, S., Lovell, C.A.K.: The Measurement of Efficiency of Production. Kluwer Academic Publisher, Dordrecht (1985)
6. Banker, R.D., Charnes, A., Cooper, W.W.: Models for estimation of technical and scale inefficiencies in data envelopment analysis. Manage. Sci. **30**, 1078–1092 (1984)
7. Mergoni, A., Emrouznejad, A., De Witte, C.H.: Fifty years of data envelopment analysis. Eur. J. Oper. Res. **326**(3), 389–412 (2025). https://doi.org/10.1016/j.ejor.2024.12.049
8. O'Neill, L., Rauner, M., Heidenberger, K., Kraus, M.: A cross-national comparison and taxonomy of DEA-based hospital efficiency studies. Socioecon. Plann. Sci. **42**, 158–189 (2008)
9. Hollingsworth, B.: The measurement of efficiency and productivity of health care delivery. Health Econ. **17**(10), 1107–1128 (2008)

10. Wang, Z., Nguyen, B.H., Zelenyuk, V.: Performance analysis of hospitals in Australia and its peers: a systematic and critical review. J. Prod. Anal. **62**, 139–173 (2024)
11. Kirigia, J.M., Emrouznejad, A., Vaz, R.G., Bastiene, H., Padayachy, J.: A comparative assessment of performance and productivity of health centres in Seychelles. Int. J. Product. Perform. Manag. **57**(1), 72–92 (2008)
12. Kohl, S., Schoenfelder, J., Fügener, A., Brunner, J.O.: The use of Data Envelopment Analysis (DEA) in healthcare with a focus on hospitals. Health Care Manag. Sci. **22**, 245–286 (2019)
13. Lotfi, F.H., Amirteimoori, A., Moghaddas, Z., Vaez-Ghasemi, M.: Cost efficiency measurement with price uncertainty: a data envelopment analysis. Math. Sci. **14**, 387–396 (2020)
14. Athanassopoulos, A., Gounaris, C.: Assessing the technical and allocative efficiency of hospital operations in Greece and its resource allocation implications'. Eur. J. Oper. Res. **133**(2), 199–218 (2001)
15. Johannessen, K.A., Kittelsen, S.A.C., Hagen, T.P.: Assessing physician productivity following Norwegian hospital reform: a panel and data envelopment analysis. Soc. Sci. Med. **175**, 117–126 (2017)
16. Puig-Junoy, J.: Partitioning input cost efficiency into its allocative and technical components: an empirical DEA application to hospitals. Socioecon. Plann. Sci. **34**(3), 199–218 (2000)
17. Nedelea, I.C., Fannin, J.M.: Testing for cost efficiency differences between two groups of rural hospitals. Int. J. Healthc. Manag. **10**(1), 57–65 (2017)
18. Gannon, B.: Testing for variation in technical efficiency of hospitals in Ireland. Econ. Soc. Rev. Econ. Soc. Stud. **36**(3), 273–294 (2005)
19. Devitt, N., Zieba, M., Dineen, D.: Technical efficiency in Irish public hospitals: a multi-output distance function SFA approach. Int. J. Comput. Econ. Econometr. **14**(2), 122–150 (2024)
20. Moradi-Motlagh, A., Emrouznejad, A.: The origins and development of statistical approaches in non-parametric frontier models: a survey of the first two decades of scholarly literature (1998–2020). Ann. Oper. Res. **318**, 713–741 (2022)
21. Ni Luasa, S., Dineen, D., Zieba, M.: Technical and scale efficiency in public and private Irish nursing homes – a bootstrap DEA approach. Health Care Manag. Sci. **21**, 326–347 (2018)
22. Simar, L., Wilson, P.W.: Estimation and inference in two-stage, semi parametric models of productive efficiency. J. Econometr. **136**(1), 31–64 (2007)
23. Simar, L., Wilson, P.W.: Two-stage DEA: caveat emptor. J. Prod. Anal. **36**(2), 205–218 (2011)
24. Coelli, T.J., Rao, D., O'Donnell, C.J., Battese, G.E.: An Introduction to Efficiency and Productivity Analysis, 2nd edn. Springer, USA (2005)
25. Daraio, C., Simar, L., Wilson, W.P.: Central limit theorems for conditional efficiency measures and tests of the 'separability' condition in non-parametric, two-stage models of production. Economet. J. **21**, 170–191 (2018)
26. Kellett, J., O'Keeffe, S.T.: The Model 2 hospital: role and challenges. Ir. Med. J. **105**(9), 294 (2008)

Identifying Input Excess and Output Shortfall in the Indian Health Systems Using Non-radial Slack Based Data Development Analysis

Nutan Shashi Tigga[1(✉)], Pinak Sarkar[2], and Nachiket Mor[3]

[1] Jindal School of Banking and Finance, O P Jindal Global University, Sonipat, Haryana 131029, India
tigga.nutan@gmail.com

[2] Economics and Public Policy, Lal Bahadur Shastri Institute of Management, Delhi 110075, India

[3] The Banyan Academy of Leadership in Mental Health, 6Th Main Road West, Mogappair, Chennai, Tamil Nadu 600037, India
nachiket@nachiketmor.net

Abstract. We analyse the efficiency of India's health system by assessing input excesses and output shortfalls, specifically focusing on government health expenditure, health personnel, and public hospital against the infant survival rate as output. Using time-series data from 2014-15 to 2019-20 of 21 states, our non-radial, non-oriented, Slack-Based Measure Data Envelopment Analysis (DEA) model yields an average efficiency score of 84%, highlighting significant potential for improvement. Bihar showed efficiency despite resource constraints, while Himachal Pradesh remained inefficient. These findings suggest reallocating resources to health services in underperforming states and encourage collaboration with the private sector in states like Kerala and West Bengal, which have higher private expenditure. This study uniquely employs slack-based DEA to compare India's health system during the specified period.

Keywords: Efficiency · Slack-based Measure Data Envelopment Analysis · government health expenditure · health system · Infant Survival Rate

JEL Classification:: I100 · I140 · I150 · I180 · C610 · C670

1 Introduction

Health system resources play a crucial role in improving the health of people in poor and developing countries. Greater progress can be achieved through the efficient utilising of these resources, especially where they are limited. India, one of the emerging economies, has made significant advancements in health over years. For instance, life expectancy at birth rose from 60.3 years in 1991–95 (RBI, 2017) to 70 years in 2022 (ORGI, 2021). However, there is substantial variation in health outcomes across Indian states. A striking example is the over tenfold difference in the Maternal Mortality Ratio (MMR) between

A. Emrouznejad et al. (Eds.): DEA 2024, LNOR, pp. 336–349, 2025.
https://doi.org/10.1007/978-3-031-98177-7_23

Assam (195 per 100,000 live births) and Kerala (19 per 1,00,000 live births) during the period of 2018–20 (ORGI, 2022). Additionally, disparities in health financing is also considerably high across the states in India. In 2019–20, Kerala had a per capita total health expenditure (THE) of Rs. 10,607, while Bihar's health expenditure was as low as Rs. 1,588 in according to per the National Health Accounts estimates (NHSRC, 2023). Given the significant differences in health system inputs and outputs, examining the relative efficiencies across the Indian states has become crucial. Measuring efficiency allows us to assess how well the health system is performing in achieving its ultimate goal: improving the population's health by providing optimal healthcare services. Mor and Shukla (2023) have raised concerns that governments in many states may already be investing enough to deliver Universal Health Coverage (UHC) to their residents but are unable to do so due to high levels of inefficiency in how these funds are utilized.

Studies measuring the efficiency of health systems are abundant, both globally and nationally. In India, research often examines the country as a whole (Shetty and Pakkala, 2010; De et al., 2012; Tigga and Mishra, 2015; Tigga & Sarkar, 2025) or focuses on specific districts within a state (Bhat et al., 2001; Ram Jat and Sebastian 2013; Purohit, 2008 & 2016). These studies have employed both parametric and non-parametric techniques. Data envelopment analysis (DEA) is a non-parametric approach, is the most widely used technique for assessing efficiency due to its simplicity, in contrast to stochastic frontier analysis (SFA), which is a parametric approach. It is considered as one of the important tool for performance evaluation (Mergoni et al. 2024).

DEA has Widley been used for assessment of health centres and hospitals (Kirigia, et al, 2008). Despite its popularity, one significant drawback of DEA is its incapability to directly account for input excesses or output shortfalls, known as input or output slacks. This limitation can lead to biased calculations influenced by radial and angular choices (Sun et al., 2023). Additionally, DEA is quite sensitive to extreme values and outliers, making it is essential to detect and correct these outliers to ensure accurate results (Simar, 2003). To address this limitation, Tone (2001) proposed a slack-based measure (SBM) that simultaneously considers input or output slacks. A hospital may seem efficient under traditional DEA if it uses fewer resources while still having unused capacity. SBM-DEA can identify these excess resources and provide more accurate scores. SBM model has been widely used to assess the efficiency of hospitals, clinics, and healthcare system across the globe (Emrouznejad et al. 2025). In India, there is limited research in the assessing the health system using this approach. Mogha et al. (2014 & 2015) used a similar methodology to assess the performance of hospitals, although, their analysis was limited to hospitals in the districts of Uttarakhand.

This study evaluates the health system in India across 21 states for the period from 2014–15 to 2019–20 using the SBM-DEA method. To our knowledge, no previous research has employed this method to assess the relative efficiency of the health system in these twenty-one states during the specified time frame. This analysis will help identify opportunities for hospitals to reduce inputs without compromising outputs or to increase outputs without incurring additional resources.

2 Literature Review

Efficiency measurements in health systems using DEA have been approached from various dimensions: efficiency measurements across national health systems, such as in Asian countries (Ahmed et al., 2019); across regions of one country, such as states in India (Shetty and Pakkala, 2010; De et al., 2012); and hospitals in a region, such as hospitals or health centres across districts of a state (Bhat et al., 2001; Dash et al., 2008; Ram Jat and Sebastian, 2013).

For measuring health system efficiency across 50 Asian countries, Ahmed et al. (2019) provided an understanding of the comparative performance of the countries. They also identified the determining factors associated with the efficiency score. Using DEA and censored Tobit regression and smoothed bootstrap models, the study selected one input variable, two output variables, and four explanatory variables. They found that most of the efficient countries belonged to the high-income group, and only one was from the lower-middle-income group. Furthermore, the study found that population density, bed density, and primary education completion rate significantly influenced health system efficiency.

In India, a few studies that measure the efficiency of the Indian health system cover all the states. Most have focused on one state or district of a state. Dash et al. (2008) assessed the efficiency of the use of health resources across the districts of Tamil Nadu. The study selected two outputs, infant mortality rate and life expectancy at birth, and five inputs, real per capita GDP, urbanisation rate, literacy rate, and number of beds and doctors. The study concluded that districts with good and moderate health outcomes and only about 20% of the districts were using their resources efficiently. Furthermore, the study indicated that inefficient districts could reduce their inputs by 13% and 10% to become efficient. Another study by Ram Jet and Sebastian (2013) evaluated the technical efficiency (TE) of public district hospitals in Madhya Pradesh. The study included three input and eight output indicators focusing on maternal services. The study concluded that the sampled district hospitals could produce the same number of outputs even after reducing inputs by 10%. Furthermore, they found that 50% of the hospitals were operating below the optimal level and that efficiency could be improved by better resource management.

Shetty and Pakkala (2010) analysed the efficiency of health systems across 19 states using two outputs, IMR and life expectancy, and four inputs, per capita health expenditure, population below the poverty line, the density of health centres, and the literacy rate. They used the non-proportional range directional model (NP-RDM) of data envelopment analysis (DEA), a measure that handles negative data. The study found that two sets of states are inefficient: one, where the states use the inputs inefficiently, and the other, which has inadequate health resources. De et al. (2012) assessed and compared the efficiency of the health system across 15 states during all the Five-Year Plan periods. The study selected two outputs, the under-five survival rate and female life expectancy at birth, and three input indicators, numbers of doctors, nurses, and hospital beds. They found that only two states, Kerala and Haryana, were efficient throughout the period and concluded the influence of other non-discretionary factors on the output levels. In addition, factors such as female literacy, poverty level, institutional delivery, and full immunisation of children explained the inefficiency of the health system across these states. Tigga and Mishra (2015) analysed the health systems of 27 states using two outputs, infant survival rates and institutional deliveries, and two inputs, health centres and health workers. The study found that of the total number of states, only six were efficient, and almost all the inefficient states used more than the required amount of inputs to achieve the existing output levels. Dutta et al. (2014) analysed the efficiency of secondary-level government-run hospitals in West Bengal using the DEA approach. The input variables in the study were labour, infrastructure and capital, and the output variable was quality-adjusted outpatients and inpatients. The results showed that the hospitals could produce 37% more output with the same input resources. The authors further assert that the gaps in planning, implementation, and managerial challenges were the principal reasons for the inefficiencies in the speciality hospitals.

Most of the studies discussed above used traditional DEA methods. One of the criticisms faced by these methods is their assumption of radiality. They fail to directly consider the input excesses or output shortfalls (input/output slacks). Then, Charnes et al. (1985) developed an additive DEA model to simultaneously account for possible input reduction and output augmentation. Subsequently, a slack-based measure (SBM) of efficiency was developed by Tone (2001) to account for both scalar measures and input/output slacks simultaneously. In the Indian context, only one study was found using SBM-DEA to report the input excess and output shortfall in the health system. Mogha et al. (2014 & 2015) measured the relative efficiencies of public sector hospitals in Uttarakhand using the slack-based measure DEA (SBM-DEA). The study used the number of beds, doctors and non-medical employees as inputs, and in-door and out-door patients and minor and major surgeries as outputs. Their findings suggest that most of the hospitals have a potential scope to increase the output levels using less input than the existing level. They further concluded that the input and output slack indicated an opportunity for a significant reduction in all the inputs and outputs to reach the frontier.

The present study intends to contribute to the literature in the following ways: first, it computes the efficiency of the health system using SBM-DEA to account for input excess and output shortfalls; second, it focuses on a group of twenty-one states in India; and third, it estimates the efficiency of health systems across these states for a period of six years.

3 Methods and Data

3.1 Slack-Based Measure

The traditional DEA models proposed by Charnes, Cooper and Rhodes (CCR) and Banker, Charnes and Cooper (BCC) models have been widely used to measure efficiency across various sectors due to their advantages. However, one of the main drawbacks of these traditional DEA models is that in an input-oriented DEA, a common factor radially contracts all inputs, and all outputs are expanded radially to their efficient level by a common expansion factor in an output-oriented DEA. Therefore, it fails to directly consider the input excesses or output shortfalls (input/output slacks). Accordingly, Charnes et al. (1985) developed an additive DEA model that could simultaneously account for and measure possible input decreases and output increases; however, this model lacked a scalar measure in the range of [0,1]. Subsequently, a slack-based measure (SBM) of efficiency was developed by Tone (2001) to account for both scalar measures and input/output slacks simultaneously. The primary purpose of the SBM is to project the DMU (Decision-Making Units) to the furthest point on the efficient frontier, with the objective function to be minimised by finding the maximum slacks (Tone, 2001); thus, it is a non-radial model in principle. Therefore, the non-radial SBM-DEA is more appropriate than traditional DEA models since the traditional and radial approaches do not consider slacks (Cooper et al., 2007). Zhou et al. (2006) found that the SBM-DEA model has a higher discriminatory power than traditional DEA models. In the non-radial approach, inputs and outputs are not impelled to improve uniformly (Lozano and Gutiérrez, 2011). Another advantage of using SBM-DEA is that the inefficiency indicator for each variable can be identified to increase the efficiency of the DMU being studied (Zhang and Kim, 2014). The non-radial, non-oriented SBM model is provided below:

Unlike the traditional DEA models, the SBM considers how much input waste can be reduced for a given output level and how much output can be increased for a given input level. The SBM-DEA model is shown in Eq. (1). The symbol ρ represents the efficiency of DMU_k, which ranges between 0 and 1.

$$min\rho = \frac{1-\frac{1}{m}\sum_{i=1}^{m} s_i^- / x_{ik}}{1+\frac{1}{q}\sum_{r=1}^{q} s_r^+ / y_{rk}}$$

$$s.t.\begin{cases} \sum_{j=1}^{n} \lambda_j x_{ij} + s_i^- = x_{ik}, i = 1, \ldots.m \\ \sum_{j=1}^{n} \lambda_j y_{rj} - s_r^+ = y_{rk}, r = 1, \ldots.s \\ \lambda_j \ge 0, j = 1, \ldots n \\ s_i^- \ge 0 \\ s_r^+ \ge 0 \end{cases} \tag{1}$$

where ρ is the health system efficiency, with $0 < \rho \le 1$. x and y are represent the observed values of DMU inputs and outputs, respectively; s − and s + are the input and output slacks for the DMU under evaluation; and λ is the weight coefficient of the reference DMU. The DMU is efficient if ρ = 1, which implies that $s_i^- = 0$ and $s_r^+ = 0$. If ρ<1, then the DMU is inefficient and requires eliminating the slack in inputs and bad outputs and augmenting the shortfalls in good outputs.

3.2 Data

In DEA, the selection of input and output variables is crucial. In the present study the inputs and outputs are derived from the previous studies as discussed in the literature review. Three input indicators- government health expenditure (GHE) (per capita in Rs.), health personnel (per 10,000), and government hospital beds (per 100,000) are considered as inputs, and infant survival rate (ISR) as output. All indicators were from the financial year 2014–15 to 2019–20. The choice of the number of states and time framework is determined by the availability of the data.

In DEA techniques, output is measured so that 'more is better'. Here, the original output is the infant mortality rate (IMR), which is a negative- output or undesirable output; hence, 'less is better'. Since both desirable and undesirable outputs are generated in a production process, various approaches have been developed. One is the transformation of conventional DEA models to make all negative data positive by adding a scalar value to the negative variables (Ali & Seiford, 1990; Lovell & Pastor, 1995; Seiford & Zhu, 2002).

Following previous studies (De et al., 2012), the IMR is transformed into a desirable/positive output using the following formula:

$$ISR = \frac{1000 - IMR}{IMR} \tag{2}$$

The data for the present study have been compiled from various sources, such as the National Health Accounts, National Health Profile and Handbook of Statistics on Indian States by the RBI. GHE includes health expenditure by all government agencies of the Union government and respective State governments, State Other Departments, Local Bodies, and quasi-governmental organisations and donors in case funds are channelled through government organisations. Health personnel include the number per 10,000 population of allopathic doctors, nurses, and midwives in the public sector. Hospital Beds include beds per 100,000 population in government hospitals in rural and urban areas and Community Health Centres (CHCs).

4 Results

4.1 Descriptive Statistics

Table 1 shows the descriptive statistics of the state-level data from 2014–15 to 2019–20. The GHE of the 21 states averaged Rs. 899 per capita for 2014–15, which increased to Rs. 1759 in 2019–20. The number of health personnel per 10,000 population was 24 in 2014–15, which rose to 29 in 2019–20. The output, i.e., mean ISR, was 33 in 2014–15 and later increased to 48 in 2019–20. Variations in the inputs and outputs across the states can be seen from the standard deviations in Table 1.

4.2 Health System Efficiency Performance

Table 2 summarises the overall technical efficiency estimated using the non-radial, non-oriented slack-based measure data envelopment analysis (SBM-DEA) model, as discussed in Sect. 3. The DMU is efficient if the efficiency scores are equal to 1, which implies that all the slack values are equal to 0. If the efficiency scores are less than 1, it implies that the DMU is inefficient, and efficiency can be achieved by eliminating slack in inputs and outputs. In other words, those DMUs will require input reduction or output augmentation to reach the production frontier. The health system efficiency (HSE) across the states from 2014–15 to 2019–20 is estimated by ρ in Eq. 2, which includes slack values of all input and output variables. The slack variable calculates the potential reduction or augmentation in input and output. Table 2 shows the results of the HSE for

Table 1. Summary statistics of input-output variables

	ISR	GHE Per Capita in Rs	HP (per 10,000)	Beds (per 1,00,000)		ISR	GHE Per Capita in Rs	HP (per 10,000)	Beds (per 1,00,000)
2014–15					**2017–18**				
Mean	33	899	24	64	Mean	41	1429	27	64
Std. Dev	15	374	22	37	Std. Dev	26	549	23	39
Min	19	338	2	10	Min	20	556	2	10
Max	82	2016	75	139	Max	142	3177	87	170
2015–16					**2018–19**				
Mean	36	1144	25	58	Mean	45	1601	27	70
Std. Dev	18	454	23	31	Std. Dev	30	620	24	49
Min	20	425	2	10	Min	21	674	2	24
Max	99	2316	79	131	Max	166	3604	91	202
2016–17					**2019–20**				
Mean	37	1290	26	66	Mean	48	1759	29	69
Std. Dev	18	522	23	40	Std. Dev	31	658	24	50
Min	20	504	2	10	Min	22	701	2	14
Max	99	2816	83	171	Max	166	3829	94	20

21 states of India from 2014–15 to 2019–20. The overall average HSE from 2014–15 to 2019–20 indicates that out of 21 states, only Bihar was found to be relatively efficient, with a score equal to 1. The HSE of the others ranged from 0.162 to 0.903 (average = 0.429), suggesting considerable potential for improvement (83.84%). Kerala (0.903) ranked first, and Himachal Pradesh ranked last (0.162) among the inefficient states. The other inefficient states with comparatively better HSE scores than the mean were Jammu and Kashmir, Jharkhand, Maharashtra and West Bengal. These four states together had a 66.35% potential to improve their HSE score.

Table 2. Health system efficiency (HSE) scores based on the slack-based measure (SBM) across states

DMU	2014–2015	2015–2016	2016–2017	2017–2018	2018–2019	2019–2020	Mean
Bihar	1	1	1	1	1	1	1
Kerala	1	1	0.421	1	1	1	0.903
Jammu and Kashmir	1	0.539	0.540	0.534	1	1	0.769

(continued)

Table 2. (*continued*)

DMU	2014–2015	2015–2016	2016–2017	2017–2018	2018–2019	2019–2020	Mean
Jharkhand	1	0.740	0.706	0.606	0.703	0.673	0.738
Maharashtra	0.449	0.509	0.484	0.427	1	1	0.645
West Bengal	0.500	0.431	0.247	0.232	0.597	1	0.501
Punjab	0.425	0.424	0.366	0.392	0.458	0.433	0.416
Uttarakhand	0.420	0.413	0.398	0.367	0.446	0.429	0.412
Telangana	0.415	0.405	0.360	0.307	0.438	0.424	0.392
Tamil Nadu	0.382	0.398	0.395	0.347	0.363	0.401	0.381
Gujarat	0.278	0.272	0.252	0.295	0.397	0.402	0.316
Haryana	0.323	0.328	0.284	0.260	0.375	0.316	0.314
Uttar Pradesh	0.331	0.288	0.336	0.270	0.339	0.277	0.307
Chhattisgarh	0.311	0.304	0.274	0.216	0.330	0.301	0.289
Karnataka	0.271	0.275	0.236	0.238	0.314	0.334	0.278
Andhra Pradesh	0.335	0.269	0.257	0.239	0.233	0.225	0.260
Madhya Pradesh	0.251	0.254	0.245	0.200	0.288	0.272	0.252
Odisha	0.239	0.248	0.215	0.200	0.291	0.272	0.244
Assam	0.266	0.229	0.216	0.181	0.211	0.208	0.219
Rajasthan	0.189	0.211	0.226	0.175	0.232	0.246	0.213
Himachal Pradesh	0.149	0.157	0.150	0.154	0.177	0.183	0.162
Mean	0.454	0.414	0.362	0.364	0.485	0.495	0.429

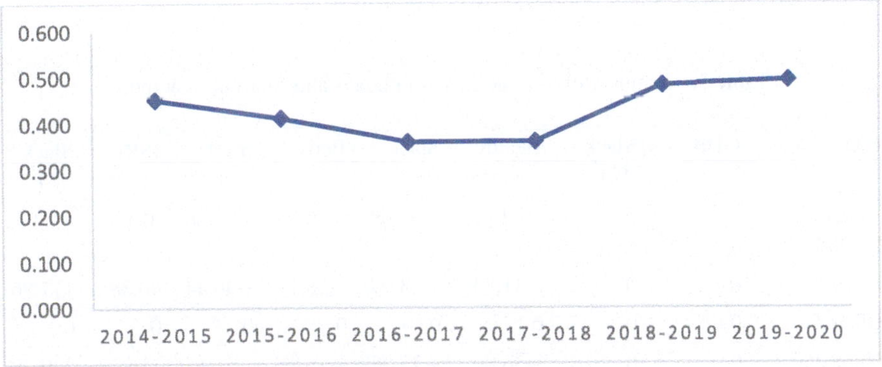

Fig. 1. Average health system efficiency across years

Aggregating the performance of 21 states from 2014–15 to 2019–20, Fig. 1 shows the HSE by year. The HSE showed significant variation over the years. The average score ranged between 0.362 and 0.495. There was a decreasing trend in the HSE from 2014–2015 to 2016–17. Post 2017–18, the HSE scores witnessed a significant increase.

4.3 Analysis of Slack Variables

Most of the states are underperforming, which indicates that there is great potential to improve the HSE. Hence, examining the slack values of input and output variables is imperative. The estimated slack values and the associated percentages are shown in Table 3. The negative values indicate the percentage of potential reduction in inputs, and the positive values indicate the potential increase in the output. As seen in Table 2, only one state, i.e., Bihar, is relatively efficient compared to the other states and has 0 slacks. The remaining states require reducing one or more inputs and augmenting the output to become efficient. States ranked lowest among the inefficient states have extremely high input and output slacks. This suggests that to increase efficiency, these states will have to lower the input levels by the estimated amount of input values. For example, Himachal Pradesh, ranked last among the inefficient states, has excess health personnel, beds, and shortfalls in ISR. The state could increase the output, i.e., ISR, by approximately 250% per year while reducing its health personnel and beds by 77.38% and 51.13%, respectively.

Similarly, Rajasthan could increase its output shortfall by 177.38% by decreasing its health expenditure by 0.41% and health personnel and beds by 88.63% and 33.69%, respectively. The other inefficient states, such as Madhya Pradesh, Odisha, and Assam, also show substantial waste in inputs and high shortfalls in output. Kerala, ranked first among the list of inefficient states, could increase ISR by 2.46% by reducing its health personnel and beds by 14.71% and 9.97%, respectively. On average, the states could increase ISR by 88.31% while reducing GHE by 6.3%, health personnel by 59.71%, and bed density by 39.55%. The overall mean slacks show that health personnel have the highest percentage of slack, followed by bed density.

Table 3. Summary of average excess in inputs and shortfall in outputs

DMU	GHE	Slack (%)	Health personal	Slack (%)	Beds	Slack (%)	ISR	Slack (%)
Andhra Pradesh	0	0	−67.87	−94.09	−50.34	−60.06	32.66	99.05
Assam	0	0	−11.83	−73.99	−28.17	−46.44	40.88	177.86
Bihar	0	0	0	0	0	0	0	0
Himachal Pradesh	0	0	−36.80	−77.38	−85.85	−51.13	116.25	249.78
Karnataka	0	0	−40.42	−88.75	−58.34	−60.33	34.54	81.24
Kerala	0	0	−12.52	−14.71	−10.87	−9.97	3.09	2.46
Maharashtra	0	0	−7.13	−43.28	−27.91	−52.29	9.50	17.76
Punjab	0	0	−31.13	−87.54	−23.96	−41.90	17.70	36.51

(continued)

Table 3. *(continued)*

DMU	GHE	Slack (%)	Health personal	Slack (%)	Beds	Slack (%)	ISR	Slack (%)
Tamil Nadu	0	0	−40.65	−87.97	−60.74	−58.47	21.37	33.91
Uttarakhand	0	0	−11.77	−76.48	−59.77	−68.29	11.17	25.56
Rajasthan	−5.32	−0.41	−36.90	−88.63	−18.44	−33.69	45.77	177.38
Jharkhand	−9.49	−1.21	−0.15	−5.22	−8.67	−26.39	7.82	22.94
Madhya Pradesh	−27.84	−3.06	−16.77	−83.83	−12.64	−34.23	28.28	138.67
Telangana	−25.04	−3.24	−3.52	−55.87	−10.05	−33.03	18.59	81.20
Odisha	−78.49	−7.08	−28.59	−88.34	−13.02	−31.76	33.17	140.19
Gujarat	−197.41	−13.43	−19.60	−80.29	−8.86	−20.77	36.08	102.99
Haryana	−221.39	−15.43	−16.22	−78.14	−4.52	−12.24	35.58	111.47
Chhattisgarh	−224.32	−17.00	−5.77	−58.18	−6.57	−18.77	36.95	151.80
Uttar Pradesh	−308.43	−17.73	−3.13	−38.38	−35.67	−46.90	45.77	143.07
Jammu and Kashmir	−388.93	−25.98	−0.04	−1.40	−30.85	−50.44	1.24	2.74
West Bengal	−495.41	−27.97	−1.93	−31.46	−64.05	−73.54	21.05	57.86
Mean	−94.38	−6.31	−18.70	−59.71	−29.49	−39.55	28.45	88.31

5 Discussion

The present study uses the slack-based measure proposed by Tone (2001) to estimate the efficiency of the health system across 21 states in India. Traditional DEA models fail to capture the slacks directly (input excess and output shortfalls); hence, many alternative approaches have been introduced to compute them. The SBM-DEA used in this study assesses the input excess and output shortfall, also called slacks. These show the reductions and augmentations that need to be made by inefficient units.

The overall average score across the states from 2014–15 to 2019–20 indicates an 84% potential for improvement. Of the 21 states, only one (Bihar) remained efficient, and Himachal Pradesh remained inefficient during the entire period. Kerala was efficient throughout the years except in 2016–17. However, the annual average for Kerala was 0.903, ranking first among the list of inefficient states, indicating an improvement of 9.7%. Kerala reported the highest per capita OOPE, i.e., Rs. 7206 in 2019–20, compared to other states. Moreover, the share of OOPE in THE was 67% against the share of GHE, which was lowest across the states, i.e., 24.4%, which is significantly lower than Bihar (44.1%). This could be one of the reasons why Kerala turned inefficient, as GHE is used as one of the inputs (NHSRC, 2023). However, Bihar's emergence as an efficient state is surprising since its output variables are not comparable to those of Kerala. The U5MR in Kerala was 8 per 1000 live births, while it was 30 per 1000 live births in Bihar, as per the latest SRS estimates (ORGI, 2022). Earlier studies measuring the efficiency of the health system have found similar results (Shetty and Pakkala, 2010; De et al., 2012; Tigga and Mishra, 2015). The reasons cited for states with poor outcomes to emerge as efficient could be due to the presence of other nondiscretionary factors that could influence the output levels (De et al., 2012; Kathuria and Sankar, 2005). Additionally, these states utilise low input levels compared to other states. For example, Bihar uses the lowest amount of GHE and health personnel to produce the current output levels compared to the rest of the Indian states. The GHE in Bihar was Rs. 701 per capita in 2019–20 compared to Himachal Pradesh, which was Rs. 3829 per capita. Shetty and Pakkala (2010) also found similar conclusions where states such as Assam, Odisha, Rajasthan, Jharkhand, and Bihar, were efficient. Alternatively, states such as Andhra Pradesh, Tamil Nadu, and Himachal Pradesh turned inefficient despite having significantly higher health resources at their disbursal and good health outcomes. Shetty and Pakalla (2010) and De et al. (2012) found similar results in their study.

The potential reduction and augmentations in input and output variables indicate the scope of the improvement in the inefficient states. Of the 21 states, 15 had an HSE score below the mean (average = 0.429). These inefficient states show a significant amount of input access and output shortfall. For example, Himachal Pradesh could increase the output, i.e., ISR, by 250% per year while reducing its health personnel and beds by 77.38% and 51.13%, respectively. Similarly, Rajasthan could increase its output by 177.38% while decreasing its health expenditure by 0.41% and health personnel and beds by 88.63% and 33.69%, respectively. Mogha et al. (2014) also found similar increases in the output levels to reach the frontier.

Kerala, which topped the list of inefficient states, can improve its ISR, albeit marginally, by 2.46% while reducing its government health personnel by 14.71% and hospital beds by 9.97%. The state has no excess of GHE. One probable reason for this could be that the share of private expenditure (59%) to THE in Kerala (OOP) is significantly higher than that of GHE (32.5%) (NHSRC, 2024). Hence, it indicates a greater dependence on the private sector than the public sector. The state's dependence on the private sector is also reflected on the proportion of total institutional deliveries conducted in the state, which was as high as 99.8%, but of which merely 34% were conducted in a public facility (IIPS and ICF, 2021). This suggests that since the private sector is doing much of the work on the ISR front, the state does not need to continue to invest in the same area and could benefit from redirecting its resources towards primary care to better address its emerging non-communicable disease burden. It could even start to direct more of its resources towards state-sponsored insurance schemes so that its residents do not have to pay out-of-pocket when they visit the private sector and slowly shrink the size of the public sector, which is, in any case, not getting much use.

In a state such as Rajasthan, the results may indicate gross inefficiencies in how its health personnel and hospital beds are deployed. We already know that at the current levels of ISR, further improvements will require much better access to emergency care (Mor, 2023). Continuing to add hospital beds and personnel without specialists who can offer emergency care will likely be a waste of resources since it will not result in a concomitant reduction in ISR.

A state like Bihar, on the other hand, spends so little on healthcare both from the government and on an out-of-pocket basis that any increases in GHE end up showing immediate gains in health outcomes. However, as the experience of other states such as Jharkhand, Madhya Pradesh, and Uttar Pradesh shows, if Bihar does not augment its specialist capacity very quickly, it will end up in the same position as these other states. Further input increases will not lead to any improvements in ISR relative to what is already being delivered through generalised economic growth in the state, and the state may achieve faster reductions in ISR by directing these resources into direct benefits-transfer schemes.

6 Conclusion

Input excess and output shortfall in a health system is a serious challenge, particularly in resource-constrained settings. With limited resources to meet the health needs of a growing population, efficient utilisation is the only solution. Many have concluded that an additional increase in resources may not necessarily lead to good health outcomes (Mor, 2022; Mor & Shukla, 2023). In such a situation, assessing health system efficiency, which can indicate the areas that require intervention, is imperative. The present study has attempted to analyse the health system using a non-radial measure to calculate the excesses and shortfalls. Such an analysis could assist in taking focused measures to improve the efficiency of the health system. We found that Bihar, one of the Empowered Action Groups (EAGs), characterised by poor socioeconomic and health indicators, has emerged as an efficient state despite having lower health expenditure and poor health outcomes. Alternatively, states such as Himachal Pradesh and Tamil Nadu, which have significantly higher levels of health expenditure and good outcomes, are inefficient. Thus, better outcomes could be expected if additional investments could be made in states such as Bihar, which are already getting the most they can out of their meagre current resources. Reallocating resources from less efficient states attaining good outcomes to more efficient states with poor outcomes may also be a solution to ensuring an overall efficient utilisation of resources. In addition, states like Kerala and West Bengal, which have a higher share of private expenditure in THE, may be better served by recognising this reality and working alongside the private sector to improve overall health outcomes in the state instead of wasting scarce resources by adding unnecessary doctors and hospital beds within the public sector.

References

Ahmed, S., et al.: Measuring the efficiency of health systems in Asia: a data envelopment analysis. BMJ Open **9**(3), e022155 (2019)

Bhat, R., Verma, B.B., Reuben, E.: Hospital efficiency: an empirical analysis of district hospitals and grant-in-aid hospitals in Gujarat. J. Health Manag. **3**(2), 167–197 (2001)

Charnes, A., Cooper, W.W., Golany, B., Seiford, L., Stutz, J.: Foundations of data envelopment analysis for Pareto-Koopmans efficient empirical production functions. J. Econom. **30**(1–2), 91–107 (1985)

Cooper, W.W., Seiford, L.M., Tone, K.: Data envelopment analysis: a comprehensive text with models, applications, references and DEA-solver software, vol. 2, p. 489. Springer, New York (2007)

De, P., Dhar, A., Bhattacharya, B.N.: Efficiency of health care system in India: an inter-state analysis using DEA approach. Soc. Work Publ. Health **27**(5), 482–506 (2012)

Dash, U., Vaishnavi, S.D., Muraleedharan, V.R.: Technical efficiency in the use of health care resources: a case study of Tamil Nadu. Indian Econ. Rev. **43**(1), 69–82 (2008)

Dutta, A., Bandyopadhyay, S., Ghose, A.: Measurement and determinants of public hospital efficiency in West Bengal India. J. Asian Publ. Policy **7**(3), 231–244 (2014)

Emrouznejad, A., Brzezickib, Ł., Lu, C.: The development and evolution of slacks-based measure models in data envelopment analysis: a comprehensive review of the literature. J. Econ. Surv., 1–20 (2025). https://doi.org/10.1111/joes.12682

International Institute for Population Sciences (IIPS) and ICF. National Family Health Survey (NFHS-5), India, 2019–21: Jharkhand. Mumbai: IIPS (2021)

Kathuria, V., Sankar, D.: Inter-state disparities in health outcomes in rural India: an analysis using a stochastic production frontier approach. Dev. Policy Rev. **23**(2), 145–163 (2005)

Kirigia, J. M., Emrouznejad, A., Gama Vaz, R., Bastiene, H., Padayachy, J.: A comparative assessment of performance and productivity of health centres in Seychelles. Int. J. Prod. Perform. Manage. **57**(1), 72–92 (2008)

Lovell, C.K., Pastor, J.T.: Units invariant and translation invariant DEA models. Oper. Res. Lett. **18**(3), 147–151 (1995)

Lozano, S., Gutiérrez, E.: Slacks-based measure of efficiency of airports with airplanes delays as undesirable outputs. Comput. Oper. Res. **38**(1), 131–139 (2011)

Mergoni, A., Emrouznejad, A., De Witte, K.: Fifty years of data envelopment analysis. Eur. J. Oper. Res., **326**(3), 389–412 (2025). https://doi.org/10.1016/j.ejor.2024.12.049

Mogha, S.K., Yadav, S.P., Singh, S.P.: New slack model based efficiency assessment of public sector hospitals of Uttarakhand: State of India. Int. J. Syst. Assur. Eng. Manage. **5**, 32–42 (2014)

Mogha, S.K., Yadav, S.P., Singh, S.P.: Slack based measure of efficiencies of public sector hospitals in Uttarakhand (India). Benchmark.: Int. J. **22**(7), 1229–1246 (2015)

Mor, N.: Lessons for developing countries from outlier country health systems. Front. Publ. Health **10** (2022)

Mor, N.: Commentary: developing technical support and strategic dialogue at the country level to achieve primary health care-based health systems beyond the COVID-19 era. Front. Publ. Health **11**, 1212271 (2023)

Mor, N., Shukla, S.K.: Estimating funds required for UHC within Indian States. Lancet Reg. Health - Southeast Asia **13**(6), 14 (2023)

National Health System Resource Center (NHSRC). National Health Accounts Estimates for India (2019–20). Ministry of Health and Family Welfare, Government of India, New Delhi (2023)

National Health System Resource Center (NHSRC). National Health Accounts Estimates for India (2021–22). New Delhi: Ministry of Health and Family Welfare, Government of India (2024)

Office of the Registrar General and Census Commissioner of India (ORGI). SRS based abridged life tables, 2016–20. Sample Registration System. Ministry of Home Affairs, Govt. of India, New Delhi (2022)

Office of the Registrar General and Census Commissioner of India (ORGI). Special Bulletin on Maternal Mortality in India 2018–20. Sample Registration System. Ministry of Home Affairs, Govt. of India, New Delhi (2022)

Purohit, B.C.: Efficiency of the health care system: a sub-state level analysis for West Bengal (India). Rev. Urban Reg. Dev. Stud. **20**(3), 212–225 (2008)

Purohit, B.C.: Health care system efficiency: a sub-state level analysis for Orissa (India). Rev. Urban Reg. Dev. Stud. **28**(1), 55–74 (2016)

Reserve Bank of India (RBI). Handbook of statistics on Indian economy, Reserve Bank of India, Mumbai (2017)

Ram Jat, T., San Sebastian, M.: Technical efficiency of public district hospitals in Madhya Pradesh, India: a data envelopment analysis. Glob. Health Action **6**(1), 21742 (2013)

Seiford, L.M., Zhu, J.: Modeling undesirable factors in efficiency evaluation. Eur. J. Oper. Res. **142**(1), 16–20 (2002)

Shetty, U., Pakkala, T.P.M.: Technical efficiencies of healthcare system in major states of India: an application of NP-RDM of DEA formulation. J. Health Manag. **12**(4), 501–518 (2010)

Simar, L.: Detecting outliers in frontier models: a simple approach. J. Prod. Anal. **20**, 391–424 (2003)

Sun, M., Ye, Y., Zhang, G., Shang, X., Xue, Y.: Healthcare services efficiency and its intrinsic drivers in China: based on the three-stage super-efficiency SBM model. BMC Health Serv. Res. **23**(1), 811 (2023)

Tigga, N.S., Mishra, U.S.: On measuring technical efficiency of the health system in India: an application of data envelopment analysis. J. Health Manag. **17**(3), 285–298 (2015)

Tigga, N.S., Sarkar, P.: Has the efficiency and productivity of the health system in India improved during post-policy period? Application of the bootstrap data envelopment analysis and Malmquist productivity index. Appl. Econ. **57**(11), 1239–1256 (2025)

Tone, K.: A slacks-based measure of efficiency in data envelopment analysis. Eur. J. Oper. Res. **130**(3), 498–509 (2001)

Zhang, N., Kim, J.D.: Measuring sustainability by energy efficiency analysis for Korean power companies: a sequential slacks-based efficiency measure. Sustainability **6**(3), 1414–1426 (2014)

Zhou, P.A.B.W., Ang, B.W., Poh, K.L.: Slacks-based efficiency measures for modeling environmental performance. Ecol. Econ. **60**(1), 111–118 (2006)

Efficiency and Productivity of Public Health Expenditures in Reducing Child Mortality in Emerging Market Economies

Pinak Sarkar[1](✉) and Nutan Shashi Tigga[2]

[1] Economics and Public Policy, Lal Bahadur Shastri Institute of Management, Delhi 11/07 Dwarka Sector 11, Near Metro Station, New Delhi, Delhi 110075, India
pinaksarkar@lbsim.ac.in

[2] Jindal School of Banking and Finance, O P Jindal University, Sonipat, Haryana 131029, India
nutan.tigga@jgu.edu.in

Abstract. In this study, we measure the efficiency and productivity of public health expenditures in improving child mortality outcomes in ten major emerging economies of the world. We use bootstrap data envelopment analysis (DEA) and the bootstrap Malmquist productivity index (MPI) to measure the progress of efficiency and productivity from 2010–2020. Furthermore, we test the statistical significance of the progress in efficiency and productivity using the Wilcoxon signed-rank test. For robustness of the efficiency and productivity estimates, we report the confidence intervals and test the statistical significance of the scores across countries. The average bias-corrected overall technical efficiency score of 0.3381 suggests that if inefficiency could be eliminated, child survival could be improved by 66% in these countries. The MPI results suggest that over the study period, on average, productivity improved for all these nations except for Saudi Arabia and South Korea. At the policy level, the study suggested that an increase in health expenditure is not sufficient to improve health outcomes unless efficiency and productivity are achieved.

Keywords: Health expenditure · bootstrap data envelopment analysis (DEA) · bootstrap Malmquist productivity index (MPI) · emerging economies · under-five mortality rate · neonatal mortality rate

1 Introduction

Emerging market economies are witnessing economic growth and rapid transition from low-income to middle-income status. These economies are at various stages of economic development and face various challenges such as economic volatility, infrastructural constraints, political instability, inadequate funding, technological bottlenecks, etc. to mention a few. Their health system also faces challenges in providing adequate healthcare services and achieving desired health outcomes. The public health spending in emerging economies were about half of that of advanced countries (Grigoli and Kapsoli 2017). And low levels of health expenditure are cited as one of the reasons for poor health

A. Emrouznejad et al. (Eds.): DEA 2024, LNOR, pp. 350–365, 2025.
https://doi.org/10.1007/978-3-031-98177-7_24

outcomes across the developing world (Barenberg et al. 2017; Novignon and Lawanson 2017; Akinlo and Sulola 2019). Moreover, the level of disparities in public health care spending within emerging economies is stark, for example, public health spending was as high as 10.30% of GDP in Brazil and as low as 2.96% of GDP in India during 2020 (Fig. 1). However, in regard to health outcomes, such as child mortality indicators, most of these countries, such as Brazil, China, India, and Vietnam, experienced substantial improvements. Significant reductions in the neonatal mortality rate (NMR) and under-five mortality rate (U5MR) were observed during 2010–2020 (Fig. 2 & Fig. 3). The reduction in child mortality rates can be seen as an exemplary achievement by these countries, as these are also the Sustainable Development Goals (SDG-3.2), where the target is to reduce NMR to 12 per 1,000 live births and U5MR to 25 per 1,000 live births by 2030.

Studies have shown that healthcare expenditure is an essential component for reducing the mortality of children, such as infants and neonates (Novignon and Lawanson 2017; Kiross et al. 2020). However, a mere increase in public health expenditure alone may not be sufficient to improve health outcomes since, beyond inadequate public spending, the associated inefficiencies pose a greater challenge. The International Monetary Fund (2019) indicated that a substantial proportion of spending is wasted due to the misallocation of resources, poor quality of public services, and corruption, leading to an estimation of additional spending of 10% of GDP in low- and middle-income countries (LMICs) and 2% of GDP in emerging market economics (Cristóbal et al. 2021). Previous research indicates significant inefficiencies in public health spending; for example, inefficiency in public health expenditure was observed across G7 countries (Verhoeven, Gunnarsson and Carcillo 2007), the European Union and emerging markets (Afonso et al. 2010), Europe and Central Asia (Hsu 2014), emerging and developing economies (Grigoli and Kapsoli 2017) and emerging countries. Such inefficiency leads to a waste of resources (Grigoli and Ley 2012) and hinders much-needed improvements in health outcomes in resource-constrained regions. Since there has been a rapid increase in public health expenditures in recent decades across countries, both in absolute figures and as a share of GDP, the role of the efficiency of health expenditures in achieving health outcomes needs to be examined (Papanicolas et al. 2018).

In this context, the present study attempts to explore the efficiency and productivity of public health expenditures in improving child survival outcomes across the ten largest emerging economies, *viz.*, Brazil, Russia, India, China, Egypt, Indonesia, Mexico, South Korea, Saudi Arabia, and Turkey. The selection of these countries was based on the ten big emerging markets (BEM) category. Some countries in this category such as Argentina, Poland and South Africa were excluded due to data unavailability and hence counties such as Egypt, Russia and Saudi Arabia were included. Here, we use bootstrap data envelopment analysis and the bootstrap Malmquist productivity index to measure the efficiency and productivity of public health expenditures in reducing the NMR and the U5MR. Our study departs from earlier studies using DEA and MPI, as we test the robustness of these nonparametric approaches by using bootstrap (2000 repetitions) and confidence intervals. Moreover, the statistical significance of the scores is also analyzed to make meaningful inferences.

2 Literature on the Efficiency and Productivity of Health Expenditures

Empirical studies have sought to analyze the relationships between health expenditure efficiency, productivity, and child health outcomes using parametric approach, *viz.* Stochastic frontier analysis (SFA) and a nonparametric approach, *viz.* Data envelopment analysis (DEA) (Hollingsworth 2003: Kirigia et al. 2008; Selamzade & Ersoy, 2024; Tigga & Sarkar, 2025; Mergoni *et al.* 2025). One of the first parametric studies to assess the efficiency of health expenditure was by Evans *et al.* (2000). They analyzed a panel dataset of 191 countries from 1993 to 1997 by using a fixed-effects panel data estimator and corrected ordinary least squares (COLS), using health expenditure and average years of schooling as the input variables and the log of DALEs as the output measure. The study concluded that health outcomes could be improved either by increasing the efficiency of the health system, i.e., moving toward the frontier, or by increasing health expenditures, i.e., moving along the expansion path. Following their work, Jarasuriya and Woodon (2003) estimated the efficiency of health and education provision across 76 countries from 1990 to 1998 by using the SFA approach. They considered net primary enrollment and life expectancy as outputs and three inputs: per capita GDP, spending per capita, and the adult literacy rate. The authors concluded that increased spending is not a guarantee to obtain better education or health outcomes. There are various disadvantages of using SFA to assess efficiency, one of which is that only one output or aggregate output can be used at a time. In the health sector, multiple outputs are produced simultaneously using the same set of inputs; hence, using SFA may inevitably provide misleading results.

Gupta and Verhoeven (2001) assessed the efficiency of education and health expenditures in 85 countries in Africa, Asia, and the Western Hemisphere by using the free disposal hull (FDH) approach. They found that African countries were inefficient relative to their Asian and Western Hemisphere peers and that the level of inefficiency and government expenditure were positively correlated. In addition to FDH, another widely used nonparametric approach is data envelopment analysis (DEA). Some studies have used the two approaches simultaneously to assess the efficiency of health expenditure. Afonso and St. Aubyn (2005) applied both DEA and FDH to evaluate the efficiency of expenditures in education and health across OECD countries. They estimated efficient frontiers using inputs such as expenditure, instruction time in hours, number of teachers, beds, medical technology, and health employment. The output for education was measured by reading, math, and science literacy, whereas infant mortality and life expectancy were used as health outputs. They found that in general, the DEA and FDH results were identical, with efficient countries in DEA being a subset of those identified as efficient under FDH. Grosskopf et al. (2006) tried to measure how efficiently health expenditures are translated into better health outcomes. They observed that better health outcomes in developed countries were not necessarily due to a large share of public expenditure on health. Rayp and De Sijpe (2007) estimated the efficiency of government expenditures for 52 developing countries. They used DEA to measure efficiency, taking government expenditure per capita (PPP) as an input measure to explain efficiency. They observed that countries with a high youth population, high illiteracy rate, and low private health spending will have difficulty registering good health and education outcomes. Hsu (2014) evaluated the performance of health expenditure across 46 countries in Europe and Central Asia using three approaches: returns to scale, slack-based measure (SBM) and super SBM. They found that countries could improve output by 1.2% using existing levels of input. Ouertani et al. (2018) assessed the efficiency of government expenditures in Saudi Arabia, they concluded that unemployment and broad money have a negative impact on government expenditures for health and infrastructure. Liu et al. (2019) measured health expenditure efficiency using the combination of a super slack-based measure (SBM) model with the Malmquist productivity index (MPI) in rural China from 2007 to 2016. This study used two inputs—per capita healthcare expenditure and total health expenditure—and multiple outputs—health institution outputs, health technical personnel outputs, health facility outputs, and the utilization rate of health resource outputs. The study revealed that despite an increase in health expenditure in rural areas, particularly in the western region, there was no increase in efficiency. A recent study by Guo et al. (2021) estimated the total factor productivity of the Chinese government's health care expenditure after the 2009 health reform. The authors concluded that optimizing government expenditure along with an increase in government health investment would improve their efficiency score.

Most of the studies discussed above have concluded that a higher level of public expenditure does not necessarily lead to greater efficiency; hence, measures such as optimizing health expenditures and reallocating health resources should be focused upon for the efficient utilization of resources.

3 Materials and Methods

The study uses data from 10 emerging economies for 2010–2020. The set of selected countries is based on the availability of the data for the desired period. The data for analysis was obtained from the World Bank Development Indicators for ten emerging economics.

3.1 Data: Inputs and outputs

The selection of inputs and outputs for the study is drawn from the literature. We used public health expenditure as the input since it is a vital indicator of a country's health system (Grigoli and Kapsoli 2007; Cetin and Bahce 2016). As outputs, we used two widely used child outcomes, the neonatal mortality rate (NMR) and the under-five mortality rate (U5MR). In the DEA model, outputs are measured in a manner such that 'more is better'. However, in the case of the two outputs selected, the same implication is not feasible, as both are undesirable outputs/negative variables. Hence, a transformation of NMR and U5MR was used. Using the following formula (De et al. 2012), the two outputs are converted:

$$\text{Neonatal Survival Rate (NSR)} = \frac{1000 - NMR}{NMR}$$

$$\text{Under} - 5\ \text{Survival Rate (U5SR)} = \frac{1000 - \text{U5MR}}{\text{U5MR}}$$

3.2 Methods

3.2.1 Data Envelopment Analysis

It is a mathematical programming technique that is based on linear programming (LP) and measures the relative performance of a group of organizational units known as decision-making units (DMUs). Within the DEA framework, the Charnes, Cooper, and Rhodes (CCR) model (Charnes et al. 1978) and the Banker, Charnes, and Cooper (BCC) model (Banker et al. 1984) are extensively used. It is important to mention that Banker's pioneering research has significantly contributed to the DEA methodologies, which have addressed fundamental challenges related to scale efficiency, statistical robustness, and the effects of contextual variables on performance (Emrouznejad et al. 2025).

The technical efficiency scores derived from the CCR model are the overall technical efficiency (OTE) scores, and those derived from the BCC model are the pure technical efficiency (PTE) scores and scale efficiency (SE) scores. Thus, the SE for each DMU is the ratio of the OTE score to the PTE score, i.e.,

$$SE = OTE/PTE$$

Additionally, the DEA model could be either input-oriented or output-oriented. In the input-oriented model, the inputs used are minimized to obtain a certain amount of output, while in the output-oriented model, the outputs are maximized with a given amount of inputs. Another important concept in the production function is the returns to scale (RTS), which explains the long-run relationship between inputs and outputs. The original CCR model assumed constant returns to scale (CRS) and was criticized for being a restrictive approach, as it is less likely that CRS would hold in many realistic scenarios. Alternatively, the BCC model allowed for variable returns to scale (VRS).

The output-oriented model is an appropriate model for the present study, with the objective of maximizing health outcomes given the fixed level of inputs. Additionally, in a health system, the input levels are usually fixed in the short run; hence, reducing them may not be feasible (Evans et al. 2001; Jacobs et al. 2006). Hence, the output-oriented model is more realistic and reflects the real-world situation (Cheng et al. 2015).

We assume a set of DMU_j ($j = 1, 2,\ldots n$) to be evaluated and define $(x_{1j}.\ldots x_{mj})$ as the input vector of DMU_j with input weight vectors $(v_1\ldots,v_m)$ and $(y_{1j}\ldots,y_{qj})$ as the output vector with output weight vectors $(u_1\ldots,v_q)$. Assume that each DMU_j consumes a x_{ij} amount of input i to produce a y_{rj} amount of output r and that the input and output of DMU_k ($k = 1\ldots,n$) to be evaluated are $(x_{1k}\ldots,x_{mk})$ and $(y_{1k}\ldots,y_{qk})$, where $x_{ik} \geq 0$ and $y_{rk} \geq 0$.

Let $\mu_r = t_{ur}$ and $v_i = t_{vi}$,

where $t = (\sum_{i=1}^{m} v_i x_{ik})^{-1}$.

The output-oriented BCC-DEA model is as follows:

Maximize $\sum_{i=1}^{m} v_i x_{ik} + v_0)$

Subject to

$$\begin{cases} \sum_{r=1}^{q} \mu_r y_{rj} + \sum_{i=1}^{m} v_i x_{ij} + v_0 \leq 0(\mathrm{j} = 1, \ldots, \mathrm{n}), \\ \sum_{r=1}^{q} \mu_r y_{rk} = 1, \\ \mu_r \geq 0(\mathrm{r} = 1, \ldots, \mathrm{q}), v_i \geq 0(\mathrm{i} = 1, \ldots, \mathrm{m}), v_0 \in \mathbf{R} \end{cases}$$

For a set of *n* DMUs, a standard DEA model is solved *n* times, i.e., one for each DMU. Efficiency scores equal to 1 indicate 'efficiency', while scores less than 1 indicate 'inefficiency'.

We choose DEA over other efficiency techniques due to its suitability in the health sector. Gong and Sickles (1992) found that SFA is a better approach than DEA for efficiency analysis if the technology and inefficiency distributions closely follow what was used in the data generation process. However, if the underlying technology and efficiency distributions are unknown, DEA is a better tool than SFA. Moreover, in the presence of heteroscedasticity, DEA-based estimators are sure to provide better results (Banker et al. 2004). Additionally, for datasets composed of smaller samples, DEA is appropriate (Resti, 2001).

3.3 Malmquist Productivity Index (MPI)

It is a nonparametric approach based on DEA and measures changes in productivity over time and can be decomposed into efficiency and technology changes. To measure productivity, we assume that a firm produces output y^{t} using input x^{t} and that the production frontier H^t that models the transformation of inputs $x^t \in \mathbb{R}_1^N$ into outputs $y^t \in \mathbb{R}_1^N$ at time *t* is as follows:

$$H^t = \left[(x^t, y^t) : x^t can\ producey^t\right]$$

The Malmquist productivity index, defined as the geometric mean of the two indexes, is as follows (Färe *et al.* 1994):

$$M_0^t = \left(x^t, y^t, x^{t+1}, y^{t+1}\right) = \left[\frac{D_0^t\left(x^{t+1}, y^{t+1}\right)}{D_0^t(x^t, y^t)} \frac{D_0^{t+1}\left(x^{t+1}, y^{t+1}\right)}{D_0^{t+1}(x^t, y^t)}\right]^{1/2} \quad (1)$$

The first ratio compares the data from periods t and t + 1 relative to production possibilities existing in period t, and the second compares the performance of the same data relative to production possibilities existing in period t + 1. If this index is greater than unity, it implies improvement in productivity between periods t and t + 1. In addition, values less than unity mean that productivity has declined.

The MPI can be further decomposed into two components as follows (Färe *et al.* 1989, 1992):

$$M_0^t = \left(x^t, y^t, x^{t+1}, y^{t+1}\right) = \frac{D_0^{t+1}\left(x^{t+1}, y^{t+1}\right)}{D_0^t(x^t, y^t)}\left[\frac{D_0^t\left(x^{t+1}, y^{t+1}\right)}{D_0^{t+1}\left(x^{t+1}, y^{t+1}\right)} \frac{D_0^t\left(x^t, y^t\right)}{D_0^{t+1}(x^t, y^t)}\right]^{1/2} \quad (2)$$

In Eq. (2), the component, i.e., $\frac{D_0^{t+1}(x^{t+1},y^{t+1})}{D_0^t(x^t,y^t)}$ measures the change in the Farrell technical efficiency between years t and t + 1. The second component, i.e., $\left[\frac{D_0^t(x^{t+1},y^{t+1})}{D_0^{t+1}(x^{t+1},y^{t+1})}\frac{D_0^t(x^t,y^t)}{D_0^{t+1}(x^t,y^t)}\right]$, measures the change in technical progress as measured by the shift in the frontier between the two periods evaluated at the input level x^{t+1} and at the input level realized at x^t. For both components, values greater than unity indicate improvement, and values less than unity suggest decline. The two terms in Eq. (2) are efficiency change (ECH) and technological change (TCH). Thus, M = ECH x TCH. The ECH indicates how close (or farther away) a firm is to the "best practice" frontier over time and measures the shift toward the best practice frontier. The TCH shows how much the frontier shifts and measures the movement of the production frontier. An MPI value greater than 1 indicates growth in productivity, while a value less than 1 indicates a decline in productivity between the periods. A value of 1 denotes stagnation in productivity.

ECH can be further decomposed into PEC and scale efficiency change (SEC) under variable returns to scale (VRS).

$$\text{ECH} = \frac{D_0^{t+1}(x^{t+1}, y^{t+1})}{D_0^t(x^t, y^t)} = \frac{D_{VRS}^{t+1}(x^{t+1}, y^{t+1})}{D_{VRS}^t(x^t, y^t)} \frac{\frac{D_{CRS}^{t+1}(x^{t+1},y^{t+1})}{D_{VRS}^t(x^t,y^t)}}{\frac{D_{CRS}^t(x^t,y^t)}{D_{VRS}^t(x^t,y^t)}} \tag{3}$$

Thus, ECH = PEC x SEC

The first part of the equation indicates the change in efficiency relative to the true VRS frontier, i.e., the proportional reduction in input usage if efficiently utilized, and the second part shows the extent to which the distance from the scale-efficient point on the VRS frontier (relative to the constant returns to scale (CRS) frontier) has changed, i.e., the reduction in input usage that occurs in the presence of CRS (scale efficiency). A PEC value greater than 1 implies efficiency gain, i.e., the firm is closer to the VRS frontier in period t, and if the PEC value is less than 1, the opposite holds true. Likewise, an SEC greater than 1 implies that the firm has become more scale-efficient between the two periods.

3.4 Bootstrapping Technique

One of the greatest drawbacks of nonparametric approaches such as DEA and MPI is their inability to account for any random noise. These are point estimates for productivity and do not provide any information on the uncertainty in the firm-specific estimates. Bootstrap techniques are one way to correct this and obtain measures of statistical precision in the estimates. The rationale behind the bootstrap procedure is to repeat the true data generation process by randomly drawing (B times) a sample from the original sample. Introduced by Efron (1979) and Efron and Tibshirani (1993), this technique has been widely used for dealing with sampling variability. However, the bootstrap approach frontier models were first used by Simar (1992), and its development for nonparametric envelopment estimators was later introduced by Simar and Wilson (1998 & 1999). Other available bootstrap approaches include those of Atkinson and Wilson (1995) and

Lothgren and Tambour (1999). We use the bootstrap method introduced by Simar and Wilson (1998 & 1999) to estimate the bias and confidence intervals for the efficiency and productivity scores using 2000 repetitions.

4 Results

4.1 Descriptive Statistics

Figure 1 shows the current health expenditure (% GDP) during 2010–2020. Among the emerging economies, Brazil has the highest CHE, which was 7.95% in 2010 and further increased to 10.31% in 2020.

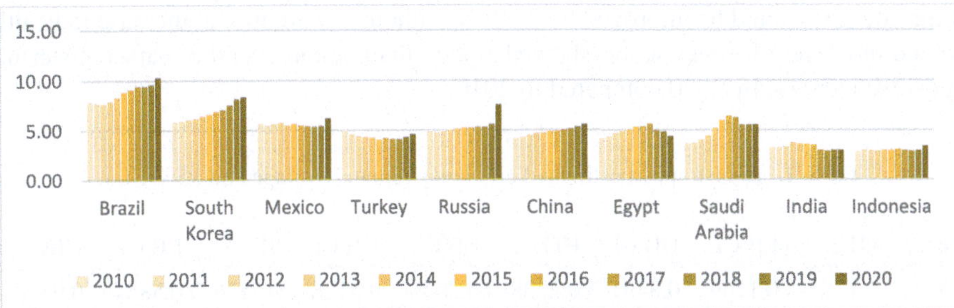

Fig. 1. Current Health Expenditure (% GDP)

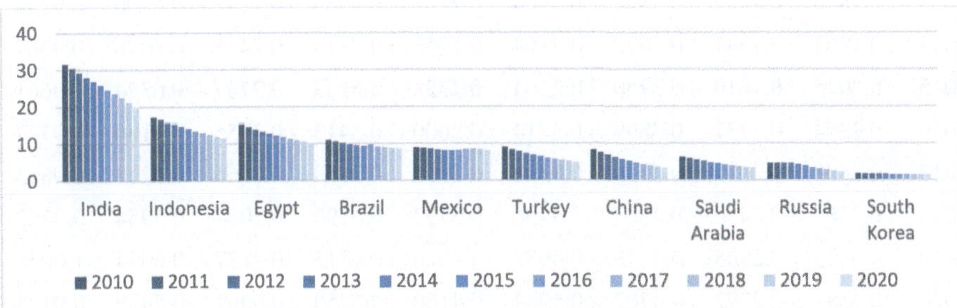

Fig. 2. Neonatal mortality rate (per 1,000 live births)

Figure 2 shows that the NMR has decreased across emerging economies in the last decade. Countries such as India, where the NMR was very high, i.e., 31.80 (per 1,000 live births) in 2010, experienced a remarkable reduction to 20.20 in 2022, indicating the constant improvement made by the country. Figure 3 shows that the U5MR has decreased across emerging economies in the last decade.

4.2 Results of Bootstrap DEA

The bootstrap efficiency scores of the ten emerging economies are shown in Table 1. The average bias-corrected OTE estimate was 0.2474, which suggests that the performance

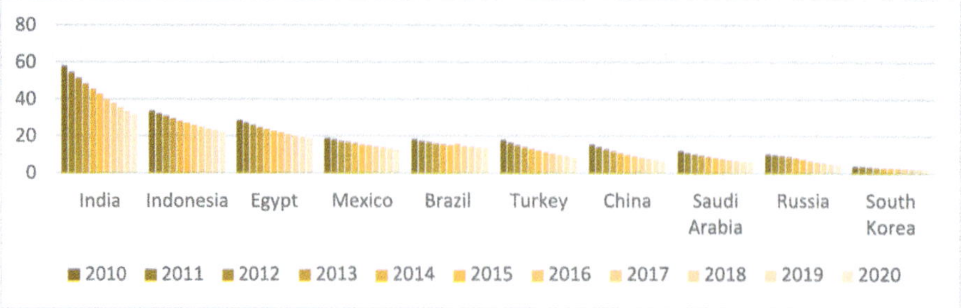

Fig. 3. Mortality rate, under-five (per 1,000 live births)

of the countries could be improved by 75.26% without any additional increase in health expenditure. The efficiency scores showed modest fluctuations over the years, registering a positive increase by 2020 compared to 2010.

Table 1. Bootstrapped estimated efficiency (annual average and 95% CIs)

Year	OTE	LB CI	UB CI	PTE	LB CI	UB CI	SE	LB CI	UBCI
2010	0.1777	0.1269	0.3491	0.2238	0.2041	0.5126	0.7436	0.5854	0.9620
2011	0.1932	0.1369	0.3598	0.2697	0.2341	0.5476	0.7009	0.5534	0.9147
2012	0.1882	0.1372	0.3574	0.2329	0.2215	0.5202	0.7239	0.5828	0.9311
2013	0.1880	0.1429	0.3596	0.1904	0.2101	0.4983	0.7648	0.6200	0.9794
2014	0.1961	0.1547	0.3678	0.2164	0.2256	0.5110	0.7465	0.6076	0.9367
2015	0.2045	0.1610	0.3750	0.2264	0.2328	0.5153	0.7747	0.6324	0.9661
2016	0.2374	0.1737	0.3899	0.2714	0.2600	0.5413	0.7555	0.6180	0.9727
2017	0.2800	0.2019	0.4154	0.3394	0.2994	0.5909	0.7466	0.6213	0.9885
2018	0.3289	0.2450	0.4527	0.4496	0.3792	0.6790	0.7087	0.5954	0.9702
2019	0.3895	0.2905	0.4878	0.4953	0.4026	0.6815	0.7177	0.6064	1.0048
2020	0.3381	0.2772	0.4702	0.5054	0.4160	0.6850	0.6967	0.5956	0.9125
Mean	0.2474	0.1862	0.3986	0.3110	0.2805	0.5712	0.7345	0.6017	0.9581

Overall technical efficiency (OTE), pure technical efficiency (PTEC), scale efficiency (SE), lower bound confidence interval (LB CI) and upper bound confidence interval (UB CI).

Using the Wilcoxon signed-rank test, proposed by Wilcoxon (1945), we test whether the increase in efficiency over the years was statistically significant. The Wilcoxon signed-rank test is a non-parametric statistical test used to compare two related samples, determining whether their population mean ranks differ. Table 2 presents a summary of the tests conducted on the annual efficiency scores between the two periods. Although the OTE scores improved between 2010 and 2015, the difference was not statistically significant. The performance of the countries differed significantly between 2015 and 2020 (p value < 0.05). Overall, the performance of the countries significantly differed between 2010 and 2020 (p-value < 0.05). The PTE showed a similar trend—efficiency performance significantly differed between 2015 and 2020 and between 2010 and 2020 (p-value < 0.05). The changes in SE were not statistically significant.

Table 2. Wilcoxon signed rank test on the bootstrap efficiency scores

Comparison	OTE	PTE	SE
	p value	p value	p value
2010 vs 2015	0.075	0.508	0.508
2015 vs 2020	0.002*	0.009*	0.047
2010 vs 2020	0.002*	0.022*	0.285

* $p < 0.05$

Table 3. Bootstrapping MPI and its components (annual average and 95% CI)

Year	MPI	LB	UB	TEC	LB	UB	TC	LB	UB	SEC	LB	UB	PTEC
2011	1.0420	0.9933	1.0948	1.1015	0.9720	1.2783	0.9483	0.8306	1.0820	0.9359	0.8177	1.0428	1.1769
2012	1.0320	0.9577	1.1134	0.9434	0.7703	1.1391	1.0936	0.9009	1.3492	1.0451	0.8851	1.2456	0.9028
2013	1.0160	0.9365	1.1033	0.9312	0.7235	1.1232	1.0919	0.8869	1.4247	1.0720	0.9036	1.2915	0.8687
2014	0.9923	0.9408	1.0454	1.0604	0.8416	1.3238	0.9365	0.7524	1.1701	0.9789	0.8268	1.1773	1.0833
2015	1.0450	0.9680	1.1342	1.0163	0.8522	1.2394	1.0274	0.8295	1.2519	1.0425	0.8970	1.2061	0.9749
2016	0.9927	0.9532	1.0416	1.0798	0.9169	1.3164	0.9198	0.7552	1.0829	0.9709	0.8409	1.1207	1.1121
2017	1.2395	1.1361	1.3618	1.1423	0.9109	1.4538	1.0613	0.8756	1.3434	0.9803	0.8162	1.1992	1.1653
2018	1.2220	1.0723	1.4530	1.2330	0.9378	1.7757	0.9881	0.7436	1.3043	0.9431	0.7375	1.3402	1.3075
2019	0.9569	0.8839	1.0501	1.0361	0.7759	1.4085	0.9206	0.7399	1.1840	1.0328	0.7629	1.4218	1.0032
2020	0.9311	0.8438	1.0319	1.0356	0.7302	1.5911	0.9071	0.6707	1.2057	0.9620	0.7077	1.4108	1.0764
Mean	1.0469	0.9686	1.1429	1.0580	0.8431	1.3649	0.9894	0.7985	1.2398	0.9964	0.8195	1.2459	1.0671

Malmquist productivity index (MPI), technical efficiency change (TEC), technological change (TC), scale efficiency change (SEC), pure technical efficiency change (PTEC), lower bound (LB) and upper bound (UB)

Table 4. Wilcoxon signed rank test on the bootstrap MPI

Comparison	MPI	TEC	TC	SEC	PTEC
	p value	p value	p value	p value	p value
2011 vs 2015	0.07	0.05	0.80	-0.46	0.03*
2015 vs 2020	0.33	0.88	0.33	0.11	0.33
2011 vs 2020	0.28	0.58	0.06	0.65	0.80

* $p < 0.05$

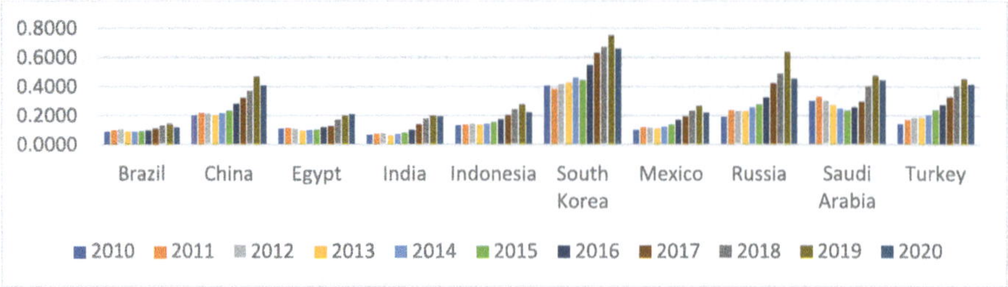

Fig. 4. OTE scores across emerging economies over time

The OTE scores across the countries show variation (Fig. 4); however, to conclude, with certainty that these countries actually differ in efficiency levels, we use confidence intervals. We use the CIs to test the *null hypothesis* that the two countries are equally efficient and the *alternative hypothesis* that they are not equally efficient. For example, the OTE scores of China and the Republic of Korea are 0.2891 and 0.5303, respectively; however, we cannot conclude with certainty that the two countries are not equally efficient because their CIs overlap. Alternatively, the OTE scores of India and China are 0.1158 and 0.2891, respectively, and their CIs do not overlap. Therefore, we reject the null hypothesis that these two countries are equally efficient.

4.3 Trends in Productivity Change

The average productivity growth across 10 emerging economies during 2010–20 was 1.047, which indicate that productivity had progressed at an average rate of 4.7% per year. This progress was purely due to an increase in the TEC (technical efficiency change); however, the TC (technological change) declined. Here, the average TEC, which indicates how far the emerging economies in aggregate have moved to or from the efficiency frontier over the years, has a value of 1.058, suggesting that on average, technical efficiency has increased by 5.8%. Alternatively, the mean TC was reported to be 0.989, indicating a decrease in technology use of 1.1% to improve U5MR and NMR (Table 3).

The productivity of the countries has shown changes over the years; however, to test whether these changes were significant, we used the Wilcoxon signed-rank test on the bootstrap MPI. The *null hypothesis* is that the bias-corrected MPIs between two years have the same distribution, and the *alternative hypothesis* states that the MPIs between two years have different distribution. The test summary depicted in Table 4 indicates that no change in productivity is noted between the periods for MPI and its components. However, for PTEC, the change between 2011 and 2015 was significant (p-value $<$ 0.05).

Overall, out of the ten countries, eight exhibited progress, and two exhibited regress. In three out of eight countries, productivity growth was explained by improvements in both technical efficiency and technological innovation. In the remaining five countries, productivity growth was due to either of them (Fig. 5).

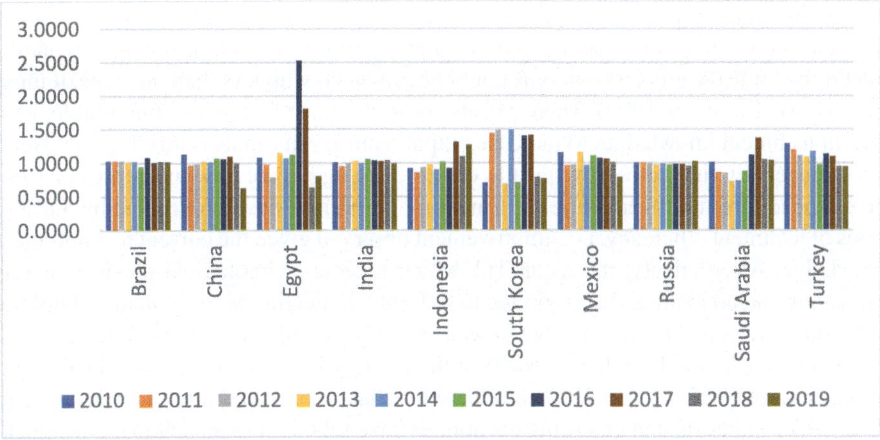

Fig. 5. MPI scores across emerging economies over the years

To examine whether the changes in productivity and its components across the countries were significant, CIs were obtained to test the null hypothesis of insignificant productivity changes. Overall, only two countries indicate significant productivity progress Brazil and Turkiye are the two countries that show significant progress in productivity.

5 Discussion

Efficiency in public health expenditure refers to the maximization of health outcomes for a given level of spending. The average bootstrap OTE was 0.3381, suggesting that if inefficiency could be eliminated, child mortality outcomes could improve by 66% in these countries. South Korea had the highest health expenditure efficiency, with an OTE of 0.6636, whereas Brazil and India had the lowest levels of efficiency. One important observation is that Brazil had the highest level of current health expenditure (% of GDP), i.e., 10.31%, during 2020 but had the lowest OTE score of 0.1250. The high inefficiency has resulted in poor health outcomes, i.e., the U5MR in Brazil was 14.70 (per 1000 live births) and the NMR was 8.70 (per 1000 live births), both of which are high compared to the emerging economy average. This suggests that a higher level of budget allocation does not guarantee greater efficiency; rather, the optimal use of resources, funds, manpower, and reduced waste determines efficiency. Mills (2014) suggested that in developing countries, inefficiencies arise due to shortages, weak management, and poor distribution of resources. The International Monetary Fund (2019) also indicated that a major part of spending is wasted due to the misallocation of resources, poor quality of public services, etc., leading to the estimation of additional spending of 2% of GDP in emerging market economics (Cristóbal et al. 2021).

The bootstrap MPI results indicate that the average productivity growth in the emerging economies from 2010 to 2020 was 1.047, which indicates that productivity had progressed at an average rate of 4.7% per year. This progress was purely due to an increase in the TEC (technical efficiency); however, the TC (technological change) regressed. A technically efficient position is achieved when the maximum possible improvement in outcome is obtained from a set of resource inputs. An intervention is technically inefficient if the same (or greater) outcome could be produced with less than one type of input (Palmer and Torgerson 1999). There can also be a situation wherein it is impossible, with current technical knowledge, to increase output from given inputs (Farrell 1957). Here, the average TEC indicates how far the countries in aggregate have moved to or from the efficiency frontier over the years. Emerging economies have progressed by 5.8% in terms of technical efficiency, i.e., improvement observed given the current technological knowledge. Alternatively, the mean TC, which reflects technological change such as innovation and adaptation, was reported to be 0.989, indicating regression in technology adoption by 1.1%, which might have slowed or restricted improvement in child mortality outcomes such as the U5MR and NMR in these countries. Notably, during the decadal period, except for 2012 and 2013, productivity change occurred due to an increase in TEC, which suggests that emerging economies have relied more on achieving technical efficiency rather than improving TC, i.e., technological innovation in improving child mortality outcomes. Technological innovation and new technology adaptation in health systems require massive investments, and emerging economies often face challenges in allocating sufficient resources and funding to their health systems. Thus, it is quite likely that these countries mainly focus on utilizing their given level of technology to improve health outcomes. At the disaggregate level, of the ten countries, eight exhibited productivity progress, and two exhibited regression. In three out of eight countries, productivity growth was explained by improvements in both technical efficiency and technological innovation. These were Brazil, India and Indonesia. The remaining five countries, i.e., Russia, China, Egypt, Mexico and Turkey, also experienced productivity growth, which was attributed to either technical efficiency progress or technological progress. On the other hand, South Korea and Saudi Arabia experienced a productivity regression, which can be explained by a reduction in technological change.

Overall, countries that have witnessed very low health expenditure efficiency, i.e., low OTE scores, such as Brazil, India, and Indonesia, have witnessed MPI scores greater than 1, implying productivity progress. On the other hand, countries such as South Korea and Saudi Arabia, which have relatively high OTE scores, have experienced productivity decreases, i.e., MPIs less than 1. This should not be seen as contradictory because throughout the decadal period, these two countries witnessed very low levels of NMR (Fig. 2) and U5MR (Fig. 3). Since the relative reductions in NMR and U5MR by these countries are also very small, productivity might have regressed, as productivity involves achieving more health outputs for the same level of inputs. On the other hand, countries such as India and Indonesia, despite having very low OTEs, have witnessed a rapid decline in these health indicators (Fig. 2 and Fig. 3). Hence, they have witnessed productivity progress.

6 Conclusion

Efficiency in public health expenditure is crucial for maximizing health outcomes, yet many emerging economies struggle with inefficiencies despite significant investments. The findings in this paper reveal that eliminating inefficiencies could substantially reduce child mortality rates. Notably, it is observed that having the highest health expenditure and increased spending alone does not necessarily ensure better health outcomes. Instead, focus should be on optimal resource utilization, effective management, and strategic allocation of funds. Furthermore, the MPI results suggest that productivity growth in emerging economies over the past decade was primarily driven by technical efficiency improvements rather than technological advancements. While countries like Brazil, India, and Indonesia experienced productivity gains despite low efficiency scores, South Korea and Saudi Arabia, which had higher efficiency, faced productivity declines that resulted in minimal reductions in child mortality rates. This paradox underscores the complex relationship between efficiency, productivity, and health outcomes. To sustain progress, emerging economies must balance technical efficiency improvements with investments in technological innovation. Policymakers should focus on strengthening healthcare systems through better governance, innovation, and equitable resource distribution which will be essential in achieving long-term improvements in public health outcomes.

References

Afonso, A., Aubyn, M.S.: Nonparametric approaches to education and health efficiency in OECD countries. J. Appl. Econ. **8**(2), 227–246 (2005). https://doi.org/10.1080/15140326.2005.12040626

Afonso, A., Schuknecht, L., Tanzi, V.: Public sector efficiency: evidence for new EU member states and emerging markets. Appl. Econ. **42**(17), 2147–2164 (2010)

Akinlo, A.E., Sulola, A.O.: Health care expenditure and infant mortality in sub-Saharan Africa. J. Policy Model. **41**(1), 168–178 (2019). https://doi.org/10.1016/j.jpolmod.2018.09.001

Atkinson, S.E., Wilson, P.W.: Comparing mean efficiency and productivity scores from small samples: a bootstrap methodology. J. Prod. Anal. **6**(2), 137–152 (1995). https://doi.org/10.1007/bf01073408

Banker, R.D., Charnes, A., Cooper, W.W.: Some models for estimating technical and scale Inefficiencies in data envelopment analysis. Manag. Sci. **30**(9), 1078–1092 (1984). https://www.jstor.org/stable/2631725

Banker, R.D., Chang, H., Cooper, W.W.: A simulation study of DEA and parametric frontier models in the presence of heteroscedasticity. Eur. J. Oper. Res. **153**(3), 624–640 (2004). https://doi.org/10.1016/S0377-2217(02)00699-9

Barenberg, A.J., Basu, D., Soylu, C.: The effect of public health expenditure on infant mortality: evidence from a panel of Indian states, 1983–1984 to 2011–2012. J. Dev. Stud. **53**(10), 1765–1784 (2017). https://doi.org/10.1080/00220388.2016.1241384

Cetin, V.R., Bahce, S.: Measuring the efficiency of health systems of OECD countries by data envelopment analysis. Appl. Econ. **48**(37), 3497–3507 (2016)

Charnes, A., Cooper, W.W., Rhodes, E.: Measuring the efficiency of decision-making units. Eur. J. Oper. Res. **2**(6), 429–444 (1978). https://doi.org/10.1016/0377-2217(78)90138-8

Cheng, Z., et al.: Technical efficiency and productivity of Chinese county hospitals: an exploratory study in Henan province, China. BMJ Open **5**, e007267 (2015). https://doi.org/10.1136/bmjopen-2014-007267

Cristóbal, J., et al.: Unraveling the links between public spending and sustainable development goals: insights from data envelopment analysis. Sci. Total Environ. **786**, 147459 (2021). https://doi.org/10.1016/j.scitotenv.2021.147459

De, P., Dhar, A., Bhattacharya, B.N.: Efficiency of health care system in India: an interstate analysis using DEA approach. Soc. Work Publ. Health **27**(5), 482–506 (2012). https://doi.org/10.1080/19371918.2012.672261

Efron, B.: Bootstrap methods: another look at the jackknife. Ann. Stat. **7**(1), 1–16 (1979). https://doi.org/10.1214/aos/1176344552

Efron, B., Tibshirani, R.J.: An Introduction to the Bootstrap. Chapman & Hall, London (1993)

Emrouznejad, A., Podinovski, V., Charles, V., Lu, C., Moradi-Motlagh, A.: Rajiv Banker's lasting impact on data envelopment analysis. Ann. Oper. Res. **351**, 1225–1264 (2025). https://doi.org/10.1007/s10479-025-06473-3

Evans, D.B., Tandon, A., Murray, C.J.L., Lauer, J.A.: Comparative efficiency of national health systems: cross-national econometric analysis. BMJ **323**, 307–310 (2001). https://doi.org/10.1136/bmj.323.7308.307

Farrell, M.J.: The measurement of productive efficiency. J. R. Stat. Soc. Ser. A Stat. Soc. **120**(3), 253–281 (1957)

Fa¨re, R., Grosskopf, S., Valdmanis, V.: Capacity, competition and efficiency in hospitals: a nonparametric approach. J. Prod. Anal. **1**, 123-138 (1989). https://doi.org/10.1007/BF00157792

Färe, R., Grosskopf, S., Lindgren, B., Roos, P.: Productivity changes in Swedish pharamacies 1980–1989: a non-parametric Malmquist approach. J. Prod. Anal. **3**(1), 85–101 (1992)

Fa¨re, R., Grosskopf, S., Lindgren, B., Roos, P.: Productivity developments in Swedish hospitals: a Malmquist Output Index approach. In: Charnes, A., Cooper, W.W., Lewin, A.Y., Seiford, M.L. (eds), Data Envelopment Analysis: Theory, Methodology and Applications, Kluwer Academic, Boston, MA, pp. 253–272 (1994)

Gong, B.H., Sickles, R.C.: Finite sample evidence on the performance of stochastic frontiers and data envelopment analysis using panel data. J. Econometrics **51**(1–2), 259–284 (1992). https://doi.org/10.1016/0304-4076(92)90038-S

Grigoli, F., Ley, E.: Quality of government and living standards: Adjusting for the efficiency of public spending (IMF Working Paper No. 12/182). International Monetary Fund, Washington, DC (2012)

Grigoli, F., Kapsoli, J.: Waste not, want not: the efficiency of health expenditure in emerging and developing economies. Rev. Dev. Econ. **22**(1), 384–403 (2018). https://doi.org/10.1111/rode.12346

Grosskopf, S., Self, S., Zaim, O.: Estimating the efficiency of the system of healthcare financing in achieving better health. Appl. Econ. **38**(13), 1477–1488 (2006). https://doi.org/10.1080/00036840500424798

Guo, X., Zhang, J., Xu, Z., Cong, X., Zhu, Z.: The efficiency of provincial government health care expenditure after China's new health care reform. PLoS One **16**(10), e0258274 (2021). https://doi.org/10.1371/journal.pone.0258274

Gupta, S., Verhoeven, M.: The efficiency of government expenditure experiences in Africa. J. Policy Model. **23**(4), 433–467 (2001). https://doi.org/10.1016/S0161-8938(00)00036-3

Hollingsworth, B.: Non-parametric and parametric applications measuring efficiency in health care. Health Care Manag. Sci. **6**(4), 203–218 (2003). https://doi.org/10.1023/A:1026255523228

Hsu, Y.C.: Efficiency in government health spending: a super slacks-based model. Qual. Quant. **48**(1), 111–126 (2014). https://doi.org/10.1007/s11135-012-9753-9

International Monetary Fund (IMF). Fiscal policy and development: human, social, and physical investment for the SDGs. IMF Staff Discussion Note SDN/19/03 (2019)

Jacobs, R., Smith, P.C., Street, A.: Measuring Efficiency in Health Care: Analytic Techniques and Health Policy. Cambridge University Press, United Kingdom (2006)

Jayasuriya, R., Wodon, Q.: Measuring and explaining country efficiency in improving health and education indicators, MPRA Paper 11183, University Library of Munich, Germany (2003)

Kirigia, J.M., Emrouznejad, A., Vaz, R.G., Bastiene, H., Padayachy, J.: A comparative assessment of performance and productivity of health centres in Seychelles. Int. J. Product. Perform. Manag. **57**(1), 72–92 (2008)

Kiross, G.T., et al.: The effects of health expenditure on infant mortality in sub-Saharan Africa: evidence from panel data analysis. Health Econom. Rev. **10**(5) (2020). https://doi.org/10.1186/s13561-020-00262-3

Liu, W., Xia, Y., Hou, J.: Health expenditure efficiency in rural China using the super-SBM model and the Malmquist productivity index. Int. J. Equity Health **18**, 1–13 (2019). https://doi.org/10.1186/s12939-019-1003-5

Lothgren, M., Tambour, M.: Bootstrapping the data envelopment analysis Malmquist productivity index. Appl. Econ. **31**(4), 417–425 (1999). https://doi.org/10.1080/000368499324129

Mergoni, A.; Emrouznejad, A.; De Witte, K. (2024) Fifty years of Data Envelopment Analysis, Eur. J. Oper. Res. **326**(3), 389–412 (2025). https://doi.org/10.1016/j.ejor.2024.12.049

Mills, A.: Health care systems in low-and middle-income countries. N. Engl. J. Med. **370**(6), 552–557 (2014). https://doi.org/10.1056/NEJMra1110897

Novignon, J., Lawanson, A.O.: Health expenditure and child health outcomes in Sub-Saharan Africa. Afr. Rev. Econom. Finan. **9**(1), 96–121 (2017)

Palmer, S., Torgerson, D.J.: Definitions of efficiency. BMJ **318**(7191), 1136 (1999). https://doi.org/10.1136/bmj.318.7191.1136

Papanicolas, I., Woskie, L.R., Jha, A.K.: Health care spending in the United States and other high-income countries. Jama **319**(10), 1024–1039 (2018)

Ouertani, M.N., Naifar, N., Ben Haddad, H.: Assessing government spending efficiency and explaining inefficiency scores: DEA-bootstrap analysis in the case of Saudi Arabia. Cogent Econom. Finan. **6**(1), 1493666 (2018). https://doi.org/10.1080/23322039.2018.1493666

Rayp, G., Van De Sijpe, N.: Measuring and explaining government efficiency in developing countries. J. Dev. Stud. **43**(2), 360–381 (2007). https://doi.org/10.1080/00220380601125230

Resti, A.: Efficiency measurement for multi-product industries: a comparison of classic and recent techniques based on simulated data. Eur. J. Oper. Res. **121**(3), 559–578 (2000). https://doi.org/10.1016/S0377-2217(99)00054-5

Selamzade, F., Ersoy, Y.: Health Efficiency Measurement of Turkish Provinces Using DEA, TOPSIS and EDAS methods. In: Advances in the Theory and Applications of Performance Measurement and Management. DEA45 2023. Lecture Notes in Operations Research. Springer, Cham (2024). https://doi.org/10.1007/978-3-031-61597-9_15

Simar, L.: Estimating efficiencies from frontier models with panel data: a comparison of parametric, non-parametric and semiparametric methods with bootstrapping. J. Prod. Anal. **3**(1–2), 171–203 (1992). https://doi.org/10.1007/BF00158775

Simar, L., Wilson, P.W.: Sensitivity analysis of efficiency scores: how to bootstrap in nonparametric frontier models. Manag. Sci. **44**(1), 49–61 (1998). https://www.jstor.org/stable/2634426

Simar, L., Wilson, P.W.: Estimating and bootstrapping Malmquist indices. Eur. J. Oper. Res. **115**(3), 459–471 (1999). https://doi.org/10.1016/S0377-2217(97)00450-5

Tigga, N.S., Sarkar, P.: Has the efficiency and productivity of the health system in India improved during post-policy period? Application of the bootstrap data envelopment analysis and Malmquist productivity index. Appl. Econ. **57**(11), 1239–1256 (2025)

Verhoeven, M., Gunnarsson, V., Carcillo, S.: Education and health in G7 countries: achieving better outcomes with less spending (IMF Working Paper No. 07/263). Washington, DC International Monetary Fund (2007)

An Investigation and Evaluation of Efficiency in Primary and Secondary Healthcare System of India Using DEA

Qaiser Farooq Dar[1,4](✉), Gulbadin Farooq Dar[2], and Ashiq Hussain Bhat[3]

[1] ICMR-National Institute of Virology, Pune - North Zone, Jammu-181221 Nagrota, Jammu & Kashmir, India
qaiserdea@gmail.com

[2] Department of Statistics, St Joseph's University, Bengaluru 560027, Karnataka, India
gulbadinstst.pu@gmail.com

[3] ICMR Rajendra Memorial Research Institute of Medical Sciences, Patna 800007, India
ashiqhb14@gmail.com

[4] Faculty of Mathematical & Information Sciences, Academy of Scientific and Innovative Research (AcSIR), Ghaziabad, India

Abstract. The rising costs of public healthcare have made efficient operation increasingly important. Evidence-based evaluations are essential for policymakers. Assessing efficiency in primary and secondary care systems is complex due to multiple input and output indicators. Data Envelopment Analysis (DEA), a non-parametric method, is used to evaluate the relative efficiency of healthcare facilities, considering inputs like financial, human, infrastructure, and material resources, and outputs such as clinical outcomes, service efficiency, patient satisfaction, and access to care.

Using secondary data from healthcare centers across India, both input- and output-oriented DEA models were applied. Results showed significant variability in efficiency, with benchmarks identified for improvement. Key determinants of efficiency include financial management, staffing, and infrastructure quality. The study offers policy recommendations to enhance healthcare delivery and emphasizes the need for continuous efficiency monitoring.

Keywords: Data Envelopments Analysis · Slack-Based Measure (SBM) · Public Healthcare Efficiency

1 Introduction

India boasts one of the largest healthcare systems globally, catering to over 1.4 billion people through a mix of public and private services. The public healthcare system, managed jointly by the central and state governments, serves as the backbone of healthcare delivery for rural and tribal populations. It provides free or low-cost services through an extensive network of primary healthcare centers (PHCs), community health centers (CHCs), district hospitals, and specialized tertiary care facilities in urban areas. This system aims to ensure equitable access to essential health services for all sections of society, especially the underserved.

A. Emrouznejad et al. (Eds.): DEA 2024, LNOR, pp. 366–377, 2025.
https://doi.org/10.1007/978-3-031-98177-7_25

On the other hand, the private healthcare sector plays a significant role in delivering secondary and tertiary care, accounting for nearly 70% of healthcare services in India. While private hospitals and clinics are renowned for offering high-quality care, especially in urban centers, their accessibility is often limited to those who can afford it. Rural areas, in contrast, face challenges such as inadequate infrastructure, high out-of-pocket costs, and limited availability of private healthcare options.

The stark divide between public and private healthcare highlights the challenges of ensuring equitable healthcare access in India. Public institutions grapple with resource constraints and operational inefficiencies, while private providers remain inaccessible to large sections of the population. Bridging this gap requires policy reforms, resource optimization, and enhanced public-private collaboration.

India's healthcare system reflects the country's socio-economic diversity and underscores the urgent need for sustainable interventions to address its disparities. Strengthening the public healthcare network while improving private sector accountability is key to achieving better health outcomes for all.

The challenges in assessing efficiency arise due to the complexity of healthcare operations, which involve multiple inputs such as medical personnel, infrastructure, and financial resources, and outputs like clinical outcomes and patient satisfaction. Data Envelopment Analysis (DEA) offers a powerful tool to evaluate the relative efficiency of healthcare facilities by comparing the performance of Decision-Making Units (DMUs) based on the relationship between inputs and outputs. The public healthcare system in India faces significant challenges, with poor service excellence being a critical concern. Among the various factors contributing to this issue, the effective utilization of resources stands out as one of the most important. Public healthcare facilities often operate with limited resources, such as medical staff, hospital infrastructure, and equipment. When these resources are optimized and effectively managed, the efficiency of healthcare services can improve, bringing about positive transformations in service delivery and patient outcomes.

In this context, the measurement of technical, pure technical, scale, and mix efficiency is vital. These metrics provide a comprehensive evaluation of how well healthcare facilities utilize their resources to deliver services. Technical efficiency examines overall resource usage, while pure technical efficiency focuses on managerial efficiency in using inputs. Scale efficiency assesses whether facilities operate at an optimal size, and mix efficiency evaluates the combination of inputs to achieve desired outputs.

Such efficiency assessments not only measure performance but also help identify sources of inefficiency within healthcare institutions. By pinpointing these areas, policymakers and administrators can implement targeted interventions, optimize resource allocation, and ultimately enhance the quality and accessibility of public healthcare services across the country.

A comprehensive evaluation of India's state-wise public healthcare systems is essential to optimize their performance. By measuring technical, pure technical, and scale efficiency using the DEA model, states can be ranked based on their efficiency levels. This ranking highlight top-performing states while identifying underperforming ones, enabling targeted interventions. The super-efficiency DEA model approach further facilitates a nuanced understanding of performance disparities, while inefficiency sources in lagging states can be pinpointed for remedial action. Setting benchmark targets based on best practices helps inefficient systems aspire toward reaching the efficient frontier and achieving improved outcomes.

To enhance efficiency, it is vital to assess the factors influencing public healthcare performance, such as resource allocation, governance, and service delivery. This analysis supports the formulation of targeted policy recommendations aimed at addressing inefficiencies and improving underperforming systems. By adopting evidence-based strategies, states can optimize their resource utilization and infrastructure while ensuring equitable access to quality healthcare. Such an approach not only bridges performance gaps across states but also strengthens the overall public healthcare system, driving better health outcomes for India's diverse and growing population.

2 Literature Review

Data Envelopment Analysis (DEA) is a widely used non-parametric method for assessing the efficiency of decision-making units (DMUs). DEA employs linear programming techniques to establish an efficiency frontier, categorizing DMUs into efficient and inefficient groups. The efficient DMUs are placed on the frontier, while the inefficient ones are assessed relative to it. The foundational models of DEA, the CCR (Charnes, Cooper, and Rhodes) and BCC (Banker, Charnes, and Cooper) models, are radial measures that evaluate efficiency by focusing on inputs and outputs [1–3]. These models can be solved using input, output, or mixed orientations, depending on whether fixed or variable levels of inputs and outputs are analyzed [4]. To address the limitations of radial measures, non-oriented and non-radial models like the Slack-Based Measure (SBM) (Emrouznejad, et al., 2025) [25] were introduced. SBM, proposed by Tone (2001), focuses on excess inputs and output shortfalls, providing a more detailed evaluation by incorporating the concepts of weak and strong efficiency [5]. Additionally, to rank DMUs with identical efficiency scores, the concept of super-efficiency was developed, which extends DEA's ability to differentiate among efficient DMUs [6–8]. DEA method logies such as window analysis further enable the study of efficiency changes over time [7, 9 10], while fuzzy DEA can be applied when input and output variables are uncertain or imprecise [11, 12]. Network DEA is an extension of the traditional DEA methodology that is specifically designed to evaluate systems or DMUs that have a multi-stage or multi-component structure [13].

DEA is widely applied across various fields to evaluate efficiency due to its flexibility in handling multiple inputs and outputs. In the banking sector, DEA has been used to assess the efficiency of central banks by analyzing their resource utilization and performance in achieving economic stability and growth [14, 15]. In the biopharmaceutical industry, it evaluates company efficiency in research and development, production, and market performance [16]. DEA is also applied in sports analytics, such as ranking English Premier League (EPL) strikers based on inputs like attempts and minutes played against outputs such as goals scored and assists, providing an objective performance evaluation [17–19]. Additionally, it has been used in the Indian cement industry to analyze production efficiency, focusing on factors like energy usage, production capacity, and output, helping identify operational inefficiencies and areas for improvement [20]. These examples highlight DEA's versatility and effectiveness in benchmarking and optimizing performance across diverse sectors.

DEA has been also extensively applied in healthcare systems worldwide to evaluate efficiency, particularly in hospital settings. Numerous studies in developed countries like the USA and the UK have demonstrated DEA's utility in assessing hospital performance [21, 22]. Similarly, developing countries such as Brazil and South Africa have used DEA to evaluate healthcare systems' efficiency at primary and secondary levels. For instance, a cross-sectional study in Florianopolis, Brazil, assessed the performance of primary healthcare centers using DEA [23]. The methodology was also applied in India, where Mandal and Madheswaran (2010) determined the technical efficiency of district hospitals in Madhya Pradesh [24]. These studies highlight the versatility of DEA in identifying inefficiencies and providing actionable recommendations for healthcare improvement. However, most DEA applications in India have been confined to hospital efficiency, leaving a significant gap in research on public healthcare efficiency at the primary and secondary levels [25].

India's public healthcare system, particularly at the primary and secondary levels, serves as the backbone for delivering healthcare services to rural and underserved populations. However, resource constraints and inefficiencies significantly limit its performance. To effectively manage and improve the public healthcare system, objective and accurate efficiency measurement is crucial. DEA provides a robust framework for such assessments by comparing the performance of public healthcare facilities against the best-performing units [26]. Despite its global popularity, DEA applications in Indian public healthcare remain limited. Addressing this gap, the current study aims to leverage DEA methodologies to evaluate the efficiency of India's primary and secondary healthcare systems, identify inefficiencies, and recommend strategies for improvement. This research holds the potential to guide policymakers in optimizing resource allocation and improving healthcare delivery [27, 28].

3 Research Methodology

This study based on the evaluation and estimation of Efficiency in Primary and Secondary Healthcare system of India by using a non-parametric approach called DEA. It is a linear programming-based technique for estimating the relative efficiency of organizational units together with multiple inputs and outputs. The mathematical formulations of DEA are based on two orientations (input and output). In this study, we are using output orientation in case of both constant and variable returns to scale. The non-oriented and non-radial measure of efficiency SBM is also used in this study, and super efficiency SBM approach is used for the ranking of DMUs. The mathematical formulation of DEA with the assumption of CRS was given by Charnes et al. (1978). Let x_{ij} and y_{rj} denote the $i^{th}(i = 1, 2, 3, \ldots, m)$ input and r^{th} $(r = 1, 2, 3, \ldots, s)$ output of $j^{th} - DMU(j = 1, 2, 3, \ldots n)$ respectively. The efficiency of kth DMU is denoted by

θ_K^* and standard envelopment form of input-orientation CCR-DEA model as;

$$\begin{aligned}
&\theta_k^* = Minimise\theta_k \\
&Subject\ to \\
&\sum_{j=1}^{n} y_{rj}\lambda_j - s_r^+ = y_{rk};\ \forall r = 1, 2, 3, \ldots, s \\
&\sum_{j=1}^{n} x_{ij}\lambda_j + s_i^- = \theta_k x_{ik};\ \forall i = 1, 2, 3, \ldots, m \\
&s_r^+ \geq 0, s_i^- \geq 0, and\ \lambda_j \geq 0;\ \forall j = 1, 2, 3, \ldots, n.
\end{aligned} \tag{1}$$

In order to get valid returns on the scale, Banker, Charnes and Cooper (1984) have extended DEA to the case of VRS by adding one more constraint known as convexity constraint in CCR envelopment model (Emrouznejad, et al., 2025)

$$\begin{aligned}
&\theta_k^* = Minimise\ \theta_k \\
&Subject\ \ to \\
&\sum_{j=1}^{n} y_{rj}\lambda_j - s_r^+ = y_{rk};\ \forall r = 1, 2, 3, ..., s \\
&\sum_{j=1}^{n} x_{ij}\lambda_j + s_i^+ = \theta_k x_{ik};\ \forall i = 1, 2, 3, ..., m. \\
&\sum_{j=1}^{n} \lambda_j = 1;\ \forall j = 1, 2, 3, ..., n \\
&s_r^+ \geq 0, s_i^- \geq 0\, and\, \lambda_j \geq 0.
\end{aligned} \tag{2}$$

where x_{ik} and y_{rk} are the inputs and outputs of kth DMU (which is under evaluation).

The input excesses and output shortfalls are denoted by s_i^-; $i = 1, 2, 3, \ldots m$ and s_r^+; $r = 1, 2, 3, \ldots s$ are also known as input and output slacks. The DMU which is under evaluation is said to be efficient if and only if $\theta_k^* = 1$ and all the slacks must be zero i.e. $s_r^+ = 0$, $s_i^- = 0$. If $\theta_k^* = 1$, but one the slack is non-zero. Then DMU under evaluation is said to be week efficient otherwise if $\theta_k^* < 1$, then the DMU which is under evaluation is said to be inefficient.

CCR efficiency is the combination of purely technical and scale efficiencies and whereas BCC efficiency is only pure technical efficiency excluding scale efficiency. Thus, the radial difference between the CCR frontier and BCC frontier is called as scale efficiency. Thus, we have an equation for scale efficiency of $k^{th}\ DMU$ as;

$$Scale\ Efficiency\ (SE) = \frac{CCR\ Efficiency\ (\theta_{k-CCR}^*)}{BCC\ Efficiency\ (\theta_{k-BCC}^*)} \tag{3}$$

Thus any DMU with both CCR and BCC efficiency can be scale efficient and are called as most productive scale size (MPSS).

3.1 Slack-Based measure (SBM) of efficiency in DEA

A non-oriented and non-radial measure of efficiency, which deals with input excesses and output shortfall of DMU is called the slack-based measure of efficiency was given by Tone (2001) [5]. The mathematical formulation is defined as x_{ij} and y_{rj} denotes the $i^{th}(i = 1, 2, 3, \ldots, m)$ input and r^{th} $(r = 1, 2, 3, \ldots, s)$ output of $j^{th} - DMU(j = 1, 2, 3, \ldots n)$ respectively. It is assumed that the data set is known and strictly positive. The production possibility set P of DMU_k is defined as:

$$P = \left\{(x_{ik}\,, y_{rk})/\, x_{ik} \geq \lambda_j x_{ij}\,;\; y_{rk} \leq \lambda_j y_{rj}\,,\; \forall\, \lambda_j \geq 0;\; j = 1, 2, 3, ..., n\right\} \forall\, i\,, r \quad (4)$$

Let ρ be the inefficiency rate of DMU_0 assessing the m-inputs and s-outputs is defined:

$$\rho_k = \left[\frac{1}{m}\sum_{i=1}^{m}(\frac{x_{ik} - s_i^-}{x_{ik}})\right] * \left[\frac{1}{s}\sum_{r=1}^{s}(\frac{y_{rk} - s_r^+}{y_{rk}})\right]^{-1}$$

The interpretation of non-oriented and non-radial DEA technique SBM is minimizing the above inefficiency rate directly on the base of slacks, subject to production possibility set P and standard mathematical form is given below:

$$Min\ \rho_k^* = \frac{1 - \frac{1}{m}\sum_{i=1}^{m}\frac{s_i^-}{x_{ik}}}{1 + \frac{1}{s}\sum_{r=1}^{s}\frac{s_r^+}{y_{rk}}}$$

Subject to

$$\sum_{j=1}^{n} \lambda_j x_{ij} + s_i^- = x_{ik}\ ;\ i = 1\,,\,2\,,\,3\,,\,\ldots,\,m. \quad (5)$$

$$\sum_{j=1}^{n} \lambda_j y_{rj} - s_r^+ = y_{rk}\ ;\ r = 1\,,\,2\,,\,3\,,\,\ldots.,\,s.$$

$$\lambda_j \geq 0;\ s_i^- \geq 0;\ s_r^+ \geq 0;\ j = 1\,,2\,,3\,,\ldots n.$$

where x_{ik} *and* y_{rk} are the inputs and outputs of the DMU_k under evaluation. $s_i^- \geq 0;\ i = 1\,,\,2\,,\,3\,,\,\ldots\,,\,m$ *and* $s_r^+ \geq 0;\ r = 1\,,\,2\,,\,3\,,\,\ldots,s$ are the input excesses and output shortfalls, also referred to as slacks. The converted linear form of SMB-model by Charnes-Cooper transformation is as given below;

$$Min\ \tau_k^* = t - \frac{1}{m}\sum_{i=1}^{m}\frac{S_i^-}{x_{ik}}$$

$$Subject\ to\quad t + \frac{1}{s}\sum_{r=1}^{s}\frac{S_r^+}{y_{rk}} = 1$$

$$\sum_{j=1}^{n}\wedge_j x_{ij} + S_i^- = tx_{ik}\ ;\ i = 1, 2, 3, \ldots, m. \tag{6}$$

$$\sum_{j=1}^{n}\wedge_j y_{rj} - S_r^+ = ty_{rk}\ ;\ r = 1, 2, 3, \ldots, s.$$

$$S_i^- \geq 0;\ S_r^+ \geq 0;\ t \geq 0\, and\ \wedge_j \geq 0;\ j\, 1, 2, 3, \ldots, n.$$

Let an optimal solution of the mathematical model (6) be $(\tau_k^*, t^*, \wedge^*, S^{-*}, S^{+*})$. Then we have an optimal solution of SBM- model is defined as:

$$\rho_k^* = \tau_k^*;\ \lambda^* = \wedge^*/t^*\ ;\ s^{-*} = S^{-*}/t^*\ and\ s^{+*} = S^{+*}/t^* \tag{7}$$

On the base of an optimal solution (iii), we decide whether DMU_k which is under evaluation is efficient or inefficient depends on the definition· In the case of SBM modal of VRS, can be expressed by adding the convexity constraint into the linear programming problem (6).

3.2 The Slack-Based Measure of Super Efficiency

Let us assume k^{th} DMU is SBM-efficient, i.e. $\rho_k^* = 1$ are using $x_i;\ (i = 1, 2, 3, \ldots, m)$ and $y_r;\ (1, 2, 3, \ldots, s)$ respectively. Then the mathematical formulation of slack-based super-efficiency of (x_k, y_k) as the optimal objective function value ρ_k^* of the following model:

$$Min\ \tau_k^* = t - \frac{1}{m}\sum_{i=1}^{m}\frac{S_i^-}{x_{ik}}$$

$$Subject\ to\quad t + \frac{1}{s}\sum_{r=1}^{s}\frac{S_r^+}{y_{rk}} = 1$$

$$\sum_{j=1, J\neq k}^{n}\wedge_j x_{ij} + S_i^- = tx_{ik}\ ;\ i = 1, 2, 3, \ldots, m. \tag{8}$$

$$\sum_{j=1, J\neq k}^{n}\wedge_j y_{rj} - S_r^+ = ty_{rk}\ ;\ r = 1, 2, 3, \ldots, s.$$

$$\sum_{j=1, j\neq k}^{n}\wedge_j = t;\ \forall t \geq 0;\ j = 1, 2, 3, \ldots, n.$$

$$S_i^- \geq 0;\ S_r^+ \geq 0\ and\ \wedge_j \geq 0;\ j = 1, 2, 3, \ldots, n.$$

The super-efficiency in powerful approach for rank the DEA efficient DMUs obtained SMB, it without solving the SBM model; one can rank efficient DMUs by solving the model (3.3.1) for each efficient DMU. Fore efficient DMUs have super-efficiency score greater than or equal to unity, while inefficient DMUs have super-efficiency score less than unity.

4 Data Collection and Analysis

In this section, the DEA methodology is applied to assess the efficiency of 36 entities, including 28 states and 8 union territories of India, for the financial year 2021–22. The data used for this analysis has been collected from various sources, including the Rural Health Mission, annual reports, and other national health-related publications. These data sets were meticulously compiled and organized using Excel spreadsheets for initial processing. For the actual DEA analysis, specialized software such as DEA Frontier, PIM-DEA, and Stata were employed to conduct the efficiency assessments. This multi-software approach ensures a robust analysis by applying various DEA models to evaluate and compare the efficiency of each entity across different healthcare parameters, including inputs such as medical personnel, infrastructure, and financial resources, and outputs like healthcare outcomes and service delivery performance.

The components of healthcare infrastructure are critical elements that support the functioning and effectiveness of a healthcare system. These components include both physical and non-physical aspects that contribute to providing quality healthcare services. Some main components are as given:

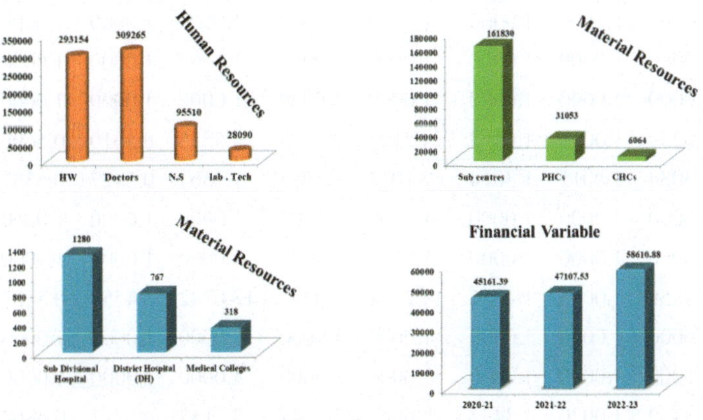

5 Results and Discussion

This section analyses the public healthcare efficiency results of 36 entities, including 28 states and 8 union territories of India, for the financial year 2021–22 obtained of by using the DEA methodology. We use both radial and non-radial measures of efficiency to obtain pure technical, technical, scale and super efficiency scores of 36 entities for the year 2021–22 as listed in Table 1. The slack-based super-efficiency model is applied for rank the DMUs.

Table 1. Efficiency Scores of 36 States and Union Territories of India for the Financial Year 2021–22.

DMUs	Envelopment Model			Non-Radial Model			Slack Based Model		
	CRS	VRS	SE	CRS	VRS	SE	CRS	VRS	SE
ST1	1.1420	1.0000	1.1420	4.1504	1.0000	4.1504	0.6377	1.0000	0.6377
ST2	1.0000	1.0000	1.0000	1.0000	1.0000	1.0000	1.0000	1.0000	1.0000
ST3	1.5403	1.0018	1.5375	5.2551	1.1728	4.4808	0.5268	0.6650	0.7922
ST4	3.9939	1.0000	3.9939	9.2742	1.0000	9.2742	0.3702	1.0000	0.3702
ST5	1.5485	1.0000	1.5485	6.1286	1.0000	6.1286	0.6251	1.0000	0.6251
ST6	1.0000	1.0000	1.0000	1.0000	1.0000	1.0000	1.0000	1.0000	1.0000
ST7	1.0000	1.0000	1.0000	1.0000	1.0000	1.0000	1.0000	1.0000	1.0000
ST8	1.2944	1.0010	1.2932	3.2651	1.0836	3.0132	0.5383	0.7108	0.7572
ST9	1.8038	1.0022	1.7999	3.8412	1.4222	2.7009	0.3709	0.3725	0.9958
ST10	1.0000	1.0000	1.0000	1.0000	1.0000	1.0000	1.0000	1.0000	1.0000
ST11	1.0418	1.0016	1.0401	3.8282	1.0692	3.5804	0.6337	0.6760	0.9374
ST12	2.3984	1.0033	2.3904	8.1104	1.4869	5.4546	0.3805	0.4221	0.9015
ST13	1.0895	1.0000	1.0895	3.1944	1.0000	3.1944	0.6704	1.0000	0.6704
ST14	1.0000	1.0000	1.0000	1.0000	1.0000	1.0000	1.0000	1.0000	1.0000
ST15	1.0017	1.0017	1.0000	1.5138	1.5138	1.0000	0.5226	0.5226	1.0000
ST16	1.0000	1.0000	1.0000	1.0000	1.0000	1.0000	1.0000	1.0000	1.0000
ST17	1.0000	1.0000	1.0000	1.0000	1.0000	1.0000	1.0000	1.0000	1.0000
ST18	1.0000	1.0000	1.0000	1.0000	1.0000	1.0000	1.0000	1.0000	1.0000
ST19	1.6359	1.0003	1.6353	5.2107	1.1443	4.5535	0.5610	0.5696	0.9849
ST20	1.3034	1.0015	1.3014	2.1012	1.2683	1.6567	0.5427	0.5381	1.0084
ST21	1.0000	1.0000	1.0000	1.0000	1.0000	1.0000	1.0000	1.0000	1.0000
ST22	1.0000	1.0000	1.0000	1.0000	1.0000	1.0000	1.0000	1.0000	1.0000
ST23	1.4958	1.0005	1.4951	14.2874	1.0962	13.0342	0.4559	0.5167	0.8823
ST24	1.0000	1.0000	1.0000	1.0000	1.0000	1.0000	1.0000	1.0000	1.0000
ST25	1.0000	1.0000	1.0000	1.0000	1.0000	1.0000	1.0000	1.0000	1.0000
ST26	1.3459	1.0020	1.3432	2.6663	1.2047	2.2132	0.4717	0.4862	0.9702
ST27	1.0000	1.0000	1.0000	1.0000	1.0000	1.0000	1.0000	1.0000	1.0000
ST28	1.0197	1.0000	1.0197	2.2700	1.0000	2.2700	0.5052	1.0000	0.5052
UT1	1.0000	1.0000	1.0000	1.0000	1.0000	1.0000	1.0000	1.0000	1.0000
UT2	1.0000	1.0000	1.0000	1.0000	1.0000	1.0000	1.0000	1.0000	1.0000

(continued)

Table 1. *(continued)*

DMUs	Envelopment Model			Non-Radial Model			Slack Based Model		
	CRS	VRS	SE	CRS	VRS	SE	CRS	VRS	SE
UT3	1.0000	1.0000	1.0000	1.0000	1.0000	1.0000	1.0000	1.0000	1.0000
UT4	1.0000	1.0000	1.0000	1.0000	1.0000	1.0000	1.0000	1.0000	1.0000
UT5	1.0021	1.0021	1.0000	1.1844	1.1844	1.0000	0.5004	0.5004	1.0000
UT6	1.0000	1.0000	1.0000	1.0000	1.0000	1.0000	1.0000	1.0000	1.0000
UT7	1.0000	1.0000	1.0000	1.0000	1.0000	1.0000	1.0000	1.0000	1.0000
UT8	1.0000	1.0000	1.0000	1.0000	1.0000	1.0000	1.0000	1.0000	1.0000

The results presented in the table show the efficiency scores of 36 entities, consisting of 28 states and 8 union territories (UTs) of India, evaluated using the DEA methodology for the financial year 2021–22. These efficiency scores were derived using three different models: the Envelopment Model (CRS and VRS), the Non-Radial Model, and the Slack-Based Model (SBM). Each model provides insights into the efficiency of the decision-making units (DMUs), considering varying assumptions about scale efficiency and input-output orientations. From the results, several key observations can be made. ST2, ST6, ST7, ST10, ST14, ST16, ST17, ST18, ST21, ST22, UT1, UT2, UT3, UT4, UT6, UT7, and UT8 have an efficiency score of 1.0000 across all models, indicating that these entities are efficient and operate at their optimal levels of resource utilization. These states and union territories, therefore, represent the benchmark in terms of efficiency. Conversely, entities such as ST1, ST3, ST4, ST5, ST8, ST9, ST12, ST19, ST23, ST26, and ST28 show inefficiencies in one or more models. For example, ST1 displays efficiency scores of 1.1420 under the CRS model and 4.1504 under the VRS model, indicating inefficiencies in both technical and scale efficiency. Similarly, ST4 has an efficiency score of 3.9939 under the CRS model, signaling a significant gap in its performance relative to its peers. The Non-Radial Model and SBM reveal that these entities need further improvements in terms of input utilization and output maximization to reach efficiency benchmarks. The introduction of the Super-Efficiency Model and SBM allows for a deeper analysis of the inefficiencies within these states and union territories. The SBM model is useful in identifying the specific excesses in inputs and shortfalls in outputs, providing a clearer understanding of the areas where improvements are necessary. ST12 and ST23, with scores of 0.3805 and 0.4559 under SBM, reflect considerable inefficiency, particularly in managing their resources.

6 Conclusions

References

1. Charnes, A., Cooper, W.W., Rhodes, E.: Measuring the efficiency of decision-making units. Eur. J. Oper. Res. **2**(6), 429–444 (1978)

2. Banker, R.D., Charnes, A., Cooper, W.W.: Some models for estimating technical and scale inefficiencies in data envelopment analysis. Manage. Sci. **30**(9), 1078–1092 (1984)
3. Mergoni, A., Emrouznejad, A., De Witte, K.: Fifty years of data envelopment analysis. Eur. J. Oper. Res. **326**(3), 389–412 (2025). https://doi.org/10.1016/j.ejor.2024.12.049
4. Farooq, D.Q., Padi, T.R., Tali, A.M.: Mixed input and output orientations of data envelopment analysis with linear fractional programming and least distance measures. Statist., Optim. Inf. Comput. **4**(4), 326–341 (2016)
5. Tone, K.: A slacks-based measure of efficiency in data envelopment analysis. Eur. J. Oper. Res. **130**(3), 498–509 (2001)
6. Andersen, P., Petersen, N.C.: A procedure for ranking efficient units in data envelopment analysis. Manage. Sci. **39**(10), 1261–1264 (1993)
7. Tone, K.: A slacks-based measure of super-efficiency in data envelopment analysis. Eur. J. Oper. Res. **143**(1), 32–41 (2002)
8. Dar, Q.F., Pad, T.R., Tali, A.M., Hamid, Y., Danish, F.: Data envelopment analysis with sensitive analysis and super-efficiency in Indian banking sector. Int. J. Data Envelopment Anal. **5**(2), 1193–1206 (2017)
9. Färe, R., Grosskopf, S., Norris, M., Zhang, Z.: Productivity growth, technical progress, and efficiency change in industrialized countries. Am. Econ. Rev. **84**(1), 66–83 (1994)
10. Emrouznejad, A., Yang, G.L.: A survey and analysis of the first 40 years of scholarly literature in DEA: 1978–2016. Socioecon. Plann. Sci. **61**, 4–8 (2018)
11. Lertworasirikul, S., Fang, S.C., Joines, J.A., Nuttle, H.L.: Fuzzy data envelopment analysis (DEA): a possibility approach. Fuzzy Sets Syst. **139**(2), 379–394 (2003)
12. Dar, Q.F., Hyo, A.Y., Dar, G.F., Bhat, S.A., Tali, A.M., Bhat, Y.H.: Fuzzy data envelopment analysis with SBM using?-level fuzzy approach. Pak. J. Stat. Oper. Res., 481–501 (2019)
13. Dar, Q.F., Ahn, Y.H., Dar, G.F., Ma, J.H.: Data envelopment analysis in three-stage network structure with slack based measure (SBM). Stat., Optim. Inf. Comput. **10**(2), 643–657 (2022)
14. Dar, Q.F., Young-Hyo, A., Dar, G.F.: Evaluation and investigation: the determinants of central banking efficiency. RAIRO-Oper. Res. **55**(2), 481–493 (2021)
15. Dar, Q.F., Ahn, Y.H., Dar, G.F.: The impact of international trade on central bank efficiency: an application of DEA and Tobit regression analysis. Stat., Optim. Inf. Comput. **9**(1), 223–240 (2021)
16. Hyo, A.Y., Farooq, D.Q., Ma, J.H.: Efficiency Evaluation of Korean Biopharmaceutical Companies: an application data envelopment analysis with negative and missing data. J. Prod. Res. **41**(3), 29–37 (2023)
17. Dar, Q.F., Zahoor, U.H.: Assessing efficiency and ranking of English premier league strikers using data envelopment analysis. 상품학연구**40**(6), 67–74 (2022)
18. Bhat, N.A., Dar, Q.F.: Ranking of football players by DEA-super efficiency model: an evidence from English premier league 2016/17 Season. Turk. Online J. Qual. Inq. **12**(4) (2021)
19. Sultana, D., Dar, QF.: A comprehensive review of data envelopment analysis (DEA) in sports. J. Sports Econom. Manag. **9**, 82–109 (2019)
20. Muhammad, A., Rao, T., Farooq, Q.: DEA window analysis with slack-based measure of Efficiency in Indian cement industry. Stat., Optim. Inf. Comput. **6**(2), 292–302 (2018)
21. Ozcan, Y.A.: Health Care Benchmarking and Performance Evaluation. Springer (2014)
22. Jacobs, R., Smith, P.C., Street, A.: Measuring Efficiency in Health Care: Analytic Techniques and Health Policy. Cambridge University Press (2006)
23. Marques, R.C., Wanke, P.: Assessing efficiency in Brazilian primary health care: evidence from a two-stage DEA analysis. Health Care Manag. Sci. **20**(4), 448–458 (2017)
24. Mandal, S., Madheswaran, S.: Benchmarking the performance of district hospitals in Madhya Pradesh, India. IUP J. Publ. Finan. **8**(3–4), 33–51 (2010)

25. World Health Organization. World Health Report 2000: Health Systems: Improving Performance. WHO(2000)
26. Pandey, A., Chaudhary, P.: Efficiency and productivity analysis of the Indian public health sector. Int. J. Health Econ. Policy **3**(1), 10–20 (2018)
27. Ghosh, S.: Assessing public healthcare performance using DEA: evidence from India. J. Health Manag. **23**(1), 5–16 (2021)
28. Ministry of Health and Family Welfare (MoHFW), Government of India. Annual Report on Health Sector (2022)
29. Emrouznejad, A., Brzezickib, Ł., Lu, C.: The development and evolution of slacks-based measure models in data envelopment analysis: a comprehensive review of the literature. J. Econ. Surv., 1–20 (2025). https://doi.org/10.1111/joes.12682
30. Emrouznejad, A., Podinovski, V., Lu, C., Charles, V., Moradi-Motlagh, A.: Rajiv banker's lasting impact on data envelopment analysis. Ann. Oper. Res. **351**, 1225–1264 (2025). https://doi.org/10.1007/s10479-025-06473-3

Strategies to Enhance the Mathematical Proficiency of Underperforming High Schools Using Data Envelopment Analysis

Niloofar Fallah and Mohsen Rostamy-Malkhalifeh(✉)

Department of Mathematics, Science and Research Branch, Islamic Azad University, Tehran, Iran
niloochl1@yahoo.com

Abstract. Education has long been a foundational element of societal advancement, with primary and secondary education playing a pivotal role in shaping cognitive and social outcomes. As educational systems face increasing complexity—marked by data abundance, external pressures, and resource constraints—evaluating institutional performance requires scientifically grounded, data-driven methodologies.

This study applies a two-stage network data envelopment analysis (NDEA) model to assess the efficiency of 45 high schools in Portugal. The model decomposes school operations into two interconnected stages: resource allocation and educational output generation, with a particular emphasis on mathematical proficiency. The analysis integrates the Anderson–Peterson (AP) super-efficiency method to provide a comprehensive ranking of schools, even among those deemed technically efficient under standard DEA. Findings indicate that only two schools achieved full efficiency across both stages, serving as performance benchmarks. The remaining schools displayed varying degrees of inefficiency, particularly in translating internal processes into strong mathematics outcomes. These insights highlight the importance of targeted interventions at both resource and process levels.

The study concludes with practical recommendations for policymakers, including data-informed resource planning, investment in math-specific programs, and institutional benchmarking. The proposed model offers a robust framework for improving educational performance and guiding strategic initiatives in underperforming schools.

Keywords: Network data envelopment analysis · Performance evaluation · Schools in Portugal

1 Introduction

In the 21st century, education has assumed a central role in preparing individuals to navigate an increasingly interconnected and rapidly evolving global environment. Technological innovation, globalization, and digitalization have transformed not only how societies

A. Emrouznejad et al. (Eds.): DEA 2024, LNOR, pp. 378–396, 2025.
https://doi.org/10.1007/978-3-031-98177-7_26

function but also how educational institutions operate and are evaluated. Amidst these transformations, education systems are required to adapt swiftly to shifting demands, expanding access, and continuously improving outcomes. Particularly, secondary education plays a pivotal role in developing essential cognitive and analytical skills—most notably, mathematical proficiency—which are critical to success in higher education, employment, and informed citizenship [1–3].

Despite its societal importance, evaluating the performance of schools remains a complex and often contested endeavor. Traditional models of assessment, which tend to emphasize financial inputs or output-based metrics such as graduation rates, often neglect the internal mechanisms and contextual factors that influence educational effectiveness [4, 5]. This is particularly true in non-profit public educational institutions, where the core mission is not revenue generation, but knowledge production and student development. As a result, a growing body of research calls for more multidimensional and process-oriented frameworks for assessing school performance [6, 7].

One promising approach to address this complexity is Data Envelopment Analysis (DEA)—a non-parametric method that enables the comparative assessment of multiple decision-making units (DMUs), such as schools, by analyzing their ability to transform inputs into outputs [8]. DEA allows for the inclusion of multiple heterogeneous variables, making it suitable for educational settings. However, conventional DEA models typically treat each school as a "black box," without examining the internal stages through which resources are processed into academic outcomes. This limits the explanatory power of the analysis and obscures potential areas for pedagogical improvement [9].

To overcome these limitations, scholars have increasingly turned to network DEA (NDEA) and two-stage DEA models, which allow for the decomposition of the overall production process into interconnected sub-stages [10]. In the context of education, this means evaluating how schools use resources not just to produce outcomes, but also to improve internal processes like instructional quality, student engagement, and organizational climate [11, 12]. These intermediate factors often serve as critical precursors to success in final academic metrics, especially in domains such as mathematics.

Building upon these methodological advancements, the current study proposes an integrated two-stage NDEA model combined with the Anderson–Peterson (AP) super-efficiency framework [13]. This hybrid model enables both the identification of relatively efficient and inefficient schools, and the ranking of units that lie on the efficiency frontier. By applying this model to 45 public secondary schools in Portugal, this study aims to assess the efficiency of educational resource utilization and the effectiveness of instructional and organizational processes—with a specific focus on mathematical proficiency.

The analysis considers both quantitative indicators—such as the number of teachers, classroom availability, and student math performance—and qualitative variables—such as availability of enrichment programs and instructional strategies. Using GAMS software and the CPLEX solver, the model processes these variables to compute efficiency scores for each stage, as well as overall super-efficiency rankings [14]. By integrating intermediate variables, the analysis provides a more granular understanding of where inefficiencies occur, whether in resource inputs, process management, or learning outcomes.

This research contributes to the educational performance literature by offering a robust, multi-stage framework for school evaluation, one that accounts for both operational and pedagogical dimensions. The findings not only identify high-performing schools that may serve as benchmarks, but also offer actionable insights for underperforming schools [15–17]. Based on the analysis, the study develops strategic recommendations to enhance mathematical achievement through better resource planning, improved internal instructional practices, and the replication of best practices from efficient peer institutions.

The remainder of this paper is structured as follows: Section 2 presents a review of the relevant literature on performance evaluation in education and the application of DEA models. Section 3 describes the methodological framework and modeling approach. Section 4 provides an overview of the dataset and presents the empirical results. Section 5 discusses the findings and their policy implications. Finally, Sect. 6 concludes with a summary and offers directions for future research.

2 Literature review

Over the past two decades, Data Envelopment Analysis (DEA) has emerged as a prominent non-parametric method for assessing efficiency across various domains, including education. Researchers have utilized DEA to uncover inefficiencies in educational institutions and to provide insights into optimal resource allocation and performance enhancement strategies [1].

A recent comprehensive review by Kikuchi [2] synthesizes empirical applications of DEA in the education sector, illustrating its versatility in identifying performance gaps and benchmarking best practices across educational settings. These studies underscore DEA's ability to handle multiple inputs and outputs, making it suitable for analyzing complex systems like schools.

In support of a multi-method evaluation framework, Hale [3] compares quantitative and qualitative approaches to educational performance assessment. The study advocates for mixed-method designs to capture the multifaceted nature of school effectiveness, arguing that DEA can serve as a valuable component within such frameworks.

Expanding the scope of DEA applications, Lin [4] investigates the link between resource allocation and performance outcomes in non-profit educational settings. Utilizing DEA, the author demonstrates that efficient use of resources significantly correlates with improved educational results, particularly in underfunded environments.

Henriques [5] introduces a value-oriented DEA model that incorporates non-discretionary variables to assess the performance of 159 secondary schools in Ecuador. The study leverages data from the International Student Assessment Program for Development and emphasizes the influence of infrastructural quality, teacher-student ratios, and socio-cultural factors on academic outcomes. The findings suggest that schools performing below the regional mean in Latin America should redirect resources toward mathematics instruction as a strategic priority.

The effectiveness of educational policy reforms is further examined in a longitudinal study by Kuntz [6], focusing on 643 secondary schools in Greece over the period 2000–2017. Using a multilevel DEA framework, the study reveals substantial efficiency gaps among schools, particularly within specific NUTS2 regions. While reforms were intended to improve school performance, their actual impact appeared minimal. Factors such as the presence of science laboratories, class size, and regional GDP per capita emerged as significant determinants of school efficiency.

Despite the study's emphasis on secondary education, certain references such as Sang [7], which analyzes research productivity among university graduates, appear misaligned with the thematic focus. Although Sang's study is methodologically relevant, its scope pertains more directly to higher education and thus offers limited applicability to secondary school efficiency assessments.

Furthermore, conceptual models such as the "education investment model" and "research-driven model" are introduced in several studies without sufficient contextual clarity or proper citation. The absence of foundational references for these concepts may hinder coherence and limit their interpretability within the literature review. Moreover, terms such as productivity, efficiency, and effectiveness are often used interchangeably, despite representing distinct dimensions of institutional performance. Clarifying these distinctions is essential for analytic precision and methodological rigor.

In summary, the reviewed literature demonstrates the adaptability and relevance of DEA models to various educational contexts, particularly in evaluating multi-input, multi-output systems. Nonetheless, the review underscores the importance of consistent terminology, conceptual clarity, and thematic alignment to enhance the robustness, coherence, and practical value of DEA-based evaluations in secondary education.

3 Methodology

The current study adopts a two-stage network Data Envelopment Analysis (DEA) framework to evaluate the performance efficiency of decision-making units (DMUs), namely high schools. The model structure builds upon the foundational works of Cao and Huang (2009), Zha and Liang (2010), and the integrated DEA structure proposed by Izadikhah (2018). The goal is to conduct an in-depth decomposition of overall efficiency by modeling each school as a network of sub-processes, consisting of a Leader Stage and a Follower Stage.

Stage One: Preliminary Evaluation (Leader Model).

In the first phase, schools (DMUs) are evaluated based on preliminary or less influential indicators, such as infrastructure or resource availability. These indicators influence internal processes and indirectly affect educational outcomes. The mathematical structure employs an input-oriented CCR model: Based on this model:

- x_{ij}: Quantity of input i used by school j
- y_{rj}: Output r produced by school j
- z_{dj}: Intermediate measure (e.g., classroom activity, lab use)
- v_i, u_r, w_d: Weights assigned to inputs, outputs, and intermediates
- θ_p: Efficiency score for school p

Equation (1) aims to maximize intermediate outputs while ensuring efficiency constraints based on inputs and outputs across all DMUs.

$$\theta_p = max \sum_{d=1}^{D} w_d z_{dp} \tag{1}$$

s.t

$$\sum_{i=1}^{m} v_i x_{ip} = 1,$$

$$\sum_{r=1}^{s} u_r y_{rj} - \sum_{i=1}^{m} v_i x_{ij} \leq 0, \forall j = 1, \ldots, n;$$

$$\sum_{d=1}^{D} w_d z_{dj} - \sum_{i=1}^{m} v_i x_{ij} \leq 0, \forall j = 1, \ldots, n;$$

$$\sum_{r=1}^{s} u_r y_{rj} - \sum_{d=1}^{D} w_d z_{dj} \leq 0 \forall j = 1, \ldots, n;$$

$$u_r, v_i, w_d \geq \varepsilon, \text{ r} = 1, \ldots, \text{ s}; \ i = 1, \ldots, \text{ m}; \text{ d} = 1, \ldots, \text{ D}.$$

Consequently, after assessing the primary phase of DMU, the second stage involves the application of the Za and Liang technique to appraise the ultimate screening. This will encompass the utilization of an assessment model as presented below:

Stage Two: Final Screening (Follower Model)

Following initial screening, schools are evaluated using outcome-based performance indicators such as student achievement, particularly in mathematics. Zha and Liang's approach introduces input allocation factors (α_i) that distribute inputs between the two stages. Equation (2) models this dual-stage interaction, maximizing the product of stage-specific efficiencies.

The terms - αi xij and xij ((1–αi)) denote input shares directed to the first and second stages, respectively.

Zha and Liang (2010) presented a technique for creating a model of the typical flow within a two-step evaluation process, in which inputs are determined in a somewhat indeterminate manner across one or two stages that are linked through a collaborative association. To depict the aforementioned processes, an allocation factor αi was attributed to each input value Xij of a decision-making unit (DMU) j.

$$\theta_P = \max \sum_{d=1}^{D} w_d z_{dp} \times \sum_{r=1}^{s} u_r y_{rp} \tag{2}$$

s. t.

$$\sum_{i=1}^{m} \pi_i x_{ip} = 1,$$

$$\delta\left[\sum_{i=1}^{m}(v_i - \pi_i)x_{ip} + \sum_{d=1}^{D} w_d z_{dp}\right] = 1,$$

$$\sum_{r=1}^{s} u_r y_{rj} - \delta\left[\sum_{i=1}^{m}(v_i - \pi_i)x_{ij} + \sum_{d=1}^{D} w_d z_{dj}\right] \le 0,$$

$$\mathrm{j} = 1, \ldots, \mathrm{n};$$

$$\sum_{d=1}^{D} w_d z_{dj} - \sum_{i=1}^{m} \pi_i x_{ij} \le 0,$$

$$\mathrm{j} = 1, \ldots, \mathrm{n};$$

$$v_i \ge \pi_i \ge 0, \forall\, \mathrm{i} = 1, \ldots, \mathrm{m}.$$

$$u_r, w_d \ge \varepsilon,\ \mathrm{r} = 1, \ldots, \mathrm{s};\ \mathrm{d} = 1, \ldots, \mathrm{D}.$$

The present study employs a method for efficiency evaluation that involves the utilization of data envelopment analysis in a dual-stage process, specifically through the dissociation of CaO and Huang as well as Zha and Liang. The assessment of leader and follower models is conducted within the framework of the two-stage models for efficiency evaluation.

in his work, Izadikhah (2018) introduced an integrated two-stage approach for The evaluation of a DMU. This is done based on the input-oriented CCR approach, which is reflected in the final model proposed by the author that addressed the evaluation of a DMU in the pharmacy industry.

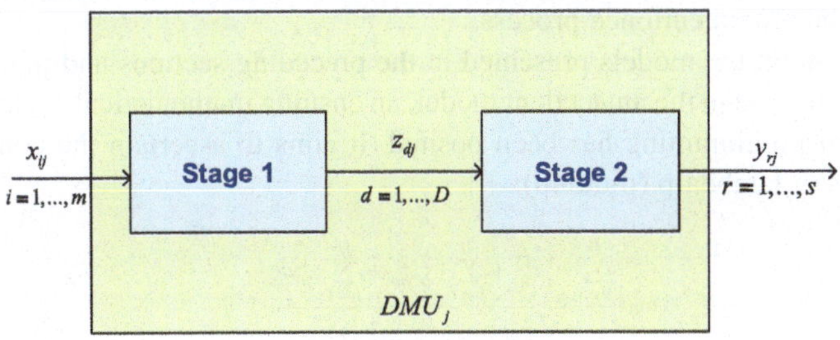

$$\theta_{2p}^{L} = \max \sum_{r=1}^{R} u_r^2 y_{rp}^2 \tag{3}$$

s. t.

$$\sum_{k=1}^{K} v_k^3 x_{kp}^3 + \sum_{t=1}^{T} v_t^2(1-\alpha_t)x_{tp}^2 + \sum_{d=1}^{D} w_d^1 z_{dp}^1 + \sum_{b=1}^{B} w_b^2 z_{bp}^2 = 1,$$

$$\sum_{r=1}^{R} u_r^2 y_{rj}^2 - \sum_{k=1}^{K} v_k^3 x_{kj}^3 - \sum_{t=1}^{T} \left(v_t^2 - \mu_t^2\right) x_{tj}^2 - \sum_{d=1}^{D} w_d^1 z_{dj}^1 - \sum_{b=1}^{B} \varphi_b^2 z_{bj}^2 \leq 0,$$

$$\forall j = 1, \ldots, n,$$

$$\sum_{d=1}^{D} w_d^1 z_{dp}^1 + \sum_{b=1}^{B} w_b^2 z_{bp}^2 + \sum_{h=1}^{H} u_h^1 y_{hp}^1 - \theta_{1P}^{U*} \left(\sum_{i=1}^{I} v_i^1 x_{ip}^1 + \sum_{t=1}^{T} \mu_t^2 x_{tp}^2 \right) = 0,$$

$$\sum_{d=1}^{D} w_d^1 z_{dj}^1 + \sum_{b=1}^{B} w_b^2 z_{bj}^2 + \sum_{h=1}^{H} u_h^1 y_{hj}^1 - \sum_{i=1}^{I} v_i^1 x_{ij}^1 - \sum_{t=1}^{T} \mu_t^2 x_{tj}^2 \leq 0,$$

$$j = 1, \ldots, n,$$

$$w_d^1, w_b^2, u_r^2, v_k^3, v_t^2, u_h^1, v_j^1, \mu_t^2, \varphi_b^2 \geq 0, d = 1, \ldots, D;$$

$$b = 1, \ldots, B; r = 1, \ldots, s; k = 1, \ldots, K;$$

$$h = 1, \ldots, H; i = 1, \ldots, I; t = 1, \ldots, T.$$

$$0 \leq \mu_t^2 \leq v_t^2, t = 1, \ldots, T,$$

$$0 \leq \varphi_b^2 \leq w_b^2, b = 1, \ldots, B.$$

The current study examines a two-stage production scheme that involves cooperative productivity, where each decision-making unit (DMU) is composed of two distinct subsets that are serially connected and possess specific characteristics. The objective is to evaluate the aforementioned process.

Drawing upon the models presented in the preceding sections and the decomposition method utilized in the antecedent model, an ensuing mathematical construct involving fractional programming has been posited. It aims to ascertain the comprehensive effectiveness of lambda-p (p-dump):

$$\theta_p = \max \frac{\sum_{d=1}^{D} w_d^1 z_{dp}^1 + \sum_{b=1}^{B} w_b^2 z_{bp}^2 + \sum_{h=1}^{H} u_h^1 y_{hp}^1}{\sum_{i=1}^{I} v_i^1 x_{ip}^1 + \sum_{t=1}^{T} v_t^2 (\alpha_t x_{tp}^2)} \times \frac{\sum_{r=1}^{R} u_r^2 y_{rp}^2}{\sum_{k=1}^{k} v_k^3 x_{kp}^3 + \sum_{t=1}^{T} v_t^2 (1-\alpha_t) x_{tp}^2 + \sum_{d=1}^{D} w_d^1 z_{dp}^1 + \sum_{b=1}^{B} w_b^2 (\beta_b z_{bp}^2)} \quad (4)$$

S.t.

$$\frac{\sum_{d=1}^{D} w_d^1 z_{dj}^1 + \sum_{b=1}^{B} w_b^2 z_{bj}^2 + \sum_{h=1}^{H} u_h^1 y_{hj}^1}{\sum_{i=1}^{I} v_i^1 x_{ij}^1 + \sum_{t=1}^{T} v_t^2 (\alpha_t x_{tj}^2)} \leq 1, \forall j = 1, \ldots, n,$$

$$\frac{\sum_{r=1}^{R} u_r^2 y_{rj}^2}{\sum_{k=1}^{k} v_k^3 x_{kj}^3 + \sum_{t=1}^{T} v_t^2 (1-\alpha_t) x_{tj}^2 + \sum_{d=1}^{D} w_d^1 z_{dj}^1 + \sum_{b=1}^{B} w_b^2 (\beta_b z_{bj}^2)} \le 1, \forall j = 1, \ldots, n,$$

$$w_d^1, w_b^2, u_r^2, v_k^3, v_t^2, v_i^1, u_h^1 \ge 0, \ \forall d = 1, \ldots, D; b = 1, B; r = 1, s;$$
$$k = 1, K; t = 1, \ T; i = 1, \ldots, I; h = 1, \ldots, H.$$

$$0 \le \alpha_t \le 1, \forall t = 1, \ldots, T,$$

$$0 \le \beta_b \le 1, \forall b = 1, \ldots, B.$$

Using Charnes-cooper deformation and settings (β_b=$w_b^2 \varphi_b^2$)($v_t^2 \alpha_t$= μ_t^2), , the model (4) can be converted into the subsequent model:

$$\theta_p = \max\left(\sum_{d=1}^{D} w_d^1 z_{dp}^1 + \sum_{b=1}^{B} w_b^2 z_{bp}^2 + \sum_{h=1}^{H} u_h^1 y_{hp}^1\right) \times \left(\sum_{r=1}^{R} u_r^2 y_{rp}^2\right) \tag{5}$$

s.t.

$$\sum_{i=1}^{I} v_i^1 x_{ip}^1 - \sum_{t=1}^{T} \mu_t^2 x_{tp}^2 = 1,$$

$$\sum_{k=1}^{k} v_k^3 x_{kp}^3 + \sum_{t=1}^{T} (v_t^2 - \mu_t^2) x_{tp}^2 + \sum_{d=1}^{D} w_d^1 z_{dp}^1 + \sum_{b=1}^{B} \varphi_b^2 z_{bp}^2 = 1,$$

$$\sum_{d=1}^{D} w_d^1 z_{dj}^1 + \sum_{b=1}^{B} w_b^2 z_{bj}^2 + \sum_{h=1}^{H} u_h^1 y_{hj}^1 - \sum_{i=1}^{I} v_i^2 x_{ij}^2 - \sum_{t=1}^{T} \mu_t^2 x_{tj}^2 \le 0, \forall j = 1, \ldots, n,$$

$$\sum_{r=1}^{R} u_r^2 y_{rj}^2 - \sum_{k=1}^{k} v_k^3 x_{kj}^3 - \sum_{t=1}^{T} (v_t^2 - \mu_t^2) x_{tj}^2 - \sum_{d=1}^{D} w_d^1 z_{dj}^1 - \sum_{b=1}^{B} \varphi_b^2 z_{bj}^2 \le 0, \forall j = 1, \ldots, n,$$

$$w_d^1, \ w_b^2, \ u_r^2, \ v_k^3, \ v_t^2, \ v_i^1, \ u_h^1 \ge 0, \quad d = 1, D; b = 1, B; r = 1, \ s;$$
$$k = 1, \ldots, K; t = 1, \ldots, T; i = 1, I; h = 1, H.$$

$$0 \le \varphi_b^2 \le w_b^2, b = 1, \ldots, B.$$

$$0 \le \mu_t^2 \le v_t^2, t = 1, \ldots, T,$$

Model (5) represents a nonlinear programming problem, for which the following can be ascertained:

$$\theta_{1p}^L \le \sum_{d=1}^{D} W_d^1 Z_{dp}^1 + \sum_{b=1}^{B} w_b^2 z_{bp}^2 + \sum_{h=1}^{H} u_h^1 y_{hp}^1 \le \theta_{1p}^U and \tag{6}$$

$$\theta_{2p}^L \le \sum_{r=1}^{R} u_r^2 y_{rp}^2 \le \theta_{2p}^U.$$

Given the adjustments made to the aforementioned model by Izadikhah (2018), the assessment of DMUs via a two-step methodology is presented as follows.

$$\theta_p = \max \sum_{r=1}^{R} \tilde{u}_r^2 y_{rp}^2 \tag{7}$$

s. t.

$$\sum_{i=1}^{I} \tilde{v}_i^1 x_{ip}^1 + \sum_{t=1}^{T} \overset{\sim 2}{\mu_t} x_{tp}^2 = \lambda,$$

$$\sum_{k=1}^{K} \tilde{v}_k^3 x_{kp}^3 + \sum_{t=1}^{T} \left(\tilde{v}_t^2 - \overset{\sim 2}{\mu_t} \right) x_{tp}^2 + \sum_{d=1}^{D} \tilde{w}_d^1 Z_{dp}^1 + \sum_{b=1}^{B} \overset{\sim 2}{\varphi_b} z_{bp}^2 = \lambda,$$

$$\sum_{d=1}^{D} \tilde{w}_d^1 Z_{dp}^1 + \sum_{b=1}^{B} \tilde{w}_b^2 z_{bp}^2 + \sum_{h=1}^{H} \tilde{u}_h^1 y_{hp}^1 - \sum_{i=1}^{I} \tilde{v}_i^1 x_{ip}^1 - \sum_{t=1}^{T} \overset{\sim 2}{\mu_t} x_{tp}^2 \le 0,$$

$$j = 1, \ldots, n,$$

$$\sum_{r=1}^{R} \tilde{u}_r^2 y_{rp}^2 - \sum_{k=1}^{K} \tilde{v}_k^3 x_{kp}^3 - \sum_{t=1}^{T} \left(\tilde{v}_t^2 - \overset{\sim 2}{\mu_t} \right) x_{tp}^2 - \sum_{d=1}^{D} \tilde{w}_d^1 Z_{dp}^1 - \sum_{b=1}^{B} \overset{\sim 2}{\varphi_b} z_{bp}^2 \le 0,$$

$$\forall j = 1, \ldots, n,$$

$$\lambda = \sum_{d=1}^{D} W_d^1 Z_{dp}^1 + \sum_{b=1}^{B} w_b^2 z_{bp}^2 + \sum_{h=1}^{H} u_h^1 y_{hp}^1,$$

$$\tilde{w}_d^1, \tilde{w}_b^2, \tilde{u}_h^1, \tilde{v}_i^1, \tilde{v}_t^2, \tilde{u}_h^1 \ge 0, \qquad d = 1, \ \ldots, D; b = 1, \ldots, B; r = 1, \ \ldots s;$$
$$k = 1, \ldots, K; t = 1, \ \ldots, \ T; i = 1, \ \ldots, I; h = 1, \ldots, H$$

$$0 \le \overset{\sim 2}{\varphi_b} \le \tilde{w}_b^2, b = 1, \ldots, B.$$

$$0 \le \overset{\sim 2}{\mu_t} \le \tilde{v}_t^2, t = 1, \ldots, T.$$

$$\theta_{1p}^L \le \lambda \le \theta_{1p}^U$$

According to the models presented by Anderson and Peterson, the following information can be inferred.

$$\max \theta \tag{8}$$

S. T

$$\sum_{i=1 i \neq p}^{n} \lambda_i X_i \leq \theta \leq X_p$$

$$\sum_{j=1 j \neq p}^{n} \lambda_j Y_j \geq Y_p$$

Model Integration and Definitions
The variables are clarified as follows:

- x, y: Actual input and output indicators (e.g., number of teachers, test scores).
- v, u: Weight variables optimized through linear programming.
- z: Intermediate outputs (process-based indicators).
- μ, φ: Auxiliary variables derived from transformation of non-linear expressions.
- λ: Normalization constant across stages.

In the final model (Eq. 9), the efficiency of each DMU is computed by integrating both stages through a multiplicative formulation. This approach enables decomposition of performance into leadership (resource management) and followership (learning outcomes), thereby capturing the educational dynamics holistically.

$$\theta_p = \max \sum_{r=1}^{R} \tilde{u}_r^2 y_{rp}^2 \tag{9}$$

s. t.

$$\sum_{i=1}^{I} \tilde{v}_i^1 x_{ip}^1 + \sum_{t=1}^{T} \overset{\sim 2}{\mu}_t x_{tp}^2 = \lambda,$$

$$\sum_{k=1}^{K} \tilde{v}_k^3 x_{kp}^3 + \sum_{t=1}^{T} \left(\tilde{v}_t^2 - \tilde{\mu}_t^2\right) x_{tp}^2 + \sum_{d=1}^{D} \tilde{w}_d^1 Z_{dp}^1 + \sum_{b=1}^{B} \tilde{\varphi}_b^2 z_{bp}^2 = \lambda,$$

$$\sum_{d=1}^{D} \tilde{w}_d^1 Z_{dp}^1 + \sum_{b=1}^{B} \tilde{w}_b^2 z_{bp}^2 + \sum_{h=1}^{H} \tilde{u}_h^1 y_{hp}^1 \leq \lambda$$

$$j = 1, \ldots, n,$$

$$\sum_{d=1}^{D} \tilde{w}_d^1 Z_{dp}^1 + \sum_{b=1}^{B} \tilde{w}_b^2 z_{bp}^2 + \sum_{h=1}^{H} \tilde{u}_h^1 y_{hp}^1 - \sum_{i=1}^{I} \tilde{v}_i^1 x_{ip}^1 - \sum_{t=1}^{T} \overset{\sim 2}{\mu}_t x_{tp}^2 \leq 0,$$

$$\forall j = 1, \ldots, n,$$

$$\sum_{r=1}^{R} \tilde{u}_r^2 y_{rp}^2 - \sum_{k=1}^{K} \tilde{v}_k^3 x_{kp}^3 - \sum_{t=1}^{T} \left(\tilde{v}_t^2 - \overset{\sim 2}{\mu}_t\right) x_{tp}^2 - \sum_{d=1}^{D} \tilde{w}_d^1 Z_{dp}^1 - \sum_{b=1}^{B} \overset{\sim 2}{\varphi}_b z_{bp}^2 \leq 0,$$

$$, \forall j = 1, \ldots, n,$$

$$\lambda = \sum_{d=1}^{D} W_d^1 Z_{dp}^1 + \sum_{b=1}^{B} w_b^2 z_{bp}^2 + \sum_{h=1}^{H} u_h^1 y_{hp}^1$$

$$\tilde{w}_d^1, \tilde{w}_b^2, \tilde{u}_h^1, \tilde{v}_i^1, \tilde{v}_t^2, \tilde{u}_h^1 \geq 0, \qquad d = 1, \ \ldots, \ D; b = 1, \ldots, B; r = 1, \ \ldots s;$$
$$k = 1, \ldots, K; t = 1, \ \ldots, \ T; i = 1, \ \ldots, I; h = 1, \ldots, H$$

$$0 \leq \overset{\sim 2}{\varphi_b} \leq \tilde{w}_b^2, b = 1, \ldots, B.$$

$$0 \leq \overset{\sim 2}{\mu_t} \leq \tilde{v}_t^2, t = 1, \ldots, T.$$

$$\theta_{1p}^L \leq \lambda$$

Measurement of Quantitative and Qualitative Variables

Quantitative variables include class size, number of faculty, number of math laboratories, and mathematics test scores. Qualitative variables—such as teaching quality and student satisfaction—were converted into numerical values using a 5-point Likert scale, based on expert evaluations. These are treated within the DEA structure as ordinal inputs or outputs, normalized across schools.

4 Results

In accordance with the two-step model of input and output evaluation applied to every decision-making unit (DMU), the input and output elements assessed In each stage are as delineated below (Table 1):

Table 1. Input and Output Indicators

Row	Output indicators	Input Indicators
1	Number of graduated students	Total number of incoming students
2	Satisfaction level of students regarding intelligence education	Annual training budget
3	Satisfaction level of parents regarding intelligent education	Number of extracurricular classes
4	Intelligent education ranking	Number of school personnel
5		The allocation of training time beyond normally scheduled hours

Table 2. Collected information from 45 schools

dmu	X_1^1	X_1^2	X_1^3	X_2^1	X_2^2	X_2^3	Z_1^1	Z_1^2	Z_1^3	Y_1^1	Y_1^2
dmu1	2	8	14	14	4	35	570	70	3501	96	12
dmu2	3	5	15	15	0	30	700	95	2403	120	15
dmu3	4	8	20	20	3	40	850	128	2010	210	38
dmu4	1	4	10	10	0	20	510	800	1500	1050	14
dmu5	4	5	12	12	0	25	900	90	1709	120	10
dmu6	3	6	16	16	0	15	600	97	1200	150	18
dmu7	3	7	17	17	2	40	1200	3100	1870	4200	40
dmu8	1	8	21	21	1	15	850	65	3000	95	9
dmu9	4	6	16	16	4	30	980	110	3500	150	18
dmu10	3	5	13	13	1	20	780	78	870	110	18
dmu11	1	10	30	30	5	40	1150	700	4250	1120	45
dmu12	4	10	35	35	4	40	1000	750	4330	1200	35
dmu13	2	5	12	12	1	15	600	55	975	85	6
dmu14	2	6	18	18	0	15	620	78	760	105	17
dmu15	4	6	16	16	1	15	790	117	3760	220	29
dmu16	5	6	22	22	1	20	820	120	2760	250	19
dmu17	4	7	36	36	0	10	510	56	680	78	7
dmu18	5	5	25	25	0	10	710	95	3540	130	11
dmu19	4	8	31	31	2	15	830	220	2570	300	30
dmu20	3	7	22	22	1	20	890	90	3490	140	25
dmu21	5	4	14	14	1	25	910	180	1680	370	28
dmu22	1	3	13	13	1	30	654	97	1350	160	19
dmu23	5	4	8	8	0	15	450	45	755	68	4
dmu24	4	6	40	40	2	35	1250	150	3980	320	32
dmu25	3	7	15	15	2	35	783	520	1130	650	35
dmu26	4	5	6	6	1	15	594	88	877	171	9
dmu27	4	6	14	14	4	35	664	150	1360	210	18
dmu28	4	5	11	11	1	15	725	190	985	260	11
dmu29	2	14	62	62	7	55	1950	3000	3555	5100	64
dmu30	5	5	13	13	1	25	840	150	1890	260	25
dmu31	1	5	40	40	1	30	836	900	1655	1350	31
dmu32	3	4	15	15	1	15	550	780	2433	960	10

(continued)

Table 2. (*continued*)

dmu	$X_1{}^1$	$X_1{}^2$	$X_1{}^3$	$X_2{}^1$	$X_2{}^2$	$X_2{}^3$	$Z_1{}^1$	$Z_1{}^2$	$Z_1{}^3$	$Y_1{}^1$	$Y_1{}^2$
dmu33	2	5	11	11	2	15	840	150	2310	210	6
dmu34	4	5	17	17	3	45	452	220	2840	320	16
dmu35	5	6	16	16	1	15	865	145	3120	180	20
dmu36	5	7	45	45	3	40	1275	590	3110	880	38
dmu37	4	4	10	10	0	25	657	700	2380	850	19
dmu38	2	5	12	12	1	15	856	450	1604	620	16
dmu39	5	4	11	11	1	35	654	670	1308	960	22
dmu40	2	6	50	50	3	20	1002	225	3650	420	13
dmu41	5	6	13	13	1	35	700	270	988	350	10
dmu42	1	7	18	18	2	25	1100	2500	1240	4670	101
dmu43	2	5	14	14	2	15	678	70	876	250	11
dmu44	1	5	13	13	0	20	856	256	1238	380	17
dmu45	5	6	14	14	6	60	2450	5220	3910	7250	210

Explanation of Table 2: Variables and DEA Model Structure
Table 2 summarizes the dataset collected from 45 secondary schools and is structured to support a comprehensive two-stage Data Envelopment Analysis (DEA) model. The table categorizes each variable based on its function within the analytical framework—inputs, intermediate variables, and outputs—allowing for a nuanced evaluation of school performance with a focus on mathematical proficiency.

Input Parameters
The input variables, denoted as X11, X12, and X13, capture core educational resources allocated to each school. Specifically:

- **X11**: Number of qualified teaching staff,
- **X12**: Number of classrooms available,
- **X13**: Annual educational budget.

These variables represent the foundational capacity of each institution and serve as the initial drivers of the educational production process. It is important to note that greater resource availability does not inherently indicate higher efficiency, especially if resource utilization is suboptimal.

Intermediate Variables
The intermediate variables include **X21, X22, X23**, and **Z11, Z12, Z13**, which reflect internal processes and operational dynamics that mediate the relationship between resource inputs and educational outcomes:

- **X21**: Number of enrolled students,

Table 3. Anderson's method of analysis and evaluation of the two-stage model

DMU	TET(U)1J	TET(L)2J	TET(U)2J	TET(L)1J	E1	E2	TETA	RANK	TETA-AP
1	1	0.577	0.577	0.00343	0.69543	0.11876	0.08259	38	0.08259
2	0.79651	0.532	0.886	0.00121	0.66412	0.62448	0.41473	8	0.41473
3	0.45897	1	1	0.18477	0.40344	0.34676	0.1399	28	0.1399
4	0.87367	0.36	0.944	0.00005	0.75772	0.93942	0.71182	3	0.71182
5	0.52849	0.43	0.728	0.02138	0.51255	0.54317	0.2784	14	0.2784
6	0.3641	0.755	1	0.3641	0.3641	0.94823	0.34525	10	0.34525
7	0.78364	0.368	0.405	0.68214	0.68214	0.40123	0.27369	15	0.27369
8	0.95069	0.466	0.495	0.00019	0.7284	0.1375	0.10016	34	0.10016
9	0.93782	0.587	0.591	0.0246	0.66762	0.13955	0.09317	36	0.09317
10	0.48224	1	1	0.10837	0.48224	0.30575	0.14745	25	0.14745
11	1	0.713	0.74	0.00013	0.99985	0.44554	0.44547	6	0.44547
12	0.79918	0.603	0.604	0.00013	0.55744	0.23701	0.13212	29	0.13212
13	0.49819	0.367	0.389	0.02222	0.47863	0.11407	0.0546	44	0.0546
14	0.45629	1	1	0.43021	0.4529	0.94846	0.42956	7	0.42956
15	1	0.938	1	0.1485	0.57793	0.3803	0.21978	17	0.21978
16	0.69129	0.533	0.659	0.10123	0.54512	0.24152	0.13166	30	0.13166
17	0.23269	0.501	0.606	0.00034	0.23269	0.55404	0.12892	31	0.12892
18	1	0.39	0.917	0.00022	0.62846	0.5374	0.33774	11	0.33774
19	0.55255	0.765	1	0.12703	0.46595	0.38535	0.17955	21	0.17955
20	0.91214	0.936	1	0.03137	0.58587	0.34998	0.20504	18	0.20504
21	0.6342	0.938	1	0.14722	0.55706	0.43226	0.2408	16	0.2408
22	1	0.754	0.894	0.00062	0.89288	0.34566	0.30863	13	0.30863
23	0.33786	0.299	0.463	0.00003	0.30634	0.38342	0.11745	32	0.11745
24	1	0.718	0.785	0.00004	0.65279	0.27322	0.17836	22	0.17836
25	0.45128	0.966	0.967	0.00006	0.43649	0.4248	0.18542	19	0.18542
26	0.56564	0.683	0.683	0.02326	0.33266	0.1943	0.06464	41	0.06464
27	0.37848	0.647	0.647	0.13326	0.36976	0.17352	0.06416	42	0.06416
28	0.37657	0.481	0.545	0.06655	0.36609	0.19801	0.07249	40	0.07249
29	0.93774	0.434	0.439	0.69567	0.93774	0.33472	0.31388	12	0.31388
30	0.57498	0.923	0.978	0.00036	0.49768	0.36196	0.18014	20	0.18014
31	0.99271	0.568	0.631	0.60446	0.80399	0.50104	0.40283	9	0.40283
32	0.93409	0.25	0.258	0.28415	0.60421	0.14244	0.08606	37	0.08606
33	0.92017	0.253	0.262	0.00029	0.70432	0.08111	0.05713	43	0.05713
34	0.86436	0.678	0.678	0.13414	0.72717	0.10852	0.07891	39	0.07891
35	0.79356	0.626	0.727	0.00002	0.556	0.25282	0.14057	27	0.14057
36	0.67417	0.577	0.646	0.19399	0.55398	0.26301	0.1457	26	0.1457
37	0.90801	0.455	1	0.88793	0.88793	0.55418	0.49208	5	0.49208
38	0.73356	0.443	0.519	0.00053	0.68286	0.23949	0.16354	24	0.16354
39	0.49505	0.533	0.592	0.00007	0.43718	0.37622	0.16447	23	0.16447
40	1	0.28	0.351	0.00005	0.78634	0.13786	0.10841	33	0.10841
41	0.30764	0.387	0.398	0.00028	0.28671	0.18007	0.05163	45	0.05163
42	1	1	1	1	1	1	1	2	1.4483
43	0.54085	0.617	0.667	0.00008	0.54085	0.17891	0.09677	35	0.09677
44	0.98218	0.583	1	0.6979	0.94824	0.72702	0.68939	4	0.68939
45	1	1	1	1	1	1	1	1	2.67248

- **X22**: Availability of specialized mathematics enrichment programs (with some schools reporting zero, indicating lack of provision),
- **X23**: Student-to-teacher ratio,

- **Z11–Z13**: Metrics related to instructional practices and infrastructure utilization, such as math lab hours, ICT tool integration, and participation in remedial sessions.

These variables are crucial in understanding how schools transform raw inputs into pedagogical outputs, especially those that directly or indirectly support mathematical skill development. Their inclusion in the model ensures that internal processes are evaluated as performance contributors in their own right.

Output Parameters
The output variables—**Y11** and **Y12**—capture the ultimate educational outcomes:

- **Y11**: Average standardized mathematics test scores,
- **Y12**: Percentage of students achieving proficiency in national mathematics benchmarks.

These outputs align directly with the central theme of the study: evaluating how effectively schools enhance mathematical proficiency. DEA examines the transformation of inputs and intermediate processes into these measurable academic results.

Performance Evaluation Framework
The DEA model assesses each school's relative efficiency by comparing its ability to convert inputs into desirable outputs through the mediating influence of intermediate variables. Schools that achieve high mathematical outcomes with proportionally fewer resources are considered more efficient. Conversely, those with abundant inputs but subpar outputs are flagged for potential performance improvement.

Analytical Considerations
Particular attention is given to variable **X22**, which shows several zero values, reflecting the absence of math-specific enrichment programs in some schools. These values were retained in the dataset to authentically capture disparities in program implementation. The DEA framework employed (input-oriented BCC model) can accommodate such zero values without compromising model validity.

Overall, the structure and content of Table 2 enable a robust, process-oriented evaluation of educational efficiency. By integrating both quantitative and qualitative indicators—particularly those tied to mathematical instruction and outcomes—the model supports a policy-relevant analysis of resource utilization and performance enhancement across schools. This approach moves beyond simplistic input-output analysis and provides actionable insights for improving the mathematical proficiency of underperforming educational institutions.

Based on the two-stage data envelopment analysis model, an assessment of the efficiency of 45 schools was conducted through a two-stage evaluation process based on the data gathered. The first and second stages of this evaluation are detailed below (Table 3).

Based on an evaluation and analysis of 45 schools utilizing input and output indicators, findings indicate that two schools have exhibited efficiency according to the prescribed indicators. Conversely, it was observed that the remaining schools demonstrated a substantial deviation from the efficiency boundary, attributable to inadequate education.

The evaluation of the initial stage has demonstrated that nine schools exhibit an appropriate level of efficiency according to the first-stage indicators of efficiency. Furthermore, based on the second stage of efficiency analysis, it has been determined that five schools have demonstrated efficiency, while other institutions have failed to meet the efficiency center's standards due to deficiencies in their input and output processes. To rectify this situation, these educational institutions must take action to improve their performance.

Finally, according to the evaluations conducted through the intersectional integration model and Anderson model, a total of 45 schools were assessed and ranked based on their performance on input and output indices. The evaluation revealed that school number 45 attained the highest rank, obtaining an intersection efficiency weight of 2.67248, while school number 42 secured a rank of second place with an intersectional weight of 1.4483. Anderson's approach to ranking inefficient schools is applied to other institutions as presented in Table 2.

Interpretation of Anderson–Peterson Super-Efficiency Scores

The table presents a detailed evaluation of 45 decision-making units (DMUs), corresponding to secondary schools, using a two-stage DEA model integrated with the Anderson–Peterson (AP) super-efficiency approach. Each row includes:

- TET(U)1J / TET(L)1J: Upper and lower bounds of efficiency in stage 1 (resource allocation, i.e., leadership stage).
- TET(U)2J / TET(L)2J: Upper and lower bounds of efficiency in stage 2 (educational output, i.e., follower stage).
- E1, E2: Stage-specific efficiency scores.
- TETA: Overall two-stage efficiency score (E1 $\times$ E2).
- RANK: Rank based on TETA.
- TETA-AP: Anderson–Peterson super-efficiency score.

General Interpretation of TETA and Ranking

The TETA column reflects the composite efficiency of each school, combining the effects of resource utilization (E1) and output achievement (E2). Higher TETA values suggest schools that not only utilize resources effectively but also deliver strong educational results, especially in mathematical proficiency.

- Top performers (e.g., DMU11, DMU4, DMU15) display balanced efficiency in both stages.
- Low-performing units (e.g., DMU41, DMU13) show inefficiencies either in converting inputs or in translating intermediate processes into outcomes.

Understanding TETA-AP and Scores Above 1 (DMU42 and DMU45).

Unlike standard DEA scores that are capped at 1, the Anderson–Peterson (TETA-AP) model allows for efficiency scores exceeding 1, specifically to differentiate among DMUs that are on the efficient frontier (i.e., those scoring TETA = 1).

- DMU42 (TETA-AP = 1.4483) and DMU45 (TETA-AP = 2.67248) achieved perfect efficiency in both stages (TET = 1), indicating they lie on the DEA efficiency frontier.

- The TETA-AP > 1 implies these schools not only perform efficiently but outperform others on the frontier in terms of input-output proportionality.
- This super-efficiency means they can serve as benchmark institutions, capable of maintaining outputs even with reduced inputs compared to other efficient peers.

These values do not indicate a modeling error but instead reflect the robustness of performance and allow the differentiation of top-tier schools for policy recommendation and strategic replication.

Additional Insights.

- Schools like DMU4 (TETA = 0.71182) and DMU44 (TETA = 0.68939) also stand out despite not being super-efficient (TETA < 1) because of their strong performance across both stages, especially in translating intermediate instructional quality into measurable math achievements.
- Schools such as DMU1 (TETA = 0.08259) and DMU41 (TETA = 0.05163) are among the lowest ranked, indicating significant room for improvement, either in their resource allocation or in the effectiveness of their educational processes.

5 Conclusion and Strategic Implications

This study aimed to address the pressing issue of low mathematical proficiency in secondary schools by introducing an advanced analytical model combining two-stage network Data Envelopment Analysis (DEA) with the Anderson–Peterson super-efficiency method. The proposed framework allowed for a nuanced decomposition of school performance into two distinct yet interconnected phases: resource utilization and educational output generation. This methodological innovation enables the identification of both holistic and stage-specific inefficiencies in school operations.

Applying this model to a dataset of 45 secondary schools revealed notable disparities in institutional performance. Only two schools (DMU42 and DMU45) attained full efficiency across both stages and were further distinguished through super-efficiency scores above unity (TETA-AP > 1). These units exemplify ideal performance benchmarks, as they not only optimized their inputs but also achieved superior learning outcomes in mathematics.

Conversely, the majority of schools failed to translate their available resources into commensurate academic outputs, particularly in mathematics. Many exhibited average or low efficiency scores, with some schools ranking at the bottom despite comparable or even greater input allocations. This pattern signals deeper systemic or operational issues such as ineffective instructional practices, poor resource management, or lack of targeted academic support.

The analysis highlighted the critical role of intermediate variables—such as math lab engagement (Z11), teacher-student interaction, and supplemental learning programs (X22)—in shaping final academic outputs. Schools that excelled in internal processes demonstrated stronger overall efficiency, suggesting that educational inputs must be complemented by process-level interventions to yield desired learning outcomes.

In light of these findings, a number of strategic recommendations are proposed to improve the mathematical proficiency of underperforming schools. First, resource allocation must be data-driven and strategically aligned with curricular priorities in mathematics. Second, investing in teacher training, curriculum design, and math-specific infrastructure can enhance the quality of intermediate processes.

Underperforming schools should be encouraged to adopt benchmarking practices by studying the operational and pedagogical strategies of super-efficient schools such as DMU42 and DMU45. These institutions can serve as practical models for designing context-sensitive reforms in schools operating under similar socioeconomic and policy constraints.

Furthermore, the utilization of DEA-based evaluation frameworks in routine educational assessment can empower administrators and policymakers with actionable insights. Periodic performance audits based on this methodology will support early detection of inefficiencies and facilitate the implementation of targeted improvement plans tailored to the unique needs of each school.

Future research could extend this model by incorporating fuzzy logic techniques (e.g., triangular or trapezoidal fuzzy DEA) to account for uncertainty in qualitative judgments and stakeholder inputs. This would further enhance the robustness and applicability of the evaluation framework, particularly in diverse educational contexts where subjectivity and variability are inherent.

References

1. Muniz, R.D.F., Andriola, W.B., Muniz, S.M.M., Thomaz, A.C.F.: The use of data envelopment analysis (DEA) to estimate the educational efficiency of Brazilian schools. J. Appl. Res. Ind. Eng. **11**(1), 93–102 (2024)
2. Kikuchi, K., Suzuki, S., Nijkamp, P.: Bullying among pupils at school and a country's educational system: an efficiency evaluation of educational performance in Europe by means of an extended data envelopment analysis. Soc. Ind. Res., 1–32 (2024)
3. Salas-Velasco, M.: Evaluation of undergraduate academic programs through data envelopment analysis and time-to-degree estimates at Spanish public universities. Socioecon. Plann. Sci. **93**, 101878 (2024)
4. Hale, D.: A causal-comparative quantitative study to examine school funding equity under House Bill 3 (Doctoral dissertation, Texas A&M University-Commerce) (2024)
5. Lin, Y.C., Yu, M.M.: Performance evaluation of compulsory education system in Taiwan: a modified dynamic network data envelopment analysis approach. Stud. Educ. Eval. **78**, 101280 (2023)
6. Henriques, C.O., Chavez, J.M., Gouveia, M.C., Marcenaro-Gutierrez, O.D.: Efficiency of secondary schools in Ecuador: A value based DEA approach. Socio-Econ. Plann. Sci., 101226 (2022)
7. Kounetas, K.E., Kaisari, M.: Secondary schools' efficiency performance and reforms in Greece: a bootstrap DEA and multilevel approach (2022). Available at SSRN 4005907
8. Mergoni, A., Emrouznejad, A., De Witte, K.: Fifty years of data envelopment analysis. Eur. J. Oper. Res. **326**(3), 389–412 (2025). https://doi.org/10.1016/j.ejor.2024.12.049
9. Gao, J., Liu, Y.: Research on input-output efficiency of junior high school—a case study of DEA-OLS model. In Journal of Physics: Conference Series, vol. 1955, no. 1, p. 012121. IOP Publishing (2021)
10. Povedano, R., Salgado Junior, A.P., Souza Junior, M.A.A.D., Rebehy, P.C.P.W.: Indicators and goals for school performance evaluation: a two-stage DEA. Ensaio: Avaliação e Políticas Públicas em Educação **29**, 899–933 (2021)

11. Delprato, M., Antequera, G.: Public and private school efficiency and equity in Latin America: new evidence based on PISA for development. Int. J. Educ. Dev. **84**, 102404 (2021)
12. Dixit, V., Singh, J., Sardar, S.: Efficiency of subnational government expenditure towards elementary education and nutrition: bootstrap data envelopment analysis for India. J. Public Aff. **24**(1), e2884 (2024)
13. Lou, Y.Y., et al.: A parallel data envelopment analysis and Malmquist productivity index model of virtual frontier for evaluating scientific and technological innovation efficiency at universities. Dec. Analytics J. **10**, 100384 (2024)
14. Zhang, G., Guo, W., Xiong, X., Guan, Z.: A hybrid approach combining data envelopment analysis and recurrent neural network for predicting the efficiency of research institutions. Expert Syst. Appl. **238**, 122150 (2024)
15. Khrapatyi, S., Tokarieva, K., Hlushchenko, O., Paramonova, O., Lvova, I.: Research on performance evaluation of higher vocational education informatization based on data envelopment analysis. STEM Educ. **4**(1), 51–70 (2024)
16. Dzulkarnain, S.N.Z.H., Nawawi, M.K.M., Kashim, R., Dzulkarnain, S.N.H.H., Dzulkarnain, S.M.H.: Efficiency measurement of a public university: a slack-based data envelopment analysis approach. Semarak Int. J. Appl. Sci. Eng. Technol. **1**(1), 18–26 (2024)
17. Izadikhah, M., Saen, R.F.: Assessing sustainability of supply chains by chance-constrained two-stage DEA model in the presence of undesirable factors. Comput. Oper. Res. **100**, 343–367 (2018)

Managerial Ability Assessment of Turkish State Universities with DEA

Ayhan Gölcükcü(✉)

İstanbul Beykent University, İstanbul, Turkey
ayhangolcukcu@beykent.edu.tr

Abstract. In today's technological era, the need for physical resources and the use of physical resources have decreased extremely compared to the past, especially in universities based on mental craftsmanship. On the other hand, as access to information has reached the same level even in the most remote areas thanks to technology, academic competence has turned into a process completely dominated by the academics' own initiatives and the institutional culture they operate in. In this context, corporate culture is a phenomenon that can be considered as a function of managerial ability. Consequently, in this study, it is aimed to examine in the context of managerial ability, the state universities in Turkey, under the conditions of equality by technology, transform their existing financial and academic human resources into trained academics, market-oriented workforce and academic influence through academic publications and citations. For this purpose, managerial ability evaluation with DEA initiated by Demerjian et al. will be applied to Turkish State-Owned Universities'.

Keywords: DEA · Managerial Ability · Universities

1 Introduction

Data Envelopment Analysis (DEA) is a conventional tool in efficiency measurement which is classically used by management for measuring the performance of employees or sections, departments or branches of the firm or organization. However, in previous decades a new role has begun to be voiced; managerial ability. Although one of the first to use the concept of "Managerial Ability" was Bennet [1] with his work in 1925, and the early studies that addressed this concept in connection with DEA were made by Gim´enez & Thieme [2], Majid & Maulana [3], Mlambo & Ncube [4], Bergantino & Musso [5], the subject gained popularity after the publication of Demerjian et al. [6], which is the cornerstone. They used an immense data set for measuring the managerial ability of CEOs' gathered from various sources which cover 177,512 firm-year observations. In their two-stage process they initially measure the firm performance by DEA in the first stage and later in the second stage they evaluate the CEOs' performance from the residuals of Tobit regression which employes DEA results as dependent variable and some contextual variables as independent variables.

Furthermore, other fundamental studies on this subject are Demerjian et al. [7], Banker et al. [8] Demerjian & Owens [9], Demerjian et al. [10], Demerjian et al. [11],

A. Emrouznejad et al. (Eds.): DEA 2024, LNOR, pp. 397–412, 2025.
https://doi.org/10.1007/978-3-031-98177-7_27

Banker & Park's three stage approach [12], Banker et al. [13], which reveal the statistical foundations of Managerial Ability, can be mentioned. Meanwhile, it is convenient to refer here as base studies by Banker et al. [14], who mentioned that management performance is an unobservable outcome, and Banker & Natarajan [15], who investigated the use of contextual variables in DEA.

Despite the above major studies, there are numerous studies also exist. These studies are mentioned in the subsequent second section titled; An Overview About Managerial Ability Under DEA as a review of literature.

In this study, it is aimed to measure the managerial ability of Turkish State-Owned Universities. DEA will be restored to its original form and the managerial ability of Turkish State-Owned Universities without profit measure will be remarked. The aim and novelty of this study will be discussed extensively as the first study which concerns managerial ability of not-for-profit state-owned universities. It will also be associated with quality assurance and accreditation efforts in Turkish Universities in the third section.

The scope, data and variables are given as a subsection under the fourth section titled; Method. Later, the managerial ability of Turkish State-Owned Universities will be evaluated with the criteria mentioned in the data and variables subsections.

In the final last section, conclusion about evaluated DEA results will be made and some recommendations will be given with future directions including a discussion whether the criteria are really a measure in terms of managerial ability.

2 An Overview About Managerial Ability Under DEA Context as a Review of Literature

Managerial Ability has long been studied using different methods as indicated in [1]. It was about farming and proposed an index for measuring managerial ability. It is extremely far away from DEA. Porter & Schully [16] tried to measure the managerial efficiency around sport and tried to find an empirical estimates of managerial efficiency in baseball by employing Cobb-Douglas production function. This study was also not a DEA study, even though it was a little closer to DEA and used similar concepts. Banker et al. [14] described a principal agent model of the relationship between shareholders and a manager. They approached the subject from an accounting point of view and their contribution was that managerial ability is a phenomenon that cannot be measured directly by sector variables but can be assessed indirectly by some other means.

In two parallel studies, Kiemo & Kamau [17] and Mlambo & Ncube [4] tried to measure managerial ability in the field of banking, and in both studies, the Panzer-Rosse H statistic was used to measure managerial ability. Further financial applications of measuring managerial ability have been seen in insurance industry [18–21], tax related issues [22–25] and earning management related with stock market [26–32]. Moreover, Amirteimoori studied the subject by working with different researchers on its applications in forestry, health and banking [33–35]. Sinha is another leading author of managerial ability who worked as a colleague of Professor Banker [20, 21], or private researcher [36] or with his associates [37, 38].

Essentially, DEA emerged to measure the performance of an education program in 1978 [39], which was revealed more clearly with the study titled "Evaluating program and managerial efficiency: an application of data envelopment analysis to program follow through" in 1981 [40], which was the follow-up to the initial study [39]. However, although this study was aimed at measuring management performance, it was still far from managerial ability. Similarly, Johnson & Ruggiero [41] decomposed Malmquist productivity index to analyze educational production of Ohio school districts, Singh et al. [42] proposed A multi-objective optimization approach for DEA models in a fuzzy environment.

Regrettably, vast majority of the valuable literature of managerial ability under DEA context is, as mentioned, financial in nature, including studies on educational performance [43, 44].

In literature, there are also studies like 45.Sowiyah et al. [45] which measures the influence of managerial ability and teacher performance out of DEA context using a descriptive verification method with a cluster random sampling.

Finally, Anggraini & Sholihin [46] made a review about managerial ability which includes 120 studies published between the dates of 1974–2020 and discussed the trend under the areas of topic, research setting, theory & methodology and data analysis. However, the studies mentioned in this review cover not only DEA but also different methods.

2.1 Novelty of the Study

As mentioned in the previous section managerial ability applications under DEA context are mainly related to the financial sector and tried to maximize earnings directly or indirectly or the studies concerning managerial ability are out of DEA context. In this study, DEA will be returned to its original form and the managerial ability of Turkish State-Owned Universities without profit measure will be remarked.

The university enrolment system in Türkiye is carried out through a post-secondary exam, and candidates gain the right to enroll in a university through central placement in line with the choices they make according to the score they receive from this exam. Here, the quality of the universities significantly affects the candidates' choices. In this context, in order to measure the quality of the universities and provide quality assurance, universities in Türkiye are rated and accredited by primarily YÖKAK (Turkish Higher Education Quality Council -THEQC) and some other various autonomous evaluation organizations, based on certain criteria. The approaches towards accredited universities could be a more tolerant regarding the number of students, opening of new departments or different state support. However, these efforts are quite time-consuming, create an additional burden on the main duties of university personnel and may disrupt their main work.

Essentially, this study will attempt to reveal the level and model of the relationship through the alternative hypothesis that "there is an association between the applied performance measures applied and the performance of universities". Our null hypothesis is naturally that there is no association between the applied performance measures and the performance of universities. In this regard, a two-stage model will be applied. In the first stage, efficiency scores will be evaluated by DEA models using the variables of academic activities of universities. Later in the second stage, measures used to evaluate the university performance are regressed with DEA scores. Since these measures are contextual and, due to the nature of performance measurement and DEA, the estimates of the latent dependent variable performance scores should have a lover and upper limit, a double-censored Tobit regression analysis will be applied in the second stage.

Therefore, in this study, it is not only aimed to measure the managerial ability of state-owned universities but also it will be questioned the success of measures used to evaluate the performance of universities in Türkiye.

3 Method

3.1 The Scope and Data of the Study: Turkish State-Owned Universities

In Turkey, universities are divided into two groups according to their sources of income. The first group includes state-owned universities whose budgets are formed by funds transferred by the state and are determined by the budget law announced every year. The second group includes foundation universities whose budgets are dependent on the founding foundation and partial tuition fee payments received from students. Since the budgets of foundation universities are their own budgets and these budgets are not disclosed to the public, these universities are excluded from the scope of study.

The Council of Higher Education (YÖK) announces non-financial data on universities, namely the number of students and academic staff, so these are compiled from the website of the Higher Education Information Management System of YÖK [47], and financial data were compiled from the budget laws announced on the website of the Strategy and Budget Presidency of the Republic of Turkey [48]. The data for 2023 was used in this study, since the data for 2024 was not announced by YOK. In addition, some statistics evaluated by the Higher Education Quality Board [49] were examined within the scope of this study. In budget law year 2023 there exist 127 universities as Decision Makin Units (DMUs).

This study is based on publicly available data. Main financial source of state-owned universities in Turkey are as mentioned the budget transferred by central government and this was used as the input variable. However, data on how this resource is spent, although it has an impact on managerial ability, could not be obtained because it is not disclosed to the public. This is a limitation in this study. In addition, since there are a limited number of universities, a limited number of variables could be used. Increasing the amount of data through panel data was not possible due to the degeneration of the data structure in the pandemic conditions. This is another limitation of our study.

3.2 Variables

In this study, variables were collected in three categories in line with the managerial ability literature and Demerjian's [6] study. Input and output variables of the DEA analysis and contextual variables for managerial ability. The contextual variables were also examined in two groups. The variables used are presented in Table 1.

Table 1. The Variables.

Description (Code)	Type of Variable	Model
Total number of academic staff (Staff)	Input	M1-M5
Total number of students (Student)	Input	M1-M5
Budget (Budget)	Input	M1-M5
Number of full professors (Prof)	Output	M1-M5
Number of research assistants (Assist)	Output	M1-M5
Number of publications indexed by WoS (WoS)	Output	M1-M5
Number of associate degree students (D1)	Output	M1
Number of undergraduate students (D2)	Output	M2
Number of MSc students (D3)	Output	M3
Number of PhD students (D4)	Output	M4
Total number of postgraduate (MSc + PhD) students (D5)	Output	M5
Number of disadvantaged students (Disadv)	Contextual (Group 1)	M1-M5
Number of foreign students (Foreign)	Contextual (Group 1)	M1-M5
Number of outgoing students (Outgoing)	Contextual (Group 1)	M1-M5
Number of incoming students (Incoming)	Contextual (Group 1)	M1-M5
Number of citations (Overall 2023) (Citation)	Contextual (Group 1)	M1-M5
Percentage of achievement of goals related to education and training activities in the institution's strategic plan (Edu)	Contextual (Group 2)	
Percentage of achievement of goals related to research activities in the institution's strategic plan (Resc)	Contextual (Group 2)	
Percentage of achievement of goals related to administrative activities in the institution's strategic plan (Admin)	Contextual (Group 2)	
Percentage of achievement of goals related to community service activities in the institution's strategic plan (Public)	Contextual (Group 2)	
The satisfaction rate of the business world regarding the competencies of graduates (B_World)	Contextual (Group 2)	

3.3 Assumptions on Variables

In the sense that, universities are black boxes which converts the efforts of academic staff with the help of administrative structure of the university to graduated students and academic outcome. There are 4 + 1 kinds of resources in universities.

- Physical resources (Campuses, Buildings and assets within)
- Financial resources
- Staff (Academic & Administrative)
- Students
- Administrative Culture

In the sense that universities are processes that use financial/physical sources, staff and students as a resource and yield products. Therefore, the total number of staff, the total number of students and the budget were taken as input. Since physical resources are managed with long-term planning and are under the control of the state depending on the investment budget, they were not used as input in this study. In addition, since organizational culture and management competence are parallel to each other and we are trying to measure this, they were not used as input.

The resulting products of universities are individuals who continue their academic careers and reach the final point, namely professorship, research assistants who have the opportunity to leave the university as researchers. Meanwhile, graduates are disconnected from the scope of the university's managerial ability, the number of graduates was not used in this study. Instead of graduates, students at different levels of education are used as another output with an idea that the product of a university is its students, who are candidates for graduation. In addition, academical studies are another product of academic life, for instance the number of publications indexed by web of science (WoS) are taken as output.

Students can graduate from the university at 4 different levels (associate degree, undergraduate degree, master's degree and doctorate) in Türkiye; each of these levels taken as output in separate different models additionally, the total of postgraduate students (Master's + Doctorate) was used as additional separate output in an additional model and the data was analyzed in a total of 5 models namely M1-M5. Due to limited data all.

Number of disadvantaged students, Number of foreign students, the number of students who went through exchange programs and the number of students who came through exchange programs and the number of citations in WoS database are first group contextual variables of second stage measurement.

The Higher Education Quality Board evaluates universities in different areas. Second group of contextual variables selected from these areas; Percentage of achievement of goals related to education and training activities in the institution's strategic plan, Percentage of achievement of goals related to research activities in the institution's strategic plan, Percentage of achievement of goals related to administrative activities in the institution's strategic plan, Percentage of achievement of goals related to community service activities in the institution's strategic plan, The satisfaction rate of the business world regarding the competencies of graduates variables were also evaluated as the second group of contextual variables.

3.4 DEA Models

As indicated in previous sections, especially state-owned universities are non-profit organizations in Türkiye so when the budget is used with the number of academic staff and number of students as input, budget minimizing models are useless. Therefore, instead of input minimization, output maximization is much more beneficial. Hence, the output-oriented CCR and BCC models were used for the first phase of this study. Objective functions of these models for first stage DEA analysis can be given in ratio form as follows respectively.

$$\min q = \frac{v_1 Staff_0 + v_2 Student_0 + v_3 Budget_0}{\mu_r y_{r0} \mu_1 Prof_0 + \mu_2 Asist_0 + \mu_3 WoS_0 + \mu_4 D1_0} \tag{1}$$

$$\min q = \frac{v_1 Staff_0 + v_2 Student_0 + v_3 Budget_0 + u_0}{\mu_r y_{r0} \mu_1 Prof_0 + \mu_2 Asist_0 + \mu_3 WoS_0 + \mu_4 D1_0} \tag{2}$$

Each of the above models is solved for every DMUs. Total of 10 DEA models are evaluated. Some statistics about these models are presented in Table 2.

Table 2. Some Statistics About Model Results

DEA Type	Statistics	M1	M2	M3	M4	M5
CCR	Number of Efficient DMUs	18	22	16	18	16
	Mean efficiency score	0.803941	0.801732	0.764848	0.737769	0.754466
	Least Efficient	0.445715	0.405465	0.394342	0.386378	0.392698
BCC	Number of Efficient DMUs	28	31	27	27	27
	Mean efficiency score	0.829428	0.823582	0.792295	0.762978	0.781138
	Least Efficient	0.572045	0.406597	0.453244	0.388865	0.419395

In the second stage of the analysis, Tobit regression is used as in Demerjian et al. [6] for each model and Group 1 contextual variables. Tobit regression was not applied between the second group of contextual variables and DEA results because the correlation between these variables and DEA scores was both very low and insignificant. Correlation betweenDEA scores and contextual variables are given in Table 3 & 4.

Table 3. Correlation between DEA scores and group 1 contextual variables*

DEA Model	Contextual Variable	M1	M2	M3	M4	M5
CCR	(Disadv)	–0.174 (0.037)	–0.138 (0.078)	–0.084 (0.193)	–0.042 (0.333)	–0.08 (0.208)
	(Foreign)	–0.082 (0.201)	–0.219 (0.012)	–0.155 (0.055)	–0.117 (0.115)	–0.158 (0.052)
	(Outgoing)	–0.271 (0.002)	–0.323 (0)	–0.318 (0)	–0.339 (0)	–0.331 (0)

(continued)

Table 3. (*continued*)

DEA Model	Contextual Variable	M1	M2	M3	M4	M5
	(Incoming)	–0.227 (0.009)	–0.253 (0.004)	–0.274 (0.002)	–0.283 (0.002)	–0.279 (0.002)
	(Citation)	–0.523 (0)	–0.536 (0)	–0.577 (0)	–0.582 (0)	–0.584 (0)
BCC	(Disadv)	–0.166 (0.044)	–0.128 (0.094)	–0.077 (0.215)	–0.035 (0.358)	–0.069 (0.239)
	(Foreign)	–0.075 (0.221)	–0.224 (0.01)	–0.115 (0.119)	–0.097 (0.159)	–0.12 (0.11)
	(Outgoing)	–0.273 (0.002)	–0.304 (0.001)	–0.292 (0.001)	–0.318 (0)	–0.307 (0.001)
	(Incoming)	–0.238 (0.007)	–0.249 (0.005)	–0.261 (0.003)	–0.271 (0.002)	–0.268 (0.003)
	(Citation)	–0.503 (0)	–0.507 (0)	–0.525 (0)	–0.542 (0)	–0.536 (0)

* Values in parentheses are significance levels.

Table 4. Correlation between DEA scores and group 2 contextual variables*

DEA Model	Contextual Variable	M1	M2	M3	M4	M5
CCR	(Edu)	0.095 (0.29)	0.061 (0.494)	0.023 (0.794)	0.085 (0.345)	0.037 (0.682)
	(Resc)	0.076 (0.396)	–0.012 (0.898)	–0.032 (0.724)	0.012 (0.897)	–0.019 (0.832)
	(Admin)	0.025 (0.779)	0.079 (0.38)	0.075 (0.402)	0.081 (0.368)	0.073 (0.417)
	(Public)	0.027 (0.766)	–0.025 (0.781)	–0.021 (0.814)	0.003 (0.971)	–0.022 (0.806)
	(B_World)	–0.032 (0.719)	–0.222 (0.012)	–0.152 (0.088)	–0.168 (0.059)	–0.155 (0.082)
BCC	(Edu)	0.063 (0.48)	0.083 (0.351)	0.034 (0.705)	0.099 (0.268)	0.048 (0.594)
	(Resc)	0.085 (0.344)	0.068 (0.445)	0.054 (0.547)	0.085 (0.344)	0.067 (0.455)
	(Admin)	0.004 (0.965)	0.094 (0.291)	0.071 (0.425)	0.088 (0.326)	0.08 (0.373)

(*continued*)

Table 4. *(continued)*

DEA Model	Contextual Variable	M1	M2	M3	M4	M5
	(Public)	0.009 (0.918)	0 (0.996)	–0.002 (0.982)	0.02 (0.821)	0.002 (0.983)
	(B_World)	–0.049 (0.588)	–0.232 (0.009)	–0.156 (0.08)	–0.174 (0.051)	–0.163 (0.068)

* Values in parentheses are significance levels.

The significant correlations in Table 3 are only between "The satisfaction rate of the business world regarding the competencies of graduates" and M2. This indicates that there is a low but significant correlation between DEA results and the business world. Unfortunately, this is not enough for performing Tobit regression analysis.

3.5 Tobit Model

The Tobit model is named after James Tobin's early work [50], and a detailed compilation and classification on the subject was made by Amemiya [51]. Details are also available in Amemiya [52, 53], Schnedler [54] and Hayashi [55]. Meanwhile, studies in which the Tobit model was used as the second step after DEA efficiency evaluation in the evaluation of managerial ability are quite few. Matranga et al. [56] which assesses the managerial ability of Italian hospitals, Chuah and Foong [57] studied the effect of familiness and foreignness of the CEOs in Malaysia. Park et al. [23] has addressed tax avoidance in Korea, while Lee and Yoon [25] issued the subject of tax planning as a part of Trade-off between tax and nontax costs. On the other hand, Park and Yung [27] also mentioned the effect of managerial ability on future stock price crash risk. Savitha et al. [19] focused on life microinsurance companies in India while Banna et al. [58] worked on microfinance institutions of Latin America. The theme of Oh [26] was the industrial concentration and investment efficiency and Zaman et al. [59] pointed out the managerial efficiency and corporate leverage policy in Pakistan.

As stated by Park and Jung [27] and Zaman et al. [59], DEA scores are limited to the range of [0,1], but when these limits are intended to be used for Tobit regression, it is meaningless since an efficiency score of 0 will be obtained if and only if the outputs are all 0. When we take the upper limit as 1, we ignore the efficiency of the efficient units, so the lower limits for Tobit regression were obtained by subtracting ε (ε was taken as 10^{-9}) from the least efficient score of each model given in Table 2, and double censoring was performed by taking $1+\varepsilon$ for the upper limit.

Due to the diagnostic results of correlations only Tobit regression analysis performed between DEA scores and Group 1 contextual variables. The employed Tobit model as an equation is as follows.

$$\begin{aligned} DEA\ Score_i &= \beta_0 + \beta_1 \text{Disadv}_i + \beta_2 \text{Foreign}_i + \beta_3 \text{Outgoing}_i \\ &\quad + \beta_4 \text{Incoming}_i + \beta_5 B\_\text{Citation}_i + \in_i \end{aligned} \tag{3}$$

Tobit Regression results for CCR and BCC models are given in Table 5 & Table 6 respectively.

Table 5. Tobit Regression results for CCR models

Models	Variables	Coef.	Std. Err.	t	P > t	
CCR-M1						
Number of obs: 127	constant	0.7408373	0.0155473	47.65	0	*
Log likelihood: 101.784	Disadv	0.0000133	0.00000416	3.2	0.002	*
LR chi2(5): 47.65	Foreign	−0.0000172	0.00000591	−2.92	0.004	*
Prob > chi2: 0	Outgoing	0.0000923	0.0002233	0.41	0.68	
Pseudo R2: −0.3056	Incoming	−0.0000171	0.0001019	−0.17	0.867	
	Citation	0.0000518	0.00000873	5.93	0	*
CCR-M2						
Number of obs: 127	constant	0.7153868	0.0188579	37.94	0	*
Log likelihood: 77.268	Disadv	0.00000754	0.00000504	1.5	0.137	
LR chi2(5): 43.18	Foreign	−0.00000675	0.00000717	−0.94	0.348	
Prob > chi2: 0	Outgoing	0.0003238	0.0002708	1.2	0.234	
Pseudo R2: −0.3878	Incoming	0.0000614	0.0001236	0.5	0.62	
	Citation	0.0000506	0.0000106	4.78	0	*
CCR-M3						
Number of obs: 127	constant	0.6667759	0.0200486	33.26	0	*
Log likelihood: 67.585	Disadv	0.0000051	0.00000535	0.95	0.343	
LR chi2(5): 48.85	Foreign	−0.00000984	0.00000762	−1.29	0.199	
Prob > chi2: 0	Outgoing	0.0002798	0.0002878	0.97	0.333	
Pseudo R2: −0.5659	Incoming	0.0001159	0.0001313	0.88	0.379	
	Citation	0.0000609	0.0000113	5.41	0	*
CCR-M4						
Number of obs: 127	constant	0.6311874	0.0212585	29.69	0	*
Log likelihood: 60.303	Disadv	0.00000409	0.00000568	0.72	0.473	
LR chi2(5): 52.89	Foreign	−0.000013	0.00000808	−1.61	0.11	
Prob > chi2: 0	Outgoing	0.0003178	0.0003051	1.04	0.3	
Pseudo R2: −0.781	Incoming	0.0001238	0.0001392	0.89	0.376	
	Citation	0.0000689	0.0000119	5.77	0	*

(*continued*)

Table 5. (*continued*)

Models	Variables	Coef.	Std. Err.	t	P > t	
CCR-M5						
Number of obs: 127	constant	0.6517079	0.0204741	31.83	0	*
Log likelihood: 64.983	Disadv	0.00000484	0.00000547	0.88	0.378	
LR chi2(5): 50.68	Foreign	–0.00001	0.00000778	−1.29	0.2	
Prob > chi2: 0	Outgoing	0.000303	0.0002938	1.03	0.305	
Pseudo R2: −0.6392	Incoming	0.0001159	0.0001341	0.86	0.389	
	Citation	0.0000636	0.0000115	5.54	0	*
* Represents significant values						

Table 6. Tobit Regression results for BCC models

Models	Variables	Coef.	Std. Err.	t	P > t	
BCC-M1						
Number of obs: 127	constant	0.782667	0.0157816	49.59	0	*
Log likelihood: 98.424	Disadv	0.0000146	0.0000043	3.39	0.001	*
LR chi2(5): 38.33	Foreign	–0.0000202	0.00000623	–3.24	0.002	*
Prob > chi2: 0	Outgoing	0.0001344	0.0002262	0.59	0.554	
Pseudo R2: −0.2418	Incoming	0.0000423	0.0001033	0.41	0.683	
	Citation	0.0000424	0.00000888	4.78	0	*
BCC-M2						
Number of obs: 127	constant	0.7453307	0.0193615	38.5	0	*
Log likelihood: 73.920	Disadv	0.00000537	0.00000518	1.04	0.301	
LR chi2(5): 33.87	Foreign	–0.00000349	0.00000736	–0.47	0.637	
Prob > chi2: 0	Outgoing	0.0003027	0.0002781	1.09	0.279	
Pseudo R2: −0.2972	Incoming	0.0000913	0.0001269	0.72	0.473	
	Citation	0.0000425	0.0000109	3.91	0	*
BCC-M3						
Number of obs: 127	constant	0.7137132	0.0210316	33.94	0	*
Log likelihood: 63.411	Disadv	0.00000663	0.00000562	1.18	0.241	
LR chi2(5): 35.62	Foreign	–0.0000129	0.000008	–1.61	0.11	

(*continued*)

Table 6. *(continued)*

Models	Variables	Coef.	Std. Err.	t	P > t	
Prob > chi2: 0	Outgoing	0.0002646	0.0003021	0.88	0.383	
Pseudo R2: −0.3906	Incoming	0.0001357	0.0001378	0.98	0.327	
	Citation	0.0000521	0.0000118	4.41	0	*
BCC-M4						
Number of obs: 127	constant	0.6760198	0.0228915	29.53	0	*
Log likelihood: 52.645	Disadv	0.00000487	0.00000612	0.8	0.428	
LR chi2(5): 37.69	Foreign	–0.0000149	0.0000087	–1.71	0.09	
Prob > chi2: 0	Outgoing	0.0003174	0.0003288	0.97	0.336	
Pseudo R2: −0.5576	Incoming	0.0001585	0.00015	1.06	0.293	
	Citation	0.0000584	0.0000129	4.54	0	*
BCC-M5						
Number of obs: 127	constant	0.6980827	0.0219749	31.77	0	*
Log likelihood: 56.086	Disadv	0.00000607	0.00000587	1.03	0.303	
LR chi2(5): 35.82	Foreign	–0.0000129	0.00000835	–1.55	0.124	
Prob > chi2: 0	Outgoing	0.0003009	0.0003154	0.95	0.342	
Pseudo R2: −0.469	Incoming	0.0001464	0.0001439	1.02	0.311	
	Citation	0.0000539	0.0000123	4.37	0	*

* Represents significant values

4 Conclusion, Recommendations, and Future Directions

The correlations between the results of the defined DEA models and the contextual variables are quite explanatory. Curiously, all correlations regarding the contextual variables of group 1 predict a negative relationship. While this correlation is low for outgoing and incoming students within the scope of exchange programs, it is at a medium level and significant in terms of number of citations. The correlation and its significance vary from model to model in terms of disadvantaged students and foreign students. While the presence of foreign students at the undergraduate level had a low-level, significant negative effect on DEA scores, the correlations observed at other levels were not found to be significant.

There is confusion in the correlations regarding the Group 2 contextual variables. The effects of the contextual variables vary in terms of sign according to the level of education, being negative in some and positive in others. However, the correlations are only low, significant and negative in the context of the variable "The satisfaction rate of the business world regarding the competencies of graduates (B_World)" for bachelor's and master's degrees in both the CCR and BCC models.

According to Table 5 and Table 6, it is seen that the Tobit regression results give similar results for both CCR and BCC models. Except for the M1model, in the M2-M5 models, only the constant term and citation count variables were found to be significant in the model and from coefficient value it is seen that they influenced DEA scores, albeit at a very low level. On the other hand, in the M1 model, in addition to the constant term and citation count, the number of disadvantaged students had a very low level of positive effect on the DEA performance of universities and the number of foreign students had a very low level of negative effect.

The contextual variables used in this study are a selection of variables used in the performance evaluation of universities, and since it was seen that their effects on performance were not as expected in terms of the obtained results, the calculated managerial ability results will not be given. It would be useful to consider the existence of different variables affecting the performance of universities and to guide future studies in this regard.

This study had aimed to evaluate the managerial ability of universities through existing measures, but at this point, the main conclusion of this study is that the current measures used to evaluate performance do not actually measure performance well, the measures claimed to provide performance do not actually provide desired performance, on the contrary, they cause waste of time and resources that could be used for real performance production, and therefore new methods and measures should be developed for performance measurement.

References

1. Bennett, M.K.: A method of measuring managerial ability in farming. J. Farm Econ. **7**(3), 347–358 (1925)
2. V Giménez, D.P., Thieme, C.: Technical efficiency, managerial efficiency and objective-setting in the educational system: an international comparison. J. Oper. Res. Soc. **58**(8), 996–1007 (2007)
3. Majid, M.S.A., Maulana, H.: Assessing performance of mutual funds in Indonesia. J. Econ. Coop. Dev. **31**(4), 49–76 (2010)
4. Mlambo, K., Ncube, M.: Competition and efficiency in the banking sector in South Africa. Afr. Dev. Rev. **23**(1), 4–15 (2011)
5. Bergantino, A.S., Musso, E.: The role of external factors versus managerial ability in determining seaports relative efficiency: an input-by-input analysis through a multi-step approach on a panel of Southern European ports. Marit. Econ. Logist. **13**(2), 121–141 (2011)
6. Demerjian, P., Lev, B., McVay, S.: Quantifying managerial ability: a new measure and validity tests. Manag. Sci. **58**(7), 1229–1248 (2012)
7. Demerjian, P.R., Lev, B., Lewis, M.F., McVay, S.E.: Managerial ability and earnings quality. Account. Rev. **88**(2), 463–498 (2013)
8. Banker, R.D., Darrough, M.N., Huang, R., Plehn-Dujowich, J.M.: The relation between CEO compensation and past performance. Account. Rev. **88**(1), 1–30 (2013)
9. Demerjian, P.R., Owens, E.L.: Measuring the probability of financial covenant violation in private debt contracts. J. Account. Econ. **61**(2–3), 433–447 (2016)
10. Demerjian, P., Lewis-Western, M., McVay, S.: How does intentional earnings smoothing vary with managerial ability? J. Acc. Audit. Financ. **35**(2), 406–437 (2020)
11. Demerjian, P., Donovan, J., Lewis-Western, M.F.: Income smoothing and the usefulness of earnings for monitoring in debt contracting. Contemp. Account. Res. **37**(2), 857–884 (2020)
12. Banker, R.D., Park, H.U.: Three-stage approach to analyze managerial ability. Data Envel. Anal. J. **5**(1), 27–54 (2021)

13. Banker, R., Park, H.U., Sahoo, B.: A Statistical Foundation for The Measurement of Managerial Ability (No. 111832). University Library of Munich, Germany (2022)
14. Banker, R.D., Datar, S.M., Maindiratta, A.: Unobservable outcomes and multiattribute preferences in the evaluation of managerial performance. Contemp. Account. Res. **5**(1), 96–124 (1988)
15. Banker, R.D., Natarajan, R.: Evaluating contextual variables affecting productivity using data envelopment analysis. Oper. Res. **56**(1), 48–58 (2008)
16. Porter, P.K., Scully, G.W.: Measuring managerial efficiency: the case of baseball. Southern Econ. J. 642–650 (1982)
17. Kiemo, S., Kamau, A.: Banking Sector Competition and Intermediation Efficiency in Kenya. Afr. Dev. Rev. **33**(4), 648–661 (2021)
18. Nourani, M., Devadason, E.S., Kweh, Q.L., Lu, W.M.: Business excellence: the managerial and value-creation efficiencies of the insurance companies. Total Qual. Manag. Bus. Excell. **28**(7–8), 879–896 (2017)
19. Savitha, B., Banerjee, S., Shetty, A.: Product diversification versus technical efficiency of conglomerate life microinsurance companies: evidence from India. Geneva Pap. Risk Insur.-Issues Pract. **44**, 527–547 (2019)
20. Banker, R.D., Amirteimoori, A., Sinha, R.P.: An integrated data envelopment analysis and generalized additive model for assessing managerial ability with application to the insurance industry. Decis. Anal. J. **4**, 100115 (2022)
21. Banker, R.D., Amirteimoori, A., Allahviranloo, T., Sinha, R.P.: Performance analysis and managerial ability in the general insurance market: a study of India and Iran. Inf. Technol. Manag. **25**(1), 19–31 (2024)
22. Ko, C.Y., Park, J., Jung, H.: Does the ability of managers affect the firm's tax avoidance? International information institute (Tokyo). Information **17**(7), 2961 (2014)
23. Park, J., Ko, C.Y., Jung, H., Lee, Y.S.: Managerial ability and tax avoidance: evidence from Korea. Asia-Pac. J. Account. Econ. **23**(4), 449–477 (2016)
24. Akbari, F., Salehi, M., Bagherpour Vlashani, M.A.: The effect of managerial ability on tax avoidance by classical and Bayesian econometrics in multilevel models: evidence of Iran. Int. J. Emerg. Mark. **13**(6), 1656–1678 (2018)
25. Lee, K., Yoon, S.: Managerial ability and tax planning: trade-off between tax and nontax costs. Sustainability **12**(1), 370 (2020)
26. Oh, H.T.: Managerial ability, industrial concentration and investment efficiency. Int. J. Appl. Bus. Econ. Res. **15**(14), 353–359 (2017)
27. Park, S.Y., Jung, H.: The effect of managerial ability on future stock price crash risk: evidence from Korea. Sustainability **9**(12), 2334 (2017)
28. Cui, H., Leung, S.C.M.: The long-run performance of acquiring firms in mergers and acquisitions: does managerial ability matter? J. Contemp. Account. Econ. **16**(1), 100185 (2020)
29. Salehi, M., Mousavi Shiri, M., Hossini, S.Z.: The relationship between managerial ability, earnings management and internal control quality on audit fees in Iran. Int. J. Product. Perform. Manag. **69**(4), 685–703 (2020)
30. Simamora, A.J.: Crime rate, real earnings management and managerial ability. Corp. Gov. Int. J. Bus. Soc. **22**(2), 405–423 (2022)
31. Putra, A.A.: Managerial ability and informative earnings management: the role of CEO-commissioner relationship and board independence. Corp. Gov. Int. J. Bus. Soc. **23**(4), 742–765 (2023)
32. Bazrafshan, A., Makarem, N., Hesarzadeh, R., Salmanabbood, W.: Managerial ability, earnings quality and ISIS: evidence from Iraq. Int. J. Emerg. Mark. **18**(9), 3085–3115 (2023)

33. Amirteimoori, A., Banker, R.D., Zadmirzaei, M., Susaeta, A.: Managerial ability and productivity growth in the European forest sector. Eur. J. Forest Res. **142**(5), 1029–1043 (2023)
34. Amirteimoori, A., Safarpour, S., Kordrostami, S., Khoshandam, L.: Performance and managerial ability analysis in health sector: a data envelopment analysis approach. In: Allahviranloo, T., Hosseinzadeh Lotfi, F., Moghaddas, Z., Vaez-Ghasemi, M. (eds.) Decision Making in Healthcare Systems, pp. 373–390. Springer, Cham (2024)
35. Amirteimoori, A., Kou, G.: Technical efficiency and managerial ability in two-stage production processes with undesirable products: a case on Asian banks. Socioecon. Plann. Sci. **96**, 102081 (2024)
36. Sinha, R.P.: The Impact of Contextual Variables on Life Insurer Performance. Metamorphosis, 09726225241282448 (2024)
37. Sinha, R.P., Vaisi, B., Edalatpanah, S.A.: Does managerial ability promote firm efficiency? An evidence from Indian banking industry. J. Appl. Res. Ind. Eng. **11**(4), 518–535 (2024)
38. Sinha, R.P., Vaisi, B.: Managerial ability in Indian life insurance companies: a comparison based on DEA and DEAGP. In: Edalatpanah, S., Hosseinzadeh Lotfi, F., Kerstens, K., Wanke, P. (eds.) Analytical Decision Making and Data Envelopment Analysis. Infosys Science Foundation Series. Springer, Singapore (2024)
39. Charnes, A., Cooper, W.W., Rhodes, E.: Measuring the efficiency of decision-making units. Eur. J. Oper. Res. **2**(6), 429–444 (1978). https://doi.org/10.1016/0377-2217(78)90138-8
40. Charnes, A., Cooper, W.W., Rhodes, E.: Evaluating program and managerial efficiency: an application of data envelopment analysis to program follow through. Manag. Sci. **27**(6), 607–730 (1981). CCR 1981. https://doi.org/10.1287/Mnsc.27.6.668
41. Johnson, A.L., Ruggiero, J.: Nonparametric measurement of productivity and efficiency in education. Ann. Oper. Res. **221**, 197–210 (2014)
42. Singh, A.P., Yadav, S.P., Singh, S.K.: A multi-objective optimization approach for DEA models in a fuzzy environment. Soft. Comput. **26**(6), 2901–2912 (2022)
43. Johnes, J., Portela, M., Thanassoulis, E.: Efficiency in education. J. Oper. Res. Soc. **68**(4), 331–338 (2017). https://doi.org/10.1057/S41274-016-0109-Z
44. Thanassoulis, E., Dey, P.K., Petridis, K., Goniadis, I., Georgiou, A.C.: Evaluating higher education teaching performance using combined analytic hierarchy process and data envelopment analysis. J. Oper. Res. Soc. **68**(4), 431–445 (2017). https://doi.org/10.1057/S41274-016-0165-4
45. Sowiyah, S., Fitriyanti, Z., Natalia, V.E.D.: The influence of principal managerial ability and teacher performance toward senior high school quality. Int. J. Educ. Stud. Soc. Sci. (IJESSS) **2**(1), 8–17 (2022)
46. Anggraini, P.G., Sholihin, M.: What do we know about managerial ability? A systematic literature review. Manag. Rev. Q. **73**(1), 1–30 (2023)
47. Higher Education Information Management System Homepage. https://Istatistik.Yok.Gov.Tr/. Accessed 31 Oct 2024
48. Strategy And Budget Presidency of The Republic of Türkiye, The Page of Central Government Budget Laws. https://Www.Sbb.Gov.Tr/Merkezi-Yonetim-Butce-Kanunlari/. Accessed 31 Oct 2024
49. Higher Education Quality Board, Indicator Values Report Page. https://Yokak.Gov.Tr/Raporlar/Indicatorvaluesreport. Accessed 31 Oct 2024
50. Tobin, J.: Estimation of relationships for limited dependent variables. Econometrica: J. Econom. Soc. 24–36 (1958)
51. Amemiya, T.: Tobit models: a survey. J. Econom. **24**(1–2), 3–61 (1984)
52. Amemiya, T.: Regression analysis when the dependent variable is truncated normal. Econometrica: J. Econ. Soc. 997–1016 (1973)

53. Amemiya, T.: Advanced Econometrics. Harvard University Press. Cambridge, Massachusetts (1985)
54. Schnedler, W.: Likelihood estimation for censored random vectors. Economet. Rev. **24**(2), 195–217 (2005)
55. Hayashi, F.: Econometrics. Princeton University Press. Princeton, New Jersey (2011)
56. Matranga, D., et al.: Evaluating the effect of organization and context on technical efficiency: a second-stage DEA analysis of Italian hospitals. Epidemiol. Biostat. Public Health **11**(1), e8785 (2014)
57. Chuah, S.F., Foong, S.S.: Managerial ability and firm performance in Malaysia: do familiness and foreignness of the CEOs Matter? Rev. Pac. Basin Financ. Mark. Policies **22**(03), 1950017 (2019)
58. Banna, H., Rana, M.S., Ismail, I., Ismail, N.: Quantifying the managerial ability of microfinance institutions: evidence from Latin America. J. Int. Dev. **31**(7), 578–600 (2019)
59. Zaman, Q., U., Hassan, M.K., Arshad, H.M., Sultan, S., Ashraf, A.: Managerial efficiency and corporate leverage policy in Pakistan. Asian Acad. Manag. J. Account. Financ. **16**(2), 25–46 (2020)

Benchmarking in Education: Data Envelopment Analysis Approaches for Solving Bi-Objective Problems

Dovilė Stumbrienė[1(✉)], José L. Ruiz[2], and Inmaculada Sirvent[2]

[1] Institute of Data Science and Digital Technologies, Vilnius University, Vilnius, Lithuania
dovile.stumbriene@mif.vu.lt

[2] Centro de Investigación Operativa, Universidad Miguel Hernández, Alicante, Spain
{jlruiz,isirvent}@umh.es

Abstract. Policy frameworks of the European Commission have been developed to benchmark and monitor progress toward a common vision. The complexity of frameworks raised the need to add a second objective within the Data Envelopment Analysis (DEA). We proposed two benchmarking approaches through a bi-objective DEA model. The first model sets goal-adjusted targets and identifies strategies for improving national performance toward the EU-level goals. The model solves a bi-objective problem that imposes two objectives of closeness, between the actual performances and targets (effort) and between targets and EU-level goals (adjustment). The second model incorporates information on gender equality and finds targets that close the gender gap and improve performance. The model also solves a bi-objective problem that imposes two objectives of closeness. One objective is concerned with the effort that needs to be made for the achievement of the targets set, while the other objective is to minimize the gap between the targets set for male and female performance. This paper summarizes the proposed models and highlights the benefit of using bi-objective DEA approaches in the context of European policy frameworks. The proposed approaches have been applied in education, and two illustrative examples are discussed. However, these models can be applied in other areas facing similar challenges at the country or organization level.

Keywords: Benchmarking · Data envelopment analysis · Target setting · Education · Equality · Gender gap · European countries

1 Introduction

European education governance by numbers, by comparison, by "what works" relies on policy tools such as standardization and benchmarking (Ydesen and Brøgger, 2022; Robertson, 2022; Brøgger, 2018). The European Commission, as the centre of the EU governance system, is responsible for developing indicators for benchmarking, ensuring periodic monitoring and evaluation, overseeing member states' progress with the frameworks, as well as providing recommendations and guidelines (Egeberg and Trondal, 2017). The use of benchmarking, indicators, and monitoring prompted a shift in the

A. Emrouznejad et al. (Eds.): DEA 2024, LNOR, pp. 413–425, 2025.
https://doi.org/10.1007/978-3-031-98177-7_28

Commission's role from its right to take the initiative to its right to evaluate and monitor (Radaelli, 2013). This paper discusses two policy frameworks of the European Commission: the strategic framework towards a European Education Area (EEA strategic framework) and the Regional Gender Equality Monitor (RGEM framework).

EEA strategic framework is the outcome of the continuous cooperation of EU Member States in education, which sets out the vision towards the European Education Area (Council of the European Union, 2021). Seven internationally comparable indicators and EU-level goals have been set to monitor the progress towards achieving the vision by 2030. While, the RGEM framework was developed with the aim to measure gender equality at the regional level within the EU (Norlén et al., 2019). This framework is built on seven domains, one of which is *Knowledge*, supported by four indicators. So, the EEA strategic framework aims to improve performance in education and achieve defined EU-level goals. The RGEM framework also aims to improve performance in education, but differently from the EEA strategic framework, it seeks to ensure gender equality.

Benchmarking, as a strategic tool, is used to assess performance and aims to improve it (Elmuti and Kathawala, 1997). Performance measuring, in comparison with other respective units (peers), identifies the best practices and indicates how to accomplish better results (targets) is possible with benchmarking. Benchmarking enables not only quantifying and monitoring performance, guiding policy decisions, but also assisting the move towards a more analytically rigorous and data-driven approach to decision-making. Data Envelopment Analysis (DEA), introduced by Charnes et al. (1978), is a benchmarking approach that evaluates the performance of decision-making units (DMUs) and provides targets that direct the improvement towards best practices. The continuous evolution of DEA methodology within the last five decades is the key to the wide application range (Mergoni et al., 2024). The education field is not the exception, a large number of papers have assessed the performance of European education systems using DEA approaches (see De Witte and López-Torres (2017) for a review and the most recent papers: Stumbrienė et al. (2024a, b), Camanho et al. (2023), Segovia-Gonzalez and Contreras (2023), Dominguez-Gil et al. (2022), Stumbrienė et al. (2022)).

In this paper, we do not focus on measuring efficiency (as common in education benchmarking studies, constructing composite indicators), but we pay special attention to setting targets. In DEA, the performance of DMUs can be evaluated by setting targets on the frontier of the attainable set determined by efficient DMUs. It is common to impose one objective that minimizes the distance to the efficient frontier and sets the targets. In this way, we have the closest targets that minimize the gap between actual performances and best practices and realize the improvement with as little effort as possible. However, as a highly complex system, the education sector can require more than one objective under analysis to be optimized simultaneously.

As mentioned before, the EEA strategic framework has two aims – one is to improve performance in education, and the other is to achieve defined EU-level goals; the RGEM framework also aims to improve performance in education and, moreover, to ensure gender equality. In light of this, we need to incorporate, within the DEA benchmarking, information on EU-level goals in the case of the EEA strategic framework and information on the gender gap in the case of the RGEM framework. DEA allows us to deal with these issues by adding a second objective to the benchmarking model together with the usual criterion of minimizing the effort to achieve targets (improving performance toward best practices). We proposed two benchmarking approaches through a bi-objective DEA model.

The model proposed by Stumbrienė et al. (2024a), which deals with the EEA strategic framework, sets targets on the frontier of the attainable set by considering two objectives – minimizing the distance between actual outputs and targets as well as minimizing the distance between targets and EU-level goals. Another model proposed by Stumbrienė et al. (2024b), which is in line with the RGEM framework, sets targets for the performance indicators seeking to minimize differences between males and females while at the same time minimizing the differences between actual outputs and targets. In both models, the series of targets can be provided by varying the importance attached to each of the two objectives. In this way, through an optimization model, we can improve performance and simultaneously direct efforts at the EU-level goals or close the gender gap. Other bi-objective DEA models formulated for benchmarking seek to close the gap between actual performances, targets, peers or goals (see, e.g., Ruiz and Sirvent (2022a), Ruiz and Sirvent (2022b), Ruiz and Sirvent (2019)).

The present paper is focused on summarizing two proposed bi-objective DEA approaches in the context of European policy frameworks, which raised the need to add a second objective within the DEA benchmarking. The structure of the rest of this paper is as follows. In Sect. 2, we briefly summarize the proposed models. In Sect. 3, we demonstrate two examples of setting targets through the bi-objective DEA models and highlight the benefit of quantitative policy guidance. We provide a concluding discussion in Sect. 4.

2 Methods

The discussion of the two policy frameworks of the European Commission here is based on the analysis of several indicators, say $y_1, \ldots, y_s$, which reflect the performance of countries/regions (obviously, the set of indicators selected for the analysis is different in each framework). For that reason, we consider the "attainable set" (AS) below proposed in Liu et al. (2011) for the benchmarking of those "decision-making units" (DMUs), which is a bounded closed convex set constructed from the observed data by maintaining some assumptions such as convexity. Liu et al. (2011) develop a theoretical framework for the so-called DEA models without explicit inputs (WEI), which can be used in the evaluation of performance (effectiveness) without regard to any production process.

$$\mathrm{AS} = \{\mathrm{Y} \geq 0 / \ \mathrm{Y} \leq \sum_{j=1}^{n} \lambda_j \mathrm{Y}_j, \ \sum_{j=1}^{n} \lambda_j = 1, \ \lambda_j \geq 0\} \tag{1}$$

The AS thus collects the set of alternative plans of performance that can be considered for benchmarking. In particular, the frontier of the AS, $\partial(\mathrm{AS})$, which can be seen as a best-practice frontier, is especially relevant because the projections of the DMUs to $\partial(\mathrm{AS})$ determine the targets that guide the actions needed to improve performance towards best practices.

When planning improvements in general, the main objective is to close the gap between actual performances and best practices. If we denote by $\hat{Y}_0$ the projection of a given DMU_0 onto $\partial(AS)$, then the benchmarking model below provides the closest targets, i.e., those that require from DMU_0 the least possible effort for their achievement

$$\begin{aligned} & \text{Min } d\{Y_0, \hat{Y}_0\} \\ & \text{s.t.}: \\ & \qquad \hat{Y}_0 \in \partial(AS) \end{aligned} \tag{2}$$

where d is a distance in the metric space $\mathbb{R}^s$.

The model below provides an operational formulation of (2)

$$\begin{aligned} \text{Min} \quad & \sum_{r=1}^{s} \frac{(\hat{y}_{r0} - y_{r0})}{y_{r0}} \\ \text{s.t.}: \quad & \\ & \sum_{j \in E} \lambda_j y_{rj} = \hat{y}_{r0} & r = 1,\ldots,s \\ & \hat{y}_{r0} \geq y_{r0} & r = 1,\ldots,s \\ & \sum_{j \in E} \lambda_j = 1 & \\ & -v + \sum_{r=1}^{s} u_r y_{rj} + d_j = 0 & j \in E \\ & v, u_r \geq 1 & r = 1,\ldots,s \\ & \lambda_j d_j = 0 & j \in E \\ & \lambda_j, d_j \geq 0 & j \in E \end{aligned} \tag{3}$$

where E is the set of effective DMUs. We note that the restrictions in (3) ensure projections onto the frontier of the AS (see Aparicio et al. (2007)).

As stated in the introduction, the European Commission's policy frameworks have other objectives besides improving the performance of countries/regions. In the case of the EEA strategic framework, the performance of countries should also be directed towards achieving defined EU-level goals. Denote by $G = (g_1, \ldots, g_s)$ those goals. Then, we seek to set targets by closing the gap both with actual performances and with goals. The following bi-objective model provides those targets

$$\begin{aligned} & \text{Min } d\{Y_0, \hat{Y}_0\} \\ & \qquad d\{\hat{Y}_0, G\} \\ & \text{s.t.}: \\ & \qquad \hat{Y}_0 \in \partial(AS), \end{aligned} \tag{4}$$

In the RGEM framework, we have the objective of closing the gender gap. The information on indicators is disaggregated by gender, y_r^M, y_r^F, $r = 1, \ldots, s$, that is, we have data on the indicators both for males and females. Targets thus result from the following bi-objective benchmarking model

$$\begin{aligned} &\text{Min } d\{Y_0, \hat{Y}_0\} \\ &\quad\;\; d\left\{\hat{Y}_0^M, \hat{Y}_0^F\right\} \\ &\text{s.t. :} \\ &\quad\;\; \hat{Y}_0 \in \partial(AS) \end{aligned} \tag{5}$$

In order to solve bi-objective models (4) and (5), we follow an approach based on a linear scalarization that depends on a parameter α, $0 \leq \alpha \leq 1$, so that we eventually solve a model having as single objective minimizing $\alpha\, d\{Y_0, \hat{Y}_0\} + (1-\alpha)d\{\hat{Y}_0, G\}$, for (4), and minimizing $\alpha\, d\{Y_0, \hat{Y}_0\} + (1-\alpha)d\left\{\hat{Y}_0^M, \hat{Y}_0^F\right\}$, for (5).

An operative formulation for (4) would be then:

Min $\alpha \sum_{r=1}^{s} \frac{|\hat{y}_{r0}-y_{r0}|}{y_{r0}} + (1-\alpha) \sum_{r=1}^{s} \frac{|\hat{y}_{r0}-g_r|}{y_{r0}}$, subject to the same constraints as in (3), except $\hat{y}_{r0} \geq y_{r0}$, $r = 1, \ldots, s$, since in this case dominance of the projection is not required.

Analogously, model (5) can be operationalized as:

Min $\alpha \sum_{r=1}^{s} \left[(\hat{y}_{r0}^M - y_{r0}^M) + (\hat{y}_{r0}^F - y_{r0}^F)\right] + (1-\alpha) \sum_{r=1}^{s} \left|\hat{y}_{r0}^M - \hat{y}_{r0}^F\right|$, subject to the same constraints as in (3), where both the indicators and their corresponding projections are disaggregated by gender. That is: $\hat{y}_{r0}^M = \sum_{j\in E} \lambda_j y_{rj}^M$, $\hat{y}_{r0}^F = \sum_{j\in E} \lambda_j y_{rj}^F$, $\hat{y}_{r0}^M \geq y_{r0}^M$, $\hat{y}_{r0}^F \geq y_{r0}^F$, $r = 1, \ldots, s$. For details of these formulations, see Stumbrienė et al. (2024a, b).

This approach provides non-dominated solutions. Through the specification of α, we may adjust the importance that is attached to each of the two objectives. The specification $\alpha = 1$ leads to the closest targets, i.e., targets that are set seeking only to minimize the effort for their achievement. As α decreases, the models set targets that are progressively closer to the EU-level goals for the EEA strategic framework or that progressively close the gender gap for RGEM strategic framework. Eventually, a series of targets are generated that may offer managers alternatives for decision-making when planning improvements.

3 Case Studies

The empirical analysis reported in this section uses seven indicators defined in the EEA strategic framework and four indicators presented in the *Knowledge* domain of the RGEM framework. The description of the indicators used in this study is provided in Table 1 and Table 2. The data were extracted from the Eurostat and OECD databases for 29 European countries in the case of the EEA strategic framework and for 30 European countries at the regional level for the RGEM framework. The data correspond to the year 2022, except for early childhood education and care, to the year 2021. For applying models (4) and (5), some indicators were adjusted by subtracting the values observed in each DMU from 100% to ensure that higher values of all indicators correspond to better performance (see Table 1 and Table 2). The descriptive statistics of the indicators are provided in Table 3 and Table 4.

Table 1. Description of EEA strategic framework indicators used in the empirical analysis.

Variable	Description of indicator	Adjusted	EU-level goals
Low_math	Underachievement in maths: the share of 15-year olds who fail to reach Level 2 on the PISA scale for mathematics	+	Below 15
Low_reading	Underachievement in reading: the share of 15-year olds who fail to reach Level 2 on the PISA scale for reading	+	Below 15
Low_science	Underachievement in science: the share of 15-year olds who fail to reach Level 2 on the PISA scale for science	+	Below 15
Early_child	Early childhood education and care: the percentage of children between age 3 and each country's starting age for compulsory primary education participating in early childhood education and care	–	Above 96
Early_leavers	Early leavers from education: the percentage of the population aged 18–24 with at most lower secondary educational attainment and no longer in formal or non-formal education or training	+	Below 9
Tertiary	Tertiary education attainment: the percentage of the population aged 25–34 with tertiary educational attainment	–	Above 45
Adults_learning	Participation adults in learning: the percentage of the population aged 25–64 who have participated in formal or non-formal education and training during the last 12 months	–	Above 47

As discussed above, models (4) and (5) evaluate performance by minimizing efforts to achieve best practices while at the same time dealing with the second objective – closing the gap between targets and EU-level goals in the case of the EEA strategic framework and closing the gender gap in the case of the RGEM framework. When $\alpha = 1$, the models set the closest targets (least possible effort), but the second objective is ignored. The opposite holds for $\alpha = 0$; the models ignore the objective of effort to achieve best practices and seek to set targets (on the frontier of the AS) by closing the gap between targets and EU-level goals (EEA strategic framework) and closing the gender gap (RGEM framework). As α increases from 0, the targets get closer to the actual data since the objective of closeness with actual performances is attached more importance and when $\alpha = 0.5$, both objectives reach the same importance. Next, the illustrative examples of Czechia and Latvia are discussed. Focusing on performance improvement towards the EU-level goals, Czechia can be an example of that; due to having space for improvements in almost all areas (Czechia had achieved only one out of seven analysed EU-level goals). Latvia, with one of the largest total gender gaps, can

Table 2. Description of the RGEM framework (*Knowledge* domain) indicators used in the empirical analysis.

Variable	Description of indicator	Adjusted
Tertiary	Tertiary educational attainment: the percentage of the population aged 25–64 with tertiary educational attainment	–
Adults_learning	Participation adults in learning: the percentage of the population aged 25–64 who have participated in formal or non-formal education and training during the last 4 weeks	–
Early_leavers	Early leavers from education: the percentage of the population aged 18–24 with at most lower secondary educational attainment and no longer in formal or non-formal education or training	+
NEET_rate	Neither in employment nor in education: the percentage of the population aged 15–29 neither in employment nor in education and training	+

Table 3. Descriptive statistics of the EEA strategic framework indicators for 29 European countries.

	Low_math	Low_reading	Low_science	Early_child	Early_leavers	Tertiary	Adults_learning
Average	29.7	27.9	25.1	88.2	8.2	45.4	45.8
St. Dev	9.6	10.4	9.5	10.5	3.4	9.8	13.9
Min	15	11.4	10.1	49.8	2.3	24.7	16.6
Max	53.6	60.6	51.8	100	15.6	62.3	73.9

Table 4. Descriptive statistics of the RGEM framework (*Knowledge* domain) indicators according to gender for 93 European regions.

	Tertiary		Adults_learning		Early_leavers		NEET_rate	
	M	F	M	F	M	F	M	F
Average	32.0	38.5	11.1	13.6	88.9	92.2	89.9	87.4
St. Dev	9.7	10.8	6.4	8.7	5.3	3.7	4.5	5.6
Min	12.5	15.6	1.3	1.5	64.7	81.6	71.2	68.6
Max	52.4	59.4	29.9	44.5	97.1	98.4	96.5	95.8

illustrate the example of how to improve performance and simultaneously direct towards gender equality.

3.1 The Case Study of Czechia: How to Improve Performance Towards the EU-Level Goals

Czechia can be characterised as standing somewhere in the middle according to the declared indicators in the EEA strategic framework – quite far from top-performing countries such as Ireland, Estonia, Sweden or Denmark, but also quite ahead of countries such as Bulgaria, Romania, Greece or Cyprus. Czechia had achieved one out of seven analysed EU-level goals; other indicators should be improved by approximately 10 percentage points or less, i.e., there is no dominant low-performing area. Different scenarios of how to improve performance towards the EU-level goals are presented in Table 5, which reports the targets provided by model (4) for the extreme cases ($\alpha = 1$ and $\alpha = 0$), the intermediate case ($\alpha = 0.5$), and, additionally, actual data of Czechia and the EU-level goals.

When $\alpha = 1$, we have the usual scenario in benchmarking, minimizing the gap between actual performances and best practices and realize the improvement with as little effort as possible. In this scenario, model (4) provides the closest targets to the actual data of Czechia and ignores the second objective of closing the gap between targets and EU-level goals. It can be noted that in this scenario, the targets for *underachievement in maths, reading and science* together with *early leavers from education* are very close to the actual data of Czechia, while the efforts should be directed mainly toward the performance in *tertiary educational attainment, early childhood education and care,* and *participation of adults in learning*. As α decreases, the targets are more in line with EU-level goals at the expense of increasing the efforts to reach the targets, and when $\alpha = 0.5$, both objectives reach the same importance.

Table 5. Targets for Czechia.

	Low_math	Low_reading	Low_science	Early_child	Early_leavers	Tertiary	Adults_learning
Data	74.5	78.7	80.1	85.4	93.8	34.6	45.7
α=1	74.5	78.3	79.0	98.8	93.7	54.4	52.1
α=0.5	85.0	86.2	89.9	91.5	89.2	43.9	48.1
α=0	84.8	86.3	89.6	91.8	89.6	45.0	48.5
Goal	85.0	85.0	85.0	96.0	91.0	45.0	47.0

The radar chart in Fig. 1 reports the gap between the actual performance of Czechia (the zero line, black) and the targets associated with $\alpha = 1$ and $\alpha = 0.5$. Points above the black line indicate that the targets have not been achieved, while points below indicate that the targets have been already achieved. It can be noted that improvements should be directed in different ways depending on the value of α. In the usual scenario when $\alpha = 1$, as mentioned above, Czechia should direct improvements mainly toward the performance in *tertiary educational attainment, early childhood education and care,* and *participation of adults in learning*. However, the EEA strategic framework has two aims – one is to improve performance in education, and the other is to achieve defined EU-level goals; in the usual scenario, we impose only one objective and EU-level goals are ignored. In the scenario when both objectives have the same importance ($\alpha = 0.5$), improvements should be redirected – less efforts to improve *tertiary educational attainment, early childhood education and care,* and *participation of adults in learning*, but also focus on improving *underachievement in maths, reading and science.*

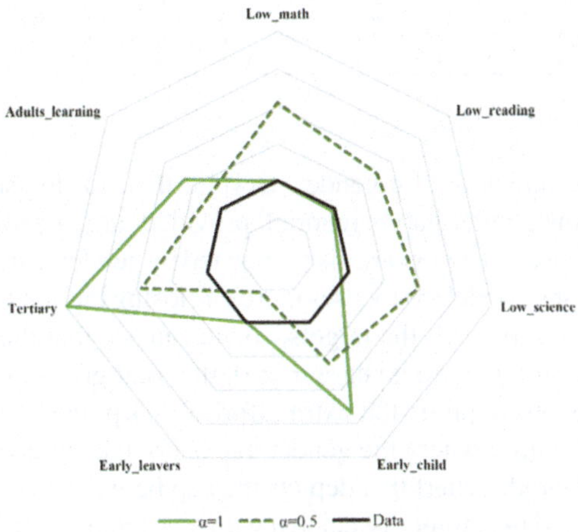

Fig. 1. Benchmarking of Czechia.

As we showed in the case of Czechia, the strategy of improving national performance differs depending on the task we have. Suppose we want to improve performance in education based on declared indicators. In this case, we can use the DEA model to minimize the distance between actual performances and best practices. But if we have declared indicators together with defined goals, we need to incorporate the second objective, in order to improve performance and simultaneously direct efforts at the EU-level goals, i.e. minimizing the distance between targets and EU-level goals.

3.2 The Case Study of Latvia: How to Improve Performance Towards Gender Equality in Education

Latvia has one of the largest total gender gaps, together with Iceland, Sweden, Estonia, Denmark, Lithuania, and some regions in Poland, Finland, and Spain. Different pathways to improve the performance of Latvia and close the gender gap are presented in Table 6, which reports the targets provided by the model (5) for $\alpha = 1$, $\alpha = 0.5$, $\alpha = 0$ and actual data for Latvia. Additionally, the gender gap, together with the total gender gap and total effort, are reported for each scenario. The gender gap for each indicator is calculated as the difference between male and female performance. In this way, a negative gender gap reflects that females, on average, outperform males. The total gender gap is calculated as the sum of the absolute value of the gender gap for each indicator. It can be noted that the highest gender gap is observed in *tertiary educational attainment* due to the low performance of males compared with females, indicating a large scope for improvement in this indicator for Latvia.

When $\alpha = 1$, on the one hand, we have the usual scenario to improve performance with as little effort as possible, but on the other hand, this scenario is not usual due

Table 6. Targets for Latvia.

	Tertiary			Adults_learning			Early_leavers			NEET_rate			Total	Total
	M	F	Gap	M	F	Gap	M	F	Gap	M	F	Gap	gap	effort
Data	29.4	48.9	-19.5	6.6	12.5	-5.9	90.7	96.0	-5.3	89.0	88.4	0.6	31.3	
α=1	39.9	48.9	-9.0	10.0	12.5	-2.5	95.8	97.4	-1.5	91.3	88.9	2.4	15.4	23.3
α=0.5	42.6	48.9	-6.3	10.8	12.5	-1.7	96.0	96.0	0.0	91.0	90.0	1.1	9.0	26.4
α=0	46.4	48.9	-2.5	11.2	12.5	-1.3	95.5	96.0	-0.5	91.5	91.2	0.3	4.7	31.7

to disaggregating the indicators by gender. Despite that, in this scenario, the second objective of closing the gender gap is ignored, provided targets reduce the total gender gap almost twice because indicators are disaggregated by gender instead of using national averages. As α decreases, model (5) focuses more on closing the gender gap at the expense of increasing the efforts to reach the targets, so we can see that the total gender gap is reduced even further. In the extreme case $\alpha = 0$, the total gender gap is reduced up to 4.7 percentage points but requires the extra effort of 8.4 points (from 23.3 to 31.7) as compared with the situation where the gender gap is not taken into account.

Figure 2 shows the radar chart that depicts the gap between the actual performances (the zero line, black) and the targets associated with $\alpha = 1$ and $\alpha = 0.5$, disaggregating by gender (the yellow line for males and the blue line for females). It can be noted that Latvia has to improve performance the most among males in *tertiary educational attainment*; also, minor improvements should be directed towards the performance of males in *early leavers from education* and *participation of adults in learning*. The noticeable difference between targets provided in these scenarios ($\alpha = 1$ and $\alpha = 0.5$) is in *tertiary educational attainment*, where the gender gap is the widest.

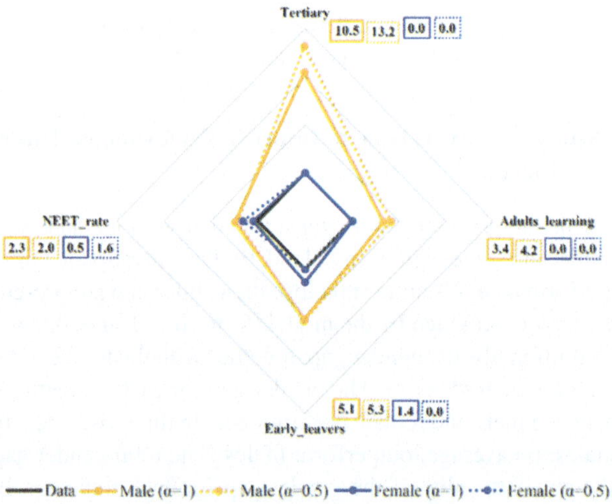

Fig. 2. Benchmarking of Latvia.

As we showed in the case of Latvia, a more accurate benchmarking analysis towards gender equality in education can be performed with disaggregated indicators by gender.

If the gender gap is wide, it is useful to add the second objective of closing it with the usual criterion of improving performance toward best practices. In this way, the targets are more directed towards ensuring gender equality.

4 Conclusions

This study highlights the need for a bi-objective DEA approach in the context of European policy frameworks and demonstrates how incorporating a second objective within the DEA benchmarking can enhance the analysis and benefit quantitative policy guidance. The proposed bi-objective DEA approaches expand the common practice in benchmarking in education in several ways. First of all, instead of evaluating performance as the degree to which the desirable goals are met, i.e. measure the "success" in achieving the goals, the proposed approaches focus on guiding improvement by identifying targets. In addition, the complexity of the European Commission's policy frameworks raised the need to incorporate a second objective to the benchmarking model together with the usual criterion of improving performance toward best practices. This led to more comprehensive improvement guidance.

Empirically, we showed how bi-objective DEA approaches can be used to improve the national performance of Czechia towards the EU-level goals defined in the EEA strategic framework and how to improve the performance of Latvia towards gender equality in education based on the RGEM framework.

First of all, the results point out different ways for improvement depending on the objective imposed in the model. One strategy for improvement is obtained when the aim is to improve performance based on set indicators, while another is obtained by incorporating the aim to achieve defined goals simultaneously to improve national performance. This supposes the importance of using the bi-objective DEA approach in cases where there are two objectives.

The results also show that in terms of improving performance towards equality in education, disaggregated indicators led to more accurate directions to improve performance towards equality (in this case, gender equality; it can also be based on other social categories, such as social class, ethnicity, race, etc.). If the gender gap is wide, it is not enough to perform an analysis with disaggregated indicators; the objective of closing the (gender) gap can be incorporated to get targets that are more directed towards ensuring equality.

In conclusion, as DEA has mostly been used for evaluating performance in education benchmarking studies, improving performance by setting targets is not as common practice as in other areas, such as transport, energy, agriculture or manufacturing. Education, as a social area, is based on broader objectives, which can go beyond improving performance against set indicators, and include broader objectives such as "education for all", gender equality, etc., which are reflected in European policy frameworks. An emphasis on more encompassing objectives than just improving performance is particularly relevant for social areas such as education and incorporating the second objective within the DEA benchmarking can be one step ahead to more accurate performance improvement by setting targets. The bi-objective DEA approaches presented in this study are in line with the policy frameworks of the European Commission but can also be applied in other areas facing similar challenges of solving bi-objective problems.

Acknowledgments. This work is financed by the Portuguese Foundation for Science and Technology within the project eduBEST – Education Systems Benchmarking with Frontier Techniques, with reference 2022.08686.PTDC, and through the grants, PID2019-105952GB-I00 and PID2021-122344NB-I00, funded by Ministerio de Ciencia e Innovación/Agencia Estatal de Investigación (MCIN/AEI/https://doi.org/10.13039/501100011033 "ERDF A way of making Europe"), and PROMETEO/2021/063, funded by Conselleria d'Innovació, Universitat, Ciencia i Societat Digital.

References

Aparicio, J., Ruiz, J.L., Sirvent, I.: Closest targets and minimum distance to the Pareto-efficient frontier in DEA. J. Prod. Anal. **28**(3), 209–218 (2007)

Brøgger, K.: The performative power of (non)human agency assemblages of soft governance. Int. J. Qual. Stud. Educ. **31**(5), 353–366 (2018)

Camanho, A.S., Stumbriene, D., Barbosa, F., Jakaitiene, A.: The assessment of performance trends and convergence in education and training systems of European countries. Eur. J. Oper. Res. **305**(1), 356–372 (2023)

Charnes, A., Cooper, W., Rhodes, E.: Measuring the efficiency of decision making units. Eur. J. Oper. Res. **2**(6), 429–444 (1978)

Council of the European Union: Council Resolution on a strategic framework for European cooperation in education and training towards the European Education Area and beyond (2021–2030) Brussels, 2021/C 66/01 (2021)

De Witte, K., López-Torres, L.: Efficiency in education: a review of literature and a way forward. J. Oper. Res. Soc. **68**(4), 339–363 (2017)

Dominguez-Gil, C., Segovia-Gonzalez, M.M., Contreras, I.: A multiplicative composite indicator to evaluate educational systems in OECD countries. Compare J. Comp. Int. Educ. **52**(8), 1296–1313 (2022)

Egeberg, M., Trondal, J.: Researching European Union agencies: what have we learnt (and where do we go from here)? J. Common Market Stud. **55**(4), 675–690 (2017)

Elmuti, D., Kathawala, Y.: An overview of benchmarking process: a tool for continuous improvement and competitive advantage. Benchmarking Qual. Manag. Technol. **4**(4), 229–243 (1997)

Liu, W.B., Zhang, D.Q., Meng, W., Li, X.X., Xu, F.: A study of DEA models without explicit inputs. Omega **39**(5), 472–480 (2011)

Mergoni, A., Emrouznejad, A., De Witte, K.: Fifty years of data envelopment analysis. Eur. J. Operational Res. **326**(3), 389–412 (2025). https://doi.org/10.1016/j.ejor.2024.12.049

Norlén, H., Papadimitriou, E., Dijkstra, L.: The regional gender equality monitor. Publications Office of the European Union, EUR 29679 EN (2019)

Radaelli, C.M.: Measuring policy learning: regulatory impact assessment in Europe. In: Zito, A. (eds.) Learning and Governance in the EU Policy Making Process, pp. 43–62. Routledge (2013)

Robertson, S.L.: Guardians of the future: International organisations, anticipatory governance and education. Glob. Soc. **36**(2), 188–205 (2022)

Ruiz, J.L., Sirvent, I.: Benchmarking within a DEA framework: setting the closest targets and identifying peer groups with the most similar performances. Int. Trans. Oper. Res. **29**, 554–573 (2022)

Ruiz, J.L., Sirvent, I.: Identifying suitable benchmarks in the way towards achieving targets using data envelopment analysis. Int. Trans. Oper. Res. **29**, 1749–1768 (2022)

Ruiz, J.L., Sirvent, I.: Performance evaluation through DEA benchmarking adjusted to goals. Omega **87**, 150–157 (2019)

Segovia-Gonzalez, M.M., Contreras, I.: A composite indicator to compare the performance of male and female students in educational systems. Soc. Indic. Res. **165**(1), 181–212 (2023)

Stumbrienė, D., Ruiz, J.L., Sirvent, I.: Rethinking EU-level goals: different strategies to improve national performance toward the European Education Area objectives. Int. Trans. Oper. Res. 1–28 (2024a)

Stumbrienė, D., Ruiz, J.L., Sirvent, I. Towards gender equality in education: Different strategies to improve subnational performance of European countries using data envelopment analysis. Socio-Econ. Plan. Sci. 102138 (2024b)

Stumbrienė, D., Želvys, R., Žilinskas, J., Dukynaitė, R., Jakaitienė, A.: Efficiency and effectiveness analysis based on educational inclusion and fairness of European countries. Socioecon. Plann. Sci. **82**, 101293 (2022)

Ydesen, C., Brøgger, K.: The crafting of a European education space and Europeanization: the role of the EU and the OECD. In: Coutaz, R., Paoli, S. (eds.) Building Europe through Education: Actors, Spaces and Pedagogies in a Historical Perspective. Routledge (2022)

Does Private Tuition Affect Efficiency of Government and Government Aided Schools? An Evidence of Higher Secondary Schools Based in Kolkata

Arpita Ghose[1(✉)] and Sangita Choudhury[2]

[1] Department of Economics, Jadavpur University, Kolkata, India
arpitaju@gmail.com

[2] Department of Applied Economics, Maulana Abul Kalam Azad University of Technology, Kolkata, West Bengal, India
sangita.choudhury@makautwb.ac.in

Abstract. This paper contributes to the literature by estimating yet unexplored role of private tuition in explaining output oriented technical efficiency levels (TEF_{out}) of higher secondary stage (HS), as estimated applying BCC (1984) DEA model, using primary survey data for 25 government and government aided schools in Kolkata, India, collected using stratified random sampling method, after controlling for the other determinants of TEF_{out}, considering private tuition taken by the students as an endogenous variables and also measuring the variation of TEF_{out} not accounted by private tuition. The results show 56% sample schools are inefficient at HS; hence more of output can be produced using the existing resources; TEF_{out} ranges between 0.652 and 1.000. A positive and significant influence of 'percentage of private tuition taking students' in determining the TEF_{out} is evident. TEF_{out} variation not explained by private tuition is very less, ranging only from 0.40% and 8.64% with 3.72% as mean, which is a serious concern. The findings also reflect, the positive role of less discussed factors on TEF_{out} at HS namely: (i) Government targeted policy variable, like proportion of the girl student receiving 'Kanyashree' (an initiative taken by the Government of West Bengal to improve the life and the status of the girls by helping economically backward families with cash so that the marriage of their girl child before eighteen years is not being arranged, because of socio economic and financial problem), (ii) the amount of government fund received by the schools (excluding salary payments), reflecting a non-linear relationship with TEF_{out}, but with positive marginal effect, implying an increase in government fund will increase TEF_{out}, (iii) role of Headmaster/mistress (HM) in school management, like (a) interaction of HM with teachers & students and (b) interaction frequency with state department of education, while taking any policy decision for the school and (iv) School infrastructure related variables, like number of smart classroom in the school and whether librarian of the school is on duty. Suitable policies are suggested to enhance TEF_{out} at HS of the schools and to reduce the dependence on private tuition.

Keywords: Efficiency Analysis of Education · Microeconomics · Data Envelopment Analysis · Cross-Sectional Models · Stratified Random Sampling

A. Emrouznejad et al. (Eds.): DEA 2024, LNOR, pp. 426–448, 2025.
https://doi.org/10.1007/978-3-031-98177-7_29

JEL CODE: I21 · B21 · C61 · C31 · C83

1 Introduction

For India, sufficient educational achievement has been made with respect to literacy rate. Indian literacy rate according to 2011 census stands at 74% compared to 12% at the end of British rule in 1947. Coming to the Indian school education, among the four levels; primary, upper primary, secondary and higher secondary; higher secondary is a stepping stone to higher education level that equips and empowers students with skills required for entering the job market. Gross enrollment ratio (GER) at Indian higher secondary level (HS) is 27% in 2019–20.

Thus, given this achievement in GER at HS, it is worthwhile to map GER at HS to the other performance indicator like efficiency. Basic question is whether the different schools at HS are technically efficient in the sense that whether maximum output given the amount of resources are being generated at HS. In tune with Hulten (1996), highlighting the role of inefficiency imposing penalty on economic growth, one may argue, it may be possible that there is some extent of inefficiency, as a result of which India may be unable to produce maximum educational output producible and may be forgoing a part of economic growth. A related question is that, what are the major factors influencing technical efficiency (TE) of different schools?

There are two concepts of technical efficiency, (i) output oriented TE, focusing on maximizing the output, keeping the inputs level constant and (ii) input oriented TE, which focuses on determining how much a firm can reduce its inputs while maintaining the same output. This paper discus output oriented TE and its determinants.

In the Indian context, given the four levels of education at school level, efficiency studies is much needed for HS; as it is a stepping stone to higher education level that equips and empowers students with skills, which in turn will foster accumulation of human capital. Studying school efficiency at HS is crucial for maximizing educational outcomes and resource allocation, leading to better learning environments and societal benefits. Analysing school TE is important for (i) **Optimizing Resource Allocation i.e.,** by understanding how schools use resources, policymakers and educators can identify areas where funding and other resources could be better allocated to improve student outcomes, (ii) **Improving Student Learning i.e.,** efficient schools are more likely to provide effective instruction, leading to better student achievement and development, While quality of education is important, TE assessment complements these efforts by ensuring that resources are used effectively to achieve best possible outcomes, (iii) **Identifying Best Practice, i.e.,** identifying efficient schools can help to identify effective teaching methods, curriculum design, and school management practices that can be replicated in other schools, (iv) **Promoting Equity i.e.,** TE studies can help to identify disparities in educational outcomes and resources, allowing for targeted interventions to ensure equitable access to quality education; one can identify the schools which are not operating efficiently, so that one can formulate appropriate policy to boost up efficiency of these schools, (v) **Social Cohesion, i.e.,** efficient and equitable education systems can lead to higher levels of social cohesion, (vi) **Focus on Student Development, i.e.,** School

TE research can help to understand how schools can better support student development, both academically and non-academically, (vii) **Data-Driven Decision Making, i.e.,** TE studies provide data that can inform evidence-based decision-making, leading to more effective policies and practices, (viii) **Accountability, i.e.** TE analysis can help to hold schools accountable for their performance and resource utilization (ix) **Long-Term Economic Growth, i.e.,** a highly efficient education system can contribute to a skilled workforce and a more productive economy by providing accumulation of human capital. Further, the role of efficiency on economic growth is highlighted by Hulten (1996), as argued above.

The positive role of human capital in economic growth is supported by existing literature, both theoretically and empirically. (Lucas (1988); Romer (1990) among others).It is the level of human capital, not its rate of accumulation, which influences the rate of productivity growth (Romer (1990)). Growth is positively correlated with the initial level of human capital. Mankiw, Romer and Weil (1992) modelled extension of Solow model incorporating human capital as a separate input and predicts that, other things being equal, a country with higher amount of human capital should have a higher level of per capita income. The education and the creation of human capital was responsible for both the differences in labour productivity and in overall levels of technology (Barro (1991)). Having established the importance of human capital in economic development, the role of expansion of education can be justified. Education can be viewed as an investment in human capital and thus the development of a strong and efficient education system is urgently needed. In this connection, the existence of some literature establishing the relationship between education and economic growth empirically (Gundlach (2001); Hanushek & Woessmann (2010); Marquez-Ramos & Mourelle (2019); Valero (2021); Ziberi, Rexha, Ibraimi, & Avdiaj (2022); DSouza & Lobo (2024)) strengthens importance of education and thereby human capital on economic growth and thus provides justification for efficiency study on education, as generation of efficiency can promote skill and human capital and in turn foster economic growth.

Several literatures exist on measurement of technical efficiency scores (TE) using Data Envelopment Analysis (DEA), and its determinants, that requires no prior specification of functional form of the criterion function (Charnes, Cooper and Rhodes (CCR) (1978, 1981); Banker, Charnes and Cooper (1984), among others). In this context, it will also be worthwhile to highlight the significant contribution of Mergoni, et al. (2025), offering analysis of DEA development within the last fifty years. Some literatures on efficiency of the school/education (i) around the globe are Bessent and Bessent (1980); Bonesronning and Rattsq (1994); Hanushek (1994, 1996); Levin (1997); Andrew, Duncombe and Yiner (2002); Bardly and Taylor (2002); Driscoll, Halcousis, Svorny (2003); Kruger (2003); Newman et al. (2006); Adkins and Moomaw (2007); Leithwood, Jantzi (2009); Arshad (2014); Rocha et al. (2014); Johnes et al. (2017); Khotvaneh et al. (2017); Hashmi and Bradley (2018); Masci, Witte and Agasisti (2018); Schiltz (2018); Sotiriadis et al. (2018); Agasisti and Zoido (2019); Javornik and Klemencic Mirazchiyski (2023); Júnior et al. (2024); Hsiao and See (2024); (ii) for India are Kingdon (1996); Sankar (2007); Tyagi et al. (2009); Sengupta and Pal (2010, 2012);Purohit (2015); Mahapatra (2015); Ghose (2014, 2017); Singh, Pant and Goel (2018), among others.

However, most of the Indian school level studies are based on secondary data and the studies using primary survey data are lacking. Further, for a developing country like India, of the school is expected to be positively influenced by the availability of fund from the government and also by the government policy variables. Sengupta and Pal (2010, 2012) considered deprivation aspects; social aspects and policy aspects which may affect technical efficiency, but the favourable effect of the policy variables on it has not been obtained. The limitation of Sengupta and Pal is that they have used composite index for determinant analysis. But to have a definite policy for improvement, the role of each of the items under the broad indicators like infrastructure, school characteristics, policy variable etc. are needed to be analyzed. It may be possible that one variable under the broad indicator is significant while the other is not. This differential effect cannot be captured under composite index comprising of different categories. Further, TE of the schools may be influenced by the government policy for some specific targeted group, which was not there in Sengupta and Pal (2010, 2012).

The perusal of the literature thus identifies some research gaps. None of the issues has been explored yet as stated in following objectives of the present paper: First of all, given the fact that there is a dearth in the literature for Indian HS education system dealing with school level study on TE using primary survey data, the present paper is involved with measurement and determinants of output oriented Technical Efficiency (TEF_{out}) at HS, applying BCC DEA Model, using primary survey data-for 25 government and government aided schools, in Kolkata, India, collected using stratified random sampling method. The reasons for the choice of public sector schools in Kolkata is that the charges for government sector schools are very low and it is necessary to investigate the TEF_{out} of such low charges schools. This study basically addresses the question whether different government and government aided schools in Kolkata are technically efficient at HS, in the sense that they are generating maximum school output given the existing resources available at school, i.e. by measuring TEF_{out} and considers how TEF_{out}, can be improved. **The role of primary survey data can be highlighted from the fact that** the use of such data helps us to collect school level information on students taking private tuition, some important infrastructural variables like existence of smart classroom, whether Liberians are on duty, percentage of students getting kanyashree, (an initiative taken by the Government of West Bengal (GoWB) to improve the life and the status of the girls by helping economically backward families with cash so that the marriage of their girl child before eighteen years is not being arranged, because of socio economic and financial problem), information on the role of headmaster/mistress (HM), i.e. before taking any policy decision whether (i) HM interacts with teachers and students and (ii) visit state education department; the interaction of HM with teachers and students and the frequency of visiting state government education departments by the HM matters, are taken as determinants. These data cannot be availed from the secondary data sources. In fact, the estimated results will highlight the role of these important variables in explaining TEF_{out}, which were unexplored yet. Further, the primary survey data help us to collect school specific inputs and outputs variables, needed for estimation purposes as listed below, which cannot be obtained from secondary sources. **Secondly**, it has been observed that most of the students take private tuition. According to State Council of Education Research and Training Report (SCERT), in West Bengal 89% of male secondary and higher secondary students avail private tuition. In this context it turns out that TEF_{out} as achieved by the schools does not reflect school's true performance. Thus it is necessary to measure the extent of variation of TEF_{out} not accounted by the private tuition. This paper adds to the literature by measuring such unexplored variation of TEF_{out} not accounted by private tuition, after controlling for the other determinants of TEF_{out}, like role of headmaster/ mistress /principal in school management while framing policies, characteristics of head of the school, characteristics of the teachers, school infrastructure, school characteristics and student composition related variables, student's characteristics and socio-economic background of the students and student's perception about different attributes for higher secondary level of education; and also considering private tuition taken by the students as an endogenous variable. Among these included control variables, for finding out TEF_{out} not accounted by the private tuition, only the significant control variables will be considered. **Thirdly,** the paper intends to find out the role of HM in school management in explaining TEF_{out} of the school., i.e., whether taking any policy decisions (i) HM interacts with teachers and students, and such interaction is an important determinants of TEF_{out} and (ii) frequency of visiting state government education departments by the HM matters in explaining TEF_{out}. **Fourthly,** in case of the state West Bengal, India, there is an initiative on the part of GoWB, popularly known as Kanyashree Prakalpa, as discussed above. Thus it will be interesting to look into the yet unexplored impact of this targeted government policy variable on TEF_{out} of the school, i.e., whether the provision of Kanyashree has been able to boost up TEF_{out} of the schools. **Finally,** in the era of new technology advancement, the existence of smart classroom plays an important positive role for teaching methods. To keep the track with the advanced technique, the GoWB is providing smart classroom to different schools with support from the West Bengal Electronics Industry Development Corporation limited and has introduced e-learning and smart classrooms in schools, providing financial assistance for smart classrooms under the "ICT and Digital initiatives" components of Samagra Shiksha Mission. Thus it is important to trace out yet unexplored impact of smart classroom on TEFout of the schools.

Given this back drop, the contributions of the paper are: **first of all,** the paper highlights that not all sample schools are efficient. The estimated results show 56% sample schools are inefficient at HS; hence more of output can be produced using the existing resources; TEF_{out} ranges between 0.652 and 1.000. **Secondly,** it measures the variation of TEF_{out} not accounted by the percentage of students taking private tuition, the measures yet unexplored in the literature. A positive and significant influence of 'percentage of private tuition taking students' in determining the TEF_{out} is evident. Since percentage of students taking private tuition is not an exogenous variable, rather determined within the system, in the first stage, the estimated value of percentage of students taking private tuition is obtained from a regression of percentage number of students taking private tuition on different explanatory variables explaining it and then in the second stage TEF_{out} is explained in terms of the estimated value of the percentage of students taking private tuition along with other factors, and accordingly TEF_{out} variation not explained by percentage of students taking private tuition is obtained. The results support that such variation is very less, ranging only from 0.40% to 8.64% with 3.72% as mean, which is a serious concern. Such result in turn highlight that TEF_{out} **of the**

schools is mainly explained by the percentage of students taking private tuition, and hence the importance of private tuition in explaining TEF_{out} Hence there is a need to reduce private tuition. Some strategies are provided to reduce the dependence of private tuition in the concluding section of the paper. **Thirdly,** the findings reflect, positive role of Headmaster/mistress (HM) in school management is supported by the fact that TEF_{out} will positively depend on (a) the interaction of HM with teachers & students and (b) interaction frequency with state department of education, while taking any policy decision for the school, the role which has not supported in the literature yet. **Fourthly,** the paper supports the positive role of less discussed factors on TEF_{out} at HS namely: (i) Government targeted policy variable, like proportion of the girl students receiving 'Kanyashree', as discussed above and (ii) the amount of GoWB fund received by the schools (excluding salary payments), reflecting a non-linear relationship with TEF_{out}, but with positive marginal effect, implying an increase in government fund will raise TEF_{out}. **Finally,** supporting the unexplored role of school infrastructure related variables, like number of smart classroom in the school, highlighting the role of new technology development in fostering TEF_{out} of the school and also whether librarian of the school is on duty will have a positive effect on TEF_{out}.

The paper departs from the literature in the sense that while finding out the determinants, a composite index was not constructed and used it as the single determinant of TEF_{out} as argued above. Suitable policies are suggested to enhance TEF_{out} at HS and to reduce the dependence on private tuition.

Rest of the paper unfolds as follows. Section 2 discusses the methodology, determinants of TE and the data. Estimated results are presented in Sect. 3. Section 4 represents summary and conclusions and recommends some policy suggestions to boost up TEF_{out} at HS.

2 Methodology of Estimation, Determinants of Technical Efficiency and Data Sources

2.1 Methodology of Estimation of Technical Efficiency

Like other production system, Educational outcome can be viewed as a production process, where inputs of labor, capital, and organization are applied to the relatively "unfinished" child, and an output- pupil achievement results. The objective is to generate the maximum possible achievement using a given amount of school resources. Following Ray (2004) an educational production function may be defined as Y = A f (Xa,..., Xm), where Y = some measure of school output—for example, a score on a scholastic achievement; Xa,....Xm = variables measuring the school inputs; typically include the amount and quality of teaching services. The term A can be thought of as a shift parameter, being function of (i) I_1, I_2, ...,I_n showing the favourable infrastructure variables, (ii) PR_1,PR_2, PR_q showing poor infrastructure variables, (iii) E_1, E_2, ..., E_P showing environmental influences of learning outside the school, e.g., the parents' educational attainment and or atmosphere at home, (iv) PO_1, PO_2, ..., PO_k showing policy variables. There are two components of efficiency: technical efficiency (TE) and allocative efficiency (AE). TE of a Decision Making Unit (DMU) can be measured by either (i) output-oriented (TEF_{out}) approach showing the maximum output quantities that can be proportionately increased without altering input quantities or (ii) input-oriented (TEF_{inp}) approach representing the maximum amount of input quantities, which can be proportionately reduced without altering output production. TEF_{out} of the DMU is the comparison of actual output with the maximum producible quantity from the observed input. The TEF_{inp} measure of DMU is the comparison of minimum input required with the actual input in use.

AE reflects the ability of a DMU to use the inputs in optimal proportions, given their respective price. *The paper is concerned with the measurement and the determinants of TE, in particular* TEF_{out}. TE measurement involves two-step. First, a benchmark production function, known as frontier, is constructed which is supposed to be perfectly efficient. Secondly, comparison of the observed performance of DMU with the benchmark is the basic approach of estimating TEF_{out}.

2.2 Methodology for Finding Output Oriented Technical Efficiency Score and the Data

Given the actual input output bundle, the paper estimates TEF_{out} by constructing the frontier using Banker, Charnes and Cooper (BCC) (1984)VRS DEA Model, after making a number of fairly general assumptions about the technology namely, (i) all actually observed input-output combinations are feasible, (ii) the production possibility set is convex, (iii) inputs and outputs are freely disposable.

Each of the existing N DMUs is producing 'g' outputs using 'h' inputs. The DMU t uses input bundle $x^t = (x_{1t} \ldots . x_{ht})$ and produces the output bundle $y^t = \left(y_{1t} \ldots . y_{gt}\right)$. The production possibility set under VRS is

$$T^{VRS} = \left\{(x, y) : x \geq \sum_{j=1}^{N} \lambda_j x^j;\; y \geq \sum_{j=1}^{N} \lambda_j y^j;\; \sum_{j=1}^{N} \lambda_j = 1;\; \lambda_j \geq 0;\; (j = 1, \ldots N)\right\} \quad (1)$$

TEF_{out} of any DMU t requires the solution of the following LP problem:

$$\max \emptyset$$

$$\text{Subject to} \quad \sum_{j=1}^{N} \lambda_j y_{rj} \geq \emptyset y_{rt}; \quad (r = 1,2,\ldots g);$$

$$\sum_{j=1}^{N} \lambda_j x_{ij} \leq \emptyset x_{it} \quad (i = 1,2,\ldots h);$$

$$\phi \text{ free}, \quad \lambda_j \geq 0 \quad (j = 1,2\ldots N)$$

$$\sum_{j=1}^{N} \lambda_j = 1 \tag{2}$$

TEF_{out} of DMU t can be determined by using Eq. (3).

$$TEF^{t}_{out} = TEF^{t}_{out}\left(x^t y^t\right) = \frac{1}{\varnothing^*} \tag{3}$$

where $\varnothing^*$ is the solution of (2) showing the maximum value of φ. The maximum output bundle producible from input bundle x^t is y^* and is dined as $y^* = \varnothing^* y^t$ (See Ray (2004) for an excellent overview).

2.3 Input and Output Variables for Technical Efficiency Measurement of HS Using School Level Data

The educational production function can be distinguished from the standard micro-economic production function since (i) to represent the educational output, suitable measures are needed, as it is not a tangible product and (ii) the unavailability of market prices of input/output requires estimation of shadow prices. Generally, HS course in Kolkata, West Bengal comprises of three streams Science, Arts and Commerce; but most of the sample HS schools did not offer commerce. Hence, Science and Arts streams are taken together and are used for analyzing TEF_{out} of the HS level of schools in the overall sense. For the sample schools, studying language group is compulsory both for science and Arts students. Accordingly, six output variables are considered: (1) highest marks obtained in HS board examination by the school students (for representing best quality achievements by the school students), (2) percentage of students scoring first division marks (for representing the quality achievements by the school students), (3) average marks obtained in Language group (for representing average quality performance of the school students in the language group), (4)average marks obtained in Mathematics(for representing average quality performance of the school students in Mathematics, actually both Science and Arts students may opt for Mathematics), (5) average marks obtained in Science group (for representing average quality performance of the school students in the science group; this is relevant for science group only) (6) average marks obtained in Arts group(for representing average quality performance of the school students in the Arts group; this is relevant for Arts group only). The four considered input variables are: (1) classroom-student ratio, it is considered as a crucial input variable because it directly impacts student's outcomes, teacher workload and the quality of instruction, ultimately influencing students learning and academic achievement. (2) teacher-pupil ratio, it also directly impacts student learning, teacher workload and overall school quality, influencing factors like academic achievement, student engagement and teacher effectiveness. (3) percentage of teachers with qualification master degree and above, representing the quality of teacher input and (4) per student monthly expenditure incurred by the school. It is an important input variable as it provides insight into resource allocation, funding disparities, and potential impacts on student's outcomes. It reveals potential disparities in funding, allowing for identification of schools with limited resources and those with sufficient funding, which can lead to unequal opportunities for the students.

2.4 The Determinant Variables of TEF_{out}

The first stage estimation of TEF_{out} is followed by the second stage regression for determinants.

The considered determinants are both from schools and students' perspective and are classified under **(a)Characteristics of teachers:** It is captured by: (i) average service experience of full-time teachers (in years) (Z1), (ii)proportion of part-timers (Z2), (iii) average service experience of part-time teachers (Z3). Usually, schools mitigate the shortage of full-time teachers by recruiting part-time teachers and hence examination of the role of part-time teachers in promoting TEF_{out} is needed. **(b) School infrastructure and administration related variables**: These are represented by (i) proportion of vacant teaching post (Z4), (ii) proportion of vacant non-teaching post (Z5), (iii) whether the Headmaster/Headmistress (HM) is on duty (Z6), (iv) whether there is smart classroom (Z7), (v) number of smart classrooms (Z8), (vi) whether the librarian of the school is on duty (Z9) and (vii) whether all laboratory staffs are on duty (Z10). **(c) School characteristics and student composition related variables:** In the light of the fact that both state and central government of India is consistently taking policies for incentivizing backward section of students for bringing them into the formal education system, it is relevant to test whether the larger proportion of backward students attending education system as well as some school characteristics improve the TEF_{out} at HS. For this the included variables are: (i) proportion of reserve category students (Z11), (ii) proportion of physically disable/PD students (Z12), (iii) proportion of students taking vocational courses (Z13), (iv) percentage of students with English proficiency (Z14),(v) class size (Z15) (defined as the number of sections in the class), (vi) proportion of reserve category teachers (Z16) and (vii) Ratio of male to female teachers (Z17).The 'proportion' mentioning variables relating to students are considered for entire set of the students in the class. **(d) Policy variables:** Here the effects of the following variables are considered: (i) proportion of needy students getting free studentship (Z18), (ii) Amount of government fund received by the school, excluding the salary payment of non-teaching and teaching staffs(Z19) and (iii) proportion of girl students who got Kanyashree (Z20).('Kanyashree

Prakalpa', is mentioned in Introduction). Although 'Z11' is included under determinant category relating to student composition, however, it can also be treated as a variable reflecting government's social inclusion policy. **(e) School managerial role of Headmaster/Headmistress (HM):** Here the considered variables are the interactions of the HM while making a decision with: (i) teachers and students (Z21), (ii) parents (Z22), (iii) interaction with other school's HM (Z23) and (iv) frequency of interaction with State Education Department by HM(Z24). **(f) Characteristics of the head of the school (HM):** Here the included variables are: (i) gender of HM (Z25), (ii) age of HM (Z26), (iii) experience of HM(in years) (Z27), (iv)level of Education of HM (Z28) and (v)salary of HM (gross) (Z29). **(g) Socio-economic background of the students and student's characteristics:** Here the included variables are: (i) mother's academic qualification (Z30), (ii) father's academic qualification (Z31), (iii) monthly parental income (Z32), (iv) percentage of student taking private tuition (Z33), (v) number of subjects learning from private tutors (Z34), (vi)number of hours of study at home (Z35), (vii) percentage of students having text books at home (Z36), (viii) percentage of students consulting reference book (Z37), (ix) travel time to school each way (Z38), (x) percentage of students working for earning money(Z39), (xi) percentage of younger siblings per student (Z40), (xii) percentage of elder siblings per student (Z41) and (xiii) percentage of students who give service at home (Z42). **(h) Quality of school attributes based on student's perception:** The included student's perception variables are student's perception about: (i) regularity of class (Z43), (ii) quality of teaching (Z44), (iii) quality of Laboratory (Z45), (iv) quality of drinking water (Z46) and (v) school's sanitation facility (Z47).

Since objective is to test individual effect of the explanatory variable on TEF_{out} at HS, any composite index (comprising of the broader categories) is not constructed.

2.5 The Process of Data Collection

Primary survey data are collected through directly filling up the questionnaires from 25 Kolkata-based government and government aided schools for the year 2019–2020. The dataset is comprised of **(i) School Survey:** the information relating to the broader category of the determinants (a) to (f) as mentioned above are collected through questioning the appropriate school administration about HS. **(ii) Student Survey:** The information about the above mentioned determinants (g) and (h) are collected through direct questioning the students regarding their socio-economic background and also on the topic of their perception about teaching, academic and infrastructural environment of the school.

Sample Selection Procedure: To select the sample schools, first the complete list of Government and Government aided schools in Kolkata, classified into 23 different circles, published by the Department of Education, GoWB is collected and treated as 'population'. Considering each of 23 circles as relevant stratum, the stratified random sampling procedure is followed for selecting the school. The population proportion of schools in each stratum is computed. Using a random number table and assumption of equality between population proportion and sample proportion, the sample schools are selected. Suppose, for the i-th stratum, the population proportion of the school turns Pi. Then assuming the equality between the population proportion and the sample proportion, from i-th stratum, **25 x Pi** number of schools are to be selected[1].

However, after selecting the sample schools, for each of the sample school, a student survey is conducted by selecting randomly eight students from arts and eight students from science stream (due to absence of commerce stream in several surveyed schools) at HS. Randomly selected eight students from each stream comprise of four general

[1] Contextually, it may be mentioned that the use of Stratified random sampling has the following advantages (i) It allows to collect more accurate data: This method of data collection can allow for more accurate information. Instead of simply collecting a general consensus from a large group, you can break the data down to provide more in-depth detail. This may result in a better understanding about the sample group as one can draw better comparison. (ii) It allows to collect more diverse data: Having multiple subgroups within the sample population also allows one to collect a more diverse range of data. During a sample survey, one can ask questions to the entire population, which can yield some results. But, if one asks the same questions to specific subgroups, a variety of more detailed and diverse answers may emerge. (iii) The method is more manageable: Having subgroups within the population one can also make the data, and the work of collecting the data, more manageable. Breaking down the sample population allows one to divide the work throughout different days or time slots. This also allows to begin the analysis of one set of data while another person is surveying a different group. This can lead to quicker results from the survey. (iv)More cost effective: If one uses stratified sampling, it can be a more cost-effective method of conducting a survey. This is because one may be able to use a smaller sample size by creating the subgroups. When using this method, one can get more information from the individual groups, which means that it may not be required the overall population to be as large. (v) Prevents sample bias: Stratified sampling allows researchers to examine their sample and build groups of participants who are free of bias. While breaking the larger population down into subgroups, the people conducting the experiment or survey can help to ensure that each group has proper representation. Building groups with the express purpose of avoiding bias, while still maintaining the accuracy of the study, can help to make sure that the results one receive are accurate and trustworthy.Although the stratified random sampling method has many advantages, it may be relevant to point out some disadvantages of this methods. (i) The sampling technique lacks versatility: One limitation of the stratified sampling method is that it lacks versatility. This method only works for studies that require sample populations and surveys. It can be difficult to use this method when conducting a non-statistical study. While the method is extremely effective when working with statistical data, it can be difficult to apply subgroups and sample populations to a research project that lacks the need for a wide range of statistics. However, this limitation is not applicable to the present paper, as the paper deals with statistical data analysis. (ii) Some cases it may be difficult to do data analysis: With more and more subgroups, there also comes a larger input of information. The information can be specific and intricate, and it can take a long time to analyze. While having detailed data can be an advantage, the time it takes to analyze that data is a disadvantage. Analyzing data from multiple subgroups and from the entire sample population can take a significant amount of their time and energy. Hence while forming the number of strata the investigator should keep these limitations in mind. One needs to keep the number of strata within manageable limit. (iii) The method requires more planning: Creating a study using a stratified sampling method can require a significant amount of planning. While setting up the study may require more planning, the benefits of using this method are well worth the extra effort. The advanced planning is much needed for construction of appropriate strata because poorly defined strata or non-representative sample within the strata can introduce bias.

category and four reserve category students. More precisely, from class XI of arts stream, two general and two reserve category (i.e. four) students are selected and from class XII of the same stream two general and two reserve category (i.e. four) students are also selected. The same number as well as the same composition of student selection also holds for class XI and class XII of science stream. The student sample includes male, female also.

2.6 Definition and Description of the Variables

For **School infrastructure & administration related variable,** information on some variables are dichotomous in nature. For example, 'Z6','Z7', 'Z9' and 'Z10'are representing'1' value for presence and '0' value for absence of the respective variable. Considering **school characteristics & student-composition related variables**, 'Z11' at HS, representing reserved category students, (including arts and science) incorporating all the backward social categories, i.e. Scheduled Tribe (ST), Scheduled Caste (SC) and Other Backward Class (OBC). Since medium of instruction, in all the surveyed schools, is Bengali, proficiency of English may positively affect TEF_{out} of the school. Information on the variable 'Z14 about English Proficiency of the students, is collected from the English teacher (including arts and science). The class size variable represents the number of sections in a particular class of the school.

The study considers two **teacher's characteristics related variables,** 'Z1' and 'Z3'which are counted in terms of years.

The policy variables, 'Z18' is represented by the proportion of free studentship receiving students at HS (including arts and science) of the school. The school receives two types of funds; namely, Government fund and Samagra Shiksha Mission fund. It is standard practice to use a portion of Samagra Shiksha Mission fund for student' s purposes and hence this is used for computation of the input variable, reflecting 'per student expenditure incurred' for the estimation of TEF_{out}. To investigate the effect of government fund on TEF_{out}, the amount of fund received from GoWB by the school excluding the salary payment of non-teaching & teaching staffs, is considered as the possible determinant of TEF_{out}. The variable 'Z20' represents the proportion of Kanyashree, considering both arts and science streams of HS of the school. **School managerial role of Headmaster/ Headmistress (HM) related variables,** are 'yes'/'no' type. For example, 'Z21', 'Z22' and 'Z23' are dummy variables representing '1' value for 'yes' and '0' value for 'no'. Since government schools are liable for interacting with state education department, frequency of interaction seems crucial for this study. The information on the variable 'Z24' is categorical in nature; representing '1' value for 'one-time interaction in six months', '2' value for 'one-time interaction in a month' and '3' value for 'weekly interaction' with state education department.

Characteristics of HM related variables, 'Z25'is dichotomous, representing '0' value for female HM and '1' value for male HM.'Z26' and 'Z27' are two variables, noted in years. The variable 'Z28' represents the number of years spent for studying (only UG = 15, UG + B.ED = 15 + 2 = 17,only PG = 17,UG + BP.ED = 15 + 3 = 18, PG + B.ED = 17 + 2 = 19, DOUBLE PG = 17 + 2 = 19, PG + BP.ED = 17 + 3 = 20, PG + B.ED + M.ED = 17 + 2 + 2 = 21, Ph.D. = 17 + 5 = 22,Ph.D. +

B.ED = 22 + 2 = 24, PGBT + DOUBLE PG + Ph.D. = 1 + 19 + 5 = 25).'Z29', the gross salary variable is expressed in rupees. Considering **students' characteristics & socio-economic background related variables**, both 'Z30' and 'Z31' variables are represented by the number of years spent for studying, i.e. '0' value refers to illiteracy, '4' represents literacy upto class IV, '8' value denotes literacy up to class VIII, '10' indicates literacy up to class X, '12' value stands for literacy up to class XII, '15' refers to literacy up to graduation, '17' denotes literacy up to post graduation. Mean value of the corresponding income range of the parents is represented as 'Z32'. The variables, 'Z34','Z35' and 'Z38' represent average value of responses by the respondent students in the class corresponding to each variable. **Student's perception variable about the quality of school attributes** are categorical in nature. Based on the student's opinion about 'teaching Quality', 'Class Regularity', 'Quality of laboratory', 'Sanitation facility' and 'Quality of drinking water' of the school, categorization is made for each of the mentioned variables as: '1' value stands for excellent quality, '2' indicates good quality, '3' represents 'not a bad quality' and '4' denotes poor quality of the attribute.

3 Empirical Findings

While collecting students' data it is noticed that majority of the students undertake private tuition, which may inflate TEF_{out} of the schools. Hence, it would be appropriate to measure the extent of the variation of TEF_{out} of the schools in absence of private tuition, taken by the students, which are unexplored yet.

The present study estimates this performing the undermentioned steps: **Step I:** A regression is carried out considering TEF_{out} of HS as dependent variable and Z33 as independent variable along with other possible determinants. The result confirms statistical significance of Z33 variable, after controlling for other determinants of $TEF_{out,}$ **Step II:** In fact Z33 is not an exogenously determined explanatory variable. Rather it is in turn determined by several other explanatory variables in the system for HS. Therefore, the determinants of Z33are identified initially in the step II and the estimated value of Z33 (Z33hat2) is obtained. **Step III:** In step III, a regression is conducted using TEF_{out} as explained variable and estimated Z33(Z33hat2) along with the other determinants of TEF_{out} as explanatory variables. This step provides the final result regarding the determinants affecting TEF_{out} and supports highly statistically significant role of 'Z33hat2' in determining TEF_{out}. **Step IV:** A regression is carried out taking into account TEF_{out} of HS as explained variable and only estimated Z33(i.e. Z33hat2) as explanatory variable. The residual series for this regression is obtained where the residual actually reflects that extent of TEF_{out} variation, of HS, which is not influenced by'Z33'. Such type of measure of TEF_{out} variation, of HS was unexplored yet.

3.1 Measure of TEF_{out} of HS of Schools and the Extent of the Variation of that TEF_{out} not Explained by Private Tuition

The Measure of TEF_{out} at HS of schools obtained by applying BCC(1984) DEA model using VRS and the extent of the variation of that TEF_{out} not explained by private tuition, is represented in Table 1.

Table 1. TEF_{out} estimates for Higher Secondary Stage of sample schools and the extent of the variation of that TEF_{out} not explained by private tuition

Serial No	Higher Secondary (HS)Schools	TEF_{out} Scores	Measure of total TEF_{out} variation which is not explained by Z33
1	HS School 1	1.000	0.0351119
2	HS School 2	1.000	0.0113506
3	HS School 3	1.000	0.0546421
4	HS School 4	0.973	0.0394412
5	HS School 5	0.873	0.0429601
6	HS School 6	1.000	0.0503452
7	HS School 7	1.000	0.0392209
8	HS School 8	1.000	0.0114496
9	HS School 9	1.000	0.0282235
10	HS School 10	0.907	0.0612875
11	HS School 11	0.933	0.044573
12	HS School 12	1.000	0.0287091
13	HS School 13	0.954	0.0045111
14	HS School 14	0.982	0.0291011
15	HS School 15	0.892	0.0744493
16	HS School 16	0.991	0.0864397
17	HS School 17	0.866	0.0101241
18	HS School 18	1.000	0.0258512
19	HS School 19	0.818	0.0249646
20	HS School 20	1.000	0.0587875
21	HS School 21	0.652	0.0415968
22	HS School 22	1.000	0.0182931
23	HS School 23	0.945	0.0040876
24	HS School 24	0.889	0.0537179
25	HS School 25	0.893	0.0497179
Mean TEF_{out}		0.943	0.0371583

Source: Author's estimation with primary survey data for the year 2019–2020

The results reflect that 56% sample schools are inefficient considering HS; hence these schools can produce more output using the existing resources. Estimated TEF_{out} of HS lie between 0.652 and 1.000. Variation of TEF_{out} among the different schools is also evident. For developing better idea about estimated TEF_{out} at HS, the following graph is attached (Fig. 1).

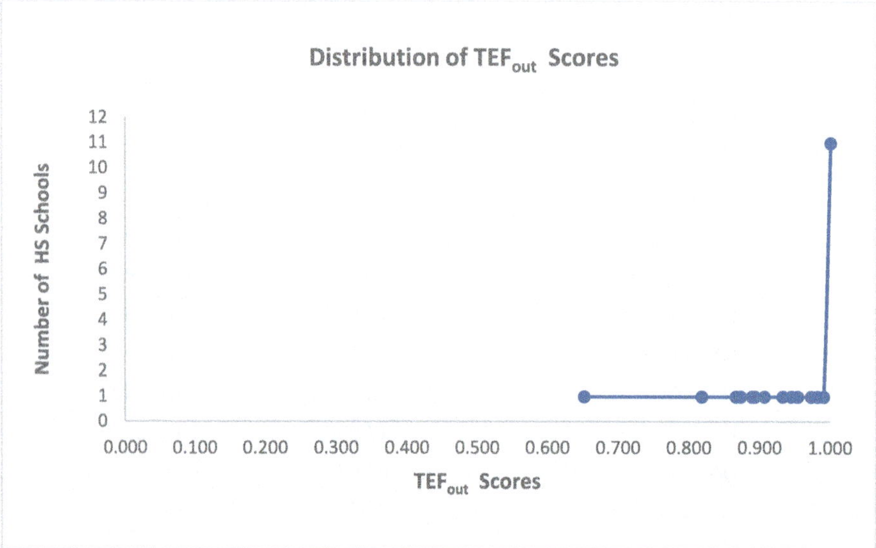

Fig. 1. Distribution curve for TEF_{out} at HS level. Source: Author's estimation with primary survey data for the year 2019–2020

Table 1 also shows that the number of schools showing above the average level of TEF_{out} is16. The result also reflects that the major part of variation in TEF_{out} of HS is explained by Z33 and the part of variation, not explained by Z33, is very less, it ranges between 0.40% and 8.64% with mean value 3.72%.

3.2 Determinants of TEF_{out} at HS Level of Schools

Determination of the determinants of TEF_{out} at HS, is a twostep process. **In first stage** the paper considers Z33 as an endogenous variable; the determinants of Z33 and hence estimated value of Z33(Z33hat2) is obtained. **In the second stage,** the determinants of TEF_{out} is obtained using Z33hat2 as an explanatory variable, after controlling for other explanatory variables (Table 2).

The OLS regression, reported in Table 2, is tested for heteroskedasticity applying Breusch–Pagan test and the result indicates acceptance of 'homoskedasticity' as null hypothesis since chi^2appears as 3.57 with Prob $>$ chi^2 $= 0.0588$; showing a higher p-value compared to the threshold value '0.05'.

Table 2 shows highly statistically significant positive relationship between TEF_{out} of HS and the estimated value of Z33 (with p value 0.0000), showing the role of private tuition in explaining TEF_{out} of HS. A common approach of defining tutoring provisions is the use of the emblem "shadow education", signifying that tutoring centers "shadow" formal schools. The sample observations of this paper suggest that majority of students

Table 2. Determinants of TEF_{out} of HS in school using 'estimated Z33' as explanatory variable

Variables	Coefficient	t-value	p-value	Adj R-squared = 0.8447
Z33hat2	.0187878	5.24	0.000	
Z9	.0463905	2.29	0.037	
Z25	.0302317	1.58	0.134	
Z19	-1.65e-07	-2.01	0.063	
M16	6.03e-13	1.85	0.083	
Z20	.004583	1.96	0.069	
M9	.0171668	2.08	0.055	
Z8	.0482924	3.67	0.002	
C	-.9266207	-2.65	0.018	

Source: Author's computation with primary survey data for the year 2019-2020

undertake private tuition. The ways in which the private tutoring aligns with that of formal schooling reveals the socio-educational ingrained ness of private tutoring within the mainstream schooling for the present sample. It may have a strong implication of broader educational policies as one may argue that by following these "shadowing" practices along with the mainstream structure, private tutoring in turn reproduces social inequality. (Gupta (2021). Thus, there should be a definite government educational policies to curb the incidence of private tuition. Some of these are highlighted in the concluding section. At the same time the previously less discussed undermentioned variables are important in explaining TEF_{out} of HS, which are significant at less than 5 or 10% level of significance, like (i) **HM's school managerial role:** An interaction variableM9, representing interaction between two variables i.e. Z21 and Z24 (M9 = Z24*Z21) representing HM's managerial role, is appearing as positive and significant, supporting the positive school managerial role of HM in promoting TEF_{out} of HS. The significance of interaction variable 'Z24*Z21' implies that given the value of Z24, TEF_{out} level of the school is dependent on Z21 and vice versa (ii) **School Infrastructure:** Here 'Z8' and 'Z9'variables are positively and significantly affecting schools TEF_{out} of HS, indicating that: (a) The school with better infrastructural facility in the form of more smart classrooms will obtain higher TEF_{out} at HS and (b) The HS school with librarian, present on duty will produce more TEF_{out} at HS (iii) **Policy Variables:** Here (a) 'Z20'shows positive significance in determining school's TEF_{out} of HS, implying positive role of Kanyashree in promoting TEF_{out} and (b) 'Z19' confirms the existence of non-linear U-shaped relationship between TEF_{out} as the coefficient of the variable turns negatively significant and the coefficient of the square term i.e. 'M16'turns positively significant, implying that after a minimum level referred as the threshold level of government fund (Z19), TEF_{out} increases with increase in'Z19'. However, the sample mean of 'Z19' shows a greater value than the threshold level value of 'Z19', after which the positive effect of 'Z19' on TEF_{out} will be felt and hence the system is at the rising segment of the curve, with positive marginal effect of 'Z19' in determination of TEF_{out} at HS. The **goodness** of fit of the model is also high. In this connection, it will be important to graphically represent the distribution curves of the significant determinants of TEF_{out} as following figures (Figs. 2, 3, 4,5 and 6):

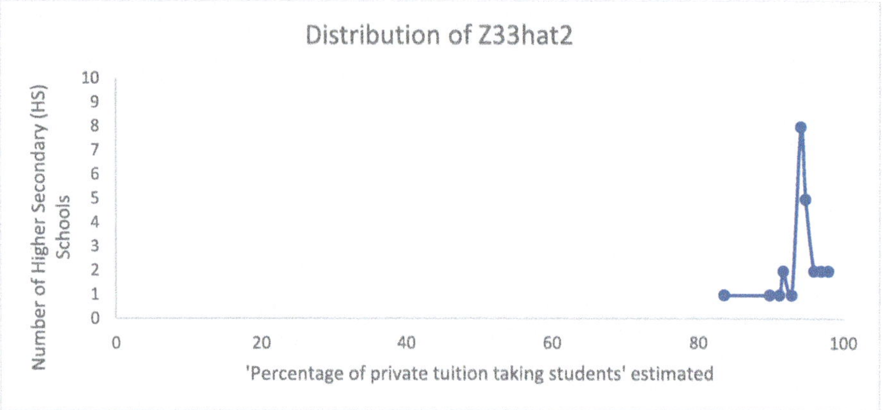

Fig. 2. Distribution curve for Z33hat2 at HS level. Source: Author's estimation with primary survey data for the year 2019–2020

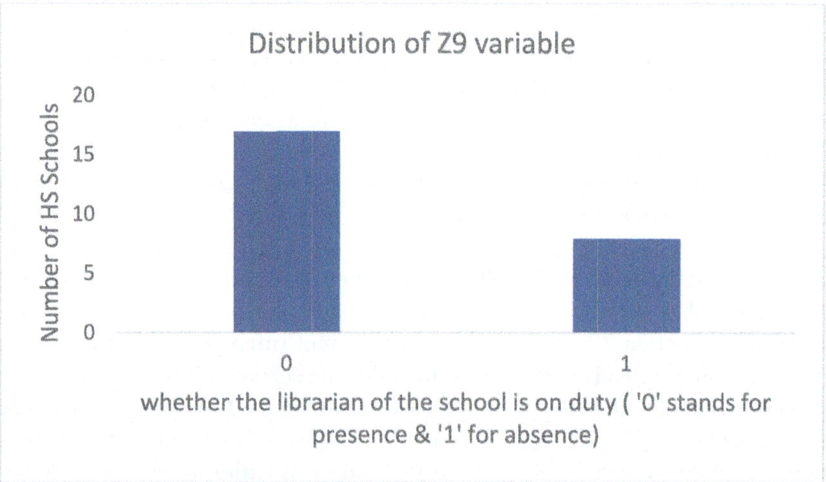

Fig. 3. Distribution of Z9 at HS level. Source: Author's estimation with primary survey data for the year 2019–2020

Relevantly, it may be interesting to point out that Z33 is significantly influenced by the following variables (i) positively by the variable 'Z4' revealing the fact that if the proportion of vacant to total teaching post increases in the school then more students will prefer taking private tuition, (ii) negatively by the variable suggesting that the students who work to earn money, may find lack of time for attending private tuition (iii) negatively by the variable 'Z3', implying that the part time teachers with more years of experience, are likely to offer better academic benefits to the students and hence it is less likely that the students join private tuition classes as an alternative. (iv) negatively by 'Z36',indicating

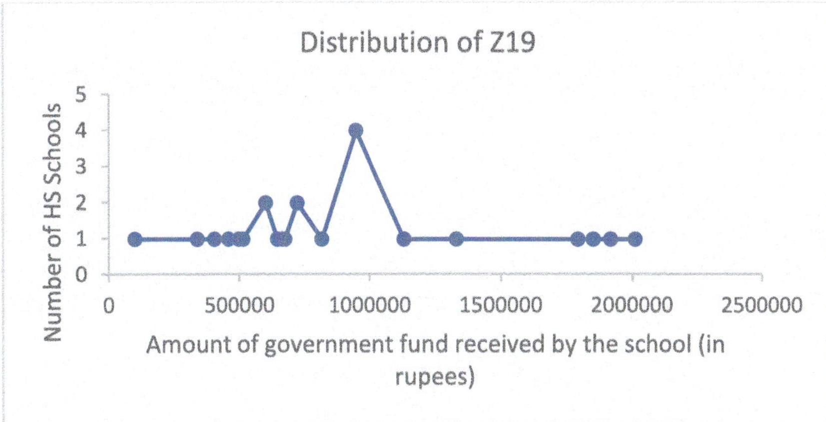

Fig. 4. Distribution curve for Z19 at HS level. Source: Author's estimation with primary survey data for the year 2019–2020

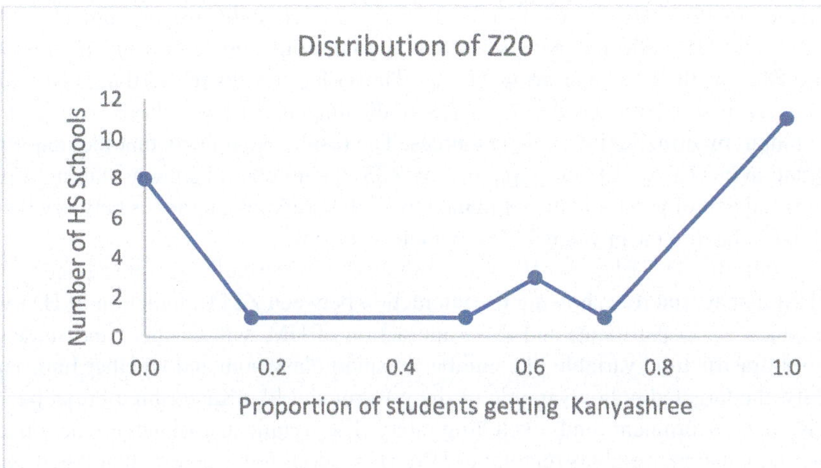

Fig. 5. Distribution curve for Z20 at HS level. Source: Author's estimation with primary survey data for the year 2019–2020

that the students having all the text books at home, may read and acquire knowledge on their own and therefore their chance of taking private tuition seems less.

4 Summary and Policy Suggestions

The paper contributes to the literature by establishing yet unexplored role of private tuition in explaining TEF_{out} of the schools at HS, after controlling for its the other determinants and finding out that part of the variation of TEF_{out}, not explained by private tuition, considering number of students taking private tuition as an endogenous variable.

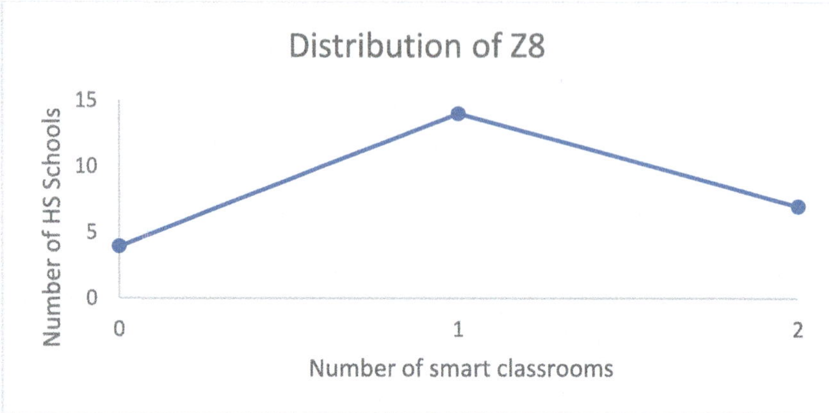

Fig. 6. Distribution curve for Z8 at HS level. Source: Author's estimation with primary survey data for the year 2019–2020

The paper uses primary survey data from 25 government aided & government schools in Kolkata for HS, collected using stratified random sample procedure and applies BCC (1984) DEA method for measuring TEF_{out}. The TEF_{out} results reveal that 56% sample schools are inefficient considering at HS; indicating that these schools can produce more output by utilizing the same resources. The result also reflects that the major part of variation in TEF_{out} of HS is explained by Z33 (the number of students taking private tuition) and part of variation, not explained by Z33, is very less, it ranges between 0.40% and 8.64%, having mean value 3.72%, which is serious.

Apart from Z33, TEF_{out} for HS is also influenced by the earlier less discussed factors like HM managerial role showing the interaction between Z24 (interaction of HM with state education Department) and Z21 (interaction of HM with teachers and students), school infrastructural variable like number of smart classroom and whether librarian is on duty, the targeted policy variable of the government like Kanyashree Prakalpa, and provision of government funds excluding salary. The significance of the interaction term of two explanatory variables relating to HM Managerial role suggests that the effect of one variable is not constant, rather it will depend on the value of the other.

Thus, to increase technical efficiency there is a need to increase the number of smart classroom, hold teachers-student meeting while taking any policy decision and visit state education Department by the HM, ensure that the librarian is on duty. Further there is also a need to continue the targeted policy of the government and provision of government fund to the school.

To reduce private tuition focus is to strengthen school infrastructure, invest in better school building, libraries and technologies to create a conducive learning environment; to increase teacher quality and resources and promoting a holistic education system that moves beyond exam-centric learning. At the same time attention should be given to make smaller class size, as smaller class size allows teachers to provide more individualized attention to students addressing their specific needs. It is also essential that schools have sufficient teaching materials, equipment sand resources to support effective teaching and learning. In this context it will be helpful to distribute free textbook to the students. It is also needed to recruit qualified teachers and implement rigorous teacher training programmes, to offer competitive salaries to attract and retain good quality talented teachers. At the same time, it is also needed to promote overall level of parent's education, specially mother's education, so that the students can get learning support from their home. Further, regular parent's teacher meeting and communication between parents and teachers will be helpful for addressing and solving the problems.

The paper considers the problems of West Bengal only. However, the problem of private tuition is there for the other states of the country as well. For example, a substantial number of students from different states of India takes private tuition at HS (Tripura 87%., Bihar (67.2%), Orissa (63.4%), Manipur (84.7%)). (Source: The 75th round of the National Sample Survey (NSS) on Household Social Consumption on Education in India (2017–18), Government of India) Some other existing study suggests a high percentage of students ((67%) in Kerala, in all subjects; (55.30%) in Maharastra for three subjects; and also a high percentage of students from Uttar Pradesh in more than three subjects receive private tuition (K. Sujatha (2021),) Hence these policy recommendations to curtail the dependence of private tuition will also be applicable to the other states as well.

Acknowledgement. This paper is a part of the project funded by Planning and Development Unit (NITI Aayog), Jadavpur University, Kolkata which is gratefully acknowledged. The authors are extremely grateful to Samagra Shiksha Mission for giving the necessary permission and making arrangements for collecting the data. The authors are indebted to the comments made by Professor Indrani Chakraborty, Director, Institute of Development Studies Kolkata and Professor Achin Chakraborty, Institute of Development Studies Kolkata, for their comments on the earlier draft. The authors sincerely acknowledges the help made by Ms. Sampita Neogi, for collection and compilation of the data and grateful to the participants of the online Seminar on presentations of ongoing projects of the Planning and Development Unit, Jadavpur University, held on October 2, 2020 for their valuable comments. However, usual disclaimer applies.

References

Adkins, L.C., Moomaw Ronald, L.: Analyzing the technical efficiency of school districts in Oklahoma. J. Econ. Insight Missouri Valley Econ. Assoc. **33**(2), 41–61 (2007)

Agasisti, T., Zoido, P.: The efficiency of schools in developing countries, analysed through PISA 2012 data. Socioecon. Plann. Sci. **68**, 100711 (2019). https://doi.org/10.1016/j.seps.2019.05.002

Arshad, M.N.M.: Efficiency of secondary education in selected OIC countries. Glob. Educ. Rev. **1**(4), 53–75 (2014)

Banker, R.D., Charnes, A., Cooper, W.W.: Some models for estimating technical and scale inefficiencies in data envelopment analysis. Manag. Sci. **30**(9), 1078–1092 (1984)

Bessent, A.M., Bessent, E.W.: Determining the comparative efficiency of schools through Data Envelopment Analysis. Educ. Adm. Q. **16**, 57–75 (1980)

Bonesronning, H., Rattsq, J.: Efficiency variation among Norwegian high schools: consequences of equalization policy. Econ. Educ. Rev. **3**(4), 289–304 (1994)

Charnes, A., Cooper, W.W., Rhodes, E.: Measuring the efficiency of decision making units. Eur. J. Oper. Res. **2**(6), 429–444 (1978)

Charnes, A., Cooper, W.W., Rhodes, E.: Evaluating program and managerial efficiency: an application of data envelopment analysis to program follow through. Manag. Sci. **27**(6), 668–697 (1981)

Driscoll, D., Halcoussis, D., Svorny, S.: School district size and student performance. Econ. Educ. Rev. **22**, 193–201 (2003)

D'Souza, O.M., Lobo, S.S.: An empirical analysis of higher education and economic growth in India. J. Inform. Educ. Res. **4**(3), 1464–1471 (2024)

Rocha, F., Oliveira, P.D., Duarte, J., Gadelha, S.R.D.B., Pereira, L.F.V.N.: Can education targets be met without increasing public spending? An analysis for Brazilian municipalities. In: Emrouznejad , A, Banker , R and Doraisamy, S. M. (eds.), Recent Developments in Data Envelopment Analysis and its Applications: Proceedings of the 12th International Conference of DEA, June 2014, Kuala Lumpur, Malaysia, November 2014. ISBN 978-1-85449-487-0. DEA2014-Proceedings_250323_133115 (3).pdf

Ghose, A.: Interstate variation of efficiency of elementary education in India: a meta frontier and group frontier analysis. In: Emrouznejad, A., Banker, R., Doraisamy, S. M.(Eds.), Recent Developments in Data Envelopment Analysis and its Applications, Proceedings of the 12th International Conference of DEA, June 2014, Kuala Lumpur, Malaysia, November 2014. ISBN: 978 1 85449 487 0, DEA2014-Proceedings_250323_133115 (3).pdf i

Ghose, A.: Efficiency of elementary education in India, empirical evidence using a nonparametric Data Envelopment Approach. Springer (2017)

Gundlach, E.: Education and economic development: an empirical perspective. J. Econ. Dev. **26**(1), 37–60 (2001)

Gupta, A.: Exposing the "shadow": an empirical scrutiny of the "shadowing process" of private tutoring in India. Educ. Rev. **75**(3), 391–410 (2021)

Hanushek, E.A.: School resources and student performance. In: Burtless, G. (ed.), Does money matter? The Effect of School Resources on Student Achievement and Adult Success, pp. 43–73. The Brookings Institution (1996)

Hanushek, E.A.: Some simple analytics of school quality (NBER Working Paper No. 10229). National Bureau of Economic Research (1994). https://www.nber.org/papers/w10229

Hanushek, E.A., Woessmann, L.: Education and economic growth. In Brewer, D.J., Patrick J.P.J. (eds.), Economics of Education, pp. 60–67. Elsevier (2010)

Hashmi, M.H., Bradley, S.: The efficiency of the educational institutes in Pakistan. In: Data Envelopment Analysis and Performance Measurement: Recent Developments: Proceedings of the DEA40: International Conference of Data Envelopment Analysis. Aston Business School, UK (2018). ISBN: 978 1 85449 438 2

Hsiao, B., See, K.F.: Maximizing efficiency and mobility in education funding: a three-stage centralized data envelopment analysis model. In: Emrouznejad, A., Thanassoulis, E., Toloo, M. (eds.) Advances in the Theory and Applications of Performance Measurement and Management. DEA45 2023. LNOR, pp. 261–277. Springer, Cham (2024). https://doi.org/10.1007/978-3-031-61597-9_20

Hulten, C.R.: Infrastructure capital and economic growth: how well you use it may be more important than how much you have, (NBER Working paper, 5847) (1996)

Javornik, Š, Klemencic Mirazchiyski, E.: Factors contributing to school effectiveness: a systematic literature review. Eur. J. Investig. Health, Psychol. Educ. **13**(10), 2095–2111 (2023)

Johnes, J., Portela, M., Thanassoulis, E.: Efficiency in education. J. Oper. Res. Soc. **68**(4), 331–338 (2017)

Júnior, M.A.A.S., Junior, A.P.S., Rebehy, P.C.P.W., Marinzeck-Santos, Y., Medeiros, L.G.: DEA as a tool for policy formulation in the public basic education: an application in Brazilian schools. In: Emrouznejad, A., Thanassoulis, E., Toloo, M. (eds.) Advances in the Theory and Applications of Performance Measurement and Management. DEA45 2023. LNOR, pp. 304–314. Springer, Cham (2024). https://doi.org/10.1007/978-3-031-61597-9_23

Khotvaneh, N., Alipour, V., Lebadi, Z.: Study the efficiency and productivity of high schools at education district 4 of Karaj (Iran). Int. J. Bus. Manag. Invent. **6**(2), 64–68 (2017). ISSN (Online): 2319 – 8028, ISSN (Print): 2319 – 801X. www.ijbmi.org

Kingdon, G.: The quality and efficiency of private and public education: a case study of urban India. Oxford Bull. Econ. Stat. **58**(1), 57–82 (1996)

K. Sujatha (2021). Private tuilions in India: trends and issues, Revue internationale d édu-cation de Sèvres. https://journals.openedition.org/ries/3913?gathStatIcon=true&lang=en10.4000/ries.3913

Leithwood, K., Jantzi, D.: A review of empirical evidence about school size effects: a policy perspective. Rev. Educ. Res. **79**(1), 464–490 (2009)

Levin, H.M.: Raising school productivity: an x-efficiency approach. Econ. Educ. Rev. **16**(3), 303–311 (1997)

Lucas, R.E., Jr.: On the mechanics of economic development. J. Monet. Econ. **22**(1), 3–42 (1988)

Mankiw, N.G., Romer, D., Weil, D.N.: A contribution to the empirics of economic growth. Q. J. Econ. **107**(2), 407–437 (1992)

Marquez-Ramos, L., Mourelle, E.: Education and economic growth: an empirical analysis of nonlinearities. Appl. Econ. Anal. **27**(79), 21–45 (2019). Emereld

Masci, C., De Witte, K., Agasisti, T.: The influence of school size, principal characteristics and school management practices on educational performance: an efficiency analysis of Italian students attending middle schools. Socio-Econ. Plan. Sci. **61**, 52–69 (2018)

Mergoni, A., Emrouznejad, A., De Witte, K.: Fifty years of data envelopment analysis. Eur. J. Oper. Res. **326**(3), 389–412 (2025). https://doi.org/10.1016/j.ejor.2024.12.049

Purohit, B.C.: Efficiency of education sector: the case of Rajasthan state (India). Madras School of Economics Working paper, 121/2015 (2015)

Ray, S.C.: Data envelopment Analysis: Theory and Techniques for Economics and Operations Research. Cambridge University Press, Cambridge (2004)

Romer, P.M.: Endogenous technological change. J. Polit. Econ. **98**(5), S71–S102 (1990)

Sankar, D.: Education system performance among Indian states: A public expenditure efficiency analysis using Linear Programming Methods. South Asia Human Development Unit (SASHD) of the World Bank (2007)

Schiltz, F.: Efficiency of education provision in least developed countries. In: Data Envelopment Analysis and Performance Measurement: Recent Developments: Proceedings of the DEA40: International Conference of Data Envelopment Analysis. Aston Business School, UK (2018). ISBN: 978 1 85449 438 2

Singh, N., Pant M., Goel, A.: ANN embedded data envelopment analysis approach for measuring the efficiency of state boards in India. Int. J. Syst. Assur. Eng. Manag. **9**(5), 1092–1106 (2018). Springer, The Society for Reliability, Engineering Quality and Operations Management (SRE-QOM), India, and Division of Operation and Maintenance, Lulea University of Technology, Sweden

Sotiriadis, D., Menexes, G., Tsamadias, C.: Investigating the efficiency of senior secondary schools: evidence from schools in the Greek region of Central Macedonia. Int. J. Bus. Econ. Sci. Appl. Res. (IJBESAR) **11**(2), 36–43 (2018)

Sengupta, A., Pal, N.P.: Primary education in India: delivery and outcome-a district level analysis based on DISE data. J. Educ. Plan. Adm. **24**(1), 5–21 (2010)

Sengupta, A., Pal, N.P.: Assessing the primary schools—A multi-dimensional approach: a school level analysis based on Indian data. Int. J. Educ. Dev. **32**(2), 264–272 (2012)

Tyagi, P., Yadav, S.P., Singh, S.P.: Efficiency analysis of schools using DEA: A case study of Uttar Pradesh state in India (2009)
Valero, A.: Education and economic growth (Discussion Paper No. 1764). Centre for Economic Performance, UK (2021)
Ziberi, B.F., Rexha, D., Ibraimi, X., Avdiaj, B.: Empirical analysis of the impact of education on economic growth. Economies **10**(4), 89. MDPI (2022)

Efficiency Analysis of Indian Pharmaceutical MSMEs: A Post-COVID Era

Varun Mahajan(✉)

Economics and Public Policy Area, Indian Institute of Management Tiruchirappalli, Tiruchirappalli, India
varun@iimtrichy.ac.in

Abstract. Indian pharmaceuticals, mainly led by Micro, Small and Medium Enterprises (MSMEs), play a critical role in the global health environment and constitute a major presence in the generic medicine and vaccine segments. It has thrown open severe vulnerabilities within these very enterprises but has also made them indispensable for riding on the escalating global demand surges. Data envelopment analysis (DEA) has been applied to study this pandemic in relation to the operational efficiency of Indian MSMEs in the pharmaceutical industry based on a sample of 399 firms. The DEA efficiency scores show that most of the firms under study have a high degree of inefficiency and are influenced majorly by the resource allocation of different firms, scale of production, and production techniques. The results indicate that micro-firms have been uniformly trending upward toward higher levels of efficiency than small and medium-sized firms, for which support systems could be administered to ensure the efficient oversight of overhead costs, regulatory compliance, and innovation. Further, the results show that adequate resource management and governmental support can make these firms more resilient and competitive. The paper advances insights that can give adequate policy guidance to policymakers and industry stakeholders on how the Indian pharmaceutical MSMEs could be enhanced in their future competitiveness.

Keywords: MSMEs · DEA · Efficiency · Pharmaceutical industry · India

1 Introduction

MSMEs contribute to the major share of the Indian pharmaceutical industry, and such units are thus highly interlinked with the international health sector, especially concerning generic drugs and vaccines. They contribute to more than 60% of the total output of the Indian pharmaceutical industry, now estimated at around $50 billion in 2024 (MSME, 2023–24). This further made the pharmaceutical industry an ally in facing global public health issues. The Indian pharmaceutical sector's MSMEs have responded effectively to the unprecedented surge in demand for basic drugs and health products. However, the pandemic has thrown up tremendous operational challenges and inefficiencies within these enterprises. Hence, understanding pharmaceutical MSMEs and efforts made to improve efficiency would become the benchmark for sustained growth and competitiveness and their contribution towards the larger healthcare ecosystem when the industry

A. Emrouznejad et al. (Eds.): DEA 2024, LNOR, pp. 449–459, 2025.
https://doi.org/10.1007/978-3-031-98177-7_30

starts to move towards recovery and resilience in the post-COVID-19 era (GoI: Annual Report, 2023–24).

This study reflects on the impact the COVID-19 pandemic has had on the efficiency of Indian pharmaceutical MSMEs. This study aims to find challenges and opportunities that MSMEs have faced post-pandemic. Based on the application of DEA regarding efficiency measurement, the general discussion considers operational, financial, and technological dimensions for identifying bottlenecks related to efficiency issues and what drives them. These aspects are expected to shed more light on matters that could help policymakers, industry players, and MSME managers formulate strategies to heighten resilience and optimum resource allocations for better industry performance in the face of future health crises.

2 Review of Literature

DEA has emerged as a powerful analytical tool in operational research and has contributed to advancements in performance evaluation methodologies its evaluation within the last 50 years (Mergoni et al., 2025). There has been exponential growth in the number of publications related to the theory and applications of DEA (Emrouznejad and Yang, 2018). In recent years, several researchers have researched the performance of the Indian small-scale industrial sector by using advanced non-parametric methods, most notably DEA. The relative efficiency of the Indian small-scale industrial sector has been analysed using non-parametric methods, notably mentioned by Chanda (2016) and Singh and Mallaiah (2022). These studies hint that DEA offers a useful tool for benchmarking and improving the operational efficiency of small-scale industries in India in support of policymakers in formulating targeted interventions for sectoral growth and sustainability. Further, recently, some studies (Gupta and Kumar, 2023; Sonkariya and Yadav, 2023) have analysed different models, such as Intuitionistic Fuzzy revenue models in DEA and extended version of Tone and Tsutsui (2009) network SBM model in different industries (Emrouznejad, Brzezicki and Chu., 2025) (Table 1).

A thorough review of the available literature reveals that most studies have been carried out on the efficiency and productivity of the Indian manufacturing sector. Still, a visible gap exists in research studies focused on MSMEs of the Indian pharmaceutical industry (IPI). This is an essential sector since it significantly contributes to the global pharmaceutical market and the Indian economy. The literature tends to be dominated by discussions of larger and better-capitalized companies, which overlooks smaller firms' issues and opportunities. Thus, our study attempts to fill this gap by being the first systematic investigation of MSMEs in the IPI. MSMEs are highly important for the pharmaceutical sector as they bring innovation, employment, and growth to the industry's economy. Strikingly, more comprehensive research on such enterprises needs to be conducted, particularly concerning their efficiency and productivity. Through the robust analytic framework of DEA, this paper shall gain insight into the operational efficiencies of Indian pharma MSMEs in the aftermath of COVID-19 and provide data grounding in its support for enhancing efficiency in this critical sector.

Table 1. Summary of Recent Empirical Studies on Efficiency and Productivity of the MSME in India

Studies	Period	Data Source	Methodology	Findings
Kaur and Bawa (2004)	1980–81, 1990–91 and 2000–01	Annual Survey of Industries (ASI)	DEA	The study analysed Inter-state efficiency in Indian manufacturing sector the study concluded with some suggestions for states having relatively inefficient manufacturing sector
Kumar (2014)	1973 to 2012	Annual Reports and censuses of MSME of the Government of India, various issues of economic surveys of the Government of India	Multiple Regression (OLS)	Pre- and post-liberalization MSME performance was compared in the research. The regression results found that a weak dependence on independent variables causes Low output
Chanda (2016)	2006–07	Fourth All India Census of registered MSMEs and Annual Survey of Industries (ASI)	Data Envelopment Analysis	The study found that most MSMEs could have been more technically efficient and less productive than their large scale counterparts
Singh (2017)	1981–82 to 2014–15	Development Commissioner MSME, Handbook of Statistics of Indian Economy published by RBI	Stochastic Frontier Approach	The technological efficiency of MSME varied throughout time, from 1981–1982 to 2014–2015. The study also shows that GDP growth rates and the proportion of MSME exports to total exports are essential for minimising technical inefficiency

(*continued*)

Table 1. *(continued)*

Studies	Period	Data Source	Methodology	Findings
Majeed et.al. (2020)	2017–18	ASI data collected by MoSPI	Stochastic Frontier Approach	This article compares the efficiencies in Northern Indian regions and finds that industrialisation in both regions is labour-intensive
Singh and Mallaiah (2022)	2011	Fourth All India Census of MSMEs	Data Envelopment Analysis	A state-by-state analysis calculates the relative technical efficiencies of MSMEs operating in Indian states and union territories. They concluded that registered MSMEs are operating inefficiently. However, in most states, unregistered MSMEs were doing better in terms of efficiency
Sinha et al. (2022)	2006–17	Prowess of CMIE	Fixed and Random Effects Models	The paper analysed the productivity of Small-scale firms and compared it with medium-scale firms. Firms' performance could be achieved by higher capital intensity and internalisation
Kathuria & Natarajan (2025)	2010–2022	Prowess of CMIE	Econometric models	Econometric analysis does not provide convincing evidence of digitalisation and government policy in improving the productivity of MSMEs

Source: Author's own compilations from different studies

3 Methodology

Appropriate instruments are needed to assess Indian MSMEs' efficiency. Efficiency can be evaluated using non-parametric or parametric techniques. Data envelopment analysis, a non-parametric technique, was used in the study to determine efficiency. Due to DEA's many advantages over the conventional method, researchers have been using it more and more frequently in recent years to assess Decision-Making Unit (DMU) performance. For instance, it does not require that a functional form be imposed pre-specified, and multiple inputs and outputs are easily handled. It offers clear points of comparison by identifying efficient peers and easily accommodates variable returns to scale. These features make DEA ideal for complex production environments and managerial benchmarking, especially when statistical noise modelling is less critical Coelli et al., (2005).

Charnes, Cooper, and Rhodes (1978) (CCR) developed this method in 1978 using the constant returns to scale (CRS), and Banker, Charnes, and Cooper (1984) (BCC) updated it in 1984 using the variable returns to scale (VRS). Since then, it has been widely utilised to assess relative efficiency in many sectors. Since its inception, this methodological framework has been extensively employed to evaluate relative efficiency across a diverse range of sectors. To gain a deeper understanding of these contributions, particularly the influential role played by Rajiv Banker, Emrouznejad, Podinovski, et al. (2025) conducted a comprehensive analysis that elucidates Banker's substantial impact on the field of Data Envelopment Analysis (DEA).

Input Oriented CCR (envelopment) model	Input Oriented BCC (envelopment) model
$Min\ Z_k = \theta_k - \varepsilon\left(\sum_{i=1}^{m} s_{ik}^{+} + \sum_{j=1}^{s} s_{jk}^{-}\right)$ (1)	$Min\ Z_k = \theta_k - \varepsilon\left(\sum_{i=1}^{m} s_{ik}^{+} + \sum_{j=1}^{s} s_{jk}^{-}\right)$ (2)
Subject to:	Subject to:
$\sum_{r=1}^{n}\lambda_{rk}y_{ir} - s_{ik}^{+} = y_{ik} \quad \forall i = 1,\ldots\ldots m,$	$\sum_{r=1}^{n}\lambda_{rk}y_{ir} - s_{ik}^{+} = y_{ik} \quad \forall i = 1,\ldots\ldots m,$
$\sum_{r=1}^{n}\lambda_{rk}x_{ji} + s_{jk}^{-} = \theta_k x_{jk} \quad \forall j = 1,\ldots\ldots s,$	$\sum_{r=1}^{n}\lambda_{rk}x_{ji} + s_{jk}^{-} = \theta_k x_{jk} \quad \forall j = 1,\ldots\ldots s,$
$\lambda_{rk} \ge 0 \quad \forall r = 1,2,\ldots\ldots,n,$	$\sum_{r=1}^{n}\lambda_{rk} = 1 \quad \forall r = 1,2,\ldots\ldots,n,$
Is unrestricted in sign	θ_k Is unrestricted in sign and
$s_{jk}^{-}, s_{ik}^{+} \ge 0 \quad \forall j = 1,2,\ldots\ldots s,\ i = 1,2,\ldots\ldots m,$	$\lambda_{rk}, s_{jk}^{-}, s_{ik}^{+} \ge 0 \quad \forall r, j, i,$

Input-oriented CCR and BCC efficiency of kth DMU.

Where s_{ik}^{+} is slack in i^{th} output of the kth DMU; s_{jk}^{-} is slack in the jth input of the kth DMU. $\lambda'_{rk}s$ are the dual variables, known as intensity variables. θ_k (Scalar) is the (proportional) reduction applied simultaneously to all inputs and results in a radial movement towards the envelopment surface.

4 Data Sources and Selection of Firms

Primary databases "PROWESS" of Centre for Monitoring Indian Economy (CMIE) and "Capitaline" offer information on several manufacturing companies, including those in the pharmaceutical industry. Even though these sources only offer financial data based on balance sheets for individual businesses and do not provide information about physical data like the number of employees or output units, etc., due to the limitations of other databases like the Annual Survey of Industry (ASI), the Prowess database was used to select pertinent cross-sectional data for 399 MSME pharmaceutical firms for which data were available for the years 2023–2024.

The output variable in this research is net sales revenue, which reflects the firm's total income from its operations. The input variables are as follows: cost of raw materials, which incorporates costs for necessary supplies; employee compensation, which includes wages, salaries, and benefits; selling and distribution costs, which reflect costs associated with marketing, advertising, and distribution activities; power, fuel, and water charges, which incorporate utility costs important to the production and operation of a business; and capital, representing investments in both fixed assets and working capital.

5 Results

Table 2 reports descriptive statistics on the input and output variables. The minimum and maximum values found in the table show consistent and high variation in MSME enterprises. The average net sales revenue per company is Rs. 911.10 million, while the standard deviation is Rs.1738.36 million. The size range for Indian pharmaceutical MSMEs can be inferred from the smallest sales revenue Rs. 0.20 million and the highest sales revenue as Rs 17680.90 million. As can be interpreted from Table 2, Indian MSME pharma firms have low levels of fixed assets, and the intensity of working capital is high. Smaller companies would, of course, turn towards outsourcing and contract manufacturing to reduce the requirement to fund major capital investments in fixed assets that include structures and equipment. Their focus on generics and formulations reduces the appetite for large capital infrastructure (Mahajan, 2020).

The values of correlation coefficients were calculated to find out the correlation between inputs and output of all the sample firms. It is observed that the output variable has a statistically significant correlation with the input variables. The value of adjusted R^2 is 0.94, suggesting that the five input variables together explain 94% of variations in net sales revenue. The F-statistics is found to be statistically significant at a 1% level. These two tests suggest that a best fit exists between explanatory and explained variables.

DEA evaluates firms to derive the production frontier. Those firms with an Overall Technical Efficiency (OTE) of 1 are considered efficient and lie on the efficiency frontier. In contrast, the firms with a score below one rank less efficient than the frontier. Levels of efficiency of less than 1 suggest considerable scope to reduce inputs without a loss of current output levels. Since the CCR model works on the basis of CRS, in which the scale size of the DMU is not considered, it is relevant to assess pure technical efficiency. Therefore, in order to find out whether inefficiency in any firm is due to inefficient production operations or due to unfavourable conditions due to the size of

Table 2. Descriptive Statistics of Inputs and Output (2023–24)

Statistics	Net Sales Revenue	Raw materials, stores & spares	Power, fuel & water charges	Compensation to employees	Selling & distribution expenses	Capital
Mean	911.099	377.440	20.825	114.783	59.290	167.201
Standard Deviation	1738.362	566.840	36.808	276.826	196.755	286.907
Minimum	0.200	0.100	0.100	0.200	0.100	0.500
Maximum	17680.900	4963.800	506.300	3123.300	2108.500	5026.200

Note: Figures in millions of rupees, NS: net sales revenue, RM: raw materials, stores & spares, PFW: power, fuel and water charges, CE: compensation to employees, SD: selling & distribution expenses, C: Capital
Source: CMIE Prowess (https://prowess.cmie.com/)

the firm, the BCC model has also been applied. Pure technical efficiency (PTE) refers to the proportion of technical efficiency which is attributed to the efficient conversion of inputs into output, given the scale size. Scale Efficiency (SE) measures the impact of scale–size on the efficiency of a firm. It is measured as the ratio of CCR efficiency to BCC efficiency. If the value of the SE score is one, then the firm is apparently operating at optimal scale. If the value is less than one, then the firm appears to be either small or big relative to its optimum scale size.

The analysis reveals that out of 399 firms, 30 (7.5%) are relatively efficient, i.e., OTE score = 1, whereas the remaining 369 firms (92.5%) are relatively inefficient, i.e., OTE score < 1. The firms on the efficient frontier form the "reference set", i.e., these firms can set an example of best operating practices for the remaining 369 inefficient firms to emulate. Figure 1 shows the distribution of the firms by the measure of OTE, PTE and SE. It can be easily noticed that 30 firms are on the Most Productive Scale Size (MPSS) (OTE = PTE = SE = 1), 65 firms are PTE efficient (PTE = 1), and 39 firms are scale efficient (SE = 1). Figure 1 further reveals that 222 firms have SE scores between 0.91 and 0.99. In total, 43 firms are in the CRS stage (10.8%), 228 firms are operating in the IRS stage, and the remaining 128 are operating in the domain of DRS. Most of the firms are under the IRS; therefore, an increase in size would increase the efficiency. Technological progress, economies of scale, and strategic market expansion generally contribute to increasing returns to scale. Firms have to innovate constantly and manage the cost of innovation to retain them in the long run. Applying DEA analysis shows that all firms under study have a high degree of inefficiency and are influenced majorly by the resource allocation of different firms, scale of production, and production techniques.

Applying DEA analysis, the results show high inefficiency for all firms under study, which is primarily influenced by different firms' resource allocation, scale of production, and production techniques. A very small proportion of firms appear on the efficient frontier, and it simply implies that those are already at optimal use of resources and

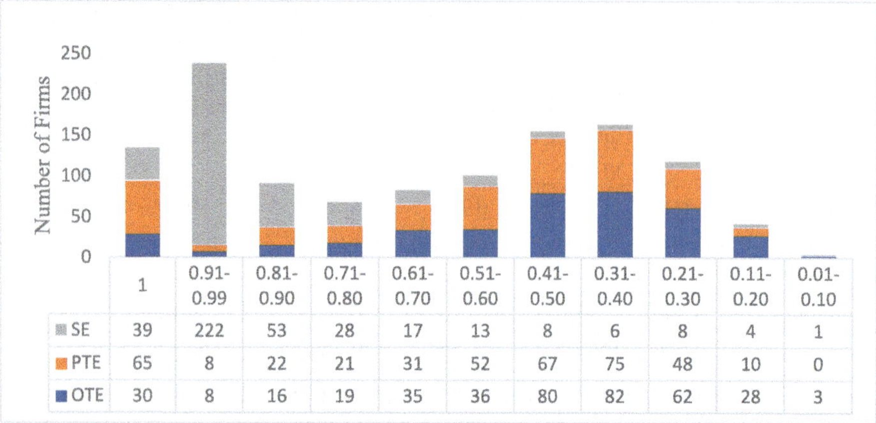

Fig. 1. Frequency Distribution OTE, PTE and SE

operating at the optimum scale through best management practices, superior technology, or better processes. That 92.5% of firms score below 1 in OTE attests to the fact that most firms have widespread inefficiencies, and most are struggling with excess inputs, suboptimal scale, or technical inefficiencies in converting inputs into outputs. The categorisation of firms into PTE and SE tells exactly where the firms go wrong. Other firms can be quite resource-efficient but must operate optimally, while others are scale-efficient but have technical problems. The distribution across the different returns to scale stages—CRS, IRS, and DRS—indicates that most firms are either too small to exploit economies of scale (IRS) or too large and experiencing diseconomies of scale (DRS), meaning scale adjustments could improve overall efficiency. These results highlight the relevance of differentiated strategies for different companies, thus underlining the operational improvement need from firms characterised by technical inefficiency and scale adjustment need to others affected by inefficient scale.

Table 3. Descriptive Statistics of OTE, PTE, SE and RTS of MSME Pharma Firms

Particulars	Micro			Small			Medium		
	OTE	PTE	SE	OTE	PTE	SE	OTE	PTE	SE
Mean	0.554	0.834	0.667	0.492	0.565	0.880	0.450	0.529	0.893
Standard Deviation	0.284	0.190	0.290	0.204	0.225	0.150	0.201	0.247	0.183
Minimum	0.108	0.450	0.108	0.189	0.244	0.263	0.198	0.201	0.263
Maximum	1	1	1	1	1	1	1	1	1
No. of efficient firms	6	20	6	14	23	17	10	22	16
No. of firms on RTS	CRS	IRS	DRS	CRS	IRS	DRS	CRS	IRS	DRS
	6	36	0	19	112	33	18	80	95
N	42			164			193		

As Table 3 shows, in the overall sample of 399 firms, 42, 164, and 193 are Micro, Small and Medium pharma firms, respectively. These firms have been classified by the MSME Act 2020. The mean OTE scores of Micro are observed to be higher than those of Small and Medium pharma firms. The mean OTE score for the Micro firm is 0.554, 0.492 for Small firms and 0.450 for Medium firms. Regarding PTE scores, Micro pharma firms have substantially higher efficiency scores than small and medium firms. This clearly shows that in the case of PTE scores, when the size of the firms is considered, micro firms are around 50% more efficient than other firms. Interestingly, medium firms have shown least OTE (0.450) and PTE (0.529) among all the categories of firms. The result in Table 3 indicates that Medium firms have the scope for a 55% reduction in their inputs without altering the output in the case of CRS technology and a 47.1% reduction in the case of VRS technology. Similarly, in the case of PTE scores, Micro have the highest mean efficiency of around 83.4%, whereas Small and Medium firms have 56.5% and 52.5% efficiency scores, respectively. In the Micro category, six firms are OTE efficient, 20 are PTE efficient, and six are scale efficient. Micro pharma firms are more efficient than small and medium firms in CRS and VRS technology.

Indian small and medium-sized pharmaceutical companies are generally more efficient than small and medium-sized businesses worldwide due to their operational flexibility, relatively low overhead costs, and personal customer relationships. This is because the processes are very agile and speedy, and the firm is nimble enough to address changes in the marketplace and adhere to regulatory requirements. Business processes are simple, and bureaucracy is minimal. As a result, administrative expenses are lowered, and better resource utilisation occurs. Moreover, micro firms tend to have a strong relationship with their customers. This is because the services are customised and receive direct customer feedback, resulting in higher customer satisfaction and continuous improvement. All these factors enable the micro firm to become more agile and cost-competitive than its larger counterparts.

6 Conclusions and Policy Implications

MSME pharmaceutical sectors are an integral part of not only the health economies of India but also worldwide, especially in the area of generics and vaccine production. COVID-19 has exposed a couple of glaring vulnerabilities to those houses, which have adversely rendered them weak to impact their survivability. To that effect, this study taking MSME pharmaceutical companies at a very granular level using DEA of 399 brings out results that are very highly significant in the context of the performance of the sector during the pandemic and its aftermath. From the results, it was found that most of the firms were operating under the IRS and had significant inefficiencies, indicating that technological advancement, economies of scale, and market expansion normally increase returns to scale. The study first indicates evidence that micro-enterprises broadly surpass small and medium enterprises with regard to both OTE and PTE. Therefore, the conclusions outlined above imply that operational adaptable and cost-effective micro firms are able to maintain relatively high-efficiency levels despite their own weak financial foundation. Contrary to that, small and medium enterprises face more challenges in dealing with overhead cost control, compliance with regulations, and innovation promotion. It also underscored that responsiveness to regulatory guidelines and increased scrutiny by international regulatory authorities sharply impacted the issue. The mounting pressure to achieve quality compliance posed by it to MSMEs involved huge investments in technology and infrastructure, an expenditure many firms require greater capital for, especially for Small and Medium-sized enterprises. It has been observed that limited access to developed manufacturing technologies, in combination with a scarcity of skilled labour, intensifies the financial problems, rendering the sector ineffective from the point of innovation and product diversification.

The practical and policy implications of the study are diverse and significant and provide critical guidance for the improvement of the Indian pharmaceutical MSME sector. From the policy perspective, findings suggest the need for government support for competitiveness and resilience in this sector of MSMEs. Policies supporting financial assistance for technology change, regulatory compliance, and upskilling of workers would be appropriate. Moreover, to present a competitive advantage in the global market, government support can make advanced manufacturing technologies and international standards more accessible. In addition, this study puts forward the necessity for policy measures related to innovation by supporting R&D activities for MSMEs, which would enable these firms to diversify their product portfolios and reduce their dependence on generics.

References

Banker, R.D., Charnes, A., Cooper, W.W.: Some models for estimating technical and scale inefficiencies in data envelopment analysis. Manag. Sci. **30**(9), 1078–1092 (1984)

Chanda, A.: Relative technical efficiency of the micro, small and medium enterprises of India: recent evidences using data envelopment analysis. Indian Econ. J. **63**(4), 668–688 (2016)

Charnes, A., Cooper, W.W., Rhodes, E.: Measuring the efficiency of decision making units. Eur. J. Oper. Res. **2**(6), 429–444 (1978)

Coelli, T.J., Rao, D.S.P., O'donnell, C.J., Battese, G.E.: An introduction to efficiency and productivity analysis. Springer science & business media (2005)

Emrouznejad, A., Yang, G.L.: A survey and analysis of the first 40 years of scholarly literature in DEA: 1978–2016. Socioecon. Plann. Sci. **61**, 4–8 (2018)

Emrouznejad, A., Brzezicki, Ł., Lu, C.: The development and evolution of slacks-based measure models in data envelopment analysis: a comprehensive review of the literature. J. Econ. Surv. 1–28 (2025). https://doi.org/10.1111.joes.12682

Emrouznejad, A., Podinovski, V., Charles, V., Lu, C., Moradi-Motlagh, A.: Rajiv Banker's lasting impact on data envelopment analysis. Ann. Oper. Res. **351**, 1225–1264 (2025). https://doi.org/10.1007/s10479-025-06473-3

Government of India: Annual Report (2023–24), Department of Pharmaceuticals, Ministry of Chemicals and Fertilizers

Gupta, M., Kumar, S.: Does income diversification affect bank efficiency, effectiveness and performance? Evidence from Indian banking industry. In: International Conference on Data Envelopment Analysis, pp. 165–185. Springer Nature Switzerland, Cham (2023)

Kathuria, V., Natarajan, R.R.: COVID-19, digitalization, and firm productivity: a tale of MSMEs in two industries. In: FDI, MSMEs, Digitalization, and Green Industrialization: Challenges, Opportunities and Policy Lessons for India, pp. 131–160. Springer Nature Singapore, Singapore (2025)

Kaur, K., Bawa, R.S.: Efficiency of Indian manufacturing sector–an inter-state analysis. In: Data Envelopment Analysis and Performance Management, Birmingham, UK: 44th International Conference on DEA, pp. 135–139 (2004)

Kumar, P.: An empirical study on performance of Indian MSME. Asia-Pac. J. Manag. Res. Innov. **10**(4), 367–375 (2014)

Mahajan, V.: Is productivity of Indian pharmaceutical industry affected with the introduction of product patent act? Indian Growth Dev. Rev. **13**(1), 227–258 (2020)

Majeed, M., Khan, J.I., Mushtaq, S.O.: An analysis of MSME's efficiency: a study of two comparable northern Indian regions. SEDME (Small Enterp. Dev. Manag. Ext. J.) **47**(3), 250–260 (2020)

Mergoni, A., Emrouznejad, A., De Witte, K.: Fifty years of data envelopment analysis. Eur. J. Oper. Res. **326**(3), 389–412 (2025). https://doi.org/10.1016/j.ejor.2024.12.049

MSME (Ministry of Micro, Small and Medium Enterprises) (2023–24), Annual Report, Government of India, New Delhi

Singh, S.: Measurement of technical efficiency of MSME sector in India: an application of stochastic frontier. Int. J. Oper. Res.-Nepal **6**(1), 47–57 (2017)

Singh, S., Mallaiah, L.C.: An economic analysis of technical efficiency of MSMEs in India. SEDME (Small Enterp. Dev. Manag. Ext. J.) **49**(1), 30–40 (2022)

Sinha, A.K., Mishra, A.K., Rl, M., Prabhudesai, R.: Examining the determinants of small firms' performance in India. Int. J. Product. Perform. Manag. **71**(6), 2496–2533 (2022)

Sonkariya, A., Yadav, S.P.: Development of intuitionistic fuzzy revenue efficiency models in DEA: an application to the Indian public sector banks. In: International Conference on Data Envelopment Analysis, pp. 137–149. Springer Nature Switzerland, Cham (2023)

Tone, K., Tsutsui, M.: Network DEA: a slacks-based measure approach. Eur. J. Oper. Res. **197**(1), 243–252 (2009)

Performance Mapping of Startups in India Using Data Envelopment Analysis

Pratigya Kwatra[1](✉) and Gokulananda Patel[2](✉)

[1] IILM University, Greater Noida, India
pratigya.kwatra@iilm.edu
[2] Birla Institute of Management Technology, Greater Noida, India
gn.patel@bimtech.ac.in

Abstract. Performance measurement of start-ups is an emerging field of research. Relative efficiencies of thirty-nine Indian start-ups that became unicorns in the year 2022 were assessed using the DEA slack-based model. The inputs used in the DEA model consisted of human, financial, and technological capital indicators, while the output indicators were primarily financial parameters. Eight Decision Making Units achieved relative efficiency of 1 and 13 DMUs displayed efficiency of more than 50%. The results indicate that start-ups in India are operating at a lower efficiency. The performance of unicorns in India has been studied for the first time in this paper. The study is original in terms of indicators used in the DEA model to predict start-up performance and the application of slack-based DEA model for analyzing start-up efficiency.

Keywords: Start-up · Performance measurement · DEA · Slack-based model · Relative efficiency

1 Introduction

Start-ups[1] are important drivers of an economy's growth, owing to their contribution to the national income and employment growth (Lumpkin and Dess, 1996, Zapata-Guerrero et al., 2020). Start-ups are significant for their contribution to developing innovative technology-based products and solutions (Goswami et al., 2023). India has seen a rise in new entrepreneurial ventures over the last decade, owing to funding opportunities, a rising aspirational young population, advancements in technology and the evolving vocation-based education system. The start-up eco-system which includes venture capitalists, angel investors, Start-up India - a Government platform, Department for Promotion of Industry and Internal Trade (DPIIT), has been the driving force behind rising innovative ventures. (David et al., 2021).

The performance of start-ups is of extensive concern globally and in India, due to two key reasons. First, the start-up needs to ensure its own sustainability as most start-ups do not sustain themselves in the long run (Reynolds, 1987), (Krishna et al., 2016).

[1] Start-up is a new enterprise, which has not completed 10 years of operation and is innovating new products or services-based solutions.

A. Emrouznejad et al. (Eds.): DEA 2024, LNOR, pp. 460–473, 2025.
https://doi.org/10.1007/978-3-031-98177-7_31

Second, due to the numerous stakeholders involved in the ecosystem, it is accountable to its investors, funding agencies, and other stakeholders. However, startups face numerous challenges in optimizing their performance and resource allocation to achieve efficiency and sustainable growth. (Preethi et al., 2023).

Factors responsible for success as well as failure of a start-up have been studied by many researchers (Pathania & Tanwar, 2024), (Díaz-Santamaría & Bulchand-Gidumal, 2021), (Krishna et al., 2016). The literature includes studies exploring the reasons for start-up failure (Pathania & Tanwar, 2024), predicting start-up success (Krishna et al., 2016), whether start-up will fail, marginally survive or take a high growth path (Cooper et al., 1994). However, empirical research regarding key performance parameters for a startup and the current efficiency of the start-ups is still in the exploratory phase.

There are three key research questions that we aim to answer in this paper. First, what key indicators can be used to analyze the performance of Indian start-ups? Second, is there a method that can evaluate start-up performance that can accommodate multiple parameters? Third, how are the current start-ups in India performing?

To answer these research questions, our paper is organized in the following manner: a literature review on startup performance which answers our first research question, followed by a section on methodology answering our second research question, followed by results and discussion where we answer our third research question and finally conclude the paper.

2 Review of Literature

The literature talks about end performance parameters for a start-up and the pre-requisites needed as a recipe which predict the success of a start-up. The end parameters throw light on the actual performance of a start-up which include mostly financial indicators. The other spectrum includes literature on parameters that predict a start-up's performance such as the initial human and financial capital of the start-up (Cooper et al., 1994), the role of entrepreneurial efficiency (Grežo & Hanák, 2024) (Haber & Reichel, 2007), (Davidsson & Honig, 2003), (Pathania & Tanwar, 2024), role of technology capital (Yu et al., 2021), (Tekic et al., 2024) to be important for a new enterprise to be successful. Along similar lines, Zang et al. identified five dimensions to assess new start-ups which included Operations, innovation, financial and financing dimensions along with public sentiment (Zhang et al., 2022). The adoption of different knowledge management practices (i.e., acquisition, documentation, creation, transfer, and application) was found to have a positive impact on the performance of global start-ups. (Battisti et al., 2022).

Following the literature, we shed light on the three most researched performance parameters in the context of start-ups.

2.1 Financial Performance

Traditionally, financial performance has been used as a parameter to analyze a venture's performance. Financial indicators that are used to define performance include financial revenue and gross loan portfolio (Chauhan, 2021), P/E ratio, and breakeven point (Preethi et al., 2023).

Initial financial capital has also been used as an important factor to predict the success of a start-up (Cooper et al., 1994) How much capital a start-up can raise in the form of seed capital or venture capital in the initial stage or during its operations and growth. A study of technology-based start-ups identified a start-up as successful if it secures funding and can earn significant revenue (Díaz-Santamaría & Bulchand-Gidumal, 2021).

2.2 Human Capital as an Indicator of Start-Up Performance

Literature on start-up performance considers the human capital as an important indicator and can predict how a start-up will perform in the long run. Inefficiencies in the management have emerged as a recurring reason for start-up failure (Pathania & Tanwar, 2024). Management of a start-up is linked directly to the experience, technical ability and management capability of the entrepreneur and his team. These characteristics of an entrepreneur have been termed Entrepreneurial Orientation (EO) (Lumpkin & Dess, 1996). The entrepreneur's expertise is quintessential when it comes to making important decisions for a new venture such as incubating a start-up, experience, and credibility in seeking external funding, setting up a team of experts, or risk-taking ability. Expertise of the entrepreneur has been identified as the key criteria used by venture capitalists while making the funding decision (Macmillan et al., 1985). Adoption of knowledge management practices, an extension of entrepreneurial expertise, has also been proven to have a positive impact on the financial performance of start-ups (Battisti et al., 2022).

Indicators of human capital studied in literature include initial human capital of the venture, number of partners in a venture and their industry know-how (Cooper et al., 1994), characteristics of the entrepreneur (Sandberg & Hofer, 1987), entrepreneur's experience and personality (Macmillan et al., 1985), entrepreneur's education (Robinson & Sexton, 1994), entrepreneur's experience and experience of his team (Grežo & Hanák, 2024).

While initial human capital has been used to predict start-up success, growth in human capital has been used as a start-up performance indicator. The amount of employment generated which is indicative of the growth of the organization has been considered as a success indicator for a start-up (Hackett & Dilts, 2004) (Zapata-Guerrero et al., 2020).

2.3 Technological Capital as an Indicator of Start-Up Performance

Technological capital, is termed as physical capital of an organization. The term technological capital has been used to imply the initial technology of the startup, research and development undertaken by the start-up, innovations in the form of new product development, new process development, or design development. Investment in innovation or technology has become imminent for any organization due to the dynamic and fast-growing external environment (Yu et al., 2021). Technology driven start-ups tend to procure funding quickly as compared to non-technology driven start-ups as they attract high valuations (Tekic et al., 2024) (Preethi et al., 2023) and add to the competitiveness of the enterprise (Guan et al., 2006). Technology-led business model innovation allowed unicorn companies to steer through crises such as Covid-19. (Rodrigues & Noronha, 2021) Lack of technology-led innovations or investments in technology has been connected to the valley of death, which refers to the challenges faced during the early developmental phases of a start-up (Gbadegeshin et al., 2022).

Indicators used to depict start-up's technological capital in literature include the expenditure on Research and development and labor involved (Yu et al., 2021), innovation rate or number of new products launched (Guan et al., 2006), years taken to become unicorns (Harshit Kumar Taneja, 2024).

3 DEA: A Tool to Predict Start-Up Performance

As can be seen from the literature, performance of start-ups is dependent on multiple variables. There is a need to approach the performance of new start-ups in a multi-dimensional way (Reynolds, 1987). Data Envelopment Analysis (DEA) is a non-parametric approach involving multiple inputs and multiple outputs which are run in a mathematical linear programming model to give relative efficiencies of all decision-making units (DMUs) (Charnes et al., 1978). Multi-factor performance evaluation using DEA measures efficiency in the range from 0 to 1 and helps in identifying performance targets for the DMUs under study. The DEA methodology does not require prior weights of inputs and outputs (Charnes et al., 1978). The DEA methodology has been undergoing constant innovations which make it relevant to wide range of applications (Mergoni et al., 2024) (Mishra et al., 2024).

Previously, DEA has been used to analyze the performance of Chinese listed companies and predict early warning of Financial Crisis (Zhang et al., 2022), the performance of incubators (Zapata-Guerrero et al., 2020), assess the diffusion of innovation amongst EU member states (Anderson & Stejskal, 2019), to assess the performance of new ventures (Macmillan et al., 1985), to evaluate the technological innovation capability and competitiveness of start-ups (Guan et al., 2006) (Preethi et al., 2023), decode complex start-up valuation (Üzgün, 2023), to study Indian Public Sector Banks (Sonkariya & Yadav, 2024) and many others.

3.1 DEA Slack Based Model

In this paper, we have used the slack-based model of DEA. The Slack Based Model (SBM) was introduced by Tone in 2001 (Tone, 2001). The SBM model can deal with inputs/outputs individually and hence provides non-uniform input/output factor efficiencies (Tone & Tsutsui, 2010). The DEA SBM model is helpful in determining the input surpluses and output deficiencies which can be applied to the inefficient DMUs for optimization. The SBM model has undergone various improvisations to accommodate negative data, imprecise data, dealing with undesired outputs (Emrouznejad et al., 2025), (Lozano, 2016), to take into account the contribution of linking activities which contribute to producing the output (Tone & Tsutsui, 2009). Tone et al. provided the SupSBM-efficiency score for strong efficient DMUs, revealing more information about the efficient DMUs (Tone et al., 2020).

Previously, DEA SBM has been used to evaluate the efficiency of enterprises (combined heat and power plants) (Pokushko et al., 2023), performance optimization of hospitals (Arya & Yadav, 2018), banks and bank branches (Shi et al., 2023) (Lozano,

2016) and educational institutions (Johnes & and Tone, 2017) where the decision making units could account for inefficiencies in specific areas. Also, Mahmoudi et al. (2019) have developed comprehensive performance evaluation of banking branches using three-stage SBM model. In this paper, the SBM model is applied to evaluate the relative efficiencies of start-ups as it gives a direction to the start-ups to improve their relative efficiency. In case of start-ups, the inputs and outputs are non-radial in nature and therefore, required change in resources (inputs) will be a desirable knowledge for these companies which work towards creating the desired outputs.

The SBM mathematical model (Cooper et al., 2006) used in the paper is as follows:

$$\min \rho = \frac{\left(1 - \frac{1}{m}\sum_{i=1}^{m}(s_i^- / s_{iq}\right)}{\left(1 + \frac{1}{r}\sum_{i=1}^{r}(s_i^+ / y_{iq}\right)}$$

Subject to the following conditions:

$$\sum\nolimits_{j=1}^{n} x_{ij\lambda_j} + s_i^- = x_{iq}, \text{i} = 1, 2, \ldots, \text{m}.$$

$$\sum\nolimits_{j=1}^{n} y_{ij\lambda_j} - s_i^+ = y_{iq}, \text{i} = 1, 2, \ldots, \text{r}.$$

$$\lambda_m,\ s_i^-,\ s_i^+ \geq 0.$$

where, ρ represents the efficiency value of DMU (x0, y0).

s_i^- represents the redundancy of the ith input,

s_i^+ represents the deficiency of the rth output.

λ is the adjustment matrix

4 Data and Methodology

We attempt to assess the performance of thirty-nine Indian start-ups, founded between 2015 and 2022, which entered the unicorn[2] category by the year 2022[3]. While India has birthed lakhs of start-ups in the last decade, only a few start-ups have survived and fewer have been able to reach the unicorn status. Hence, it is important to analyze the performance of these unicorn start-ups which act as role models for other non-unicorn start-ups. Unicorn start-ups are a relatively new phenomenon, they have not received much attention in the academic fraternity (Rodrigues & Noronha, 2023). Also, the availability of data is a challenge in India, especially when it comes to the emerging world of start-ups. However, reaching the unicorn category is a unique phenomenon that warrants relatively more data or information being available in the public domain.

The DEA methodology requires the identification of appropriate inputs and outputs. A study that applied DEA to evaluate innovative EU member states used total human

[2] Unicorns are those technology based start-ups that have reached a valuation of one billion dollars.

[3] The year 2022 has been chosen as it is at a safe distance from the year 2020 which was the Covid year, the year of many disruptions and financial upheavals.

resource, finance, and total innovation expenditure as inputs while employment and sales were employed as outputs in the DEA model (Anderson & Stejskal, 2019). Another study that evaluated Mexican incubators using DEA identified the number of technical staff of the incubator, financial resources, the total area of the incubator, financing opportunities, access to research centers, access to physical and virtual technological resources, academic level of tutors as inputs, while outputs were number of companies graduated, the growth of employment and survival to the first year of operation. (Zapata-Guerrero et al., 2020).

Table 1 gives an essence of inputs and output variables used in the start-up DEA literature. There is a mix of indicators from the finance, people, and innovation dimensions being used as inputs and financial and technology or RnD based indicators being employed as outputs.

Table 1. Input and output variables of the efficiency model in Startup Literature

Inputs	Outputs	References
• No of employees	• Financial Revenue • Gross Loan portfolio	Chauhan, P. 2021(Chauhan, 2021)
• Number of events announced by listed companies in China	• Shareholder comments positive sentiment index	Zhang et al., 2022(Zhang et al., 2022)
• Manufacturing ability • Cost	• Revenue • Delivery rate	Preethi et al., 2023(Preethi et al., 2023)
• Total number of resources • Number of square meters	• Annual rate of graduates • Annual survival rate new companies	Zapata Gurero et al., 2020(Zapata-Guerrero et al., 2020)
• new doctorate graduates, • lifelong learning, • opportunity-driven entrepreneurship, • R&D expenditure in the public sector, • venture capital investments, • R&D expenditure in the business sector • non-R&D innovation expenditure	• employment of knowledge • intensive service • sales of developed or newly produced products	Anderson and Stejskal, 2019 (Anderson & Stejskal, 2019)
• Learning • R&D • Manufacturing • Marketing • Organizing • Resource	• Market share • Sales growth • Export rate • Profit growth • Productivity • New product rate	(Guan et al., 2006)

(continued)

Table 1. *(continued)*

Inputs	Outputs	References
• Number of RnD personnel working hours • New Product Development Expenditure • RnD Expenditure	• Sale of new products • Number of patents	(Lu et al., 2017)
• RnD Expense • Rn D labor	• Patents • New Products	(Yu et al., 2021)

Table 2 shows the input and output variables used in the DEA SBM model applied in our paper. It is worth noting that input and output variables used in the paper have been selected after careful reference to literature and based on the availability of data in the Indian context.

Table 2. Input and Output Variables used for DEA SBM model

Dimension	Input Variable	Output Variable
Human Capital	Number of employees in the year 2022–23	
Financial Capital	Initial investment raised by the startup to initiate operations	Profit/Loss for the year 2022
	Number of rounds of investment raised	Valuation in the year 2022
Technology Capital	No of years taken to enter unicorn category has been taken as a proxy to investment done in technology	

Inputs Variables

Backed by literature, three performance parameters have been used in our study, namely, financial, human, and technological capital. Other parameters used in literature include industry structure, strategy of the start-up (Sandberg & Hofer, 1987), and running out of money (Goswami et al., 2023) among others have not been used in our study as these parameters are qualitative while the DEA model requires numerical data.

Human Capital: The literature points out that the capabilities of the entrepreneur and his team are important factors in making a start-up a success. However, it was difficult to quantify the capabilities of the start-up founder(s), hence the number of employees (size of the team) has been taken as the proxy for human capital due to the key reason of availability of data.

Financial Capital: Initial capital raised by the start-up at the start of operations and a number of total investment rounds raised by the start-up have been taken to represent financial capital of the start-up as input variables.

Technology Capital: There is a lot of reference to technology-based start-ups in literature as highlighted in Sect. 2 of the paper. However, total investment in technology or RnD has not been disclosed by all start-ups, and hence cannot be used. However, the author theorize that the number of years taken to reach the unicorn category is directly proportional to the investment done in technology by the start-up. Therefore, we have taken that as one of our key input variables.

Output Variables

Financial indicators have been used as output indicators. While the literature suggests technology based and human capital based indicators as important outputs of a start-up, due to lack of availability of data on indicators such as number of patents or new products launched, we restricted to financial indicators as key output variables.

Profits earned by the start-up in the year 2022: One of the key output indicators considered is the profits reported by the start-ups. DEA requires input and output data on a comparable scale, hence data needs to be normalized. (Preethi et al., 2023) There were cases where the company had reported a loss for the year 2022. Since the DEA model used for this study does not process negative data values, the axis of the profit/loss values was shifted by subtracting all the values by 800, so that all values become positive.

Valuation of Start-up: Another output indicator considered is the valuation of the start-up in the year 2022 which represents the total market value of the start-up as evaluated by various market investors and experts. Valuation of a start-up was declared an important factor in predicting success of a start-up by Krishna et al. (Krishna et al., 2016).

5 Results and Analysis

The key logic of applying DEA for start-up evaluation is that startups could strategically utilize their resources to maximize efficiency by understanding the relationship between the performance determinants (Preethi et al., 2023).

Descriptive statistics of our data can be seen in Table 3. There were a maximum of 10 employees and a minimum of 3 employees working with the start-ups in our data. The start-ups raised initial capital of $5.4 million on average, raised investment from 2.8 rounds on average, and reached the unicorn category in a minimum of 2 years from the inception year and a maximum of 16 years. The average profit earned by the start-ups was $7.1 million with an average valuation of $2.6 billion.

Table 4 gives the efficiencies of all DMUs and the slacks for each DMU. A total of 8 DMUs achieved an efficiency of 1. Average efficiency achieved was 0.48. There are thirteen DMUs which have efficiency of more than 50%. The result indicates that start-ups in India need to work on enhancing their efficiency levels. It is those DMUs that have less than 50% relative efficiency which need high level of introspection in terms of input and output corrections.

It is observed that four DMUs have an efficiency of less than 30%, DMU 1, 13, 17, and 24. A common observation of all these DMUs is that they have employed a large number of employees, have raised a large amount of capital, and took a long tenure (average of 8.75 years) to become a unicorn as compared to other start-ups. These DMUs have an average profit of 5.8 mn and an average valuation of 1.7 mn, which is lower than the average. These companies have exploited a large amount of resources but have not been able to become highly profitable or highly valued companies and therefore their efficiencies are the lowest among the peer group.

Table 3. Descriptive analysis of Input and Output Variables

	D-Input	D-Input2	D-Input3	D-Input4	D-Ouput1	D-Output2
	no of employee	capital raised (mn)	no of rounds of investment	no of years to become unicorn	Profit (mn)	Valuation (mn)
average	5.6	5.4	2.8	6.4	7.1	2.6
sd	1.9	4.8	2.3	2.7	1.5	2.1
median	5.0	3.9	2.0	6.0	7.7	1.7
mode	5.0	2.5	1.0	5.0	7.7	1.1
max	10.0	25.0	10.0	16.0	8.4	12.0
min	3.0	0.5	1.0	2.0	1.0	1.0

The DEA SBM model has an advantage of suggesting manipulations in inputs and outputs, if undertaken by DMUs can enhance efficiency levels.

Table 4. DMU Efficiency and slacks from DEA SBM

DMU	efficiency	slack. Input 1	slack. Input2	slack. Input3	slacks. Input4	slacks. Ouput1	Slacks Output2
1	0.26	1.38	0	2.17	5.03	0	3.71
2	0.64	0	0	4.11	2.20	0	0.29
3	1	0	0	0	0	0	0
4	1	0	0	0	0	0	0
5	0.39	0	0	1.45	0.68	0	3.51
6	1	0	0	0	0	0	0
7	0.51	0	0.38	0	0	0	1.81
8	1	0	0	0	0	0	0
9	0.86	0	0	0.21	0	0	0.38
10	0.79	0	0.95	0	0	0	1.08
11	0.41	0	0.19	3.54	0	0	3.50
12	0.44	0.45	0	0.73	1.91	0	3.18
13	0.26	0	0	4.99	0	0	4.25

(*continued*)

Table 4. *(continued)*

DMU	efficiency	slack. Input 1	slack. Input2	slack. Input3	slacks. Input4	slacks. Ouput1	Slacks Output2
14	0.54	0.15	7.81	1.57	0	0	2.55
15	0.41	0	0.45	5.06	0.34	0	3.92
16	0.36	1.14	0	1.21	0.63	0	2.18
17	0.10	0	5.61	1.24	5.72	7.05	0
18	0.66	0	0	0.64	0	0	0.91
19	0.45	0.0009	0	0.11	1.77	0	5.48
20	1	0	0	0	0	0	0
21	0.44	0	0.91	1	2	0.051	0.75
22	1	0	0	0	0	0	0
23	0.57	0	3.96	0.45	5.43	1.54	0
24	0.22	0	0	2.79	6.29	0	6.43
25	0.72	0	0	0.39	0.91	0	0.44
26	1	0	0	0	0	0	0
27	1	0	0	0	0	0	0
28	1	0	0	0	0	0	0
29	0.60	0	0	1.04	0	0	0.93
30	0.52	0	0	0.21	0	0	1.96
31	0.50	0	5.86	0.30	0	0	3.49
32	0.52	3.15	1.18	0.05	0.06	0	0.90
33	1	0	0	0	0	0	0
34	0.36	1.49	0	3.27	5.61	0	4.29
35	0.41	0	7.11	0.08	3.33	0	6.02
36	1	0	0	0	0	0	0
37	0.50	0	5.69	1.06	1.63	2.49	0
38	0.34	2.73	1.05	0	1.92	0	8.89
39	0.39	2.36	14.06	0.38	0	0	8.51

As can be seen in Table 4, for inefficient DMUs, the DEA SBM model has given slack for all or selective inputs and outputs. But for DMUs whose efficiency is 1, the slack is zero for the inputs and outputs. The direct application of the DEA result is that it can be recommended to the start-up owners that reduction in inputs by the precise amount given by the model and increasing the outputs can make them more efficient. For instance, DMU 1 has an efficiency of 33%. According to the DEA results, the DMU is suggested to reduce the first input which is the number of employees from 8 to 1.034, no change is suggested in input 2 which is the funds raised by the start-up, input 3 the number of rounds investment is raised remains same or can be reduced from 3 to 2.2 and input 4 that is the number of years taken to reach the current status of a unicorn from 9 to 5.3, which could be interpreted as with better technology, the start-up can reduce the number of years it is investing in to turn into unicorn which will add to the efficiency of the company. At the same time, the DMU can extend its output 2 valuation from 1.4 to 1.87. With these slack adjustments in inputs and outputs, the DMU can reach the required relative efficiency of 1.

It is also observed in the findings that the model has suggested fewer changes in input 1 which is the number of employees working with the start-up and more changes in input 2 which is initial capital raised or used and input 3 which is the number of investment rounds. Start-ups are not employing more than the required number of employees which are the key cost centers for any organization. However, they need to look into the fact that more output can be obtained with less capital, less number of investment rounds and less number of years to reach unicorn status. This can only be achieved through business model correction. Similarly, the model suggests less correction with the first output which is profit and more corrections in the valuation of start-ups. The implication is towards tweaking the model towards long-term sustainability which will help in raising the confidence of investors and raise valuation.

While it is important for new ventures to innovate their business models or else they face the threat of further reducing their performance or even death (Reynolds, 1987), all start-ups may or may not be open to the idea of drastic business model changes. However, the unicorns can respond dynamically to market needs as they can adapt their business model for making specific digital innovations as solutions (Rodrigues & Noronha, 2023). Innovative startups, especially unicorns, are considered to be better prepared to deal with market dynamics as they seek to adapt and identify better entrepreneurial opportunities (or performance) through internal restructuring or business model adjustment (Linnenluecke, 2017).

Other than business model innovation, knowledge management which includes building knowledge in form of innovative products, systems, patents etc., preserving and passing it to the successors or using it as a revenue stream, has been found to have significant impact on performance of start-ups (Battisti et al., 2022). Knowledge management can be said to be a bi-product of both investment in technology as well as entrepreneurial expertise.

6 Conclusion and Way Forward

The paper approaches the performance measurement of start-ups in India from the lens of multi-dimensional DEA slack-based model. The slack based model is helpful in identifying the modifications needed in inputs and outputs for start-ups to attain higher efficiency. This is an attempt to experiment with new models in start-up performance measurement research. An important contribution of the paper comes in the form of new input and output variables that are used in the DEA model. The findings of the paper can be practically implemented by practitioners in terms of input and output adjustments that can be done to their enterprises to enhance start-up performance. Adjusting inputs and outputs can lead to remarkable business model innovations which are critical for the success of any venture. Also, companies exhibiting high relative efficiency were found to be profit-making companies as well.

Availability of data in India is a challenge and it plays an important role in identifying inputs and outputs indicators in the DEA model. The research gives way to further refinement of DEA model in terms of weights restrictions and the choice of inputs and outputs variables that can enhance the results as well as practical application of results. Another directive for future research is to work around undesired outputs in case of start-up performance.

References

Anderson and Stejskal. Diffusion efficiency of innovation among EU member states: a data envelopment analysis. Economies **7**(2), 34 (2019). https://doi.org/10.3390/economies7020034

Arya, A., Yadav, S.P.: Development of intuitionistic fuzzy super-efficiency slack based measure with an application to health sector. Comput. Ind. Eng. **115**, 368–380 (2018). https://doi.org/10.1016/j.cie.2017.11.028

Battisti, E., Alfiero, S., Quaglia, R., Yahiaoui, D.: Financial performance and global start-ups: the impact of knowledge management practices. J. Int. Manag. **28**(4), 100938 (2022). https://doi.org/10.1016/j.intman.2022.100938

Charnes, A., Cooper, W.W., Rhodes, E.: Measuring the efficiency of decision making units. Eur. J. Oper. Res. **2**(6), 429–444 (1978). https://doi.org/10.1016/0377-2217(78)90138-8

Chauhan, S.: Measuring financial efficiency and ranking of Indian MFIs: an analysis using DEA vs PCA (2021)

Cooper, A.C., Gimeno-Gascon, F.J., Woo, C.Y.: Initial human and financial capital as predictors of new venture performance. J. Bus. Ventur. **9**(5), 371–395 (1994). https://doi.org/10.1016/0883-9026(94)90013-2

Cooper, W.W., Seiford, L.M., Tone, K.: Data Envelopment Analysis: A Comprehensive Text with Models, Applications, References and DEA-Solver Software (2nd ed. 2007 edition). Springer-Verlag New York Inc (2006)

David, D., Gopalan, S., Ramachandran, S.: The startup environment and funding activity in India. In: Taghizadeh-Hesary, F., Yoshino, N., Kim, C.J., Morgan, P.J., Yoon, D. World Scientific Series in Finance, vol. 17, pp. 193–232. World Scientific (2021). https://doi.org/10.1142/9789811235825_0007

Davidsson, P., Honig, B.: The role of social and human capital among nascent entrepreneurs. J. Bus. Ventur. **18**(3), 301–331 (2003). https://doi.org/10.1016/S0883-9026(02)00097-6

Díaz-Santamaría, C., Bulchand-Gidumal, J.: Econometric estimation of the factors that influence startup success. Sustainability **13**(4), 4 (2021). https://doi.org/10.3390/su13042242

Emrouznejad, A., Brzezicki, Ł., Lu, C.: Development and evolution of slacks-based measure models in data envelopment analysis: a comprehensive review of the literature. J. Econ. Surv. (2025). https://doi.org/10.1111/joes.12682

Gbadegeshin, S.A., et al.: Overcoming the valley of death: a new model for high technology startups. Sustain. Futures **4**, 100077 (2022). https://doi.org/10.1016/j.sftr.2022.100077

Goswami, N., Murti, A.B., Dwivedi, R.: Why do Indian startups fail? A narrative analysis of key business stakeholders. Indian Growth Dev. Rev. **16**(2), 141–157 (2023). https://doi.org/10.1108/IGDR-11-2022-0136

Grežo, M., Hanák, R.: Entrepreneurial experience and venture success: a comprehensive meta-analysis of performance determinants. J. Entrep. Manag. Innov. **20**(1), Article 1 (2024). https://doi.org/10.7341/20242011

Guan, J.C., Yam, R.C.M., Mok, C.K., Ma, N.: A study of the relationship between competitiveness and technological innovation capability based on DEA models. Eur. J. Oper. Res. **170**(3), 971–986 (2006). https://doi.org/10.1016/j.ejor.2004.07.054

Haber, S., Reichel, A.: The cumulative nature of the entrepreneurial process: the contribution of human capital, planning and environment resources to small venture performance. J. Bus. Ventur. **22**(1), 119–145 (2007). https://doi.org/10.1016/j.jbusvent.2005.09.005

Hackett, S.M., Dilts, D.M.: A systematic review of business incubation research. J. Technol. Transf. **29**(1), 55–82 (2004). https://doi.org/10.1023/B:JOTT.0000011181.11952.0f

Taneja, H.K.: Exploring the success of Indian unicorns: a study of growth trends and economic impact. Int. J. Sci. Res. Arch. **12**(2), 537–545 (2024). https://doi.org/10.30574/ijsra.2024.12.2.1259

Johnes, G., Tone, K.: The efficiency of higher education institutions in England revisited: comparing alternative measures. Tert. Educ. Manag. **23**(3), 191–205 (2017). https://doi.org/10.1080/13583883.2016.1203457

Krishna, A., Agrawal, A., Choudhary, A.: Predicting the outcome of Startups: less failure, more success. In: 2016 IEEE 16th International Conference on Data Mining Workshops (ICDMW), pp. 798–805 (2016). https://doi.org/10.1109/ICDMW.2016.0118

Linnenluecke, M.K.: Resilience in business and management research: a review of influential publications and a research agenda. Int. J. Manag. Rev. **19**(1), 4–30 (2017). https://doi.org/10.1111/ijmr.12076

Lozano, S.: Slacks-based inefficiency approach for general networks with bad outputs: an application to the banking sector. Omega **60**, 73–84 (2016). https://doi.org/10.1016/j.omega.2015.02.012

Lu, J., Jia, G., Zong, H., He, W.: The technological input-output efficiency of high-technology enterprises in china based on the DEA method. MATEC Web Conf.erences **100**, 05031 (2017). https://doi.org/10.1051/matecconf/201710005031

Lumpkin, G.T., Dess, G.G.: Clarifying the entrepreneurial orientation construct and linking it to performance. Acad. Manag. Rev. **21**(1), 135–172 (1996). https://doi.org/10.2307/258632

Macmillan, I.C., Siegel, R., Narasimha, P.N.S.: Criteria used by venture capitalists to evaluate new venture proposals. J. Bus. Ventur. **1**(1), 119–128 (1985). https://doi.org/10.1016/0883-9026(85)90011-4

Mahmoudabadi, M.Z., Emrouznejad, A.: Comprehensive performance evaluation of banking branches: a three-stage slacks-based measure (SBM) data envelopment analysis. Int. Rev. Econ. Financ. **64**, 359–376 (2019)

Mergoni, A., Emrouznejad, A., De Witte, K.: Fifty years of data envelopment analysis. Eur. J. Oper. Res. **326**(3), 389–412 (2025). https://doi.org/10.1016/j.ejor.2024.12.049

Mishra, S., Singh, A., Sufia.: Environmental performance of coal-based thermal generation plants: a two-stage DEA model with undesirable outputs. In: Emrouznejad, A., Thanassoulis, E., Toloo, M. (eds.), Advances in the Theory and Applications of Performance Measurement and Management, pp. 151–163. Springer (2024). https://doi.org/10.1007/978-3-031-61597-9_13

Pathania, A., Tanwar, S.: Decoding startup failures in Indian startups: insights from interpretive structural modeling and cross-impact matrix multiplication applied to classification. J. Entrep. Manag. Innov. **20**(2), Article 2 (2024). https://doi.org/10.7341/20242025

Pokushko, M., Medina-Bulo, I., Kuzmich, R., Pokushko, R., Karaseva, M., Dubovik, N.: Slack based model for enterprises' efficiency improvement. 351–356 (2023). https://doi.org/10.15405/epct.23021.43

Preethi, K.N., et al.: Enhancing Startup efficiency: multivariate DEA for performance recognition and resource optimization in a dynamic business landscape. Int. J. Adv. Comput. Sci. Appl. **14**(8) (2023). https://doi.org/10.14569/IJACSA.2023.0140869

Reynolds, P.D.: New firms: societal contribution versus survival potential. J. Bus. Ventur. **2**(3), 231–246 (1987). https://doi.org/10.1016/0883-9026(87)90011-5

Robinson, P.B., Sexton, E.A.: The effect of education and experience on self-employment success. J. Bus. Ventur. **9**(2), 141–156 (1994). https://doi.org/10.1016/0883-9026(94)90006-X

Rodrigues, C.D., de Noronha, M.E.S.: What companies can learn from unicorn startups to overcome the COVID-19 crisis. Innov. Manag. Rev. **20**(3), 211–226 (2021). https://doi.org/10.1108/INMR-01-2021-0011

Rodrigues, C.D., Noronha, M.E.S.D.: What companies can learn from unicorn startups to overcome the COVID-19 crisis. Innov. Manag. Rev. **20**(3), 211–226 (2023). https://doi.org/10.1108/INMR-01-2021-0011

Sandberg, W.R., Hofer, C.W.: Improving new venture performance: the role of strategy, industry structure, and the entrepreneur. J. Bus. Ventur. **2**(1), 5–28 (1987). https://doi.org/10.1016/0883-9026(87)90016-4

Shi, X., Wang, L., Emrouznejad, A.: Performance evaluation of Chinese commercial banks by an improved slacks-based DEA model. Socioecon. Plann. Sci. **90**, 101702 (2023). https://doi.org/10.1016/j.seps.2023.101702

Sonkariya, A., Yadav, S.P.: Development of intuitionistic fuzzy revenue efficiency models in DEA: an application to the Indian public sector banks. In: Emrouznejad, A., Thanassoulis, E., Toloo, M. (eds.), Advances in the Theory and Applications of Performance Measurement and Management, pp. 137–149. Springer, Switzerland (2024). https://doi.org/10.1007/978-3-031-61597-9_12

Tekic, Z., Chemikhin, A., Malyy, M.: Market-first or technology-first? Exploring unicorns' pathways to extreme valuations. Technol. Soc. **77**, 102526 (2024). https://doi.org/10.1016/j.techsoc.2024.102526

Tone, K.: A slacks-based measure of efficiency in data envelopment analysis. European Journal of Operational Research, **130**(3), 498-509 (2001)

Tone, K., Toloo, M., Izadikhah, M.: A modified slacks-based measure of efficiency in data envelopment analysis. Eur. J. Oper. Res. **287**(2), 560–571 (2020). https://doi.org/10.1016/j.ejor.2020.04.019

Tone, K., Tsutsui, M.: Network DEA: a slacks-based measure approach. Eur. J. Oper. Res. **197**(1), 243–252 (2009). https://doi.org/10.1016/j.ejor.2008.05.027

Tone, K., Tsutsui, M.: Dynamic DEA: a slacks-based measure approach. Omega **38**(3), 145–156 (2010). https://doi.org/10.1016/j.omega.2009.07.003

Üzgün, Y.O.: Startup valuation with data envelopment analysis. In: Derindere Köseoğlu S. (ed.), A Practical Guide for Startup Valuation: An Analytic Approach, pp. 275–296. Springer (2023). https://doi.org/10.1007/978-3-031-35291-1_13

Yu, A., Shi, Y., You, J., Zhu, J.: Innovation performance evaluation for high-tech companies using a dynamic network data envelopment analysis approach. Eur. J. Oper. Res. **292**(1), 199–212 (2021). https://doi.org/10.1016/j.ejor.2020.10.011

Zapata-Guerrero, F.T., Ayup, J., Mayer-Granados, E.L., Charles-Coll, J.: Incubator efficiency vs survival of start-ups. RAUSP Manag. J. **55**(4), 511–530 (2020). https://doi.org/10.1108/RAUSP-04-2019-0063

Zhang, Z., Xiao, Y., Fu, Z., Zhong, K., Niu, H.: A study on early warnings of financial crisis of Chinese listed companies based on DEA–SVM model. Mathematics **10**(12), 2142 (2022). https://doi.org/10.3390/math10122142

Evaluating Performance Efficiency in the Malaysian Telecommunications Industry: A Data Envelopment Analysis Approach

Arumugam G Sithamparam(✉) and Senthil Kumar

Asia Pacific University, Kuala Lumpur, Malaysia
{arumugam.gs,senthil.kumar}@apu.edu.my

Abstract. This purpose of this study is to evaluate the efficiency of four leading operators in Malaysia's telecommunications sector, namely Telekom Malaysia, CelcomDigi, TIME dotCom, and Maxis Bhd—amidst growing competition and technological advancements, such as the transition to 5G. Using the Data Envelopment Analysis (DEA, the constant and variable returns-to-scale analysis were done to assess the operators' operational efficiency from 2018 to 2023, a period marked by intense market competition, merger, and the rollout of next-generation networks. Given the limited number of primary decision-making units (DMUs) in this study, DEA window analysis was applied to enhance discriminatory power for firms that show variance in the efficiency performance, enabling more nuanced temporal comparisons of efficiency. The findings indicate that efficiency gains were realized by firms following strategic mergers, although some firms incurred short-term reductions in pure technical efficiency. The analysis indicated that, while capital and operational investments are vital for maintaining competitiveness, excessive expenditure without proportional revenue growth may hinder the performance of EBITDA. Furthermore, this study highlights the positive impact of scale adjustments in maximizing efficiency, underscoring the role of consolidation in Malaysia's telecommunications landscape. Future research is suggested to explore productivity dynamics through the Malmquist index to understand shifts in efficiency frontiers over time as the industry continues to navigate the challenges of digital transformation and infrastructure investments.

Keywords: Data envelopment analysis · Telecommunications · Window analysis

1 Introduction

Malaysia's telecommunications industry has transformed significantly since the Communications and Multimedia Act (CMA) of 1998, which restructured the sector by separating its regulatory and service roles. This led to the establishment of the Malaysian Communications and Multimedia Commission (MCMC) and the privatization of Telekom Malaysia, fostering a competitive environment that attracted private players. According to the International Telecommunication Union's (ITU) 2023 ICT Development Index, Malaysia scored 94.5 out of 100, ranking 15th globally and 3rd

A. Emrouznejad et al. (Eds.): DEA 2024, LNOR, pp. 474–491, 2025.
https://doi.org/10.1007/978-3-031-98177-7_32

in the ASEAN region, following Singapore and Brunei (International Telecommunication Union [ITU], 2023). As of 2023, Malaysia's Information and Communication Technology (ICT) sector, encompasses publishing, broadcasting, telecommunications, and computer services. The GDP contribution in the third quarter (Q3) 2024 is 3.7%, a decline from Q3 2023 and Q3 2022 which were 3.8 and 4.5 respectively (DoSM, 2024). Advancements in mobile technology, from 3G to 4G, and now 5G, have driven Malaysia's digital landscape. The shift to 4G has resulted in faster data speeds and mobile applications, while the 2021 rollout of 5G, managed by Digital Nasional Berhad (DNB), has introduced ultra-fast speeds essential for supporting IoT and smart city applications, enhancing sectors such as healthcare and manufacturing. The merger of Celcom and Digi in 2022 to form CelcomDigi is another key development in creating a more extensive network to provide broader coverage and improved services. The International Telecommunication Union's ICT Development Index (IDI) 2023 by the International Telecommunication Union (ITU) highlights Malaysia's significant progress in ICT infrastructure and affordability, reflecting strong alignment with national digital transformation goals. The report ranks Malaysia favorably in mobile and fixed-broadband subscriptions, marking advancements in universal connectivity. The report (International Telecommunication Union [ITU], 2023) emphasizes Malaysia's advancements in universal connectivity, with high rankings in mobile and fixed-broadband subscriptions. This achievement aligns with Malaysia's national digital transformation goals, as the country continues to enhance both ICT infrastructure and service affordability, solidifying its role in the digital economy.

These developments reflect Malaysia's ongoing commitment to enhance its telecommunications infrastructure and ensure that its citizens and businesses can leverage modern communication technologies. However, it emphasizes areas for improvement, such as high-speed connectivity expansion and ICT skills development which are essential to further enhance meaningful connectivity across demographics (International Telecommunication Union, (ITU) 2023). These insights underscore Malaysia's commitment to digital inclusivity and its efforts toward comprehensive digital literacy. These developments reflect Malaysia's ongoing commitment to enhance its telecommunications infrastructure and ensure that its citizens and businesses can leverage modern communication technologies.

The industry remains central to Malaysia's economic development goals, including the digital transformation ambitions under the MyDIGITAL Blueprint (https://www.mydigital.gov.my/).

Figure 1 and Fig. 2 illustrate the growth trends of mobile and fixed broadband in Malaysia, reflecting the sector's expanding role in the country's digital landscape. Figure 1 shows a steady rise in mobile broadband subscriptions, increasing from 28.53 million in 2016 to 45.34 million in 2023, with slight moderation during 2018–2020 and accelerated growth post-2020. This upward trend indicates the growing reliance on the mobile internet for both personal and business use, particularly as digital adoption surged after the COVID-19 period. Figure 2 highlights a similar trend for fixed broadband, with the penetration rate increasing from 39% in Q1 2021 to 50.6% in Q4 2023. This consistent quarterly (Q) increase signifies the expansion of broadband accessibility as telcos strive to reach more households and businesses.

The period from 2021 to 2023 was marked by two significant industry developments: the merger of Celcom and Digi and the rollout of 5G, which collectively fueled the growth seen in both mobile and fixed broadband adoption. These shifts underscore the need for telcos to enhance their operational efficiency, particularly when facing rising infrastructure and service demands. The merger allows for resource consolidation and reduced operational redundancies, whereas 5G requires careful capital expenditure (CAPEX) planning to optimize network expansion. As telcos expand broadband coverage and adopt new technologies, efficiency in operating expenditure (OPEX) and strategic investments are crucial for meeting growing customer expectations and maintaining a competitive edge in a dynamic market.

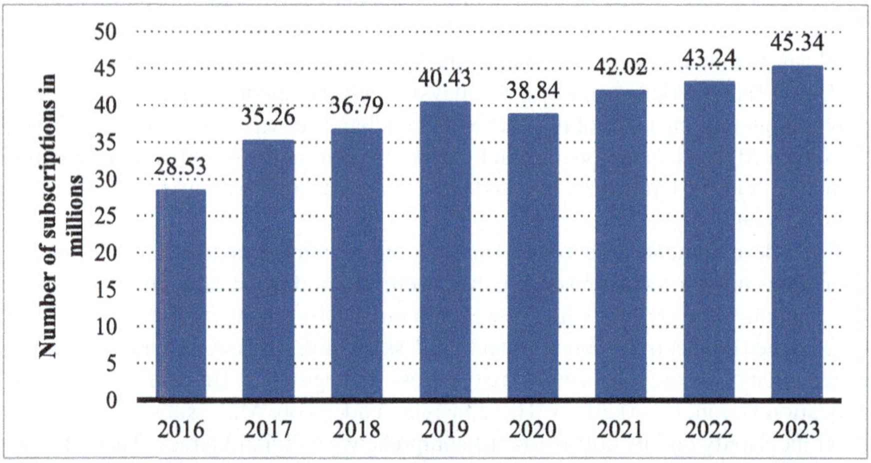

(Source: (Malaysian Communications and Multimedia Commission [MCMC], 2024))

Fig. 1. Number of mobile broadband subscriptions in Malaysia from 2016 to 2023 (in millions).

The remainder of this paper is organized as follows. Section 2 reviews the application of DEA in the telecom sector. Section 3 provides an overview of the DEA methodology and its extension to a window analysis. Section 4 details the data and highlights the key milestones in the history of these four firms. Section 5 presents and discusses the study's empirical results. Section 6 concludes the study and suggests potential directions for future research.

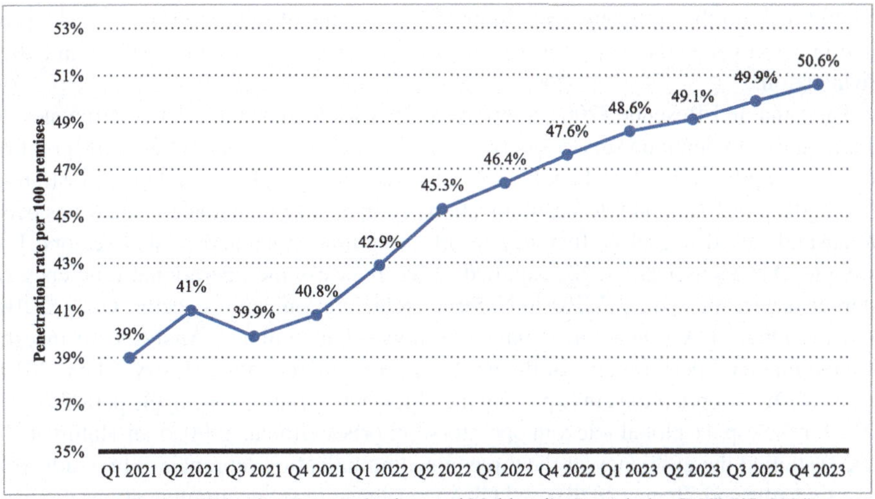

(Source: Malaysian Communications and Multimedia Commission [MCMC], 2024)

Fig. 2. Fixed broadband penetration rate in Malaysia Q1 2021-Q4 2023.

2 Literature Review

Given the current situation, enhancing efficiency has become critically important for telecommunications operators aiming to increase their profit margins. Among the various techniques for measuring efficiency, the Data Envelopment Analysis (DEA), a method introduced by Charnes et al., (1978) to assess the relative efficiency of decision-making units (DMUs) that utilize multiple inputs to generate multiple outputs. A key consideration when applying DEA is that the number of DMUs should be at least twice the sum of the inputs and outputs. Otherwise, the model may yield many units that appear to be relatively efficient, thereby reducing its discriminating power. However, since there are only four primary operators in this study, DEA window analysis (as developed by Charnes et al., 1985) is implemented to expand the number of DMUs. This approach also allows the evaluation of each DMU's performance across different periods. Additionally, due to acquisitions among operators that alter scale size, the BCC model (discussed later) is used to assess the impact of these acquisitions on scale efficiency.

Although Data Envelopment Analysis (DEA) and its extensions have been widely used over the past two decades, DEA window analysis remains relatively uncommon in the literature. Carbone (2000) applied DEA window analysis to identify efficiency trends over time in semiconductor manufacturing. Similarly, Sueyoshi and Aoki (2001) analyzed the performance of Japanese postal services from 1983 to 1997 by combining DEA window analysis with the Malmquist index. Asmild et al., (2004) combined DEA window analysis with the Malmquist index to assess productivity changes in Canada's banking industry from 1981 to 2000. In assessing the efficiency of the telecommunications industry, the use of the DEA approach is limited. The extant literature search has some studies including Sueyoshi (1998) who explored various aspects of Nippon Telegraph and Telephone's performance, including whether the Japanese government should break up the company, examining its economies of scale, returns to scale, production activities, and cost performance, and comparing performance before and after privatization.

FacAnha and Resende (2004) employed DEA to analyze the relative efficiency of Brazilian local telephone services between 1998 and 2002. Cooper et al., (2001) introduced an imprecise DEA to handle imprecise data and weight restrictions, illustrating a method using a Korean telecommunications company. In recent years, DEA has been increasingly used to study efficiency in telecommunications and related sectors. For example, Ajebesin et al., (2024) applied DEA to assess the operational efficiency of internet service providers (ISPs) in Nigeria's telecommunications sector. Goto (2010) used the DEA- DA (Data Envelopment Analysis–Discriminant Analysis) to investigate the financial performance of the world telecommunications industry. Chen (2019) extended DEA applications in exploring the efficiency variation during the period 2012–2017 for the top 41 global telecom operators in Forbes Global, 2000. Tsai-Halim et al., (2023) studied the efficiency of the three telecommunications companies in Saudi Arabia. Oredegbe and Zhang (2020) used DEA to evaluate the telecommunications industry efficiency in 19 countries categorized into high income countries (HICs) and middle income countries (MICs). The extant literature search did not reveal any studies on Malaysia's telecommunications industry efficiency performance using the DEA approach. Hence, the DEA technique and the use of Window analysis in analyzing the Malaysian telecommunications firms' efficiency is likely the first of its kind.

3 Methodology

Data Envelopment Analysis (DEA) is a non-parametric model, meaning it does not require a predefined production function to handle the observed data. Let us consider a set of n decision-making units (DMUs). For DMU k, let (where r = 1,…,s) represent the level of the rth output, and (where i = 1,…,m) represent the level of the ith input. To assess the efficiency of DMU k, Charnes et al. (1978) proposed the following model.

$$
\begin{aligned}
&\min \quad \theta \\
&\textit{subject to } \theta_{x_{ik}} - \sum_{j=1}^{n} \lambda_j x_{ij} \geq 0, i = 1, \ldots, m, \\
&\sum_{j=1}^{n} \lambda_j y_{rj} \geq y_{rk}, r = 1, \ldots, s, \\
&\lambda_j \geq 0, j = 1, \ldots, n, \quad \theta \textit{ unrestricted in sign}.
\end{aligned}
\tag{1}
$$

The optimal efficiency score, denoted as *, falls within the range. When = 1, a DMU is considered technically efficient and is on the efficiency frontier, along with the set of most efficient units. The inefficient DMUs are situated below this frontier, with their observed data being enveloped by it. In the DEA, the efficiency of the DMUs is measured against the DMUs positioned on the frontier, which are the best performing units. Introduced by Charnes et al. (1978), this approach is commonly known as the CCR model, characterized by the assumption of constant returns to scale (CRS) in the production function. Consequently, the efficiency score derived from the CCR model is referred to as technical efficiency (TE) to differentiate it from other efficiency measures.

3.1 The BCC Model

To account for the variable returns-to-scale (VRS), Banker, Charnes, and Cooper (1984) extended the original CCR model (Model 1) to create what is now known as the BCC model. Unlike the CCR model, which assumes constant returns-to-scale (CRS), the BCC model allows for variations in efficiency owing to changes in the scale of operations. This flexibility makes the BCC model particularly useful for analyzing situations in which increasing or decreasing the scale of inputs may affect the output efficiency.

$$\begin{aligned} &Min \quad \theta \\ &Subject\ to\ \theta_{x_{ik}} - \sum_{i=1}^{n} \lambda_j x_{ij} \geq 0, i = 1, \ldots, m, \\ &\sum_{i=1}^{n} \lambda_j y_{rj} \geq y_{rk}, r = 1, \ldots, s, \\ &\lambda_j \geq 0, j = 1, \ldots, n, \quad \theta\ unrestricted\ in\ sign. \end{aligned} \tag{2}$$

In the BCC model, similar to the technical efficiency (TE) in the CCR model, the objective value is referred to as the pure technical efficiency (PTE). By adding the constraint the BCC model limits its feasible region to a subset of that in the CCR model, allowing for variable returns-to-scale. This means that PTE is always equal to or greater than TE, as the PTE isolation efficiency is based solely on resource utilization, disregarding the influence of scale. A low PTE indicates inefficient resource management by a Decision-Making Unit (DMU) within its environment. The BCC model enables the decomposition of TE into PTE and scale efficiency (SE), where SE is calculated as SE = TE/PTE. The SE measures the impact of the scale size on a DMU's efficiency. If a DMU exhibits high PTE but low TE, resulting in low SE, this suggests that while the DMU uses its resources effectively (high PTE), it does not operate at an optimal scale. In such cases, adjusting the scale, either increasing or decreasing it, could enhance the overall efficiency by aligning the DMU with the most productive scale size. This decomposition helps to identify whether inefficiencies stem from internal resource management (PTE) or from scale-related factors (SE), providing valuable insights for optimizing operational size and resource utilization.

DEA Window Analysis: DEA window analysis extends the concept of moving averages to assess the efficiency trends of Decision-Making Units (DMUs) over time. By treating each DMU within a specific window as a separate entity, this approach enables the comparison of a DMU's efficiency across different periods, thus increasing the discriminatory power of the DEA when limited DMUs are available. Choosing an appropriate window width is important: it should be small enough to avoid unfair comparisons across different time conditions but large enough to provide an adequate sample size for meaningful analysis, as recommended by Asmild et al. (2004). Unlike traditional methods that analyze all DMUs regardless of their efficiency scores, this study applies DEA window analysis selectively, focusing only on DMUs that show variations in efficiency over time. This targeted approach provides clearer insights into the efficiency dynamics of units in which performance fluctuates, allowing for a more focused examination of the factors influencing these changes.

As described by Asmild et al., (2004), consider a set of N Decision-Making Units (DMUs) (where $n = 1, \ldots$), each using r inputs to produce s outputs, observed across T periods (where $t = 1,\ldots,T$). Let *DMUnt* denote the observation of DMU n during period t, with an associated.

$$\text{input vector } X_n^t = \begin{pmatrix} x_n^{1t} \\ \vdots \\ x_n^{rt} \end{pmatrix} \text{ and output vector } Y_n^t = \begin{pmatrix} y_n^{1t} \\ \vdots \\ y_n^{rt} \end{pmatrix}.$$

If the window begins at time k (where $1 \le k \le T$) and has a width of w (where $1 \le w \le T$), then the input and output data within this window can be represented in matrix form. These matrices capture the inputs and outputs for each DMU over a specified time window, allowing the analysis of efficiency across multiple periods. By structuring the data in this way, DEA window analysis can evaluate a DMU's performance not only within a single period, but also relative to its behavior across a range of periods within the selected window. This setup enables a more dynamic analysis of efficiency trends over time, and is especially useful when observing changes in performance or adjustments in resource utilization across different time frames. The matrices are expressed as follows:

$$X_{kw} = \begin{bmatrix} x_1^k & x_2^k & \cdots & x_N^k \\ x_1^{k+1} & x_2^{k+1} & \cdots & x_N^{k+1} \\ \vdots & \vdots & \ddots & \vdots \\ x_1^{k+w} & x_2^{k+w} & \cdots & x_N^{k+w} \end{bmatrix}, Y_{kw} = \begin{bmatrix} y_1^k & y_2^k & \cdots & y_N^k \\ y_1^{k+1} & y_2^{k+1} & \cdots & y_N^{k+1} \\ \vdots & \vdots & \ddots & \vdots \\ y_1^{k+w} & y_2^{k+w} & \cdots & y_N^{k+w} \end{bmatrix}.$$

By substituting the inputs and outputs of into Model (1) or Model (2), the efficiency results for DEA window analysis can be obtained. This process involves evaluating the efficiency of each DMU within a specified window by applying the DEA model to its observed inputs and outputs for each period in the window. By repeating this analysis across different time windows, the DEA window analysis allows us to track how each DMU's efficiency evolves over time. This approach is especially valuable for identifying trends and fluctuations in performance because it enables comparisons of a DMU's efficiency in different periods within the same window. Consequently, DEA window analysis provides a more comprehensive view of efficiency dynamics, highlighting both short and long-term trends in resource utilization and output production across multiple time frames.

4 Data and Milestones of Telecommunication Firms

Before detailing the data, the key biographical milestones for each of the three telecommunication firms are presented below. The significant milestones for each operator are outlined as follows.

Telekom Malaysia (TM) was established in 1984 and quickly became a major force in Malaysia's telecommunication landscape. TM was listed on the Kuala Lumpur Stock Exchange in 1990; after the 2008 restructuring, it focused solely on fixed-line, broadband, and data services. With its UniFi broadband service launched in 2010, TM achieved the largest broadband market share in Malaysia, accounting for approximately 40% of the fixed broadband market by 2023 (Twimbit, 2023). TM's role as a key provider in the 5G rollout through its partnership with Digital Nasional Berhad (DNB) has further solidified its position in Malaysia's telecommunications infrastructure (Bernama, 2023).

CelcomDigi was formed through the merger of Celcom (established in 1988) and Digi (founded in 1995) in 2022, creating Malaysia's largest mobile network operator. This merger boosted CelcomDigi's market position, resulting in approximately 45% of its mobile market share in 2023 (Twimbit, 2023). With extensive network coverage and active participation in Malaysia's 5G rollout, CelcomDigi continues to dominate the mobile market, with a wide array of prepaid and postpaid services (Bernama, 2023).

TIME dotCom (TIME) was established in 1996 and has become a key player in Malaysia's fiber and broadband infrastructure market. It was listed on the Kuala Lumpur Stock Exchange in 2001, and by 2010, TIME had expanded its focus to high-speed broadband for businesses and urban residential areas. As of 2023, TIME holds approximately 15% of the fixed broadband market share, primarily targeting urban and enterprise markets, with a strong focus on data centers and cloud services (Twimbit, 2023).

Maxis Bhd, founded in 1995, quickly became a leader in both the mobile and broadband segments. Maxis launched Malaysia's first 3G service in 2002 and LTE (4G) in 2013, reinforcing its position as a provider of high-speed mobile Internet. As of 2023, Maxis accounts for approximately 25% of the mobile market share and 20% of the fixed broadband market share (Twimbit, 2023). Maxis has been actively involved in the 5G rollout and has focused on digital transformation and IoT solutions for businesses (Bernama, 2023).

The primary data source for this analysis is the London Stock Exchange Group (LSEG) database, formerly known as Refinitiv, rather than publicly available financial statements. The LSEG is a global provider of comprehensive financial and economic data known for its breadth, accuracy, and reliability, making it invaluable for data-driven research. Based on data availability and previous studies, assets, operating costs, and capital and labor expenditures were chosen as input variables, whereas revenues, ebitda, and assets were selected as output variables. Table 1 presents the descriptive statistics for these variables. To validate the selection of inputs and outputs, Spearman correlation coefficient analysis was conducted to test for isotonicity, confirming that increases in inputs do not reduce outputs. As indicated in Table 2, all variables were positively correlated, supporting their suitability for the efficiency analysis.

The four major Malaysian telecommunications companies—Telekom Malaysia (TM), CelcomDigi, TIME dotCom, and Maxis Bhd—serve distinct customer bases, impacting the suitability of the subscriber count as an output variable for efficiency analysis (Twimbit, 2023; Malaysian Communications and Multimedia Commission, 2023). TM, for example, serves a mix of residential, SME, and large corporate clients, with high-capacity services for corporate and government clients driving much of its revenue, making the subscriber count less indicative of its efficiency (TM Annual Report, 2023). CelcomDigi, the largest mobile network in Malaysia, serves primarily individual and SME users but has a lower revenue per user compared to firms with more corporate clients (CelcomDigi Annual Report, 2023). TIME dotCom focuses on high-demand urban and corporate clients, including data centers, leading to fewer subscribers overall, but higher revenue per user, meaning that subscriber counts undervalue their efficiency (TIME Annual Report, 2023). Maxis, with a mix of individual and corporate clients, also derive significant revenue from enterprise solutions such as IoT and digital services, which are not fully captured by subscriber numbers (Maxis Annual Report, 2023). Internationally, Verizon in the U.S. and Vodafone in Europe illustrate a similar pattern, where a strong corporate client base yields high average revenue per user (ARPU), which is not represented by subscriber count (Verizon Annual Report, 2023; Vodafone Annual Report, 2023). Tsai et al., (2006) used the EBITDA margin as an output variable implying it (EBITDA) a more balanced comparison, focusing on profitability and operational efficiency across telcos, It is justified to argue that EBITDA as an output variable which captures revenue potential and operational effectiveness independent of customer composition, offers a fair assessment of efficiency across telcos.

5 Discussion of the Efficiency and Empirical Findings

A DEA Window Analysis was employed to examine the efficiency trends of DMUs over time (AlKhars et al., 2022). Each DMU within a window is treated as a distinct entity, allowing comparisons of its efficiency across different periods. This method enhances the discriminating power, particularly when analyzing a limited number of DMUs. The window should be small enough to avoid unfair comparisons, but large enough to ensure adequate representation (Asmild et al., 2004). As such the use of the analysis in this study is justified.

The number of windows (Cooper et al., 2007) is calculated as follows:

Number of windows, $w = k - p + 1 = 6 - 3 + 1 = 4$.

Where:

- $k =$ total periods $= 6$
- $p =$ window width $= 3$

Based on this calculation, there were four windows, each spanning three years. This resulted in 48 data points, as shown below:

Number of "different" firms or data points $= n \times p \times w = 4 \times 3 \times 4 = 48$.

While there is no theory to determine the window size, a width of three years has been used in the past studies carried out by Jia and Yuan (2017) and Hemmasi et al., (2020). A review by Maidamisa et al., (2012) also indicates that many studies have used a window width of three years.

5.1 Analysis

Stata (v. 18) was used to compute the descriptive statistics (Table 1) and Spearman's correlation (Table 2).

Table 1 presents the descriptive statistics for the selected inputs (OPEX, CAPEX, LABEX) and outputs (REVENUE, EBITDA, ASSETS). The monetary data is measured in millions of Ringgit Malaysia. Significant variations in firm expenditures and financial performance can be noted. Firms could have similar operational cost structures as indicated by the mean of the operating expenses (OPEX), which has relatively low dispersion (standard deviation of RM 350 million). On the other hand, greater variability is reflected from the capital expenditures (CAPEX) and labor expenditures (LABEX), where the standard deviations are RM701 million and RM 935 million, respectively. This could imply different investment strategies and workforce structures across firms. The revenue output variable shows a significant differences in firm sizes and market performance. The EBITDA variable has relatively low variation, which could mean that most firms maintain a similar level of operational profitability despite differences in expenditures. The standard deviation for total assets is high, indicating that some firms hold significantly larger asset bases than others. The DEA method is deemed suitable given notable difference in values across all variables.

Table 1. Descriptive Statistics of Inputs and Outputs

Variable	Mean	Std. Dev	Min	Max
Input				
OPEX	5,850	350	681	10,700
CAPEX	1,180	701	248	2,660
LABEX	964	935	199	2,620
Output				
REVENUE	7,460	4,140	983	12,700
EBITDA	.422875	.0542829	.289	.523
ASSETS	16,000	9,870	3,130	37,000

Data in millions of Ringgit Malaysia.

Table 2. Spearman Correlation Coefficients of Inputs and Output

	OPEX	CAPEX	LABEX	REVENUE*	EBITDA*	ASSETS*
OPEX	1.0000					
CAPEX	0.9600	1.0000				
LABEX	0.9243	0.8870	1.0000			
REVENUE*	0.9843	0.9548	0.9183	1.0000		
EBIDTA*	−0.6203	−0.5763	−0.7451	−0.5807	1.0000	
ASSETS*	0.9365	0.8670	0.8826	0.9270	−0.5724	1.0000

* Outputs

After selecting the input and output variables, this study employs the Spearman correlation coefficient analysis to test for isotonicity. The choice of input and output variables is based on its appropriateness to measure the efficiency of firms focused on commercial objectives. Saxena et al. (2015) considered revenue, operating profit (commonly known as EBIT (Earnings Before Interest and Tax) or EBITDA (Earnings Before Interest Tax Depreciation and Amortization)) and EPS (Earnings Per Share), while the input variables were cost of material, employee benefit expenses (LABEX) and capital employed. Abdel-Halim et al., (2023) studied capital expenditure and operating expenses as inputs with revenue as one of the outputs. Tsai et al., (2006) had total assets, CAPEX and number of employees as inputs, and revenue, EBITDA and EBIT as outputs. Morita and Avkiran (2009), studied total assets as an output variable.

As shown in Table 2, the Spearman correlation analysis reveals essential insights into the relationships between inputs and Operating Expenses (OPEX), Capital Expenditure (CAPEX), and Labor Expenses (LABEX)—and outputs, namely Revenue, EBITDA, and Assets, in the telecommunications industry. OPEX shows a strong positive correlation with revenue (0.9843), suggesting that operational spending is integral in driving revenue growth. In telecom, a high OPEX, which includes network maintenance, customer support, and service enhancements, is crucial for maintaining service quality and expanding market reach, which directly impacts revenue. However, OPEX has a negative correlation with EBITDA (-0.6203), implying that, while these expenses boost revenue, they simultaneously pressure profitability. EBITDA is computed by subtracting Cost of Goods Sold (COGS) and OPEX from total revenue. Consequently, when the other things are unchanged, a higher OPEX leads to lower EBITDA.

This non-isotonic relationship reflects the fact that in capital-intensive industries, excessive operating costs can erode EBITDA margins if not matched by equivalent revenue gains. Similarly, CAPEX is positively correlated with revenue (0.9548), indicating that capital investments are necessary for revenue growth. CAPEX funds infrastructure expansion, technological upgrades, and network enhancement, which are the essential components of service quality and customer acquisition. However, CAPEX also negatively correlates with EBITDA (−0.5763), indicating that while CAPEX drives long-term growth, it can reduce profitability in the short term if revenue gains do not immediately offset costs. LABEX is also strongly positively correlated with Revenue (0.9183), as labor expenses support customer service, technical maintenance, and field operations, which are crucial for high-quality services. However, LABEX's negative correlation with EBITDA (−0.7451) further emphasizes the challenge of balancing labor costs with profitability, as excessive labor expenses can reduce margins. The strong positive correlations of Assets with all three inputs (OPEX: 0.9365, CAPEX: 0.8670, LABEX: 0.8826) demonstrate that high operational, capital, and labor expenses contribute to asset growth, reinforcing the capital-intensive nature of telecoms. This analysis highlights the need for telecom companies to strategically manage expenditures across operations, capital investments, and labor to achieve revenue growth while preserving profitability. The non-isotonic relationships with EBITDA suggest that, while these costs are essential for sustaining revenue, careful cost control is crucial for maintaining efficiency and profitability, making DEA analysis valuable for assessing how well companies balance these competing demands. In Data Envelopment Analysis (DEA), the assumption of isotonicity, where increasing inputs should not decrease outputs, is commonly applied, as it reflects the expectation that more resources lead to higher output levels. However, a negative correlation between inputs, such as Operating Expenses (OPEX), Capital Expenditure (CAPEX), Labor Expenses (LABEX), and EBITDA (a measure of profitability), is considered non-isotonic. This non-isotonic relationship is valid in certain industries, including telecommunications, where increased expenditures often lead to lower profitability margins, if not managed efficiently. In industries with high operating and capital expenses, such as telecommunications, rising costs may support revenue, but can simultaneously reduce profitability (EBITDA) if expenses outweigh revenue gains. Studies show that this non-isotonic relationship is acceptable in DEA because it reflects real economic behavior where high costs can erode margins. The negative correlation between expenses and EBITDA provides insights into a firm's cost management efficiency, showing that efficiency improvements are achieved not only by increasing revenue but also by effectively managing costs.

High technical efficiency (TE) signifies that a DMU effectively converts inputs into outputs. The TE is calculated under the assumption of constant returns to scale, meaning that the output changes proportionally with input changes.

Table 3. Technical Efficiency over 2018–2023

Unit name (DMU)	2023	2022	2021	2020	2019	2018
CelcomDigi Bhd	100.00%	100.00%	100.00%	100.00%	100.00%	100.00%
Maxis Bhd	100.00%	97.50%	100.00%	100.00%	100.00%	100.00%
TIME dotCom Bhd	100.00%	100.00%	100.00%	100.00%	100.00%	100.00%
Telekom Malaysia Bhd	93.60%	88.80%	95.90%	100.00%	100.00%	79.00%

The data in Table 3 show that CelcomDigi Bhd and TIME dotcom Bhd operate at the maximum relative efficiency. In contrast, Maxis and Telekom Malaysia Bhd had lower efficiency scores, with TM exhibiting more variability than their peers. The lower efficiency scores for TM and Maxis can be attributed to their relatively higher capital, operating, and labor expenses. Additionally, these two companies have higher asset levels. It appears that these companies incur significant expenses on operations, capital, and labor without a corresponding increase in revenue, which has also led to lower EBITDA margins.

Compared with the CRS assumption, the VRS assumption (BCC Model) offers a more nuanced perspective by accounting for the size of operations when computing efficiencies. Table 4 presents the pure technical efficiency scores of the sample companies, providing a more realistic assessment.

Table 4. Pure Technical Efficiency 2018–2023

Unit name	2023	2022	2021	2020	2019	2018
CelcomDigi Bhd	100.00%	100.00%	100.00%	100.00%	100.00%	100.00%
Maxis Bhd	100.00%	100.00%	100.00%	100.00%	100.00%	100.00%
TIME dotCom Bhd	100.00%	100.00%	100.00%	100.00%	100.00%	100.00%
Telekom Malaysia Bhd	96.00%	100.00%	100.00%	100.00%	100.00%	100.00%

For instance, TM has lower revenue in 2023 than CelcomDigi. TM can reduce labor expenses and focus on improving revenues.

It can be noted in Table 4, CelcomDigi Bhd, Maxis Bhd, and TIME dotCom Bhd achieved 100% scores for the Pure Technical Efficiency (PTE) for 2018–2023 period. This means these firms operated at the highest efficiency levels in utilizing their inputs to generate outputs during the studied six years. This may not be same for Telekom Malaysia Bhd (TM). The firm relented in its efficiency score to achieve a relatively lower score of 96% in 2023. This implies that TM has room for improvement in optimizing its resources efficiently. A possible reason could be TM achieved a lower revenue in 2023 compared to its competitors. Lower efficiencies suggest that firms can benefit from scaling up or down.

Scale efficiency scores help determine whether a company operates on an appropriate scale. Table 5 presents the scale efficiency scores, showing that Maxis had a slightly lower scale efficiency in 2022.

Table 5. Scale Efficiency 2018–2023

Unit name	2023	2022	2021	2020	2019	2018
CelcomDigi Bhd	100.00%	100.00%	100.00%	100.00%	100.00%	100.00%
Maxis Bhd	100.00%	97.50%	100.00%	100.00%	100.00%	100.00%
TIME dotCom Bhd	100.00%	100.00%	100.00%	100.00%	100.00%	100.00%
Telekom Malaysia Bhd	97.50%	88.80%	95.90%	100.00%	100.00%	79.00%

TM exhibited higher variability, with significant decreases in 2022 and 2018. Lower revenues, coupled with relatively higher capital, labor, and operating expenses, likely contributed to lower efficiencies. To achieve higher efficiencies, it is crucial for TM to increase revenue and reduce these significantly higher expenses.

Window Analysis: The efficiency scores of CelcomDigi Bhd and TIME Dotcom Bhd were nearly perfect, and the Maxis (VRS) efficiency scores were also almost perfect. Therefore, window analysis of such data is unlikely to provide new insights. Consequently, only Maxis Bhd (CRS and SE) and Telekom Malaysia Bhd (CRS, VRS, and SE), which exhibited variations during the study period, were selected for this analysis. The variation in Maxis' efficiency scores is minimal (refer Table 6), with a slight dip observed in 2022 from the unitary DEA analysis. The window analysis also indicates a reduction in technical efficiency, likely due to increased capital and labor expenses. Despite higher revenue, Maxis could not maintain 100% efficiency. In contrast, Telekom Malaysia showed a clear declining trend. TM's expenses are the highest among its peers, yet its revenue is lower than that of CelcomDigi, indicating the inefficient utilization of inputs. The pure technical efficiency of TM, as shown in Table 7, reflects this inefficiency. Maxis scores were consistently 100%, suggesting that further analysis may not provide new insights. However, the window analysis reveals a realistic and better understanding of the scale of operations, highlighting a reduction in pure technical efficiency.

Table 6. DEA Window Analysis – Technical Efficiency

TM - CRS	2023	2022	2021	2020	2019	2018	Average
2021–2023	93.60%	88.80%	95.90%				92.77%
2020–2022		88.80%	95.90%	100.00%			94.90%
2019–2021			95.90%	100.00%	100.00%		98.63%
2018–2020				100.00%	100.00%	79.00%	93.00%
Average	93.60%	88.80%	95.90%	100.00%	100.00%	79.00%	94.83%
Maxis-CRS	2023	2022	2021	2020	2019	2018	Average
2021–2023	100.00%	97.50%	100.00%				99.17%
2020–2022		97.50%	100.00%	100.00%			99.17%
2019–2021			100.00%	100.00%	100.00%		100.00%
2018–2020				100.00%	100.00%	100.00%	100.00%
Average	100.00%	97.50%	100.00%	100.00%	100.00%	100.00%	99.58%

Table 7. DEA Window Analysis – Pure Technical Efficiency

TM - VRS	2023	2022	2021	2020	2019	2018	Average
2021–2023	96.00%	100.00%	100.00%				98.67%
2020–2022		100.00%	100.00%	100.00%			100.00%
2019–2021			100.00%	100.00%	100.00%		100.00%
2018–2020				100.00%	100.00%	100.00%	100.00%
Average	96.00%	100.00%	100.00%	100.00%	100.00%	100.00%	99.67%

The scale efficiencies (window analysis) of the TM and Maxis are presented in Table 8.

The data indicate that the TM's scale efficiency of the TM is relatively low and shows a declining trend. This decline can be attributed to the technical efficiency and its influencing factors. In contrast, the decline in Maxis' scale efficiency was minimal. Significant fluctuation in the scale efficiency of TM is noted, from a low of 79% in 2018 to 97.5% in 2023. The fluctuations could indicate a lack in managing the resources efficiently, perhaps due to high CAPEX, LABEX and infrastructure costs. Maxis consistently operates at near-optimal scale efficiency throughout the study period. This suggests Maxis is managing its costs and operations effectively. The Celcom-Digi merger (2022) probably resulted in intensified competition which led to competing firms adapting their efficiency strategies. As the 5G and fiber broadband rollout progresses, telecommunications firms must align infrastructure investments with revenue potential to maintain long-term efficiency. TM's gradual recovery in scale efficiency suggests improving financial strategies, but sustaining this requires continued optimization in OPEX and capital utilization to compete effectively in the industry (Fig. 3).

Table 8. DEA Window Analysis – Scale Efficiency

TM - SE	2023	2022	2021	2020	2019	2018	Average
2021–2023	97.50%	88.80%	95.90%				94.07%
2020–2022		88.80%	95.90%	100.00%			94.90%
2019–2021			95.90%	100.00%	100.00%		98.63%
2018–2020				100.00%	100.00%	79.00%	93.00%
Average	97.50%	88.80%	95.90%	100.00%	100.00%	79.00%	95.15%
Maxis-SE	2023	2022	2021	2020	2019	2018	Average
2021–2023	100.00%	97.50%	100.00%				99.17%
2020–2022		97.50%	100.00%	100.00%			99.17%
2019–2021			100.00%	100.00%	100.00%		100.00%
2018–2020				100.00%	100.00%	100.00%	100.00%
Average	100.00%	97.50%	100.00%	100.00%	100.00%	100.00%	99.58%

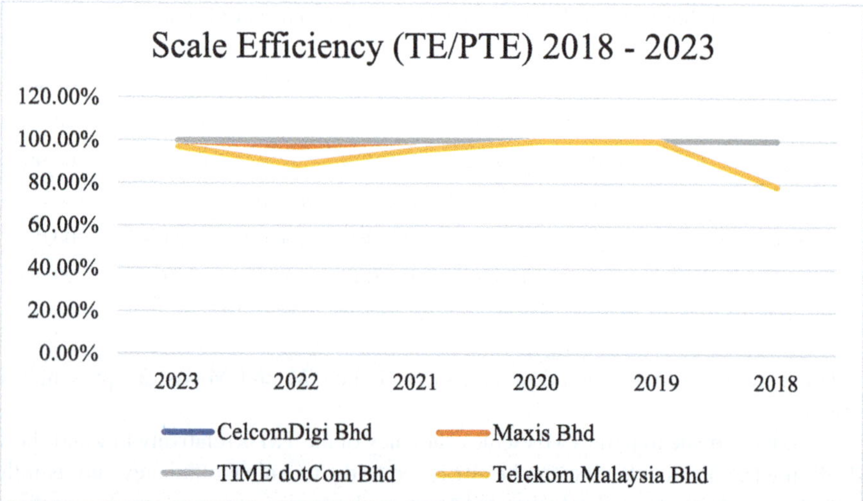

Fig. 3. Scale Efficiency 2018 – 2023

Overall, the technical efficiency of CelcomDigi and TIME dotcom was nearly perfect, with similar trends observed for pure technical efficiency and scale efficiency. In contrast, TM has the lowest technical and scale efficiency among peer groups, indicating inefficient input conversion. Reducing input expenses could help improve TM efficiency. Maxis efficiencies are slightly lower, particularly in terms of technical and scale efficiency, likely because of relatively higher capital expenses. Rationalizing capital expenditure could benefit Maxis.

6 Conclusion

This study applied the DEA model and DEA window analysis to assess the efficiency of Malaysia's four major telecommunications companies—Telekom Malaysia (TM), CelcomDigi, TIME dotCom, and Maxis Bhd—over recent years characterized by intense competition, technological advancements, and shifting market dynamics. The industry has seen major developments, including the merger of Celcom and Digi, and the roll-out of 5G, which have reshaped the competitive landscape. Given the impact of these changes on firm size and operational complexity, the BCC model was employed to capture scale efficiency adjustments. To address the limited number of Decision-Making Units (DMUs), this study used window analysis with quarterly data, with each window covering eight quarters, thus increasing the number of DMUs and enhancing the discriminative power in the efficiency assessment.

By employing both the CCR and BCC models, this study calculated technical efficiency, pure technical efficiency, and scale efficiency for each company. Three key findings emerged from the analysis. First, recent mergers and acquisitions within the sector, such as the Celcom-Digi merger, positively impacted scale efficiency by allowing these firms to leverage a larger operational base, although this initially led to some challenges in pure technical efficiency as integration processes occurred. Second, strategic adjustments, particularly in cost control and capital expenditure management, helped firms maintain relatively stable scale efficiency across periods despite variations in market conditions. Third, Telekom Malaysia, the largest government-linked entity, exhibited relatively lower efficiency levels across all measures, indicating potential areas for improvement in its resource utilization and suggesting that public-sector efficiency reforms may still be relevant in a privatizing industry landscape.

It is common to find that financial variables such as OPEX, CAPEX and revenue are used in DEA models (Asmild et al. 2004; Cooper et al. 2001). However, such financial metrics may not fully capture the operational efficiency of firms. The use of qualitative variables- service quality, customer satisfaction, regulatory compliance and innovation, which are critical in the telecommunications industry (Sueyoshi & Aoki, 2001), could lend a strong support in measuring efficiency performance wholistically. Additionally, the financial variables could often portend a short-term efficiency, while brand value which is associated with a long-term investment may not be immediately captured in the analysis (Jia & Yuan, 2017). For further research, it is worth exploring the potential application of the Malmquist Productivity Index to evaluate productivity changes over time. This would allow for the analysis of whether each company is moving closer to or further from the efficient frontier as market conditions evolve, or whether the frontier itself is expanding or contracting due to technological advancements and market pressures.

In conclusion, increasing competition and rapid technological evolution in the telecommunications industry necessitate ongoing efficiency improvements for both private and government-linked companies. This study illustrates that DEA is a valuable tool for efficiency analysis as it does not require a predefined production function and enables the separation of pure technical efficiency from scale efficiency. Pure technical efficiency reflects a firm's ability to optimize resource use under external constraints, whereas scale efficiency is crucial in an era of frequent mergers and acquisitions. By applying DEA models under both constant and variable returns to scale, this study suggests that while acquisitions and expansions may improve scale efficiency, they may also temporarily hinder pure technical efficiency as companies integrate resources and adapt to their new scale.

Reference:s

Abdel-Halim, A., Al Khars, M., Alnasser, A.: Evaluating the technical efficiency of Saudi Arabian telecommunications companies using window DEA and AHP techniques. Int. J. Islam. Middle East. Financ. Manag.Financ. Manag. **16**(6), 1306–1325 (2023)

Ajibesin, A.A., Akinmoyede, Y., Thangavel, S.K., Salahudeen, R.: Exploring productivity by evaluating ISP efficiency in Nigeria's telecommunications sector. In: Ajibesin, A., Vajjhala, N. (eds.), Data Envelopment Analysis (DEA) Methods for Maximizing Efficiency, pp. 261–282. IGI Global (2024)

AlKhars, M.A., Alnasser, A.H., AlFaraj, T.: A survey of DEA window analysis applications. Processes **10**(1836) (2022)

Asmild, M., Paradi, J.C., Aggarwall, V., Schaffnit, C.: Combining DEA window analysis with the Malmquist index to measure productivity change in Canada's banking industry (1981–2000). J. Prod. Anal. **21**(1), 67–89 (2004)

Banker, R.D., Charnes, A., Cooper, W.W.: Some models for estimating technical and scale inefficiencies in data envelopment analysis. Manage. Sci. **30**(9), 1078–1092 (1984)

Bernama.: ICT sector contributes 23.5% to Malaysia's GDP in 2023, https://www.mida.gov.my/mida-news/ict-e-commerce-generated-rm427-7b-in-2023-contributing-23-5-to-malaysias-gdp/, Accessed 05 Nov 2024

Carbone, T.A.: Measuring efficiency of semiconductor manufacturing operations using data envelopment analysis (DEA). In: IEEE/SEMI Advanced Semiconductor Manufacturing Conference, pp. 56–62. IEEE (2000)

Charnes, A., Cooper, W.W., Rhodes, E.: Measuring the efficiency of decision making units. Eur. J. Oper. Res.Oper. Res. **2**, 429–444 (1978)

Charnes, A., Clark, T., Cooper, W.W., Golany, B.: A developmental study of data envelopment analysis in measuring the efficiency of maintenance units in the U.S. Air Forces. Ann. Oper. Res. **2**(1), 95–112 (1985)

CelcomDigi Berhad. Annual Report 2023 (2023). https://celcomdigi.listedcompany.com/

Chen, C.-M.: Evaluating the efficiency change and productivity progress of the top global telecom operators since OTT's prevalence. Telecomm. Policy **43**(5), 1–24 (2019)

Cooper, W.W., Park, K.S., Yu, G.: Imprecise data envelopment analysis: handling imprecise data and weight restrictions. Eur. J. Oper. Res.Oper. Res. **129**(2), 338–347 (2001)

FacAnha, L.O., Resende, M.: Price cap regulation, incentives and quality: the case of Brazilian telecommunications. Int. J. Prod. Econ. **92**(2), 133–144 (2004)

Hemmasi, A., Talaeipour, M., Khademi-Eslam, H., Farzipoor Sean, R., Pourmousa, S.H.: Using DEA window analysis for performance evaluation of Iranian wood panels industry. Afr. J. Agric. Res. **6**(7), 1802–1806 (2020)

Goto, M.: Financial performance analysis of US and world telecommunications companies: Importance of Information Technology in the telecommunications industry after the AT&T breakup and the NTT divestiture. Decis. Support. Syst.. Support. Syst. **48**, 447–456 (2010)

International Telecommunication Union. (2023). ICT Development Index 2023: Malaysia's global digital performance ranking, https://trc.gov.jo/EchoBusV3.0/SystemAssets/%D8%AA%D9%82%D8%B1%D9%8A%D8%B1%202023_compressed, Accessed 05 Nov 2024

Jia, T., Yuan, H.: The application of DEA (Data Envelopment Analysis) window analysis in the assessment of influence on operational efficiencies after the establishment of branched hospitals. BMC Health Serv. Res. **17**(265), 1–8 (2017)

Maidamisa, A.A., Ahmad, R., Ismail, A.A.: A comparative analysis of window width selection technique in data envelopment analysis. Int. J. Comput. Appl.Comput. Appl. **41**(5), 21–26 (2012)

Malaysian communications and multimedia commission. (March 1, 2024). Number of mobile broadband subscriptions in Malaysia from 2016 to 2023 (in millions) In: Statista. Retrieved November 01, 2024. https://www.statista.com/statistics/1051148/malaysia-mobile-broadcast-subscriptions/

Malaysian Communications and Multimedia Commission. (2023). Industry Performance Report 2023. https://www.mcmc.gov.my/en/resources/industry/industry-performance-report

Maxis Berhad. (2023). Annual Report 2023. https://maxis.listedcompany.com/ar23/

Morita, H., Avkiran, N.K.: Selecting Inputs and outputs in data envelopment analysis by designing statistical experiments. J. Oper. Res. Soc. Japan **52**(2), 163–173 (2009)

Oredegbe, A., Zhang, Y.: Telecommunications industry efficiency: a comparative analysis of high and middle-income countries. Telecommun. Policy **44**(5), 1–11 (2020)

Saxena, P., Saxena, R.P., Sehgal.: Efficiency evaluation of the energy companies in CNX 500 Index of the NSE, India using data envelopment analysis. Benchmarking: Int. J. **23**(1), 113–126 (2015)

Sueyoshi, T., Aoki, S.: A use of a nonparametric statistic for DEA frontier shift: the Kruskal and Wallis rank test. Omega **29**(1), 1–18 (2001)

Sueyoshi, T.: Privatization of Nippon telegraph and telegram: was it a good policy decision? Eur. J. Oper. Res.Oper. Res. **107**(1), 45–61 (1998)

Tsai, H.-C., Chen, C.-M., Tzeng, G.-H.: The comparative productivity efficiency for global telecoms. Int. J. Prod. Econ. **103**(2), 509–526 (2006)

TIME dotCom Berhad.: Annual Report 2023, https://www.time.com.my/wp-content/uploads/pdf/annual-reports/Time-Annual-Report-2023, Accessed 06 Nov 2024

TM Berhad.: Annual Report 2023, https://tm.com.my/iar2023/, Accessed 06 Nov 2024

Twimbit.: Malaysia telecoms update 2023: Market insights and key players, https://cdn.twimbit.com/uploads/2023/11/09141131/State-of-malaysia-telco-2023, Accessed 06 Nov 2024

Verizon Communications Inc.: Annual Report 2023. https://www.verizon.com/about/investors/annual-report, Accessed 06 Nov 2024

Vodafone Group Plc.:Annual Report 2023. https://investors.vodafone.com/~/media/files/v/vodafone-ir/documents/performance/financial-results/2023/vodafone-fy23-annual-report, Accessed 06 Nov 2024

Evaluating Efficiency and Total Factor Productivity of Top Five Tourism Countries Using Malmquist DEA

Suneet Singh[1(✉)], Saurabh Pratap[1], and Snehal Laud[2]

[1] Industrial Management, Department of Mechanical Engineering, IIT (BHU) Varanasi, Varanasi, India
Suneetsingh.rs.mec21@itbhu.ac.in

[2] Department of Industrial and Production Engineering, Shri G. S. Institute of Technology and Science, Indore, MP, India

Abstract. The present work examines the efficiency and Total Factor Productivity (TFP) changes in the tourism sector of the top five countries with the highest international visitors in 2022: France, Spain, USA, Italy, and Turkey. Focusing on the period from 2019 to 2022, the study employs an output-oriented Malmquist Data Envelopment Analysis (DEA) under the Variable Return to Scale (VRS) assumption to evaluate efficiency and TFP changes. The analysis uses international tourism receipts as the output and the number of people employed in the tourism sector and the number of international visitors as inputs, with data sourced from Statista and the World Travel and Tourism Council (WTTC). The results indicate a significant decline in TFP across all countries in 2020 due to the COVID-19 pandemic, except for the USA, which saw a 5.8% increase in TFP by reducing employment in the sector. The findings highlight the pandemic's impact on the tourism sector's efficiency and offer insights for managing inputs during future crises.

Keywords: Malmquist Data Envelopment Analysis · Technical efficiency · Total factor productivity · Travel and Tourism

1 Introduction

"The travel and tourism industry, it's just a huge part of our economy." (Karen Hughes).

The quote by Karen Hughes clearly states the role of travel & tourism sector in a countries' economy (You, 2020). Tourism sector acts as the backbone of an economy for many countries. Tourism generates significant revenue, jobs, and thereby stimulates growth of various sectors, as the sector consists of many activities sus as; food, transport, accommodation, entertainment, and many others (Baloch et al., 2023). A report by the World Economic Outlook (WEO) highlights that countries with a larger share of their GDP contributed by tourism sector tend to be more economically resilient. The analysis

A. Emrouznejad et al. (Eds.): DEA 2024, LNOR, pp. 492–503, 2025.
https://doi.org/10.1007/978-3-031-98177-7_33

of Covid-19 pandemic demonstrated this clearly, as these countries were able to bounce back economically faster than those with less reliance on tourism (UN Tourism, 2023). In 2023, the travel and tourism sector experienced significant growth, adding 27 million new jobs - a 9.1% increase from 2022. This follows the industry's strong rebound in 2022, highlighting its ongoing recovery and expansion. Additionally, spending by international tourists jumped by 33.1% in 2023 (World Travel & Tourism Council, 2023). Moreover, the global economy is expected to accelerate by 2.9% in 2024, adding an impressive $11.1 trillion to the world's total economic output (World Travel & Tourism Council, 2024).

Region-specific research on contribution of tourism sector in the Bali and Indonesia's economy, clearly suggests how tourism sector can play a vital role in contributing towards nation's economy. The research examines how a country's economy can be impacted when the tourism sector is represented as a tertiary sector versus when it is represented specifically through hotels, trade, and restaurants. When represented as a tertiary sector, economic growth was projected to increase from 65.28% to 68.28% between 2010 and 2014, with employment rising from 48.96% in 2010 to 51.81% in 2014. In contrast, when represented as the trade, hotel, and restaurant sector, Bali's economy grew from 27.82% to 31.35% during the same period. Similarly, the no. of employees in this sector accelerated from 26.24% to 27.64%. The work sheds light on the importance of emphasizing and investing in the tourism sector, demonstrating its significant contribution to economic growth (Antara & Sumarniasih, 2017).

However, the COVID-19 the global pandemic completely disrupted the field, causing millions to lose their jobs (Bakar & Rosbi, 2020). As per the World Travel & Tourism Council, 63 million people lost their jobs in the year 2020 Worldwide, whereas significant growth in the sector can be observed as the figure in 2021 got reduced to roughly 43 million, and further got reduced to 39 million in 2022 (World Travel & Tourism Council, 2021). As the sector is of great importance towards the economy and employment of the nation, it is essential to study the growth and decline within the sector. However, it is not always necessary that the boom in the tourism sector will affect the economy positively; for instance, an investigation was performed on the economic factors which drive Chinese tourists to Norway using ARDL and descriptive data analysis. The results of the work showed, how tourism sector grew within Norway but weakened the local currency (Xie & Tveterås, 2020).

On the basis of above discussion, this research focuses on top five economies namely, France, Spain, USA, Italy, and Turkey, and observes how their tourism sector got affected during the COVID-19 pandemic. The countries were selected on the basis of the highest number of international visitors in 2022. To comprehensively assess the sector's performance and recovery, the study spans the years 2019 to 2022, capturing both the pre-pandemic stability, the severe downturn in 2020, and the subsequent recovery trends. The current study utilises Output-Oriented Malmquist Data Envelopment Analysis (M-DEA) approach to find out the Technical Efficiency Change (TEC) and Total Factor Productivity (TFP) change of the five economies for their Inbound tourism sector. The study is conducted from the year 2019 to 2022. Malmquist DEA (M-DEA) approach helps to compare year on year growth or decline within the sector. Each country was treated as Decision Making Unit (DMU) to observe the change. The aim of employing

M-DEA over DEA is to determine the changes in efficiency over time, it allows the assessment of productivity growth or decline across multiple periods. The study uses number of international visitors and number of employs employed within the tourism sector as inputs and the international tourism receipts for each respective country for their Inbound tourism as output. The selection of input and output variables aligns with prior studies in tourism efficiency analysis. The number of employees in the tourism sector is a key input, as highlighted by Soysal-Kurt (2017), Barišić & Cvetkoska (2020), Wu et al., (2011) and Prorok et al. (2019). The number of international arrivals is another critical input, supported by Bayrak & Bahar (2018) as it reflects tourism demand. International tourism receipts are used as the output, consistent with Soysal-Kurt (2017), as they capture the sector's economic contribution. The output-oriented DEA model specifically aims to maximize outputs while keeping the inputs unchanged. The objective of selecting output-oriented model is to determine how much international tourism receipts an economy can increase by keeping its international visitors and number of people employed within the sector constant.

The study is organized as follows: The 2nd section reviews previous works in this field. The 3rd section details the methodology used. The 4th section discusses the results. Finally, the 5th section concludes the study.

2 Literature Review

The current section is devoted to explore previous studies conducted using Data Envelopment Analysis (DEA) and travel & tourism and economic growth of an economy associated with it. In the year 2019, Barisic & Cvetkoska, (2019) evaluated the efficiency of travel and tourism in contributing to GDP and employment using Data Envelopment Analysis (DEA). They examined 28 EU countries in 2017, using internal tourism consumption and capital investment as inputs, and tourism's contribution to GDP and employment as outputs. Their findings offer insights for policymakers to enhance tourism strategies. A study by Higuerey et al., (2020) evaluated the efficiency of Ecuador's hotel industry using DEA on 147 hotels (2013–2017). It analysed revenue as output and personnel, assets, and consumption as inputs. Its results have shown that third-class hotels are most efficient, while those in tourist-rich areas optimize resources better, improving productivity. Subsequently, Narmatov et al., (2021) reviewed 350 tourism-related DEA articles, categorizing them by industry and methodology. They found a focus on hotels and Asia/Europe, highlighting emerging DEA models, sustainability, and research gaps. Wu & Lin, (2022) applied a dynamic DEA model to assess cultural tourism efficiency in Asian destinations. Cambodia, China, Hong Kong, and Singapore performed best. The studies' results highlight resource reallocation's impact on efficiency and the importance of cultural resource integration. Chen, (2022) assessed the operational efficiency of tourism factories before and after COVID-19 using the DEA approach. He developed the PADME model to assess non-financial factors and analyzes financial efficiency using turnover, net profit, assets, R&D expenses, and employees. Findings from 12 listed Taiwanese companies offer managerial and academic insights. While this study examined the operational efficiency of tourism factories in Taiwan using the Product, Aesthetic, Digitalization, Management and Experience model and DEA approach, our research focuses on the impact of COVID-19 on the tourism sector in France, Spain, the USA, Italy, and Turkey. We employ an Output-Oriented Malmquist DEA to analyze efficiency changes. Simultaneously, Wu & Lin, (2022) used dynamic DEA model for assessing 14 tourist destinations in Asia and considered each destination as a DMU (Fuchs M and F Zach, 2004). In this study, five countries are considered as DMUs, and employed Malmquist DEA to observe changes in their efficiency scores and change in Total Factor Productivity (TFP) over a period of time of four years. The Malmquist productivity index was introduced by Färe et al., (1994), to analyse the productivity growth in 17 Organisations for Economic Co-operation and Development (OECD) countries over the time span of 1979- 1988. They decomposed the Malmquist Productivity Index into two parts: changes in technology and changes in efficiency. The Malmquist DEA approach has also been applied to specific regions until now. Hereafter, Bire, (2021) applied Malmquist DEA in the Nusa Tenggara Timur (NTT) Province of Indonesia to assess the performance of regions in NTT. Malmquist DEA has been used to other sectors such as agriculture sector and various industries to compare their performance. Subsequently, Chaubey et al., (2022) evaluated the performance of agricultural sector across 31 states and Union territories of India from 2012 to 2017. The states and Union territories were ranked on the basis of efficiency scores. Simultaneously, Wang et al., (2022) applied a 2-stage DEA method to assess the operations related efficiency of Vietnam's ports. The study used top 18 Vietnam's Sea port for determining their operational efficiency. The method integrated the Malmquist and resample technique to assess the performance efficiency of the Vietnamese ports. The DEA resampling was applied for forecasting the performance of seaports performance based on the efficiency scores for the next 3 years. The objective of combining Malmquist productivity index was to analyse the total factor productivity changes through technological and technical change based on the selected inputs.

Until now, similar approaches have been used either exclusively with the DEA method or have focused on tourist destinations within a particular country (Emrouzne jad et al., 2025). The authors are not aware of any studies that compare the efficiency and productivity of inbound tourism among countries with the highest number of international tourists. Therefore, this study tries to address this recess by determining the technical efficiency changes and TFP changes for the top five countries with the highest number of international tourists.

3 Methodology

In DEA namely two approaches are used, input centered approach, & output centered approach. The efficiency is evaluated by comparing the input levels of each DMU against a constructed frontier of best practice (Mergoni et al., 2024; Glavin, 2024; Zhao, 2024). The most efficient DMU lies on the frontier, indicating that it uses the least number of inputs to produce its outputs compared to other DMU. The efficiency of each DMU is assessed by measuring how far it is from the efficiency frontier.

Technical efficiency (TE) for a DMU is generally calculated by taking the ratio of the weighted sum of its outputs to the weighted sum of its inputs. This helps determine how well a DMU is using its resources compared to the best possible performance (Nandy & Singh, 2020).

$$\mathrm{TE} = \frac{\sum_{n=1}^{s} u_n v_{ni}}{\sum_{m=1}^{p} x_m y_{mi}} \tag{1}$$

Here, u_n represent the weights of outputs and x_j represents the weights of inputs. The output quantities are denoted as v_n (n = 1, 2, 3.... s) And the input quantities are denoted as y_j (m = 1, 2, 3...p). i denotes the DMU (i = 1, 2, 3 .... l).

The constant return to scale model is expressed as;

$$max_{\theta,\lambda}\theta \tag{2}$$

Subject to,

$$-y_j + Y\lambda \geq 0 \tag{3}$$

$$\Theta\, x_j - \mathrm{X}\,\lambda \geq 0 \tag{4}$$

$$\lambda \geq 0 \tag{5}$$

In this context, Θ denotes the efficiency of a DMU, and λ is a vector of constants with dimensions N × 1. The output & input matrices are denoted by Y and X, respectively, with y_j representing the output vector and x_j the input vector for each DMU. Θ must be less than or equal to 1, and a DMU is considered to be on the efficient frontier when θ equals 1 (Charnes et al., 1978; Banker et al., 1984; Coelli et al., 2005). Total technical efficiency can be computed via. The above presented model (2).

The Variable Return to Scale (VRS) model is expressed as;

$$max_{\theta,\lambda}\theta \tag{6}$$

Subject to,

$$-y_j + Y\lambda \geq 0 \tag{7}$$

$$\Theta\, x_j - X\lambda \geq 0 \tag{8}$$

$$\mathrm{N}\ 1'\lambda = 1 \tag{9}$$

$$\lambda \geq 0 \tag{10}$$

Here, N1 is a vector of ones with dimensions N × 1. To compute the pure technical efficiency (PTE), Eq. 9 is being introduced with convexity constraint. The relationship between TTE, PTE & Scale efficiency (SE) is as follows; TTE = PTE × SE (Wang et al., 2017; Erena et al., 2021; Yildirim et al., 2019).

The output-focused DEA method, in conjunction with the Malmquist productivity index, is employed to assess the technical performance of the five countries. Malmquist (1953) introduced the Malmquist productivity index, and it is utilised to quantify the changes in productivity over a period of time. Färe et al. (1992) synthesised Farrell's (1957) concepts on measuring productive efficiency with Caves et al.'s (1982) methodologies for measuring productivity to create the Malmquist productivity index. This index is calculated directly from input and output data using DEA.

Nikolić & Miljković (2023) define an output-oriented Malmquist productivity change metric as:

$$m_0(y_{t+1}, x_{t+1}, y_t, x_t) = \left[\frac{d_o^t(x_{t+1}, y_{t+1})}{d_o^t(x_t, y_t)} \times \frac{d_o^{t+1}(x_{t+1}, y_{t+1})}{d_o^{t+1}(x_t, y_t)}\right]^{1/2} \quad (11)$$

The productivity of production point (x_{t+1}, y_{t+1}) in relation to $(x_{t,}y_t)$ production point is represented by Eq. (11).

Further Eq. (11) may be decomposed as the following;

$$m_0(y_{t+1}, x_{t+1}, y_t, x_t) = \frac{d_o^{t+1}(x_{t+1}, y_{t+1})}{d_o^t(x_t, y_t)} \times \left[\frac{d_o^t(x_{t+1}, y_{t+1})}{d_o^{t+1}(x_{t+1}, y_{t+1})} \times \frac{d_o^t(x_t, y_t)}{d_o^{t+1}(x_t, y_t)}\right]^{1/2} \quad (12)$$

The variation in the technical efficiency between t + 1 period and period t, is defined in the following way,

$$TE\Delta = \frac{d_o^{t+1}(y_{t+1}, x_{t+1})}{d_o^t(y_t, x_t)} \quad (13)$$

Technological change can be written as:

$$T\Delta = \left[\frac{d_o^t(y_{t+1}, x_{t+1})}{d_o^{t+1}(y_{t+1}, x_{t+1})} \times \frac{d_o^t(y_t, x_t)}{d_o^{t+1}(y_t, x_t)}\right]^{1/2} \quad (14)$$

Technical efficiency change in Malmquist Data Envelopment Analysis (DEA) refers to a DMU's ability to maximize output relative to its inputs over time. Pure efficiency change measures a DMU's shift towards or away from the efficiency frontier, reflecting operational improvements or declines.

$$\text{EC} = \frac{D_{t+1}(y_{t+1,}x_{t+1})}{D_t(y_{t,}x_t)} \quad (15)$$

where, $D_t(y_{t,}x_t)$ represents the distance function at time t, and $D_{t+1}(y_{t+1,}x_{t+1})$ represents the distance function at time t + 1.

To be the most efficient DMU, the DMU must have a technical efficiency score of 1 in both t and t + 1 time period, which demonstrates that the DMU is lying on the efficiency frontier in both time periods. The DMU must not have any slacks in inputs as well as outputs, ensuring that there can be decrease in the input or increase in output without affecting the efficiency score. The total factor productivity (TFP) change must be equal to greater than 1, as the TFP change is measured by the Malmquist productivity index (MPI). If a DMU fulfils all these criteria then it may be considered to be the most efficient DMU (Färe et al., 1994; Shah et al., 2024).

Technological efficiency measures the extent to which a firm uses its resources to produce the highest possible output, based on the best practice frontier, considering technological advancements. It is captured by the distance of the firm's production to the frontier of best practice (Färe et al., 1994). The Malmquist Index can be formulated as follows:

$$\text{Malmquist Index} = \text{Technical efficiency change} \times \text{Techno logical change}.$$

Where, Technical efficiency is the change in a DMU'S efficiency relative to the frontier, whereas the technological change reflects the movement of the frontier, capturing technological progress or regress. So, the TFP is the aggregate measure of the productivity, while Malmquist Index decomposes productivity growth into technical efficiency changes and technological advancements.

4 Results and Discussion

The study uses DEAP software version 2.1 for the measurement of the efficiencies and Total Factor Productivity. The study analyses efficiency changes and Total Factor Productivity (TFP) changes to distinguish between improvements in a DMU's operational performance and the impact of technological progress on productivity. There is no absolute calculation of efficiency or productivity, both the efficiency and productivity are demonstrated in terms of change while taking the previous year as base. Since the year 2019 does not have any previous year data to serve as base, the study will show the efficiency and productivity changes from the year 2020 as the year 2019 will serve as the base to compute the efficiency change (EC) and productivity change (Table 1).

Table 1. Results for the year 2020 with 2019 as the base year

Country	Technical EC	Technological EC	Pure EC	Scale EC	TFP change
France	1.176	0.571	1.000	1.176	0.672
Spain	0.815	0.587	1.000	0.815	0.478
USA	1.000	1.058	1.000	1.000	1.058
Italy	1.037	0.635	0.938	1.105	0.658
Turkey	1.083	0.681	1.000	1.083	0.737

For the year 2020, Total factor productivity for the 4 countries namely; France, Spain, Italy, and Turkey, declined while the total factor productivity for USA got increased by

5.8%. The change in TFP is clearly driven by a 5.8% improvement in its technological efficiency. The decline in TFP is attributed to a reduction in input, specifically the number of employees in the tourism sector, which dropped from 17.49 million in 2019 to 9.75 million in 2020. The decrease in the efficiencies and productivity are due to the COVID-19 impact. The TFP change for Spain declined by 52.2% followed by Italy whose TFP declined by 34.2%.

In 2021, as the effects of COVID-19 began to ease compared to the previous year, more people started traveling again, leading to an increase in the number of visitors and higher tourism revenue (Table 2).

Table 2. Result for the year 2021 with 2020 as the base year

Country	Technical EC	Technological EC	Pure EC	Scale EC	TFP change
France	1.000	1.152	1.000	1.000	1.152
Spain	1.227	1.146	1.000	1.227	1.406
USA	1.000	0.878	1.000	1.000	0.878
Italy	1.010	1.149	1.154	0.875	1.161
Turkey	1.233	1.135	1.000	1.233	1.399

The TFP for each country accelerated in the year 2021 when compared with the year 2020 except for USA which had highest TFP change in the year 2020. The TFP change for USA declined by 12.2%, while the TFP change for Spain increased by 40.6%. As the change in technical efficiency for France and USA is equal to 1, that means that both these countries have neither improved nor worsened in their ability to convert inputs into output relative to the production frontier. As Spain, Italy, and Turkey have technical efficiency change > 1, it means, they have improved their efficiency and have become more efficient in converting inputs into output.

In the year 2022, as countries began to open their borders for more tourists, most of the countries saw an increase in the number of tourists, 93.2 million tourists visited France which is 2.29 million more than the number of tourists it had in 2019. But the tourism receipts still lacked by 3.8 US billion dollars, in spite of having a greater number of visitors (Table 3).

As of 2021, The total factor productivity for all the countries increased except for USA, whose TFP decreased by 2.6%. Spain increased its TFP by 36.7%, this change is due to the change of 36.7% in its technological efficiency. While, the TFP decline in USA is also due to the technological efficiency decline of 2.6%. As in both the cases the pure efficiency as well as the scale efficiency has remained constant. As the technical efficiency for France and Turkey is <1, it denotes that they have become inefficient compared to last year. This means either the DMU is using more quantity of input to produce same quantity of output or is producing less amount of output with same amount of input.

Table 3. Result for the year 2022 with 2021 as the base year

Country	TEC	Technological EC	Pure EC	Scale EC	TFP change
France	0.881	1.462	1.000	0.881	1.288
Spain	1.000	1.367	1.000	1.000	1.367
USA	1.000	0.974	1.000	1.000	0.974
Italy	1.025	1.218	1.000	1.025	1.248
Turkey	0.891	1.198	0.879	1.014	1.068

5 Conclusion, Limitations and Future Scope

The present work aims to find the efficiency changes & TFP changes in the tourism sector of the top most five countries with the greatest number of international visitors for the year 2022, which were France, Spain, USA, Italy, and Turkey. The study focused on a four-year period from 2019 to 2022 and evaluated the efficiency and TFP changes using M- DEA. The purpose of using Malmquist DEA is its ability to function with a no. of outputs & inputs, and it focuses on the relative efficiency of DMUs comparing their performance and efficiency with the most efficient DMU which lies on the efficiency frontier. The results of the study comprise of 3 tables for the year 2020, 2021, and 2022 showcasing the efficiency and the total factor productivity changes for each year while taking the efficiencies of the previous year as base. The efficiency and TFP changes for 2019 are not evaluated as the data for the year 2018 was not available to serve as base for the year 2019. The paper uses a single output and double input Malmquist DEA model, with considering international tourism receipts for respective countries as the output, and number of people employed in tourism sector and the number of international visitors for each year as the inputs. The results indicate a significant decline in TFP across all countries in 2020 due to the COVID-19 pandemic, except for the USA, which saw a 5.8% increase in TFP by reducing employment in the sector. These findings highlight the pandemic's impact on the tourism sector's efficiency and offer insights for managing inputs during future crises. Further, the study can be used for future untoward incidents, and can provide an insight on how the inputs can be managed at the time of such incidents.

Using just single output and two input variables to measure the efficiency and TFP of the tourism sector in different countries provides a basic view. However, including more input and output variables would give a more detailed and accurate picture of the entire sector. Upcoming research can be carried further with increasing the number of input and output variables. Also, a greater number of countries can be used as DMUs, this study only uses top five countries with highest number of international visitors in 2022. Furthermore, the Malmquist DEA can be combined with machine learning techniques or other advanced DEA techniques such as fuzzy DEA can be used on the same idea of measuring the efficiency of the tourism sector of countries to provide them with an approach on how to manage the inputs in such untoward incidents.

References

Antara, M., Sumarniasih, M.S.: Role of tourism in economy of Bali and Indonesia. J. Tourism Hospitality Manag. **5**(2) (2017). https://doi.org/10.15640/jthm.v5n2a4

Bakar, N.A., Rosbi, S.: Effect of coronavirus disease (COVID-19) to tourism industry. Int. J. Adv. Eng. Res. Sci. **7**(4), 189–193 (2020). https://doi.org/10.22161/IJAERS.74.23

Banker, R.D., Charnes, A., Cooper, W.W.: Some models for estimating technical and scale inefficiencies in data envelopment analysis. Manage. Sci. **30**(9), 1078–1092 (1984)

Barišić, P., Cvetkoska, V.: Analyzing the efficiency of travel and tourism in the European Union. In: Advances in Operational Research in the Balkans: XIII Balkan Conference on Operational Research, pp. 167–186. Springer (2020)

Bayrak, R., Bahar, O.: Economic efficiency analysis of tourism sector in OECD countries: an Emprical study with DEA. Int. J. Econ. Adm. Stud. (20), 83–100 (2018). https://doi.org/10.18092/ulikidince.277725

Bire, R.B.: Mapping destination competitiveness in Indonesia's Nusa Tenggara Timur (NTT) province: a Malmquist–data envelopment analysis approach. Reg. Sci. Policy Pract. **13**(3), 820–835 (2021). https://doi.org/10.1111/RSP3.12420

Caves, D.W., Christensen, L.R., Diewert, W.E.: The economic theory of index numbers and the measurement of input, output, and productivity. Econometrica **50**(6), 1393 (1982). https://doi.org/10.2307/1913388

Charnes, A., Cooper, W.W., Rhodes, E.: Measuring the efficiency of decision making units. Eur. J. Oper. Res. **2**(6), 429–444 (1978)

Chaubey, V., Sunil Sharanappa, D., Kumar Mohanta, K., Narayan Mishra, V., Narayan Mishra, L.: Efficiency and productivity analysis of the Indian agriculture sector based on the Malmquist-DEA technique. Univ. J. Agric. Res. **10**(4), 331–343 (2022). https://doi.org/10.13189/ujar.2022.100402

Chen, T.A.: Business performance evaluation for tourism factory: using DEA approach and Delphi method. Sustainability **14**(15), 9209 (2022)

Coelli, T.J., Prasada Rao, D.S., O'Donnell, C.J., Battese, G.E.: An introduction to efficiency and productivity analysis. An Introduction to Efficiency and Productivity Analysis, 1–349 (2005). https://doi.org/10.1007/B136381/COVER

Emrouznejad, A., Podinovski, V., Charles, V., Lu, C., Moradi-Motlagh, A.: Rajiv Banker's lasting impact on data envelopment analysis. Ann. Oper. Res. **351**, 1225–1264 (2025). https://doi.org/10.1007/s10479-025-06473-3

Erena, O.T., Kalko, M.M., Debele, S.A.: Technical efficiency, technological progress and productivity growth of large and medium manufacturing industries in Ethiopia: a data envelopment analysis. Cogent Econ. Finan. **9**(1), 1997160 (2021)

Färe, R., Grosskopf, S., Lindgren, B., Roos, P.: Productivity changes in Swedish pharamacies 1980–1989: a non-parametric Malmquist approach. J. Prod. Anal. **3**(1–2), 85–101 (1992). https://doi.org/10.1007/BF00158770/METRICS

Färe, R., Grosskopf, S., Norris, M., Zhang, Z.: Productivity Growth. Tech. Prog. Effi. Change Ind. Countries. **84**(1), 66–83 (1994)

Farrell, M.J.: The measurement of productive efficiency. J. Royal Stat. Soc. Ser. A (General) **120**(3), 253–281 (1957). https://doi.org/10.2307/2343100

Fuchs, M., Zach, F.: On the usefulness of data envelopment analysis for strategy development: a tourism destination case study. In: Data envelopment analysis and performance management, Proceedings of the 4th International Conference of DEA (DEA2004), Birmingham, UK, vol. 266 (2004). ISBN: 0 902683 73 X

Glavin, M.P.V.: Remembering Jon A. Chilingerian. In: Advances in the Theory and Applications of Performance Measurement and Management. DEA45 2023. Lecture Notes in Operations Research. Springer, Cham (2024). https://doi.org/10.1007/978-3-031-61597-9_1

Higuerey, A., Viñan-Merecí, C., Malo-Montoya, Z., Martínez-Fernández, V.A.: Data envelopment analysis (DEA) for measuring the efficiency of the hotel industry in Ecuador. Sustainability **12**(4), 1590 (2020)

Malmquist, S.: Index numbers and indifference surfaces. Trabajos de Estadistica **4**(2), 209–242 (1953). https://doi.org/10.1007/BF03006863/METRICS

Mergoni, A., Emrouznejad, A., De Witte, K.: Fifty years of data envelopment analysis. Eur. J. Oper. Res. **326**(3), 389–412 (2025). https://doi.org/10.1016/j.ejor.2024.12.049

Nandy, A., Singh, P.K.: Farm efficiency estimation using a hybrid approach of machine-learning and data envelopment analysis: evidence from rural eastern India. J. Clean. Prod. **267**, 122106 (2020). https://doi.org/10.1016/J.JCLEPRO.2020.122106

Nikolić, I., Miljković, M.: Assessing the productivity of Serbian manufacturing industry with Malmquist DEA indices. Industrija **51**(2), 29–47 (2023). https://doi.org/10.5937/INDUSTRIJA51-48617

Nurmatov, R., Lopez, X.L.F., Millan, P.P.C.: Tourism, hospitality, and DEA: where do we come from and where do we go? Int. J. Hosp. Manag. **95**, 102883 (2021)

Prorok, V., Šerić, N., Peronja, I.: Analysis of overall and pure technical efficiency of tourism in Europe. Trans. Maritime Sci. **8**(02), 219–229 (2019)

Radovanov, B., Dudic, B., Gregus, M., Marcikic Horvat, A., Karovic, V.: Using a two-stage DEA model to measure tourism potentials of EU countries and Western Balkan countries: an approach to sustainable development. Sustainability **12**(12), 4903 (2020)

Shah, W.U., Zhu, N.: Impact of China's agricultural water policy reforms on water usage efficiency and total factor productivity change. In: Advances in the Theory and Applications of Performance Measurement and Management. DEA45 2023, LNOR. Springer, Cham (2024). https://doi.org/10.1007/978-3-031-61597-9_21

Soysal-Kurt, H.: Measuring tourism efficiency of European countries by using data envelopment analysis. Eur. Sci. J. **13**(10), 31–49 (2017). https://doi.org/10.19044/esj.2017.v13n10p31

UN Tourism. Tourism's importance for growth highlighted in world economic outlook report (2023). https://www.unwto.org/news/tourism-s-importance-for-growth-highlighted-in-world-economic-outlook-report

Wang, C.N., et al.: A two-stage DEA approach to measure operational efficiency in Vietnam's port industry. Mathematics **10**(9), 1385 (2022). https://doi.org/10.3390/MATH10091385

Wang, Y., Shi, L., Zhang, H., Sun, S.: A data envelopment analysis of agricultural technical efficiency of Northwest Arid Areas in China. Front. Agric. Sci. Eng. **4**(2), 195–207 (2017)

World Travel & Tourism Council. WTTC Research Reveals Travel & Tourism's Slow Recovery Is Hitting Jobs and Growth Worldwide (2021). https://wttc.org/news-article/wttc-research-reveals-travel-and-tourism-slow-recovery-is-hitting-jobs-and-growth-worldwide

World Travel & Tourism Council. Economic Impact Research (2023). https://wttc.org/research/economic-impact

World Travel & Tourism Council. Travel & Tourism set to Break All Records in 2024, reveals WTTC (2024). https://wttc.org/news-article/travel-and-tourism-set-to-break-all-records-in-2024-reveals-wttc

Wu, J., Tsai, H., Zhou, Z.: Improving efficiency in international tourist hotels in Taipei using a non-radial DEA model. Int. J. Contemp. Hosp. Manag. **23**(1), 66–83 (2011)

Wu, Y.C., Lin, S.W.: Efficiency evaluation of Asia's cultural tourism using a dynamic DEA approach. Socioecon. Plann. Sci. **84**, 101426 (2022). https://doi.org/10.1016/J.SEPS.2022.101426

Wu, Y.C., Lin, S.W.: Efficiency evaluation of Asia's cultural tourism using a dynamic DEA approach. Socioecon. Plann. Sci. **84**, 101426 (2022)

Xie, J., Tveterås, S.: Economic drivers for the Chinese tourists. Scand. J. Hosp. Tour. **20**(2), 110–125 (2020)

Yildirim, C., Kacak, H., Yildirim, S., Kavuncubasi, S.: Comprehensive efficiency assessment of Turkish teaching hospitals: technical, Pure technical and scale efficiencies with data envelopment analysis. J. Appl. Bus. Econ. **21**(3) (2019)

You, Q.: The travel and tourism industry, It's just a huge part of our economy. -Karen Hughes: Notebook with Unique Golden Marble Touchl Travel l Journal and Notebook l 120 Pages 6'x9'. Independently Published (2020). https://books.google.co.in/books?id=8ax-zQEACAAJ

Zhao, Y.: Empirical estimation of the production frontier. In: Advances in the Theory and Applications of Performance Measurement and Management. DEA45 2023, LNOR, Springer, Cham (2024). https://doi.org/10.1007/978-3-031-61597-9_6

Efficiency Assessment of Container Terminals: The Case of Southwest Asia's Ports

Reza Babazadeh[1], Amna Al Raisi[2], and Mehdi A. Kamran[2(✉)]

[1] Department of Industrial Engineering, Faculty of Engineering, Urmia University, Urmia, West Azerbaijan, Iran

[2] Department of Logistics, Tourism, and Services Management, German University of Technology in Oman, PC 130, P. O. Box 1816, Muscat, Oman

mehdi.kamran@gutech.edu.om

Abstract. As global trade expands and maritime transport becomes increasingly essential, the efficiency of container terminals plays a critical role in facilitating seamless trade flow and supporting economic growth. This research examines the efficiency of container terminals in Southwest Asia using the Super Slack-Based Measure Data Envelopment Analysis (SBM-DEA) methodology. The study focuses on the operational performance of twenty container terminals across nine countries. The findings reveal significant disparities in efficiency levels among these terminals, with efficient ports serving as benchmarks for others to enhance their performance. These insights offer valuable implications for port authorities, shipping companies, and policymakers, contributing to a deeper understanding of maritime logistics and trade dynamics in the region.

Keywords: DEA · SBM-DEA · Container Port · Efficiency · Prioritizing

1 Introduction

Container terminals are pivotal to the global movement of goods, acting as essential hubs in international trade. Over 80% of the world's goods are transported by sea, with this percentage even higher in many developing countries (UNCTAD, 2021). The significance of maritime transport has grown with increasing globalization, placing pressure on ports to operate efficiently to meet the demands of the international market (Talley, 2006). According to UNCTAD (2021), over 80% of global trade by volume is carried by sea, with this figure exceeding 90% for many developing regions. The report highlights that containerized trade volumes grew by 3.2% in 2021, reflecting the resilience of maritime transport despite global disruptions.

The West Asia region, particularly Southwest Asia, is strategically positioned on key maritime routes connecting Europe, Asia, and Africa. Ports in countries such as Iran, KSA, UAE, Oman, and Qatar are integral to the region's economy, supporting trade flows between the East and the West. As these ports play a central role in the global maritime industry, improving their operational efficiency has become a priority for enhancing the region's economic competitiveness.

A. Emrouznejad et al. (Eds.): DEA 2024, LNOR, pp. 504–519, 2025.
https://doi.org/10.1007/978-3-031-98177-7_34

The ports in Southwest Asia operate under unique conditions, including environmental challenges, intense regional competition, and varying levels of technological adoption. As such, traditional methods of performance evaluation may not fully capture the complexities of port operations in this region. Therefore, a customized approach is necessary to assess port efficiency more accurately.

This study uses the Super Slack-Based Measure (SBM) (Emrouznejad et al., 2025) Data Envelopment Analysis (DEA) model to evaluate the performance of container terminals in Southwest Asia. This methodology not only addresses input and output slack but also provides a full ranking of decision-making units (DMUs), offering a more nuanced view of port efficiency compared to traditional DEA models (Mergoni et al., 2025). By identifying strengths and weaknesses across the region's ports, this research aims to provide actionable insights that can drive improvements in logistics and maritime operations.

The efficiency of ports has been assessed from various perspectives, including user preferences and selection criteria. In this context, factors influencing port attractiveness and user choices are analyzed using multinomial logistic regression (Niavis & Tsekeris, 2012). Key performance indicators commonly employed include efficiency, productivity, utility, and expediency (Suárez-Alemán et al., 2016). A substantial body of literature exists on port efficiency assessment worldwide. Various input variables, such as the number of quay cranes and the area and length of the selected port along the straddles, are utilized to evaluate container throughput as an efficiency indicator (Cullinane et al., 2005).

DEA is a versatile and widely applicable method for evaluating container terminal efficiency. Notteboom and Rodrigue (2008) highlight DEA as an essential tool for benchmarking and comparing terminal performance. The application of DEA in maritime studies has proven effective in assessing container terminal efficiency, providing insights into productivity, resource utilization, and potential improvements (Tongzon, 2005; Cullinane et al., 2006). Research indicates that DEA is highly effective in evaluating port and terminal efficiency within maritime logistics. For instance, Barros et al. (2016) demonstrate its efficacy in assessing container terminals in Latin America. Additionally, studies by Tone and Tsutsui (2009) illustrate DEA's adaptability for evaluating container terminal performance across diverse global contexts. Lee (2021) has evaluated the applications of SBM-DEA and super SBM-DEA models in the presence of non-positive data. Undoubtedly, DEA is a reliable method for assessing container terminal efficiency across a wide range of geographical and operational contexts.

The findings of this research will benefit port authorities, shipping companies, logistics providers, and policymakers involved in the maritime industry. By providing insights into the factors influencing container terminal efficiency in the Southwest Asia, it contributes to a deeper understanding of maritime logistics and trade dynamics (Ducruet & Notteboom, 2023). Stakeholders from various sectors can leverage the results of this study. This paper employs the super SBM-DEA model to rank terminal ports in the Southwest Asia region. This method offers several advantages over traditional DEA models, including enhanced discrimination power for fully ranking DMUs, consideration of input slack, the ability to handle non-radial measures, flexibility with variable returns to scale, and robust handling of multiple inputs and outputs. This study addresses gaps in the literature by focusing on Southwest Asia, a region with limited research on container terminal efficiency despite its strategic importance. Unlike prior studies that primarily use traditional DEA models, we apply the Super SBM-DEA model to provide a more detailed ranking of efficiency. Additionally, the unique dataset covering twenty terminals across nine countries enhances the relevance and contribution of this research.

2 Problem Statement and Methodology

This study employs the Super Slack-Based Measure Data Envelopment Analysis (SBM-DEA) model to compare the efficiency of container ports in Southwest Asia, particularly those located along the Persian Gulf, Gulf of Oman, Red Sea, Gulf of Aden, and the Red Sea. Figure 1 illustrates the geographical distribution of the seaports in Southwest Asia. The goal is to assess the operational efficiency of these container terminal ports by examining various input and output indicators.

Southwest Asia is a region of strategic maritime importance due to its location on critical shipping routes that connect Europe, Asia, and Africa. The region's ports are characterized by diverse institutional and operational frameworks, influenced by varying economic policies, trade agreements, and levels of investment in infrastructure. For instance, the UAE boasts highly modernized ports, such as Jebel Ali and Khorfakkan, which are equipped with state-of-the-art technologies and serve as regional transshipment hubs. These ports have benefited from significant government investments and partnerships with leading global operators.

In contrast, ports in countries such as Yemen and Iraq operate under more challenging conditions, including political instability and limited infrastructure. For example, Aden Port in Yemen faces constraints due to ongoing conflict, which has hindered its ability to expand and modernize its facilities. Similarly, Iraq's Umm Qasr Port, despite being one of the country's most significant container terminals, struggles with outdated infrastructure and operational inefficiencies due to historical underinvestment.

Moreover, the institutional arrangements governing the ports in this region also vary significantly. While some ports, such as Sohar in Oman, operate under public-private partnerships with international operators, others, like Bandar Abbas in Iran, are primarily managed by state-owned enterprises. This diversity in governance and operational models adds a layer of complexity to assessing the efficiency of these ports.

Key performance indicators (KPIs) used in this analysis include quay length, number of berths, quay cranes, reefer plugs, yard equipment, terminal area, terminal capacity, and throughput. Data for the twenty container terminals from nine countries were gathered from various sources, including port websites, academic publications, Google Maps measuring tools, and prior research. The dataset was verified and cross-checked to ensure accuracy.

Efficiency levels of container terminals are influenced by factors such as quay length, number of berths, yard equipment, and technology adoption (Cullinane et al., 2005; Suárez-Alemán et al., 2016). Ports along major trade routes, like Jebel Ali, benefit from economies of scale, while conflict-affected ports, such as Aden, face operational disruptions (Ducruet & Notteboom, 2023). Governance models, workforce productivity, and public-private partnerships also play critical roles, as seen in Sohar Port's collaborative structure (Barros et al., 2016; Yu & Hsu, 2020). These insights highlight key factors shaping terminal efficiency across regions.

Due to data availability challenges, some values were adjusted based on general published information. For instance, Jebel Ali Container Terminal in the UAE comprises four separate terminals, and the data provided for this study represents the combined values of these four. Similarly, throughput for the Khorfakkan and Sharjah terminals in the UAE was reported as a single value, which was divided equally between the two terminals for analysis purposes. The same approach was applied for the Shuwaikh and Shuaiba terminals in Kuwait.

Table 1 presents the gathered data for the ports included in this case study. The selection of input and output variables was based on their importance in capturing the operational performance of container terminals, as highlighted in existing literature (Cullinane et al., 2005; Suárez-Alemán et al., 2016). The input variables, such as quay length, number of berths, quay cranes, reefer plugs, yard equipment, terminal area, and terminal capacity, were chosen to reflect the key resources and infrastructure of the ports. Throughput was selected as the main output variable to measure the effectiveness of resource utilization. Quay length across terminals ranges from 388 m (Bushehr) to 9,860 m (Jebel Ali), with an average of 2,095 m. Terminal area varies significantly, with a range of 18 hectares (Sharjah) to 510 hectares (Jebel Ali), while throughput spans from 106,290 TEUs (Bushehr) to 15,370,000 TEUs (Jebel Ali). This variability highlights the operational diversity among the container terminals analyzed.

Fig. 1. Southwest Asia Countries

Table 1. Data of the case study

Country	Port	Quay length (m)	No. of berths	No. of quay cranes	Maximum draft	No. of reefer plugs	No. of yard equipment	Terminal area (ha)	Terminal capacity (TEU[1])	Throughput (TEU)
Oman	Sohar	970	3	13	18	1200	295	68	2400000	838712
	Duqm	1130	1	4	18	600	31	34	1700000	200000
	Salalah	2197	6	20	18	1372	252	87	5000000	4512308
Yemen	Aden	710	2	7	16	269	119	45	900000	465000
	Hodeidah	500	2	4	10	130	167	28	713000	299548
Iraq	Umm Qasr	600	4	7	14	1208	121	40	300000	4000000
UAE	Khorfakkan	1880	6	20	16	1000	362	70	5000000	1375143
	Jebel Ali	9860	31	102	18	2582	1136	510	41400000	15370000
	Sharjah	740	1	2	13	364	48	18	690000	1375143
	Abu Dhabi	3295	9	33	19	3400	138	28	7800000	4910000
Kuwait	Shuwaih	1200	3	6	10	128	49	26	850000	492408
	Shuaiba	880	4	7	14	80	39	31	350000	492408
KSA	Jubail	1000	4	6	14	800	200	57	1800000	244718
	King Abdullah	2200	6	20	18	3000	137	120	25000000	2910000
	Jeddah Islamic Port	4800	17	35	18	6100	267	195	7000000	4738002
	Dammam Port	2380	4	15	16	2500	212	100	3400000	2038787
Qatar	Hamad	2400	2	15	17	1250	35	75	7500000	1543591
Bahrain	Khalifa Bin Salman	1800	1	4	15	896	118	110	1100000	464010
Iran	Bushehr	388	2	2	10	500	63	25	320000	106290
	Bandar Abdas	2834	9	18	14	414	73	85	270000	4500000

[1]twenty-foot equivalent units

2.1 Mathematical Formulation of the SBM-DEA Model

The Super SBM-DEA model is employed to overcome the limitations of traditional SBM models, such as partial ranking of decision-making units (DMUs). The SBM model is particularly effective for evaluating efficiency in situations where inputs and outputs are asymmetric. It allows for the simultaneous consideration of input reduction and output enhancement, distinguishing it from other DEA models.

- **SBM Model**

The foundation of this study is based on the SBM model. The SBM model is one of the DEA techniques specifically designed to evaluate the efficiency of DMUs. This model offers several advantages due to its special features. This developed non-radial model with non-linear value data helps to evaluate efficiency in situations where inputs and outputs are asymmetric (Deng and Song, 2016).

To solve the problem, Let i represent the number of DMUs, j, f stand for inputs and outputs, respectively, X the input set, and Y the output set with $X = (x_{rs}) \in R^{j*i}, X > 0$ and $Y = (y_{rs}) \in R^{f*i}, Y > 0$.

Thus, the production possibilities set (PPS), is defined in the form of Eq. (1):

$$PPS = \{(x, y)|x \geq X\gamma, y \leq Y\gamma, \gamma \geq 0\}, \quad (1)$$

where γ is a non-negative vector in R^i in above relation.

In this way, relations (2) and (3) help us to define a certain DMU (x_0, y_0):

$$x_0 = X\gamma + s^-, \quad (2)$$

$$y_0 = Y\gamma - s^+ \quad (3)$$

where γ, s^- and $s^+ \geq 0$ in R^i. In these equations, s^- shows the excess of inputs and s^+ shows the outputs shortfall and are called slacks. Also, due to the non-negativity of $X(X > 0)$ and $\gamma \geq 0$, we can conclude that:

$$x_0 \geq s^-. \quad (4)$$

Here is how δ index is defined using s^- and s^+:

$$\delta = \frac{1 - \frac{1}{j}\sum_{r=1}^{j}\frac{s_r^-}{x_{r0}}}{1 + \frac{1}{f}\sum_{r=1}^{f}\frac{s_r^+}{y_{r0}}} \quad (5)$$

The properties (a) units invariant and (b) monotone decreasing in input/output slacks are satisfied by δ, as can be verified. Additionally, it holds from (4):

$$0 < \delta \leq 1 \quad (6)$$

In this regard, the following SBM fractional model is formulated to calculate the efficiency of a unique DMU (x_0, y_0) as follows:

$$min\delta = \frac{1 - \frac{1}{j}\sum_{r=1}^{j}\frac{s_r^-}{x_{r0}}}{1 + \frac{1}{f}\sum_{r=1}^{f}\frac{s_r^+}{y_{r0}}}$$

Subject to:

$$\begin{aligned} x_0 &= X\gamma + s^- \\ y_0 &= Y\gamma - s^+ \\ \gamma, s^+, s^- &\geq 0 \end{aligned} \quad (7)$$

The existing model can be converted into a linear program using the Charnes-Cooper transformation (Charnes et al., 1978).

If we call the optimal solution of the SBM model $(\delta^*, \gamma^*, s^{-*}, s^{+*})$ it will be an efficient DMU that This condition is valid only if there is no surplus in the inputs and no shortage in the outputs. In other words: $\delta^* = 1, s^{-*} = s^{+*} = 0$.

- **Super SBM Model**

One of the primary challenges in efficiency evaluation models is the simultaneous assignment of identical scores to multiple DMUs, leading to difficulties in differentiating their rankings. To address this issue, super-efficiency models have been proposed. These models offer several advantages over basic models, particularly in DMU ranking and performance assessment. According to Yu and Hsu (2020), super-efficiency models, such as the unified additive super-efficiency model, enable direct calculations of efficiencies and slacks, allowing for clearer distinctions between efficient DMUs. These models provide ratings greater than 1, which helps differentiate the efficiency levels of units relative to one another. They outperform standard models in robustness and accuracy, particularly in ranking efficient DMUs more effectively (Xie et al., 2021). Furthermore, they address the limitations of traditional DEA models by accommodating integer-valued inputs and outputs (Yu & Hsu, 2020).

Assuming that DMU (X_O, Y_O) is identified as an efficient DMU in terms of SBM model $(\delta^* = 1)$, we can now discuss super-efficiency models:

We define a new set of production possibilities. This production possibility set, denoted as $PPS\backslash(x_0, y_0)$, is defined to as covering (X, Y) except for (x_O, y_O) as follows:

$$PPS\backslash(x_0, y_0) = \{(\overline{x}, \overline{y})|\overline{x} \geq \sum_{s=1,\neq 0}^{i} \gamma_s x_s, \overline{y} \leq \sum_{s=1,\neq 0}^{i} \gamma_s y_s, \overline{y} \geq 0, \gamma \geq 0\} \tag{8}$$

Moreover, it is necessary to define a subset of the existing set $(PPS\backslash(x_0, y_0)$. This subset is created assuming that X and $Y > 0$, $\overline{PPS}\backslash(x_0, y_0)$ is not vacant. This subset is called $\overline{PPS}\backslash(x_0, y_0)$ and is shown as follows:

$$\overline{PPS}\backslash(x_0, y_0) = PPS\backslash(x_0, y_0) \cap \{\overline{x} \geq x_0 \; and \; \overline{y} \leq y_0\} \tag{9}$$

The ϑ distance according to the Eq. (10) was considered as the weighted distance d_I, from (X_O, Y_O) to $(\overline{X}, \overline{Y}) \in \overline{PPS}\backslash(X_0, Y_0)$:

$$\vartheta = \frac{\frac{1}{j}\sum_{r=1}^{j} \frac{\overline{x}_r}{x_{r0}}}{\frac{1}{f}\sum_{k=1}^{f} \frac{\overline{y}_k}{y_{k0}}} \tag{10}$$

According to Eq. (9) for the defined ϑ distance:

i. ϑ is always greater than one $(\vartheta \geq 1)$.

ii. ϑ is only one if the removal of efficient DMU (X_O, Y_O) that $(X_0, Y_0) \in \overline{PPS}\backslash(X_0, Y_0)$ has no effect on the main production possibility set (PPS).

This index has the following interpretation. The numerator is the average expansion rate of the point $(\overline{X}, \overline{Y}) \in \overline{PPS}\backslash(x_0, y_0)$ from x_O to $\overline{x}(\geq x_0)$, since it is a weighted d_I distance from x_O to $\overline{x}$. It is a half-value decrease rate between y_O and $\overline{y}(\leq y_0)$ of $(\overline{x}, \overline{y}) \in \overline{PPS}\backslash(x_0, y_0)$ since the denominator is a weighted distance between y_O and $\overline{y}$, that called d_I. The closeness of y_0 to $\overline{y}$ is the result of the large denominator of the fraction. One way to think of its opposite is as a distance index between y_O and $\overline{y}$. Therefore, ϑ is a product of two indices: the input and output space's distance. Both indices are dimension-less.

Based on the above, the value of the optimal objective function of Eq. (11), which is equal to ϑ^*, is introduced as the super-efficiency value:

$$\vartheta^* = \min \vartheta = \frac{\frac{1}{j}\sum_{r=1}^{j} \frac{\overline{x}_r}{x_{r0}}}{\frac{1}{f}\sum_{k=1}^{f} \frac{\overline{y}_k}{y_{k0}}}$$

Subject to:

$$\begin{aligned} &\overline{x} \geq \sum_{s=1,\neq 0}^{i} \gamma_s x_s \\ &\overline{y} \leq \sum_{s=1,\neq 0}^{i} \gamma_s y_s \\ &\overline{x} \geq x_0, \overline{y} \leq y_0, \overline{y} \geq 0, \gamma \geq 0 \end{aligned} \quad (11)$$

3 Computational Results

Table 2 presents the efficiency scores and rankings obtained from both the Super SBM-DEA and traditional SBM-DEA models. According to the results of the Super SBM-DEA model, the three top-performing ports are Umm Qasr, Bandar Abbas, and Jebel Ali, with efficiency scores of 1.78, 1.44, and 1.21, respectively. These ports demonstrate superior operational efficiency and can serve as benchmarks for other ports in the region.

Notably, ports such as Abu Dhabi, Sharjah, and Hamad rank fourth, fifth, and sixth, respectively, with only minor differences in their efficiency scores compared to the top three ports. These small differences suggest that with slight improvements in key operational areas, these ports could significantly enhance their efficiency.

In contrast, the remaining ports exhibit varying degrees of inefficiency, with larger discrepancies in their scores. Ports such as Duqm, Jubail, and Bushehr which rank toward the bottom, highlight the need for targeted interventions to improve their performance.

The Super SBM-DEA model provides a more granular ranking of efficient ports compared to the traditional SBM-DEA, which tends to cluster multiple ports with the same efficiency score. This enhanced discrimination allows for better insights into operational inefficiencies and areas for improvement across the region's container terminals.

Table 2. The obtained efficiency scores and rankings

DMU	Super SBM-DEA		Traditional SBM-DEA	
	Score	Rank	Score	Rank
Sohar	0.141485	17	0.1415	17
Duqm	8.96E-02	18	0.0896	18

(continued)

Table 2. (*continued*)

DMU	Super SBM-DEA		Traditional SBM-DEA	
	Score	Rank	Score	Rank
Salalah	0.556697	7	0.5567	7
Aden	0.16638	15	0.1664	15
Hodeidah	0.170519	14	0.1705	14
Umm Qasr	1.780009	1	1	1
Khorfakkan	0.188416	13	0.1884	13
Jebel Ali	1.207068	3	1	1
Sharjah	1.089914	5	1	1
Abu Dhabi	1.107718	4	1	1
Shuwaih	0.247198	12	0.2472	12
Shuaiba	0.268116	10	0.2681	10
Jubail	5.16E-02	19	0.0516	19
King Abdullah	0.336322	8	0.3363	8
Jeddah Islamic Port	0.331256	9	0.3313	9
Dammam Port	0.263511	11	0.2635	11
Hamad	1.018897	6	1	1
Khalifa Bin Salman	0.14539	16	0.1454	16
Bushehr	5.09E-02	20	0.0509	20
Bandar Abdas	1.440292	2	1	1

3.1 Verification of the Results of Super SBM-DEA

Verification and validation of the super SBM-DEA model is necessary before its application in real-world practices. To do so, the non-parametric spearman's rank correlation method (Sheskin, 2000) is applied. The method analyses the positive correlation between the two sets of ranks obtained by the super SBM-DEA and SBM-DEA models using the measure $\rho = 1 - \frac{6\sum d_i^2}{n(n^2-1)}$. Where, d_i is difference between the ranks of each method and n stands for the total number of DMUs. Null hypothesis $H_0 : r_s = 0$ is tested against alternative hypothesis $H_1 : r_s > 0$ as follows:

- $\boldsymbol{H_0}$. *There is no correlation between the ranks obtained by the super SBM-DEA model and the SBM-DEA model.*
- $\boldsymbol{H_1}$. *There is a positive correlation between the ranks obtained by the super SBM-DEA model and the SBM-DEA model.*

Since the test's p-value $= 0.000$ is close to zero, H_0 is rejected. Stated otherwise, a high and robust correlation (0.981) is seen between the ranks by the super SBM-DEA and super SBM-DEA models. Discrimination power of the super SBM-DEA could be observed in Table 2. While SBM-DEA leads to partial ranking of some DMUs like DMUs 8,9,10, 17, and 20, the super SBM-DEA gives full ranking of DMUs.

For a better understanding, Fig. 2 shows the ranking results obtained from the super SBM model.

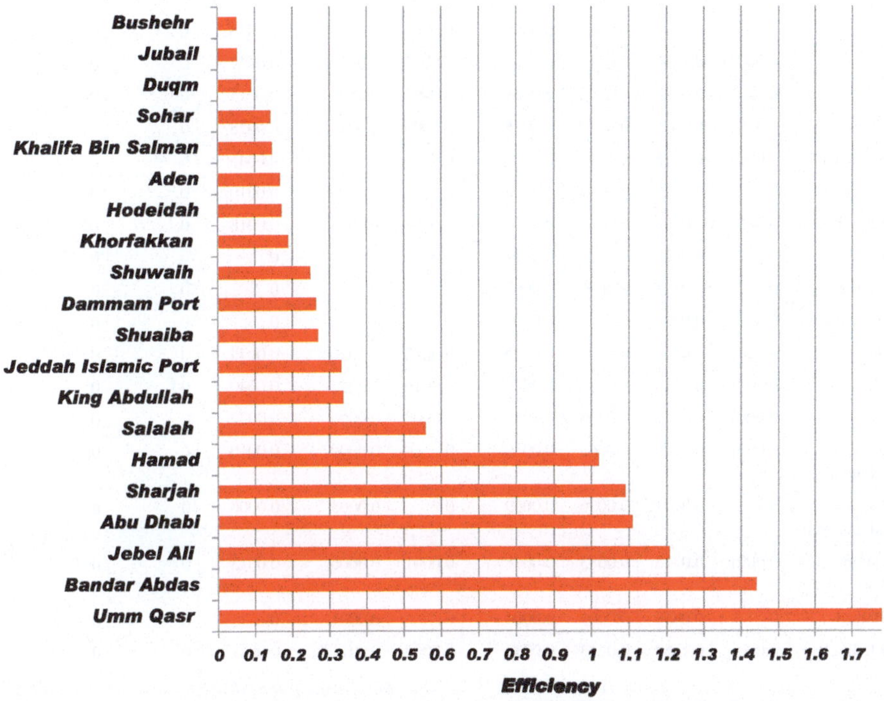

Fig. 2. Ranking of units based on the Super SBM method

Table 3 presents the weights assigned to the input and output variables. Notably, the maximum draft received the highest weight of 0.91, indicating its paramount importance among the variables. Following this, the number of berths and quay cranes were allocated weights of 0.75 and 0.27, respectively, placing them in the next two ranks of importance. Consequently, these three indicators are critical due to their significant weights.

In contrast, the output indices—throughput and terminal capacity—are assigned weights close to zero, making them the least important relative to the other variables. This suggests that inefficient units can enhance their overall efficiency more effectively by focusing on improving these key indicators.

Tables 4 and 5 outline the changes necessary for inefficient DMUs to achieve efficiency. Specifically, these tables detail the required adjustments for each variable and the ideal values for each index. In Table 4, negative numbers indicate the percentage decrease needed for a specific index, while positive numbers reflect the required percentage increase. Table 5 lists the ideal values corresponding to these percentages.

These targets provide concrete targets for inefficient DMUs. For instance, if a port has an inefficient quay length or insufficient yard equipment, adjustments to these input

Table 3. Weights obtained for variables by Super SBM method

DMU	Input								Output
	Quay length	No. of berths	No. of quay cranes	Maximum draft	No. of reefer plugs	No. of yard equipment	Terminal area	Terminal capacity	Throughput
Sohar	0.0001	0.0417	0.0096	0.0069	0.0001	0.0004	0.0018	0	0
Duqm	0.0001	0.1250	0.0313	0.0069	0.0002	0.0040	0.0037	0	0
Salalah	0.0001	0.0208	0.0063	0.0069	0.0001	0.0005	0.0014	0	0
Aden	0.0002	0.0625	0.0179	0.0078	0.0005	0.0011	0.0028	0	0
Hodeidah	0.0003	0.0625	0.0313	0.0125	0.0010	0.0007	0.0045	0	0
Umm Qasr	0	0.0010	0.0002	0.7868	0	0.0002	0.0019	0	0
Khorfakkan	0.0001	0.0208	0.0063	0.0078	0.0001	0.0003	0.0018	0	0
Jebel Ali	0	0.0001	0	0.0007	0	0	0	0	0
Sharjah	0	0.0123	0.0006	0.0005	0	0	0	0	0
Abu Dhabi	0	0.0004	0	0.0004	0	0	0	0	0
Shuwaih	0.0001	0.0417	0.0208	0.0125	0.0010	0.0026	0.0048	0	0
Shuaiba	0.0001	0.0313	0.0179	0.0089	0.0016	0.0032	0.0040	0	0
Jubail	0.0001	0.0313	0.0208	0.0089	0.0002	0.0006	0.0022	0	0
King Abdullah	0.0001	0.0208	0.0063	0.0069	0	0.0009	0.0010	0	0
Jeddah Islamic Port	0	0.0074	0.0036	0.0069	0	0.0005	0.0006	0	0
Dammam Port	0.0001	0.0313	0.0083	0.0078	0.0001	0.0006	0.0013	0	0
Hamad	0	0.0490	0.0001	0.0004	0	0.0001	0	0	0
Khalifa Bin Salman	0.0001	0.1250	0.0313	0.0083	0.0001	0.0011	0.0011	0	0
Bushehr	0.0003	0.0625	0.0625	0.0125	0.0003	0.0020	0.0050	0	0
Bandar Abdas	0	0.0004	0.0001	0.0005	0	0	0	0	0
Average	0.0001	0.0374	0.0137	0.0456	0.0003	0.0009	0.0019	0	0
Sum	0.0017	0.7477	0.2749	0.9111	0.0052	0.0188	0.0381	0	0
Max	0.0003	0.1250	0.0625	0.7868	0.0016	0.0040	0.0050	0	0
Min	0	0.0001	0	0.0004	0	0	0	0	0

variables can help it move toward the efficient frontier. The differences between the current values and the efficient target values for each input and output are presented. By implementing these changes, the DMU can reach the efficient frontier. For example, Salalah Port needs to reduce its number of reefer plugs by 0.01 percent (as shown in Table 4) to achieve a target value of 1362.72 (as indicated in Table 5) and attain efficiency.

In the Super SBM model, inherently efficient DMUs can be further distinguished by their super efficiency scores. In such cases, predictions may not be necessary since the DMU is already performing above the efficiency frontier. However, there are still opportunities for improvement beyond this frontier. For example, Umm Qasr, despite being ranked first and attaining a super efficiency score greater than one, received recommendations for further performance enhancements. Specifically, it is suggested that the terminal area index be increased by 0.87% to a value of 74.77 for improved performance.

Table 4. The percentage of the need to change indicators based on the Super SBM model

DMU	Input								Output
	Quay length	No. of berths	No. of quay cranes	Maximum draft	No. of reefer plugs	No. of yard equipment	Terminal area	Terminal capacity	Throughput
Sohar	-0.87	-0.72	-0.89	-0.84	-0.79	-0.91	-0.88	-0.97	0.00
Duqm	-0.97	-0.80	-0.91	-0.96	-0.90	-0.80	-0.94	-0.99	0.00
Salalah	-0.69	-0.25	-0.61	-0.12	-0.01	-0.46	-0.48	-0.93	0.00
Aden	-0.90	-0.77	-0.88	-0.90	-0.48	-0.88	-0.90	-0.96	0.00
Hodeidah	-0.91	-0.85	-0.87	-0.90	-0.30	-0.95	-0.89	-0.97	0.00
Umm Qasr	3.18	0.96	1.24	-0.05	-0.68	-0.44	0.87	0.00	0.00
Khorfakkan	-0.89	-0.77	-0.88	-0.70	-0.58	-0.89	-0.80	-0.98	0.00
Jebel Ali	-0.02	-0.01	-0.40	1.66	-0.45	-0.78	-0.43	-0.98	0.00
Sharjah	-0.72	0.38	0.20	-0.62	0.14	-0.13	-0.24	-0.85	0.00
Abu Dhabi	-0.78	-0.45	-0.74	-0.07	-0.56	0.08	0.79	-0.95	0.00
Shuwaih	-0.74	-0.67	-0.67	-0.85	-0.65	-0.84	-0.64	-0.97	0.00
Shuaiba	-0.65	-0.75	-0.72	-0.89	-0.43	-0.80	-0.70	-0.92	0.00
Jubail	-0.96	-0.94	-0.93	-0.94	-0.91	-0.96	-0.96	-0.99	0.00
King Abdullah	-0.80	-0.52	-0.75	-0.43	-0.71	-0.36	-0.76	-0.99	0.00
Jeddah Islamic Port	-0.85	-0.72	-0.76	-0.08	-0.77	-0.46	-0.76	-0.95	0.00
Dammam Port	-0.87	-0.49	-0.76	-0.55	-0.75	-0.71	-0.80	-0.96	0.00
Hamad	-0.81	0.00	-0.75	-0.69	-0.70	0.15	-0.74	-0.99	0.00
Khalifa Bin Salman	-0.96	-0.54	-0.80	-0.89	-0.84	-0.88	-0.96	-0.97	0.00
Bushehr	-0.96	-0.95	-0.91	-0.96	-0.94	-0.95	-0.96	-0.98	0.00
Bandar Abdas	-0.76	-0.50	-0.56	0.13	2.28	0.86	-0.47	0.25	0.00

To streamline the analysis, we summarize key findings from Tables 3, 4, and 5. The results highlight that maximum draft, number of berths, and quay cranes have the highest influence on efficiency scores. Ports with lower efficiency, such as Duqm and Bushehr, need to optimize these factors to improve performance. Additionally, inefficient terminals require reductions in excess capacity while enhancing key operational resources to reach the efficiency frontier.

The efficiency scores reveal significant variations among ports. Umm Qasr (1.78), Bandar Abbas (1.44), and Jebel Ali (1.21) rank as the most efficient terminals, benefiting from strategic locations, high throughput, and advanced infrastructure. These ports

Table 5. The projection (ideal value) of indicators based on the Super SBM model

DMU	Input								Output
	Quay length	No. of berths	No. of quay cranes	Maximum draft	No. of reefer plugs	No. of yard equipment	Terminal area	Terminal capacity	Throughput
Sohar	125.81	0.84	1.47	2.94	253.29	25.37	8.39	62903	838712
Duqm	30.00	0.20	0.35	0.70	60.40	6.05	2.00	15000	200000
Salalah	676.85	4.51	7.90	15.79	1362.72	136.50	45.12	338423	4512308
Aden	69.75	0.47	0.81	1.63	140.43	14.07	4.65	34875	465000
Hodeidah	44.93	0.30	0.52	1.05	90.46	9.06	3.00	22466	299548
Umm Qasr	2506.66	7.83	15.65	13.26	391.46	67.43	74.77	300000	4000000
Khorfakkan	206.27	1.38	2.41	4.81	415.29	41.60	13.75	103135	1375143
Jebel Ali	9679.68	30.74	61.48	47.82	1414.04	249.34	290.32	922200	15370000
Sharjah	206.27	1.38	2.41	4.81	415.29	41.60	13.75	103135	1375143
Abu Dhabi	736.50	4.91	8.59	17.19	1482.82	148.53	49.10	368250	4910000
Shuwaih	310.11	0.98	1.97	1.53	45.30	7.99	9.30	29544	492407
Shuaiba	310.11	0.98	1.97	1.53	45.30	7.99	9.30	29544	492407
Jubail	36.71	0.24	0.43	0.86	73.90	7.40	2.45	18353	244718
King Abdullah	436.50	2.91	5.09	10.19	878.82	88.03	29.10	218250	2910000
Jeddah Islamic Port	710.70	4.74	8.29	16.58	1430.88	143.32	47.38	355350	4738002
Dammam Port	305.82	2.04	3.57	7.14	615.71	61.67	20.39	152909	2038787
Hamad	450.51	2.00	3.73	5.23	370.32	40.29	19.49	108923	1543591
Khalifa Bin Salman	69.60	0.46	0.81	1.62	140.13	14.04	4.64	34800	464010
Bushehr	15.94	0.11	0.19	0.37	32.10	3.22	1.06	7971	106290
Bandar Abdas	675.00	4.50	7.88	15.75	1359.00	136.13	45	337500	4500000

serve as benchmarks for others aiming to improve operational performance. Medium-efficiency ports, such as Abu Dhabi (1.11) and Sharjah (1.09), demonstrate competitive performance but can further optimize resource utilization. In contrast, inefficient ports like Duqm (0.0896) and Bushehr (0.0509) suffer from outdated infrastructure and lower trade volumes, necessitating targeted investments and operational reforms. A strong correlation between efficiency and maximum draft ($\rho = 0.78$) suggests that deeper ports accommodate larger vessels, enhancing efficiency. Ports with higher berth and quay crane counts also tend to be more efficient, highlighting the importance of infrastructure capacity.

To examine the relationship between efficiency and operational scale, we classified ports into: High-throughput (>3M TEUs), Medium-throughput (1M–3M TEUs), Low-throughput (<1M TEUs). High-throughput ports (e.g., Jebel Ali, Umm Qasr) generally exhibit higher efficiency, while medium-throughput ports show variability due to factors like governance and technology. Spearman's correlation analysis reveals a strong positive correlation between efficiency and maximum draft (ρ = 0.78) and a moderate correlation with berths (ρ = 0.61). This classification provides benchmarks for improving terminal performance.

The following managerial insights could be extracted from the results:

- **Benchmarking Best Practices:** Ports with consistently higher efficiency levels than others include Umm Qasr, Bandar Abbas, and Jebel Ali. Underperforming terminals in the area should use these ports as a benchmark. To increase efficiency, port authorities can benefit from their operational strategies, resource management techniques, and infrastructure use.
- **Targeted Efficiency Improvements:** The study identifies important factors that have a major impact on terminal efficiency, including quay length, berth count, and maximum draft. To maximize terminal operations, managers should give priority to these areas of development, especially at ports with many indicators of inefficiency like Duqm, Jubail, and Bushehr.
- **Resource Allocation:** According to the data, ports with a lot of quay cranes, berths, and larger terminal areas typically have superior performance. To have a more balanced and effective throughput, strategic investments in these resources should be made, particularly for ports that are falling behind.
- **Policy and Infrastructure Planning:** When organizing port expansions and infrastructure upgrades, policymakers ought to take these findings into account. Ports like Abu Dhabi and Sharjah, which have significant geographic advantages, might benefit from strategic investment to increase their competitiveness in international trade routes.

The efficiency analysis in this study adopts an input-oriented approach, as the primary focus is on minimizing the use of resources while maintaining or enhancing output levels. This orientation is particularly relevant for container terminals, where improving resource utilization—such as quay length, yard equipment, and terminal capacity—is often prioritized over increasing throughput. The analysis highlights how efficient terminals, such as Umm Qasr and Jebel Ali, utilize their resources optimally compared to less efficient terminals like Duqm and Bushehr. This provides valuable insights into specific areas where underperforming terminals can improve their efficiency, such as optimizing quay utilization and increasing yard equipment efficiency.

The high correlation (ρ = 0.981) between SBM-DEA and Super SBM-DEA confirms consistency in rankings, with Super SBM-DEA providing better discrimination. Umm Qasr, Bandar Abbas, and Jebel Ali emerge as super-efficient due to their strategic locations, high throughput, and advanced infrastructure. In contrast, lower-ranked ports like Duqm and Bushehr struggle due to limited capacity and outdated facilities, emphasizing the need for investment and modernization.

4 Conclusion

This study provides a comprehensive evaluation of the operational efficiency of container terminals in Southwest Asia using the Super SBM-DEA model. The findings reveal substantial variation in performance across the region's ports, with Umm Qasr, Bandar

Abbas, and Jebel Ali emerging as the most efficient terminals. These high-performing ports can serve as benchmarks for others, offering operational strategies and resource management practices that can be emulated.

The key factors influencing port efficiency in this region include quay length, the number of berths, and the availability of quay cranes. Ports with inefficiencies, such as Duqm and Jubail, should focus on enhancing these critical areas to improve their operational performance. Moreover, strategic investments in infrastructure and technology, particularly in underperforming terminals, could lead to significant efficiency gains. The analysis highlights Duqm as one of the least efficient ports, primarily due to limited berths and quay cranes. To enhance its performance, investments in increasing the number of berths and acquiring additional quay cranes are recommended. These upgrades would improve its capacity to handle larger vessels and boost throughput. Additionally, modernizing yard equipment and adopting advanced terminal operating systems could streamline operations and reduce inefficiencies.

Regulatory frameworks play a crucial role in port efficiency. Ports in the UAE and Oman benefit from private sector involvement, while state-controlled ports like Bandar Abbas face investment constraints. Price regulations help stabilize trade but may reduce efficiency incentives. Ports with competitive regulatory environments, such as King Abdullah Port, show better performance. Future DEA studies should incorporate institutional factors to provide a more comprehensive efficiency analysis.

Adjustments in the data, such as consolidating terminal values (e.g., combining throughput for Jebel Ali's four terminals or splitting throughput for Sharjah and Khorfakkan), may impact the comparative efficiency rankings. While these adjustments ensure consistency in analysis, they could overestimate or underestimate efficiency levels depending on the accuracy of the division or aggregation. For instance, combining terminal data for Jebel Ali reflects its overall capacity but may obscure operational inefficiencies within individual terminals. These limitations were acknowledged to ensure the robustness of the results.

The results underscore the importance of customized performance evaluation methodologies, such as the Super SBM-DEA, which are better suited to the unique competitive and environmental challenges faced by ports in Southwest Asia. By addressing these inefficiencies, the region's ports can improve their competitiveness on international maritime routes, contributing to broader economic development.

The findings of this study are not only relevant to port authorities but also to policymakers, shipping companies, and logistics providers who seek to enhance the efficiency of maritime operations in this strategically important region.

References

Barros, C.P., Chen, Z., Wanke, P.: Efficiency in Chinese seaports: 2002–2012. Maritime Econ. Logist. **18**, 295–316 (2016)

Charnes, A., Cooper, W.W., Rhodes, E.: Measuring the efficiency of decision making units. Eur. J. Oper. Res. **2**(6), 429–444 (1978). https://doi.org/10.1016/0377-2217(78)90138-8

Cullinane, K., Wang, T.F., Song, D.W., Ji, P.: The technical efficiency of container ports: comparing data envelopment analysis and stochastic frontier analysis. Transp. Res. Part A: Policy Pract. **40**(4), 354–374 (2006)

Cullinane, K., Song, D.-W., Wang, T.: The application of mathematical programming approaches to estimating container port production efficiency. J. Prod. Anal. (2005)

Ducruet, C., Notteboom, T.: Port Systems in Global Competition: Spatial-economic Perspectives on the Co-development of Seaports. Routledge, Taylor & Francis Group (2023)

Deng, G., Li, L., Song, Y.: Provincial water use efficiency measurement and factor analysis in China: based on SBM-DEA model. Ecol. Ind. **69**, 12–18 (2016)

Emrouznejad, A., Brzezickib, Ł., Lu, C.: The development and evolution of slacks-based measure models in data envelopment analysis: a comprehensive review of the literature. J. Econ. Surv. 1–28 (2025). https://doi.org/10.1111/jos.12682

Lee, H.S.: Slacks-based measures of efficiency and super-efficiency in presence of nonpositive data. Omega **103**, 102395 (2021)

Mergoni, A., Emrouznejad, A., De Witte, K. Fifty years of data envelopment analysis. Eur. J. Oper. Res. **326**(3), 389–412 (2025). https://doi.org/10.1016/j.ejor.2024.12.049

Niavis, S., Tsekeris, T.: Ranking and causes of inefficiency of container seaports in south-Eastern Europe - European transport research review. SpringerOpen, 13 June 2012. https://doi.org/10.1007/s12544-012-0080-y

Notteboom, T., Rodrigue, J.-P.: The future of containerization: perspectives from maritime and inland freight distribution. Geo J. (2008)

Suárez-Alemán, A., Sarriera, J.M., Serebrisky, T., Trujillo, L.: When it comes to container port efficiency, are all developing regions equal? Transp. Res. Part A: Policy Pract. **86**, 56–77 (2016)

Talley, W.K.: An economic theory of the port. Res. Transp. Econ. **16**, 43–65 (2006)

Tone, K., Tsutsui, M.: Network DEA: a slacks-based measure approach. Eur.opean J. Oper. Res. (2009)

Tongzon, J.L.: ASEAN-China free trade area: a bane or boon for ASEAN countries? World Econ. **28**, 191–210 (2005). https://doi.org/10.1111/j.1467-9701.2005.00643.x

Yu, S.H., Hsu, C.W.: A unified extension of super-efficiency in additive data envelopment analysis with integer-valued inputs and outputs: an application to a municipal bus system. Ann. Oper. Res. **287**(1), 515–535 (2020)

Xie, Q., Zhang, L.L., Shang, H., Emrouznejad, A., Li, Y.: Evaluating performance of super-efficiency models in ranking efficient decision-making units based on Monte Carlo simulations. Ann. Oper. Res. **305**, 273–323 (2021)

UNCTAD (United Nations Conference on Trade and Development). Review of Maritime Transport 2021. Geneva: United Nations (2021). https://unctad.org/system/files/official-document/rmt2021_en.pdf

Sheskin, D.: Handbook of Parametric and Nonparametric Statistical Procedures. CRC Press, Florida (2000)

Energy Efficiency Evaluation Based on DEA: Case Study of Adriatic Container Ports

Mozhgan Mansouri Kaleibar and Evelin Krmac(✉)

Faculty of Maritime Studies and Transport, University of Ljubljana, Pot pomorscakov 4, 6320 Portoroz, Slovenia
{mozhgan.mansouri.kaleibar,evelin.krmac}@fpp.uni-lj.si

Abstract. The supply and correct utilisation of energy is one of the most important issues influencing the economic progress of countries. Maritime transport also consumes part of the energy. Assessing its performance in terms of energy consumption and its environmental impact seems to be important. The Adriatic region plays a central role in maritime transport and container ports are important hubs in the global supply chain network. However, the energy consumption of these ports raises environmental issues and operational costs. To address this issue, Data Envelopment Analysis (DEA) methods are used to assess the energy efficiency of the Adriatic container ports.

This paper presents a comprehensive analysis of the energy efficiency of Adriatic container ports using improved DEA methods. This study considers various input and output parameters in terms of energy consumption, operational performance and environmental impact. By using improved DEA models, the relative efficiency of container ports is evaluated and the most efficient container ports are identified. These results provide valuable perspectives for decision-makers, port authorities and stakeholders to improve energy efficiency, promote sustainability and increase the overall performance of container ports in the Adriatic region.

Keywords: Data envelopment analysis (DEA) · Container port · Adriatic ports · Energy efficiency · Energy consumption

1 Introduction

The effective utilisation of energy resources is essential for sustainable economic development around the world. In the maritime transport sector, which is a cornerstone of international trade, improving energy efficiency has become increasingly important due to its significant impact on environmental sustainability and operational costs. Container ports, which serve as important hubs in the maritime logistics network, are at the forefront of this issue. These ports not only drive international trade, but also play a crucial role in the economic progress of the region. However, their operations tend to be energy-intensive, leading to notable environmental and financial challenges [21].

The Adriatic region, strategically located between Central and Eastern Europe, has a network of container ports that are an essential part of the global supply chain. These

A. Emrouznejad et al. (Eds.): DEA 2024, LNOR, pp. 520–528, 2025.
https://doi.org/10.1007/978-3-031-98177-7_35

ports are essential for the economic integration of the region with Europe and beyond. However, increasing energy demand and associated environmental concerns emphasise the need for a thorough assessment of the energy efficiency of these ports. Insight into the energy performance of these ports is essential to uncover opportunities to reduce energy consumption, minimise environmental impact, and increase economic competitiveness.

Data Envelopment Analysis (DEA) has established itself as an effective method for assessing efficiency in various industries, including maritime transport [4, 16]. DEA provides a comprehensive framework for evaluating the relative efficiency of decision-making units (DMUs) by incorporating multiple inputs and outputs. In this study, DEA is used to assess the energy efficiency of the Adriatic container ports in order to gain a deeper understanding of their operational performance. By using advanced DEA modelling, the study examines key factors such as energy consumption, environmental effects and operational outcomes to determine the relative efficiency of these ports.

The results of this study aim to provide port authorities, policy makers and other stakeholders with practical insights to drive energy efficiency improvements. By identifying the most efficient ports, the study creates benchmarks and best practises that can be adopted by others to improve their performance. Ultimately, this study contributes to a broader discussion on sustainable maritime transport and the role of ports in promoting environmental responsibility and economic growth in the Adriatic region.

2 Literature Review

Research on the assessment of ports using mathematical models has developed considerably since the introduction of these models and their unique properties [13]. This area of research is particularly important for assessment of energy efficiency in seaports as it has a profound impact on environmental sustainability and operational costs. Ports are large energy consumers and use significant amounts of electricity for activities such as cargo handling, lighting and logistics [1]. Improving energy efficiency in these facilities is critical to reducing environmental impacts and improving financial outcomes [10, 14].

DEA has become a widely accepted tool for assessing port efficiency by evaluating multiple inputs and outputs [4]. Its ability to measure relative efficiency makes it particularly effective for complex systems such as ports [5]. For example, in studies by Barros and Athanassiou (2004) and Cullinane and Wang (2007), DEA was used to investigate the technical efficiency and scale efficiency of container ports, where significant differences in efficiency were found between different locations [2, 6].

Recent research has increasingly focussed on environmental efficiency or eco-efficiency within the port sector. Chang (2013) utilised DEA to investigate how environmental factors are incorporated into port efficiency assessment [3]. Similarly, Guimarães et al. (2014) used DEA to assess the environmental performance of Brazilian container terminals and emphasised the need to consider environmental outputs in efficiency assessments [11].

Quintanoet al. (2020, 2021) used non-radial DEA models to evaluate eco-efficiency in the port sector and assess the eco-efficiency of economic activities in seaport. Their application of advanced DEA techniques to identify unit segments and evaluate environmental performance has provided a more detailed understanding of port efficiency [18, 19].

Kong and Liu (2021) investigated sustainable port cities by incorporating the coupling of coordination and environmental efficiency into their DEA models. Their research emphasises the need to strike a balancing between economic development and environmental stewardship, which is consistent with the overall goals of sustainable port management [12].

Additionally, DEA has been applied to assess the port efficiency in various national contexts. For example, Prakash et al. (2016) have analysed the efficiency of major Indian ports from 2007–08 to 2013–14, highlighting the variations in the performance of different ports and the need for targeted improvements. This national study complements the broader global perspective on port efficiency by providing specific insights into the Indian context [17]. Mansouri Kaleibar and Krmac (2024) utilized a generalized fuzzy SBM model to assess the efficiency of Adriatic ports, addressing uncertainties and identifying key inefficiency factors to enhance port performance and trade competitiveness [15].

Although extensive research has been conducted on port efficiency and environmental performance using DEA, there is a notable lack of studies that specifically address the Adriatic region. The Adriatic Sea, with its dense network of container ports that are crucial for European maritime logistics, has not yet been thoroughly analysed in terms of energy efficiency. This study attempts to fill this gap by using advanced DEA modelling to assess the energy efficiency of container ports in the Adriatic.

3 Research Methodology

The aim of this study is to assess the energy efficiency of container ports in the Adriatic region by applying the ND-SBM-UO (Non-Discretionary Slack-Based Measure with Undesirable Outputs) model. This DEA model was selected due to its ability to account for both positive and negative outputs, which allows for a thorough assessment of port performance.

3.1 DEA Model

The ND-SBM-UO model, an extended version of the DEA's Slack-Based Measure (SBM) model [15, 20], is well suited for assessing efficiency when both positive outputs, such as freight throughput, and negative outputs, such as energy consumption and emissions, are present. This model is particularly beneficial for this study as it takes into account non-discretionary inputs, i.e., factors that cannot be controlled by the decision making units (DMUs), such as water depth, which is critical to port operations but is inherently fixed. By including both undesirable outputs and non-discretionary inputs, the ND-SBM-UO model provides a comprehensive and nuanced analysis of port performance that captures the full range of environmental and operational efficiency.

The ND-SBM-UO is defined as follows:

$$
\begin{aligned}
&\tau^* = Mint - \frac{1}{m}\sum\nolimits_{i\in}^{D} \frac{s_i^-}{x_{io}} \\
&s.tt + \frac{1}{s_1 + s_2}\left(\sum_{r=1}^{s_1} \frac{s_r^g}{y_{ro}^g} + \sum_{r=1}^{s_2} \frac{s_r^b}{y_{ro}^b}\right) = 1, \\
&\sum_{j=1}^{n} \Lambda_j x_{ij} + s_{iD}^- = tx_{io}, i \in D, \\
&\sum_{j=1}^{n} \Lambda_j z_{ij} + s_{iND}^- = tz_{io}, i \in ND, \\
&\sum_{j=1}^{n} \Lambda_j y_{rj} - s_r^g = ty_{ro}^g, r = 1, \ldots, s_1, \\
&\sum_{j=1}^{n} \Lambda_j y_{rj} + s_r^b = ty_{ro}^b, r = 1, \ldots, s_2, \\
&\Lambda_j \geq 0, s_{iD}^- \geq 0, s_{iND}^- \geq 0, i = 1, \ldots, m, j = 1, \ldots, n, \\
&s_r^g \geq 0, s_r^b \geq 0, r = 1, \ldots, s, t > 0.
\end{aligned}
\tag{1}
$$

We assume that there are n DMUs. X and Y are the input and output variables, respectively, and m, s_1 and s_2 are the number of inputs, desirable outputs and undesirable output variables, respectively where, x_{ij} = ith input of jth DMU and z_{ij} = ith uncontrollable input of jth DMU. Note that only the controllable input is included in the target. A DMU is ND-SBM-UO efficient if and only if $\tau = 1$.

4 Case Study

The purpose of this case study is to evaluate the energy efficiency of the Adriatic container ports using the ND-SBM-UO DEA model to assess relative performance. As container ports play a crucial role in both regional and international trade, their energy consumption has a significant impact on environmental sustainability and economic competitiveness. The analysis focusses on a selected group of Adriatic ports and considers different input and output variables to determine their operational efficiency.

4.1 Inputs and Outputs

The study focusses on container ports in the Adriatic region. These ports are the port of Trieste (Italy), the port of Venice (Italy), the port of Ravenna (Italy), the port of Ancona (Italy), the port of Koper (Slovenia), the port of Rijeka (Croatia), the port of Ploce (Croatia) and the port of Bar (Montenegro).

For the purposes of this study, the dataset consists of observations for eight container ports in the Adriatic region. The data covers the period from 2021 to 2023 and is divided into three main groups of variables.

Table 1. Descriptive statistics for Adriatic container ports (2021–2023)

Variable	Year	Min	Max	Mean	SD
Energy consumption (MWh/year)	2021	4,100	82,000	39,860	30,366
	2022	4,140	82,843	39,607	30,457
	2023	4,200	84,000	40,063	30,469
Number of employees	2021	300	1,300	850.00	369.44
	2022	300	1,300	850.00	369.44
	2023	300	1,300	850.00	369.44
Level of connection	2021	7.0	9.0	7.88	0.63
	2022	7.0	9.0	7.88	0.63
	2023	7.0	9.0	7.88	0.63
Water depth (m)	2021	10.5	17.9	11.76	3.16
	2022	10.5	17.9	11.76	3.16
	2023	10.5	17.9	11.76	3.16
Throughput (TEU/year)	2021	21,526	997,574	438,644	391,163
	2022	21,729	1,017,788	434,849	398,249
	2023	21,934	1,066,093	424,013	390,752
Emissions (kg CO2-eq/year)	2021	857,200	22,234,200	10,243,250	8,654,883
	2022	851,600	25,455,179	10,525,840	9,147,010
	2023	857,200	25,500,000	10,570,062	9,153,393

Input variables: These represent the resources used by the ports, which include: Energy consumption: Measured in MWh/year, this variable indicates the total energy consumed by the port for its operations.

- *Number of employees*: The total number of employees working at the terminal and contributing to the port's operational activities.
- *Level of connection with hinterland*: This qualitative measure reflects the extent to which the port is integrated with the surrounding logistics infrastructure, including road and rail connections.
- *Water depth*: A critical factor for port operations, indicating the depth of water available to accommodate larger vessels.
- *Desired output variables*: These represent the productive outputs of the ports and focus on:
- *Container throughput*: Measured in TEUs (Twenty-foot Equivalent Units) per year, this variable represents the total volume of containers handled by each port and reflects operational performance.
- *Undesired output variables*: These variables represent the negative impacts associated with port operations:

- *Emission*: Measured in kg CO2-eq/year, this variable indicates the level of greenhouse gas emissions generated by port activities, particularly energy consumption.

The data used in this analysis comes from reliable databases and reports, including the Findaport database [9], the annual reports of the Adriatic ports and the European Statistical Office (Eurostat) [8]. Descriptive statistics for these variables across all eight ports for the period from 2021 to 2023 are presented in Table 1 and provide an overview of the data used in the efficiency assessment.

4.2 Results and Discussion

This section analyses the energy efficiency, throughput and emissions of the Adriatic container ports, focusing on eight ports: Koper, Trieste, Rijeka, Venice, Ravenna, Ancona, Ploce and Bar. The analysis covers the years 2021 to 2023.

One input variable, water depth, is considered uncontrollable and serves as an external constraint that affects the performance of the assessed DMUs. Since water depth cannot be controlled by the terminals themselves, the model proposed in this study is designed to treat this variable as a constant. Therefore, the aim of the model is to evaluate the container terminals considering this uncontrollable input, focusing on reducing the input and increasing the output, excluding the desired input.

The energy efficiency results for the container terminals are summarised in Table 2.

Table 2. Energy efficiency results

Ports	2023	Rank	2022	Rank	2021	Rank
Koper	1	1	1		1	
Trieste	0.88084	2	1		0.83613	2
Rijeka	0.73049	3	0.78312	2	1	
Venice	0.44209	6	0.42864	5	0.38123	5
Ravenna	0.46018	5	0.45131	4	0.39632	4
Ancona	0.72325	4	0.71646	3	0.65226	3
Ploce	0.15995	8	0.17242	7	0.15102	7
Bar	0.19869	7	0.21877	6	0.25939	6

The analysis showed that the port of Koper is the only port to achieve full efficiency consistently from 2021 to 2023. This indicates that Koper has successfully rationalised its operations while effectively managing energy consumption and emissions. The stability of its efficiency scores suggests that Koper has adopted best practice in energy management, positioning it as a benchmark for other ports in the region.

The Port of Trieste showed a solid performance, with efficiency scores fluctuating around 0.88084 in 2023 and reaching full efficiency in 2022, resulting in a consistent second place. This reflects a generally strong but fluctuating efficiency, indicating the need for continuous improvement to maintain its competitive advantage.

The Port of Rijeka, while showing a downward trend in efficiency, managed to maintain a third place in 2023 with a score of 0.73049, having achieved full efficiency in 2021. Despite this decline, Rijeka remains comparatively efficient compared to other regional ports.

The port of Ancona improved to fourth place with an efficiency score of 0.72325 in 2023, showing a positive development in its energy balance. The ports of Venice and Ravenna, on the other hand, were slightly less efficient, ranking fifth and sixth respectively.

The ports of Ploce and Bar have the lowest efficiency values of 0.15995 and 0.19869 respectively in 2023. These low scores indicate that these ports face significant challenges in terms of energy consumption and operational efficiency, emphasising the need for the immediate introduction of energy saving initiatives and operational improvements.

The extent of connectivity with the hinterland has a significant impact on operational efficiency. Ports that are well connected to transport networks are better able to optimise their logistics and minimise energy waste, thereby improving their overall performance. Therefore, improving connectivity should be a strategic priority for less efficient ports that want to improve their energy performance.

These findings emphasise the need to tackle energy consumption and emissions in the maritime transport industry. As environmental sustainability comes under increasing scrutiny, ports need to focus on minimising their energy footprint and reducing greenhouse gas emissions. The results reveal a pressing need for comprehensive environmental strategies tailored to the specific conditions of individual port.

A key component of this analysis is the link between energy consumption and emissions. Higher energy consumption is often associated with higher greenhouse gas emissions. The results show that ports with higher throughput generally produce higher emissions, which emphasises the importance of implementing effective energy management strategies to reduce environmental impacts.

The Port of Koper, for example, achieves full efficiency with maintaining relatively low emissions, likely due to its effective operational practises and investments in cleaner technologies. In contrast, ports such as Ploce and Bar, which have lower energy efficiency, have significantly higher emissions due to their inefficient use of energy and operational practises.

Ports such as Koper and Trieste effectively maximise their container throughput while minimising energy consumption and emissions. This efficiency shows that optimising operations not only increases throughput, but also leads to a reduction in emissions per unit of energy used.

On the other hand, ports with poor energy efficiency tend to have a negative emissions throughput, i.e., they generate more emissions per TEU handled. This inefficiency increases operating costs and raises significant environmental concerns. To improve logistics and reduce energy waste, it is essential for these ports to improve their connectivity to the hinterland, which could ultimately lead to better performance results.

5 Conclusion

This study highlights the remarkable differences in energy efficiency between Adriatic container ports, which have a significant impact on operational sustainability and environmental impact. While some ports, such as Koper, have achieved impressive levels of efficiency, others face significant challenges that require prompt action. The results of this analysis not only enrich the academic discussion on port efficiency, but also provide practical recommendations for stakeholders seeking to improve energy efficiency in the Adriatic maritime transport sector.

The results provide valuable insights for policy makers and port authorities. Learning from the best practises of efficient ports such as Koper can be helpful for strategic improvements across the region. Initiatives that focus on investing in energy-efficient technologies, implementing thorough training programmes for port staff and promoting better cooperation with logistics partners can significantly improve energy efficiency.

In addition, the data reveals differences in energy efficiency between the ports, leading to an investigation into the underlying causes. A benchmarking analysis comparing the Adriatic ports with similar ports worldwide could provide a broader perspective on their performance. Furthermore, examining historical data could clarify whether the efficiency scores from 2021 to 2023 reflect actual improvements or declines compared to previous years.

While this study provides important insights into the energy efficiency of Adriatic container ports, it is important to note some limitations. The analysis is based on a three-year data span, which may not fully capture long-term trends or fluctuations. Additionally, inconsistent data availability across ports may hamper comparability efforts.

Future research could further investigate the factors contributing to efficiency differences, including the impact of new technologies and best practises on port operations. Sensitivity analysis could also help to identify which input parameters have the greatest impact on efficiency scores, enabling targeted action to be taken.

References

1. Acciaro, M., McKinnon, A.C., Roso, V.: The role of ports in the logistics chain: an energy perspective. Transp. Res. Part D: Transp. Environ. **30**, 143–154 (2014). https://doi.org/10.1016/j.trd.2014.05.008
2. Barros, C.P., Athanassiou, M.: Measuring the performance of the Portuguese port system: an application of data envelopment analysis. Marit. Policy Manag. **31**(4), 355–367 (2004). https://doi.org/10.1080/0308883042000290522
3. Chang, Y.T.: A study of the environmental efficiency of ports using data envelopment analysis. Sustainability **5**(8), 3595–3608 (2013). https://doi.org/10.3390/su5083595
4. Cooper, W.W., Seiford, L.M., Tone, K.: Data Envelopment Analysis: A Comprehensive Text with Models, Applications, References and DEA-Solver Software. Springer, Heidelberg (2007)
5. Cullinane, K., Wang, T.F., Hsu, Y.: The efficiency of container ports: a comparative study of the UK and the USA. Marit. Policy Manag. **31**(3), 221–236 (2004). https://doi.org/10.1080/0308883042000251767

6. Cullinane, K., Wang, T.F.: The efficiency of European container ports: a cross-sectional data envelopment analysis. Transp. Rev. **27**(5), 579–601 (2007). https://doi.org/10.1080/01441640701208653
7. Emrouznejad, A., Brzezickib, Ł., Lu, C.: The development and evolution of slacks-based measure models in data envelopment analysis: a comprehensive review of the literature. J. Econ. Surv. 1–20 (2025). https://doi.org/10.1111/jos.12682
8. European Statistical Office, epp.eurostat.ec.europa.eu
9. Findaport, https://www.findaport.com/
10. Gritsenko, D., Krause, S.: Energy efficiency in ports: a conceptual framework and empirical evidence. Int. J. Shipping Transp. Logist. **11**(6), 731–751 (2019). https://doi.org/10.1504/IJSTL.2019.102563
11. Guimarães, J.C., Leal Junior, A.B., de Almada Garcia, D.: Assessing the environmental performance of Brazilian container terminals: an application of data envelopment analysis. Environ Sci Policy **40**, 69–77 (2014). https://doi.org/10.1016/j.envsci.2014.02.008
12. Kong, L., Liu, W.: Exploring sustainable port cities: a data envelopment analysis approach. Sustain. Cities Soc. **70**, 102915 (2021). https://doi.org/10.1016/j.scs.2021.102915
13. Krmac, E., Mansouri Kaleibar, M.: A comprehensive review of data envelopment analysis (DEA) methodology in port efficiency evaluation. Maritime Econ. Logist. 25(4), 817–881 (2023). https://doi.org/10.1057/s41278-022-00239-5
14. Lam, J.S.L., Notteboom, T.: The impact of the introduction of eco-efficiency measures on the efficiency of container terminals. Transp. Res. Part D: Transp. Environ. **29**, 1–11 (2014). https://doi.org/10.1016/j.trd.2014.04.003
15. Mansouri Kaleibar, M., Krmac, E.: A generalized DEA approach to performance assessment: case study of northern Adriatic ports. In: Advances in the Theory and Applications of Performance Measurement and Management, pp. 123–136 (2024)
16. Mergoni, A., Emrouznejad, A., De Witte, K.: Fifty years of data envelopment analysis. Eur. J. Oper. Res. **326**(3), 389–412 (2025). https://doi.org/10.1016/j.ejor.2024.12.049
17. Prakash, R., Dhingra, C., Raju, R.: Assessing the efficiency of Indian ports using data envelopment analysis. Transp. Policy **50**, 120–128 (2016). https://doi.org/10.1016/j.tranpol.2016.03.002
18. Quintano, C., Mazzocchi, C., Rocca, L.: Eco-efficiency in the port sector: an analysis using non-radial data envelopment analysis models. Resour. Conserv. Recycl. **163**, 105118 (2020). https://doi.org/10.1016/j.resconrec.2020.105118
19. Quintano, C., Mazzocchi, C., Rocca, L.: Assessing eco-efficiency in seaport economic activities: a non-radial DEA approach. Environ. Sci. Pollut. Res. **28**(14), 17467–17480 (2021). https://doi.org/10.1007/s11356-021-12973
20. Tone, K.: A slacks-based measure of efficiency in data envelopment analysis. Eur. J. Oper. Res.Oper. Res. **130**(3), 498–509 (2001). https://doi.org/10.1016/S0377-2217(00)00153-2
21. UNCTAD: review of maritime transport. https://unctad.org/publication/review-maritime-transport-2023

Analysis of Technical Efficiency of Major Indian Ports Using Markov Chain Montecarlo Based Stochastic Frontier Analysis and Malmquist Index

Jayanta Kumar Das[1], Deepankar Sinha[2](✉), and Sujit Mukherjee[3]

[1] Indian Institute of Social Welfare and Business Management, Kolkata, WB 700073, India
jayanta@iiswbm.edu

[2] Indian Institute of Foreign Trade, Kolkata, WB 700100, India
dsinha@iift.edu

[3] MCKV Institute of Engineering, 243, GT Road North, Liluah, Howrah, West Bengal 711204, India
sujit.mukherjee@mckvie.edu.in

Abstract. In the last two decades, the Indian government has implemented several port reform programs to increase the efficiency of the major Indian ports. Operational efficiency of the major Indian ports over the period 2011–2023 has been evaluated by framing three research questions: Have port reforms improved port operational efficiency (RQ1)? What are the directions of Technical Efficiency change (Catch-up) and Technological change (Frontier Shift) (RQ2)? And whether the government reforms have an equal or partial impact on ports (RQ3). Most efficiency studies of Indian ports use data envelope analysis (DEA) as a tool. A major drawback of DEA is its inability to consider deviations in output from its frontier due to random exogenous shocks. DEA does not consider two crucial characteristics of the port industry: heterogeneity in production technology, the time-varying nature of operational efficiency, and the influence of random shocks which are uncontrollable by the port authorities. This paper suggests using Markov Chain Monte Carlo-based Stochastic Frontier Analysis to overcome the drawbacks of DEA. Malmquist productivity index is used for further analysis. The analysis indicates that the central government has been able to increase the overall productivity of all the major ports (except Chennai port) homogenously through technological and infrastructure upgrades at a fast pace. However, the major ports are still struggling to keep up with this pace and have been unable to upgrade their technical and operational efficiency as expected.

Keywords: Bayesian Stochastic Frontier Analysis · Seaport Technical Efficiency · Seaport Productivity · Markov Chain Monte-Carlo based Stochastic Frontier Analysis · Indian ports Technical Efficiency · Government Reforms

A. Emrouznejad et al. (Eds.): DEA 2024, LNOR, pp. 529–552, 2025.
https://doi.org/10.1007/978-3-031-98177-7_36

1 Introduction

Seaports are a crucial node in international supply chains, and its performance impacts the global business. Nayak et al. (2022) showed that operational efficiency, physical infrastructure, and socio-economic dimensions considerably impact ports. However, it also concludes that certain latent factors influence port cargo throughput. Other studies also establish the multidimensionality of port performance (Chakrabartty & Sinha, 2024). Hence, performance measures and benchmarking of ports are not complete unless all factors are identified and measured. Stochastic frontier analysis (SFA) serves as a method to determine the impact of both controllable and uncontrollable factors (random errors) (Põldaru & Roots, 2004).

Ports alone cannot affect the desired infrastructure as its performance depends on transportation connectivity, warehouses, dry ports and similar facilities to improve its performance. The ports' efficiency also depends on the effectiveness of associated regulatory stakeholders, namely the customs department and participating government agencies. Studies show that governments play an essential role in improving the logistics performance of their country. Government reforms need to focus on technological readiness, infrastructure, and process improvements, amongst other factors (Ekici et al., 2019). However, the measurement of the effectiveness of such reforms merits consideration.

India has over 200 ports across its coastline, and 12 of them are classified as "major" ports under the aegis of the central government. The share of these ports has declined since 2000 (Sinha & Roy Chowdhury, 2019) compared to the ports under the state governments, and reforms such as port privatization have not positively impacted port performance homogenously (Dasgupta and Sinha, 2016). So far, very insignificant studies have attempted to identify the combined impact of controllable and un- controllable factors on port's performance.

This paper aims to determine the efficiency of major Indian ports and identify the effectiveness of the Indian Government's reforms on their country ports' performance.

The Indian government has implemented several port reform programs over the last two decades to increase the performance of Indian ports. In 2005, the GOI introduced the National Maritime Development Programme (NMDP), intending to expand port capacity and modernize the existing port infrastructure through dredging operations to increase channel and berth depth, constructing new berths, installing high-capacity modern cargo-handling equipment, and improving hinterland road and rail connectivity.

The Indian Maritime Agenda (IMA) 2010–2020, apart from the aims of the NMDP reform program, focused on increasing port performance and coastal and inland waterways shipping. The focus was also on finding ways to corporatize port trusts later, implement the landlord port model, and incorporate competition in the port sector by effectively reforming the existing regulatory mechanisms and encouraging private-public partnerships. Through the Major Port Authorities Act (2015), the government granted greater autonomy to port trusts to fulfil the aims of the Indian Maritime Agenda (2010–2020).

In 2014, Project UNNATI (https://www.pib.gov.in/newsite/PrintRelease.aspx?relid=177280) was launched by the Ministry of Shipping to – evaluate and benchmark the operational and financial performance of the major ports with the best international ports and Indian private ports for identifying the problem and opportunity areas of the major ports.

In 2016, the central government launched the Sagarmala program (MoS, 2016) with several initiatives, like setting up manufacturing clusters around ports and coastal economic zones, institutional improvements to public-private partnerships (BOT: build, operate, and transfer mechanisms), and the development of the coastal community to facilitate more jobs.

In a competitive environment, a port must be efficient in terms of operations. According to Monteiro (2018), the Efficiency evaluation of ports is necessary as it helps monitor the progress of the port sector and, thereby, is core to port policy consideration. Many researchers from time to time (for example, Prakash, Rao, & Anupama, 2013; Abhilash & Gopinathan, 2013; Rajasekar & Deo, 2014) have tried to understand the effect of reformations on operational efficiency and performance aspects of the major ports of India. However, studies in the period, especially from 2013 onwards (Monteiro, 2018; Iyer & Nanyam, 2021; Adler et al., 2021), are much fewer. According to Nanyam and Jha (2023), the choice of study period is crucial. Accordingly, in this paper, the entire period of 2011–12 to 2021–23 has been divided into two time-period windows – 2011–12 to 2015–16 and 2016–17 to 2021–23 (to capture the influence of several programs launched by the GoI, especially the Unnati project in 2014 and Sagarmala project, 2016) to answer three research question RQ1: Have port reforms improved port operational efficiency? RQ2: What are the directions of technical efficiency change (Catch-up) and Technological change (Frontier Shift) RQ3: Do government reforms have an equal or partial impact on ports?

The paper tries to model the production process of major Indian ports through Markov Chain Monte Carlo (MCMC) based Stochastic Frontier Analysis (SFA) and to address both the research questions, uses the Malmquist productivity index (MPI), which is a product of two components- Technical efficiency change (Catch up) and technological change (Frontier shift).

Data Envelopment Analysis (DEA), a powerful non-parametric tool (Mergoni, 2024) for efficiency computation first introduced by Charnes et al. (1978), has been widely used to measure the performance of firms in terms of efficiency in several areas like the healthcare sector (Pavananunt K, 2004; Jung et al., 2023), Banking sector (e.g., Paradi & Zhu, 2013; Sonkariya & Yadav, 2024), Agricultural sector (e.g., Emrouznejad et.al.,2023) and more recently in measuring environmental performance of coal based thermal power plants (Mishra et al., 2024) and in higher education sector (Klumpp, 2018). DEA was initially developed as a pure nonparametric tool that considers data deterministic and does not account for measurement errors, sample noise, and specification errors in data (Olesen & Petersen, 2016). Banker ("Stochastic Data Envelopment Analysis," Working Paper, Carnegie Mellon University, 1988) first extended the traditional DEA model and introduced the stochastic DEA model. Stochastic DEA uses a linear programming technique and incorporates random variables to represent uncertainty in the input and output data (Amirteimoori et al., 2021). Instead of a single efficiency score, Stochastic DEA provides a range or probability distribution of efficiency scores. In 1998, Simar and Wilson introduced "Bootstrap DEA" (a form of Stochastic DEA), which uses a resampling technique to add statistical inference and confidence intervals to the deterministic results of traditional DEA. Bootstrap DEA has recently drawn the attention of the research community (Samad & Armstrong, 2022; Diallo et al., 2022).

In addition to the research in non-parametric DEA and its various extensions, there has been research in the area of parametric methods for finding the efficiency of firms using econometric approaches like Stochastic Frontier Analysis (Aigner, Lovell & Schmidt, 1977; Meeusen & Van Den Broeck 1977). SFA is parametric as it requires a functional form of the frontier and is thus parametric. The classical SFA is an ordinary least square-based SFA with limitations; hence, Bayesian SFA using Markov Chain Monte Carlo Simulation was introduced.

Technical efficiency refers to the production of maximum output by a firm given some resources as inputs. In the last few decades, there have been numerous studies on evaluating the efficiency of ports and in-depth analysis using several input variables and output variables using mathematical techniques like Data Envelopment Analysis (non-parametric method) and Stochastic Frontier Analysis (parametric econometric method) in ports, worldwide.

Stochastic frontier analysis (SFA) is a valuable tool as it separates inefficiency from random stocks or measurement errors (Liu et al. 2020). Most efficiency studies of Indian ports have been conducted using Data Envelopment Analysis. Papers on evaluating the efficiency of decision-making units using Stochastic Frontier Analysis (SFA) in the Indian port sector are practically missing. Besides, traditional SFA has certain limitations, and the need to ensure the inefficiency distribution is positively skewed merits consideration.

Tongzon and Heng (2005) applied the stochastic frontier production function to panel data of selected container terminals worldwide to investigate the quantitative relationship between port ownership structure and efficiency. Cullinane and Song (2006) have estimated the relative technical efficiency of some European container ports using a stochastic frontier model with the frontier in log-linear Cobb-Douglas production form. The result shows that the size of a port or terminal is closely correlated with its efficiency. Krljan et al. (2021) estimated the technical efficiencies using panel SFA on selected North Adriatic Ports Association (NAPA) interconnected medium- sized terminals, which are strong competitors of each other. In their model, the authors introduced control input variables into the stochastic Cobb-Douglas and the trans- logarithmic form of the stochastic frontier model. The authors conclude that NAPA terminals with the highest estimated technical efficiencies do not necessarily need to be the most productive ones, and vice versa.

However, in the above studies, the effectiveness of the country's reforms on ease of doing business, logistics performance and trade or process facilitation have not been considered.

In the Indian context, DEA has been widely used to evaluate port performance. Abhilash and Gopinathan (2013) used DEA to evaluate the performance efficiency of Cochin port relative to other major Indian ports in the period 2011–2012. The result showed that in the given period, JNPT outperformed the others, and Cochin Port was the lowest performer. Another important finding was that from 1993 to 2011, the efficiency was not correlated with the size of the Indian ports. Rajasekar and Deo (2014) performed an efficiency analysis of major Indian ports from 1993 to 2013. The results also indicated that upgrading port infrastructure and technological improvement was necessary to improve the efficiency of some inefficient ports, except JNPT, Kandla, Mormugao, and Tuticorin, which had high-efficiency scores. However, these, compared to world ports, proved to be inefficient.

Dasgupta and Sinha (2016) investigated the relative efficiency of service quality of privately managed terminals in major Indian ports, with privately managed container terminals using output-oriented data envelopment analysis (DEA). The study aimed to evaluate the effect of liberalization on the efficiencies of container terminals of "major" ports in India. The results of the study show that the efficiency of the container terminals was positively affected by privatization to a great extent. However, they depend on other factors too.

Monteiro (2018) analysed the efficiency and performance of Indian ports using the DEA and Malmquist index and showed that the increase in productivity of the Indian ports is attributed to technical change rather than technical efficiency improvement.

Adler et al. (2021) investigated the impact of cargo specialization, port ownership, inter-port competition and tariff regulation on the technical efficiency of major Indian ports. From 1995–1996 to 2013–14, the average efficiency estimate increased from around 45% to 75%. The technical efficiency improved due to a change in the ownership structure (opting for a landlord governance structure) and increased external stakeholder participation. Through their study, the authors indicated that Indian reform programmes should focus more on the modernization of major seaports. However, in the Indian context, inter-port competition has resulted in decreased efficiency of "major" ports.

Nanyam & Jha (2023) evaluated the operational efficiency and the financial and service quality efficiency of the major Indian ports separately using the DEA - Malmquist Productivity Index for the period 2013–20. Most of the ports were efficient in terms of operational and service quality efficiency; however, only six of these ports showed high performance in terms of financial aspects based on the MPI values (refer Table 1b).

This paper has five sections. The following section discusses the extant literature on the subject matter, identifying the research gaps. Section 3 establishes the stochastic frontier model used in this paper. Section 4 puts forward the results, and the last section discusses the results and concludes the findings, indicating the future scope of research.

In the following section, this paper analyses the theoretical background of Stochastic frontier analysis and comparative analysis of the advantages and disadvantages of classical DEA, Stochastic DEA, Stochastic Frontier Analysis and Bayesian Stochastic Frontier Analysis to arrive at the right method to evaluate the impact of government reforms on port performance.

2 Methodology and Data

The major drawback of classical deterministic DEA is its inability to consider deviation of output from its frontier due to random noise or random exogenous shocks. The entire deviation of output from the frontier is attributed solely to inefficiency, which also varies greatly depending on the model specification (input or output- oriented model). Moreover, if more and more firms are included in the sample, the efficiency scores may start

Table 1. List of variables chosen as inputs and outputs and mathematical/statistical techniques used for efficiency studies.

Researcher	Inputs	Outputs	Technique
Tongzon & Heng (2005)	Terminal quay length; terminal surface area	cargo traffic volume	Panel SFA
Cullinane & Song (2006)	Terminal quay length; terminal area; No. of cargo handling equipment	Container throughput volume	Panel SFA
Wang et al. (2013)	Average revenue per ton; Average total labour cost; Net investment in PP&E/employees; Cargo traffic volume; Dummy variable for Governance type	Profit per ton of Traffic	Cross-sectional SFA
Abhilash & Gopinathan. (2013).	Number of Berth; Number of cranes; Manpower	Cargo volume	DEA
Rajasekar & Deo (2014)	Number of equipment; Manpower; Number of berth Container berth length	Cargo volume	DEA-Super efficiency
Dasgupta & Sinha, (2016)	Draft Depth; Number of cranes; No. of berths	Quality of service	DEA
Monteiro (2018)	Storage area, Manpower, Number of cranes, Number of other equipment, Number of berths	Cargo volume, number of vessels handled	Malmquist DEA
Krljan et al. (2021)	Quay Length; Stacking Area; Water Depth; Number of cranes and Yard equipment	Container Throughput (TEU)	Panel SFA

(*continued*)

decreasing (Zhang & Bartels, 1998). DEA computes estimating "relative" efficiency;

Table 1. *(continued)*

Adler et al. (2021)	Capital; No. of Berths; Variable cost; Manpower; Number of cranes	No. of vessels handled; Cargo volume; Quality of service (represented by Inverse of Turn round time)	DEA
Iyer & Nanyam (2021)	Draft Depth; Quay Length; Number of Quay cranes; Number of Yard equipment; Yard area	Cargo volume	DEA
Nanyam & Jha (2023)	Number of Berths, Manpower, Number of cranes	Average output per ship berth day [Operational Efficiency] Profit in Lacs [Financial Efficiency] Quality of service (represented by Inverse of Turn round time) **[Service Quality Efficiency]**	Malmquitst DEA

thus, the measurements are only valid in a sample. When one has a few observations and more than one input and output, many DMUs will appear on the DEA frontier. The efficiency scores are only relative to the best firm, which means the computation is only valid for each sample (Bezat, 2009). Since classical DEA is a nonparametric technique, statistical hypothesis testing is difficult.

Stochastic DEA extends the classical deterministic DEA by incorporating provisions for statistical hypothesis testing. Like classical DEA, Stochastic DEA also utilizes linear programming to calculate the efficiency scores of each decision-making unit. Still, it additionally incorporates stochastic elements to represent the uncertainty in data. Stochastic DEA uses several approaches like Bootstrapping and Montecarlo simulation, and it assumes several probability distributions for the random elements in the data (Olesen & Petersen, 2016).

The key difference between Stochastic DEA and Stochastic Frontier Analysis (both classical or Bayesian SFA utilising Markov Chain Montecarlo Simulation) lies in the fact that in the case of Stochastic DEA, there is no definite functional form of production frontier, which has to be specified in case of Stochastic Frontier Analysis. Moreover, the difference lies in the treatment of noise. Stochastic DEA assumes random variation in data itself, indicating that the deviation from the frontier is considered solely inefficiency without considering the difference between random noise and actual inefficiency. In classical or Bayesian SFA, the error term contains two distinct components – random noise (modelled by normal distribution) and inefficiency (positively skewed distribution).

According to Yan et al., 2009, two characteristics of the port industry – heterogeneity in production technology and the time-varying nature of operational efficiency and influence of exogenous shocks like economic condition, business environment, competition, etc. have been ignored in efficiency studies using any flavour of DEA (deterministic or stochastic). In the case of port data, these random shocks need to be considered because they are uncontrollable by the port authorities.

It is advantageous to use Stochastic Frontier Analysis when, in addition to comparing the efficiency values of Decision-Making units, one also wants to investigate the effect of input factors contributing to inefficiency.

The most crucial limitation of classical SFA is that the maximum likelihood function does not produce reliable efficiency values in small samples, or in other words, a large number of decision-making units are required.

In this paper, the Markov Chain Monte Carlo simulation-based Bayesian stochastic frontier method has been applied using BayES software to evaluate the efficiency scores of major Indian ports and analyse the results. This method works well with small sample sizes (internally generates more samples by data augmentation technique using Gibbs sampler) and overcomes the limitation of classical SFA. Moreover, the method applies a less restrictive Bayesian inference method and offers feasible approximate alternatives based on maximum likelihood (Makieła & Mazur, 2022).

Stochastic DEA and Bayesian Stochastic Frontier Analysis were initially not used due to high computational complexity and convergence time in large data. However, in recent times, several specialized software like WINBUGS, BayES, R programming language, etc., have been introduced by the research community for researchers, which have efficient modules and can handle large amounts of data efficiently.

A production function is expressed as:

$$y_i = f(\boldsymbol{x}_i, \boldsymbol{\beta}) \quad (1)$$

where $\boldsymbol{y}_i$ is the observed output of the ith firm obtained from $\boldsymbol{x}_i$ which is a vector of inputs of a **non-stochastic** nature and $\boldsymbol{\beta}$ are unknown parameter vector to be estimated through the analysis.

A firm's technical efficiency (T.E) can be defined as **a ratio of observed output to maximum feasible output.**

If the ith firm attained a $T.\ E_i = 1$, it has attained the maximum feasible output and if $T.\ E_i \leq 1$, it implies a shortfall.

Adding T.E term in Eq. 1:

$$y_i = f(\boldsymbol{x}_i, \boldsymbol{\beta}) * T.E_i \quad (2)$$

Since the maximum technical efficiency is 1, it can be denoted as T. E = 1 − u_I where u_I denotes inefficiency.

Technical efficiency can be this denoted as $\exp\{-u_I\}$ as $\exp\{-u_I\} \approx 1 - u_I$.

In the real world, a firm's production process is often affected by Random exogenous shocks (stochastic) and inefficiency (arising from dynamic characteristics of competition, strategic policies, etc.). Adding a stochastic random component to the deterministic part of Eq. 2 makes the production frontier a "stochastic production frontier". Adding the random component as denoted by $\exp\{vI\}$, Eq. 2 can be expressed as:

$$y_i = f(\boldsymbol{x}_i, \boldsymbol{\beta}) * \exp\{-u_I\} * \exp\{v_I\} \tag{3}$$

By taking natural logarithm on both sides, Eq. 3 can be expressed in the well-known Cobb Douglas form of production function as:

$$ln\, y_i = \beta_0 + \sum_k \beta_k \ln x_i + (v_I - u_I) \tag{4}$$

The term $(v_I - u_I)$ termed in literature as a 'Composite error' (Meeusen & Van Den Broeck, 1977) is the error term ε_I and $\varepsilon_I \leq 0$. The random term **v** follows two-sided Normal distribution i.e. $v \sim N[0, \sigma^2]$ and $u \geq 0$.

Equation 4 assumes cross-sectional data of dependent and independent variables for frontier analysis. The efficiency of a firm may change with time. Equation 4, considering time to suit panel data, can be expressed as:

$$ln\, y_{it} = \beta_0 + \sum_k \beta_k \ln x_{it} + (v_{it} - u_{it}) \tag{5}$$

where $v_{it} \sim N[0, \sigma^2]$ and $u \geq 0$.

Several researchers have proposed a one-sided distribution (positively) for the inefficiency term **u** – e.g., the Half-Normal distribution (Aigner, Lovell & Schmidt 1977) and the Exponential distribution (Meeusen & Van Den Broeck, 1977).

The choice of the distribution of inefficiency greatly affects the efficiency values, and it must allow the inefficiency to be separated from the random term in composite error (Materov, 1981).

The basic assumption in classical stochastic frontier analysis is that the inefficiency distribution should be positively skewed, which implies that the composite error term in Eq. 4 must have negative skewness. A positively skewed error distribution leads to the variance of inefficiency towards zero, resulting in very high-efficiency scores of competing firms banding near one which in the research community is termed 'super efficiency'. However, a very high negative skewness in the error distribution is also problematic as it implies that only a small fraction of the firms attain a level of productivity close to the frontier (Papadopoulos & Parmeter, 2023).

Recently, many researchers (E.g. Kumbhakar & Tsionas, 2005) have focused on using the Bayesian technique introduced by Van Den Broeck et al. (1994). The Bayesian technique can be applied to a small sample size. Bayesian inference considers a preliminary inefficiency distribution to express uncertainty about unknown parameters, allows regulatory conditions to be incorporated in the model, and then works on the given data to collect evidence and represents the result in terms of probability density functions of unknown parameters (Terzi et al., 2023). Researchers frequently use a Markov Chain Monte Carlo (MCMC) method using Gibbs Sampling to estimate parameters in

Bayesian stochastic frontier analysis. The Gibbs sampling method uses the data augmentation technique and is the most frequent choice as a posterior simulator for generating draws from the posterior.

In this paper, two distributions of inefficiency terms are considered for estimating the technical efficiency of major Indian ports, namely Exponential and Half-normal.

The Data for the stochastic frontier analysis has been sourced from the Basic Port Statistics Report (https://shipmin.gov.in/transport-reseach/basic-port-statistics) published by the Ministry of Ports, Shipping and Waterways, GOI. From 2011–12 to 2021–23.

Table 2 shows the Descriptive statistics for the data.

Table 2. Descriptive statistics for input and output variables for analysis

Variables	Code	Mean	Median	Std.Dev	Minimu m	Maximu m
INPUTS						
Berth Depth (Mts)	BERTHD EPTH	11.80	12.14	1.70	8.20	14.95
Number of Berths	BERTHCT	19.43	18.00	8.15	8.20	14.95
Number of Cranes	NOC	27.34	16.00	42.75	1.00	228
Capital (Lacs)	CAPEMP LD	1612.93	1196.70	1131.29	102.56	5316
Average LoadPer Ship (MTons)	AVGLDP ERSHP	0.03	0.03	0.01	0.01	0.06
OUTPUT						
Traffic Volume	Traffic	49.65	42.06	27.66	11.74	127.1

The major drawback of DEA is its inability to consider deviation of output from its frontier due to random exogenous shocks. The entire deviation of output from the frontier is attributed to inefficiency, which also varies greatly depending on the model specification (input or output-oriented model). Moreover, if more and more firms are included in the sample, the efficiency scores may start decreasing (Zhang & Bartels, 1998). DEA computes estimating "relative" efficiency; thus, the measurements are only valid in a sample. When one has a few observations and more than one input and output, many DMUs will appear on the DEA frontier. The efficiency scores are only relative to the best firm, which means the computation is only valid for each sample (Bezat, 2009). Since DEA is a nonparametric technique, statistical hypothesis testing is difficult.

In conducting an ordinary least square-based SFA, the efficiency of firms is measured with respect to the deviation of their output from a frontier, which is a specific functional form that needs to be specified apriori. The most crucial limitation of SFA is that the maximum likelihood function does not produce reliable efficiency values in small samples, or in other words, a large number of decision-making units are required.

According to Yan et al., 2009, two characteristics of the port industry – heterogeneity in production technology and the time-varying nature of operational efficiency and influence of random shocks like economic condition, business environment, competition, etc. have been ignored in efficiency studies using DEA as DEA technology has no provision to incorporate these in the model. These random shocks need to be considered because they are uncontrollable by the port authorities.

In this paper, the Markov Chain Monte Carlo simulation-based Bayesian stochastic frontier method has been applied to evaluate the efficiency scores of major Indian ports and analyse the results. This method works well with small sample sizes (internally generates more samples by data augmentation technique using Gibbs sampler) and overcomes the limitation of classical SFA. It considers the influence of random shock for the computation of efficiency scores, thus overcoming the disadvantage of DEA. The following section explains the theoretical background of SFA in more detail.

Stochastic Frontier Analysis is a parametric method for estimating the productive efficiency of decision-making units or firms utilizing frontier production functions. The technique was first introduced in economic literature by Aigner, Lovell & Schmidt (1977) and Meeusen & Van Den Broeck (1977).

A production function is expressed as:

$$y_i = f(\boldsymbol{x}_i, \boldsymbol{\beta}) \tag{6}$$

where $\boldsymbol{y}_i$ is the observed output of the ith firm obtained from $\boldsymbol{x}_i$ which is a vector of inputs of a **non-stochastic** nature and $\boldsymbol{\beta}$ are unknown parameter vector to be estimated through the analysis.

A firm's Technical efficiency (T.E) can be defined as a **ratio of observed output to maximum feasible output.**

If the ith firm attained a $T.\,E_i = 1$, it has attained the maximum feasible output and if $T.\,E_i \leq 1$, it implies a shortfall.

Adding T.E term in Eq. 1:

$$y_i = f(\boldsymbol{x}_i, \boldsymbol{\beta}) * T.E_i \tag{7}$$

Since the maximum technical efficiency is 1, it can be denoted as $\mathrm{T.\,E} = 1 - \mathrm{u_I}$ where $\mathrm{u_I}$ denotes inefficiency.

Technical efficiency can be this denoted as $\exp\{-\mathrm{u_I}\}$ as $\exp\{-\mathrm{u_I}\} \approx 1 - \mathrm{u_I}$.

In the real world, a firm's production process is often affected by Random exogenous shocks (stochastic) and inefficiency (arising from dynamic characteristics of competition, strategic policies, etc.). Adding a stochastic random component to the deterministic part of Eq. 2 makes the production frontier a "stochastic production frontier". Adding the random component as denoted by $\exp\{\mathrm{v_I}\}$, Eq. 2 can be expressed as:

$$y_i = f(\boldsymbol{x}_i, \boldsymbol{\beta}) * \exp\{-\mathrm{u_I}\} * \exp\{\mathrm{v_I}\} \tag{8}$$

By taking natural logarithm on both sides, Eq. 3 can be expressed in the well-known Cobb Douglas form of production function as:

$$ln\, y_i = \beta_0 + \sum\nolimits_k \beta_k \ln x_i + (\mathrm{v_I} - \mathrm{u_I}) \tag{9}$$

The term ($v_I - u_I$) termed in literature as a 'Composite error' (Meeusen & Van Den Broeck, 1977) is the error term ε_I and $\varepsilon_I \leq 0$. The random term **v** follows two-sided Normal distribution i.e. v ~ N[0, σ^2] and u $\geq$ 0.

Equation 4 assumes cross-sectional data of dependent and independent variables for frontier analysis. The efficiency of a firm may change with time. Equation 4, considering time to suit panel data, can be expressed as:

$$ln\, y_{it} = \beta_0 + \sum\nolimits_k \beta_k \ln x_{it} + (v_{it} - u_{it}) \quad (10)$$

where v_{it} ~ N[0, σ^2] and u $\geq$ 0

Several researchers have proposed a one-sided distribution (positively) for the inefficiency term **u**. For example, the Half-Normal distribution was proposed by Aigner, Lovell & Schmidt (1977) and the Exponential distribution by Meeusen & Van Den Broeck (1977).

The choice of the distribution of inefficiency greatly affects the efficiency values, and it must allow the inefficiency to be separated from the random term in composite error (Materov, 1981).

The basic assumption in classical stochastic frontier analysis is that the inefficiency distribution should be positively skewed, which implies that the composite error term in Eq. 4 must have negative skewness. A positively skewed error distribution leads to the variance of inefficiency towards zero, resulting in very high-efficiency scores of competing firms banding near one which in the research community is termed 'super efficiency'. However, a very high negative skewness in the error distribution is also problematic as it implies that only a small fraction of the firms attain a level of productivity close to the frontier (Papadopoulos & Parmeter, 2023).

Recently, many researchers (E.g. Kumbhakar & Tsionas, 2005) have focused on using the Bayesian technique introduced by Van Den Broeck et al. (1994). The Bayesian technique can be applied to a small sample size. Bayesian inference considers a preliminary inefficiency distribution to express uncertainty about unknown parameters, allows regulatory conditions to be incorporated in the model, and then works on the given data to collect evidence and represents the result regarding probability density functions of unknown parameters (Terzi et al., 2023). Researchers frequently use a Markov Chain Monte Carlo (MCMC) method using Gibbs Sampling to estimate parameters in Bayesian stochastic frontier analysis. The Gibbs sampling method uses the data augmentation technique and is the most frequent choice as a posterior simulator for generating draws from the posterior.

In this paper, two distributions of inefficiency terms are considered for estimating the technical efficiency of major Indian ports, namely Exponential and Half-normal.

3 Empirical Results

From Fig. 1(a) and 1(b), it is evident that, as desired, the error kernel density for the normal-exponential and normal-half Normal models are negatively skewed to some extent. Figures 1(c) and 1(d) show that the kernel density of the inefficiency terms is positively skewed, as per the basic assumption of the stochastic frontier model.

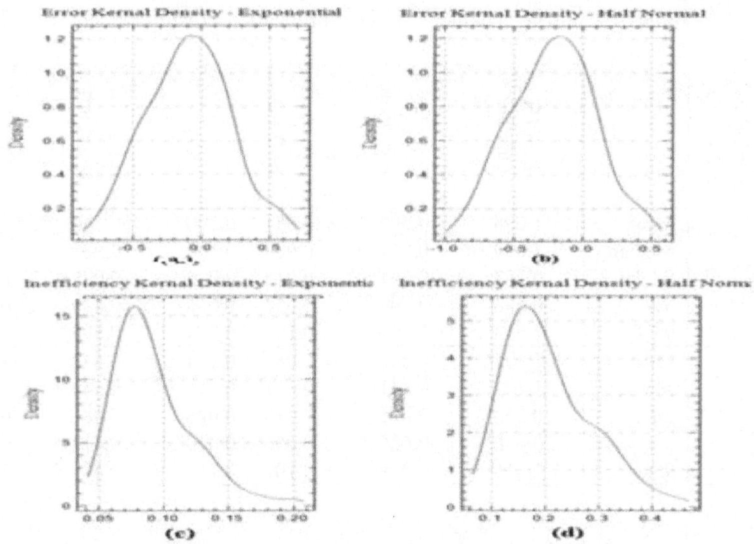

Fig. 1. The Kernel Density plots for the Error and Inefficiency terms for the models – Exponential and Half Normal.

For the models, the skewness of the error term can be estimated using:

$$Skewness\,[e] = \left(\frac{1}{n}\right) \sum\nolimits_i e_i^3 \tag{11}$$

where, e^3, may be estimated by subtracting the estimated output from the observed output.

Accordingly, the skewness of the error kernel density of the Half Normal model is estimated to be (**–0.06931**). For the Exponential model, the skewness of the error kernel density is estimated to be (**–0.02589**). In summary, the shape of the kernel density of errors and inefficiency terms are valid for both models.

3.1 Computed Coefficients and Parameters of Stochastic Frontier Models

From Table 3 and Table 4, it is evident that the coefficients (both mean and median) of BERTHDEPTH, AVGLDPERSHP, CAPEMPLD, NOC and BERTHCT and CONSTANT term and the individual parameters and error standard deviations are all statistically significant as the individual [5–95] % quantile range does not contain zero.

From Table 3, considering the median values of the coefficients of the input variables, the stochastic frontier for the Normal-Exponential model can be written as:

$$\begin{aligned} \ln(\text{Traffic}^{*}) = {} & 1.09897 * \ln(\text{BERTHDEPTH}) + 0.861444 * \ln(\text{AVGLDPERSHP}) \\ & + 0.080186 * \ln(\text{CAPEMPLD}) + 0.109782 * \ln(\text{NOC}) \\ & + 0.561114 * \ln(\text{BERTHCT}) + 1.78696 \end{aligned} \tag{12}$$

Table 3. Estimated coefficients of the Production Frontier models

Distribution Assumed for Inefficiency		BERTH DEPTH	AVGL DPERSHP	CAP EMPLD	NOC	BERTHCT	CONST
Exponential	Mean	1.098	0.861	0.081	0.200	0.560	1.790
	Median	1.099	0.861	0.080	0.190	0.561	1.786
	Sd.dev.	0.252	0.091	0.015	0.026	0.079	0.888
	5%	0.683	0.704	0.054	0.066	0.430	0.328
	95%	1.513	1.020	0.105	0.253	0.690	3.251
Half Normal	Mean	1.102	0.841	0.081	0.114	0.542	1.865
	Median	1.102	0.841	0.081	0.114	0.543	1.863
	Sd.dev.	0.253	0.095	0.015	0.026	0.081	0.880
	5%	0.656	0.686	0.056	0.070	0.412	0.421
	95%	1.515	0.998	0.105	0.157	0.673	3.314

Table 4. Descriptive statistics of the model parameters

		Mean	Median	Sd.dev	5%	95%
Exponential	tau	11.744	11.468	2.069	9.008	15.389
	lambda	12.715	10.977	6.730	5.433	25.977
	sigma_v	0.294	0.295	0.024	0.255	0.333
	sigma_u	0.092	0.091	0.045	0.039	0.184
Half Normal	tau	14.604	13.995	3.552	10.4238	20.680
	phi	14.95	14.296	5.161	7.77968	24.442
	sigma_v	0.266	0.267	0.027	0.220	0.310
	sigma_u	0.270	0.264	0.048	0.202	0.358

The stochastic frontier for this Normal-Half Normal model can be written as:

$$\begin{aligned}\ln(\text{Traffic}^*) = {} & 1.10252 * \ln(\text{BERTHDEPTH}) + 0.8409 * \ln(\text{AVGLDPERSHP}) \\ & + 0.080298 * \ln(\text{CAPEMPLD}) + 0.113848 * \ln(\text{NOC}) \\ & + 0.542652 * \ln(\text{BERTHCT}) + 1.86348 \end{aligned} \tag{13}$$

The stochastic frontier Equations [7–8] has an econometric explanation. For illustration, considering Eq. 8, it can be seen that the most important factor that positively influences the traffic is the water depth of the berth. A 1% increase in berth depth can increase the traffic volume by 1.102%. The following important influencing factor is the Average load per ship. A 1% increase in average load per ship will increase the traffic volume by 0.84%. The Average load per ship can be obtained by dividing the traffic volume by the number of vessels the port handles. The next influencing factor is berthing count, where a 1% increase in berth count can increase the traffic volume by 0.54%. A 1% increase in the number of cranes increases the traffic volume by 0.11%.

The reason for deviation from the stochastic frontier can be attributed to random or inefficiency components. This can be evaluated by the following computation proposed by Krljan et al., 2021.

$$\gamma = \frac{\sigma_u^2}{\sigma_u^2 + \sigma_v^2} \tag{14}$$

If the value of γ is close to zero, the deviation is due to random error, and if γ is close to one, the deviation is due to inefficiency. γ is computed for both the models and shown in Table 5.

Table 5. Comparison of the models considering γ value

Model	γ value	Dominant cause of deviation from production frontier
Normal- Exponential	0.10	Random error
Normal-Half Normal	0.5075	Inefficiency and Random error are equally dominant

As shown in Table 5, both random error and Inefficiency components are equally dominant in the Normal–Half Normal model. This result is realistic because, in real scenarios, port efficiency is affected by both random shock and inefficiency.

The Lewis & Raftery Log-Marginal Likelihood values and the Posterior Model Probability of the two models are compared for the final model selection. The result is shown in Table 5. The higher the Log-Marginal Likelihood value and the Posterior Model Probability (greater than the priori probability), the better the model.

Table 6. Model Comparison

POSTERIOR MODEL PROBABILITIES:

Model	Log-Marginal Likelihood	Type of log-ML Approximation	Prior Model Probability	Posterior Model Probability
Exponential	-79.6511	Lewis & Raftery	0.5	0.376582
Halfnormal	-79.1471	Lewis & Raftery	0.5	0.623418

Table 6 shows that the Normal-Half Normal Stochastic frontier model performs better with the input data. Hence, for further analysis, the Normal-Half Normal model is selected.

Table 7 shows the Technical Efficiency values obtained for each port with a normal-half-normal model between 2011–12 and 2021–23.

Table 7. Efficiency value estimated with Normal-Half Normal Production Frontier Model

Year	Chennai	Cochin	HDC	JNPT	Kamrajar	Kandla	KDS	Mormugao	Mumbai	N.Mangalore	Paradip	Tuticorin	Visakhapatnam
2011-12	0.76	0.76	0.89	0.92	0.86	0.88	0.75	0.82	0.83	0.89	0.89	0.85	0.85
2012-13	0.80	0.70	0.87	0.92	0.74	0.88	0.73	0.73	0.83	0.82	0.81	0.82	0.85
2013-14	0.78	0.73	0.87	0.90	0.83	0.85	0.73	0.70	0.86	0.82	0.83	0.79	0.84
2014-15	0.78	0.73	0.87	0.91	0.77	0.83	0.75	0.73	0.83	0.82	0.82	0.84	0.88
2015-16	0.75	0.85	0.88	0.91	0.84	0.85	0.78	0.71	0.84	0.85	0.81	0.85	0.83
2016-17	0.73	0.87	0.88	0.91	0.80	0.85	0.74	0.80	0.84	0.86	0.83	0.87	0.81
2017-18	0.73	0.87	0.88	0.91	0.80	0.85	0.75	0.81	0.85	0.86	0.86	0.82	0.81
2018-19	0.74	0.87	0.87	0.91	0.79	0.86	0.75	0.71	0.87	0.86	0.88	0.80	0.82
2019-20	0.71	0.88	0.88	0.86	0.80	0.86	0.73	0.69	0.91	0.85	0.87	0.83	0.81
2020-21	0.67	0.87	0.87	0.86	0.76	0.86	0.67	0.68	0.81	0.83	0.86	0.77	0.77
2021-23	0.786	0.78	0.81	0.90	0.77	0.83	0.64	0.64	0.80	0.81	0.84	0.71	0.81

Table 8. MCMC Diagnostics of Normal-Half Normal Model

MCMC Diagnostics: Halfnormal

dataset:	myData	total retained draws:	320000
lhs variable:	Traffic	burnin (draws/chain):	10000
# of rhs variables:	6	number of chains:	10
# of observations:	143	thinning parameter:	1
log-ML (Lewis & Raftery):	-79.1471		
log-ML (Chib & Jeliazkov):	-79.4046		

	Mean	Spectral density at 0	MCMC sd. error	Relative Numer. Eff.	Inefficiency factor
DRAFT	1.10216	0.0183362	0.000600025	0.551627	1.81282
Noofcranes	0.113778	0.000284012	7.46764e-005	0.388076	2.57682
CapitalEmployed	0.080828	5.45258e-005	3.27202e-005	0.658571	1.51844
Berthcount	0.542671	0.00439775	0.000293853	0.22957	4.35597
constant	1.86534	0.215539	0.0020572	0.571908	1.74853
AVGLDPERSHP	0.841158	0.00395024	0.000278501	0.366209	2.73068
tau	14.6041	45.4359	0.0298686	0.0442083	22.6202
phi	14.9598	97.3056	0.0437103	0.0435702	22.9515
sigma_v	0.266376	0.00214604	0.000205274	0.0564708	17.7083
sigma_u	0.270387	0.0103671	0.000451174	0.0355617	28.1201

Table 8 represents the diagnostics of the Monte Carlo Simulation carried out for the stochastic frontier model. The Markov Chain Monte Carlo simulation was carried out using ten chains (number of Markov chains running in parallel), with Burn-in (initial draws rejected per chain to reduce the selection of highly rejected correlated draws) of 10000 draws and thinning parameter 1(which means after every single draw, the next draw is rejected to reduce the correlation between immediate neighbouring draws further). Finally, the total draws retained from the posterior distribution were 3, 20, 000. The Inefficiency factor in Table 8 is relatively less (<50). The detailed theoretical and working information on MCMC is described by Chib et al. (2002). The results show that the Markov chain Montecarlo simulation using Normal-Half Normal distribution is efficient.

3.2 Computation of Malmquist Index, Frontier Shift and Catch- up for the Ports

The efficiency values generated for each port are used to compute the Malmquist Productivity Index (MPI). The Malmquist index captures the performance of ports from two perspectives - namely, the degree to which the major ports have made progress with reference to the timeline (**Catch up effect**) and the shift of the frontiers with respect to the timeline (**Frontier Shift**). The efficiency value is chosen from the Normal-Half Normal model for computing MPI.

The Malmquist Index is computed using Eq. 10–14 (Fare et al., 2016; Nanyam & Jha, 2023)

$$Catchup\ Effect_i = \frac{Efficiency\ of\ (X, Y)_i^{CP}\ w.r.t\ frontier\ of\ CP}{Efficiency\ of\ (X, Y)_i^{BP}\ w.r.t\ frontier\ of\ BP} \tag{15}$$

$$Frontier\ Shift_i = \sqrt{\frac{1}{Catchup\ Effect_i} \times \left[\frac{Efficiency\ of\ (X, Y)_i^{CP}\ w.r.t\ frontier\ of\ BP}{Efficiency\ of\ (X, Y)_i^{BP}\ w.r.t\ frontier\ of\ CP}\right]} \tag{16}$$

where BP refers to the base period, and CP refers to the current period.

$$Malmaquist\ Index_i = Catchup\ Effect_i \times Frontier\ Shift_i \tag{17}$$

where i refers to a particular port, the base period here is 2011–12, and the current year is 2021–23.

The stochastic frontier analysis has been applied to panel data. Equation 8 states the expression for the stochastic production frontier, which applies to the entire period between 2011–12 and 2021–23.

Hence, the Efficiency of the current period of the port with reference to the frontier of the base period for a port can be computed using:

$$Efficiency\ of\ (X, Y)_i^{CP}\ w.r.\ to\ frontier\ of\ base\ period = \frac{Y_{observed}^{CP}}{Y_{frontier}^{BP}} \tag{18}$$

$$Efficiency\ of\ (X, Y)_i^{BP}\ w.r.t\ frontier\ of\ CP = \frac{Y_{bserved}^{BP}}{Y_{frontier}^{CP}} \tag{19}$$

Use of Equations [10–14] with data for Chennai port in Table 9 is illustrated below: Using Eq. 13,

$$Efficiency\ of\ (X, Y)_{Chennai}^{2021-23}\ w.r.to\ frontier\ of\ 2011-12 = \frac{3.883}{4.45099}$$

Table 9. Data set for Chennai Port

Period	Frontier	Observed Traffic	Derived Efficiency
2011–12	4.45099	3.949	0.758523
2021–23	4.28201	3.883	0.785991

Using Eq. 14,

$$Efficiency\ of\ (X, Y)^{2011-12}_{Chennai}\ w.r.t\ frontier\ of\ 2021\text{–}23 = \frac{3.949}{4.28201}$$

Finally, using Eq. 11:

$$Frontier\ Shift_{Chennai} = \sqrt{\frac{4.45099}{3.883} \times \left[\frac{\frac{3.883}{4.45099}}{\frac{3.949}{4.28201}}\right]} = \mathbf{0.94957}(<\mathbf{1})$$

A similar calculation for all the other "major" ports is carried out, as shown in Table 10. The entire period of 2011–12 to 2021–23 has been divided into two time period windows: 2011–12 to 2015–16 and 2016–17 to 2021–23, with 2011–12 taken as the base period and 2016–17 as the base period for windows 2011–12 to 2015–16 and 2016–17 to 2021–23 respectively to bring out the clear picture.

4 Discussion and Conclusion

Table 10 displays the numerical values of frontier shift, catch up and MPI for the major ports in the two time windows. The negative signs of the percentage change in Frontier Shift, Catch-up, and MPI depict a downward shift of Frontier, catch up and fall in MPI. In contrast, the positive signs indicate the opposite for individual ports and answer **RQ2.** From 2011–12 to 2015–16, the technical efficiency enhancement was above average (avg. 0.984 but < 1) for Chennai, HDC, KDS, JNPT, Mumbai and Tuticorin. In terms of capacity enhancement, development was more prevalent in JNPT, Kamrajar, Kandla, Mormugao, N. Mangalore, Paradip, and Visakhapatnam than in the rest of the ports. From 2016–17 to 2021–23, the technical efficiency enhancement was above average (avg. 0.93 but < 1) for JNPT, Kamrajar, Kandla, Mumbai, N. Mangalore and Visakhapatnam. Capacity / technological development represented by frontier shift was higher than average (avg. 1.171) for Cochin, HDC, Kamrajar, KDS, Tuticorin and Visakhapatnam. The rest of the ports below the average value are also greater than one (except Chennai), indicating high-capacity addition and infrastructure development from 2016–17 to 2021–23 compared to the earlier period. Among the ports, Visakhapatnam is the only port with catch-up and frontier shift value above one, indicating growth in technological / capacity development and technical efficiency upliftment. The productivity of Visakhapatnam has increased by 2.60% in the current period compared to the last period. However, the efficiency has fallen by 0.44% (but still > 1) compared to the previous period, i.e. 2011–12 to 2015–16. From 2011–12 to 2015–16, there were seven ports

Table 10. Frontier-Shift, Catchup and Malmquist Index for two periods: 2011–12 to 2015–16 & 2016–17 to 2021–23

Port	2011–12 to 2015–16			2016–17 to 2021–23			% Change Frontier Shift	% Change Catch Up	% Change MPI
	Frontier Shift	Catch Up	MPI	Frontier Shift	Catch Up	MPI			
Chennai	1.00	0.99	1.00	0.882	1.072	0.945	−12.11	8.01	−5.07
Cochin	0.72	1.12	0.812	1.393	0.894	1.245	92.42	−20.36	53.24
HDC	0.99	0.986	0.974	1.331	0.917	1.220	34.68	−7.03	25.21
JNPT	1.04	0.989	1.028	1.108	0.991	1.098	6.59	0.21	6.82
Kamaraj	1.10	0.971	1.060	1.139	0.959	1.092	4.25	−1.20	3.00
Kandla	1.07	0.962	1.028	1.096	0.977	1.071	2.64	1.51	4.19
KDS	0.93	1.030	0.961	1.300	0.852	1.108	39.32	−17.31	15.21
Mormugao	1.23	0.860	1.058	1.255	0.792	0.994	2.08	−8.00	−6.08
Mumbai	0.96	1.017	0.977	1.068	0.950	1.014	11.15	−6.59	3.82
N.Mangalore	1.08	0.956	1.030	1.156	0.937	1.082	7.26	−1.99	5.13
Paradip	1.12	0.921	1.030	1.064	0.969	1.030	−4.97	5.23	−0.01
Tuticorin	0.98	1.009	0.990	1.406	0.867	1.219	43.18	−14.05	23.07
Visakhapatnam	1.02	0.976	1.000	1.020	1.006	1.026	-0.44	3.05	2.60
Avg	1.019	0.984	0.996	1.171	0.937	1.088	17.388	−4.500	10.087

(JNPT, Kamrajar, Kandla, Mormugao, New Mangalore, Paradip and Visakhapatnam) with a productivity index greater than one. However, in the current period, i.e. 2016–17 to 2021–23, the number of ports has increased to eleven except Chennai and Mormugao, with a fall in their productivity by 5.07% and 6.08%, respectively. A comparison of the two time periods shows that Cochin Port has achieved the highest growth percentage (53.24%) in terms of the Malmquist Index, followed by HDC, Tuticorin and KDS, respectively.

The average percentage growth in Frontier shift is 1.091 in 2011–12 to 2015–16 and 1.171 in 2016–17, greater than 1. However, the Catch-Up values for the two periods are 0.984 and 0.937 below 1. The Malmquist productive Index in the period 2011–12 to 2016–17 was below 1 (0.996), indicating that the overall performance of Indian ports was low in the said period. From 2016–17 to 2021–23, the ports have increased the Malmquist productivity index, which has gone up over 1(1.088), indicating that the major Indian ports are now performing high in productivity. The Productivity has gone by 10.087% from 2016–17 to 2021–23 compared to the earlier period. In summary, productivity growth is evident in the current period compared to the past, indicating the positive role of the port reforms measure taken by the central government in terms of capacity addition and modernization.

From Table 10, it can be further seen that the average growth rate of the Frontier Shift value is 17.388%. Frontier Shift occurs when ports expand their capacity and modernize the port. The value is nearly four times the absolute value of the percentage change in technical efficiency (4.5%). According to Munim (2020), the technical efficiency of a port may fall if there is a significant capacity increase and modernization of the port in a short period for which the port authority struggles to keep up with the pace of capacity addition. Hence, the answer to the **RQ1** is - that the central government, through their port reform programme (Unnati project, 2014; Sagarmala project, 2016) implementations, have increased the overall productivity of the ports through technological and infrastructure up-gradation; however, the major ports are struggling to keep up by upgrading their technical/operational efficiency. The technical efficiency change showed a negative growth (-4.5%) in the second period of the study compared to the first period. The Frontier Shift of Chennai port has fallen below one. All other ports have frontier shifts greater than one, indicating that the central government reform measure impacted most of the ports homogeneously, which is the answer to **RQ3.**

The result of this study validated the findings of the efficiency analysis of Indian ports by Monteiro, 2018, who analysed the efficiency and performance of Indian ports using the DEA and Malmquist index for the period 1993 to 2015 and showed that the increase in productivity of the Indian ports is attributed to technical change rather than technical efficiency improvement. This result also matches with the results of Munim, 2020.

It is evident that despite several reformation programmes undertaken by the central government (Unnati project, 2014 and Sagarmala project) in the chosen period (2011- 12 to 20121–23) that led to the upgradation of the ports in terms of technology and infrastructure at a high pace. Still, the ports have not been able to raise their technical/operational efficiency.

The Government policies should not only aim at improving the infrastructure but also direct towards process reforms. The impact of the privatization of port operations has yielded mixed results, indicating shortcomings in operationalization. A study of agreements on privatization shows that port authorities seek private operators to meet the minimum guaranteed throughput (MGT) but provide no criteria to enhance productivity and capacity over time (India Shipping News, 2024). The government projects on road infrastructure (Bharatmala), seaports (Sagarmala) and similar are implemented by individual ministries and government agencies, suggesting a silo- based approach. Instead, an integrated planning and coordination linking the infrastructure and the corresponding processes will provide the desired uniform results. Thus, an apex body may be formed to plan and deliver the reform projects and lay down the framework for the reassessment of port policies and processes to capitalize on government endeavours. At present, a digital platform to bring 16 ministries for integrated planning and coordination of infrastructure has been put in place under the umbrella of - Prime Minister Gati Sakti Master Plan for Multimodal Connectivity (PM Gati Shakti, 2024). However, this platform only discusses and assists infrastructure- related issues but does not delve into the integration of processes and the sub-policies of individual service providers such as sea ports.

Bayesian SFA is an advanced technique with limitations: the computational complexity and sensitivity to prior distributions. The correct model averaging requires proper priors of model-specific parameters and prior coherence. Interpretation of Bayesian model averaging results under incoherent priors may be difficult, and results depend on posterior model probabilities. These limitations can be overcome with real-life validation of conclusions (Makieła & Mazur, 2020).

In case of complex data sets where the underlying production function is unclear and in situations where modelling noise and inefficiency flexibilities are required, Stochastic DEA can outperform Bayesian SFA (Samad & Armstrong, 2022; Diallo et al., 2022, Amirteimoori et al., 2021). The data set used in this study has less complexities and results show consistencies with reality.

As a future research work, the individual port's efficiency results can be compared, and their processes and policies assessed to suggest a uniform framework for ports to adapt to the enhanced infrastructure and changing global scenarios. Besides, a comparison of results between Stochastic DEA, Smoothed Bootstrap and Bayesian SFA can be done to establish the right approach for assessment of Indian Port efficiency.

References

Abhilash, K.K.,Gopinathan, P.V.: Benchmarking the performance of Indian port sector by incorporating data envelopment analysis. In: International Conference on Materials for the Future-Innovative Materials, Processes, Products and Applications, pp. 206–216 (2013)

Adler, N., Hirte, G., Kumar, S., Niemeier, H.M.: The impact of specialization, ownership, competition and regulation on efficiency: a case study of Indian seaports. Marit. Econ. Logist. **24**(3), 507–536 (2021)

Aigner, D., Lovell, C., Schmidt, P.: Formulation and estimation of stochastic frontier production function models. J. Econom. **6**(1), 21–37 (1977)

Amirteimoori, A., Biresh, K.S., Vincent, C., Saber, M.: Stochastic data envelopment analysis. In: International Series in Management Science/Operations Research/International Series in Operations Research & Management Science, pp. 55–76 (2021). https://doi.org/10.1007/978-3-030-89869-4_4

Bezat, A.: Comparison of the deterministic and stochastic approaches for estimating technical efficiency on the example of non-parametric DEA and parametric SFA methods. Metody Ilościowe W Badaniach Ekonomicznych/Szkoła Główna Gospodarstwa Wiejskiego **10**(1), 20–29 (2009)

Cano-Leiva, J., Cantos-Sanchez, P., Sempere-Monerris, J.J.: The effect of privately managed terminals on the technical efficiency of the Spanish port system. Case Stud. Transp. Policy **13**, 101026 (2023)

Chakrabartty, S.N., Sinha, D.: Multi-dimensional assessment of port performance: a composite measure. J. Marit. Res. **21**(2), 235–245 (2024)

Chib, S., Nardari, F., Shephard, N.: Markov chain Monte Carlo methods for stochastic volatility models. J. Econom. **108**(2), 281–316 (2002)

Cullinane, K., Wang, T.F., Song, D.W., Ji, P.: The technical efficiency of container ports: comparing data envelopment analysis and stochastic frontier analysis. Transp. Res. Part A Policy Pract. **40**(4), 354–374 (2006)

Dasgupta, M.K., Sinha, D.: Impact of privatization of ports on relative efficiency of major ports of India. Foreign Trade Rev. **51**(3), 225–247 (2016)

Diallo, K. S., Mendy, P., Degla, G., Ndiaye, B.M.: Data envelopment analysis and bootstrap approaches for efficiency measure of the autonomous port of dakar. J. Math. Res. **14**(51) (2022). https://doi.org/10.5539/jmr.v14n4p51

Efecan, V., Temiz, Z.: Assessing the technical efficiency of container ports based on a non-monotonic inefficiency effects model. Utilities Policy **81**, 101494 (2023)

Ekici, ŞÖ., Kabak, Ö., Ülengin, F.: Improving logistics performance by reforming the pillars of global competitiveness index. Transp. Policy **81**, 197–207 (2019)

Emrouznejad, A., Petridis, K., Charles, V.: Data Envelopment Analysis with GAMS, p. 338. Springer, Cham (2023). https://doi.org/10.1007/978-3-03130701-0

Fare, R., Grosskopf, S., Norris, M., Zhang, Z.: Productivity growth, technical progress, and efficiency change in industrialized countries. **84**(1), 66–83 (2016)

Griffin, J.E., Steel, M.F.J.: Bayesian stochastic frontier analysis using WinBUGS. J. Prod. Anal. **27**(3), 163–176 (2007)

Gupta, M., Kumar, S.: Does income diversification affect bank efficiency, effectiveness and performance? Evidence from Indian banking industry. In: Emrouznejad, A., Thanassoulis, E., Toloo, M. (eds.) Advances in the Theory and Applications of Performance Measurement and Management. DEA45 2023. Lecture Notes in Operations Research. Springer, Cham (2024). https://doi.org/10.1007/978-3-031-61597-9_14

Haschka, R.E.: Endogeneity in stochastic frontier models with "wrong" skewness: copula approach without external instruments. Soc. Sci. Res. Netw. (2023)

Hlali, A.: Efficiency analysis with different models: the case of container ports. J. Mar. Sci. Res. Dev. **8**(2) (2018)

India Shipping News: Short-term contracts: A New era for Indian Port privatisation strategies - India Shipping News, https://indiashippingnews.com/short-term-contracts-a-wew-era-for-indian-port-privatisation-strategies/, (2024). Accessed 03 Feb 2025

Iyer, K.C., Nanyam, V.N.: Technical efficiency analysis of container terminals in India. Asian J. Shipp. Logist. **37**(1), 61–72 (2021)

Iyer, K.C., Nanyam, V.N.: Technical efficiency analysis of container terminals in India. Asian J. Shipp. Logist. Asian J. Shipp. Logist. **37**(1), 61–72 (2021)

Liu, J., Wang, M., Yang, L., Rahman, S., Sriboonchitta, S.: Agricultural productivity growth and its determinants in South and Southeast Asian countries. Sustainability **12**(12), 4981 (2020)

Jung, S., Son, J., Kim, C., Chung, K.: Efficiency measurement using data envelopment analysis (DEA) in public healthcare: research trends from 2017 to 2022. Processes **11**(3), 811 (2023). https://doi.org/10.3390/pr11030811

Kaleibar, M.M., Krmac, E.: A generalized DEA approach to performance assessment: case study of northern Adriatic ports. In: Advances in the Theory and Applications of Performance Measurement and Management. DEA45 2023. LNOR. Springer, Cham (2024). https://doi.org/10.1007/978-3-031-61597-9_11

Klumpp, M.: Higher education discipline-specific efficiency analysis with DEA: incorporating time lag and output selection effects for German universities. In Data Envelopment Analysis and Performance Measurement: Recent Developments: Proceedings of the DEA40: International Conference of Data Envelopment Analysis, Aston Business School, UK (2018). ISBN: 978 1 85449 438 2

Krljan, T., Grubisic, A., Hess, S., Grubisic, N.: The stochastic frontier model for technical efficiency estimation of interconnected container terminals. J. Marine Sci. Eng. **9**(5), 515 (2021)

Kumbhakar, S.C., Tsionas, E.G.: Measuring technical and allocative inefficiency in the translog cost system: a Bayesian approach. J. Econom. **126**(2), 355–384 (2005)

Makieła, K., Mazur, B.: Model uncertainty and efficiency measurement in stochastic frontier analysis with generalized errors. J. Prod. Anal. **58**(1) 35–54 (2022). https://doi.org/10.1007/s11123-022-00639-y

Makieła, K., Mazur, B.: Bayesian model averaging and prior sensitivity in stochastic frontier analysis. Econometrics **8**(2), 13 (2020)

Materov, I.S.: K probleme polnoi identifikatsii modeli stokhasticheskikh granits proizvodstva. Ekonomika I Matematicheskie Metody **17**, 784–788 (1981)

Meeusen, W., Van Den Broeck, J.: Efficiency estimation from cobb-douglas production functions with composed error. Int. Econ. Rev. **18**(2), 435 (1977)

Mergoni, A., Emrouznejad, A., De Witte, K.: Fifty years of data envelopment analysis. Eur. J. Oper. Res. **326**(3), 389–412 (2025). https://doi.org/10.1016/j.ejor.2024.12.049

Mishra, S., Singh, A., Sufia: Environmental performance of coal-based thermal generation plants: a two-stage DEA model with undesirable outputs. In: Emrouznejad, A., Thanassoulis, E., Toloo, M. (eds.) Advances in the Theory and Applications of Performance Measurement and Management. DEA45 2023, pp. 151–163. LNOR. Springer, Cham (2024). https://doi.org/10.1007/978-3-031-61597-9_13

Monteiro, J.G.R.: Measuring productivity and efficiency of seaports in India using DEA technique. Cent. Eur. Rev. Econ. Manag. **2**(3), 153 (2018)

Munim, Z.H.: Does higher technical efficiency induce a higher service level? A paradox association in the context of port operations. Asian J. Shipp. Logist. **36**(4), 157–168 (2020)

Nayak, N., Pant, P., Sarmah, S.P., Jenamani, M., Sinha, D.: A novel Index-based quantification approach for port performance measurement: a case from Indian major ports. Marit. Policy Manag. **51**(2), 174–205 (2022)

Nanyam, V., Jha, N.K.: Modeling challenges affecting the performance of major ports of India. Asian J. Shipp. Logist. **39**(3), 26–38 (2023)

Notteboom, T., Coeck, C., Van Den Broeck, J.: Measuring and explaining the relative efficiency of container terminals by means of Bayesian stochastic frontier models. Int. J. Marit. Econ. **2**(2), 83–106 (2000)

Olesen, O.B., Petersen, N.C.: Stochastic data envelopment analysis—a review. Eur. J. Oper. Res. **251**(1), 2–21 (2016). https://doi.org/10.1016/j.ejor.2015.07.058

Papadopoulos, A., Parmeter, C.F.: The wrong skewness problem in stochastic frontier analysis: a review. J. Prod. Anal. (2023)

PM Gati Shakti. https://www.india.gov.in/spotlight/pm-gati-shakti-national-master-plan-multi-modal-connectivity. Accessed 03 Feb 2024

Põldaru, R., Roots, J.: Modeling grain yield in Estonian counties: a stochastic frontier analysis approach. Data Envelopment Anal. Perform. Manag. 259 (2004)

Paradi, J.C., Zhu, H.: A survey on bank branch efficiency and performance research with data envelopment analysis. Omega **41**(1), 61–79 (2013). https://doi.org/10.1016/j.omega.2011.08.010

Pavananunt, K.: Hospital efficiency: a study of public community hospital in Thailand. In: Data envelopment analysis and performance management. Proceedings of the 4th International Conference of DEA (DEA2004), Birmingham, UK (2004). 197, ISBN: 0 902683 73X

Prakash, N.B., Rao, B.R., Anupama, T.: Major ports of India. Indian Econ. J. **60**(4), 163–172 (2013)

Rajasekar, T., Deo, M.: Is there any efficiency difference between input and output-oriented DEA models: an approach to major ports in India. J. Bus. Econ. Policy **1**(2), 18–28 (2014)

Samad, A., Armstrong, V.S.: Bootstrap-DEA management efficiency and early prediction of bank failure: evidence from 2008–2009 U.S. Bank failures. Cent. Bank Rev. **22**(3), 119–127 (2022). https://doi.org/10.1016/j.cbrev.2022.08.002

Simar, L., Wilson, P.W.: Sensitivity analysis of efficiency scores: how to bootstrap in nonparametric frontier models. Manag. Sci. **44**(1), 49–61 (1998). https://doi.org/10.1287/mnsc.44.1.49

Sinha, D., Roy Chowdhury, S.: Optimizing private and public mode of operation in major ports of India for better customer service. Indian Growth Dev. Rev. **12**(1), 2–37 (2019)

Sonkariya, A., Yadav, S.P.: Development of intuitionistic fuzzy revenue efficiency models in DEA: an application to the Indian public sector banks. In: Advances in the Theory and Applications of Performance Measurement and Management. DEA45 2023. Lecture Notes in Operations Research. Springer, Cham (2024)

Terzi, E., Aydin, S.G., Cengiz, M.A.: Stochastic energy performance evaluation using a bayesian approach. J. Math. **2023**, 1–7 (2023)

Tongzon, J., Heng, W.: Port privatization, efficiency and competitiveness: some empirical evidence from container ports (terminals). Transp. Res. Part A Policy Pract. **39**(5), 405–424 (2005)

Van Den Broeck, J., Koop, G., Osiewalski, J., Steel, M.F.: Stochastic frontier models. J. Econom. **61**(2), 273–303 (1994)

Wang, G.W., Knox, K.J., Lee, P.T.W.: A study of relative efficiency between privatized and publicly operated US ports. Marit. Policy Manag. Marit. Policy Manag. **40**(4), 351–366 (2013)

Yan, J., Sun, X., Liu, J.J.: Assessing container operator efficiency with heterogeneous and time-varying production frontiers. Transp. Res. Part B **43**(1), 172–185 (2009)

Zhang, Y., Bartels, R.: The effect of sample size on the mean efficiency in DEA with an application to electricity distribution in Australia, Sweden and New Zealand. J. Prod. Anal. **9**, 187–204 (1998)

Correction to: Efficiency of Indian Life Insurance Companies-A Cluster Based Approach

Ram Pratap Sinha and Alireza Amirteimoori

Correction to:
Chapter 21 in: A. Emrouznejad et al. (Eds.): ***Advances in the Theory and Practice of Data Envelopment Analysis*****, LNOR,**
https://doi.org/10.1007/978-3-031-98177-7_21

In the original version of Chapter 21, the author's name, Alireza Amirteimoori was misspelled. This has now been rectified and the correction has been updated.

The updated version of this chapter can be found at
https://doi.org/10.1007/978-3-031-98177-7_21

A. Emrouznejad et al. (Eds.): DEA 2024, LNOR, p. C1, 2025.
https://doi.org/10.1007/978-3-031-98177-7_37

Author Index

A. Emrouznejad et al. (Eds.): DEA 2024, LNOR, pp. 553–554, 2025.
https://doi.org/10.1007/978-3-031-98177-7

The manufacturer's authorised representative in the EU is Springer Nature Customer Service Centre GmbH, Europaplatz 3, 69115 Heidelberg, Germany. If you have any concerns regarding our products, please contact ProductSafety@springernature.com

Printed and bound by CPI Group (UK) Ltd, Croydon, CR0 4YY

30/09/2026

02225301-0014